108.75

FOREST ECOLOGY

Fourth Edition

BURTON V. BARNES
University of Michigan

DONALD R. ZAK
University of Michigan

SHIRLEY R. DENTON
Biological Research Associates
Tampa, Florida

STEPHEN H. SPURR
University of Texas at Austin

John Wiley & Sons, Inc.

Acquisitions Editor   Ellen Schatz
Marketing Manager   Catherine Beckham
Production Editor   Sandra Russell
Designer   Harold Nolan
Illustration Editor   Edward Starr

This book is printed on acid free paper.⊚
This book was set in Times Roman by Ruttle, Shaw & Wetherill and printed and bound by Hamilton Printing.
The cover was printed by Phoenix Color Corporation.

The paper in this book was manufactured by a mill whose forest management programs include sustained yield harvesting of its timberlands. Sustained yield harvesting principles ensure that the numbers of trees cut each year does not exceed the amount of new growth.

Library of Congress Cataloging-in-Publication Data
Forest ecology / Burton V. Barnes ... [et al.]. — 4th ed.
        p.   cm.
    Rev. ed. of: Forest ecology / Stephen H. Spurr, Burton V. Barnes.
    3rd ed. c1980.
    Includes bibliographical references and index.
    ISBN 0-471-30822-6 (cloth : alk. paper)
    1. Forest ecology.   I. Barnes, Burton Verne, 1930–
    II. Spurr, Stephen Hopkins.   Forest ecology.
    QK938.F6F635   1997
        577.3—dc21   97-29124
                                                                                        CIP

Printed in the United States of America
10 9 8 7 6 5 4

DEDICATION

We dedicate the 4th edition of Forest Ecology in memory of its originator: *Stephen H. Spurr (1918–1990)*: Scientist, Dynamic Teacher, Distinguished Educator, and Visionary Leader in Forestry.

Few individuals in their lifetime significantly influence their field as did Stephen Spurr, including fields of botany, geology, photogrammetry, and forest measurements, besides forest ecology. His legacy to students of forest ecology in North America and around the World is immense. Upon receiving the honorary degree of Doctor of Laws by The University of Michigan in 1988, a distinguished colleague observed: "It would be difficult to find any person who has made broader and more significant contributions toward improving the quality of civilization."

PREFACE

Forest Ecology deals with forest ecosystems—spatial and volumetric segments of the Earth—and their landforms, soils, and biota. It is designed as a textbook for people interested in forest ecosystems—either in the context of courses in forest ecology, silvics, principles of silviculture, and environmental science or as a reference for those in professional practice. Field ecologists, foresters, naturalists, wildlife ecologists, and others interested in forest ecosystems and their management, conservation, and restoration will find this book a source of basic ecological concepts and principles.

Since the appearance of the first edition in 1964, ecology has stimulated great public interest and gained scientific prominence. This increased interest and public concern presents forest ecologists with the challenge to understand ecosystem properties and processes as the basis for managing ecosystems sustainably. This book provides an understanding of the ecological relationships of individual trees and forest ecosystems.

The book has six major subdivisions. "Ecosystems at Multiple Spatial Scales" examines forest ecosystems first from the outside—the broad view of space and time first. "The Forest Tree" considers the variations among individual trees, the causes of diversity within and between species, regeneration ecology, and selected aspects of tree structure and function. "The Forest Environment" treats the physical factors of forest ecosystems—the influences of climate, light, temperature, physiography, soil, and fire on the individual forest plant and on plant communities. The concluding chapter in this part, considers the evaluation of the forest site and ecosystems using single and multiple factors. In Part 4, "Forest Communities," we consider the forest community of trees and associated plants and animals that form a key structural component of forest ecosystems—one part of the whole. "Forest Ecosystem Dynamics" brings into focus the functional relationships of the physical environment and the biota. We consider the extent to which disturbance, an ecosystem process, initiates change through time (termed succession). Chapters on carbon balance and nutrient cycling present a detailed consideration of the pattern in which carbon and plant nutrients flow within forest ecosystems and how natural and human-induced disturbances alter these patterns. In addition, the diversity of forest organisms and ecosystems is considered as well as the important field of landscape ecology. Finally, "World Forests" explores phytogeography. Here, the historical development and distribution of North American and world forests are briefly discussed.

In this edition, we have tried to retain the readable qualities of the previous editions while emphasizing forest ecosystems and their dynamics with new organization, new chapters, and new material. Although traditionally, the focus in forest ecology has been on organisms and communities, they always should be viewed as parts of landscape ecosystems. Organisms and communities come alive when viewed as integral parts of ecosystems. As such, this is why we have chosen to begin our discussion from the broad context of spatial and temporal scales and emphasis on climate, physiography, and soil followed by a consideration of forest communities.

Due to the enormous volume of ecological literature, a rigorous selection was necessary. For example, we elected not to develop chapters on forest tree physiology and hydrology/watershed management for which entire courses are devoted and detailed textbooks are available. However, water and physiological relationships are an integral part of many chapters because they influence many important ecological processes.

Although many literature citations are provided, they constitute only a small sample of the many thousands of references reviewed by the authors. Preference has been given to major English-language works. At the same time, however, the effort was continued to include a judicious selection of important earlier work, as well as representative papers written in languages other than English dealing with forest other than those in the United States.

We are indebted to many colleagues for their contributions and help in preparing the revised edition. In particular, we acknowledge reviews and other assistance by Dennis A. Albert, Stephen Arno, Kathleen Bergen, Joann Constantinides, Bruce P. Dancik, Hazel and Paul Delcourt, Melany Fisk, Gary Fowler, Jerry Franklin, J. Steve Godley, Dave Grigal, Peter Groffman, Melanie Gunn, Bill Holmes, George Host, Daniel Kashian, Mark Kubiske, Jean MacGregor, William J. Mattson, Carl Mikan, G. Mühlhäusser, Glenn Palmgren, Kurt S. Pregitzer, David Rothstein, Terry L. Sharik, Ed Trager, Wayne Walker, John Zasada, and Greg Zogg. Our thanks to Kathleen Bergen who assisted with the remote sensing section and to Douglas Pearsall who generously contributed data and assistance in preparing the diversity chapter. A special thank you to J. Stan Rowe for his reviews and for sharing his ideas and verbiage. Fuqing Han's outstanding assistance in research and other significant details in preparing this edition is greatly appreciated. In addition, we thank text readers Lenora W. Barnes, Virginia H. B. Laetz, and Donna Zak. D. R. Zak was partially supported by the National Science Foundation and the Department of Energy during the preparation of this text.

Burton V. Barnes
Donald R. Zak
Shirley R. Denton

Ann Arbor, Michigan
Tampa, Florida

CONTENTS

CHAPTER 8   LIGHT   182

CHAPTER 9   TEMPERATURE   206

CHAPTER 10   PHYSIOGRAPHY   224

CHAPTER 19   NUTRIENT CYCLING   524

CHAPTER 1

CONCEPTS OF FOREST ECOLOGY

A **forest** is a three-dimensional ecological system dominated by trees and other woody vegetation that exist in dynamic interaction with the air–earth matrix of the landscape. The forest is more than a stand of trees or a community of woody and herbaceous plants. It is a complex ecological system, or **ecosystem**, characterized by a layered structure of functional parts. **Ecology** is the study of ecological systems and their interacting abiotic and biotic components. **Forest ecology**, therefore, is concerned with the structure, composition, and function of forests as landscape ecosystems. It is concerned with the climate, physiography, and soil of diverse areas in which occur the individual trees and other organisms constituting the forest community. In forest ecology, we study forest organisms and their response to physical factors of the environment as well as functional relationships of whole ecosystems in forested landscapes. Characterized by the predominance of woody vegetation substantially taller than humans, forests are widespread on land surfaces in humid climates outside of the polar regions. It is with forests in general, and with the temperate North American forest in particular, that this book is concerned.

Forest ecosystems may be examined and studied at several levels. Most obviously, the forest may be considered first simply in terms of the trees, those plants that give the forest its characteristic aboveground appearance or **physiognomy**. Thus we think of a beech–sugar maple forest, a spruce–fir forest, or of other **forest types**, for which the naming of the predominant trees alone serves to characterize the forest ecosystem.

Another approach is to take into account the obvious interrelationships that exist between forest trees and other organisms. Certain herbs and shrubs are commonly found in beech–sugar maple forests, whereas others are found in spruce–fir or loblolly pine forests. Similar interrelationships may be demonstrated for birds, mammals, arthropods, mosses, fungi, and bacteria. Part of the forest ecosystem, therefore, is the assemblage of plants and animals living together in a **biotic community**. The **forest community**, then, is an aggregation of plants and animals living together and occupying a common area. It is thus a more organismally-complex unit than the forest type, which historically has been defined on the basis of the trees only. A **forest stand** is a sub-community consisting of trees in a local setting and possessing sufficient uniformity of species composition, age, spatial arrangement, or condition to be distinguishable from adjacent stands (Ford-Robertson, 1983).

A third approach is to focus on geographic or **landscape ecosystems**. This approach is centered conceptually and in practice on whole ecosystems—not their parts—such as trees or communities, landforms, or soils. Understanding that our primary focus should be real live chunks of earth space, i.e., landscapes and waterscapes (oceans, lakes, rivers; hereafter included as parts of landscape), we can usefully study their organisms, landforms, and soils while recognizing that each is but one part of a functioning whole. Thus we emphasize this focus on ecosystems rather than on the individual organisms and species that are parts of them.

In the past, the forest stand or the species has been the focus in natural resource fields such as forestry and wildlife. However, because of the inseparable nature of the life-giving environment and the diverse biota it supports, we are really managing whole forest ecosystems. We may usefully study ecosystem parts such as the atmosphere, landforms, soils, forest communities, and individual forest organisms. However, the interacting forest ecosystem remains our conceptual focus despite its incredible complexity. A consideration of the field of ecology from this viewpoint provides an overall perspective.

ECOLOGY

The broader the field of scientific inquiry, the more difficult it becomes to limit and define that field. Ecology, as the broadest of the life sciences, is also the most indistinct. In 1866, Ernst Haeckel proposed the term **oecology**, from the Greek *oikos* meaning home or place to live, as the fourth field of biology dealing with environmental relationships of organisms and distinguished from morphology (form), physiology (internal function), and taxonomy (likeness). Thus ecology literally means "the knowledge of home," or "home wisdom." Since its introduction the term has been applied at one time or another to almost every aspect of scientific investigation involving the relationship of one organism to another, or to the relationship of an organism to its environment (Rowe, 1989). Although Haeckel's biological view of ecology was organisms, its subject matter is now being extended to those larger wholes within which organisms exist, i.e., volumes of land–water that envelop and contain organisms (Hagen, 1992; Golley, 1993). Thus Rowe (1989, p.230) suggests:

> *Ecology is, or should be, the study of ecological systems that are home to organisms at the surface of the earth. From this larger-than-life perspective, ecology's concerns are with volumes of earth-space, each consisting of an atmospheric layer lying on an earth/water layer with organisms sandwiched at the solar-energized interfaces. These three-dimensional air/organisms/earth systems are real ecosystems—the true subjects of ecology.*

In adopting this **landscape ecosystem approach** we emphasize the two volumetric and structural–functional entities, organisms and ecosystems, whose interrelationships are considered next.

LANDSCAPE ECOSYSTEMS

The British botanist–ecologist Arthur Tansley (1935) introduced the term ecosystem in reaction to conventional terms of that time: "complex organism" and "biotic community." He wrote that the more fundamental conception is "the whole 'system,' including not only the organism-complex but also the whole complex of physical factors." He also noted that ecosystems are of the "most various kinds and sizes," and from the point of view of the ecologist, ecosystems "are the basic units of nature on the face of the earth." Tansley was a biologist and vegetation ecologist, and so his idea of ecosystem was centered on the organism (species or community) rather than geographic or landscape entities. Using the prefix *bio*, we may characterize this view as a **bioecosystem** approach. In this approach, "ecosystem" derives its meaning from particular plant or animal organisms of interest; and a nonspecific "abiotic" environment having less emphasis is added thereto. In this relatively elastic approach, every organism defines its own ecosystem, which are nearly infinite in number, non-coincident in boundaries, and difficult to study and use as a basis for management and conservation.

On the other hand, others (Troll, 1968, 1971; Rowe, 1961a) view ecosystems centered on geographic or landscape units (i.e., landscape ecosystem, and using the prefix *geo*, the **geoecosystem**) of which organisms are notable structural components. In many parts of this book, we focus on organisms, species, and communities—always remembering, however, that they are parts of local ecosystems whose functioning is, in turn, influenced by larger, regional ecosystems as described below.

The forest ecosystems we consider in this book are segments of the Ecosphere—units that may be identified and separated out at multiple scales from the landscape continua in which they occur. Figure 1.1 illustrates the relationship of volumetric systems to one another, showing several size scales from Ecosphere to cell. It is useful to think of landscapes from the large to the small (i.e., "top down")—from the one, large, complete ecological system—the **Ecosphere** (Cole, 1958) to a microsite enveloping one or more organisms. Each of the volumetric ecosystems occupies a segment of earth space for a given time span. Forest ecologists not only study (1) organisms of these systems and their aggregates as communities and populations (see Chapters 4–6, 14, 15), but also (2) the functioning of local ecosystems that involves complex interactions among organisms and their supporting environment (Chapters 8–13, 15–19), and (3) the spatial patterns of occurrence and interrelationships of entire forest ecosystems (Chapters 2, 18, 19, 21)

We view ecosystems as geographic or terrain objects—a segment of the Earth, with all that is in it and on it, including atmosphere, landforms, soils, and biota. The term ecosystem is a generic term for the volumetric units of nature that extend downward from the largest ecosystem we know, the Ecosphere, through macro-level units of continents and seas and meso-level units of regional ecosystems (major physiographic units and their included organisms); to local ecosystems (for example, local tracts of forested land); and finally to the smallest level of homogeneous layered environment with organisms enveloped in it (Figure 1.1). Thus we conceive the Ecosphere and its landscapes as ecosystems, large and small, nested within one another in a hierarchy of spatial sizes (see Chapter 2).

The layered structure of landscape ecosystems is shown in Figure 1.2, and Figure 1.3 distinguishes their basic components. Figure 1.4 serves to illustrate that these components are inseparable and interacting in complex but patterned ways. These illustrations also show that populations and communities, aggregates of organisms that co-occur in a given area, are important components of ecosystems.

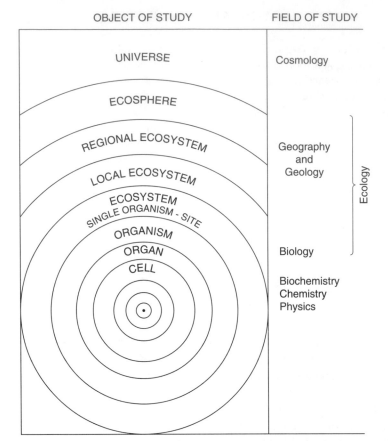

Figure 1.1.   Diagrammatic representation of objects-of-study from the most inclusive (ecosphere) to the least inclusive levels of organization (cell and organelle and molecule below it). Note that each higher level envelops the lower ones as parts of its whole. Some corresponding fields of study are also shown. Aggregates of organisms, such as populations and communities, are components of ecosystems at all scales. Like other components such as atmosphere, landform, and soil, they do not appear in this diagram of first-order objects-of-study but are shown in Figures 1.2, 1.3, and 1.4. (After Rowe, 1961a.)

Our definition of landscape ecosystem emphasizes its structurally-layered and volumetric nature (Rowe, 1961a): *"Any single perceptible ecosystem is a topographic unit, a volume of land and air plus organic contents extended areally over a particular part of the earth's surface for a certain time."* This definition, compared to more generalized ones (e.g., "a community of organisms and their physical environment interacting as an ecological unit" [Lincoln et al., 1982]), conveys the geographic/volumetric concept (geoecosystem) that is useful to field ecologists, naturalists, foresters and other land managers, and natural resource professionals. Other terms have been introduced to express the same idea but are less commonly used. The smallest indivisible landscape ecosystem was termed the ecotope by Troll (1963a, 1968) and the ecoterresa by Jenny (1980). In Canada, Hills (1952) used "total site" to express a hierarchy of geographic ecosystem units at local and regional levels.

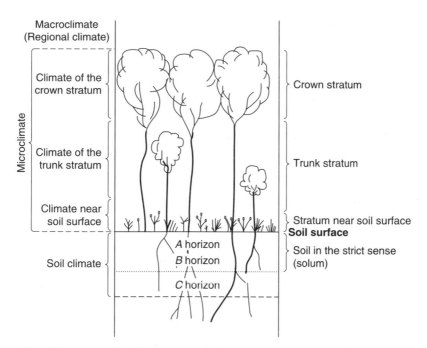

Figure 1.2. Diagrammatic illustration of the layered structure of a local landscape ecosystem. (Modified from Troll, 1963a. Reproduced by permission, Singapore Journal of Tropical Geography.)

The idea of the geoecosystem is not new, deriving from the Cowles (1901) emphasis on specific landforms, and today, it is a key feature in landscape ecology. This concept is described in Chapters 2, 13, and 21, and is emphasized in the works of many authors (Rowe and Sheard, 1981; Barnes et al., 1982; Zak et al., 1989; Jones, 1991; Host and Pregitzer, 1992; Bailey, 1995; Albert, 1995; Barnes, 1996) and in books by Matthews (1992), Huggett (1995), and Bailey (1996). Significantly, the Ecological Society of America (Christensen et al., 1996) has defined an ecosystem as: *a spatially explicit unit of the*

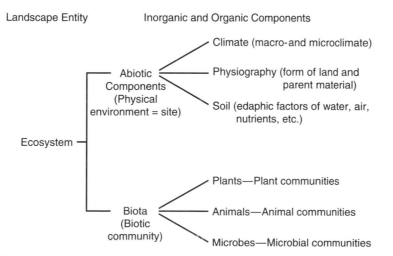

Figure 1.3. Diagrammatic illustration of abiotic and biotic components of landscape ecosystems.

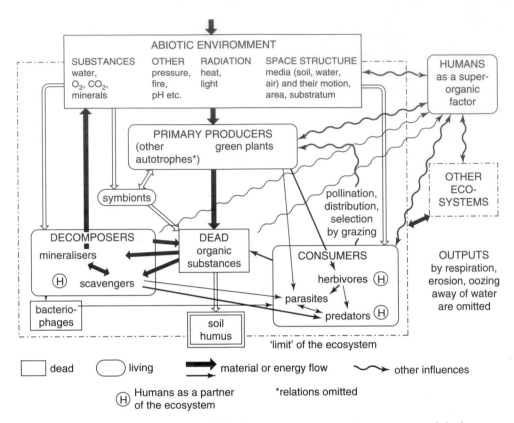

Figure 1.4.   A simplified model of a landscape ecosystem or geoecosystem and the interactions between the physical environment and the biota. (After Ellenberg, 1988. Reprinted from *Vegetation Ecology of Central Europe,* © 1988 by Cambridge University Press. Reprinted with permission of Cambridge University Press.)

Earth that includes all the organisms, along with all components of the abiotic environment within its boundaries (italics added for emphasis).

   An awareness of these ways of conceptualizing ecosystems is useful in understanding how the term ecosystem is used in the literature and applied in practice. Both practical and theoretical considerations suggest the need for a fundamental science of ecosystems focused on these basic units of nature on the face of the Earth, melding the earth sciences and biology. For studies of organisms in their immediate surroundings the biological approach is often useful. Nevertheless, for management of ecosystems, studies of forest productivity, and the conservation and restoration of forest ecosystems, the landscape ecosystem approach is eminently practical and theoretically sound. Examples of landscape ecosystems are presented later in this chapter and in Chapters 15, 17, and 21.

Landscape Ecosystem and Community

Although the term ecosystem was introduced in 1935 by Tansley, the idea of an interacting system of physical and biotic parts is very old (Lindeman, 1942; Sjörs, 1955; Major, 1969; Golley, 1993). Terms in various languages, such as taiga, heath, bald, Auewald (river floodplain forest), pampas, prairie, chaparral, Maquis, hammock, muskeg, and bog are conceived to depict vividly this concept of earth-air-organism interaction. More often than not, however, the terms emphasize a distinctive plant community and may lead one to the assumption that the ecosystem is simply an extension of community. Tansley's definition of ecosystem as organism-plus-environment leads to this view, and general definitions such as "biotic community + environment = ecosystem" reinforce this view. Applied in the landscape, this logic leads directly to vegetation classification and mapping as a way to apprehend or define ecosystems. Although indispensable for the ecologist and manager, vegetation types may or may not coincide with the fundamental climate-landform-soil-based ecosystems because of disturbance and unknown historical factors. Thus we emphasize the importance of geography and physiography within a regional climatic setting as the basis of understanding not only the vegetation types, but whole ecosystem structure and function. We use the adjectival modifier **landscape** to emphasize that ecosystems are spatial in nature and to underscore that ecosystems are not simply extensions of the biotic community. Our focus is therefore on landscape ecosystems rather than on the individual organisms and communities that are parts of them. Rowe (1989, p.229) observes:

> *Organisms do not stand on their own; they evolve and exist in the context of ecological systems that confer those properties called life. The panda is part of the mountain bamboo-forest ecosystem and can only be preserved as such. The polar bear is a vital part of the Arctic marine ecosystem and will not survive without it. Ducks are creatures born of marshes. Biology without its ecological context is dead.*

The idea of the landscape ecosystem, as a more important object-of-study than the biotic community, provides a consistent theoretical and practical basis for ecology, for ecosystem-based management, and the conservation of ecosystems and diversity of their biota.

This conceptual approach is not to say that communities are unimportant; they form the key ecosystem component whose response is indispensable in effecting ecosystem change and indicating the integrated effects of many site factors (Chapter 13). Communities and populations, however, differ basically from entities such as ecosystems, organisms, and cells because they are aggregates of individuals but *not functional systems*. Entities such as ecosystem, organism, and cell are "volumetric" units because they have structurally joined parts that form a functioning unit (Rowe, 1961a, 1992c). Communities and populations are assemblages or aggregates of spatially separated trees and understory plants that have no necessary structural connections. However, as a part of the structurally integrated ecosystem, organisms (as biotic communities) link the atmosphere and soil to form the functional system as illustrated in Figures 1.2, 1.3, and 1.4. Thus the vertical structure and species composition of communities as well as species interactions provide insights that are indispensable in understanding properties and processes of ecosystems.

Ecosystem Structure

Structure is the spatial arrangement of parts. In local forest ecosystems, multiple strata of atmosphere and soil (Figure 1.2), as well as the form of the land, influence the composition and patterns of occurrence of the biota as well as their function within ecosystems (biomass accumulation, water and nutrient cycling, successional trends). Plants are "rooted" not only in the soil strata but also in the aboveground air layers. The substantiality of the soil, with its layered structure and related physical and chemical properties, is well understood. However, the substantiality and stratification of composition, density, temperature, turbidity, and chemistry of the atmosphere have largely escaped our attention (Rowe, 1961a; Woodward, 1987). The seriousness of air pollution and forest decline in parts of the world have called particular attention to these essentially invisible layers enveloping tree crowns and entire plants and animals. Vegetation itself is also layered, i.e., structured vertically. Its physiognomy varies markedly in different regional and local ecosystem types (Chapters 6 and 8).

The landscape itself is also horizontally structured into spatial patterns of ecosystems reflecting differences in climate, geology, and physiography. The natural communities of these systems reflect limiting factors of soil water, nutrients, and disturbances and, in turn, modify the physical factors. Different ecosystem mosaics characterize mountain, plain, and river valleys due to fundamental differences their physical factors and the vegetation adapted thereto (Chapter 10).

Ecosystem Function

Local landscape ecosystems, besides having a structure of interconnected parts, are functional units characterized by many processes. Local ecosystems have characteristic functional properties that are defined by their processes. Mineral weathering, organic matter decomposition, cycling of water, nutrients and carbon, and biomass accumulation are typically considered ecosystem-level processes, i.e., part of the entire system and not restricted only to the physical or biotic parts. Other processes are often more associated with plant species, populations, or communities—photosynthesis and respiration, reproduction, regeneration, mortality, and succession. However, these are also ecosystem processes because characteristic ecosystem factors of temperature, water, nutrients, and disturbance regimes (such as fire, windstorm, and flooding) act to mediate and regulate them.

Landscape ecosystem structure and function are tightly linked to one another by the physical environment, prevailing regimes of repeating kinds of disturbances that reset succession, and the life histories of the plants and animals that comprise the biotic community. Climate, physiography, and soil provide the possibilities and set the Darwinian stage upon which organisms survive, adapt, and evolve. It is the interactions of all these factors that determine the function of ecosystems and their dynamics in time and space.

Accepting the idea expressed by Tansley (1935) that ecosystems are the basic units of nature, ecologists have concentrated enormous conceptual and mathematical resources on ecosystem functioning focused on biota. The field of systems ecology that developed over the past four decades has emphasized functioning. Furthermore, mathematical modeling and the computer caused a quantum leap in activity in ecology. However, a treatment of these fields and approaches is beyond the scope of this book. A good overview of ecosystem function is given by McIntosh (1985), and it is also treated by Hagen (1992) and Golley (1993) in books that describe the history of the ecosystem concept.

The strong emphasis placed on ecosystem functioning and its expression via box and arrow diagrams, flow charts, and computer modeling may tend to obscure the fact that ecosystems are real geographic entities. They have spatial structure and a specific history of changes that shaped their dynamic evolution as we observe them today. The spatial position of an ecosystem in relation to its surrounding neighbors also affects its functioning. The size, shape, and juxtaposition of ecosystems within a regional landscape affect the exchanges of matter and energy and thus affect the function of any individual ecosystem. Local landscape ecosystems are functional parts of regional ecosystems (Chapter 2). They contribute to the functioning of regional complexes just as parts of a given local ecosystem (climate, landform, soil, and biota) contribute to the function of an individual local ecosystem type.

Vertical and Horizontal Approaches

A great deal of forest ecology involves the study of individual plants or communities in relation to environmental factors. However, an understanding of both the function of individual landscape ecosystems (vertical approach) and their spatial arrangement (horizontal approach) and among-ecosystem exchanges and interactions is equally important, especially in the field of landscape ecology (Chapter 21).

As three-dimensional objects, landscape ecosystems can be studied both "vertically" by an internal or physiological approach and "horizontally" by an external and ecological approach (Rowe, 1961a; 1988). Troll (1971) described the two aims of a landscape ecology approach as understanding:

1. the functional interrelationships of an ecosystem at a given place through a relatively **vertical approach**, and

2. the spatial differentiation of the Earth's surface by examining the interplay of natural phenomena using a relatively **horizontal approach**.

Two landscape ecosystems in Figure 1.5 (Rowe, 1984b), one on the upper slope and one on the lower slope, illustrate characteristic ecosystem differences in hilly or mountainous terrain. The two are distinguished by a different geomorphology (convex upper slope versus concave lower slope, high versus low topographic slope position) that mediates microclimate, soil water, and nutrient availability. The dashed line is placed at an ecologically significant boundary between the two, separating them spatially on a horizontal

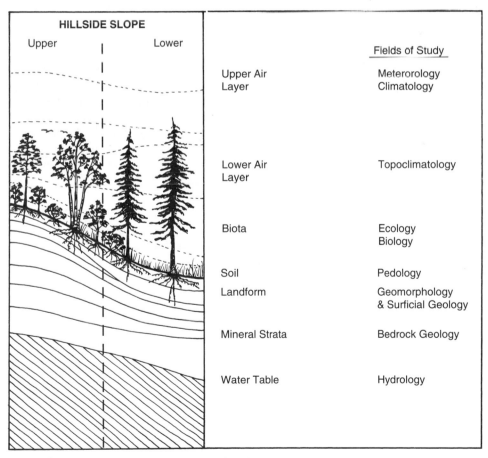

Figure 1.5. Structural profile illustrating landscape ecosystems of upper and lower slopes. Air–earth layers surround the organisms at the Earth's energized surface. The vertical line is set at an ecologically significant topographic break, dividing the upper and lower slope ecosystems. (After Rowe, 1984b.)

plane. Organisms in these spatially different ecosystems are located at the energized Earth's surface, enveloped in a layered matrix of the air and earth strata that we term environment. Also shown in Figure 1.5 are the traditional fields of study in which individuals seek to understand each of the ecosystem components. However, the forest ecologist, by using both vertical and horizontal approaches, aims to understand the integrated effects of all components.

In the vertical plane, *within each landscape ecosystem*, we can study how ecosystems function, i.e., their internal physiology. Solar radiation and precipitation enter from above; gravity affects leaf fall, water infiltration, and movement of water and nutrients through the rooting zone and into the water table. Processes of water and nutrient cycling, transpiration, soil development, and stratal accumulation of carbon into biomass operate in essentially an up-down fashion downslope and below the ground surface (Chapters 11, 18, and 19). In addition, the composition and abundance of plants, animals, and their populations in the vertical plane provide insights on plant reproduction, establishment, competition and mutualism, and succession as well as animal–plant interactions.

Complementing the vertical approach is the horizontal or ecological study of landscape ecosystems. Through this approach the adjacent ecosystems of Figure 1.5 may be contrasted. On a broad landscape scale, the mosaic of ecosystems, large and small, may be identified and mapped regionally and locally and their recurrent pattern in the landscape and interrelationships with one another can be examined. Adjacent ecosystems, especially in mountainous or hilly terrain, affect one another by the lateral exchanges of materials and energy. Water and snow, soil, organic matter, nutrients, and seeds are transported downhill; the effects on other ecosystems depend on the size, shape, and composition of the systems. Ecosystem function is affected by lateral transfers as well as by vertical, gravity-driven processes. The larger the ecosystem, the less the relative importance of horizontal transfer. In small ecosystems, the cross-boundary transfers are relatively large (Odum, 1971). Therefore, an understanding of the spatial pattern and configuration of landscape ecosystems can play an important role in the management of ecosystems and their biota.

Also, using the horizontal approach, the diversity of landscape ecosystems can be assessed (Lapin and Barnes, 1995). Such an understanding provides the spatial ecosystem framework for programs in biodiversity that seek to conserve and manage the diversity of organisms and maintain and increase populations of rare and endangered species (Chapters 20 and 21).

EXAMPLES OF LANDSCAPE ECOSYSTEMS

Ecosystems are real geographic or landscape objects, not simply box and arrow diagrams. Examples are many. Some ecosystems are easily discerned in the field, whereas others must be carved out of geographic continua. Bogs in the glaciated terrain of northern and boreal regions are easily discernible ecosystem types. They exhibit distinctive physiography, soil, and vegetation; and they recur in the landscape. Also, the distinct bottoms, or terraces, of river floodplains distinguish local ecosystem types that differ in microclimate, soil, drainage, vegetation, and their dynamics. Mountains are characterized by gradients of elevation, aspect, and slope steepness along which strikingly different ecosystems can be purposefully delimited and mapped although their boundaries are not sharply demarcated. On old, lake-bed terrain, beach ridges are easily distinguished from adjacent depressions of swampy land. Differences in water drainage and oxygen availability in these adjacent systems result in major differences in their native tree and ground–flora vegetation. Unseen are the very different processes that also distinguish these ecosystems.

A diagrammatic illustration of a topographic sequence of landscape ecosystems in northwestern Ontario (Figure 1.6) shows a continuum from mineral soil of coarse loamy till to organic matter at the base of the slope. Ecosystem types are distinguished at each of six positions in this sequence on the basis of landform, bedrock position, soil, and micro-

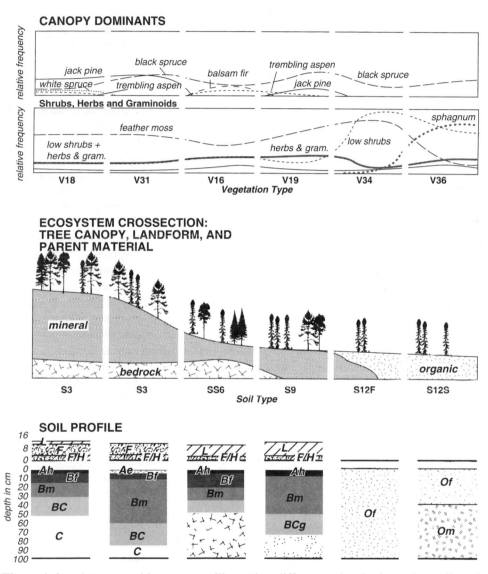

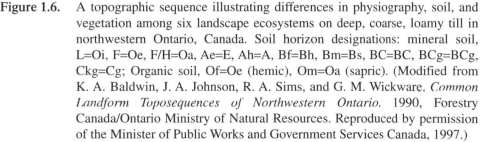

Figure 1.6.   A topographic sequence illustrating differences in physiography, soil, and vegetation among six landscape ecosystems on deep, coarse, loamy till in northwestern Ontario, Canada. Soil horizon designations: mineral soil, L=Oi, F=Oe, F/H=Oa, Ae=E, Ah=A, Bf=Bh, Bm=Bs, BC=BC, BCg=BCg, Ckg=Cg; Organic soil, Of=Oe (hemic), Om=Oa (sapric). (Modified from K. A. Baldwin, J. A. Johnson, R. A. Sims, and G. M. Wickware. *Common Landform Toposequences of Northwestern Ontario.* 1990, Forestry Canada/Ontario Ministry of Natural Resources. Reproduced by permission of the Minister of Public Works and Government Services Canada, 1997.)

climatic differences giving rise to differences in vegetation. The fire-prone ecosystem of the upper slope is dominated by jack pine. It gives way downslope to moister systems characterized by black spruce and balsam fir. On the lowermost slope position wetland ecosystems prevail and are characterized by poor drainage and low oxygen availability, limited nutrient availability in acidic, water-saturated soils, and cold micro-climate. Cold-air drainage into the lowland basin increases local frequency of potentially killing frosts throughout the growing season. The wetlands are dominated by black spruce, low shrubs, and sphagnum moss that maintain a high groundwater table at or near the soil surface. Different soil profiles are shown for each ecosystem, and microclimate changes markedly from warm, dry upland to cold, wet swamp.

Ecosystem components change in time, giving rise over the course of hundreds, thousands, or millions of years to a sequence of different landscape or waterscape ecosystems on a given place on the Earth's surface. Striking changes in the fossil record illustrate long-term changes in physiography, soil, and biota (Chapter 3). Short-term changes also occur. Such changes are most easily noted when natural or human-caused disturbances affect vegetation. A landform-based, site-specific location over 100 to 300 years may exhibit a range of different forest communities, young and old, as disturbances or lack thereof affects the biota. The composition and vertical layering of vegetation and the function of ecosystems at this site change over time. The landform and associated pattern of parent material, soil, and climate, change more slowly than the suite of species available to recolonize the disturbed site. Thus what we term the **ecosystem site type** or simply **ecosystem type** (land area supporting potentially equivalent ecosystems) can be distinguished and mapped regardless of the forest community currently present.

One such example is illustrated in Figure 1.7 where deciduous forests occur on glaciated terrain in southern Michigan. In this setting, three local *ecosystem types* are distinguished by differences in physiography (outwash and moraine landforms); soil; drainage; and overstory, understory, and ground-cover vegetation. The relatively fine-textured, silty soil on the outwash plain (type 1) supports old-growth forest dominated by white oak, whereas the drier, coarse-textured, sandy outwash soil (type 2) supports a black oak community. The fine-textured, moist, clayey soil on the rolling moraine landform (type 3) supports a community dominated by northern red oak. The moraine formerly supported a beech–sugar maple forest. However, recurrent fires through the drier white and black oak ecosystems killed the fire-sensitive mesic species of the adjacent moraine ecosystem and led to dominance of northern red oak. An occasional beech is still found, and red maple has invaded the shaded understory.

The western portion of ecosystem type 1 was clear-cut about 70 years ago, and a *cover type* markedly different from the adjacent old-growth white oak forest has formed (Figure 1.7, type 1a). The overstory tree layer of the disturbed area is now dominated by white, black, and red oaks, white ash, black maple, American elm, black cherry, and sassafras. These species either sprouted from the base of cut trees, were already present in the white oak forest understory, or seeded-in from adjacent communities following cutting. Today, we can recognize two (1a and 1b) of the many compositionally different forest communities that might occur at the site of ecosystem type 1; these easily could be mapped as two different forest cover types.

Furthermore, because the cut and uncut areas of ecosystem type 1 have different structure and function, we may also recognize two *current* ecosystems (1a and 1b) that are characteristic of ecosystem type 1. Because the cut-over area (1a) has physiography and soil like type 1b, its vegetation may gradually become similar in composition and vertical layers to those of ecosystem 1b, providing that no major changes in climate, soil, or

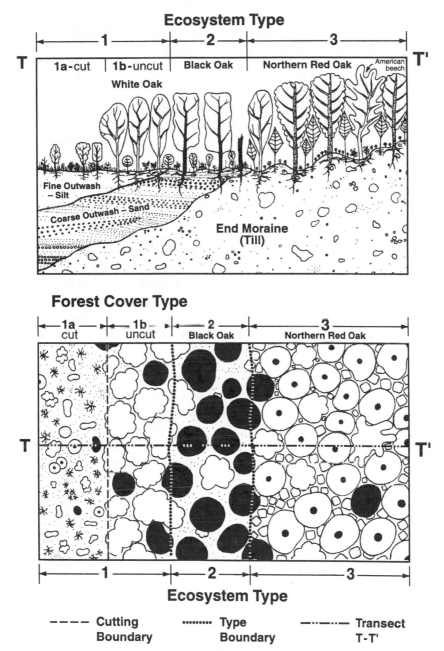

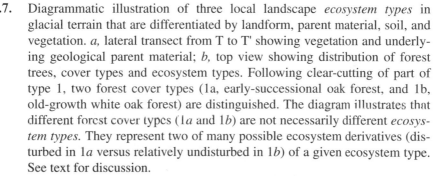

Figure 1.7. Diagrammatic illustration of three local landscape *ecosystem types* in glacial terrain that are differentiated by landform, parent material, soil, and vegetation. *a,* lateral transect from T to T' showing vegetation and underlying geological parent material; *b,* top view showing distribution of forest trees, cover types and ecosystem types. Following clear-cutting of part of type 1, two forest cover types (1a, early-successional oak forest, and 1b, old-growth white oak forest) are distinguished. The diagram illustrates that different forest cover types (1*a* and 1*b*) are not necessarily different *ecosystem types.* They represent two of many possible ecosystem derivatives (disturbed in 1*a* versus relatively undisturbed in 1*b*) of a given ecosystem type. See text for discussion.

ecosystem processes (including changed browsing pressure by herbivores) were caused by clearcutting. If a wildfire had just burned through another part of type 1, we could identify a new cover type, and as ecosystem function changed, another derivative ecosystem. Thus a given local ecosystem type, defined by relatively stable features of physiography and soil, may have a suite of disturbance-induced cover types and derivative ecosystems (Simpson et al., 1990). In Figure 1.7, the suite of ecosystems for ecosystem type 3 would be markedly different than those of either type 1 or 2 and related strongly to the different disturbance regimes of fire and windstorm as well as different physiographic and soil conditions. Therefore, mapping forest cover types is not necessarily likely to provide a useful estimate of site potential for management or conservation. However, cover types are extremely useful in management planning for wildlife, timber, water, and recreational use of existing forest communities.

It is often the conspicuous forest cover type that receives our immediate attention. However, the enormous complexity of earth space and changes in its component ecosystems through time require that major attention be directed to atmospheric, geologic, physiographic, and soil properties of forest landscapes. In summary, for every landscape, a combination of factors should be used to distinguish the pattern of local and regional landscape and waterscape ecosystems that have similar ecological potential in the long run. Sometimes it may be enough to recognize that these core units occur, or a taxonomic key may be provided. Typically however, mapped units at multiple spatial scales (Chapters 2, 13, 21) are needed to provide the basis for monitoring ecosystem change over time and for ecosystem management.

AN APPROACH TO THE STUDY OF FOREST ECOLOGY

The scope of forest ecology is in part synthesis—understanding the big picture, the ecosystem framework within which the components (atmosphere, earth, organisms) are integrated. It is also, in part, analysis—taking things apart and examining the components *but remembering that these bits and pieces are parts of wholes*. This chapter has provided an introduction to thinking in terms of wholes, the Ecosphere and its regional and local ecosystems whose components—landforms, air, water, soil, organisms—have evolved together and not separately. Based on this "top-down" perspective, we can then examine the organic and inorganic parts of ecosystems, their interactions, and the structure and function of entire ecosystems.

The sequence of presentation is: Part I, the big picture of forest ecosystems in space and time; Part II, the forest tree and its variation, life history, and structure; Part III, the forest environment; Part IV, forest communities; Part V, ecosystem dynamics; and Part VI, forests of the world. The inseparable nature of ecosystem components has been repeatedly emphasized by many authors (Herbertson, 1913, reprinted 1965; Tansley, 1935; Lindeman, 1942; Rowe, 1961a), and an explicit separation and analysis of the biotic and physical parts is not possible or desirable. Nevertheless, these divisions seem reasonable to provide both analysis and synthesis.

In Part I, the Earth and its sectoral ecosystems at multiple spatial scales are the primary objects-of-interest. Ecological literacy begins first in understanding ecosystems as large volumetric segments of the Ecosphere within which smaller volumetric units are nested (Chapter 2). Next, change in ecosystems is considered over a temporal scale of millions to hundreds of years (Chapter 3). Primary bases for this change come from geologic and physiographic studies of the land itself and determination of plant species and their communities as deduced from the paleoecological record. Here the ecological historian must play the part of the detective, for written records are few and scanty (Williams, 1989; Whitney, 1994), and precise measurements in the past are almost nonexistent. Fortunately, developments in the reconstruction of past forests through analysis of fossil pollen accumulations, in the dating of fossil organic matter through radioactive carbon dating, and in precise tree ring studies are all increasing our knowledge and understanding of the past. This analysis leads to a consideration of the present-day occupancy of regional ecosystems by gymnosperms and angiosperms and, in turn, to the examination of forest organisms themselves in Part II.

The forest tree owes its appearance, rate of growth, and size, in large part, to the environment in which it has grown throughout its life. Put in modern language, the **phenotype,** the individual as it appears in the forest, is the product of the environment acting on its **genotype**, its individual genetic constitution. An understanding of forest trees as parts of ecosystems is based, in part, upon the environment of the forest and the way it affects the biota making up the forest and their interactions. But it also depends, in part, upon the genetics of the trees themselves and the way in which their genetic heritage affects their response to the environment they live in. These genetic and environmental interactions at the individual and species level are discussed in Chapter 4 of Part II. In addition, life history features of reproduction, regeneration, as well as anatomical and physiological aspects of forest trees in relation to environment are considered (Chapters 5 and 6). Physiological processes related to photosynthesis, respiration, and growth are included chapters on light, temperature, carbon balance, and nutrient cycling (Chapters 8, 9, 18, and 19, respectively).

The **site** is the sum total of environmental factors surrounding and available to the plant at a specific geographic place. These factors, treated in Part III (Chapters 7-13) are primarily the atmospheric, physiographic, and soil components of the physical environment, but also include important influences of plants and animals (biotic factors). The growing vegetation itself affects the microclimate and soil so that the site conditions change literally as the plants themselves respond and grow.

Among atmospheric factors, solar radiation, air temperature, precipitation, humidity, wind, and carbon dioxide content all vary throughout the day, from day to day, month to month, and year to year, making up the complex we term **climate.** Of the many climatic factors, special consideration is given to **sunlight** (solar radiation) and **temperature**. **Water** is of universal importance in forest ecosystems and as such cannot be treated in a single chapter. We examine water relations and dynamics throughout the text in appropriate contexts: growth (Chapter 6), climate and plant distribution (Chapter 7), soil water (Chapter 11), carbon balance (Chapter 18), and nutrient cycling (Chapter 19).

Physiography of the terrain plays a key role in affecting not only climate but the amount and rates at which radiation and moisture are received and distributed in forest ecosystems. Below the ground surface, the supply of soil moisture and nutrients, the physical structure of the soil, microbial communities in the soil, and the nature and decomposition pattern of the organic matter, i.e., factors pertaining to the **soil–edaphic** factors, all affect the growth and development of plants. Fire, the dominant fact of forest history, is also treated here as a site factor. The study of the environmental factors and their effects on individual plants constitutes the field of **autecology.** Thus in this part, we summarize those aspects of forest autecology most pertinent to an understanding of forest ecosystems. The concluding chapter of Part III, treating site quality and its evaluation, synthesizes the foregoing material on site factors by focusing on the degree to which individual site factors and combinations of them are used to estimate the productivity of and determine management prescriptions for forest ecosystems.

Forest communities and their biota of plants and animals are considered in Part IV as integral parts of ecosystems. First, we examine the important roles of animals in affecting all phases of plant development, as well as their effects on forest communities and ecosystem processes. Forest communities are then treated, emphasizing their composition, occurrence, and the interactions (competition and mutualism) of their constituent individuals.

Forest ecosystems are ever changing, and their dynamics are treated in Part V. The physical environment and the forest community are interacting components of forest ecosystems that are operationally inseparable. Thus one ultimately deals with ecosystem function in emphasizing the interdependence and causal relationships of plants and animals in their physical environment. Disturbance and succession as ecosystem processes are treated first followed by chapters on whole-ecosystem functioning, including the carbon balance of trees and ecosystems and the dynamics of nutrients. Also included are diversity and landscape ecology—two areas of intense interest and activity that illustrate the dynamic properties of ecosystems.

Part VI brings together information on the origin and occurrence of present forest ecosystems of the world. Temperate and tropical forests are examined, emphasizing ecological contrasts among species and ecosystems.

APPLICABILITY IN ECOSYSTEM MANAGEMENT

The delineation of forest ecosystems as spatial units and an understanding of forest ecosystems along with the functioning of their interconnected parts are the objective bases of ecosystem management (Christensen et al., 1996). Management may include broad land-use decisions at the scale of regional ecosystems or manipulation of components at the local ecosystem level. Regardless of the scale, consideration is given to whole systems and their long-term sustainability and integrity (Lubchenco et al., 1991; Christensen et al., 1996)—not just use of their parts such as recreation, wildlife, or timber. Although knowledge and understanding of ecosystems for their own sake may be sufficient justification for the ecologist, forest ecology is taught in professional schools of natural resources, forestry, and environmental science because of its importance in conservation and management practices. In the past, applied forest ecology, or **silviculture,** was conceived in terms of vegetation management. Spurr (1945) defined silviculture as the theory and practice of controlling forest establishment, composition, and growth. Certainly, it was understood that the silviculturist's actions were more far-reaching than just the forest stand. Nevertheless, we explicitly emphasize that for the 21st century silviculture is the theory and practice of controlling forest ecosystem composition, structure, and function (Barnes, 1996). Even local management operations may have significant and often long-term effects on surrounding landscapes and on aquatic systems far removed from the specific site of human activity.

An understanding of the structure and function of forest ecosystems and the autecology of their organisms (i.e., forest ecology) is the basis for managing, conserving, and restoring landscape ecosystems. Never has there been a time when ecological understanding of whole systems was more important. Deforestation has occurred for centuries (Perlin, 1989), and it continues today with a myriad of ecological consequences, including habitat destruction and decreased diversity and extirpation of plants and animals; erosion, landslides, and mudflows; siltation of rivers and streams; flooding, and local climatic changes. Outside the forest, human activities associated with industrialization, such as air pollution, have resulted in local tree mortality and, in certain ecosystems, widespread death and decline of forest trees and other organisms. Increasing carbon dioxide concentrations in the atmosphere due to the burning of fossil fuels and deforestation may lead to climatic changes and concomitant changes in forest ecosystems. In the public sector, we have experienced a revolution in forestry, a new paradigm (Rowe, 1994; Kohm and Franklin, 1997) that is changing the view of the forest—from that of trees and stands with the goal of single commodities (wood production, wildlife, water) to forestry understood as sustaining ecosystems and their structure, diversity, and function. New approaches are being applied by managing human uses to ensure ecosystem sustainability. These include managing seminatural lands in perpetuity from an ecosystem perspective, ecological restoration of exploited lands, provision for old-growth ecosystems, variable retention harvesting systems, ecological management of artificial ecosystems, and regional ecosystem management to maintain ecosystem and biological diversity.

In our age, maintaining the integrity of ecosystems across the landscape, e.g., sustaining natural ecosystem processes, is more important than production of single crops, be they wood, wildlife, or water. Concerning forestry, Kohm and Franklin (1997) note: "If 20th-century forestry was about simplifying systems, producing wood, and managing at the stand level, 21st-century forestry will be defined by understanding and managing complexity, providing a wide range of ecological goods and services, and managing across broad landscapes." The landscape ecosystem approach is a new way of sensing the world, a reevaluation of the Ecosphere and of our place in it, an ethical way of living in and with it (Rowe, 1992b). The essence of the ecosystem approach is maintenance of landscapes *as complete ecosystems*, because the only way to assure the sustained yields of forests, wildlife, and water, now and in the future, is to maintain the integrity of their processes and keep their biota in a healthy state.

Suggested Readings

Bailey, R. G. 1996. *Ecosystem Geography*. (Chapters 1 and 10) Springer-Verlag, New York. 204 pp. + 2 maps.

Christensen, N. L., A. M. Bartuska, J. H. Brown, S. Carpenter, C. D'Antonio, R. Francis, J. F. Franklin, J. A. MacMahon, R. R. Noss, D. J. Parsons, C. H. Peterson, M. G. Turner, and R. G. Woodmansee. 1996. *Ecological Applications* 6:665–691.

Cole, L. C. 1958. The ecosphere. *Sci. Amer.* 198:83–92.

Golley, F. B. 1993. *A History of the Ecosystem Concept in Ecology*. Yale Univ. Press, New Haven, CT. 254 pp.

Hagen, J. B. 1992. *An Entangled Bank*. Rutgers Univ. Press. New Brunswick, NJ. 245 pp.

Kohm, K. A., and J. F. Franklin (eds). 1997. *Creating a Forestry for the 21st Century*. Island Press, Washington, D.C. 475 pp.

Perlin, J. 1991. *A Forest Journey, The Role of Wood in the Development of Civilization*. Harvard Univ. Press, Cambridge, MA. 445 pp.

Rowe, J. S. 1961. The level-of-integration concept and ecology. *Ecology* 42:420–427.

_____. 1989. The importance of conserving systems. In M. Hummel (ed.). *Endangered Spaces: The Future for Canada's Wilderness*. Key Porter Books, Ltd. Toronto.

_____. 1992. The ecosystem approach to forestland management. *For. Chron.* 68:222–224.

_____. 1994. A new paradigm for forestry. *For. Chron.* 70:565–568.

Tansley, A. G. 1935. The use and abuse of vegetational concepts and terms. *Ecology* 16:284–307.

PART 1

Ecosystems at Multiple Spatial and Temporal Scales

A basic goal of the forest ecologist is to understand the dynamics of landscape ecosystems, their structure and function. In the context of an ecology course or in solving ecological problems one should come to conceive ecosystems at several spatial and temporal scales. Gaining such understanding is a formidable task, and it has been observed that the environmental complex is unknowable and inexpressible! Nevertheless, synthesis is possible by conceiving landscapes as ecosystems and proceeding "from above," that is, from the biggest to the smallest units to understand their spatial relationships. Although local sites and stands with their familiar species appear a convenient starting point, we know that these are, in turn, affected by geological and climatic factors of higher levels within which they are embedded. Like Chinese boxes or Russian dolls stacked within one another, landscape ecosystems are nested within one another in a hierarchy of spatial sizes. Thus it is best to consider the big picture first—the comprehensive view from the outside—rather than the minutia of details as seen from inside forest ecosystems. And it is best as well to think in terms of whole ecosystems rather than their complementary parts, biotic and abiotic, that too often claim our complete attention. Thus our perspective in understanding ecosystems involves establishing a framework for studying the components—a framework of the spatial, hierarchical pattern of ecosystems of all sizes making up the ecosphere.

Similarly, a consideration of multiple temporal scales is important because components of ecosystems such as climate, soil, and vegetation change over time and because ecological processes may proceed at different time scales; from hours and days to millions of years. Therefore in succeeding chapters, before becoming immersed in the detail of organisms, sites, and their interrelationships, we wish to examine landscape ecosystems in a perspective of spatial levels and their processes at different temporal scales. In Chapter 2, we present the concept of scale and consider the hierarchy of landscape ecosystems. In Chapter 3, ecosystem change through time is considered, emphasizing the occurrence and rate of change for vegetation as a basis for the variation, life history, and ecology of individual organisms that follows in Part 2.

CHAPTER 2

LANDSCAPE ECOSYSTEMS AT MULTIPLE SPATIAL SCALES

The forest ecologist is often a landscape ecologist—a person whose domain of interest and practice necessarily encompasses landscapes of many different scales. A landscape is a heterogeneous land area composed of a group of intermeshed ecosystems, each with interacting atmosphere, earth, and biota. Landscape ecology is the study of ecosystems and of their relationship to one another, for each is set in the "environment" of its surrounding neighbors (Rowe, 1988). In other words, landscape ecology considers the spatial nature of ecosystems, focusing on their spatial patterns and functional interrelationships. In this chapter, we examine the spatial occurrence and complexity of landscape ecosystems, whereas the discipline of landscape ecology is considered in Chapter 21.

OVERVIEW OF SPATIAL AND TEMPORAL SCALES

Forest ecologists ask questions, conduct field experiments, and resolve problems at different scales of spatial resolution. The problems posed, the questions asked, and the techniques used necessarily differ from scale to scale, from very broad to fine, from regional to local. Processes occurring in landscape ecosystems at a given scale are influenced and constrained by a hierarchical set of controlling factors. For example, the effects of microclimate on forest composition and growth depend on the macroclimate of a given region. As a framework for considering scale relationships, the hierarchical model in Figures 2.1 and 2.2 illustrates four hierarchical spatial and temporal scales (Delcourt and Delcourt, 1988). Within the micro-time scale (within 500 years), the micro- and meso-spatial scales are the primary domains of interest for the landscape manager, forester, wildlife ecologist, and naturalist. Events at the micro-spatial scale take place within a spatial dimension of 1 m^2 to 10^6 m^2 (0.01 to 100 ha; or up to 1 km^2 = 0.3861 mi^2). For the meso-scale, the di-

SCALE PARADIGM

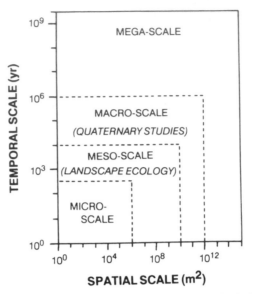

Figure 2.1. General spatial-temporal scales for the hierarchical characterization of disturbance regimes, ecosystem and biotic responses, and vegetation units. Quaternary studies at the macro-scale refer to investigations of species migrations and extinctions in the time frame of 10,000 yr to 1,000,000 yr in the Quaternary Period (see Chapter 3) on a continental scale. (After Delcourt and Delcourt, 1988.)

mensions range from 1 km^2 to 100 km^2 (10^6 to 10^{10} m^2; 100 to 10,000 ha). It is at these micro- and meso-spatial scales that seasonal patterns of precipitation and temperature, as well as longer-term weather trends and climatic fluctuations lasting from decades to centuries, affect the distribution, growth, and interactions of plants and animals. Local to widespread disturbances of relatively short duration, such as wildfire, windthrow, insect and disease outbreaks, and clear-cutting have immediate effects on ecosystem processes and community composition (Figure 2.2). These disturbances along with geomorphic processes (soil creep, debris avalanches, stream transport and deposition of sediments) affect vegetation at levels from individual plants to large forest stands. They occur on areas extending from the size of sample plots (m^2) and stream watersheds (ha) to mountain ranges (km^2).

The macro- and mega-scales (Figures 2.1 and 2.2) extend in space and time to include not only events of ecosystem change and plant migration occurring at the level of physiographic regions (macro-scale), but also plate tectonics and evolution of biota on the spatial scale of continents and the Ecosphere (mega-scale). As a yardstick, areas the size of Alaska (1,518,808 km^2 = 586,412 mi^2) or larger would be mega-scale, whereas areas the size of the state of Texas (692,409 km^2 = 267,339 mi^2) and all other American states would fall in the macro-scale.

The disturbance regimes, ecological responses, and vegetation units that are of concern to the forest ecologist occur over a range of spatial scales in the hierarchical framework shown in Figure 2.2. Disturbance events, such as fire and pathogen outbreaks, may occur at a local site or over thousands of square kilometers. Wind may uproot a single tree in old-growth forest or may affect huge land areas for much longer intervals, such as the

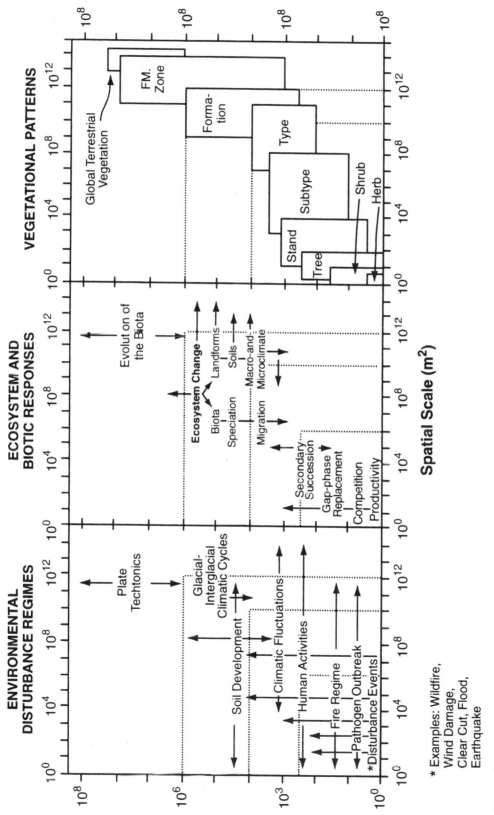

Figure 2.2. Disturbance regimes, ecosystem and biotic responses, and vegetational units viewed in the context of four spatial-temporal scales shown in Figure 2.1. (Modified from Delcourt and Delcourt, 1988.)

23

18,211 km^2 (1.8 million ha) of forest land damaged in South Carolina by hurricane Hugo in 1989 (Sheffield and Thompson, 1992). Vegetational change or succession is a good example of an ecosystem process that may be examined at different spatial and temporal scales. For example, along a continuum from fine to very broad scales in space and time, vegetation change may occur in just a few years in small tree-fall gaps, as widespread secondary succession over decades or centuries, or as evolution over millennia on a continental spatial scale. Flooding may be quite localized in its effects along the Mississippi and tributary rivers. However, the flood waters of 1993 extended into the mega scale of at least one million square kilometers. Although multiple scales of disturbance regimes, biotic responses, and vegetational units are emphasized in Figure 2.2, entire terrestrial and aquatic ecosystems themselves can be conceived, distinguished, and mapped at multiple hierarchical scales. This theme is examined in the sections that follow.

HIERARCHICAL ECOSYSTEM SCALES IN SPACE

The Ecosphere is the largest and most all-inclusive ecosystem. Nested within the Ecosphere and composing it are the volumetric segments, i.e., landscape ecosystems of several levels. Three general levels may be identified (Rowe and Sheard, 1981; Rowe, 1992b; Bailey, 1996). The **macro-level** ecosystems are those we identify as seas and continents and their major units. This level would essentially encompass the macro- to mega-scales shown in Figure 2.2. Within continents at a lower level are **meso-level** ecosystems (mesospatial scale), characterized by major physiographic features, including broad plains, rolling hills, and mountain ranges. Within these are the **micro-level** ecosystems (microspatial scale) that are visible as local landforms supporting upland forests, swamps, and lakes. Considering these ecosystem units from the "top-down," one proceeds from the more complex and heterogeneous to less complex and relatively homogeneous. The three levels cited above are arbitrary; the complexity of the landscape itself and the questions being asked determine the number of ecosystems one may wish to distinguish.

Ecosystems at each level, whether at the broad regional scale (macro- and meso-ecosystems) or at a microecosystem level, include all the essentials of climate, physiography, soil, and organisms. Ideally, we would like to delineate and map landscape ecosystems by integrating all these components at all levels. Although desirable, this formidable task has not yet been accomplished except at the local or micro level (Barnes et al., 1982; Pregitzer and Barnes, 1984; Spies and Barnes, 1985a; Simpson et al., 1990). However, single-component classifications of ecosystems by climate, vegetation, or physiography not only illustrate the spatial pattern of the respective components, but provide a starting point for understanding ecosystems at the broad scales. Despite the incompleteness of any single factor, examples of these approaches examined as follows provide an entree to the complexity of landscape ecosystems at multiple scales.

Climatic Classification

At the broadest levels a climatic classification, such as the "eco-climatic zones" of Bailey (1988; Figure 2.3), illustrates the pattern of world climates. It is one basis for distinguishing macro- and meso-level ecosystem units to which other components can be linked. Because climate strongly controls genetic differentiation of organisms and hence their distribution in space and time, its importance is unquestioned. In Figure 2.3, each of the 13 eco-climatic zones, adapted from the climatic classification of Köppen (1931) and Trewartha (1968), is characterized by a particular climatic regime. Each climatic zone essen-

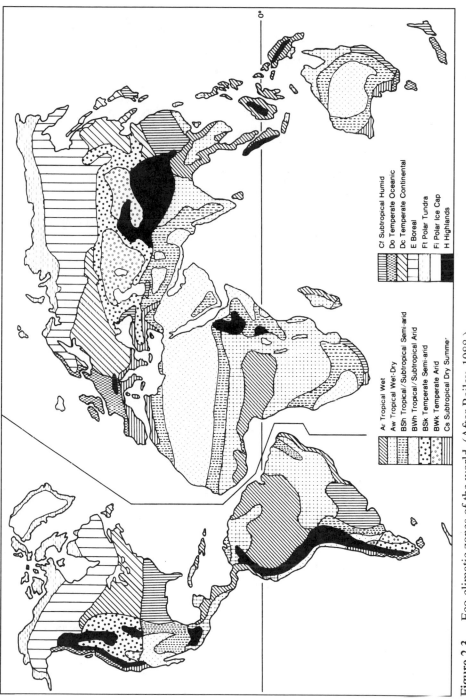

Figure 2.3. Eco-climatic zones of the world. (After Bailey, 1988.)

Table 2.1   Zonal relationships between macroclimate, soil orders, and vegetation.[1]

Eco-climatic zone[2]	Soil order[3]	Vegetation
Ar	Oxisols (Latisols)	Evergreen tropical rainforest (selva)
Aw	Oxisols (Latisols)	Tropical deciduous forest or savanna
BS	Mollisols, Aridisols (Chestnut, Brown soils, and Sierozems)	Shortgrass
BW	Aridisols (Desert)	Shrubs or sparse grasses
Cs	Entisols, Inceptisols (Mediterranean brown earths)	Sclerophyllous woodlands
Cf	Ultisols (Red and Yellow Podzolic)	Coniferous and mixed coniferous–deciduous forest
Do	Alfisols (Brown Forest and Gray-brown Podzolic)	Coniferous forest
Dc	Alfisols (Gray-Brown Podzolic)	Deciduous and mixed coniferous–deciduous forest
E	Spodosols and associated Histosols (Podzolic)	Boreal coniferous forest (taiga)
Ft	Entisols, Inceptisols, and associated Histosols (Tundra humus soils with solifluction)	Tundra vegetation (treeless)

Source: After Bailey, 1988.

[1]After Walter (1984, p. 3)

[2]Abbreviations of the zones are shown in Figure 2.3.

[3]Names in parentheses are soil orders (USDA Soil Conservation Service, 1975).

tially corresponds to a soil order and a late-successional type of vegetation (Table 2.1). In Figure 2.3, we see the climatic zones stretching across the continents. High mountains occur within many zones. These may be differentiated into a vertical stack of eco-climatic zones or sequences of altitudinal belts that are related to elevation, slope aspect, and associated climatic and soil characteristics (Figure 2.4). The sequence of altitudinal belts is not always the same, but differs according to the geographic position of the zone in which it is located. Thus around the world where high mountains occur, we observe a horizontal sequence of eco-climatic zones within which a vertical zonation of terrain ecosystems occurs as well.

In this example, climate formed the primary basis for classification of the mountainous landscape into horizontal and vertical zones at the macro level. However, gross physiographic features and vegetative cover play a major role in affecting climate both horizontally and vertically at this continental scale. At the meso-scale, Bailey (1988, 1996)

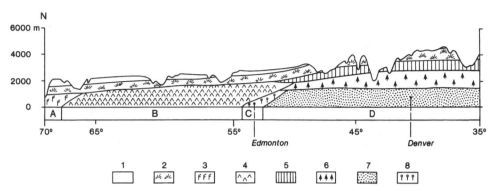

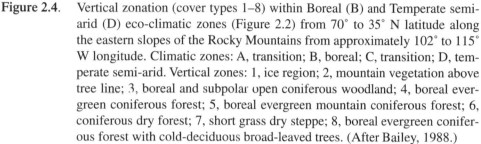

Figure 2.4. Vertical zonation (cover types 1–8) within Boreal (B) and Temperate semi-arid (D) eco-climatic zones (Figure 2.2) from 70° to 35° N latitude along the eastern slopes of the Rocky Mountains from approximately 102° to 115° W longitude. Climatic zones: A, transition; B, boreal; C, transition; D, temperate semi-arid. Vertical zones: 1, ice region; 2, mountain vegetation above tree line; 3, boreal and subpolar open coniferous woodland; 4, boreal evergreen coniferous forest; 5, boreal evergreen mountain coniferous forest; 6, coniferous dry forest; 7, short grass dry steppe; 8, boreal evergreen coniferous forest with cold-deciduous broad-leaved trees. (After Bailey, 1988.)

identifies physiography (specific landforms and their surficial form and parent material) as the key factor upon which to build the next finer, more-discerning level of landscape classification. Physiographic features can markedly affect both macro- and microclimate, both of which significantly affect the distribution and composition of vegetation. Because vegetation is fascinating in its complexity of composition and structure and is easier to describe and map than physiography or climate, it has been the traditional criterion for distinguishing broad landscape units. This approach is examined next.

Vegetation Types and Biomes

A classification of vegetation, its physiognomy (form) and species composition may serve to delineate the extent and boundaries of macro- and meso-level ecosystems. Because vegetation visibly integrates the effects of other ecosystem components, it provides a useful approach in distinguishing macro- and meso-level ecosystems. Note however that any ecosystem component taken singly—vegetation, climate, soil, landform—is an incomplete indicator of the whole.

At the very broad continental scale, macro-level ecosystems would roughly coincide with the **formations** (the macro-scale in Figure 2.1), plant community types, life zones (Merriam, 1890), biotic provinces (Dice, 1943), or **biomes** that are traditionally recognized by biogeographers to reflect macroclimatic regimes and major physiographic discontinuities. The term, formation, is applied to macro-level plant communities, but the term biome (biotic region) is more commonly used because both plants and animals are interrelated parts of these units. A general model of the world's biomes (Tallis, 1991), illustrating their strong relation to macroclimate, is shown in Figure 2.5.

Thus the world's major biotic regions, shown in Figure 2.6 (Dasman, 1959), can be used to indicate the location and extent of the Earth's macro-level ecosystems conceived as extensions of the biota (bioecosystems). Similar diagrams and descriptions of these major units are included in many textbooks that emphasize this plant-community-based

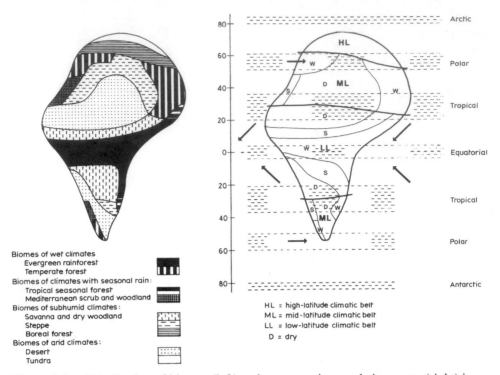

Biomes of wet climates
  Evergreen rainforest
  Temperate forest
Biomes of climates with seasonal rain:
  Tropical seasonal forest
  Mediterranean scrub and woodland
Biomes of subhumid climates:
  Savanna and dry woodland
  Steppe
  Boreal forest
Biomes of arid climates:
  Desert
  Tundra

HL = high-latitude climatic belt
ML = mid-latitude climatic belt
LL = low-latitude climatic belt
D = dry

Figure 2.5.   Distribution of biomes (left) and source regions and air masses (right) in relation to a generalized land mass. Source regions are shown shaded; arrows
show direction of prevailing winds. HL = high-latitude climatic belt; ML =
mid-latitude climatic belt; LL = low-latitude climatic belt; D = dry; W =
wet; S = seasonal precipitation. (After Tallis, 1991. Reproduced with permission from *Plant Community History* by Chapman and Hall.)

approach. Detailed descriptions of the major vegetation types shown in Figure 2.6, and
vegetation of lower levels as well, are provided by several authors (Vankat, 1979; Rowe,
1972; Barbour and Billings, 1988; Barbour and Christensen, 1993). In addition, books in
the series, Ecosystems of the World (for example, *Tropical Rain Forest Ecosystems* (Golley, 1983), *Temperate Broad-Leaved Evergreen Forests* (Ovington, 1983), *Temperate Deciduous Forests* (Röhrig and Ulrich, 1991) provide comprehensive accounts of many forest communities and ecosystems.

The use of formations or biomes at the macro-level represents a grouping of ecosystems based on similar plant **physiognomy**, the form and structure of plant communities.
Plant physiognomy is a direct response to environment, for example the temperate and
tropical rain forests of favorable environments in contrast to tundra and sclerophyllous
scrub of harsh climates. Macroclimate (general climate over a large geographical area, including macro- and mesoecosystem levels) is often perceived as the primary environmental factor controlling the form, vertical layering, and composition of these broad units.
However, physiographic features and vegetation at all scales, from mountain ranges to
local ridges and depressions, influence climate and hydrology especially where abrupt
discontinuities markedly affect the exchanges of energy and materials at the surface (Hare
and Ritchie, 1972). In Figure 2.6, we observe differences in the vegetation of the tundra,
coniferous forests, deciduous forests, grassland, and tropical rain forest. The biomes re-

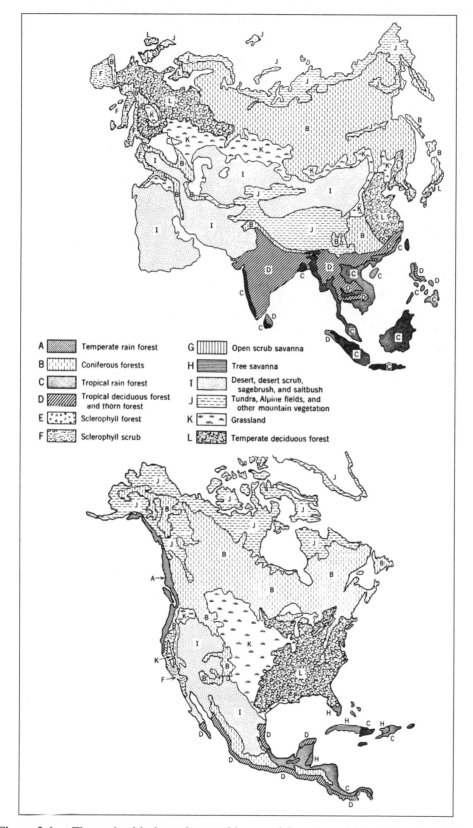

Figure 2.6. The major biotic regions or biomes of Eurasia and North America. (After Dasman, 1984. Reproduced by permission from *Environmental Conservation*, 5th ed., © by John Wiley & Sons, Inc. Reprinted by permission of John Wiley & Sons, Inc.)

flect, to a certain degree the "ecoclimatic zones" of Figure 2.3, but we also observe many differences by comparing these figures.

Physiography

The configuration of the Earth's surface and associated geologic substance or parent material, i.e., **physiography**, is perhaps the single most useful component in distinguishing landscape ecosystems at any scale. Physiography, with its relatively stable landforms, controls climatic regimes at all scales. The Earth's structures and forms mediate the fluxes of energy and materials at the Earth's surface where atmospheric and Earth layers meet. Furthermore, the vegetation often corresponds to the landform and to the soil that develops from the parent material. Thus a classification of the physiographic regions of North America (Figure 2.7) provides a useful basis for distinguishing macro- and meso-level ecosystems. Although there is reasonable correspondence among physiographic regions, climatic zones (Figure 2.3), and biotic regions (Figure 2.6), the long north–south dimension of several western regions (e.g., Pacific Border System, Interior Mountains and Plateaus System, Great Plains Province, among others) requires modification provided by the climatic and biotic viewpoints before it can appropriately be viewed as an ecosystem classification.

Within eastern North America the hierarchical physiographic classification by Fenneman (1938) provided the basis for vegetation classification by Braun (1950) and eco-region classification by Omernick (1987). Braun used Fenneman's physiographic classification to provide the landscape framework for her classification of deciduous forest regions and sections of the eastern United States (Figure 2.8). The match-up in boundaries of the broad regional geologic units with major climatic regions was also good (Bryson and Hare, 1974). For example, the macroclimatic conditions of Braun's forest regions were compared by Lindsey and Sawyer (1970) and Lindsey and Escobar (1976) using the Holdridge (1947, 1967) model of world bioclimatic formations or life zones (see Chapter 7). Eleven hundred U.S. Weather Bureau Stations were classified according to their location within eight of Braun's forest regions. Their position, based on biotemperature (sum of average monthly temperature above 0°C divided by 12), precipitation, and the ratio of potential evapotranspiration/precipitation (E/P), was plotted on a three-axis graph (Figure 2.9). For the most part the physiographically-based forest regions are well separated. A close similarity is seen between the Mixed Mesophytic and Oak–Chestnut regions and also is indicated by the similarity of canopy species composition of these regions at low and moderate elevations. This example illustrates that physiography is an excellent basis for delineating landscape ecosystems at regional scales. When coupled with the distribution of pre-settlement or near-natural vegetation, a useful classification and map may be developed.

In summary, although classifications and maps of the individual components of physiography, climate, and vegetation are useful, there is nevertheless a lack of specific correspondence of boundaries. Even at broad scales, differences should be expected, due, in part, to the different assumptions inherent in map portrayal as well as to differences in the map scale of resolution used. Landscape ecosystems are volumetric units of many interrelated components. Therefore, the delineation of regional ecosystems, or ecosystems at any spatial scale, is best accomplished by the simultaneous integration of physiographic features with those of climate, soil, and vegetation. Mapping necessarily depends on visible features such as what can be seen on air photos, as well as on the ground, i.e., landform and vegetation. Fortunately, the ecological relationships of these components provide useful insights to less observable ecosystem components: the mobile animal community, the opaque soils, the hydrologic regime, and the transparent and continuously shifting weather regimes that comprise climate.

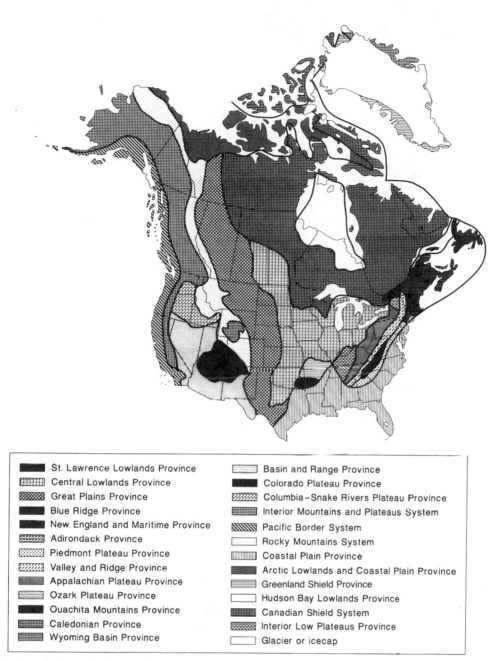

St. Lawrence Lowlands Province	Basin and Range Province
Central Lowlands Province	Colorado Plateau Province
Great Plains Province	Columbia–Snake Rivers Plateau Province
Blue Ridge Province	Interior Mountains and Plateaus System
New England and Maritime Province	Pacific Border System
Adirondack Province	Rocky Mountains System
Piedmont Plateau Province	Coastal Plain Province
Valley and Ridge Province	Arctic Lowlands and Coastal Plain Province
Appalachian Plateau Province	Greenland Shield Province
Ozark Plateau Province	Hudson Bay Lowlands Province
Ouachita Mountains Province	Canadian Shield System
Caledonian Province	Interior Low Plateaus Province
Wyoming Basin Province	Glacier or icecap

Figure 2.7. Physiographic regions of North America and their regional landforms. (After Brouillet and Whetstone, 1993. From *Flora of North America: North of Mexico,* Vol. 1: Introduction by the Flora of North America Editorial Committee; Nancy R. Morin, Convening Editor: Copyright © 1993 by the Flora of North America Association. Reprinted by permission, Oxford University Press, Inc.)

DISTINGUISHING AND MAPPING LANDSCAPE ECOSYSTEMS AT MULTIPLE SPATIAL SCALES

In progressively dividing the Ecosphere into a hierarchical series of ecosystems, we can envision a nested series of units fitting together like Chinese boxes. This geographic division of landscapes from the top-down is termed **regionalization**. Such division provides the basis for a purposive, logical grouping or **classification** of appropriately-sized ecosys-

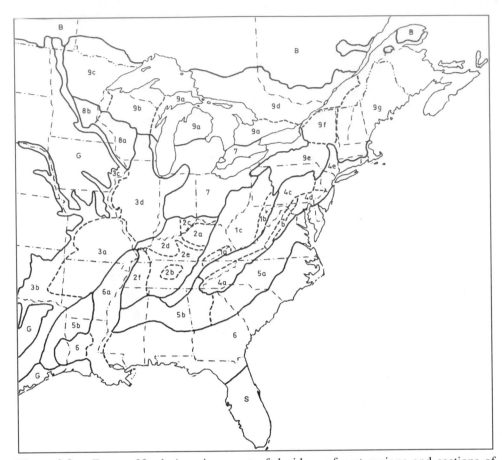

Figure 2.8. Eastern North American map of deciduous forest regions and sections of the eastern and mid-western United States. Formation, region and section names: B = boreal or spruce–fir forest formation; G = grassland or prairie formation; S = subtropical broad-leaved evergreen forest formation. Regions and sections of the deciduous forest formation: **1** = Mixed Mesophytic forest region: a, Cumberland Mountains; b, Allegheny Mountains; c, Cumberland and Allegheny Plateaus. **2** = Western Mesophytic forest region, sections: a, Bluegrass; b, Nashville Basin; c, Area of Illinoian Glaciation; d, Hill; e, Mississippian Plateau; f, Mississippi Embayment. **3** = Oak–Hickory forest region: Southern division, sections: a, Interior Highlands; b, Forest–Prairie transition area; Northern division, sections: c, Mississippi Valley; d, Prairie Peninsula. **4** = Oak–Chestnut forest region, sections: a, Southern Appalachians; b, Northern Blue Ridge; c, Ridge and Valley; d, Piedmont; e, Glaciated. **5** = Oak–Pine forest region, sections: a, Atlantic Slope; b, Gulf Slope. **6** = Southeastern Evergreen forest region, sections: a, Mississippi Alluvial Plain. **7** = Beech–Maple forest region. **8** = Maple–Basswood forest region, sections: a, Driftless; b, Big Woods. **9** = Hemlock–White Pine–Northern Hardwoods region: Great Lakes–St. Lawrence Division, sections: a, Great Lake; b, Superior Upland; c, Minnesota; d, Laurentian; Northern Appalachian Highland division, sections: e, Allegheny; f, Adirondack; g, New England. (Based on a map of physiographic provinces and sections of Fenneman (1938). After Braun, 1950.)

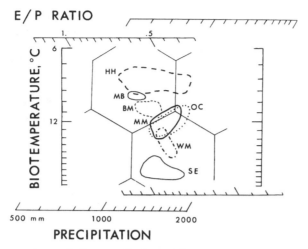

Figure 2.9. Comparison of climate of selected forest regions of Braun (1950) based on precipitation, biotemperature, and ratio of potential evapotranspiration/precipitation (E/P) of representative weather stations. Biotemperature = sum of average monthly temperature above 0° divided by 12. HH = hemlock–white pine–northern hardwoods, MB = maple–basswood; BM = beech–maple; M = mixed mesophytic; OC = oak–chestnut; OP = oak–pine; WM = western mesophytic; SE = southeastern evergreen. The upper left hexagon delimits the "Cool Temperate Moist Forest" of Holdridge (1947, 1967), the upper right bounds the "Cool Temperate Wet Forest," and the lower hexagon delimits the "Warm Temperate Moist Forest." (After Lindsey and Escobar, 1976.)

tems that have practical significance for management and conservation. Not surprisingly, no one standardized scheme will, or can be expected to, encompass the enormous diversity of ecosystem patterns around the world. However, to synthesize current thinking about ecosystem levels, we present a general overview in Table 2.2 using a hierarchical system of units proposed by Klijn and Udo de Haes (1994). The prefix "eco" is used to indicate entire ecosystems in a holistic sense. At left in the table, we show the corresponding general levels of macro-, meso-, and microecosystems that correspond reasonably well with the hierarchical spatial scales (macro-, meso-, micro-) shown in Figures 2.1 and 2.2. Also, the corresponding ecological units of the system in wide use today by the U.S. Forest Service (Avers et al., 1994) is shown in the right-hand column of Table 2.2. Actually, the number and name of ecosystem levels employed around the world varies greatly depending on the size and ecological configuration of the landscape, the purposes for which the classification is developed, and personal preferences for names of the levels.

Recall from Chapter 1 and the above discussion that a key feature of ecosystem geography is that ecosystems are nested within one another. Climate and gross physiography of a given macroecosystem affect properties and processes of each mesoecosystem and their constituent microecosystems differently than those of adjacent macroecosystems. A given management practice in one macro- or mesoecosystem will not necessarily have the same results in a different macro-or mesoecosystem. Thus recognition of ecosystem hierarchy, as shown in Table 2.2 and its expression in ecosystem maps, provides a safeguard against extrapolating management policies and practices too broadly across the landscape.

Table 2.2   Overview of hierarchical ecosystem classification at various spatial scales within the general framework of macro-, meso-, and microecosystem levels.

General Ecosystem Level[1]	Ecosystem Unit Name[2]	Mapping Scale[2]	Size of Basic Mapping Unit[2]	U.S. Forest Service Ecological Unit Name[3]
MACRO-ECOSYSTEM	Ecozone	1: >50,000,000	>62,500 km^2	Domain Division
	Ecoprovince	1: 10,000,000-50,000,000	2,500-62,500 km^2	Province
MESO-ECOSYSTEM	Ecoregion	1: 2,000,000-10,000,000	100-2,500 km^2	Section
	Ecodistrict	1: 500,000-2,000,000	625-10,000 ha	Subsection
	Ecosection	1: 100,000-500,000	25-625 ha	Landtype Association
MICRO-ECOSYSTEM	Ecoseries	1: 25,000-100,000	1.5-25 ha	Landtype and Landtype Phase
	Ecotype	1: 5,000-25,000	0.25-1.5 ha	Landtype Phase
	Eco-element	1: <5,000	<0.25 ha	Landtype Phase

Source: Modified from Klijn and Udo de Haes, 1994.

[1]Correspondence with spatial scales shown in Figures 2.1 and 2.2 (Delcourt and Delcourt, 1988): Macroecosystem level approximates mega- and macro-scales combined; Mesoecosystem level approximates meso-scale; Microecosystem level approximates micro-scale.

[2]After Klijn and Udo de Haes (1994).

[3]U.S. Forest Service National Hierarchical Framework of Ecological Units (Avers et al. 1994).

The increasing use of mapped ecosystem units at multiple scales is encouraging the shift away from managing bits and pieces of ecosystems (species, stands, soils) to a more integrated perspective of the linkages among ecosystems and their components. A consideration of how regional and local ecosystems are delineated are presented next (see also Chapter 13). Also, Robert Bailey's (1996) book, *Ecosystem Geography,* provides a detailed treatment of the theory and principles of ecosystem classification and mapping at multiple scales.

Regional Landscape Ecosystems

The process of delineating regional ecosystems, those at the macro- and mesoecosystem level, has been described in detail (Bailey, 1983, 1988, 1996; Bailey and Hogg, 1986; and Bailey et al., 1985). For the United States, broad-scale maps of ecoregions have been developed by Omernik (1987) and Bailey (1994, 1995). A detailed map and description of regional landscape ecosystems of Michigan, Minnesota, and Wisconsin is also available (Albert, 1995). In Canada, ecological regionalizations began with Halliday's (1937) macro-level forest classification, which was revised by Rowe (1972) as the *Forest Regions of Canada.* Classification at the meso level then followed with the works of Hills (1960) and Wickware and Rebec (1989) in Ontario, Loucks (1962) in the Maritimes, Demarchi et al. (1990), and MacKinnon et al. (1992) in British Columbia. Concern for a Canada-wide system led to a comprehensive and multipurpose classification of seven hierarchical levels from broad "ecozones" to site-specific "ecoelements" (Wicken, 1986).

The upper levels of the regionalization of North America (Bailey and Hogg, 1986) are shown in Figure 2.10. These three hierarchical levels, termed Domain, Division, and Province, are reasonably equivalent to the macroecosystems described above. At the broadest scale, Domains are distinguished by climate: Polar, Humid Temperate, Dry, and Humid Tropical. These correspond only at the broadest level with the "ecoclimatic zones" (Figure 2.3), the biotic regions of North America (Figures 2.6), and the physiographic regions (Figure 2.7). The four Domains differ markedly among themselves, but not surprisingly, they are characterized by great within-Domain spatial heterogeneity of climate, physiography, soil, and vegetation. For example, in the 200-series designations of the Humid Temperate Domain (Figure 2.10) we find temperate rain forests of the Pacific Northwest, sclerophyllous scrub of southern California, grasslands of the Great Plains, deciduous forests of the eastern United States, and coniferous forests of northeastern United States and adjacent Canada (Figure 2.6). The heterogeneity of ecosystem components decreases as the subdivision proceeds to Division and Province and the finer-scale levels of spatial resolution below these. Divisions are delineated on the basis of finer distinctions in climate, and Provinces are distinguished on the basis of potential natural vegetation (Küchler, 1964). In mountainous terrain, a vertical zonation of landscape ecosystems based on elevational belts and associated mountain physiography of ridges, valleys, and slope-aspect orientation of hillsides is superimposed on the geographic framework described previously (cross-hatched areas in Figure 2.10; see also Figure 2.4 and related discussion).

REGIONAL LANDSCAPE ECOSYSTEMS OF MICHIGAN

Using a multiscale, multifactor integrated approach, regional landscape ecosystems of Michigan were determined, described, and mapped at three hierarchical levels, termed Region, District, and Subdistrict (Albert et al., 1986). These mesoecosystem-scale units lie within the Province level of Bailey's ecoregions (1983) (Figure 2.10). Figure 2.11 illustrates these three levels, the first two of which (Region, District) were determined by integrating climatic and physiographic/soil classifications. Because it becomes progressively more difficult to delineate ecosystems at finer levels, especially with the drastically altered remnants of natural (or early settlement) vegetation, detailed analytical methods were employed to develop a climatic classification of Michigan (Denton and Barnes, 1988) and integrate it with physiography, soil, and vegetation. The geographic location of Michigan is unique because it consists of two large peninsulas, perpendicular to one another, that project into the Great Lakes. This positioning dramatically affects and accentuates climatic differences of both peninsulas, and the physiographic setting is reflected in the classification (Figure 2.11). The distinctiveness of warm, vegetationally diverse southern Lower Michigan (Region I, Figure 2.11) and cold Upper Michigan (Region IV, Figure 2.11) is in marked contrast to lake-moderated Regions II and III where District-level ecosystems parallel lakes Huron, Michigan, and Superior. Within Districts, major physiographic features and their associated soil and vegetation were used to identify Subdistricts. Macroclimate is more or less homogeneous within a given District and even more so within a Subdistrict.

District 1 in southeastern Michigan (approximately 56,000 km^2) provides a good example of the spatial pattern of local ecosystems distinguished at the subdistrict level. Sub-

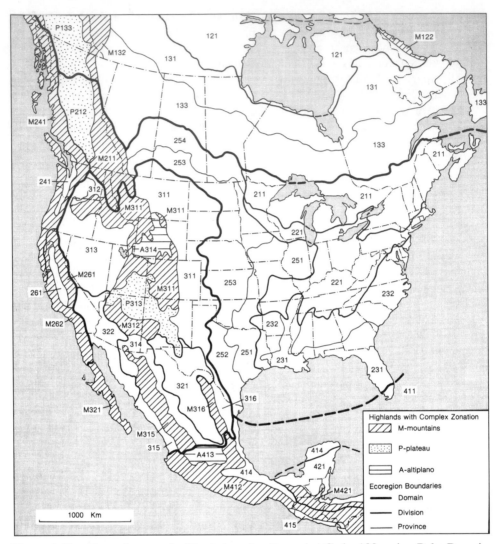

Figure 2.10.   Ecoregion map for North America. Ecoregion Code: 100-series, Polar Domain; 200-series, Humid Temperate Domain; 300-series, Dry Domain; 400-series, Humid Tropical Domain. Key to letter symbols: M = mountains; P = plateau; A = altiplano. (After Bailey and Hogg, 1986. Reprinted from *Environmental Conservation,* © 1986 by the Foundation for Environmental Conservation. Reprinted with permission of the Foundation of Environmental Conservation.)

district 1.1 (shaded area in Figure 2.11) is the Detroit metropolitan heat island that is superimposed on an extensive glacial lake plain of Subdistrict 1.2. The plain extends south into Ohio and north in Michigan where it forms major parts of Districts 5, 6, and 7 (Figure 2.11). This relatively flat plain of lacustrine clay and sandy soil is interspersed with sand beach ridges and adjacent wet depressions. In contrast to the flat lake–plain landform, Subdistrict 1.3 is characterized by large, end-moraine ridges of rolling terrain and fine to medium textured soils. Adjacent to it is Subdistrict 1.4, characterized by hilly ice-contact terrain of kettle (wet or dry depressions) and kame (steep hills) topography with distinctive microclimatic patterns very different from the other subdistricts. Within each of these subdistricts of uniform macroclimate, local landscape ecosystems and their markedly different communities occur in distinctive patterns reflecting those of the diverse landforms.

Figure 2.11. Map of regional landscape ecosystems of Michigan. Three hierarchical levels are mapped: Regions I–IV; Districts 1 to 20, and Subdistricts within many Districts. Region I = southern Lower Michigan, Districts: 1, Washtenaw; 2, Kalamazoo; 3, Allegan; 4, Ionia; 5, Huron; 6, Saginaw; 7; II, northern Lower Michigan, Districts: Arenac; 8, Highplains; 9, Newaygo; 10, Manistee; 11, Leelanau; 12, Presque Isle. III, eastern upper Michigan, Districts: 13, Mackinac; 14, Luce; IV = western upper Michigan, Districts: 15, Dickinson; 16, Michigamme; 17, Iron; 18, Bergland; 19, Ontonagon; 20, Keweenaw. (After Albert et al., 1986.)

Local Landscape Ecosystems

Up to now we have described primarily the delineation of regional ecosystem units—from Domains to Districts and Subdistricts encompassing six hierarchical levels. This process has been one of regionalization; progressive division of the landscape using gross climatic and physiographic features and associated soil and natural vegetation. However, sooner or later we reach a level, and land area, within which the *regional macroclimate is relatively homogeneous*. This key feature, together with similar gross physiography and recurrent patterns of vegetation, distinguish the **local** or **microecosystem level**. Within such an area, major landforms and local landscape ecosystem types occur in a mosaic or complex network of interrelated systems. Delineation of local ecosystems cannot proceed exactly as it did at the broader, regional levels. At these levels, lines are drawn completely around large contiguous areas (as in Figures 2.10 and 2.11) that are distinguished by major discontinuities of macroclimate, gross physiography, and vegetation. But now, within a climatically homogeneous area (Subdistrict level), one proceeds first by recognizing the kinds of **physiographic systems** or major landforms (outwash plain, mountain slope or plateau, and moraine) that recur within it. At these higher levels, landform has a profound influence on topoclimate, soil development, the biotic community, and successional trends. The local ecosystem types (microecosystems) can then be grouped appropriately by these major physiographic or landform complexes, their local landforms (such as floodplain, kame, and ground moraine) and with even more specific site factors of soil texture, nutrient availability, and microclimate (Pregitzer and Barnes, 1984; Spies and Barnes, 1985a; Archambault et al., 1990; Bailey, 1996).

In any method of regionalization of a large land area, the identification of the largest unit of relatively homogeneous macroclimate (i.e., the District level in Figure 2.11) is of the utmost importance. Within such units the extrapolation of macroclimate is of great ecological significance because plants are strongly genetically adapted to climatic factors along gradients of latitude and elevation (Chapter 4). Throughout an area where macroclimate is homogeneous, individual species and groups of plant species (Chapter 13) can be used effectively to indicate different site-factor complexes of water, nutrients, and light (Spies and Barnes, 1985b; Archambault et al., 1989). However, these same species cannot be expected to indicate the same relationships with one another or with other species in areas of different macroclimate (e.g., different Districts in Michigan).

LOCAL LANDSCAPE ECOSYSTEMS IN UPPER MICHIGAN

The complexity and patterned recurrence of local ecosystem types in glacial terrain is illustrated in Figure 2.12 for an area of relatively homogeneous macroclimate and physiography (ice-contact terrain) of the Sylvania Wilderness Area in upper Michigan. The diurnal, monthly, and yearly microclimate of many of these local ecosystems, as well as their soil and vegetation, is markedly different although the macroclimate of the Subdistrict in which they occur is homogeneous. The presence of many lakes, bogs, and coniferous swamps (types 16, 17, 18, 19) and adjacent sandy uplands (hemlock–northern hardwood types 5, 9, 10, 11) illustrates the local complexity of ice-block disintegration features that were formed in late glacial time. Many tiny (< 0.1 ha) depressions of black ash swamp (type 22) dot the entire area. A lake- and swamp-edge ecosystem (type 14), characterized by sandy soil and fire-regenerated pines and white birches, repeatedly occurs along the fire-prone margins of lakes and large swamps. In the unique shoreline landscape position, offshore winds continuously dry the site and coniferous needles of the forest floor, which ignite readily when lightning strikes the tall red and white pines. Geologic processes that

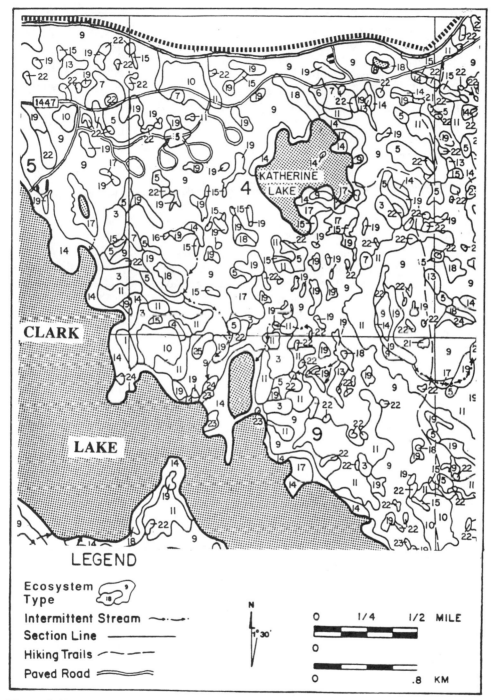

Figure 2.12. Example of the patterns of local landscape ecosystems that occur within the local level. The map illustrates local landscape ecosystem types surrounding part of Clark Lake in the north central part of the Sylvania Wilderness Area, Ottawa National Forest, Upper Michigan. Area shown mapped: 13.9 km^2 (5.4 mi^2); 1,390 ha (3,435 acres).

formed the Sylvania Wilderness have created the setting for a remarkable diversity of ecosystem sizes, shapes, and spatial patterns of wetland, upland, and lake-shore types.

It is the challenge of wildlife ecologists, foresters, and land managers to manage areas with a range of local ecosystems from single ecosystem types less than 0.1 ha to enormously complex patterns of ecosystems extending over thousands of square kilometers. Insights on how to deal with such complexity are considered in Chapter 13 and 21. In this chapter, we have considered the entire suite of ecosystems of different scales from the macroecosystem level of the North American continent to the tiny 0.1 ha microecosystem swamp—all of which are volumetric units of the Ecosphere, nested within one another and with characteristic structure and function.

In the previous sections, we have examined ecosystem diversity at different spatial scales. Ecosystems also change at a given place over time. For example, a mere 10,000 years ago in the upper Michigan area described above, the topographic form and parent material of these sites were little different from that of today. However, the climate, soil, and vegetation were quite different. And over time, ecosystems different in structure and function from those of today occupied the area illustrated in Figure 2.12 at successive times. Even stronger evidence of ecosystem change is strikingly apparent in the geologic and paleoecologic record. However, the record of vegetation change is much more complete compared to that for physiography or soil. Therefore, we can use the paleoecologic record of vegetation change as a window to provide insights into site-specific changes of landscape ecosystems over the long run. In Chapter 3, we briefly outline the change in ecosystems and vegetation that occurred in North America prior to the arrival of European settlers.

Suggested Readings

Albert, D. A. 1995. Regional landscape ecosystems of Michigan, Minnesota, and Wisconsin: a working map and classification. *USDA For. Serv. Gen. Tech. Report NC-178.* North Central For. Exp. Sta., St. Paul, MN. 250 pp. + map.

Bailey, R. G. 1988. Ecogeographic analysis. *USDA For. Serv. Misc. Publ. 1465.* Washington D.C. 18 pp.

_____. 1996. *Ecosystem Geography.* Springer-Verlag, New York. 204 pp. + 2 maps.

Bailey, R. G., and H. C. Hogg. 1986. A world ecoregions map for resource reporting. *Env. Conservation* 13:195–202.

Delcourt, H. R., and P. A. Delcourt. 1988. Quaternary landscape ecology: relevant scales in space and time. *Landscape Ecology* 2:23–44.

Denton, S. R., and B. V. Barnes. 1988. An ecological climatic classification of Michigan: a quantitative approach. *For. Sci.* 34:119–138.

Klijn, F., and H. A. Udo de Haes. 1994. A hierarchical approach to ecosystems and its implications for ecological land classification. *Landscape Ecology* 9:89–104.

Omernik, J. M. 1987. Ecoregions of the conterminous United States. *Ann. Assoc. Amer. Geog.* 77:118–125.

Rowe, J. S. 1992. The ecosystem approach to forestland management. *For. Chron.* 68:222–224.

Rowe, J. S., and J. W. Sheard. 1981. Ecological land classification: a survey approach. *Env. Manage.* 5:451–464.

Spies, T. A., and B. V. Barnes. 1985. A multi-factor ecological classification of the northern hardwood and conifer ecosystems of Sylvania Recreation Area, Upper Peninsula of Michigan. *Can. J. For. Res.* 15:949–960.

CHAPTER *3*

LONG-TERM ECOSYSTEM AND VEGETATION CHANGE

In Chapter 2, we examined the spatial relations of landscape ecosystems seen as nested geographic and volumetric segments of the Ecosphere. In the sections that follow, changes of ecosystems are considered at multiple temporal scales (Figures 2.1 and 2.2). Important changes in ecosystem properties and processes occur at fine scales—daily, seasonally, and yearly. However, more noticeable changes in landforms, soils, and biota occur in time frames of decades, centuries, and millennia. Forest ecosystem change occurring over a range of decades to centuries with resulting change in composition, structure, and biomass of vegetation is typically termed **succession** (Figure 2.2; Chapter 17). With a focus on vegetation, spatial and genetic changes over thousands and millions of years are regarded as plant migration, speciation, and evolution (Figure 2.2). These changes also occur within an ecosystem context. In this chapter, we emphasize long-term ecosystem changes, and especially distributional changes in north temperate forest trees, over the last 20,000 years as they reflect climatic, physiographic, and soil changes in landscape ecosystems.

Long-term ecosystem change is known primarily from studies of geology in formerly ice-covered areas, of plant fossils, and from the forest trees themselves. That is, plants are the surrogates, or at least among our best indicators, of landscape ecosystems at times thousands of years ago for which we have no written record of climate, physiography, and soil conditions. Whereas it is possible to track and interpret the migration of a tree species, or group of related species, little is known of the changes in regional and local physiography, climate, soil, and disturbance regimes that largely controlled the migrations of trees and associated vascular and non-vascular plants. Fortunately, however, paleoecologists have developed a remarkable overview of forest tree migration that provides insights into ecosystem change related to macroclimate.

This long-term overview emphasizes the fact that species are not postage stamps; they change genetically in response to the changing environment at a given place. Plants migrate with changing environments and may establish totally new spatial ranges. In the fossil pollen record, they leave trails of extinction at places of former dominance. The genus *Sequoia,* for example, is known from fossil records from Europe, central Asia, and North America as well as from Greenland, Spitsbergen, and the Canadian Arctic, but it is restricted today to one species with local distributions in California. Tundra and boreal species in North America no longer exist where they once did 20,000 years ago, except as high montane outliers in areas such as the Southern Appalachian Mountains. Other species have persisted in given geographic sites for several thousands of years, their populations changing genetically with changes in regional and local climate. The American beech of a given locality in eastern North America 9,000 years ago is not quite the same genetically as the beech population occupying the same place today.

Forest species differ greatly in their specific adaptations. However, groups such as conifers and deciduous angiosperms exhibit similarities in key adaptations and physiological tolerances that have determined their worldwide distributions (Chabot and Mooney, 1985). The *ecological* occurrences of these broad, biome-level groups, with respect to one another, have not changed greatly over 20,000 years or even millions of years. Thus to conclude the chapter and bridge past and present, we discuss the present-day distributions and overall adaptations of coniferous and deciduous species, as notable parts of the space–time continuum of forest ecosystems.

CHANGE BEFORE THE PLEISTOCENE AGE

Throughout millions of years, landscape ecosystems have undergone major physical changes caused by movement of continents, mountain building, global climatic changes, and continental glaciations with concomitant changes in biota. Although whole ecosystems at all scales changed, we only have records of these changes at the broadest ecosystem levels through studies of geophysical and geomorphological properties of the Earth's surface, studies of geochemical properties of sediments whereby global temperature changes are estimated by oxygen-isotope analyses of deep-ocean cores, and studies of preserved biological materials, especially plant pollen (Tallis, 1991).

In the last 65 million years, encompassing the Cenozoic Era, flowering plants (angiosperms) and mammals dominated the Earth in remarkable spatial configurations. Changes in vegetation accompanied physiographic, climatic, and soil changes at all spatial scales. Mixed conifer–broadleaf forests occurred at high latitudes (Alaska, Greenland, Siberia) 50 million years ago when tropical forests covered southern Britain (Tallis, 1991; Graham, 1993). Tree taxa ancestral to many modern-day taxa were present in these forests. However, whether these trees were associated in regional-level and local-level ecosystems in the same communities that we observe today is unknown, but highly unlikely. For example, tropical forests occupied western Kentucky and Tennessee 45 million years ago that have no exact modern analog (Graham, 1993). Considering ecosystems at the local level, with their total vegetative complements and site-specific physiographies, microclimates, and soils rather than just trees, no exact correspondences in either site conditions or communities are expected. However, if the physical components were generally similar, certain ancestral genera of the canopy and groundflora layers may have been present in combinations similar to those that exist today. Such appears to be the case for some of the wide-ranging dawn redwood forests of the Tertiary Period forests 50 million years ago and those of today in China. The dawn redwood was widely known only from the fossil record until its startling discovery in 1949 in a remote valley in the northern Hubei Province of China. It is related to the present-day redwood of the northern coast of California but characteristic of a more subtropical flora. Many genera of the Tertiary forests of what is now north-eastern Oregon (*Acer, Alnus, Ailanthus, Celtis, Fagus, Os-*

trya/Carpinus, Pinus, Quercus, Tsuga, Ulmus/Zelkova), where the dawn redwood once grew, are also found today in the valley in China where the dawn redwood presently grows (Tallis, 1991).

PLEISTOCENE GLACIATIONS

As climate became progressively colder in temperate latitudes during the Tertiary Period, ice began to accumulate as a direct consequence. Glacially-deposited materials are evident in both northern and southern hemispheres from about 10 million years ago onward (Tallis, 1991). The "Pleistocene Ice Age," beginning about 1.6 million years ago, consisted of multiple glacial-interglacial cycles (possibly as many as 18) of about 100,000 years duration. The cold glacial stage lasted approximately 90,000 years, as glacial ice sheets built up on continents, and it was followed by a short and warm interglacial stage of about 10,000 years. An overview of the past climatic variation over the last 150,000 years and expected future changes in climate are illustrated in Figure 3.1. The last continental cold-warm cycle in continental North America, the Wisconsinan Glaciation, reached its maximum extension 20,000–18,000 years ago. Thereafter, the climate warmed, and the sudden warming about 10,000 yr B.P. (Before Present) marks the beginning of the present Holocene Epoch.

Obviously, during previous "ice ages," forests were eliminated as glacial ice overrode the land in northern climes. Because these now-deglaciated lands constitute some of the most important forest areas of the earth today, it is equally obvious that their present vegetation dates from the last continental retreat of Pleistocene ice sheets. Figure 3.2 illustrates the extent of the last Pleistocene glaciation 20,000 years ago in North America. Not all the northern world, however, was under ice. A large area, Beringia, in the interior of Alaska and adjacent Yukon Territory (Figure 3.2) escaped glaciation, possibly because of low precipitation. This region apparently supported a polar tundra or Arctic steppe flora, probably because summers were too short and too cold to support forests (Hopkins, 1967; Hopkins et al., 1982). Unglaciated areas on the islands off the Northwest Pacific coast,

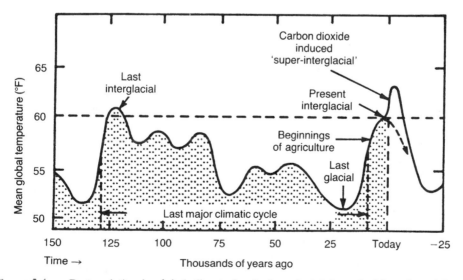

Figure 3.1. Past variation in global climate for the last glacial–interglacial cycle and future changes in climate projected for the next 25,000 years. (After Imbrie and Imbrie, 1979. Reproduced from *Ice Ages: Solving the Mystery,* by permission of Enslow Publishers.)

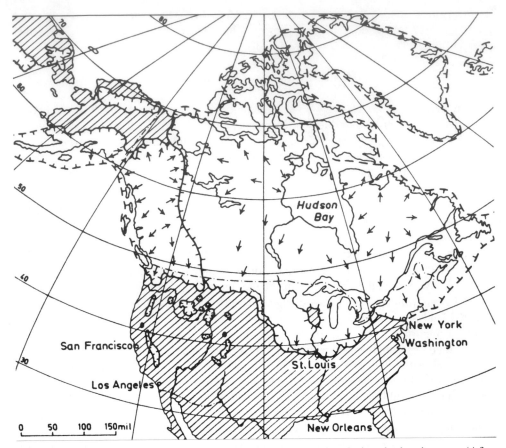

Figure 3.2.   Extent of continental glaciers in North America during the last ice age. (After Nilsson, 1983. Reprinted by permission of Kluwer Academic Publishers.)

such as the Queen Charlotte Islands of British Columbia, likely served as refugia for Sitka spruce and western hemlock during glacial maxima. Detailed considerations of ice ages, their causes and consequences, and their legacies of physiographic features for migrating vegetation are available from diverse perspectives (Delcourt and Delcourt, 1987, 1993; Pielou, 1991; Tallis, 1991; Dawson, 1992; Graham, 1993; and Hambrey, 1994).

Obviously, major changes in regional and local landscape ecosystems occurred throughout the Pleistocene ice age. New landscapes were created by the dynamics of the glaciers and through the action of waters as the ice disintegrated. These new features markedly affected climate on regional and local scales in ice-covered and adjacent lands (Figure 3.2), but their consideration is beyond our scope. Nevertheless, the significance of physiography in glacial landscapes is discussed in Chapter 10. The migration patterns of tree genera are a major source for understanding climatic change in the last 20,000 years as well as the past and current tree distributions. This period is briefly examined in the next sections, with special emphasis on vegetational change.

ECOSYSTEM AND VEGETATIONAL CHANGE SINCE THE LAST GLACIAL MAXIMUM

Instability on a grand scale characterized temperate landscapes as temperatures warmed at the end of the Wisconsinan Glaciation (Figure 3.3; Pielou, 1991; Tallis, 1991). Climatic

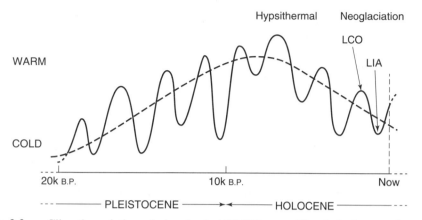

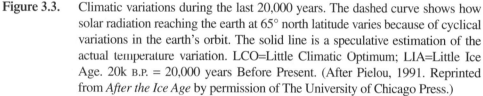

Figure 3.3. Climatic variations during the last 20,000 years. The dashed curve shows how solar radiation reaching the earth at 65° north latitude varies because of cyclical variations in the earth's orbit. The solid line is a speculative estimation of the actual temperature variation. LCO=Little Climatic Optimum; LIA=Little Ice Age. 20k B.P. = 20,000 years Before Present. (After Pielou, 1991. Reprinted from *After the Ice Age* by permission of The University of Chicago Press.)

amelioration was gradual, but as shown diagrammatically in Figure 3.3, oscillations occurred. The maximum warm period, or Hypsithermal, was followed by a cooling trend that has not been continuous. Climate moderated from A.D. 1450 to A.D. 1850 ("Little Climatic Optimum" in Figure 3.3) and cooled again in the "Little Ice Age."

As glacial ice receded, new physiographic features and their soil parent materials were uncovered and colonized by organisms. Melt waters carrying massive amounts of sediment formed new lands of outwash plains; lakes formed and reformed at different levels and reworked the lands they covered. Over the last 20,000 years the sites of landscape and waterscape ecosystems changed dramatically. However, the most conspicuous part of this change available in the prehistoric record is that of tree genera and certain species whose pollen morphology is distinctive. Thus scientists have tracked and mapped the directions and rates of their population migrations using fossil pollen and plant parts and by reconstructing the forest composition (Delcourt and Delcourt, 1987). The study of prehistorical plants is known variously as **Quaternary biogeography** or **geography** if the emphasis is on geographical distribution of the various elements of the flora, and as **Quaternary paleoecology** if environmental changes are of primary concern. Recent advances in paleoecology and past climates during the last deglaciation have been summarized for North America (Delcourt and Delcourt, 1987; Ruddiman and Wright, 1987), Europe (Huntley and Prentice, 1993), and at a global scale (Wright et al., 1993).

One of the major scientific advances of recent decades has been the unraveling of much of the complex history of the vegetation and climates of the world from the time of the last great glacial advance to the present. By bringing together archeological data and fossil records, and by using tree-ring analyses, radiocarbon dating, and in particular pollen analysis, it has become possible to spell out in considerable detail the overwhelming of forests before the advancing Pleistocene ice and the migration of forests back onto the debris-covered ice and onto the land itself as the ice melted. The irrefutable evidence for temperate regions is that vegetation has been in an almost constant state of instability and adjustment due in part to an almost continuously changing climate over the past 20,000 years and even more fundamental changes in the configuration of the land and the composition of its surficial materials (Davis, 1986). This evidence has done more than anything else to demonstrate that present vegetation patterns are closely related to events in recent geological history, of which climatic change is one important part.

The composition of forests in times past was determined by the combination of geologic processes and interrelated climatic events. In western North America, cooling and mountain building in the Tertiary Period were the major processes (Graham, 1993). However, in eastern North America, events of Pleistocene glaciation and postglacial climate have been the primary influences on ecosystem change and forest distribution and composition. In landscapes once covered by glaciers, the effects of glacial action in reworking the landforms and the geomorphic processes associated with deglaciation provided a legacy of physiographic features and associated parent materials and soil formation. These features were also important in affecting changes in ecosystem structure, function, and plant composition at regional and especially at local spatial scales (see Chapters 10 and 11).

The end of the Pleistocene glaciation was marked by a clear, worldwide climatic change about 20,000 years ago that was unidirectional in its major effects. For thousands of years afterward, the climate in the ice-free areas ameliorated and vegetation migrated onto formerly glaciated sites. A powerful approach in the difficult task of reconstructing and understanding these long-term vegetative changes has been that of pollen analysis.

Pollen Analysis

Lake sediments are rich reservoirs of fossil pollen and plant remains (Hebda, 1985). Lakes are common in glaciated areas, and the pollen grains preserved within their sediments reflect the regional character of the vegetation. In addition, the cold, wet, anaerobic environment provided by acid peat bogs constitutes a perfect preservative for hard plant parts. None is preserved better than microscopically-sized pollen and spore grains, whose hard outer coats, or exines, are extremely resistant to decay. However, trees growing on the bog surface are likely to contribute disproportionately large amounts of pollen to the local "pollen rain" of a peatland. Thus lake sediments are selected as the chief source for pollen analytical studies because they collect pollen from a broader area. In Figure 3.4, the lake is seen as central to the approach whereby recovery of pollen from the bordering forest via the lake sediments facilitates reconstruction of the sequence of species invasion and population dynamics on the surrounding uplands.

Pollen analysis, although perhaps the most useful single tool in deciphering long-term vegetational history, is potentially fraught with many dangers of misinterpretation. The pollen rain falling on a lake, open bog, or other repository in which the pollen may be embedded within sediments and preserved will contain an over-representation of species that produce abundant, wind-disseminated pollen resistant to decay, and an under-representation of species that produce pollen sparsely or whose pollens are easily decayed (e.g., *Populus* spp.) or are disseminated by insects or other means. Thus to minimize these problems, comprehensive calibrations are used and are based upon data collected across the distributional ranges of the taxa examined (Delcourt et al., 1984; Delcourt and Delcourt, 1987). Furthermore, a variety of different quantitative techniques now may be used to reconstruct past vegetation, develop maps, and interpret paleoecological results (Grimm, 1988). In the sections that follow, examples of the use of pollen analysis to reconstruct past vegetation are given for eastern and western North America.

Eastern North America

Based on radio-carbon-dated pollen diagrams for 100 localities in eastern North America, Paul and Hazel Delcourt (1981) prepared paleovegetation maps spanning the past 40,000

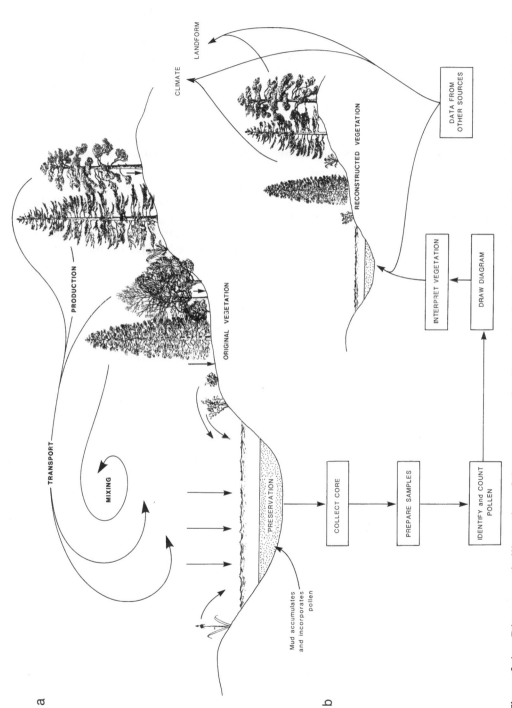

Figure 3.4. Diagrammatic illustration of the processes of pollen production, dispersal, deposition, and preservation (a) and its recovery, analysis, and interpretation of its significance (b). (After Hebda, 1985.)

years. The maps from 18,000 yr B.P. to the present (Figure 3.5) illustrate the general shift in location and areal coverage of major vegetation types with climatic amelioration. The major features of climatic history of this period were summarized by Webb et al., (1993): (1) climate changed continuously from 18,000 yr B.P. to the present, (2) a major increase in mean annual temperature occurred between 15,000 and 9,000 yr B.P., (3) the greatest seasonal contrast between cold winters and hot summers took place between 12,000 and 9,000 B.P., (4) the modern spatial gradients for July temperature and annual precipitation first appeared at 9,000 yr B.P., (5) the time of maximum warmth was about 6,000 B.P. when temperatures were not more than 1–2°C higher than today; and (6) the time of maximum dryness in the last 12,000 years was 6,000 B.P. in the Midwest and 9,000 B.P. in the northeastern United States.

Besides an overview of climatic history, it is important to note key characteristics of tree species that help us understand the migration patterns and community composition. Trees that have persisted for millions of years in temperate and boreal environments have evolved dispersal mechanisms for long-distance dispersal by adaptations to wind and water and mutualistic interactions with birds and mammals. The potential range of environments in which trees can colonize and grow is far greater than we may perceive by examining their present-day natural occurrences. Tree species persisting today have evolved a mix or balance of genetic and non-genetic mechanisms to maintain fitness in a given environment or the flexibility to change with changing conditions (Chapter 4). Mutualisms with other plants and animals are exceedingly important in expanding or restricting the occurrence of tree species. Competition may severely restrict forest species to a relatively small part, i.e., their realized habitat or niche (Chapter 4), of their potential spatial range. Finally, many forest trees have adaptations enabling them to invade ecosystems with already-established communities and maintain themselves, depending on the specific site conditions.

OVERALL MIGRATION SEQUENCE AND PATTERNS

In the eastern North America, climatic cooling during glacial times was apparently severe even to middle latitudes, and forests were displaced far south of the ice margin. During the late Wisconsinan Glaciation, 18,000 yr B.P., a boreal-like coniferous forest dominated by jack pine and spruces occurred on both sides of the Mississippi River Valley south to about 34° N latitude (Figure 3.5a). A treeless zone of tundra was situated adjacent to the ice sheet. Just as today, composition of the full-glacial boreal forest varied across the region; species dominant east of the Mississippi differed from those to the west. Boreal species such as jack pine and white spruce were displaced far to the south; the southern limit of spruce was apparently in the coastal plain between north-central Georgia and northern Florida (Watts and Stuiver, 1980). A forest of jack pine and some black spruce occurred in the Piedmont region of northern Georgia 23,000–15,000 yr B.P. (Watts, 1970). This vegetation, typical of modern forests 1,100 km to the north, indicates the severity of the climatic change (Wright, 1971). White spruce, fir, and tamarack also extended south-

Figure 3.5.   Map of vegetation change for North America. a, Map of vegetation for 18,000 yr B.P.; b, Map of vegetation for 10,000 yr B.P.; c, Map of vegetation for 5,000 yr B.P.; d, Map of vegetation for 200 yr B.P. The dots represent plant–fossil localities that provided the paleoecological evidence of the appropriate radio carbon age used to generate each paleovegetation map. (After Delcourt and Delcourt, 1981. Reprinted from the Proceedings of the 1980 Geobotany Conference, R. Romans (ed.), © 1981 Plenum Publishing Corp. Reprinted by permission of the Plenum Publishing Corp.)

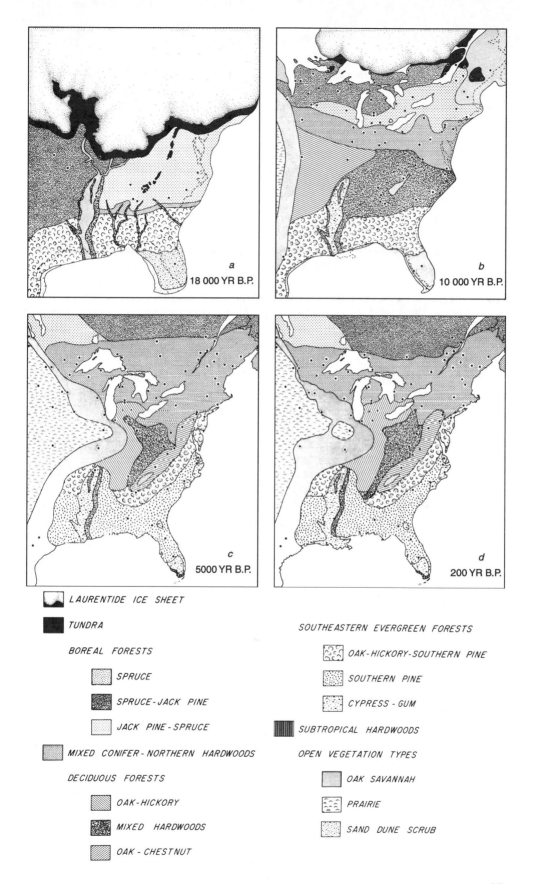

	LAURENTIDE ICE SHEET	
	TUNDRA	
	BOREAL FORESTS	SOUTHEASTERN EVERGREEN FORESTS
	SPRUCE	OAK-HICKORY-SOUTHERN PINE
	SPRUCE-JACK PINE	SOUTHERN PINE
	JACK PINE-SPRUCE	CYPRESS-GUM
	MIXED CONIFER-NORTHERN HARDWOODS	SUBTROPICAL HARDWOODS
	DECIDUOUS FORESTS	OPEN VEGETATION TYPES
	OAK-HICKORY	OAK SAVANNAH
	MIXED HARDWOODS	PRAIRIE
	OAK-CHESTNUT	SAND DUNE SCRUB

ward to 31° N latitude within the lower Mississippi Valley and its tributaries, occupying extensive sand flats of braided streams that carried glacial melt-water sediments into the Gulf of Mexico (Delcourt and Delcourt, 1977; Royall et al., 1991).

Elsewhere during full-glacial times a warm-temperate forest of oak, hickory, and southern pines persisted across the Gulf and lower Atlantic coastal plains (Figure 3.5a; Delcourt and Delcourt, 1981). A narrow transition zone of cool-temperate conifers and northern hardwoods is speculatively mapped from northern Mississippi to South Carolina between Jack Pine–Spruce Forest and Oak–Hickory–Southern Pine Forest. Forests of many different deciduous species, the Mixed Mesophytic Forest, are mapped along the loessal blufflands east of the Mississippi Valley and within protected valleys of major river systems of the Southeast. The blufflands of the eastern escarpment of the lower Mississippi Valley form a 20–30 km-wide zone of fertile, mesic habitat of calcareous silt loam soil (Caplenor et al., 1968). The loess (wind-deposited soil of silt-sized particles) is readily eroded to form steep ravines and banks that would provide protected sites for the species-rich mesophytic communities (Delcourt and Delcourt, 1979). A comparison of the geographic position of the Mixed Mesophytic Forest then with it's geographic location today (Figure 3.5a and 3.5d) illustrates that a broad vegetation type, perceived to be similar in tree species, may occur in two quite different physiographic regions and in markedly different conditions of geologic substrate, physiography, and soil. The compensation of site factors, climate and soil water and nutrients, accounts in part, for the occurrence of many of the same species in local ecosystems of such different geographic areas.

As the result of a minor climatic warming about 16,500 years ago, tree populations migrated slowly northward. Spruce–jack pine forest colonized lands formerly occupied by tundra, and conifer–northern hardwoods forest expanded north and east along what had been the southern border of the boreal forest (Watts, 1979; Delcourt, 1980; Whitehead, 1981). Following a major climatic amelioration about 12,500 yr B.P., the Mixed Mesophytic Forest expanded widely to the north and east into the Appalachian highlands (Figure 3.5b). Boreal forest then declined extensively after 10,000 yr B.P. in the southeast, and red spruce and Fraser fir populations were stranded at higher elevations in the Southern Appalachians and persist today on some of the higher montane peaks.

Major changes in vegetation also occurred in mid-Holocene times, 8,000–4,000 yr B.P. (Figure 3.5c). With increasing warmth and aridity in the Great Plains accompanying a general increase in the strength of prevailing westerly winds, oak–hickory forest, oak savanna, and prairie, in turn, shifted eastward. Their eastern limit was reached about 7,000 yr B.P. during the Hypsithermal interval (Figures 3.1 and 3.3). Fire, fanned by the westerly winds across the predominantly flat landscape, was a major factor in pushing back and maintaining the prairie–forest border. Reaching into Minnesota, Illinois, Indiana, southern Michigan, and Ohio, prairie openings apparently gradually replaced mixed pine–deciduous forests to the north and elsewhere deciduous forests and oak savannas. This **Prairie Peninsula** (Transeau, 1935; Stuckey, 1981) is evidenced by the widespread occurrence of prairie soils, remnants of oak savannas, and by the persistent fragments of the prairie until the initiation of farming by European settlers (Whitney, 1994). The change to a cooler, moister (late-interglacial) climate was slower than the rapid, early-interglacial onset of the warm period had been. Pines and temperate hardwoods gradually replaced prairies, although prairie openings probably survived in east central Minnesota until 4,000 yr B.P. (Wright, 1971). The Prairie Peninsula posed a barrier to the northward migration of upland mesophytic species such as beech, hemlock, and tuliptree. These species migrated into the midwest from the east.

During this same period (after 10,000 B.P.), mixed conifer–northern hardwoods forest moved north into southern Canada, the Mixed Mesophytic Forest species were restricted to moist coves and slopes of the Cumberland and Allegheny Plateaus, and oak–chestnut communities dominated the southern and central Appalachians. Southern pines replaced oak–hickory forests on sandy uplands of the Gulf and Atlantic coastal plains, and oak–hickory–pine forests were restricted to the Piedmont Province and the Ozark and Ouachita Mountains. Throughout this interglacial period regional physiographic features of plains, plateaus, and mountains were important in shaping the diverse vegetation types of modern regional ecosystems.

The boundaries of 200 years ago (Figure 3.5d) are similar to those at 5,000 years except for a shift south of the southern boundary of the Boreal Forest and a retreat to the west of the prairie-forest boundary. These changes were caused by a long-term cooling trend at high latitudes, together with increased precipitation across the Midwest that has occurred since about 6,000 to 5,000 yr B.P. In general, the presettlement vegetation boundaries of 200 years ago are similar to those of Braun's (1950) forest regions (Figure 2.8). This correspondence illustrates the importance of regional physiographic features in the reconstruction of the broad vegetative types whether by pollen analysis or field study of near-natural ecosystems. Macroclimate is not the only ecosystem component responsible for the distribution of vegetation in this 18,000-year sequence. Multiple geologic and soil factors, and episodic forces of disturbance such as fire, flooding, and windstorm, acted at all spatial and temporal scales to shape the distributional pattern of the vegetation of regional and local ecosystems.

ECOSYSTEM CHANGE IN THE SOUTHERN APPALACHIANS

We can gain a good perspective of ecosystem change in both time and space by examining the vertical displacement of vegetation types that Delcourt and Delcourt (1987) describe for the Mt. LeConte (elevation 2,000 m) of the Great Smoky Mountain National Park, east Tennessee to west North Carolina (Figure 3.6). At 20,000 to 16,500 yr B.P., elevations above 1,500 m in the Southern Appalachians were characterized by mean annual temperature well below 0° C, perennially frozen ground in the form of permafrost, and permanent snow packs. Intensive freeze–thaw churning of soil prevented establishment of trees, but the montane environments were suited for alpine tundra. Conifer krummholz dominated lower elevations, at and below 1,500 m.

From 16,500 to 12,500 yr B.P. (Figure 3.6) rising temperatures to near 0° C led to geomorphic processes of freeze-thaw surface heaving. Boulder fields and patterned ground features developed above 1,500 m. At high elevations, tundra and krummholz formed a patchwork cover. Boreal and cool-temperate forests became established at intermediate and lower elevations, respectively.

Climatic amelioration after 12,000 yr B.P. changed prevailing geomorphic processes as temperate stream processes replaced periglacial and colluvial ones. By 10,000 yr B.P. conifers and northern hardwoods occupied mountain summits and upper slopes, and temperate deciduous hardwoods (including the Mixed Mesophytic Forest Type) dominated the lower elevations after 10,000 yr B.P.

Today, a great diversity of ecosystems comprises the landscapes of Mt. LeConte from low to high elevation, as seen in Figure 3.6. The mountain landform with its physiographic diversity in elevation, aspect, and slope characters provides the spatially stable base for the biota. Obviously, similar mountain sites were not available to these deciduous and coniferous species 20,000 years ago in the South and Southeast where they spent full-glacial times.

Western North America

In western North America, the continental ice sheet did not extend nearly as far south as it did in eastern North America (Figure 3.2). However, species were displaced to the south, and mountain glaciation was widespread in the Rocky Mountains and other ranges south

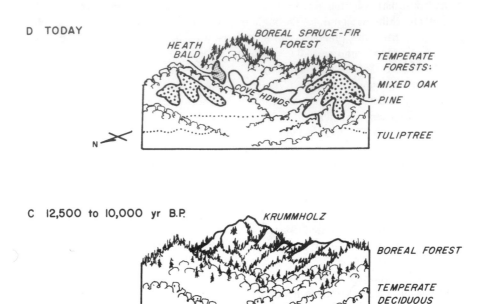

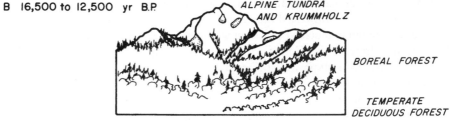

Figure 3.6. Diagrammatic illustration of vegetational changes on Mt. LeConte, Great Smoky Mountains National Park: A, full-glacial interval, 20,000–16,500 yr B.P.; B, late-glacial interval, 16,500–12,500 yr B.P.; C, early-Holocene interval, 12,500–10,000 yr B.P.; D, today. (After Delcourt and Delcourt, 1987. Reprinted from *Long-term Dynamics of the Temperate Zone,* © 1987 by Springer-Verlag New York, Inc. Reprinted by permission of Springer-Verlag New York, Inc.)

of the continental ice border. The postglacial history in the Pacific Northwest (from Oregon north to Alaska) is complicated by the fact that the Cordilleran ice radiated east and west from the north-south-oriented mountain ranges rather than simply moving southward, as in eastern North America. Also, the maritime climatic conditions originating from the Pacific Ocean moderated at least some of the postglacial climatic fluctuations found in the East. Therefore, refugia for various plants and animals, including Sitka spruce and western hemlock, along the Pacific Coast served to restock much of the glaciated terrain rather than migration from the south as in the case of the eastern part of the continent (Hansen, 1947, 1955; Heusser, 1960; 1965, 1983; Barnosky et al., 1987). As a result, temperate conifer forests were never displaced from the Pacific coast during the Pleistocene, and they ranged from the edge of the ice sheet in western Washington and British Columbia to northern California. An extensive refugium in unglaciated interior Alaska existed for trees such as spruce and birch, along with shrubs and herbs of the Arctic tundra.

The Pacific Northwest has a late-glacial and postglacial history similar to that of eastern North America. Glacial ice had begun to recede as early as 14,000 years ago and a late-glacial cold period was followed by a warmer and drier trend, that ultimately gave way to a cooler, more humid climate. Further inland in the mountains of southern British Columbia glacial recession began about 10,000 yr B.P., whereas on the prairies in Saskatchewan the recession began about 20,000 yr B.P. Following the retreat of the glaciers, a lodgepole pine parkland developed in the Pacific Northwest and northwestward as far as southeastern Alaska (MacDonald and Cwynar, 1985). Lodgepole pine reached its current most northward locality possibly less than a century ago (Cwynar and MacDonald, 1987) and migrated at an average speed of about 180 m yr^{-1} over a distance of 200 km in about 12,000 years (18.3 km per century). As the climate ameliorated, alder, birch, and more tolerant coniferous species replaced pine. Douglas-fir became the dominant species of the Pacific Northwest during the warmer and drier times. Where rainfall was heavy, western hemlock replaced pine and mountain hemlock, and Sitka spruce became well represented from British Columbia to southeastern Alaska. In the Willamette Valley, the succession following pine was to Douglas-fir and Oregon oak, and east of the Cascades grassland achieved dominance on the dry plateaus. As in the East, the approximate boundary dates of this period of maximum warmth and dryness are 8,000 to 4,000 yr B.P.

PATTERNS OF TREE GENERA AND SPECIES MIGRATIONS

The migration pathways, directions, and rates of about 20 genera of forest trees of eastern North America have been described (Bernabo and Webb, 1977; Davis, 1981; 1983a,b; Webb, 1988; Webb et al., 1993; Davis and Jacobson, 1985; Delcourt and Delcourt, 1987). Migration routes and arrival times of spruces, white pine, beech, and chestnut are illustrated in Figure 3.7. In these diagrams we see the leading edge of the species at a given time when its pollen was first consistently well represented in the pollen sequence for each site. Many trees migrated generally from southern refuges northward along a broad front as seen for the spruces in Figure 3.7*a*. In contrast, white pine (Figure 3.7*b*) and hemlock migrated west and north from refugia on the central coastal plain or on the exposed continental shelf (Davis, 1983a,b). American beech (Figure 3.7*c*) migrated northward, and after about 9,000 years ago expanded eastward through the Great Lakes region. American chestnut (Figure 3.7*d*) migrated through the Appalachian Mountains in a northeasterly direction and arrived in New England after most other tree species with which it was associated.

As one might expect, early-successional species with many tiny, wind and water dispersed seeds migrated the fastest (Godwin, 1956). For example, aspens (*Populus*) expand-

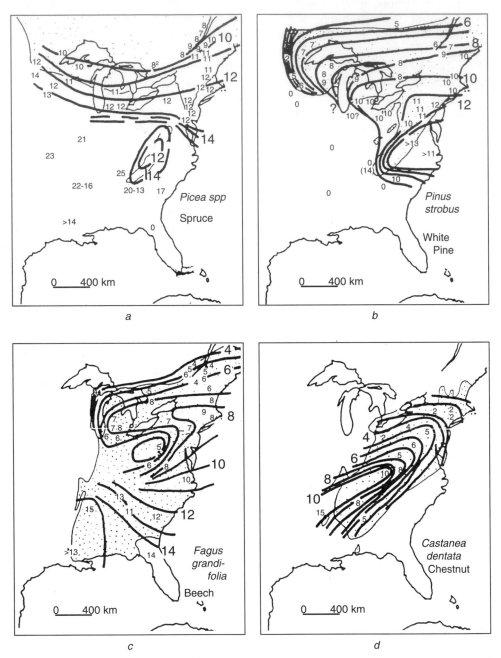

Figure 3.7.   Migration routes of tree species in late-glacial and early Holocene. Small numbers on the map indicate the time of arrival at individual sites. Contours show the leading edge of population advance at 1,000-year intervals. Shaded area is the presettlement range for the genus or species. (After Davis, 1983b. Reprinted by permission of Margaret B. Davis and the Missouri Botanical Gardens.)

ing into tundra areas exhibited a rate of advance of 544 m yr^{-1} from 14,000 to 12,000 yr B.P.The rate declined thereafter, reaching a relative minimum of 145 m yr^{-1} from 10,000 to 8,000 yr B.P. Averaged over all migration tracks from east to west, aspen advanced at a mean rate of 263 m yr^{-1}. The comparable rate for spruces was 141 m yr^{-1}, for oaks 126 m yr^{-1}, and for hickories 119 m yr^{-1}.

Also, as might be expected, heavy-seeded species, such as oaks and hickories, typically migrated more slowly, but the rate was highly variable depending on time period and migration track (Delcourt and Delcourt, 1987). In general, the dispersal rate was not an overriding factor limiting the rate of migration for boreal and temperate trees in eastern North America (Delcourt and Delcourt, 1987); rather, the availability of suitable sites for establishment and subsequent growth was the more critical limiting factor. Thus physiography and soil played important local roles in glacial–interglacial range expansion and late-Quaternary community dynamics.

Animals may aid in establishment by caching seeds and thereby, in part, providing a favorable environment for establishment. For heavy-seeded and fleshy-fruited species a mutualistic relationship with animals is, therefore, exceedingly important. Birds were important seed dispersal agents for many plant species. Such mutualism is significant in tree species of the Fagaceae, especially for oaks, American beech, and American chestnut (Johnson and Webb, 1989). In particular, the blue jay is known to disperse and cache seeds of many oak species as well as beech and chestnut. The blue jay dispersal system, may largely explain the rapid expansion of the range of oaks in late glacial time. By 10,000 yr B.P., oaks had reached southern Canada, whereas beech and chestnut migrated more slowly. The slow migration rate of chestnut may be partially explained by the large size of chestnuts and the apparent preference by jays for small nuts. The lag for beech seems to be associated with climatic limitations rather than dispersal constraints (Johnson and Webb, 1989). In addition, American beech has relatively narrow site requirements for establishment, is not favored either by fire or flooding, and due to very slow growth may not reach seed-bearing age for 40 years or more.

Long-distance fruit and seed transport by birds enables plants to reach suitable sites for establishment far ahead of the leading edge of their main advance. Similarly, strong winds can place small seeds in outlying sites far ahead of the advancing front. For example, by 12,000 years ago aspen populations had established in Maine, then a region of tundra vegetation 700 km east of the continuous range margin for aspen in the Lake States (Davis and Jacobson, 1985; Delcourt and Delcourt, 1987). In the western Great Lakes region, hemlock also advanced by long-distance dispersal of seeds, founding discontinuous populations ahead of the main front (Davis, 1987). Such dispersal by long-distance jumps is a common way for organisms to spread, especially when disturbances such as catastrophic fire and windstorm provide suitable seedbeds.

Migration Irregularities and Disturbance

At the broad regional scale we perceive wave after wave of trees and other organisms establishing and migrating over the formerly glaciated sites to reach their present areas of distribution. A continuum of change, illustrated by a series of curves depicting expanding ranges such as those in Figure 3.7, is often used to portray at broad temporal and spatial scales the general migration pathway. At time intervals of one or two thousand years, the mapped range limits appear relatively smooth and similar to organismal diffusion from areas of high concentration to areas of low concentration. Barriers or corridors, such as major rivers or mountain ranges, rarely show up on the subcontinental scale maps as they might with fine-scale resolution of migration at hundred-year intervals. Nevertheless, even at these broad time and spatial scales various irregularities from a more or less smooth advance are recognized. These include the migration of beech around or across Lake Michigan (Webb, 1986; Davis et al. 1986; Davis, 1987; Delcourt and Delcourt, 1991) and patterns of hemlock expansion in the western Great Lakes region (Davis et al., 1986; Davis, 1987). And in New England, early colonization of sites by elms are noted 300–700 years earlier than the main "explosion" in pollen values (Webb et al., 1983; Davis and Jacobson, 1985).

Disturbances such as windstorm, flooding, and especially fire, were undoubtedly very important in providing local corridors and competition-free patches in existing forests for migrating species. Catastrophic and minor disturbances also were important in changing species composition of forest ecosystems over thousands of years as late-arriving species were incorporated into existing communities. Fire provides suitable seedbeds, temporarily reduces competition, and generally favors the migration and maintenance of all boreal species, pines, oaks, and many other temperate species in a landscape mosaic. Even a species such as eastern hemlock, although perceived as fire sensitive, establishes readily on fire-prepared seedbeds. Colonizing burned areas at the same time as trembling aspen, paper birch, or white pine, hemlock grows slowly under a canopy of one or more of these species, reaching the overstory only after 100 years or more. Research by Swain (1973, 1978) revealed that short-term changes caused by a specific fire are indicated by pollen percentage increases of early-seeding aspen and birch. The percentage of white pine pollen increases only after approximately 50 years and hemlock after 100 years. The interrelationship of fire and climatic change on fire regimes and their influence on temperate forest composition has been demonstrated in several paleoecological studies (Green, 1981; Clark, 1989, 1990) and reviewed by Walker (1981), Patterson and Backman (1988), and Delcourt and Delcourt (1991).

Long-Term Change at a Given Site

In previous examples and in Figure 3.7, we have observed individual species or genera expanding over broad regional landscapes. Forest trees have been the main focus. However, what if we made landscape ecosystems the focus and, in a special time machine, were able to record whole ecosystem changes at a given site over 20,000 or more years, from melting of the ice to the present? What stunning changes we would observe, with climate and vegetation changing markedly upon the relatively stable landform, parent material, and soil. Over time we would record major changes in processes of soil development, water and nutrient cycling, and the accumulation of carbon into plant biomass and its recycling into the soil. Starting with an ice-free substrate, at various intervals in the time continuum thereafter, we could record an entire sequence of different ecosystem types, based on ecosystem composition, structure, and function. We could observe the vegetation change as a result of biotic interactions with all physical factors. At a site in northern Ohio we might record tundra vegetation 14,000 years ago, a white pine–hemlock forest at 10,000 yr B.P., a northern red oak forest at 6,000 yr B.P., a beech–sugar maple–basswood forest at 200 yr B.P., and a cornfield today! Our ecological fantasy illustrates that the sites of landscape ecosystems, landform and soil, form the relatively stable base upon which the more labile vegetative changes occur.

Actually, we can come reasonably close to observing at one place an even longer time change in an unglaciated dunal landscape of eastern Australia (Walker et al., 1981). At Cooloola, Queensland overlapping dune systems of quartz sands extend 40 km along the coast and 10 km inland. Six dune systems range in age from several thousand years to perhaps a half-million years—the result of continuous wind and water action. Although "place" has not been held exactly constant, the six systems, from outermost (youngest) to innermost (oldest) approximate differences in soil weathering, nutrient changes, resultant vegetative differences that might occur at a single place over this time span. A sequence of increasing vegetation height and coverage from the younger dune systems 1 (low grasses and shrubs) and 4 (extremely tall and layered trees and shrubs) is followed by their decrease through older dune ecosystems 5 and 6 (low shrubs and grasses).

It is not assumed that the plant community now present on a particular dune system was previously represented by the same composition on an equivalent site in the past. However, the biomass and structure of the vegetation, rather than plant composition, are likely to vary in the same relative fashion across the dune systems. These vegetation differences are related to soil development and the change in nutrient status through time in contrast to the prevailing view of vegetation change forced by climate. Initially, nutrients are taken from the soil as plants increase in size and mass and are "stored" in organic matter at increasing depths. Over the long term—up to a half-million years—nutrient loss into the drainage water occurs, and the decline in plant height and biomass through time accompanies the declining nutrient status. This example illustrates the importance of soil development, and nutrient status in particular, in affecting changes in community structure, biomass, and composition in the long term. We discuss these topics in Chapters 18 and 19. Although macroclimate is a main determinant or forcing function of ecosystem and vegetation change, geomorphic processes also change in time and interact with climate at regional and local levels. Thus vegetation change over the long term is a complex ecosystem process involving many interrelated factors.

INDEPENDENT MIGRATION AND SIMILARITY OF COMMUNITIES THROUGH TIME

Paleoecologists have made enormous advances over the years in determining the general patterns of response of a variety of tree genera to the changing conditions of the last 20,000 years. The general migration routes and rates of major tree genera and species have been determined. One viewpoint is that plant taxa have migrated independently of each other in response to climatic changes, and that plant communities arise fortuitously (Tallis, 1991). The viewpoint of independent migration, however, is based on a limited record over the last 20,000 years, restricted essentially to a few temperate dominant tree taxa. As Tallis (1991) notes, little evidence exists for the behavior of herbaceous plants of the groundflora. He writes: "It is difficult to believe that these migrated completely independently of the tree taxa." We agree, but also emphasize that intact communities don't migrate as a unit.

Some tree genera and species that we observe typically growing together today apparently did not co-occur in the past because they migrated on different tracks or at different rates. Most boreal taxa today appear to have had similar geographic ranges and had the opportunity to co-occur geographically and ecologically. However, the range expansion of some tree species, such as hemlock and beech, lagged behind that of other temperate species with which they are associated today, for example in the western Great Lakes region. Using precolonial ecosystems of 300 years ago as a baseline, ecosystem and community similarity would tend to be least in the early stages of migration and become progressively similar as these species invaded established communities.

In regard to the similarity of past and modern communities, spatial scale is an important consideration. At a broad subcontinental scale, plant formations that correspond to modern ones (spruce–jack pine, mixed conifer–northern hardwoods, mixed mesophytic, oak–hickory–southern pine) may be differentiated geographically and their migration shown over broad time intervals (Figure 3.5). Also, communities with genera in common with those of today are shown occupying successive elevational zones in the southern Appalachians (Figure 3.6) for a period of 20,000 years. However, the occurrence of similar tree genera in a geographically broad area or elevational zone should not necessarily imply that the local ecosystems and their communities of the past are necessarily the same as those of today. Generally, the similarity of communities, past and present, depends on the similarity of the physical environment and disturbance regimes of the site where the plants grow. Certain species are similar enough in their site requirements and tolerances that they respond in similar ways to disturbances of fire, flood, windstorm, and specific site conditions of soil water and nutrient availability. We would expect "fire species" combinations such as black spruce, tamarack, and leatherleaf; jack pine and red pine; eastern white pine and eastern hemlock; and loblolly and shortleaf pine to be more often associated together (with characteristic groundflora species) in fire-prone ecosystems of their respective regional ecosystems rather than growing on floodplain ecosystems with willows and cottonwoods or elms and silver maples. Thus where regional gradients or local environments of the past were similar to those of today, it is probable that certain plants and animals occurred in characteristic site-species patterns that are not unlike those of today. However, in the past if many present-day taxa in a given area did not have access to a similar range of sites, this relationship would not hold. Obviously there is no exact correspondence nor can entire intact communities migrate together because they are structurally and functionally linked to physiographic and soil features of landscape ecosystems that do not migrate.

Ecosystem complexity is so great that assessing the degree of similarity, if based on more than occurrence of tree genera, is beyond the resolution of paleoecological analysis of microscopic pollen grains and other plant remains. Based on our current understanding of communities at the broad generic level, Delcourt and Delcourt (1991, p. 3) summarize the point very well. They conclude that as continents and global climate have changed through time, the degree of correspondence in composition and structure of past communities has varied from complete dissimilarity to very close analogy when compared with today's communities.

We conclude that in many instances markedly different communities from those of today occurred where the site conditions and disturbance regimes differed as well. In other cases, certain combinations of plants, genetically and physiologically similar in their range of site requirements, occurred now as then. Thus certain genetic patterns of adaptation expressed in physiological tolerances to certain site conditions provide a link in understanding the similarity of the vegetation of past and present ecosystems. In the next section, we examine several broad relationships of vegetation and environment.

ADAPTATIONS OF CONIFERS AND ANGIOSPERMS

Today's regional and local forested landscape ecosystems of temperate North America are dominated by either gymnosperms (mainly conifers) and angiosperms, or mixtures of them. These groups occur today in distinctive geographic and ecological patterns in relation to one another (Figure 2.6) as they did throughout the last 20,000 years (Figures 3.5 and 3.6) and in the millions of years of their existence (Chapter 22). The conifers are predominantly evergreen, whereas the angiosperms are primarily broad-leaved deciduous trees. Evergreen angiosperms such as sweetbay magnolia and deciduous gymnosperms such as tamarack are relatively rare in forests and occupy specialized sites. Before examining the details of variation of individuals and populations in Chapter 4, it is useful to consider the broad picture of these groups and their overall adaptations to regional site conditions.

Conifers dominate the Boreal Forest Region (Rowe, 1972, Larsen, 1980, Elliot-Fisk, 1988), the temperate rain forests of the Pacific Northwest (Franklin, 1988), the Rocky Mountains (Peet, 1988), parts of the southeastern Coastal Plain (Braun, 1950; Christensen, 1988a), and fire-prone plains of eastern North America south of the Boreal Forests. In contrast, deciduous forests characterize much of eastern North America (Figures 2.6 and 2.8) and certain regional and local ecosystems of western North America. These include riverine ecosystems of the West, Alaska, and other parts of the Boreal Forest, oak woodlands of California (Barbour, 1988) and Oregon (Franklin and Dyrness, 1973), and trembling aspen parklands and mountain slopes of the Rocky Mountain and Intermountain regions (Moss, 1932; Barbour, 1988; Barnes, 1991).

Although deciduousness may be a primary trait distinguishing temperate angiosperms from conifers, there are several other important differences. Conifers evolved long before angiosperms, with a lineage reaching back to late Paleozoic time, over 200 million years ago. Deciduous angiosperms evolved about 125 million years ago, long after the origin of conifers, and many broad-leaved deciduous species (often termed "hardwoods") grow in less extreme conditions than conifers. Conifers tend to exclude deciduous angiosperms under certain conditions that limit photosynthesis: cold temperature; short, cool growing season; aridity; fire; high wind velocity; mid-summer drought; and low nutrient availability. Deciduous angiosperms are generally favored by long, warm, humid growing seasons, although some species are highly competitive under certain extreme environmental conditions as well.

Geographically, three broad groups of conifer-dominated ecosystems where angiosperms tend to be less common are: (1) boreal forest (jack pine, larches, spruces) and montane and subalpine Rocky Mountain forests (Douglas-fir, larches, Engelmann spruce, subalpine fir, limber and whitebark pines), (2) Pacific Northwest forest, (Douglas-fir, Sitka spruce, true firs, western hemlock, western redcedar), and (3) southeastern Coastal Plain and Piedmont forest (longleaf pine, loblolly pine, shortleaf pine). Cold temperatures, short or cool growing season, low nutrient availability, fire, and high wind velocity are most important in boreal and Rocky Mountain forest ecosystems. Pacific Northwest forest ecosystems have less extreme temperature conditions, yet summers are short and relatively cool; here, mid-summer drought is important. Also, nutrient availability is generally low. The southern pine forests are markedly different in temperature, having relatively long, humid growing seasons. Nevertheless, the pines typically grow on nutrient poor soils and in early-successional stages that are maintained by frequent fires. Details of the temperature, carbon balance, and nutrient relations of these groups in relation to their distribution are presented in Chapters 9, 18, and 19.

In this chapter, our examination of ecosystem change has focused primarily on a single physical environmental factor, climate, which is unquestionably an overriding ecological control, especially affecting species distributions of angiosperms and gymnosperms over evolutionary time scales. Obviously, it is an oversimplification to think of climate as the only factor determining the composition, structure, and function of landscape ecosystems. Nevertheless, plant species are closely integrated to their geographic ecosystems through strong genetic adaptations to climatic factors. This relationship is a dominant theme in the following chapter on tree variation and diversity. In the first three chapters, the ecosystem level has been our focus. In the next three chapters we focus on another volumetric unit, the organism, and consider organisms and their populations as integral parts of landscape ecosystems.

Suggested Readings

Clark, J. S. 1990. Fire and climate change during the last 750 years in northwestern Minnesota. *Ecol. Monogr.* 60:135–169.

Davis, M. B. 1983. Holocene vegetational history of the eastern United States. In *The Late Quaternary Environments of the United States.* Vol. 2, *The Holocene,* H. E. Wright and S. Porter (eds.). Univ. Minnesota Press, Minneapolis.

Davis, M. B. 1987. Invasions of forest communities during the Holocene: beech and hemlock in the Great Lakes region. In A. J. Gray, M. L. Crawley, and P. J. Edwards (eds.), *Colonization, Succession, and Stability.* Blackwell, London.

Delcourt, H. R., and P. A. Delcourt. 1991. *Quaternary Ecology, A Paleoecological Perspective.* Chapman and Hall, New York. 242 pp.

Delcourt, P. A., and H. R. Delcourt. 1987. *Long-Term Forest Dynamics of the Temperate Zone.* Springer-Verlag. 439 pp.

_____ 1993. Paleoclimates, paleovegetation, and paleofloras during the late Quaternary. In N. Moran (conv. ed.) *Flora of North America North of Mexico,* Vol 1, *Introduction,* Oxford Univ. Press, Oxford.

Pielou, E. C. 1991. *After the Ice Age.* Univ. Chicago Press, Chicago. 366 pp.

Tallis, J. H. 1991. *Plant Community History.* Chapman and Hall, New York. 398 pp.

Wright, H. E., Jr., J. E. Kutzbach, T. Webb III, W. F. Ruddiman, F. A. Street-Perrott, and P. J. Bartlein (eds.). 1993. *Global Climates Since the Last Glacial Maximum.* Univ. Minnesota Press. Minneapolis. 569 pp.

PART 2

The Forest Tree

In the three chapters of this Part, our focus is on forest organisms and their populations in relation to forest ecosystems, of which they are a part. In Chapter 4, the genetic differentiation of populations is seen to be closely related to the specific climatic and physiographic conditions in which plants grow. In addition, we examine the considerable within-plant variation that is important in survival and persistence of woody plant populations. The processes and factors affecting variation within and among populations and at the species level, such as hybridization and polyploidy, are also considered.

In Chapter 5, the life history of forest organisms is considered with emphasis on the regeneration of woody plants. The important processes of reproduction, germination, and establishment are discussed. In Chapter 6, the focus is on how trees grow in relation to site conditions. We examine the structural parts of woody plants, the crown and its architecture and the structure and growth of stems and roots. Also, the effects of water stress on the growth of plants are emphasized.

CHAPTER 4

FOREST TREE VARIATION

The inseparable interconnection between organism and environment is nowhere more evident than in the genetic adaptations linking these seemingly distinct ecosystem components. A forest tree, or any other organism, cannot exist without a governing biochemical control mechanism that can be passed on from generation to generation to perpetuate it. Also the organism cannot exist independently of the environment in which it occurs. It is futile to argue whether genetic or environmental factors control the form and development of an organism. Both always together determine the nature of the phenotype. Environmental factors, being generally visible and readily accessible, are the most obvious, and it is natural that most ecologists have been preoccupied with their study. Genetic factors have been fully appreciated only in recent decades, and their assessment in forest trees still lags behind their study in smaller and more tractable organisms.

The ability of organisms to live and reproduce in a given range of environments, termed **adaptedness** (Dobzhansky, 1968), has been known since the days of Aristotle. A convincing explanation was given by Darwin, although the causes of heredity variation were then unknown. Darwin's theory of evolution by natural selection is an ecological theory based upon ecological observations. It is important for the forest ecologist to consider adaptedness of forest species and their ability to adapt to changing environmental conditions.

In the present chapter, the primary focus is variation: the sources of variation; the kinds and extent of variation within and between individuals, populations, and species; and how the environment and biotic factors elicit adaptive changes in tree populations. Detailed treatments are available for the fields of physiological and population genetics (Ayala, 1982; Gottlieb and Jain, 1988; Adams et al., 1992); geographic variation in trees (Morgenstern, 1996), and forest genetics and tree improvement (Zobel and Talbert, 1984).

COMPONENTS OF PHENOTYPIC VARIATION

As indicated in Chapter 1, the ecologist may work with organisms at various levels of complexity—the individual and aggregations of individuals, such as populations and communities. The individual is the least arbitrary of these units. The genetic constitution of an individual is termed the **genotype**. We can never see a genotype because from the moment of fertilization the genotype is influenced by the plant's environment—its internal environment of cells and biochemical reactions and its external environment of light, temperature, and moisture. We see only the result—the **phenotype**—the observable properties of an organism produced by the genotype together with its environment. We may express this relationship for the entire organism, or for individual characters, by the simple formula **P = G + E + GE**. The phenotype (**P**) is the sum total of the effects of these components, the genetic information coded in the chromosomes (**G**), the environment (**E**), and the interaction of the genotype and the environment (**GE**). Often the genotype–environment interaction is small and can be disregarded. However, in certain cases it is of major importance. Through complex pathways, the genes control physiological functions, and many of these influence the morphology of the plant. The interrelationship of the genotype, environment, and plant processes of the phenotype is illustrated in Figure 4.1.

A recurrent question confronting the ecologist is the degree to which a phenotypic character is controlled by the genotype and by the environment, respectively. For example, consider a dominant tree exhibiting excellent growth in a forest plantation and an adjacent dying tree of the same species. The dominant has a superior phenotype, but to what

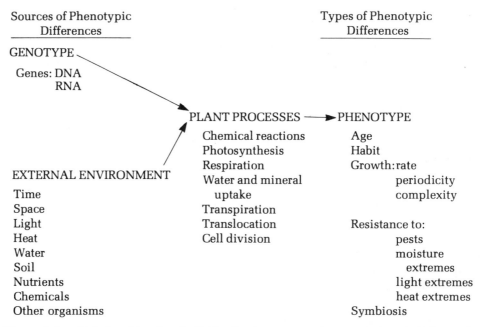

Figure 4.1.   Relationship of an individual's phenotype to its genotype and environment. Differences in plant phenotypes have their origin in the genetic constitution (genotype) of an individual and the effects of environmental factors on the genotype. Phenotypic differences of the individual become apparent as physiological processes occur in the internal environment of plant cells, tissues, and organs. (Courtesy of J. W. Hanover.)

degree is its genotype responsible, and to what degree does a favorable microenvironment or the random factor of better handling in the nursery account for the difference? If we set P in the formula equal to 100, do environment and genotype contribute equally (50–50), or is one more important than the other? If so, how much? Despite its poor phenotype, the genotype of the suppressed tree may be superior to that of the dominant—or, of course, it may not. Both components are always involved and no two phenotypes are exactly alike. The important point is the relative degree of genetic and nongenetic influence on a given character.

To determine the relative effects of genotype and environment, forest geneticists use statistical methods to compute genetic (**Vg**) and environmental (**Ve**) variance in the process of making a quantitative estimate of the strength of genetic control for a given character. Once these values have been determined, the total phenotypic variance (**Vp**) is equal to their sum (Vp = Vg + Ve). The strength of genetic control, termed **heritability** (i. e., ability of a character to be passed on to successive generations) of a character is then determined by using the ratio of genetic variance to the total phenotypic variance (heritability = Vg/Vp). If the ratio of the genetic to total variance is high (for example, 80 percent), it indicates a strong genetic control for the trait. Strong genetic control (high heritability—and therefore high likelihood of being inherited) has been shown for stem straightness, stemwood specific gravity, susceptibility to leaf rusts, and date of bud burst in spring. Traits that are highly influenced by the environment, such as height (strongly influenced by soil water and fertility) and diameter (strongly influenced by stand density) are predictably under weak genetic control, and accordingly, have low heritabilities.

Plasticity of the Phenotype

A given genotype may assume one set of characters (exhibit one phenotype) in one environment and exhibit a different phenotype in another environment. This plasticity, termed the **plasticity of the phenotype** (Bradshaw, 1965; Schlichting, 1986) is defined as the degree to which a character of a given genotype can be modified (a nongenetic change) by environmental conditions.

For herbaceous species, Bradshaw cites as plastic characters: size of vegetative parts; number of shoots, leaves, and flowers; and elongation rate of stems. Nonplastic characters include leaf shape, serration of leaf margin, and floral characteristics. There is good reason to believe these findings hold true for woody plants as well. In general, characters formed over long time periods of meristematic activity, such as stem elongation, are more subject to environmental influences and are more plastic than characters like reproductive structures that are formed rapidly, or traits such as leaf shape, whose pattern is impressed at an early stage of shoot development.

Plasticity has substantial adaptive value to plants in general and trees in particular because trees are rooted in their environment and have life spans typically longer than annuals or herbaceous perennials. An example of plasticity of adaptive significance is the rooting habit of individuals of many tree species, particularly Norway spruce, white spruce, and balsam fir. Roots develop either shallow or deep, depending on the soil environment, for example, shallow in a poorly-drained swamp versus deeply in a sandy loam upland soil. If individuals of a species exhibit high plasticity for certain establishment and growth characters, they may be able to regenerate and maintain themselves in a variety of sites. And as adults they may endure decades or centuries of fluctuating climate.

We can summarize the differences in phenotypes by examining three hypothetical situations involving individuals of a given species:

Situation A	*Situation B*	*Situation C*
P1 = G1 + E1	*P1 = G1 + E1*	*P1 = G1 + E1*
P2 = G2 + E2	*P2 = G1 + E2*	*P2 = G2 + E1*
P3 = G3 + E3	*P3 = G1 + E3*	*P3 = G3 + E1*

The phenotypes in A illustrate the typical situation in the field. All phenotypes have different genotypes, and the environments are all different enough to contribute to differences in the phenotypes. Situations B and C illustrate experimental situations where we can either hold constant the genotype (in B) or test different genotypes in a given environment (in C).

Situation B illustrates plasticity, but would have to be related to a specific characteristic. Different phenotypes of a single genotype (G1) are the result of the environmental differences; the environment has modified growth and development. In nature, the degree of plasticity of a character cannot be measured precisely because each individual typically has a different genotype (as in A). The extent of environmental modification can only be *inferred*. For example, individual genotypes of an even-aged stand in rolling terrain may occur from a dry ridge top to a moist, fertile valley. We observe a marked increase in tree height as we progress from ridge top into the valley. If it is unlikely that there are major differences among genotypes along the gradient, we may infer that environment is the major factor controlling the observed phenotypic differences in height. However, to determine precisely the plasticity for representative genotypes we would have to conduct experiments based on model C.

In situation C, we see that if the environment is the same for all individuals, phenotypic differences are due to differences among genotypes, and the amount of genetic variation can be estimated directly from the phenotypes. In practice, the environment cannot be held constant. However, we may approach this ideal by using either growth chambers or relatively uniform field test plots and a replicated experimental design. This "common garden" method is widely applied in determining genetic differences among selected individuals or populations.

SOURCES OF VARIATION

As we have seen, variation among phenotypes is attributable partly to the genotype and partly to the environment. The major sources of genetic variation are **mutation** and **recombination** of genes. Gene mutations have the effect of adding to the pool of genetic variability by increasing the number of **alleles** (the different forms of a gene) available for recombination at each locus.

Continuous or polygenic variation is typical for most characters of plants. This is due to the simultaneous segregation and interaction of many genes affecting the character and the continuous variation arising from nongenetic causes. Only a few traits are controlled by a single gene with major effects. Chlorophyll deficiency (albinism) in seedlings of species of the Pinaceae (Franklin, 1970) is one example. Also, single-gene mutations may be responsible for the marked differences in leaf morphology of closely related taxa. For example, the rare Virginia round-leaf birch of the Southern Appalachians is similar to sweet birch except for its round leaves (Sharik et al., 1990). Similarly, a single-gene mutation in a white ash population may have produced the single-leaf ash, a form of white ash but with simple adult leaves rather than compound leaves (Wagner et al., 1988).

Although mutation is the ultimate source of genetic variation, it is recombination that spreads mutations and extracts the maximum variability from them. Recombination is regarded as the major source of genetic variation of individuals in sexual systems. It distributes the raw material of variation which is then acted upon by natural selection.

The exchange of genes between different populations is termed **gene flow**; it is another source of variation. Migrants, in the form of pollen and seed, bring to a population new genetic material from other populations. When the populations involved are substantially different (such as species), the process is termed **hybridization**.

The major sources of nongenetic variation are (1) the external or physical environment (climatic, landform, soil, biotic factors) and (2) the internal environment of the plant. Factors of the physical environment that modify plants in many ways, and elicit genetic adaptations, are considered in Chapters 7 to 12. Much less appreciated is variation within an individual that is not directly related to factors of the external environment.

Although all cells of a tree have the same genetic constitution, the internal environment of the organism may affect the expression of genes and hence the traits we observe and measure. In the development of a seedling to an adult tree, striking physiological changes occur, and a series of developmental phases is recognized. Best known are the differences between the juvenile and adult phases. The characteristic features of these phases are apparently due to changes that take place in the apical meristems as they age. The most universal feature is the inability of trees in the juvenile stage to flower (Chapter 5).

Schaffalitzky de Muckadell (1959, 1962) investigated many other characters exhibiting differences in juvenile and adult phases. In European beech and oak, the brown and withered leaves (marcescent condition) are retained over winter by trees in the juvenile phase, but the leaves are deciduous in the adult phase. This feature is also observed in American beech and many oaks. Surprisingly, entire portions of the lower trunk and branches, even of very old trees, remain in the juvenile phase and retain the juvenile trait of leaf retention. Reciprocal grafting experiments of juvenile and adult branches show that the juvenile phase consistently leafs out later than the adult phase of European beech. This juvenile trait may be of adaptive value because late spring frost can pose a serious problem in young beech stands.

THE EVOLUTIONARY SEQUENCE

In evolutionary biology, the basic unit in sexual populations is not the individual, which by itself has only a limited future. Instead, it is the local interbreeding population,[1] a group of genotypes that are potentially interbreeding and that constitute the gene pool for that population. Without the heritable variation of these genotypes, natural selection cannot bring about evolutionary changes. Evolution is the cumulative change in genetic make up of populations of an organism during the course of successive generations—in simplest form a change in gene frequencies. Such changes are brought about by the guiding force of natural selection. Natural selection is defined as the differential and nonrandom reproduction of genotypes, stressing that ultimately it is the differential reproduction of genotypes, rather than just differential mortality of individuals, that brings about evolutionary change.

Of the thousands or millions of zygotes of a species on a given site that might develop to maturity and contribute offspring to the next generation, only a few survive. An example of the raw materials available to selection and the severity of selection comes from sugar maple. Curtis (1959) reported that of 6,678,400 potentially viable sugar maple seeds, 55.7 percent germinated, giving 3,673,100 seedlings per hectare. In late summer of

---

[1]A population is any group of individuals considered together because of a particular spatial, temporal, or other relationship (Heslop-Harrison, 1967).

the same year, 198,740 remained, and 2 years later only 35,380 seedlings remained alive—less than 3 percent of those germinating. He estimated that the opening resulting from the death of a mature sugar maple tree (about 6 to 7 m in diameter) would initially support about 15,000 seedlings. They would be reduced to about 150 during the first 3 years, and eventually only 1 or 2 trees would occupy the opening. Although selection acts in every phase of the plant's life, it is most effective on young seedlings and in many species eliminates all that are not well adapted to their immediate environment.

Sexual and Asexual Systems

The sexual breeding system, dominated by cross-pollination, characterizes most woody species. Evolution proceeds much faster in a sexual system where new genotypes are constantly generated. Nevertheless, most woody species have some form of asexual reproduction or **apomixis**—any means of reproduction, including vegetative propagation, which does not involve fertilization. In angiosperms, this includes sprouting from the basal part of the stem (oaks, hickories) and roots (aspens, sweetgum) following fire or cutting. Conifers rarely form sprouts (exceptions being redwood and juvenile individuals of loblolly pine), but the production of adventitious roots from branches (termed **layering**) in larch, spruces, and firs is an important mechanism for regeneration in swamp or moist upland environments. Apomixis may even occur when seeds are formed asexually, as in hawthorns. Whatever the mechanism, asexual reproduction gives the plant immediate fitness in the prevailing environment; it is a uniformity-promoting device.

A striking example is seen in the aspens of North America and Eurasia, which sucker from roots to form natural colonies termed clones. A **clone** is the aggregate of stems produced asexually from one sexually produced seedling. The clone is the typical growth habit of aspens throughout their worldwide range (Barnes, 1967), and their asexual proclivity may be a major factor in their ability to compete successfully in conifer-dominated landscapes where fire is ever present to stimulate sprouting. In parts of western North America, aspens occur in large clones, some over 40 ha in size (Kemperman and Barnes, 1976). They may live almost indefinitely by recurrent suckering, provided fire is not excluded.

GENETIC DIVERSITY OF WOODY SPECIES

We can estimate the overall genetic diversity in a species and its populations by determining the variation in allelic variants (termed **allozymes**) at a single gene locus. Gene frequencies can be determined for the allelic variants and, when computed for many loci for many populations and individuals, the genetic diversity of species and their populations may be estimated.

Based on a review of plant allozyme literature, including data from 213 woody species representing 54 genera, long-lived woody plants show higher levels of allozyme variation within their populations than other plant groups (Hamrick et al, 1992; Hamrick and Godt, 1989). Compared to other life forms, the mean genetic diversity of long-lived woody plants is 15 percent greater than that of annuals, 42 percent higher than in herbaceous perennials, and 53 percent greater than in short-lived woody species. The best predictor of levels of allozyme variation in long-lived woody species is geographic range.

Widespread and regionally distributed species maintain the most diversity, whereas endemic woody species contain the least. Species exhibiting high levels of genetic diversity are Scots pine and white spruce, whereas exceptionally low levels are found in red pine, Torrey pine, and balsam poplar. Hamrick et al. (1992) conclude that woody species are more diverse than herbaceous species because they combine the following life history and population traits that act to preserve genetic diversity: large continuous populations, long lives, large size, outcrossing breeding systems, and relatively long-distance pollen and seed dispersal. The patterns of this amazing diversity and the processes that account for it are described in the sections that follow—again emphasizing the inseparability of organism and environment.

GENECOLOGY

The foregoing review of the causes of changes in gene frequencies in populations and great allelic diversity lead to the consideration of **genecology,** a term Turesson (1923) applied to the study of variation in plant species from an ecological viewpoint. Specifically, genecology is the study of adaptive properties of any sexual population—species, subspecies, race, local interbreeding population—in relation to environment (Langlet, 1971). Turesson, like others before him, demonstrated conclusively that ecologically-correlated phenotypic variation among populations was usually genetically based rather than merely the result of environmentally induced modification of individuals. This concept has major practical implications for the land manager when introducing populations into a new environment. In tree-introduction attempts throughout the world, we have largely used a trial-and-error method, with some resounding successes and some dismal failures. Such a method is too problematical and costly as a general practice. Instead, we need to be able to predict how a given species population will perform when grown in a new environment. To do this we need to know what environmental and biotic factors elicit a genetic response, how finely populations are adapted to these factors, and the patterns of adaptation along major environmental gradients. The basis of our knowledge of genecological adaptation and examples of it are presented in the following text sections.

Comparative cultivation of seedling populations of forest trees, originating from environmentally different sites, was pioneered by Duhamel du Monceau about 1745 (Langlet, 1971), and the methods were continued and refined by other workers such a P. de Vilmorin (in the 1820s), Kienitz (1879a,b), Cieslar (1887–1907), Engler (1905–1913), among others (Langlet, 1971). The careful historical documentation by Langlet (1971) makes it clear that the concept of genecological diversity was known to forest botanists long before Turesson's work with herbaceous species. For example, Cieslar in Austria (1895, 1899) and Engler in Switzerland (1905, 1908) experimentally determined that forest trees in the Alps were genetically adapted to the climatic conditions of their respective environments. In 1895, Cieslar published evidence of a continuous gradient of juvenile height growth for Norway spruce, demonstrating its genetic adaptation to growing-season conditions grading from low to high altitude. These results served to document observations of the previous century, published in 1788, that seedlings of lowland areas proved worthless on mountain sites.

Unlike Cieslar and Engler, who used seeds for their experimental work, Turesson transplanted whole individuals from markedly different habitats, but, like his predecessors, he grew them under standard conditions in a common garden. The phenotypes Turesson observed in nature were usually different in habit of growth (procumbent or erect) and in various morphological characters. These differences were usually maintained in the garden and hence indicated genetic differences among the populations studied. Presumably, natural populations have been exposed to the factors of their respective environments for generations. Forces of natural selection have guided the genetic differentiation of each population so that it is more or less adjusted to the daily, seasonal, yearly, and even longer-term climatic and soil water fluctuations of its respective environment.

Because of the problems of preconditioning of whole plants (Rowe, 1964) and the difficulty in transplanting whole forest trees, forest scientists typically collect seeds from the desired populations, termed **provenances**, raise the seedlings in a common garden, and study the differences among the provenances. Such experiments are termed **provenance**, or **seed source**, tests. This type of testing determines (1) if there are significant genetic differences among populations in the characters chosen for study, and (2) the amount of genetic differentiation among provenances under the environmental conditions of the common garden. Provenance testing does not directly indicate what mechanisms caused the differences, although these may be inferred.

Patterns of Genecological Differentiation

A wealth of evidence has accumulated confirming that genetically-based ecological differentiation or divergence of populations, termed genecological differentiation, is a recurrent feature of plants. However, controversy arose over the pattern of differentiation—whether it is discontinuous or whether it is continuous in nature. Huxley (1938, 1939) introduced the term **cline** to designate a gradation in measurable characters, which might be continuous or discontinuous, stepped or smooth, or sloping in various ways. The term cline itself, as the definition indicates, does not mean or necessarily imply a genetically-based gradation in a character. It could refer to any gradation of phenotypic characters observed along a natural gradient, as well as a gradation that is genetically based.

Genecological differentiation is a multidimensional response of individuals of a population to their environment. Although the response is unique for each population and species, we present the following generalizations as best summarizing current understanding of differentiation in forest species.

1. The total natural range of a species, the distribution pattern (continuous, discontinuous, mosaic) of a species within this range, and the way in which the conditioning environmental factors vary are three major determinants of the differentiation pattern. If a species is distributed continuously over a wide range, particularly in latitude or elevation, it is subjected to more or less continuously varying climatic factors, and genetic variation tends to be continuous. If discontinuities occur in the species' distribution, or if the conditioning factors are discontinuous and sufficiently distinct, a discontinuous pattern may result. The variation pattern may be visualized as a series of contour lines whose spacing reflects the rate of change in the conditioning factors.

2. The results of a given genecological study tend to be related to the scale in which it is conceived and conducted (Heslop-Harrison, 1964). A wide-ranging investigation of a species along a north–south gradient may expose a clinal variation pattern that may mask other clines associated with elevation at a given latitude, or local discontinuities that may arise from marked changes in microclimate or soil drainage.

3. However continuous a cline may be, it is usually possible to show a seeming discontinuity by incomplete sampling and certain methods of data analysis (Langlet, 1959).

4. The dominant pattern of genetically-based variation is more or less continuous because the major factors of climate that elicit genetic variation are continuous; a discontinuous pattern is usually the exception. The clinal pattern has a fundamental basis in the genetic system of many forest trees favoring a high degree of recombination and outbreeding (cross-pollination) and associated features of (a) long life span of individuals, (b) relatively great site stability, (c) high and selective seedling mortality, and (d) high physiological tolerance in the adult to fluctuating environmental conditions. Discontinuous variation, being favored by inbreeding, low rates of gene recombination, and short life span in strongly fluctuating sites and communities, is more typical in herbaceous than in tree species. These and other evolutionary aspects of genecological differentiation are discussed by Heslop-Harrison (1964), Linhart (1989), Hamrick (1989), and Hamrick et al. (1992).

GENECOLOGICAL CATEGORIES

It is often useful to subdivide into classes the continuum of genetic differentiation occurring among species populations along ecological or geographic gradients. The generic term **race** is widely used to designate species populations that differ significantly in one or more morphological and physiological characters in common garden tests. Races may be defined as populations of a species that differ in the frequencies of one or more gene alleles (Dobzhansky, 1951). One may identify different kinds of races by appropriate adjectives related to scale of differentiation or eliciting factor: geographic or local race; climatic, edaphic (soil), or photoperiodic race. Races are basic elements of evolution; if genetic differences are especially great the populations may be recognized as formal taxonomic categories or taxa: variety, subspecies, or species. The term **ecotype** also is used to designate genetically different population units.

The Swedish Botanist, Göte Turesson, pioneered experimental studies of herbaceous plants in common gardens and emphasized the ecological basis of population differentiation by introducing the term ecotype to signify an ecological race. Turesson (1922a,b) defined ecotype as the product arising as a result of the genotypic response of a population to a "definite habitat" or "particular habitat," i.e., a local ecological race. However, the term ecotype has been used in many different contexts, each having a different genecological significance. The term fails to provide a useful concept for a specialized local race for which it seems best intended. We suggest use of the term, race, to distinguish units of genetic differentiation below the taxonomic rank of variety. An appropriate prefix (local, swamp, geographic) then may be used to explicitly characterize the spatial scale or key site factor. In the classic studies of intraspecific variation in native herbaceous perennials of California, for example, Clausen, Keck, and Heisey used the term ecotype in their early studies (1940), but abandoned it in favor of ecological races in general and climatic races in particular in their later publications (1958a,b).

Factors Eliciting Genecological Differentiation

Marked genetic differences in growth and other characters usually are expressed when populations are grown in latitudes or elevations substantially different from that of their native habitats. Limiting environmental factors affecting the length and nature of the growing season in the native habitat (such as mean and extreme temperatures, occurrence of early and late frosts, thermoperiod, photoperiod, and amount and periodicity of rainfall) are major selective forces affecting survival, growth, and reproduction and thereby elicit genetic differences in plant populations.

Not only are characters influenced by many factors of the physical environment, but also by plant associates and the interrelated selective pressures of insects, mammals, and birds (see Chapter 14). Coevolutionary systems have been reported for animals and reproductive traits of various woody-plant species. In studies of woody legumes in Central America, Janzen (1969) listed 31 traits that may act to eliminate or lower the destruction of seeds by bruchid beetles. The major defense mechanisms against these predators are deterrents, such as biochemical repellents (alkaloids and free amino acids), or an increase in the number of seeds to the point of predator satiation, probably requiring a decrease in seed size. These considerations emphasize the multidimensional nature of adaptation.

GROWTH CESSATION

The nature and patterns of genetic differentiation are closely related to the plant's efficient use of the growing season. In temperate zones, plants must not grow too late in the autumn or they will be damaged or killed by early autumn frosts. However, they must not cease growth too soon because longer-growing individuals may overtop and suppress them. In their native habitats, species anticipate seasonal fluctuations by responding genetically to the more reliable factors of their environment (such as day length and heat sum) than to more variable factors such as occurrence of frost. In temperate regions, response to a photoperiodic signal of shortening days of summer and autumn sets in motion a gradual and complex process of acclimation to dormancy (Chapter 9). **Photoperiodism**, the response of plants to the timing of light and darkness (usually expressed as day length), is a biological clock enabling plants to adjust their metabolism to certain seasonal fluctuations. Unlike other environmental factors, day length changes everywhere in a regular annual cycle, except at the equator where there is no change. Plants can monitor these changes with remarkable precision.

Photoperiod largely controls the entrance into dormancy of many woody plants, particularly species with northern ranges. These species are genetically adapted to a photoperiod that enables them to become dormant before the time when factors of their prevailing environment, such as freezing temperatures, become limiting. For plants in northern climates or at high elevations, early frosts in autumn and cold winters are factors that significantly affect survival. Hence a reliable mechanism, such as photoperiod, in triggering the dormancy sequence may be highly developed. For example, Figure 4.2 illustrates a very strong relationship between growth cessation and latitude for Sitka spruce and reflects a close adaptation to day length. Sitka spruce occurs in a narrow band along the Pacific coast of North America, extending over 20° of latitude from California to Alaska. This re-

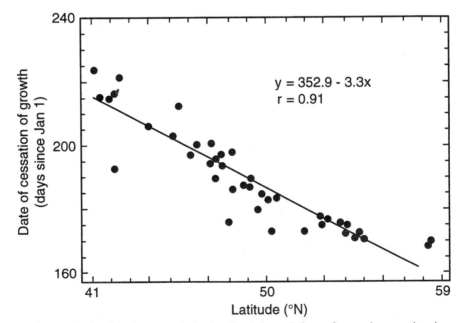

$$y = 352.9 - 3.3x$$
$$r = 0.91$$

Figure 4.2. Relationship between latitude of origin and date of growth cessation in autumn of origin for 43 Sitka spruce provenances from the Pacific coast of North America. Provenances three years old; tested in Germany; data from Kleinschmit (1978). (After Morgenstern, 1996. This excerpt is reprinted with permission of the publisher from *Geographic Variation in Forest Trees* by E. K. Morgenstern © UBC Press 1996. All rights reserved by the Publisher.)

lationship is very strong because latitude, and its associated day-length regime, characterizes very well the progressive change in relatively homogeneous climate along this coastal strip. Such a close relationship with latitude is not found where species range over mountainous terrain and encounter temperature and moisture conditions that may vary widely within any given latitude.

In almost all genecological and provenance tests, populations are grown in day-length regimes different from that prevailing in their native habitats. In the western species black cottonwood, for example, individuals of high-latitude provenances ceased height growth in June when planted at a low-latitude site near Boston, Massachusetts (Figure 4.3; Pauley and Perry, 1954; Pauley, 1958). Southerly provenances, moved north to the test site, continued height growth until September and October; some individuals ceased growth only when their terminal shoots were killed by the first severe frost. Generally, movement from the natural habitat northward, into longer days, prolongs the active period of growth and results in greater plant size of the southern populations compared to the northern populations native to that site. However, such a move, if too far north, may render plants susceptible to early frosts and may lead to injury and decreased growth or death. Movement of northern populations southward into shorter summer days shortens the active growth period in comparison to native plants at or south of the test site. In the black cottonwood example, individuals of high-latitude provenances grew only about 15 to 20 cm, whereas those from southern localities grew about 2 m (Pauley, 1958). When clones from the high latitude at the test site were given longer days by artificial light, they grew over 1.3 m (Pauley and Perry, 1954), indicating a strong influence of day length in regulating growth.

Although a significant, genetically-based, clinal response was shown in relation to latitude, the response is not simple and direct (compare with the Sitka spruce example shown in Figure 4.2) as evidenced by substantial variation among provenances from 44° to 48° (Figure 4.3). These provenances included a variety of sources sampled from the Pacific coast to western Montana and over an elevational range from sea level to 1,525 m. Although the difference in latitude is not great, there is known to be a marked difference

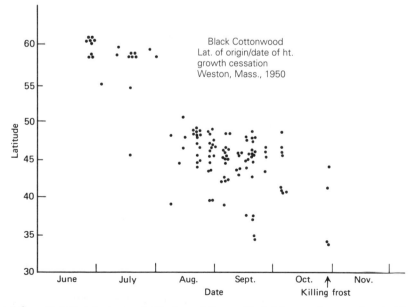

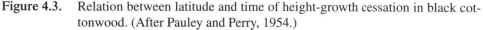

Figure 4.3.   Relation between latitude and time of height-growth cessation in black cottonwood. (After Pauley and Perry, 1954.)

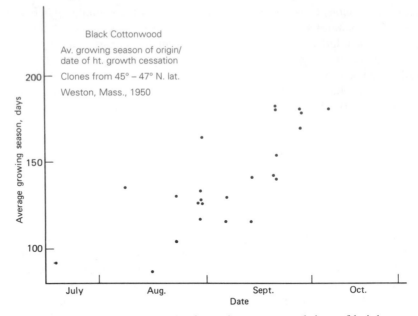

Figure 4.4. Relation between length of growing season and time of height-growth cessation in black cottonwood. (After Pauley and Perry, 1954.)

in the length of growing season among these sources due to elevation, aspect, and microsite conditions. Clinal genetic adaptation to length of growing season was found within the narrow latitudinal range of 45° to 47° (Figure 4.4) and elevation of the source probably explains much of the variation not accounted for by latitude. Populations at low and high elevations at a given latitude have growing seasons of different lengths. They become adapted to different photoperiods accordingly, and in particular to a critical day length in autumn that is important in regulating their entrance into dormancy. High-elevation populations necessarily cease growth earlier than low-elevation populations due to earlier occurrence of killing frosts. Hence they adapt to relatively longer day lengths (occurring earlier in the year) than those at low elevations.

Carrying this approach one step further, we see that a population at a high elevation may have the same length of growing season as a population at a lower elevation that occurs several degrees of latitude farther north due to the compensation of latitude for elevation. Through equivalence in length of growing season, they would have a similar photoperiodic adaptation mechanism and, if interchanged, may show only negligible differences in growth rate.

This interrelationship between elevation and latitude rarely has been recognized in genecological studies. Often, correlations of cessation of growth or plant size and latitude of source are confounded by elevational differences. To avoid this problem, Wiersma (1962) modified a formula developed for Swedish conditions, using growing season (number of days ≥6°C) to relate latitude and elevation. He reported that a displacement of 1 degree of latitude north is equivalent to a displacement of 100 m upward in altitude. Wiersma (1963) recomputed correlations of latitude of source and various characters from published papers using this adjustment and found greatly improved relationships. Similarly, Sharik and Barnes (1976) found that adjusting latitude for elevation substantially improved the correlation of latitude of origin and cessation of height growth for yellow birch and black birch populations, compared to the unadjusted relationships.

The functioning of this photoperiodic mechanism over gradients of latitude and elevation has been demonstrated experimentally for the wide-ranging Norway spruce of northern and central Europe. The critical night length that stimulates bud set is about 6 to 7 hours in southerly populations but only 2 hours in northern ones (Figure 4.5). Such a clinal pattern of variation is also evident for elevation. Austrian populations at 700 m exhibited a critical night length of 7 hours, whereas it was 5 hours for those at 1,400 m. The marked effect of growing season length at a given latitude and elevation is illustrated in the difference in critical night length for populations on the west coast of Norway contrasted with those of northern Finland (4 versus 2 hours) (Figure 4.5). Besides controlling growth cessation, photoperiod may also affect the optimum growth of plants during the growing season itself. Southern populations of both Scots and lodgepole pines grew vigorously in a 16-hour day, whereas northern populations ceased to grow; their optimum was an 18-hour day (Ekberg et al., 1979).

GROWTH RESUMPTION

Although the initiation of growth cessation for temperate trees is based on photoperiod, flowering and the resumption of vegetative growth in the spring, **flushing** or **bud burst**, once a winter chilling requirement has been satisfied, is strongly related to temperature

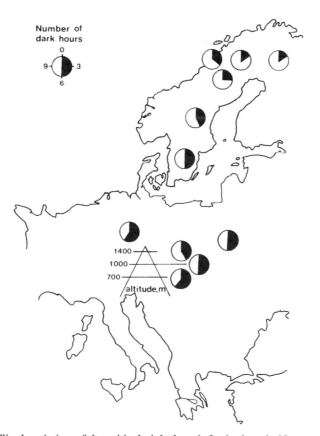

Figure 4.5. Clinal variation of the critical night length for bud-set in Norway spruce populations of different geographic origin in Fennoscandia and central Europe. Critical night length is based on the number of dark hours bringing about bud-set in 50 percent of the plants after pretreatment with continuous light. (After Ekberg et al., 1979. Reprinted by permission of Munksgaard Int. Booksellers & Publishers.)

(Sarvas, 1969; Hunter and Lechowicz, 1992; Chapter 9). Trees of northern climate are genetically adapted to initiate flushing or flowering once a certain number of "heat units," above a base or threshold temperature have accumulated (Owens et al., 1977). This relationship has been reported for many conifers (Campbell and Sugano, 1979) and for 26 hardwood species of eastern North America (Hunter and Lechowicz, 1992).

Working in northern and central Europe, Linsser (1867) discovered that the mean temperature or heat sum[2] of spring flowering for both northern and southern populations of a given species is the *same fraction* of the whole year's average heat sum for their respective locality. That is, northern (or high elevation) individuals of a species are genetically adapted to flower and flush at a lower absolute heat sum than individuals of the same species at a more southerly location where the total heat sum is greater. Northern plants in nature may flower and flush at a relatively late calendar date, because spring warming is delayed and slow. However, when their progeny are grown at warmer, lower latitudes or lower elevations, together with native populations, they tend to flush *earlier* than the native populations and vice versa when southern populations are moved north. What might be perceived as a general north–south relationship, however, only holds when frost is a significant selection force. Exceptions to this north–south relationship, where northern sources don't flush earlier (Nienstaedt, 1974), are species with southern ranges (black walnut, sweetgum, tuliptree, American sycamore) or, as in the case of Sitka spruce (Burley 1966), with a coastal, ocean-moderated distribution.

Examples of Genecological Differentiation

The woody-plant literature abounds in references citing genetic differences in physiological and morphological characters among populations of many species. Much of the information comes from provenance tests established to meet the practical objective of finding the most suitable provenances for planting in one or more localities. Many of the early tests were therefore not designed to answer genecological questions of how and why populations are adapted to their environments. Despite inadequacies, provenance studies have uncovered major adaptive responses, primarily along latitudinal and altitudinal gradients.

EASTERN NORTH AMERICAN SPECIES
The great amount of genetic variation in deciduous angiosperms has been demonstrated repeatedly in common garden tests. Photoperiodic races associated with latitude have been reported for many species (Barnes, 1991; Morgenstern, 1996) and a variety of characters, including survival, time of growth cessation and flushing, height and diameter growth, frost resistance, winter kill and crown dieback, tree form, and foliage color.

Similarly, eastern conifers have shown widespread clinal genecological differentiation, including jack pine (Mátyás and Yeatman, 1992; Morgenstern, 1996), loblolly pine (Wells and Wakely, 1966; Wells et al., 1991), eastern white pine (Wright, 1970; Morgenstern, 1996), and black and white spruce (Park and Fowler, 1988; Morgenstern, 1996). The rate of change in a clinal variation pattern is well illustrated by the genetic variation of slash pine seedlings (Figure 4.6 Squillace, 1966). This study revealed weakly defined or highly fluctuating gradients as well as distinct clinal trends. The reversal in the general cline, common to many of the 25 characters studied, is illustrated in the variation pattern of needle length (Figure 4.6). From a low of 16 cm in southernmost Florida, needle length

---

[2]A **temperature** or **heat sum** is the product of temperature above a certain base or threshold level (such as 0° or 5° C) and the time duration of that temperature. It may be expressed in degree-hours or degree-days.

Figure 4.6. The pattern of variation of needle length (cm) in seedling progenies of slash pine. (After Squillace, 1966. Reprinted with permission of the Society of American Foresters.)

of the progenies increased to its longest values, 19 to 20 cm, in south central Florida and then progressively decreased to the north.

The variation patterns of the disease, fusiform rust, as well as growth and crown form of the wide-ranging and commercially important loblolly pine of the South have been of great interest (Wells et al., 1991; Wells and Wakely, 1966). In general, sources west of the Mississippi river and in southeastern Louisiana are resistant; sources from east of this area, except those from the northern Atlantic coast, are susceptible. Climatic conditions of the past are thought to explain this relationship. The hypothetical distribution of loblolly pine and slash pine at the height of the Wisconsinan Glaciation, 18,000 years ago, is shown in Figure 4.7. The wetter conditions in the west and along the northern Atlantic coast were optimum for selection for fusiform rust resistance, whereas dry conditions in Florida gave rise to susceptible populations that migrated into present day Alabama, Georgia, and South Carolina. It was once thought that the dividing line between resistant sources in the west and susceptible ones in the east was the Mississippi River basin. However, the alluvial plain, recognized today as the Mississippi Delta did not exist. Rather, it is probable that the Desoto Canyon (Figure 4.7) east of the Mississippi, acted to isolate eastern and western populations.

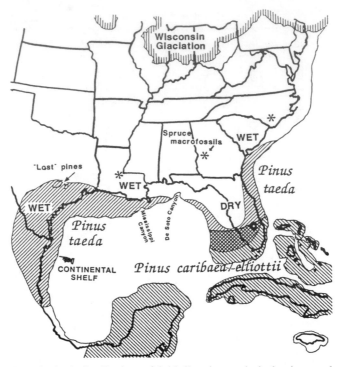

Figure 4.7.   Hypothetical distribution of loblolly pine and slash pine at the Wisconsinan glacial maximum, 18,000 years ago. Caribbean pine (*Pinus caribaea*) and Slash pine (*P. elliotii*) may have been a single species at that time and are shown as such. (After Wells et al., 1991. Reprinted by permission of the Bundesforschungsanstalt für Forst- und Holzwirtschaft, Grosshansdorf, Germany.)

SCOTS PINE

The Earth's most wide-ranging pine, spanning approximately 150° of longitude from Scotland to eastern Siberia, has been investigated in many studies at European (dating from 1745) and North American test sites. Genetic differences in such diverse characters as height growth, foliage color, stem form, rooting habit, resistance to insect attack, fruitfulness, and time of bud set have been demonstrated (Langlet, 1959; Troeger, 1960; Wright et al., 1966; Wright et al., 1967). A clinal pattern is evident for most of these characters. Portions of a cline, designated as races, ecotypes, or varieties, may be useful as the basis for selecting seed-collection zones, particularly where major differences in tree characteristics are important. American Christmas tree growers, for example, prefer Scots pine varieties from Spain and elsewhere in southern Europe, that remain green in winter, to those of northern Europe that turn yellow (Wright et al., 1966).

WIDE-RANGING WESTERN NORTH AMERICAN CONIFERS

Nowhere has the genetic response of species to heterogeneous environments been better demonstrated than in western North America. The extreme physiographic and climatic heterogeneity of landscapes throughout a wide range of latitude, longitude, and elevation from Mexico to northern British Columbia and Alberta, Canada have elicited very localized differentiation, as well as broad patterns of variation. Detailed studies of ponderosa pine, lodgepole pine, and Douglas-fir illustrate the range of differentiation within these species.

   Douglas-fir and lodgepole pine exhibit the most variation, in large part due to the diversity of sites they occupy (Critchfield, 1957, 1985; Wheeler and Critchfield, 1985; Rehfeldt, 1988). Both have coastal populations and range eastward to high elevations in the Rocky Mountains and arid interior lands. Ponderosa pine also exhibits marked variability on the broad subcontinental scale, substantial variation related to elevation in various parts of its range, and significant local differentiation. Examples of the different kinds of genetic responses of ponderosa pine and Douglas-fir to their heterogeneous environments are described next.

Ponderosa Pine

Ponderosa pine is one of the most wide-ranging pine species in North America, with races occurring from central Mexico to southern British Columbia, Canada. Ponderosa pine, classified in the Subsection Ponderosae of the genus *Pinus*, is subdivided into two varieties: var. *scopulorum* in the Rocky Mountain portion of its range and var. *ponderosa* in the western part of its range (Conkle and Critchfield, 1988). Two major races occur within var. *scopulorum* (Figure 4.8), and three geographic races occur within var. *ponderosa* (Figure 4.9) (Conkle and Critchfield, 1988).

Figure 4.8.   Eastern geographic races of ponderosa pine, excluding poorly understood populations in central Mexico. (After Conkle and Critchfield, 1988.)

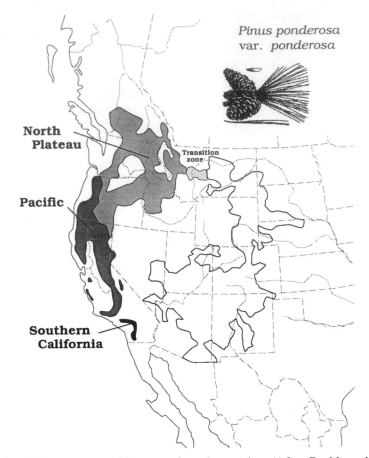

Figure 4.9. Western geographic races of ponderosa pine. (After Conkle and Critchfield, 1988.)

Evidence from morphological studies, biochemical analyses, and range-wide prove-nance testing illustrates differences among the major races. The Rocky Mountain variety was named for the compact, brushlike, bushy-tuft (scopulate) appearance of its foliage (Figure 4.8) in contrast to the open, plumelike foliage of far-western populations (Figure 4.9). In addition, the western variety is distinctive morphologically because of its general lack of 2-needle fascicles, compared to the Rocky Mountain variety (Figure 4.10). The number of needles per fascicle is influenced by tree age (fewer in young trees), climate, and site conditions (Haller, 1965). Two-needle fascicles have less surface area and fewer stomates, thereby reducing water loss, and they require less energy to produce than 3-nee-dle fascicles. These features are of survival value in harsh Rocky Mountain conditions.

Also, studies of the monoterpene components of xylem resin by Smith (1977) illus-trate marked differences among geographic races (Figure 4.11). Starting with the distinc-tive southern California race, a clockwise pattern of decrease is seen for a-pinene and limonene through the Pacific, North Plateau, and Rocky Mountain races to the Southwest-ern race. Conversely, 3-carene is negligible in southern California but present in signifi-cant amounts in the other races. Conkle and Critchfield (1988) emphasize the correspon-dence of physiographic barriers and the distinct monoterpene races and their sharp transition zones.

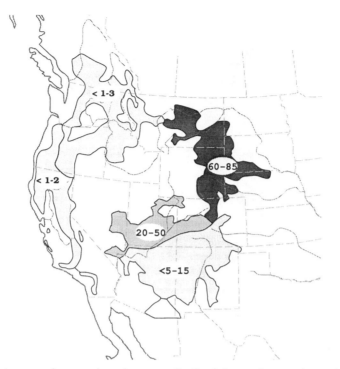

Figure 4.10. Average frequencies of two-needle fascicles on (mature/young) native ponderosa pines (Haller, 1965). Similar results were reported by Wiedman
(1939) and Read (1980) for native and plantation grown pines. (After Conkle and Critchfield, 1988.)

All these lines of evidence illustrate the enormous complexity of variation that can
be expected over such a diverse range of regional and local ecosystems that differ in
macro- and microclimate, elevation, and evolutionary history. In addition, great genetic
variation also occurs within these varieties and races, and it is strongly related to length of
the frost-free period along elevational gradients (Conkle, 1973; Read, 1980; Rehfeldt,
1986a,b, 1990).

Douglas-fir
Due to its ecological and commercial importance, the genetic differentiation of Douglas-
fir has been investigated in more detail than any other North American tree species. Two
major geographic varieties, the coastal or green type, *Pseudotsuga menziesii* var. *menziesii*, and the interior, Rocky Mountain, or blue type, *P. menziesii* var. *glauca*, span an
enormous geographic range in western North America. They thrive in heterogeneous environments in coastal and mountain areas of the Pacific Northwest and in the Rocky Mountains from Mexico to Alberta and British Columbia, Canada. Much broad-scale provenance testing has been conducted in Europe because this species is of the great value in
species-poor Europe. The most extensive provenance test involved 182 native provenances with plantings in 30 countries. Overall results showed a relatively clear separation
between coastal and interior varieties in many traits. Marked differences have been reported in frost sensitivity, phenology, morphology, and growth traits with broad and local
clinal trends associated with latitude and elevation (Kleinschmit and Bastien, 1992).
Overall, many of the populations of the coastal variety that perform best in field tests in

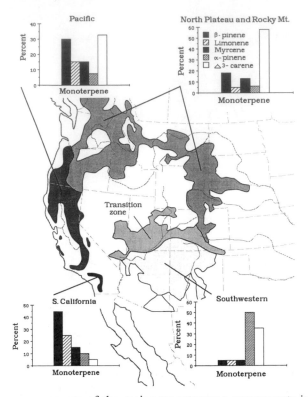

Figure 4.11. Average amounts of the major monoterpene components in xylem resin of native ponderosa pines (Smith, 1977). (After Conkle and Critchfield, 1988.)

Europe and the Pacific Northwest come from areas of high genetic diversity (Silen, 1978; Kleinschmit and Bastien, 1992). Douglas-fir not only exhibits marked genetic variation throughout its range but displays remarkable variation in morphological and physiological characters within physiographic regions of both coastal (Campbell and Sorensen, 1978; Campbell, 1986, 1991) and Rocky Mountain races (Rehfeldt, 1989).

LOCAL GENECOLOGICAL DIFFERENTIATION

In the previous discussion, we have seen that ecologically-based genetic differentiation is the rule at the macro- or range-wide scale for virtually all tree species and also mirrors the pattern of the heterogeneous environment. How finely then are populations adapted to local segments of these geographic and ecological gradients, i.e., to micro-scale differences in soil water, elevation, aspect, and soil type? Local adaptations to soil and microclimatic conditions have been demonstrated repeatedly for herbaceous species (Antonovics, 1971; Bradshaw, 1971; Snaydon and Davies, 1972, 1976). Chemical and physical properties of soils can elicit sharp discontinuities in plant distribution, and woody-plant species showing marked edaphic preferences are common. However, examples of intraspecific genetic adaptation to local soil type or to soil-water differences are rare.

Occasionally, local differentiation has been shown for populations on adjacent sites where soil water is either severely limiting or in excess. Genetic differences have been demonstrated between populations of jack pine growing in adjacent wet and dry sites (Wright et al., 1992). Those originating from the wet site had significantly more tertiary roots than those from dry sites. The greater intensity of tertiary rooting in the wet-site population probably enhances nutrient uptake in the spring when the soil is likely to be water-logged and lacking oxygen. Genetic differences were reported between a few populations of northern white-cedar from wet and dry sites (Habeck, 1958; Musselman et al., 1975) but have not been demonstrated for upland and lowland populations of black spruce (Fowler and Mullin, 1977; Boyle et al., 1990).

Slope aspect is often so different ecologically that local races are formed. Aspect differences have been demonstrated for Douglas-fir seedlings originating from north and south slopes in southern and central Oregon (Herman and Lavender, 1968; Campbell, 1979) and for Sitka spruce in southeastern Alaska (Campbell et al., 1989). Elevation and aspect are often associated with local differentiation, presumably due to marked differences in microclimate associated with these physiographic factors. For example, microevolution of the coastal race of Douglas-fir was demonstrated by Campbell (1979) in a single 6,100 ha (61 km^2) watershed in central Oregon where elevations range from 500 to 1,600 m. Seedlings of 193 trees located throughout the watershed were grown in a common garden. Genetic differentiation was surprisingly large in 14 of 16 traits, including seed germination rate, height, bud burst, bud set, and dry weight. For traits related to vegetative growth, variation appeared to be mainly associated with elevation. Estimates of maladaptation showed that none of the seedlings of a subpopulation would be adapted to locations 670 m higher or lower on the same slope.

Although it may once have been the popular assumption that vast stands of wind-pollinated trees are genetically homogeneous, it is clear today that there is significant genetic heterogeneity among local populations. Studies of local populations of lodgepole pine (Knowles and Grant, 1985), trembling aspen (Mitton and Grant, 1980), ponderosa pine (Linhart et al., 1981), and Engelmann spruce and subalpine fir (Shea, 1990) demonstrate significant genetic heterogeneity over very short spatial distances, e.g., within two hectares for ponderosa pine (Linhart, 1989). In a Colorado ponderosa pine stand, all trees occupying an area of approximately two hectares were mapped and their genetic constitution determined for seven polymorphic loci. Most trees fell into one of six spatially-definable clusters that differed from one another at one or more loci (Linhart, 1989). The patchy nature of forest disturbance and regeneration in ponderosa pine promotes this population structure. However, this fine-grain or patchy genetic heterogeneity is not necessarily related to adaptive differences among sites. Nevertheless, wherever selection pressures are strong enough and gene exchange limited, we can expect to find examples of localized ecological races. For example, in the case of Engelmann spruce and subalpine fir (Shea, 1990), there were significant differences in allele frequencies between wet and dry sites.

Factors Affecting Differentiation: Gene Flow and Selection Pressure

The extent of differentiation depends on the amount of gene flow via pollen, seeds, and other propagules between populations and on the intensity of selection. Gene flow is a cohesive force acting to keep populations from diverging. Isolating factors such as spatial distance between populations, or ecological isolation (south versus north slopes, wetlands versus uplands) act to disrupt gene flow. Selection is also important. Populations will tend to remain similar if subjected to similar selection forces but will differentiate if they are not. Over short distances gene flow is likely to be great and differentiation seems therefore unlikely. When it does occur in these situations, it is brought about chiefly by high selection pressure. First let us consider gene flow and then an example of intense selection.

Gene flow is not simply a function of how widely pollen and seeds are dispersed. Several factors, besides limited dispersal of pollen and seeds, that act to restrict gene flow are (1) limited number of breeding trees, (2) differences in flowering times of individuals, and (3) biological and ecological factors that control zygote viability and seedling establishment (genetic incompatibility, frost, drought, shade, herbivory).

It is well known that there are marked differences among species in the age of seed bearing and periodicity of seed crops. Furthermore, within a species, trees vary greatly in their reproductive capacity; some are highly fruitful, some moderately so, and others are completely barren year after year. Of the many trees in a population that could potentially exchange genes, only a few breeding trees may contribute appreciably to the next generation. For example, Schmidt (1970) found that in an 89-year-old Scots pine stand, 30 percent of the trees produced 71 percent of the female strobili and 64 percent of the male strobili.

The time of pollen release and female receptivity (phenology of flowering) is vital in pollination and is closely related to air temperature and humidity. Trees within a given population are more likely to be synchronized with one another than with trees progressively farther away. Although the timing of flowering on the average favors local gene exchange, the possibility of gene flow from distant sources definitely exists. Viable pollen may be transported many miles, but it may reach another stand too early or too late to compete effectively with local pollen. This effect becomes even more important in Scots pine and probably other pines and conifers since the capacity of the pollen chamber is limited (Sarvas, 1962), and all grains do not have an equal chance to fertilize the eggs. Of the many pollen grains reaching the micropyle of the ovule, only two, on average, have the opportunity to fertilize the eggs of each ovule, one of which eventually develops into the embryo. Therefore, pollen grains of neighboring trees may have a higher probability of achieving fertilization than those of trees at progressively more distant sites because of the greater probability of their being first to reach the micropyle.

In forest trees, the range of seed dispersal is limited, although the very small-seeded species, such as birches, hemlocks, poplars, and willows are exceptions. Bird dispersal in such groups as the Fagaceae (Johnson and Webb, 1989) and pines (Vander Wall and Balda, 1977) may extend dispersal to 20 km or more. In contrast, we know that pollen can be carried great distances (Lanner, 1966). Andersson (1963) reported that pollen was blown from Germany to southern Sweden, a distance of 72 km, and that in Sweden one year the pollen crop was so heavy that clouds of pollen were mistaken for forest fires. The pollen-dispersal distance of Scots pine is at least in the tens of kilometers, and its transfer of 600 to 700 km in 10 to 12 hours has been reported (Koski, 1970).

The many reports of widespread dispersal might lead to the conclusion that populations over large areas are prevented from diverging because of widespread gene exchange. However, from many studies it is clear that most pollen and seeds are dispersed close to the source, their frequency declining rapidly with distance (Ellstrand, 1992). Accumulating evidence indicates that for wind-pollinated species (many north temperate species) in natural stands, most individuals are pollinated and fertilized by their neighbors of the surrounding stand or from adjacent stands less than about 100 m away (Koski, 1970; Shea, 1990; Adams, 1992). Due to the great difficulty of monitoring flow and destination of pollen from individual trees and from different stands, and considering the vagaries of environmental and biological factors, we can expect no generalized answer to the question of the extent of gene exchange. At times, gene flow may be restricted to nearest neighbors, favoring inbreeding, whereas at other times gene exchange may occur over considerable distances, thus significantly affecting the gene pool of receptor populations.

It is against the prevailing level of gene flow that natural selection guides the genetic makeup of populations in a given ecosystem. Selection pressures of the particular environment play a significant role and in herbaceous species may be effective over a distance of only 2 to 4 m (Aston and Bradshaw, 1966). The effect of intense selection was strikingly demonstrated by clinal changes in glaucous and nonglaucous (green) phenotypes of *Eucalyptus urnigera* in Tasmania (Barber and Jackson, 1957). Green phenotypes are typical of low elevations and more sheltered sites. Glaucous individuals, having leaves covered with a whitish wax, become more frequent in increasingly exposed and colder environments along a gradient from low to high elevation. Clines of glaucousness are correlated with frost occurrence; the more glaucous populations occurring in the more frosty localities (Barber, 1955). The structure of wax makes it impossible for water to come in contact with the cuticle at high elevations, and thereby freezing is prevented (Hall et al., 1965). The change from green to glaucous types was essentially complete over a vertical distance of 122 to 152 m (0.8–1.6 km ground distance) in the adult populations. Glaucous seedlings may be produced from non-glaucous mother trees and vice versa, indicating gene flow via insect and bird pollinators. Nevertheless, intense selection eliminates glaucous seedlings as they mature in lower elevation forests, and at the higher elevations green seedlings are eliminated. In this case, intense selection can build up great genetic diversity over a relatively short distance, even in the face of considerable gene flow.

ECOLOGICAL CONSIDERATIONS AT THE SPECIES LEVEL

As populations change through time and radiate into new areas, they typically become increasingly diverse in response to genecological differentiation. Also, due to geographic separation and other isolating mechanisms, some populations become less able to exchange genes with other populations; they become reproductively isolated in varying degrees. An important route of speciation involving geographic isolation followed by reproductive isolation is shown in Figure 4.12 (Stebbins, 1966). The isolating factors which act

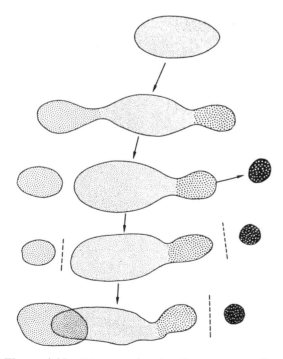

First stage.
A single population in a homogeneous environment.

Second stage.
Differentiation of environment, and migration to new environments produces racial differentiation of races and subspecies (indicated by different kinds of shading).

Third stage.
Further differentiation and migration produces geographic isolation of some races and subspecies.

Fourth stage.
Some of these isolated subspecies differentiate with respect to genic and chromosomal changes which control reproductive isolating mechanisms.

Fifth stage.
Changes in the environment permit geographically isolated populations to exist together again in the same region. They now remain distinct because of the reproductive isolating barriers which separate them, and can be recognized as good species.

Figure 4.12. Diagram showing the sequence of events which leads to the production of different races, subspecies, and species, starting with a relatively homogeneous, similar group of individuals. (Reprinted from Processes of Organic Evolution by Stebbins, G. Ledyard © 1966. Reprinted by permission of Prentice-Hall, Inc., Upper Saddle River, NJ.)

Table 4.1 Major Isolating Mechanisms Acting to Separate Plant Species

Prefertilization Mechanisms: Prevent fertilization and zygote formation.

1. Geographical separation. Populations live in different regions (allopatric).

2. Ecological separation. Populations live in the same regions (sympatric) but occupy different habitats.

3. Seasonal or temporal separation. Populations exist in the same regions and may exist in the same habitat, but they have different flowering times.

4. Ethological separation. Pollination is accomplished by specific pollinators (as in some tropical species).

5. Gametic incompatibility. Pollination may occur but gametes are incompatible before or at fertilization.

Postfertilization Mechanisms: Fertilization takes place. Hybrid zygotes are formed, but they are inviable, or give rise to weak or sterile hybrids.

6. Hybrid inviability or weakness. Zygotes are formed but are unable to germinate or become established. If established, they break down before reproductive organs are formed.

7. Hybrid sterility. Hybrids are sterile because reproductive organs develop abnormally or meiosis breaks down due to chromosome incompatibilities.

8. F_2 breakdown. F_1 hybrids are normal, vigorous, and fertile, but F_2 generation contains inviable or sterile individuals.

Source: (Reprinted from Processes of Organic Evolution by Stebbins, G. Ledyard, © 1966. Reprinted by permission of Prentice-Hall, Inc. Upper Saddle River, NJ.)

singly or, more usually, in combination, are presented in Table 4.1. The various degrees of population differentiation (Figure 4.12) have been classified in a hierarchical system. Populations exhibiting marked genetic differences in morphology and physiology are classified as species, subspecies, or varieties. These classes are formal taxonomic designations, i.e., taxa.

Species are not all of the same degree of divergence. This fact can often be noted in the degree of morphological difference and in their reproductive isolation. "Good" species, from the standpoint of reproductive isolation, are those that exist in the same geographic area (**sympatric** species) and maintain their distinctness, although individuals of each are within effective pollinating range of one another, and hybridization would be possible (Figure 4.12, fifth stage). Sympatric species that are reproductively isolated by one or more of the isolating mechanisms (i.e., "good species") remain distinct although they may hybridize and form hybrid individuals such as in loblolly pine and shortleaf pine and yellow birch and paper birch. Isolating mechanisms operating at later stages (Table 4.1, stages 6–8) often render hybrids inviable or sterile.

Where individuals of two species come together and hybridize, the species distinctions may be dissolved or "swamped out" in a flood of intermediates. In this situation, complete reproductive isolation has not yet occurred, and the populations may be more appropriately regarded as subspecies. This situation is reported from Canada where the morphologically similar white spruce and Engelmann spruce, and balsam poplar and black cottonwood, intermingle. In both cases, the recognition of subspecies has been suggested (Taylor, 1959; Brayshaw, 1965; Viereck and Foote, 1970).

Biologists recognize that no single definition of a species is entirely applicable to classify the enormous diversity of organisms nor serve the various purposes desired by different scientists. The problem of speciation has been considered concisely by Stebbins (1966), Heslop-Harrison (1967), and Solbrig (1970), and in detail by Stebbins (1950 and 1970) and Grant (1963, 1971, 1977).

Niche

Ecosystem interactions have resulted in marked genetic differentiation among species such that each occupies a different **niche**. Ecologists use the term niche in an attempt to express in one word: where, when, and how a species is genetically adapted to persist with other species in its site, its relative time of temporal dominance in the successional sequence on that site, and by its functional (physiological) adaptations. The niche of a species is the result of the multidimensional specialization of that species in space and time.

For convenience in examining the niche differentiation of woody species, we recognize three components: a spatial component (the physical site conditions to which the species is adapted), a temporal component (the relative time that a species dominates in the successional sequence of an area, for example early or late in succession), and a functional component—the physiologically-based genetic adaptations, sometimes termed natural history traits, such as number of seeds produced; dispersal time and mechanism; growth rate; and tolerance of shade, drought, fire, and flooding. A species' functional component is the particular set of genetic adaptations that enable it (1) to occupy a characteristic geographic range and local sites and (2) to dominate at a characteristic time in the course of succession on a given site. These three niche components identify where (spatial), when (temporal), and how (functional) a species competes and persists in regional and local landscape ecosystems. We use the term niche as the most concise formulation of this genetic specialization.

The **spatial component** of the niche may be illustrated by silver maple, which occupies river floodplain sites, whereas the related sugar maple thrives on upland sites. Their niche differentiation is primarily one of the different physiological adaptations of these species which make them relatively more competitive on these respective sites. The **temporal component** may be illustrated by paper birch and eastern hemlock on an upland site in the hemlock–northern hardwood forest. Paper birch is a pioneer species, and dominates the site following fire early in the succession of biota on the area. Hemlock seedlings establish simultaneously on the same site but grow slowly under the birch canopy. They dominate the site a century or more later as the birches decline and die. In this case, the species are niche differentiated, not by site conditions primarily, but by the various physiological adaptations that enable them to dominate the site at different times. The **functional component** may be illustrated by the physiological adaptations of paper birch and hemlock. Birches colonize the burned site quickly, and their seedlings grow rapidly with leaves of high photosynthetic efficiency in sites with high light levels. Hemlock seedlings are photosynthetically efficient at low light levels and, using other physiological adaptations to obtain soil water and nutrients as well, they are able to develop under the birches, replacing them in 100 to 200 years.

Hybridization

Hybridization, the crossing between individuals of populations having different adaptive gene complexes (races, subspecies, species), is frequent in natural populations of many woody-plant groups. The great number of reports of hybrids during the 20th century undoubtedly reflects the widespread disturbance of ecosystems providing open sites for their establishment. For example, major human disturbances have enabled the European white poplar (*Populus alba*) to initiate naturally occurring hybrids on three continents: the gray poplar (*P. ×canescens*) in Europe, Rouleau's poplar (*P. ×rouleauiana*) in eastern North America, and the hairy poplar (*P. ×tomentosa*) in China. In each area, the European white poplar hybridized with a native aspen. In North America, the white poplar itself was not native, but was introduced as an ornamental tree by early settlers. Similarly, the introduction of the eastern cottonwood into gardens of France and England enabled it to hybridize with the native European cottonwood and produce the highly successful black poplar hybrid (FAO, 1980). Although most natural hybrids demonstrate hybrid weakness, two of the best examples of so-called hybrid vigor in trees are clones of the European black poplar hybrid and Rouleau's poplar (Little et al., 1957; Spies and Barnes, 1981).

In the past, many hybrids were treated as normal divergent species and given binomial names, e. g., *Populus acuminata* (hybrid between narrowleaf cottonwood and Fremont cottonwood (Crawford, 1974; Eckenwalder, 1977, 1996). However, hybrids are now considered **nothospecies** (hybrid species) and may be either designated, as in the case of the hybrid between shingle oak and northern red oak, by a taxonomic formula, e.g., *Quercus imbricaria* × *Q. rubra* or as a binomial with the multiplication sign (×, signifying hybrid) placed directly before the epithet, e.g., *Quercus ×runcinata* (Wagner, 1983).

Hybridization is of major evolutionary significance, acting as an evolutionary catalyst (Stebbins, 1969, 1970). Arnold (1994) calls attention to the estimate that approximately 70 percent of all flowering plants owe their existence to past natural hybridization between different species or genera (Whitham et al., 1991). In woody species, hybrids are of major ecological and practical significance. Many hybrids are important in ecosystem management and horticulture due to rapid growth, good form, disease resistance, or frost hardiness (Duffield and Snyder, 1958; Wright, 1976; Nikles, 1970; Zobel and Talbert, 1984).

The incidence of natural hybridization today is not necessarily the same as in presettlement forests due to massive human disturbances that have affected habitats and plant populations. Such disturbances have increased greatly the likelihood of hybridization and gene flow by the creation of disturbed sites and the concomitant reduction of competition that favors establishment and survival of hybrid plants. Natural hybrids often occur in zones of contact between species (Brayton and Mooney, 1966; Remington, 1968) and in disturbed habitats. The disturbed area may be an intermediate or hybrid habitat (Anderson, 1948) where neither parent is well adapted.

An excellent example of the occurrence of hybrids in the contact zone between species is that of hybrids between the narrowleaf cottonwood and Fremont cottonwood along the Weber River in northern Utah (Whitham, 1989). The hybrids occupy a zone of overlap between the morphologically distinct parents that occur in the upper and lower elevations, respectively, along the river. An extreme concentration of leaf gall-producing aphids has been found on leaves of hybrids in a 13 km zone where the host's parents interbreed. Pure Fremont cottonwood is totally resistant to the aphid, and aphids rarely colonize leaves of pure narrowleaf cottonwood. Susceptibility to this parasite illustrates one kind of weakness that is often found in hybrids.

It has been popular to report natural hybrids, usually based on morphological characters, perhaps due to their presumed rarity, or as Wagner (1968) relates, seeking hybrids "was all part of the 'game,' and added to the thrill of the chase, like adding a new stamp to the collection." However, little detailed ecological study has been devoted to the comparative establishment of hybrids and their parents or the presumed differences between the so-called hybrid habitat and that of the parents.

Hybridization can enrich the gene pool of species by the process of introgressive hybridization or **introgression**: the gradual infiltration of germ plasm of one species into that of another as a consequence of hybridization and repeated backcrossing (Anderson, 1949, 1953). Introgression is achieved in three phases: (1) initial formation of F_1 hybrids,

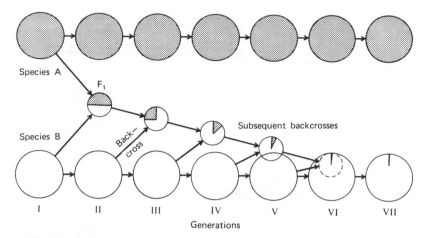

Species A

F₁

Species B

Back–cross

Subsequent backcrosses

I   II   III   IV   V   VI   VII

Generations

Figure 4.13. Diagram illustrating introgression–the interbreeding of species followed by backcrossing of some genes from one parent into at least some members of the population of the other. (After Benson, 1962, *Plant Taxonomy: Methods and Principles,* © 1962 by The Ronald Press Company. Reprinted with permission of John Wiley & Sons, Inc.)

(2) their backcrossing to one or other of the parental species, and (3) natural selection of certain favorable recombinant types (Davis and Heywood, 1963; Figure 4.13). This process is simply gene flow between species. If hybridization occurs between two closely-related species, the probability of gene flow is higher than when the species have diverged sufficiently to have well-integrated, but different, gene pools. Genes from one population will be incorporated into the gene pool of the other (regardless of rank—species, subspecies, etc.) if they improve the well-integrated harmony of the foreign gene pool. If they tend to disrupt the harmony, their frequency will be reduced; this is merely natural selection in action (Bigelow, 1965). Not surprisingly, we find frequent reports of introgression between the closely related species.

Oaks are notorious for hybridization (Little, 1979; Miller and Lamb, 1985). In California alone, more than 20 hybrids have been recognized and 11 of these are named (Pavlik et al., 1991). Introgression is, therefore, likely among the closely related oak species. Hybridization and possible backcrossing among oaks is graphically illustrated by Wagner and Schoen (1976) between shingle oak, having entire, non-lobed leaf blades, and associated northern red and black oaks that have multiple leaf lobes. The spectrum of leaf shapes of shingle oak, northern red oak, and their hybrid is shown in Figure 4.14. Some plants are possible backcrosses to northern red oak (Figure 4.14*b*) or shingle oak (Figure 4.14*e*), and introgression is suspected.

There is enormous variation in morphological characters of woody plant populations; detailed studies of parent and hybrid variation using standardized collections, chemical traits, and molecular genetic techniques are needed to estimate the extent of hybridization and gene flow. In some cases, divergence caused by intense selection, as cited by Barber and Jackson (1957), may be misinterpreted as introgression. Thus, in determining the amount of gene flow between species it is important to establish, through more than observations of morphological characters of phenotypes in nature, that (1) hybridization and backcrossing have actually taken place, and (2) increased variation of the parent species occurs outside the area of hybridization and is due to hybridization and not solely to intense selection along an environment gradient. The use of nuclear and chloroplast DNA markers is expected to bring about a better understanding of hybridization and its contribution to adaptation and speciation.

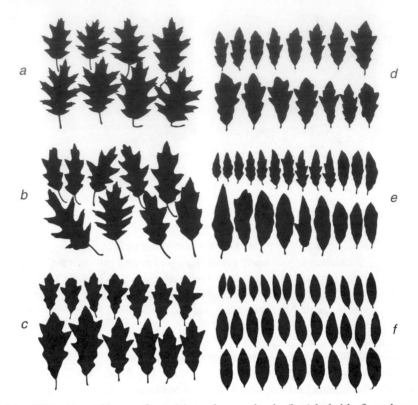

Figure 4.14. Silhouettes of leaves from (a) northern red oak, (b-e) hybrid of northern red oak and shingle oak, and (f) shingle oak. (After Wagner and Schoen, 1976.)

Variation at the species level is not only influenced by the combination of two different chromosomal sets of genes, as in hybridization, but also by the *number of similar sets* of chromosomes that individuals of a species possess—the subject of the next section.

Polyploidy

Polyploids are organisms with three or more sets of chromosomes. The ploidy level of a species (triploid = 3 sets, tetraploid = 4 sets, etc.) is measured in relation to the base or **x** chromosome number established for the genus or family, usually the lowest haploid (gametic) number for the group. For example, the base number for the genus Pinus is 12 ($x = 12$), and all pine species have 24 chromosomes ($2x = 24$, the diploid number). In contrast, the birches, with a base number of 14 chromosomes, exhibit a stunning array of polyploid levels among species. These include diploids (black birch and river birch; $2x = 28$ chromosomes), tetraploids (bog birch, paper birch, and European white birch; $4x = 56$), and a hexaploid (yellow birch; $6x = 84$).

Polyploids typically arise by hybridization of related species followed by doubling of the chromosomes of the hybrids, a process that produces a new species. Polyploids are of considerable evolutionary and ecological significance. They exhibit a wide geographical distribution, including alpine, Arctic, and tropical environments. They occur not only on a wide array of harsh sites that have cold, arid, wet, or droughty conditions but on mesic sites as well (Stebbins, 1985). The major reason for the success of polyploids is probably their greater ability to invade and colonize new or disturbed habitats relative to that of their diploid progenitors (Stebbins, 1985). These sites typically may be associated with zones of contact and hybridization between genetically differentiated diploid populations. The resultant polyploids exhibit aggressive gene combinations for colonization that were gained by hybridization and that are buffered and maintained by the polyploid condition. For example, polyploids are notable for their ability to colonize and persist on cold, harsh sites, particularly those that were formed during and following Pleistocene glaciation.

A major reason for the success of polyploids compared to related diploids under these severe conditions is that polyploidy is a uniformity-promoting mechanism. In general, polyploidy acts like a sponge, absorbing mutations but rarely expressing them. In diploids, mutations or new recombinations are more easily expressed due to the low chromosome number and fewer of each kind of chromosome. Polyploids may have four, six, or more chromosomes of each kind, and new gene combinations are not likely to produce a major change in the phenotype. Mosquin (1966) reasoned that polyploids represent an efficient buffering system, resisting the effects of natural selection on particular genes and promoting and preserving phenotypic uniformity. Thus the narrow adaptational limits of high latitude and weedy polyploids are an adaptive feature corresponding to the narrow and relatively uniform environments of boreal, Arctic, and disturbed or weedy habitats. The occurrence of polyploids is often great, as in high-latitude birches and willows, because such habitats are themselves widespread.

High levels of polyploidy have also been reported for certain tropical floras, some of which are of very ancient origin. According to Stebbins (1970), newly opened habitats were available in ancient times for evolving angiosperms, and increasing polyploidy accompanied the establishment and spread of new groups of angiosperms during the early stages of their history. The ability of polyploids to colonize newly-opened environments is apparently the common denominator of their success in diverse regions of the world, whether the time of their origin was ancient or modern. Temperate species such as American basswood and tuliptree are examples of polyploids of ancient origin that have outlived their ancestors. In contrast, a newly evolved polyploid birch, an octoploid, was recently discovered on a disturbed lake margin in southern Michigan (Barnes and Dancik, 1985). Overall, polyploidy illustrates a significant and intimate ongoing process of site–plant relationships throughout evolutionary time.

Polyploidy is rare in gymnosperms, occurring in less than five percent of the species (Delevoryas, 1980). Only three conifers are known to be polyploids (Khoshoo, 1959), the most notable being redwood, a hexaploid with 66 chromosomes. In angiosperms, at least 40 percent are estimated to be ancient or modern polyploids (Goldblatt, 1980; Lewis, 1980). Even greater numbers, at least 70 percent, occur in monocots (Goldblatt, 1980), and between 80 and 90 percent are reported for grasses (Stebbins, 1985). Some woody angiosperm genera have no polyploid species (*Populus, Juglans, Robinia*), whereas many others each have species of various ploidy levels (*Prunus, Salix, Betula, Alnus, Magnolia, Acer,* and many others). The evolutionary aspects of polyploidy are examined by Stebbins (1950, 1970, 1985) and Jackson (1976).

The Fitness–Flexibility Compromise

To survive and persist through time, a species must not only show adaptedness to its present environment, but have adaptability, that is, the potential to change. Mather (1943) expressed this compromise between fitness for the environment as it exists and flexibility that will permit further adaptive change. Flexibility is favored by variability-promoting mechanisms, such as cross-pollination and a high rate of recombination (Mosquin, 1966). Inbreeding, apomixis, polyploidy, and a low rate of recombination are uniformity-promoting devices that favor fitness in the given environment.

In most woody species, we see various mixes of uniformity- and variability-promoting devices giving plants the best of both possible worlds. Selection through time has produced a different mix and different mechanisms, depending on the particular environmental situation. The proportion of each and the nature of the mechanisms differ among species even within genera. The fitness-flexibility compromise is closely related to the limiting ecological factors of the environment that supports the species.

For example, compare trembling aspen, widely distributed in northern, glaciated, and disturbed sites, with eastern cottonwood, also wide-ranging but primarily found in lower latitudes and growing in river floodplains. Both species are primarily dioecious (male and female flowers borne on different individuals), the most effective device for ensuring cross-pollination. Thus a great amount of variation is generated in these diploid species and then widely circulated through abundant seed production and widespread wind dispersal. Aspen has the remarkable ability of vegetative propagation by root suckers, which assures genetic uniformity; the clonal growth habit is pronounced throughout its range. Fire is probably the main environmental factor that has favored this adaptation. In contrast, cottonwood rarely produced root suckers in nature (in an essentially fire-free environment), but its branches and young shoots root easily in soil. This trait may be of considerable selective advantage and of immediate fitness in river bottoms subject to periodic disturbance by flooding, whereby branches are broken off by ice and debris and deposited in new soil. Thus both species have strongly developed mechanisms, closely linked to their respective environments, that provide both fitness and flexibility.

Pines, typically cross-pollinating species, maintain a certain amount of self-pollinating ability in their breeding systems. Because fire plays a major role in establishment of pines around the world, the ability of one or a few survivors of a severe fire to colonize the site, if necessary by self-fertilization, is of major significance. Colonization is likely to be accompanied at first by an increase in the degree of inbreeding, but outbreeding will tend to be restored as the stand density increases (Bannister, 1965).

The selection pressure imposed by fire has apparently been instrumental in promoting a very high degree of self-fertility in red pine (Fowler, 1965a). The species is highly self-fertile, and unlike most other pines, seedlings resulting from self-pollination are as vigorous as those from cross-pollination (Fowler, 1965b). However, a high degree of self-fertility has been achieved, seemingly at the cost of variability, because red pine is one of the most uniform of all woody-plant species and has a relatively restricted range compared to its associates, jack pine and eastern white pine.

Suggested Readings

Adams, W. T., S. H. Strauss, D. L. Copes, and A. R. Griffin (eds.). 1992. *Population Genetics of Forest Trees*. Kluwer, Boston, MA. 420 pp.

Conkle, M. T., and W. B. Critchfield. 1988. Genetic variation and hybridization of ponderosa pine. In D. M. Baumgartner and J. E. Lotan (eds.). Symp. Proc. *Ponderosa Pine, The Species and Its Management*, Wash. State Univ., Pullman, WA.

Ekberg, I., G. Eriksson, and I. Dormling. 1979. Photoperiodic reactions in conifer species. *Holarctic Ecol.* 2:255–263.

Hamrick, J. L., M. J. W. Godt, and S. L. Sherman-Broyles. 1992. Factors influencing levels of genetic diversity in woody plant species. In W. T. Adams, S. H. Strauss, D. L. Copes, and A. R. Griffin (eds.). *Population Genetics of Forest Trees,* Kluwer, Boston, MA.

Heslop-Harrison, J. 1967. *New Concepts in Flowering-Plant Taxonomy*. Harvard Univ. Press, Cambridge, MA. 134 pp. (Ecological differentiation of populations, pp. 44–58; Geographical and reproduction isolation. pp. 59–78.)

Langlet, O. 1971. Two hundred years' genecology. *Taxon.* 20:653–721.

Morgenstsern, E. K. 1996. *Geographic Variation in Forest Trees*. Univ. British Columbia Press, Vancouver, BC. 209 pp.

Rehfeldt, G. E. 1988. Ecological genetics of Pinus contorta from the Rocky Mountains (USA): a synthesis. *Silvae Genetica* 37:131–135.

Rehfeldt, G. E. 1989. Ecological adaptations in Douglas-fir (*Pseudotsuga menziesii* var. *glauca*): a synthesis. *For. Ecol. Manage.* 28:203–215.

Stebbins, G. L. 1985. Polyploidy, hybridization, and invasion of new habitats. *Ann. Mo. Bot. Garden* 72:824–832.

Whitham, T. G. 1989. Plant hybrid zones as sinks for pests. *Science* 244:1490–1493.

CHAPTER 5

REGENERATION ECOLOGY

In Chapters 5 and 6, we consider the life history of forest trees in the context of landscape ecosystems. In this chapter, we examine woody plant regeneration, especially the environmental factors that influence this process from reproduction to establishment. In Chapter 6, we discuss selected aspects of the structure and function of shoots, crowns, and roots as related to site conditions and growth of forest trees. A treatment of tree physiology is not attempted; excellent treatments of the physiology of woody plants are available in the books of Kozlowski (1971a,b), Zimmermann and Brown (1971), Fitter and Hay (1987), Raghavendra (1991), and Kozlowski and Pallardy (1997a,b). Tree growth is succinctly described by Wilson (1984). The life history of plants has been examined from the viewpoint of plant population biology by Grime (1979, 1988) and Silvertown and Doust (1993). We examine processes of regeneration ecology first as a critical part of the forest tree's life cycle.

Physiological processes occur in cells and organs of individual plants, but they affect the growth and form of the whole plant and, furthermore, the associated populations of plants and animals. These processes are dependent on and conditioned by the physical factors of the location-specific ecosystem in which plants reproduce, establish, and grow. A schematic model of a population of individuals in Figure 5.1 illustrates major features culminating in the regeneration by sexual reproduction of plants in a given ecosystem. We observe individuals of a population recruited and established from a "bank" of seeds stored on or in the forest floor (Phases I and II). Growth in height and mass, Phase III, requires space, light, nutrients, and moisture that may be insufficient to allow vigorous growth of all individuals. Some plants may persist for many years as part of the ground cover ("stored" as seedlings or sprouts) until favorable events enable their recruitment into the understory and overstory. Some plants die (unbranched stems, T), whereas others thrive (shown by branched systems). Although potentially capable of branching indefinitely, the individuals

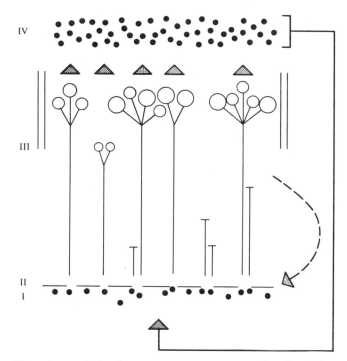

Figure 5.1. The volumetric landscape ecosystem supports the plant life cycle with trees shown as a series of repeating modular units of shoots. (After Harper, 1977, © 1977 by J. L. Harper. Reprinted by permission of Academic Press, Inc.) Phases shown:

I   the bank of seeds on and in the forest floor

II the establishment and recruitment of seedlings

III growth in height, mass, and number of modular units; vertical bars represent environmental constraints on growth; dashed line indicates the influence of the overstory on establishment and recruitment.

IV seed production and dispersal

are sooner or later restrained by physical and biotic limits of the ecosystem. The constraints of limited abiotic resources and herbivory are indicated by the vertical bars on either side of the population. Development of a plant canopy brings changes to the environment of the understory and forest floor (this feedback shown by dashed arrow) that, in turn, affect the recruitment of new individuals. Reproduction occurs in Phase IV, and seeds are dispersed to the forest floor. In this chapter, we examine several features of this plant life cycle and emphasize regeneration as a complex of ecosystem processes involving sexual and vegetative reproduction, dispersal, and establishment in relation to environmental factors.

REGENERATION

Plants maintain and expand their populations over time by the process of regeneration. Regeneration includes seed production and the maturation of seeds so that they are ready to be dispersed. In sexual reproduction, **seed production** is followed by **dispersal** of fruits and seeds, **germination**, and finally the **establishment** of seedling on the forest floor. The steps of the sexual regeneration process are as follows:

Reproduction → Fruit and seed → Active or dormant → Seed ──────→ Plant
                    dispersal   seed bank   germination   establishment

Regeneration also encompasses vegetative reproduction, whereby stems of existing plants develop to maintain and expand the forest community.

Regeneration, including both sexual and asexual reproduction, is an ecological process ensuring the development of successive generations of plants of a given landscape ecosystem. The established plants either die or undergo a process of **recruitment,** whereby certain individuals residing near the ground surface (i.e., in the ground-cover layer) reach the lower (understory) layer(s) of the forest. Finally, even fewer of these individuals of the understory are recruited into the overstory or canopy layer of the forest. Regeneration is not a property of plants alone. Factors of the physical environment, such as light, temperature, moisture, nutrients, wind, and disturbance regimes strongly influence steps in the regeneration process as do biotic factors such as herbivory, disease, and competition and mutualism with plants and animals.

The model of sexual regeneration is presented in Figure 5.2, and the individual features and processes of this model are described in the following sections. Regenerative methods of widespread occurrence in forest ecosystems are shown in Table 5.1. Their functional characteristics and site factors illustrate the interconnectedness of environment and biota and where each "strategy" is successful. Regeneration and recruitment of a new generation may occur as a single cohort (especially in V, F, and W in Table 5.1) or sporad-

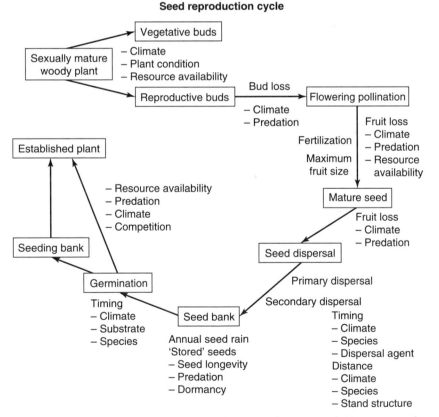

Figure 5.2. Diagrammatic illustration of the sexual regeneration process of woody plants. (After Zasada et al., 1992. Reprinted with the permission of Cambridge University Press.)

Table 5.1 Regenerative strategies of widespread occurrence in woody plants.

Regeneration method		Functional characteristics	Conditions under which strategy appears to enjoy a selective advantage
Vegetative: clone maintenance or expansion	V	New vegetative shoots remaining attached to parent plant until well established	Productive or unproductive sites subject to low or high intensities of disturbance
Active seed bank	B_a	Viable seeds that have no dormancy requirement; reside in seed bank less than a year	Sites with favorable weather or seasonally predictable disturbance by climatic or biotic factors
Dormant and persistent seed bank	B_d	Viable but dormant seeds present throughout the year; some persisting more than a year	Sites subjected to temporally unpredictable disturbance
Fire-induced opening of cones	F	Heat of fire opens cones where seeds are stored; seeds germinate immediately following favorable moisture and site conditions	Sites prone to relatively frequent, intense fires
Widely wind dispersed seeds	W	Propagules numerous and exceedingly buoyant in air, widely dispersed and often of limited viability	Sites subjected to spatially unpredictable disturbance
Locally dispersed seeds	L	Propagules few and heavy; dispersed by gravity and animals; seed buried	Sites predictable in vicinity of parent plant
Persistent juveniles	B_j	Offspring derived from an independent propagule; seedling capable of long-term persistence in a juvenile state	Sites subjected to low intensities of disturbance

Source: Modified from Grime, 1988. Reprinted from p. 378 in *Plant Evolutionary Biology* by L. D. Gottlieb and S. K. Jain (eds.), © 1988 by Chapman & Hall. Reprinted with the permission of Chapman & Hall.

ically over a period of years (especially in B_j). Seeds in the active seed bank (B_a) have no dormancy requirement (aspens, cottonwoods, many pines) and germinate readily once seedbed conditions are favorable. However, in temperate forests, seeds of many trees and shrubs lie dormant over winter (B_d) and germinate the following spring.

Sexual reproduction is the basic mode by which plants maintain their populations, adapt to changing environments, and thus persist in space and time. Male sperm and female egg cells unite to form a zygote, which is genetically different from either parent and other offspring. Vegetative or asexual reproduction (V in Table 5.1), i.e., **clone** formation, in plants is actually a growth process whereby genetically identical stems or **ramets** are derived from a sexually produced plant (the **genet**) and form a spontaneous clone. The multistemmed clone best illustrates the concept of a tree as a population of *modular units,* the shoots, where each shoot becomes a ramet of the original plant. Almost all woody plants are capable of some form of cloning (see discussion later in this chapter). By fragmenting its genotype, the plant gains growing space, water, and nutrients and eventually increases its capacity for sexual reproduction. As far as we know, a plant's vigorous asexual reproduction does not diminish its sexual ability to produce flowers and seeds.

SEXUAL REPRODUCTION

The model of sexual reproduction and regeneration (Figure 5.2) illustrates the processes involved and key factors that affect each. This set of processes is considered following an introduction of the reproduction process.

Maturation and the Ability to Flower

Individuals of forest trees progress from a *juvenile phase*, characterized by no flowering,[1] to the *adult phase* in which flowering occurs. Maturation is a term used to refer to the gradual changes in woody plant meristems that occur with increasing chronological age (Wareing, 1987). Marked differences in many characteristics, including ability to flower, leaf morphology and retention, disease resistance, rooting ability, and growth rate, characterize the juvenile phase in contrast to the adult phase of the same plant (Greenwood and Hutchison, 1993; Bonga and Aderkas, 1993). Although trees mature, and we may categorize them as juvenile or adult, maturation does not occur simultaneously in all meristems. The juvenile phase is not only characteristic of all meristems of seedlings and young trees but is found in parts of mature trees. It occurs at branch positions close to the trunk and near the bottom of mature trees, apparently where meristems remain dormant for long periods (Bonga and Aderkas, 1993). The adult phase first appears at the top of the crown as a tree gets older and increases in size, apparently where the meristems have divided the most times. Less active meristems, those remaining dormant at the lower parts of mature trees, retain their juvenile-phase characteristics.

The duration of the juvenile period varies markedly among species, from 1 to over 40 years (Owens, 1991). Fast-growing, shade-intolerant species, such as paper birch and Virginia pine, flower sooner than slow-growing, shade-tolerant trees such as American beech and hemlock. For example, the juvenile period is estimated to last 5 to 10 years in Scots pine, whereas in European beech it is 30 to 40 years (Wareing, 1959). Also, the length of the juvenile phase may be greatly influenced by the site conditions at different geographic locations (Ross et al., 1983).

Attainment of the adult phase is more closely related to tree size than to age. Usually, however, height and diameter of young plants are sufficiently correlated with age so that either size or age may be used to predict when flowering may commence. Nevertheless, small trees in the forest understory, suppressed by the overstory canopy, may never flower—even at ages of 50 to 100 or more years. For many species the attainment of a certain minimum size, rather than the number of periods of growth and dormancy, is the critical factor in attaining the adult or flowering phase. For example, European larch normally remains juvenile for 10 to 15 years but was observed to flower in just 4 years when seedlings were grown to the minimum size for flowering using warm temperatures and long days in a greenhouse (Wareing and Robinson, 1963). Flowering itself is not directly triggered by the attainment of a critical size, but through the activation of "flowering genes" by internal hormonal controls. For example, gibberellins (hormones) apparently play an important role in inducing flowering in juvenile conifers, especially for members of the Cuppressaceae (Pharis and Kuo, 1977; Ross et al., 1983).

[1]Angiosperms produce flowers, whereas conifers do not bear flowers but produce cones (strobili) that bear naked ovules. For simplicity we will use the term flowering to denote the reproductive process of both groups.

Trees in natural stands exhibit great variation in their genetic disposition to flower, particularly in number of reproductive buds produced and the ratio of female to male buds. Some trees of adult phase, even those favorably situated in the overstory canopy, do not flower or rarely flower, whereas adjacent trees are highly fruitful. Some trees that bear both male and female unisexual flowers on the same tree (**monoecious** condition) are predominantly male, whereas others are predominantly female.

Species of several genera (*Acer, Alianthus, Diospryos, Fraxinus, Gymnocladus, Maclura, Populus, Sassafras*) bear unisexual male and female flowers on different individuals (**dioecious** condition). The dioecious trait ensures outcrossing between genetically different individuals and precludes self-pollination (selfing). In trees, selfing and inbreeding generally lead to growth depression (red pine is a notable exception; see Chapter 4, p. 92). Seedlings produced by selfing or inbreeding are typically eliminated at an early stage in natural populations through competition. Angiosperms rarely produce viable self-pollinated seeds. Conifers, however, are more likely to produce viable self-pollinated seeds (especially pines), and this may be of ecological significance when isolated trees survive fire.

INCREASING SEED PRODUCTION

The most effective way to reduce the length of the juvenile phase is to grow seedlings in such a way that they attain a large size as rapidly as possible. This reduction is accomplished by using appropriate light conditions, by applying fertilizers, or using other cultural methods. The use of gibberellic acid and the judicious choice of cultural treatments will allow for early flowering in virtually any conifer species (Ross et al., 1983). However, to avoid lack of flowering in the juvenile phase, and for other reasons, the use of clones rather than seedling populations is emphasized for growing tree crops in high-yield plantations (Ahuja and Libby, 1993). Once trees are in the adult phase, they may be stimulated to increase flower and seed production by a variety of cultural methods. Many investigators have demonstrated such increases over untreated controls in a great variety of species using many different methods (Matthews, 1963; Kozlowski, 1971a).

Reproductive Cycles

The entire reproductive cycle of a forest species is closely adapted to the complex of environmental factors of the site where it grows. Many river floodplain and other wetland species (most willows, cottonwoods, elms, silver maple, river birch, red maple) flower in the early spring and disseminate seeds four to six weeks thereafter. The seeds are dispersed into moist seedbeds (e.g., recently flooded river sites and wetlands) where they germinate readily (B_a and W in Table 5.1). In these instances, the period between pollination and seed dispersal is only a few weeks, and large seeds are not developed. Instead, millions of small seeds are produced, some of which find a favorable seedbed for establishment so that a large food reserve in the seed is unnecessary.

In contrast, in most other North Temperate Zone trees, fruits and seeds develop throughout the growing season of two to four months and are disseminated in the fall or winter. The medium-size and large seeds that are produced typically lie dormant over the winter and germinate in the moist forest floor the following spring (B_d in Table 5.1). Seedlings of many of these species (upland oaks and hickories) are soon subjected to drying soils and soil-water stress. The large amount of stored food in the seed is therefore needed to develop a root system that can cope with the decreasing moisture supplies (see discussion later in this chapter). A description of the reproductive cycle of many North American forest trees is available from the USDA (1990).

In virtually all tree species, reproductive buds or primordia (earliest bud stage) are initiated during the growing season of the year before the opening of the flowers (anthesis). Usually, the flower buds are visible along the current year's shoots of conifers and in the axils of leaves of hardwoods in the fall of the year prior to flowering. The reproductive cycle of Douglas-fir, shown in Figure 5.3 (Allen and Owens, 1972), illustrates that lateral bud primordia are initiated in April along the vegetative shoot that develops inside the terminal vegetative bud (Figure 5.3 A). Some or many of these primordia may become pollen or seed cone buds, provided the internal environment is favorable. As the vegetative bud bursts, needles flush-out (Figure 5.3 B), the shoot elongates (Figure 5.3 C), and the lateral buds become visible (by late July or August) along the young shoot (Figure 5.3 D and E). These buds will enlarge and flush-out the following spring to produce new vegetative shoots or male or female cones. By October, one can determine whether the lateral buds are vegetative or reproductive. Lateral buds at the base of the young shoot tend to become pollen cones, whereas buds toward the tip of the shoot become either seed cones or vegetative shoots.

The alternative pathways of lateral bud primordia development are shown in Figure 5.4. Some bud primordia abort early, degenerate, and leave no trace. Others form bud scales and then cease to develop; they are termed **latent buds**. If the terminal bud is removed, for example by herbivory, the latent buds are usually stimulated to develop into vegetative buds. The remaining primordia develop into pollen cone buds, seed cone buds, or vegetative buds. As an example, latent buds play an important role in oaks whose foliage is killed by spring frost or defoliated by insects. An entirely new set of leaves develops from these buds.

The internal nutrition and hormonal relations in the shoot and tree largely determine the disposition of lateral bud primordia. Although the same number of primordia may be initiated in two consecutive years, the proportion of vegetative and reproductive buds may be quite different. Thus the marked variation in seed cone production in Douglas-fir, and probably many other forest trees, is the result of the proportion of primordia that develop into cones rather than variation in the number of primordia originally initiated.

In Douglas-fir, pollination and fertilization take place the same year; cones mature during the summer, and seeds are released in the fall. This cycle is typical of most conifers and many hardwoods as well. In contrast, the reproductive cycle of the genus *Pinus*, and also in the red oak group, is different because fertilization occurs 12 months after pollination, and the cycle is a full year longer.

Pollination

The timing of pollen release and female receptivity in deciduous species is closely related to the mode of pollination, whether by wind or animals. Wind-pollinated species (aspens, birches, elms, red maple) flower in the early spring before the leaves flush-out. Insect-pollinated species, such as basswood, tuliptree, and black cherry, typically flower later when the leaves are flushing. In the tropical rain forest, pollination by animals is the rule (Chapter 14). Conifers are exclusively wind pollinated. Wind pollination is promoted by the production of enormous amounts of pollen (much more than in insect-pollinated deciduous species) and the positioning of the female cone buds at the ends of shoots in the top third of the crown.

Weather conditions, especially temperature, markedly affect the shedding of pollen and the receptivity of female flowers. Flowering of female and male buds is synchronized but does not precisely coincide for female and male flowers on the same tree, thus reducing the likelihood of self-pollination. Flowering occurs rapidly in wind-pollinated species. For Scots pine in Finland, Sarvas (1962) reported that most of the pollen for individual stands is shed within three to seven days, and even over a shorter period for individual

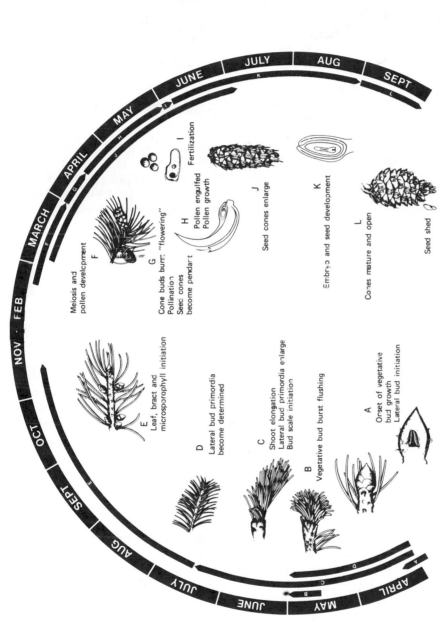

Figure 5.3. Reproductive cycle of Douglas-fir. The entire cycle extends over 17 months. Lateral buds are initiated in April and differentiate into vegetative, pollen, or seed cone buds during the ensuing 10 weeks. Pollination of the seed cones occurs the following April and the mature seeds are shed in September of the second year. The various stages are identified by letters A–L and are briefly described. The approximate length of each stage is shown by the arrows. (After Allen and Owens, 1972. Reproduced by permission, Canadian Forest Service.)

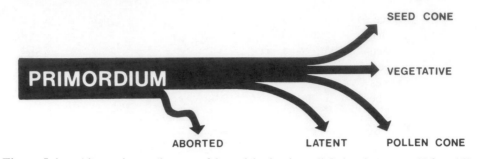

Figure 5.4.   Alternative pathways of lateral bud primordial development. (After Allen and Owens, 1972. Reproduced by permission, Canadian Forest Service.)

trees. Pollen discharge was highly correlated with high temperature and low humidity. The day of maximum shed coincided with the warmest day of the flowering season in most years. Even in unfavorable springs, pollen discharge was delayed until the occurrence of several favorable days (or even one) that were adequate for pollen shed and spreading. Rapidity, therefore, is one of the distinct advantages of wind pollination. In 14 years of monitoring pollination of Scots pine and European white birch, not once was the major part of the pollen crop destroyed by unfavorable weather.

In contrast to pollination, fertilization, and the process of ovule development may fail due to severe weather conditions. For example, regeneration failure of European linden in northwestern England is reported due to temperatures too low to permit fertilization (Pigott and Huntley, 1981).

Periodicity of Seed Crops

Abundant flowering and seed production occur irregularly in natural stands of forest trees. The cyclic nature of seed production is one of the most important traits of life history, greatly promoting the establishment of tree seedlings. Periodicity of fruiting and seed production is a pattern of repeated cycles of exceptionally high production in one year, a mast-fruiting or mast-seeding year, followed by one or more years of lesser production. Historically, **mast** refers to beechnuts, acorns, and chestnuts, the fruits of species in which this pattern is pronounced. Mast-seeding is thought to be an antipredator adaptation. In predator-prone plant species it satiates seed predators because more seeds are produced than can be consumed (Smith, 1970; Janzen, 1971; Silvertown, 1980). Wind pollination may be another reason for the evolution of mast-flowering (Norton and Kelly, 1988). Wind pollination is an undirected process compared with insect pollination. However, an economy of scale is achieved when female and male reproductive processes are in synchrony, i.e., as between a plant and its specific insect pollinators.

In general, fast-growing, shade-intolerant species (aspens, birches, cherries, cottonwoods, junipers, many pines) tend to have fewer years between unusually large seed crops than slower growing, more shade-tolerant species (beeches, firs, sugar maple, northern red and white oaks). The most remarkable flowering cycle in woody plants is that of bamboo. Bamboo clones of many species of the Indian-Asian subtropics grow vegetatively for 12 to 120 years, flower and fruit synchronously over very large areas, and then die (Janzen, 1976). According to Janzen, animals eliminated bamboo plants that were out of phase and, through such selective action, were responsible for the synchronized reproduction that is of adaptive advantage to bamboo. The massive seed crops produced are large enough to satiate all seed consumers and still provide sufficient seeds for successful regeneration in relatively open conditions.

The abundance and periodicity of flowering are not only controlled by genetic and physiological mechanisms of the plant but are strongly influenced by the external environment as well, particularly light, temperature, and moisture. In the closed forest stand, the large dominant trees that have crowns well exposed to sunlight are the primary seed producers; smaller trees with narrow and suppressed crowns yield few if any seeds. Large, open-grown trees, being well lighted throughout the crown and well supplied with water, flower more frequently and abundantly than equally large individuals of the same species in the forest stand. Such open-grown trees may produce large quantities of seed each year, provided they are well pollinated. However, isolated, open-grown conifers may produce many seeds by self-pollination, thereby decreasing the viability of filled seeds and increasing the likelihood of reduced seedling growth.

If flowering and seed production are studied over a long time, a typical cycle of seed production is evident, although some seeds may be produced each year by the dominant trees. For a stand of Scots pine in central Finland, over a 50-year period a seed production level of 100 seeds m^{-2} occurred every 5 years, and a level of 150 seeds m^{-2} occurred at 8-year intervals (Koski and Tallqvist, 1978). As shown in Figure 5.5, in western white pine, abundant cone crops occur about every three to four years (Figure 5.5; Rehfeldt et al., 1971; Eis, 1976). Cone production tends to drop markedly in years following peak production (1952, 1960, and 1963 in Figure 5.5). Individual trees tend to be on a similar cycle although there is considerable variation in inherent productive ability. For example, tree 58 is highly productive compared to tree 17; low years for tree 58 are higher than most years for tree 17. In the genus *Pinus*, good cone and seed crops typically occur at three to seven-year intervals. However, they may occur nearly every year in some species (Virginia, Monterey, jack, and lodgepole pines). A three-year cycle of good cone crops has been reported for several northwestern fir species (Franklin, 1968), and in Douglas-fir the cycle is about five years. The periodicity of flowering and the minimum seed-bearing age for most North American trees and shrubs have been summarized by the U.S. Department of Agriculture (1974; 1990).

The production of seed is governed by two major sets of factors: those that influence the initiation of flower primordia and those that act thereafter to preclude fertilization and ovule development or to cause the loss of flower buds, fruit, or cones, once they form. Although no single factor or general relationship has emerged for all species, high temperatures and dry conditions in the summer of the year before flowering and pollination often

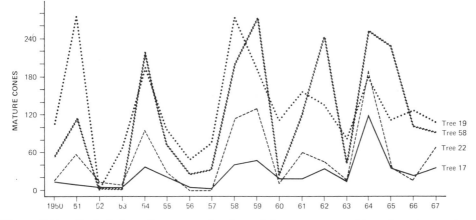

Figure 5.5. Periodicity of cone production of four western white pines in northern Idaho over 18 years. (After Rehfeldt et al., 1971. Reprinted with permission of the Society of American Foresters.)

have been associated with good flower and seed crops, as in ponderosa pine (Daubenmire, 1960), European beech (Matthews, 1955; Holmsgaard, 1962), red pine (Lester, 1967), and western white pine (Rehfeldt et al., 1971; Eis, 1976). The relationship is complex because two, and sometimes three, reproductive cycles are proceeding simultaneously (in pines and the red oak group). The magnitude of the maturing crop, because it is a sink for photosynthate and associated plant nutrients, strongly influences the number of new flower buds initiated and the development of those already formed.

Generally, the more favorable the site conditions (climate, soil water, and nutrients), the greater the flower and seed crop. For example, Sarvas (1962) found that on poor sites in southern Finland (height of dominant trees, 16 m at 100 years) fewer than 30 seeds were produced per square meter. On sites of medium fertility (height of dominants, 23 m) and high fertility (height of dominants 27 m) 60 seeds m^{-2} and 90 seeds m^{-2} were produced, respectively. Similarly, pollen yields were much higher on fertile sites (35 kg $ha^{-1)}$) than on infertile sites (9 kg ha^{-1}).

Once flower buds have been formed, various physical and biotic factors cause mortality (Figure 5.2; Sork et al., 1993). Spring frosts may kill newly formed reproductive buds or injure immature fruits and cones. High temperatures or severe drought may cause abortion of immature fruits and cones or cause a marked decrease in seed size. Low temperatures limit fertilization and ovule development. Strong winds and hail may mechanically damage the seed crop. Cone insects and seedbugs may also cause severe losses.

In a study of mast-fruiting in black, northern red, and white oaks in east-central Missouri, Sork et al. (1993) considered both weather variables and the effects of prior seed production on the current year's acorn crop. They found that the size of an acorn crop was determined by both abundance of flowers produced and their survival to mature fruit production. Furthermore, they found that the length of reproductive cycles differed among the species, although evidence exists for widespread synchronous flowering of oak species in mast years (Downs and McQuilken, 1944; Beck, 1977). These cycles in black, white, and northern red oak appeared to be two, three, and four years, respectively. They concluded that there was an inherent mast-year periodicity for each species that additionally may be affected by weather at critical times. Favorable spring temperatures during the season of fruit maturation appeared to be critical for good acorn crops of all species, whereas summer drought had a negative effect.

In general, the entire reproductive cycle is adapted to the prevailing regime of site factors for a given regional and local ecosystem, and a major departure from such conditions tends to disrupt the reproductive cycle, but only temporarily. The initiation of fruits, cones, and seeds and the factors affecting their development are considered in detail by Kozlowski (1971a). Considerable loss may occur from biotic agents after seed dispersal, and these are considered together with the benefits of animal dispersal in Chapter 14.

Effects of Reproduction on Vegetative Growth

Flowering and the production of fruit and seed crops reduce vegetative growth. In fruit trees, decades of research have demonstrated a strong decrease in shoot, cambial, and root growth with increasing fruit productivity (Kozlowski, 1971a). In forest trees, too, heavy seed crops markedly decrease both height and radial growth (Morris, 1951; Blais, 1952; Eis et al., 1965; Tappeiner, 1969). For example, in mature balsam fir, the weight of new foliage produced in an abundant cone year was only 27 percent of that in a noncone year (Morris, 1951). The developing cones or the fruits of angiosperms are major reservoirs or metabolic sinks to which photosynthetic materials (photosynthate) of the current year are allocated. Not only is current shoot and radial growth depressed, but terminal and lateral buds receive less photosynthate and nutrients so that immature shoots telescoped within them (next year's shoots) are reduced in size. Obviously, this reduced allocation, in turn, affects shoot growth the following year. For example, in heavy seed-crop years, the radial growth of 100-year-old European beech in Denmark was markedly reduced to about one-half of that in years unaffected by seed crop; it was also considerably reduced in the following 2 years (Holmsgaard, 1955). Also, Holmsgaard and Bang (1989) reported that the stem volume increment of Norway spruce was reduced 12 to 25 percent the year of an unusually large or "bumper" crop and from 4 to 9 percent in the 2 following years.

In yellow birch, heavy fruit crops can lead to crown dieback as well as reduced radial growth. For example, the enormous seed crop of 1967 in Ontario was eight times greater than the previous year, which had been regarded as a good seed year (Gross and Harnden, 1968; Gross, 1972). Fruits were produced at virtually every bud site. The nutrient demand for flowering and fruiting suppressed early leaf expansion, and leaves in heavily fruiting crown areas were reduced 75 percent in size. These small leaves led to poor shoot growth and small buds or lack of buds on the shoots. As a result, marked dieback occurred in 1968, and a new crown developed below the dead branches. Radial increment in 1967 was reduced to 47 percent of the annual average increment, and diameter growth of current shoots was less than 5 percent of that in the previous year. Negligible fruit crops were produced the two following years.

Dispersal

Seeds are dispersed by gravity, wind, water, animals, and by combinations of these agents to microsites where they may germinate and become established seedlings. The mechanism of seed dispersal is one of the functional traits of a species that enables it to compete in a particular spatial and temporal niche. In general, most tree species disperse seeds locally, with a high proportion falling within 40 to 50 m or so of the parent. However, many seeds of certain species are widely distributed by water and wind. At this extreme are pioneer trees such as willows, sycamores, cottonwoods, birches, and aspens (W in Table 5.1). Although most of the seeds of even these species may be dispersed locally, many are blown by wind or carried by water to disturbed sites where establishment is far more probable than in the locality of the parent. Seedlings of these light-seeded, shade-intolerant, short-lived pioneer species are less likely to reestablish in the same locality where their parents have matured sexually than in freshly disturbed sites some distance away. In presettlement forests, disturbances were common, and widespread dispersal was a successful means of reproducing pioneer species by sexual reproduction.

The majority of trees and shrubs of northern latitudes, however, have a relatively local dissemination pattern. Typically, the seed rain from a tree decreases exponentially with distance from the tree, and the dispersal pattern of many tree species may be characterized by the relationship shown in Figure 5.6 (Roe, 1967). The dispersal of Engelmann spruce seeds in bumper seed years from four stands into adjacent openings illustrates that most seeds fall near the parent. In these four stands, approximately 70 percent of the seeds fell within 50 m of the edge of the standing timber. For white spruce in British Columbia, at 50, 100, and 200 m from the stand, a seed rain of 19, 12, and 4 percent, respectively, were reported (Dobbs, 1976). And for the same species and distances in Alaska, only 9, 5, and 1 percent were reported (Zasada, 1985). In continuous forest stands, tree crowns of the overstory and understory intercept many seeds. Thus dispersal distances into openings may be greater than may be expected within the stand itself.

Heavy-seeded species, such as walnuts and oaks, are distributed in the vicinity of the parents, primarily by animals (L in Table 5.1). It is of selective advantage for their seeds to be dispersed relatively close (10 to 30 m) to the parent where soil-site conditions are

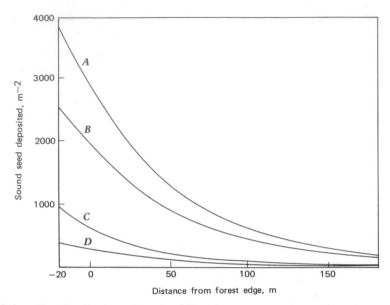

Figure 5.6.   The distribution of seeds of Engelmann spruce in good seed years from the edges of forest stands into adjacent openings. A) Togwotee Pass (1964), Teton National Forest, Wyoming; B) Falls Creek (1952), Flathead National Forest, Montana; C) Griffin Top (1964), Dixie National Forest, Utah; D) Fisher Creek (1964), Payette National Forest, Idaho. (After Harper, 1977, © 1977 by J. L. Harper. Reprinted with permission of Academic Press, Inc., data from Roe, 1967.)

likely to be similar to those where the parent tree became established. Dispersing to appropriate sites may be of such overriding importance that these species have evolved various other adaptations that allow them to establish and compete with the plants and animals of these sites. Black walnut, for example, is highly sensitive to site conditions, and is not competitive where the soil is too wet, too dry, or too shallow. Being relatively site specific, its heavy seeds are an adaptation for ensuring that they reach the ground under the trees. Squirrels and other animals then disperse the seeds to microsites away from the parent tree but similar in soil-site conditions.

Most pines have moderately heavy, winged seeds, and most seeds are wind disseminated within 50 m of the source. In longleaf pine, the maximum amount of seed per square meter falls within 10 m of the base of the tree (Buttrick, 1914). In one study, about 80 percent of the seeds caught in traps (placed at increasing distances from an isolated mature tree) fell within 40 m of the tree. Similarly, for shortleaf pine in Arkansas, Yocom (1968) reported that one-half of the seeds trapped in a forest opening fell 10 m from the edge of the stand; 85 percent fell within 50 m of the stand edge. Because pines are typically regenerated by fire, and because chances are good that fire will occur within the lifetime of the parent tree, regeneration is likely to occur in the vicinity of the parent.

Bird dispersal is the predominate means of dissemination for junipers and for the stone and pinyon pine groups. Seed wings are absent or rudimentary in stone pines (limber, whitebark, and Swiss stone, among others), and seed wings of the pinyon pines remain attached to the cone scales as the cones open. Bird and mammal dispersal is extremely important for these and many other forest species and is considered in Chapter 14.

Dispersal of seeds may occur immediately after seeds mature or periodically if seeds are stored on the tree. This depends in part on the establishment adaptations of the seedlings to conditions of different ecosystems (Zasada et al., 1992). Seeds of elms, aspens, and silver maple, all spring germinators, are dispersed over a relatively short period in the spring immediately after ripening. They germinate and establish on moist or wet sites. In marked contrast, some pine species (jack, lodgepole, Virginia, Monterey) have cones that may remain closed (termed **serotinous** cones) for short or long periods of time (F in Table 5.1; Chapter 12). These species are adapted to hot, fire-prone sites. Some cones may open periodically when temperatures are hot enough. Other cones remain closed until fire-generated heat opens them. Their seeds are then dispersed on the fire-prepared seedbed and germinate following summer rains. In most other pines, cones open shortly after ripening and seeds are rapidly dispersed over a four- to eight-week period.

The effect of weather conditions on duration of seedfall of eastern white pine is illustrated in Figure 5.7. In 1965, 75 percent of the filled seeds were shed over a 3-week period under warm and dry conditions. In contrast, cool, moist conditions of 1968 delayed and lengthened the dispersal period; approximately the same amount of seed was shed over a seven-week period.

In moist, cool climates, seeds of conifers, notably hemlocks, spruces, and firs, are typically dispersed over many months. In fir species, dispersal usually begins in September or October and continues over the winter as the cones disintegrate and release the seeds. In black spruce, cones are retained on the tree for several years in a semiserotinous state. The cones open when fire occurs or when the weather is warm and dry; they close when cool and moist conditions prevail. A continuous seed supply is thereby assured. Another adaptive advantage is that the seeds are not all on the forest floor where a wildfire would destroy them.

Seed Bank, Dormancy, and Germination

Seeds disseminated into the forest floor are stored in the seed bank until they germinate (Figure 5.8). Seeds of many forest trees arrive at the forest floor in a dormant state. They remain in the dormant seed bank until internal and external conditions are favorable for germination. Seeds of some tree species are capable of germinating immediately following dispersal, and they enter directly into the active seed bank (Figure 5.8; B_a in Table 5.1). Compared to seeds of herbs, the residence period of viable tree seeds in the dormant seed bank (B_d in Table 5.1) is relatively short; being relatively large, they are eaten by animals, and they decompose easily.

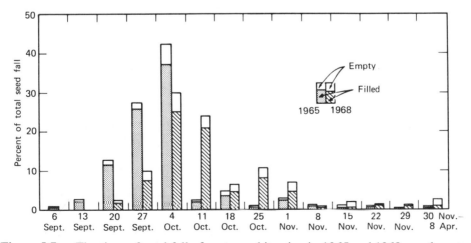

Figure 5.7. The time of seed fall of eastern white pine in 1965 and 1968, southwestern Maine. (After Graber, 1970.)

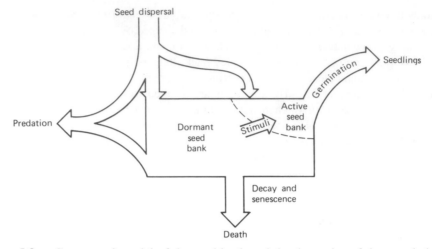

Figure 5.8.   Conceptual model of the seed bank and the dynamics of the population of seeds. (Modified from Harper, 1977, © 1977 by J. L. Harper. Reproduced by permission of Academic Press, Inc.)

Seeds of some species, particularly those disseminated in the spring (e.g., eastern cottonwood and silver maple) enter the active seed bank and germinate readily a few days or weeks after being dispersed. To germinate, viable seeds (1) imbibe water, (2) activate metabolic processes, and (3) initiate growth of the embryo. Seeds of species which are unable to satisfy these requirements (i.e., blockage of any of these processes) are said to be in a state of dormancy.

In species with nondormant seeds, the seeds are typically adapted to certain favorable germination conditions immediately following dispersal. This adaptation applies to seeds of spring-disseminated, river floodplain species. Also, seeds of many species of pines, especially the hard-pine group (including jack, lodgepole, longleaf, pitch, ponderosa, and Virginia), lack a dormancy requirement. Being strongly adapted to fire, their seeds are ready to germinate following rains on the fire-prepared seedbed. Other pines, including all pines of the soft-pine group (white and stone pines, among others), exhibit a pronounced period of seed dormancy although they also depend on fire for establishment.

During the breaking of dormancy, morphological and physiological changes must occur before the seed is capable of germinating. For seeds that overwinter in the forest floor, these changes take place slowly, and the seeds typically germinate the following spring (B_d in Table 5.1). The moist, cool conditions of the forest floor over a period of weeks or months act to decrease germination inhibitors, increase germination-promoting hormones, and create favorable internal conditions for germination. Seeds of some species, however, including basswood, white ash, and black cherry, may require much longer periods before their dormancy requirements are satisfied. Thus their seeds are stored in the dormant seed bank for varying periods and germinate one, two, or even three years after dissemination. Seeds of many shrub species (*Ribes, Rhus, Ceanothus,* and many chaparral species) may be stored for several years in the soil and germinate vigorously after fire.

ESTABLISHMENT FOLLOWING SEXUAL REPRODUCTION

Upon germination of sexually produced seeds, woody plants put out roots, stems, and leaves, and the seedlings cope with their environment. Because great numbers of seedlings perish soon after germination, the few seedlings that survive and exhibit vigorous growth are regarded as established. The period of establishment typically lasts for one to five years or more depending on the species and site conditions.

Establishment is the single most critical stage in the life history of an individual. As indicated in earlier sections, a great many adaptations have occurred that enable the species to produce and disperse seeds in such a way that many seedlings are likely to encounter favorable conditions and survive. Each species has a regeneration ecology that is closely tied to the physical and biotic factors of the sites where it typically establishes. Seedlings must not only survive under the particular physical site conditions of their seedbed, but compete with seedlings of their own species as well as plants already established in that site. These interactions, assuming natural processes and disturbance regimes are sustained, bring about the apparent order and an acceptable likeness of organization that enable us to recognize particular forest communities that are characteristic of different sites.

Although each species exhibits different establishment adaptations, two general patterns of germination and early seedling development may be contrasted: species with **epigeous** (above the soil surface) and those with **hypogeous** (below the soil surface) germination and development. In the epigeous condition, the cotyledons, often still with the pericarp (fruit wall) attached, are elevated above the surface by the elongating hypocotyl (Figure 5.9 A and B). This pattern is typical of nearly all conifers and most angiosperms. In hypogeous species (most nut-producing species, such as oaks, hickories, walnuts, and buckeyes), the cotyledons remain below the surface (remaining attached to the seedling for weeks or months) while the epicotyl grows upward and develops true leaves (Figure 5.9 C and D). Species with epigeous development store relatively little food in endosperm and the cotyledons; they rely strongly on the cotyledons for photosynthesis to stimulate early root development. Four stages in cotyledon development are recognized (Marshall and Kozlowski, 1977):

Storage. Cotyledon cells are packed with stored foods (fats, carbohydrates, proteins) and mineral nutrients. These foods and nutrients are utilized in the first days of seedling growth.

Transition. When exposed to light, a series of changes takes place: chloroplasts develop and chlorophyll synthesis begins, stomates develop, epidermal cells expand, and large intercellular spaces form in the mesophyll.

Photosynthesis. Major contributions in photosynthesis occur to further the development of the shoot and tap and lateral roots. In black locust, red maple, and American elm, appreciable photosynthesis begins four to six days after radicle emergence (Marshall and Kozlowski, 1976). Photosynthesis peaks 8 to 15 days after radicles emerge and continues for about 4 weeks.

Senescence. Dry weight decreases and some mineral nutrients are translocated back into the seedling as the cotyledon function declines.

In epigeous species, the cotyledons are extremely important to the development of seedlings during the first few weeks. Any damage to the cotyledons, such as might be caused by animals or frost, inhibits seedling growth.

In contrast, hypogeous seedlings have large fleshy cotyledons that remain below ground during seedling development and are enclosed in the pericarp (Figure 5.9 C and D). The large amount of stored food favors extensive root development prior to the development of a transpiring shoot and leaf system. The cotyledons also store considerable water and have enough food reserves to reestablish the epicotyl should it become damaged. Furthermore, being underground and inside the pericarp, the cotyledons are protected from browsing animals.

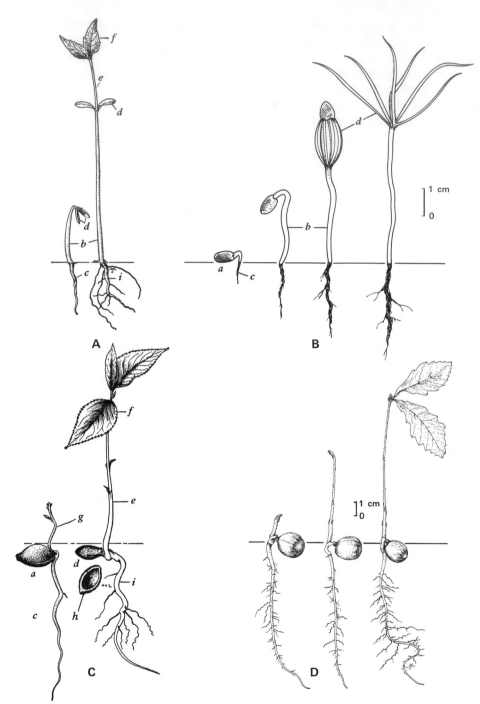

Figure 5.9. Epigeous and hypogeous seed germination and development. A) epigeous germination of pin cherry seedlings at 1 and 10 days; B) epigeous germination of red pine at 1, 2, 6, and 10 days; C) hypogeous germination of Allegheny plum seedlings at 1 and 9 days; D) hypogeous germination of bur oak seedlings at 1, 5, and 12 days. (a) seed; (b) hypocotyl; (c) radicle; (d) cotyledons; (e) epicotyl; (f) leaves; (g) plumule; (h) pericarp; (i) primary root. The plumule consists of the epicotyl and the emerging leaves. (A, C, and D after USDA, 1974; B by W. H. Wagner, Jr.)

The hypogeous system is characteristic of the large-seeded woody species whose seeds are typically buried by squirrels and rodents and thereby dispersed relatively close to the parent tree (L in Table 5.1). Many oaks and hickories grow in dry sites characterized by summer drought. Their seedlings typically establish themselves under overstory trees whose crowns shade the forest floor and whose roots compete for soil water and nutrients. Thus production of relatively few but well-provisioned seeds has proved of adaptive advantage. On mesic sites, where soil water is available throughout the growing season, mixed hardwood forests may be composed of species with both systems: bitternut hickory and northern red oak are hypogeous, and sugar maple, white ash, and beech are epigeous.

Following germination, young seedlings pass through a succulent stage during the first several weeks. At this time, tissues are soft and highly susceptible to fungal infections, damage by insect larvae and other animals, smothering, and desiccation. Seedlings of species whose young roots develop only in the uppermost soil layers are especially subject to mortality compared to those of species with deeply penetrating "tap" roots. Tissues soon begin to harden, and there follows a juvenile period when the seedling becomes increasingly hardy but still subject to mortality. Seed germination and survival of Douglas-fir seedlings throughout the most critical part of the growing season are illustrated in Figure 5.10 (Lawrence and Rediske, 1962). A major reason for the low accumulative germination and survival is that 46 percent of the seeds were destroyed by fungi and animals prior to germination; another 27 percent failed to germinate the first year. The diagram illustrates the timing of factors causing mortality and the relatively few seedlings left at the end of the first growing season.

Of particular interest to the forest ecologist are the kinds of sites and seedbeds where seedlings of different species become established. The vast majority of tree species are dependent on either fire, windthrow (uprooting of trees), or flooding to provide suitable seedbeds for their establishment. Fast-growing, light-demanding species (all pines, willows, cottonwoods, aspens, birches) are favored by catastrophic disturbances that provide open and extreme sites. The adaptations of fire-depending species are presented in Chapter 12.

Whereas the establishment of tree species is favored or not severely limited by partial shade of an overstory, few species can establish and persist in a heavily shaded understory. Similarly, although most seedlings can establish under continuously moist forest soils, only a few are able to establish and persist as soils dry out and soil-water stress becomes severe. Thus each species has a unique set of adaptations that facilitates establishment under certain physical and biotic conditions. Although it is difficult to generalize, we have divided the continuum of regeneration systems into three groups of species that differ in establishment pattern in relation site conditions and the amount and kind of disturbance.

1. Pioneer species seed into open areas following major disturbances such as fire, flooding, windstorm, and landslide. Little existing tree competition is present, and the environment is harsh (hot, cold, dry, wet, windy). Germination and growth are rapid. Roots soon penetrate deep enough into the soil, as in the case of pines, to enable the seedlings that eventually become established to withstand yearly summer drought. Floodplain species tolerate variable water levels and have the ability to generate adventitious roots from their stems if the stems are covered with water or silt.

Fire-dependent pioneer species, such as the pines of southeastern North America (longleaf, shortleaf, loblolly, and slash) typically establish in dense, even-aged stands on areas that recently have been burned. Fire prepares the seedbed in various ways (Chapter

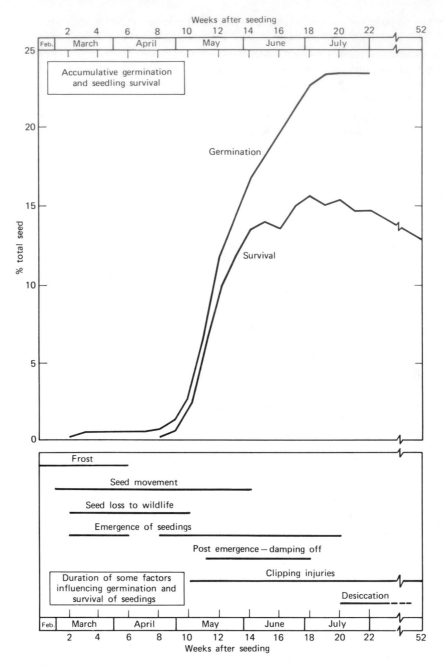

Figure 5.10. The cumulative germination and survival of Douglas-fir seeds during the first growing season. 440 Scandium[46]-tagged seeds were seeded by hand in February and their fate carefully followed. (After Lawrence and Rediske, 1962. Reprinted with the permission of the Society of American Foresters.)

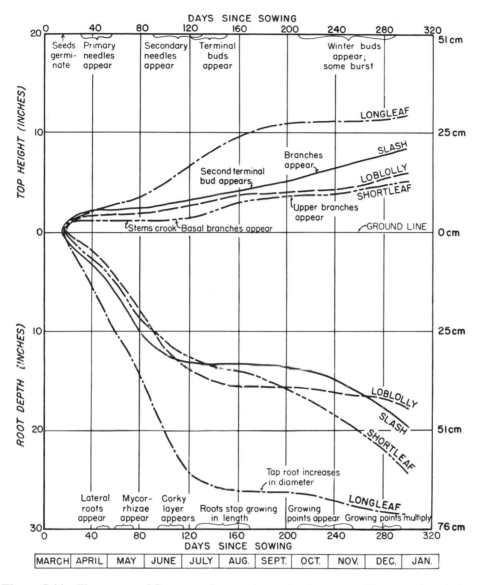

Figure 5.11. The course of first-year shoot and root development of four species of pines under favorable conditions for establishment (in a forest nursery) in Louisiana. (After Wakeley, 1954.)

12), and the seedlings develop rapidly during the first year (Figure 5.11; Wakeley, 1954). Taproot development is rapid because considerable soil water is needed to meet transpiration demands in the needles and shoots which develop rapidly in the ensuing juvenile period.

2. Gap-phase species establish under the existing forest canopy and are shade-tolerant enough, unlike pioneer species, to persist until a local disturbance enables them to penetrate a gap in the canopy (Watt, 1947; Bray, 1956; Chapter 17). White ash, black cherry, white and northern red oak, red maple, yellow birch, basswood, black walnut, slippery elm, white spruce, and to a lesser extent Douglas-fir and white pines, are representatives of this group. Individuals of any species may reach the overstory canopy via a gap, if positioned in the right place at the right time. However, gap-phase species not only possess the ability to tolerate the deeply shaded forest understory but grow quickly enough to colonize the gap before they are overtopped by individuals of competing species or the gap is filled by crowns of overstory plants.

Considerable interest has developed around the phenomenon of **gap** or **patch dynamics** in temperate and tropical forests. This emphasis illustrates the importance of fine-scale processes in succession and the importance of ecosystem-specific disturbances that create canopy gaps and below them open patches on the forest floor of different sizes and shapes. Light-requiring pioneer species are more effective in regenerating in large openings and reaching the canopy via large gaps than gap-phase species or the slow-growing, very shade tolerant species that can reach the canopy in small gaps (Chapter 17).

3. Seedlings of extremely **shade/understory-tolerant species** are able to establish in the shaded understory on moist or **mesic** sites (i.e., where soil water is available throughout the growing season) and persist for long periods (B_j in Table 5.1). They gradually penetrate the canopy as overstory trees die or windthrow provides openings for them. Sugar maple, American beech, hemlocks, true firs, and western redcedar belong in this group.

Groups 2 and 3 are similar in that seedlings are established under a forest canopy and must persist in the understory until conditions are favorable for growth into the overstory. Many different kinds of adaptations have evolved, enabling the various species to persist in the understory and respond to release depending on (1) the physical and biotic conditions of the understory, (2) the species' intrinsic growth rate, and (3) the nature of disturbance that sooner or later provides openings enabling young trees to reach the overstory. In both groups (but much more so in group 3), there is an accumulation of a few to hundreds or millions of seedlings per hectare in the understory.

Species of the two groups have different functional adaptations that enable them to tolerate and survive the physical rigor (low light and soil-water stress) and biotic hazards (herbivores and diseases) of the understory environment. Gap-phase species typically require higher light intensity to survive than the extremely shade-tolerant species, but they can better tolerate soil-water stress. In addition, the deciduous gap-phase species have the marked ability to sprout after fire or animal damage and continue to persist in the understory. In contrast, the most shade-tolerant and slow-growing species constitute group 3; as a group they require more soil water and typically dominate on mesic sites. They may accumulate large seedling populations that experience high mortality. However, they endure to shade out any seedlings that might try to establish under them. Sometimes these seedlings build up for 20 to 30 years or more. In extreme cases, understory hemlocks may be 100 or more years old. In these two groups, the trees that eventually reach the overstory have been recruited from a bank of established and persisting seedlings.

The establishment ecology of two associated species of the northern hardwood forest, yellow birch (group 2) and sugar maple (group 3), illustrates contrasting establishment strategies. Although these two species may grow literally side by side in the northern forest, their physiological adaptations and realized niche are markedly different. In contrast to pioneer species, members of the gap-phase and understory-tolerant groups become established on a forest floor that is more or less shaded by an existing overstory. Sugar maple disperses seeds in the early fall (before snowfall) during leaf drop. Although yellow birch seeds also mature in the early fall, they are gradually dispersed throughout the winter and may be blown for great distances on top of the snow (Figure 5.12*a*; Tubbs, 1965).

Sugar maple seeds germinate in early spring under the snow and in a layer of leaves where temperatures are only slightly above freezing (about 1°C) (Figure 5.12*b* and *c*). At the same time, yellow birch seeds still rest on top of the snow (Figure 5.12*a*). As the snow melts, birch seeds come to rest on the forest floor and germinate in late spring at higher temperatures (about 10°C). Radicles of maple seedlings penetrate the wet leaf mat and, following a good seed year, establish by the million, often forming a carpet of first-year seedlings on the forest floor. The tiny radicles of the birch seeds (over 50 times lighter than sugar maple seeds) are unable to penetrate the thick leaf mat, which tends to dry out rapidly. Most yellow birch seedlings soon desiccate and die. Nevertheless, some birch seeds come to rest by chance on rotting logs, moss-covered rocks, or mineral soil of mounds formed by uprooted trees. In these microsites, yellow birch seedlings may become established, provided sufficient light is available and the substrate does not dry out.

Sugar maple seedlings are highly shade tolerant and can survive the forest floor environment better than most other tree species in its geographic range. In contrast, yellow birch seedlings cannot endure heavy shade. However, at higher light levels, yellow birch inherently grows faster than sugar maple. Thus when a large gap in the canopy occurs via windstorm or disease, an appropriately-positioned yellow birch is able to outcompete the slower-growing maples for a place in the overstory canopy. Similarly, many other gap-phase species are able to colonize openings in the canopy; but they must be able to endure the rigors of the understory (low light intensity, low soil water, smothering by leaves and woody debris, and the attacks of insects, disease, and herbivores) until a gap develops. In many ecosystems of the upland hemlock-northern hardwood forest, sugar maple is able to dominate both understory and overstory because of its effective seed production, dispersal, germination, and establishment adaptations.

Post-Establishment Development

Under favorable environmental conditions, tree species exhibit rapid growth in the juvenile phase. This grand period of growth is followed by a leveling off of the growth curve and a declining rate in old age. The high mortality of the establishment period, in terms of number of stems, decreases with time. Trees increase in size and their density (number of stems per unit area) decreases as competition continues (Chapter 15). Relationships of shoot, crown, and root structure to regeneration ecology and tree growth are considered in the next chapter.

VEGETATIVE REPRODUCTION

In many situations, particularly following disturbances of various kinds, reproduction by vegetative means is more important for the survival of woody plant populations than sexual reproduction. All woody angiosperms are able to reproduce themselves vegetatively once they have become established; it is a major fitness trait (Chapter 4). Conifers are less adapted to reproduce by vegetative means. This advantage of angiosperms may be one of the major reasons for the dominance of angiosperms over conifers in abundance and diversity since the Cretaceous Period.

Vegetative reproduction enables the plant to survive and reestablish itself in place and often to expand its spatial coverage of an area. For pioneer species such as aspens, willows, sumacs, dogwoods, and many other shrub species, cloning enables these species to colonize disturbed sites quickly. Following subsequent disturbance, new ramets establish and grow rapidly due to an already-established and functioning root system—quite

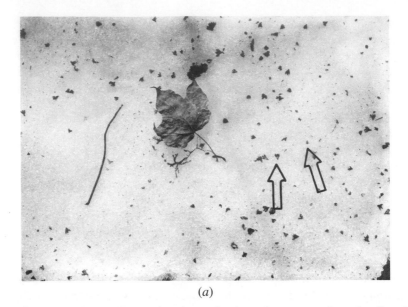

(a)

Figure 5.12. Sites of seed dispersal and germination of sugar maple and yellow birch in the northern forest. *(a)* Yellow birch bracts and seeds litter the surface of melting snow on April 30 in a northern hardwood stand on the Upper Peninsula Experiment Forest, Michigan. The overstory is composed primarily of sugar maple of seeding age, but seldom are maple seeds observed on top of or within the snow cover even after a bumper crop. Right arrow indicates a yellow birch seed; left arrow points to a bract. *(b)* Removing the snow from the exact spot shown in *(a)* to the top of the previous fall's leaf layer reveals no seeds of any species. The leaves are compressed into a soggy mat that is often partially frozen. Spring ephemerals have pushed through the mat (arrow). *(c)* When the top leaf layer shown in *(b)* is removed, the ability of sugar maple to germinate underneath a snow cover in early spring is revealed. Arrows point to germinated seeds. In those areas sampled on the Experimental Forest, the bulk of sugar maple seed was found under a layer of leaves, whereas yellow birch seeds were distributed on the top of the snow as illustrated in *(a)*. (After Tubbs, 1965; U. S. Forest Service photos.)

unlike seedling establishment. Also, once established, they may persist for long periods of time. Furthermore, on fire-prone ecosystems, clonal species resprout after fire and act to prevent erosion and to rapidly recouple the nutrient cycle.

A model of vegetative regeneration is presented in Figure 5.13 and illustrates contributions of meristems to the bud bank from several sources: crown, basal stem and root collar, and underground roots and rhizomes. Once vegetative buds are released by stimuli such as temperature, fire, wind breakage, and herbivores, etc., new shoots are formed and clonal regeneration and expansion follows (Figure 5.13). Sprouting and the rooting of stems are the primary types of vegetative reproduction common in woody plants. Sprouting i.e., development of new shoots from adventitious buds, occurs in different anatomical structures:

Basal stem: New shoots develop from dormant or adventitious buds from the basal part of the stem (i.e., at the root-collar where stem joins root) of an established plant.

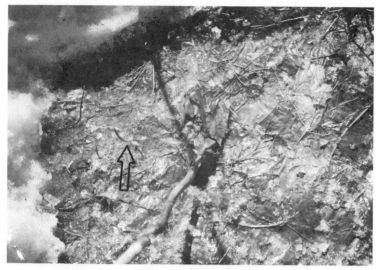

(*b*)

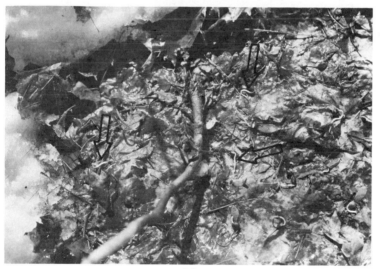

(*c*)

They form multiple-stemmed shrubs or trees such as, oaks, hickories, basswoods, ashes, walnuts, birches, alders, prickly ash (Reinartz and Popp, 1987), and species of many other genera.

Root: New shoots arise from adventitious buds on roots. Root sprouting is characteristic in clone-forming trees and shrubs such as, aspens, American beech, sassafras, sumacs, and sweetgum, among others.

Rhizome: New shoots develop from stems growing more or less horizontally underground. The distal end of a shoot becomes erect and bears leaves and flowers. This is typical of herbs; many shrubs, such as salmonberry (Tappeiner et al., 1991) and other *Rubus* species; and trees such as flowering dogwood, striped maple, Gambel oak (Tiedemann et al., 1987), and dwarf chestnut oak.

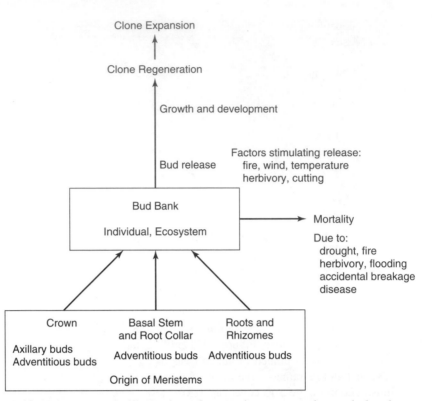

Figure 5.13. Diagrammatic illustration of vegetative regeneration and clonal expansion of woody plants. (Modified from Zasada et al., 1992. Reprinted with permission of Cambridge University Press.)

Lignotuber: New shoots arise from a buried mass of stem tissue termed the lignotuber; it is characteristic of Gambel oak and eucalypts.

Vegetative reproduction that occurs by rooting of aerial stems and subsequent shoot development include:

Reproduction via stolons or runners: Stolons are the arching branches of shrubs, such as red-osier dogwood and *Rubus* species, that take root when they come in contact with the soil surface and form new plants at that point. Runners, or procumbent stems, of plants like creeping strawberry bush and Virginia creeper take root at various points along their length as they encounter the soil surface.

Fragmentation: Branches that are broken off willow, cottonwood, and other streamside species by ice, flood debris, or wind may take root and become established after being buried in soil of the stream bank or the active floodplain.

Layering: Lower branches of boreal and northern conifers are often pressed into the soil by the weight of snow or by woody debris such as vine maple (O'Dea et al., 1995). They take root and form clones around the perimeter of the parent tree. Layering is common in spruces, firs, and larches.

Tipping: Northern white-cedar trees in swamps are often uprooted just enough by wind so that they very slowly tip over and eventually lie procumbent on the peaty surface of the swamp (Curtis, 1959). The lateral tree branches turn upright and develop into independent trees as moss and organic matter covers the old tree, and roots form at the base of each branch.

Besides maintaining individual genotypes for long periods of time in disturbance-prone ecosystems, vegetative reproduction expands the area occupied by many shrubs and trees: blueberries, dogwoods, sumacs, sassafras, beeches, aspens, sweetgum, and sycamores. Trembling aspen clones, consisting of many genetically identical ramets up to 0.1 ha in size, are very common in burned-over forests. Clones of very large size occur in the Interior West of North America. Figure 5.14 illustrates a clone in south-central Utah 43 ha in extent (Kemperman and Barnes, 1976; Barnes, 1975). This clone is reported to be one of the world's largest organisms (Grant et al., 1992; Grant, 1993).

Whereas clones of aspens and sassafras spread expansively by sprouting from roots, many forest shrubs spread by the repeated rooting of aerial stems that are pressed into the forest floor, take root, and produce new ramets. For example, clones of the shade tolerant vine maple develop in the understory of Douglas-fir forests in the Pacific Northwest by rooting of stems that are pinned to the forest floor by falling trees and branches (O'Dea et al., 1995; Figure 5.15). Establishing by seeds following fire or windstorm, vine maple individuals develop as multiple-stemmed groups. Their vinelike branches intertwine with those of young Douglas-fir trees. As the trees thin out, the maple branches lose their support, and they droop to the forest floor. In time, falling tree limbs and trunks pin the flexible maple stems to the forest floor (Figure 5.15) and facilitate the layering; 99 percent of the layered stems have been pinned by fallen trees or branches. After repeated layering events by the genet and its ramets, vine maple clones may extend over considerable area. The clone illustrated in Figure 5.15 covers approximately 210 m², whereas the crown cov-

Scale

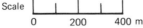

0   200   400 m

Figure 5.14. Large clone of trembling aspen, 43 ha in extent, with the boundary outlined; south-central Utah. Note the smaller clones around the large clone; the differences in tone indicate differences in fall coloration of clone foliage. (After Kemperman and Barnes, 1976. Reprinted with permission of National Research Council Canada.)

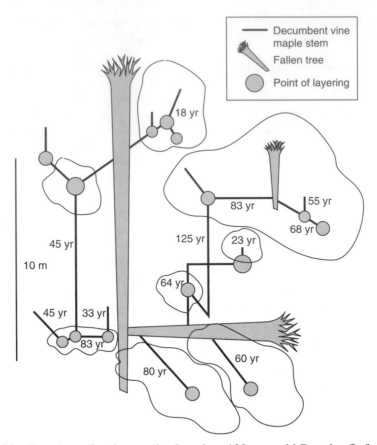

Figure 5.15. Top view of a vine maple clone in a 130-year-old Douglas-fir forest showing decumbent stems and points of layering. Aerial stems are not shown. Irregular enclosed areas indicate vine maple crown coverage. Note the large range in age among ramets of the clone. (After O'Dea et al., 1995.)

erage of all ramets was only 80 m². The 40 upright stems ranged in age from 18 to 125 years. Through this clonal process, vine maple is able to persist, potentially indefinitely, in the understories of Douglas-fir forests, perpetuated in part by the mutualistic relationship with the dominant trees themselves.

Vegetative reproduction plays a significant role in the juvenile phase following establishment of deciduous tree seedlings. Although fires or herbivores may damage or destroy a portion or all of the aboveground part of an established seedling, it can sprout back vigorously. Many species are able to withstand repeated browsing or defoliation and still resprout for many years. Sprouting ability is greatest in the juvenile phase (when it is of greatest adaptive significance) and declines as the plant matures. All deciduous trees and shrubs are capable of sprouting from roots, rhizomes, or aboveground stems.

Conifers rarely reproduce vegetatively. Even in the northern, boreal, and alpine areas where reproduction by layering is common, conifers reproduce primarily by sexual means. Only a few species are able to sprout vigorously and regenerate themselves after fire, grazing, or cutting: for example, the redwood sprouts vigorously from root-collar, stump, and stem. Also, various hard pines, including pitch pine, Virginia pine, and pond pine, sprout from dormant buds along the stem after fire or cutting (Kozlowski, 1971a). Near the upper limit of tree growth in the southern Rocky Mountains, Marr (1977) reported "tree islands" of subalpine fir and Engelmann spruce that moved along the ground by repeated layering and growth to leeward. Movement of 5 m in 11 years was common, and some clonal islands apparently have moved at least 15 m. Seedlings apparently become established in sheltered microsites, and the clones expand and colonize adjacent microsites that are inhospitable to seedling establishment.

Vegetative reproduction has several adaptive benefits for the persistence of plants. First, by spreading spontaneously below- and aboveground, an individual colonizes new space and can tap new sources of light, soil water, and nutrients that are patchily distributed. Second, in the case of non-spreading clonal species, a single genet's survival is ensured either by resprouting from its roots or protected stem following destruction of the aboveground parts of the genet or by spreading mortality risk of the genet among existing ramets that are capable of independent survival (Cook, 1979). For example, certain tropical palm species produce root-forming ramets at the base of the genet (De Steven, 1989). Because the large genet, 16 to 20 m tall, is at risk by windthrow and stem breakage from severe tropical storms, one or more of the basal ramets perpetuates the plant if the genet is killed. Overall, the clonal habit is an extremely important adaptation for the persistence and spatial dominance of woody species. However, it is sexual reproduction that provides the genetic variation for persistence in the face of marked environmental change.

Regardless of the mode or origin, sexual or vegetative, the form or architecture of tree crowns, stems, and roots, together with their growth patterns, strongly control the vegetative composition and productivity of different ecosystems. Selected aspects of tree structure and growth and their relationship to site factors are considered in the next chapter.

Suggested Readings

Allen, G. S., and J. N. Owens. 1972. The life history of Douglas-fir. Environment Canada. For. Serv., Ottawa. 139 pp.

Grubb, P. J. 1977. The maintenance of species-richness in plant communities: the importance of the regeneration niche. *Biol. Rev.* 52:107–145.

Kozlowski, T. T. 1971. *Growth and Development of Trees*, Vol. I. *Seed Germination, Ontogeny, and Shoot Growth*. Academic Press, New York. 443 pp.

Kozlowski, T. T., and S. G. Pallardy. 1997. *Physiology of Woody Plants*. 2nd ed. Academic Press, San Diego, CA. 411 pp.

Owens, J. N. 1991. Flowering and seed set. In A. S. Raghavendra (ed.), *Physiology of Trees*, John Wiley, New York.

Silvertown, J. W. 1980. The evolutionary ecology of mast seeding in trees. *Biological Journal of the Linnean Society*. 14:235–250.

U. S. Department of Agriculture. 1990. *Silvics of North America*. USDA For. Serv. Agr. Handbook 654. Washington, D.C. Vol. 1, Conifers, 675 pp; Vol 2, Hardwoods, 877 pp.

Zasada, J. C., T. L. Sharik, and M. Nygren. 1992. The reproductive process in boreal forest trees. In H. H. Shugart, R. Leemans, and G. B. Bonan (eds.), *A Systems Analysis of the Global Boreal Forest*. Cambridge Univ. Press, Cambridge.

CHAPTER 6

STRUCTURE AND GROWTH

The development of the seedling into a large tree is a complex process involving two kinds of growth: extension of each growing point forming the shoots of the crown and the roots (**primary growth**) and expansion of stem and root diameter (**secondary growth**). A combination of these growth processes, following an inherent architectural model and influenced by environmental factors, gives each species its characteristic aerial and subsurface structure and form. In this chapter, we emphasize the tree crown, stem/shoots, and roots and selected aspects of their structure and growth in relation to environmental factors, especially soil water. Physiological processes, especially growth and development of shoots and roots are considered, but a detailed treatment of tree physiology is beyond our scope. For detailed treatments of tree physiology the reader is directed to books by Larcher (1995), Kramer and Boyer (1995), and Kozlowski and Pallardy (1997a, b). For considerations of tree structure and function, the classic treatment by Büsgen and Münch (1929) is notable as well as modern works by Zimmermann and Brown (1971), Wilson (1984), and Oldeman (1990).

TREE FORM

Tree habit, form, or architecture is as incredible as it is fascinating. In tropical forests especially, the diversity of regional and local ecosystems favors many species and sites, many vertical stand layers, and the potential for rhythmic or continuing growth for long periods. Many factors such as apical control of growth, shoot types, branching orientation, timing of meristematic activity, herbivory, and shedding of shoots all affect tree architecture as well as do physical factors of light, temperature, soil water, and nutrient availability. Our treatment is necessarily limited, and readers are urged to examine more detailed treatments by Hallé et al. (1978), Wilson (1984), Oldeman (1990), Barthelemy et al. (1991), and Bell (1991).

122

The form or habit of a woody plant is a combination of its inherent architectural plan, determined genetically, and environmental influences occurring during its development. Each species has a precisely determined growth plan or architectural model adapted to the physical and biotic factors of the ecosystems where it lives.

In northern and alpine environments, trees (primarily conifers) are adapted to grow in a conical form (Figure 6.1*a*), due to strong selection pressures of snow, ice, and wind. That this trait is under strong genetic control is demonstrated by the maintenance of the conical form when spruces, larches, and firs from high altitudes are grown in parks far from their native sites. In these species, the terminal shoot or leader grows faster than the lateral branches below it in such a systematic way that a central stem and a conical crown are the result. Termed **excurrent** tree form, it is typical of northern conifers and a few deciduous trees, such as sweetgum and tuliptree. The excurrent form is an expression of strong apical control, whereby the terminal leader maintains complete control year by year over the laterals below it (Brown et al., 1967; Wilson, 1984.).

Balsam fir   Engelmann spruce

(a)

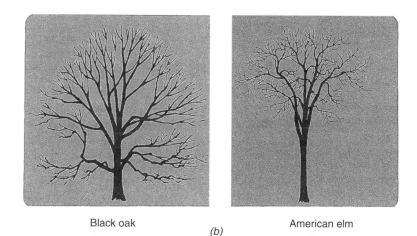

Black oak   American elm

(b)

Figure 6.1. Examples of excurrent (*a*) and decurrent or deliquescent (*b*) forms of trees. (After Hosic, 1969. Reproduced by permission from *Native Trees of Canada,* 7th ed. by R. C. Hosie; published by the Canadian Forestry Service and Fitzhenry and Whiteside Ltd, 1969.)

Not all conifers exhibit the excurrent form; the change to a more rounded and spreading form is related to climate. Pines of the southeastern United States, such as loblolly and longleaf, exhibit a strong central stem, but a more rounded crown than spruces and firs of the far north. In the arid Southwest, pinyon pines and junipers tend to have short, compact, rounded, and bushy crowns. This form is related to the high temperature and soil-water stress of their environment that favors extensive root development and compact crowns rather than tall excurrent growth that would severely expose the crown to strong, dry winds.

In many deciduous species, the lateral branches grow nearly as fast or faster than the terminal leader. The main stem may fork repeatedly, giving rise to a spreading form (Figure 6.1*b*). This **decurrent** or **deliquescent** form is typical of elms, oaks, maples, and many other species that grow in less harsh environments. In these species, apical control is lost, repeated forking occurs in the crown, and a multiple terminal leader system is often developed. This form is most pronounced in American elm and many oaks. The broad, spreading crowns may have an adaptive function in spacing out the relatively large deciduous leaves compared to the densely packed needles of many excurrent conifers. The spreading form allows light to reach the leaves, and the growth rate of the shoots acts together with branch angle, gravity, and other factors to control the spatial configuration and the density or compactness of the crown.

Within the group of decurrent angiosperms, some species that grow along rivers and streams, besides having spreading crowns, may exhibit markedly leaning trunks compared to species in closed forests on other sites. For example, Loehle (1986) reported that black willow, sycamore, and river birch had a lean from the vertical of 21 to 22°, whereas the average lean of three upland oaks was only 7°. The growth response toward light, **phototropism**, markedly affects tree form of certain species and enables these shade-intolerant species to obtain light over the river itself. However, we observe that other bottomland species, American elm, eastern cottonwood, and red ash, typically lack such a pronounced lean. Each of these species has a different kind of branching architecture to display the crown and leaves for optimum capture of light for photosynthesis.

Detailed simulations of plant growth and canopy expansion by Küppers (1994) for woody plants and by Givnish (1986) for herbs lead to the conclusion that branching patterns, as well as leaf physiology explain the competitive roles and success of co-occurring species. Photosynthesis and carbon acquisition by leaves is a relatively conservative process (Chapter 18). It requires coupling with the appropriate site-mediated trunk and crown architecture for ecological success of the plant.

Architectural Models

Generalized architectural models of trees (that apply to shrubs and herbs as well), stylistically drawn to represent the fully expressed form, were developed by Hallé and Oldeman (Hallé et al., 1978) who initially worked in the tropics. They reduced the total diversity of tree forms to 23 architectural models; selected examples are illustrated in Figure 6.2. A given species may be intermediate between two models or share features of several models, as Hibbs (1981) found for eastern hemlock.

Criteria for identifying the models included whether shoot extension is rhythmic or continuous, whether the tree is unbranched (as in some palms) or branched, whether or not branch differentiation is **orthotropic** (erect) or **plagiotropic** (horizontal), and whether flowering is terminal or lateral. Specific combinations of these characters identify the models, each of which is named for a well-known botanist. Illustrated in Figure 6.2 are several models that Hallé et al. (1978) identified for temperate plants as well as for many tropical taxa. These include the models of Leeuwenberg (sumacs, red-osier dogwood, and rosebay rhododendron), Rauh (most species of pines, red and live oak, and probably most

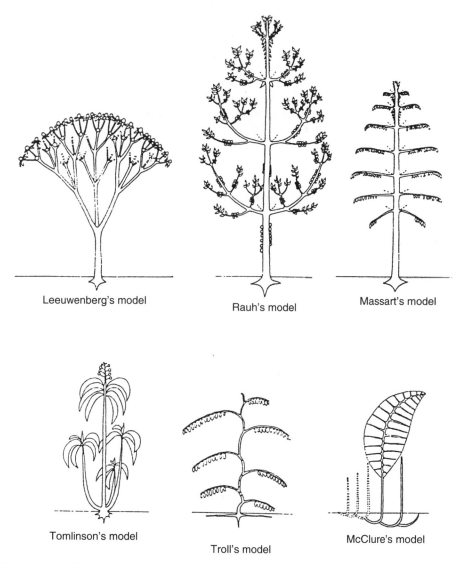

Leeuwenberg's model   Rauh's model   Massart's model

Tomlinson's model

Troll's model

McClure's model

Figure 6.2. Generalized architectural models of trees, selected from the 23 models of Hallé et al. (1978). Besides tropical species the following temperate groups or species are characterized: Leeuwenberg's model, sumacs, rosebay rhododendron; Rauh's model, most pines, red and live oaks; Massart's model, redwood and true firs; Tomlinson's model, multiple-stem palms; Troll's model, American beech, American elm; McClure's model, many bamboo species. (After Tomlinson, 1983. Reprinted with the permission of the Sigma Xi Scientific Research Society.)

oaks), Massart (redwood and true fir species such as balsam fir and European silver fir), Tomlinson (most multiple-stem palms and many sedges and grasses), Troll (American beech, American elm), and McClure (many bamboo species). The inclusion of excurrent forms of pines together with the spreading forms of oaks illustrate the broad nature of these architectural plans.

Hallé et al. (1978) observe that if readers of their book on tropical tree architecture should retain a sense of the monstrous, the fabulous, or the unreal, they should mentally just reverse the situation. How strange, they say, is the temperate tree: ". . . leafless for a large part of the year, with such marked synchrony in its development, its brief period of extension growth, its ability to flower only once each year and with its peculiar annual radial increments of growth in the wood. Here is a bizarre object indeed!"

The study of tree architecture leads to an understanding of how woody species respond to stresses and crown damage as they mature. We observe a given model most readily as seedlings or saplings because their continued development is modified by environmental stresses and accidental events. When a tree is damaged and new shoots arise, the original architectural growth plan of the species is repeated or **reiterated**. In addition to traumatic reiteration, the process (termed adaptive reiteration by Barthelemy et al., 1991) can also occur naturally in old crowns and in spontaneous vegetative reproduction from roots or stems. Thus the crown shape of woody plants is the result of the genetic process of an architectural plan, environmental modifications of it, and the reiteration of the plan following disturbances as trees mature.

The survey by Hallé et al. (1978) confirmed that the greatest diversity of woody plant forms occurs in the tropics. For example, Poore (1968) reported 374 species of trees in a 23-ha plot in lowland western Malaysia. All 23 models are represented in the tropics and many are restricted there, particularly those where shoot extension is continuous. Tomlinson (1983) suggested that the profusion of tropical forms is, in fact, due to the relative uniformity of climate found in many lowland tropical areas. Being uniformly favorable for plant growth, success of a species depends primarily on complex interactions with other plants and animals in reproduction, dispersal, and recruitment into one or more layers of the tropical forest. An ecological consideration of tree forms in relation to different climatic and successional situations in tropical and temperate forests is given by Brunig (1976).

The canopy architecture of individual species is combined in a given temperate or tropical ecosystem to produce a particular canopy structure, usually of several layers. Multiple tree and shrub layers characterize moist tropical and temperate forests (for example, see Figure 8.13), whereas dry, tropical or temperate forests may exhibit a totally different vertical canopy structure (for example, see Figure 22.3). The canopy structure directly influences the amount of leaf surface area present to capture light for photosynthesis. The interrelated factors of tree form, vertical canopy structure, and leaf area, as they influence interception of light, and photosynthesis and overall tree productivity, are discussed in Chapter 8.

SHORT AND LONG SHOOTS

The kinds of shoots and the patterns of shoot growth are important in determining tree form, reproductive and regeneration potential, and competitive ability in different ecosystems. Plants are characterized by an *open system of growth* whereby the shoot (the stem with its collection of leaves and buds) is the modular unit of construction. We examine next the kinds of shoots that make up the modular units and also the patterns of shoot growth that affect tree structure and function.

Long and short shoots are characteristic of woody plants; all trees bear long shoots, and some species have both. Short shoots exhibit little or no interleaf elongation; they may be called **dwarf shoots** in conifers and **spur shoots** in deciduous trees. The boundaries between consecutive annual shoots are marked by groups of ringlike scars left by the bud scales. Annual short-shoot growth is only a few millimeters or less so that internodes are extremely short and the annual ring scars lie virtually next to one another. For example, we collected a 28-year-old short shoot of yellow birch that measured only 7 cm long.

In contrast, the annual growth of long shoots may range from a few centimeters to several meters. Annual long-shoot growth of over 3 m is not uncommon for young eastern cottonwood trees on alluvial sites. Short shoots are characteristic of beeches, birches, maples, and rosaceous genera such as *Prunus* (cherries) and *Malus* (apples). In the gymnosperms, short shoots are characteristic of the deciduous genera *Larix*, *Pseudolarix*, and *Ginkgo*. In the pines, needle-bearing **fascicles** are modified short shoots. They are determinate, persist more than one season, but lack extension growth. Lateral short shoots are typical of temperate trees, whereas terminal short shoots are common in tropical trees. Long shoots may be either terminal or lateral in temperate or tropical forests and often occur in the upper crown. In European white birch, for example, short shoots make up over 90 percent of all shoots on tree. However, in the top of the tree, where the main extension growth occurs, long shoots account for approximately 80 percent of the shoots. Short-shoot leaves are produced at the same place every year and have an exploitive capacity as primary photosynthetic units (Bell, 1991, p. 254). For example, the broad paired, short-shoot leaves of the shrub, hobblebush, which grows in shaded understories of conifer forests of the Appalachian Mountains, are positioned to capture the maximum amount of light available.

Short shoots bear only "early" leaves that are preformed in the dormant or resting bud; they also may bear flowers, cones, or thorns. In contrast to the seasonally-determinate short shoots, long shoots can be either seasonally-determinate or indeterminate. Determinate long shoots are those preformed in the dormant bud, and they cease growth and set buds after their initial spring growth extension. The basal part of indeterminate shoots is also formed within the bud. After this portion elongates (in the early spring in temperate forests) the shoot continues to grow throughout the growing season, providing it is in a well-lit position and other site conditions are favorable. These indeterminate long shoots have an exploratory capacity in significantly extending the framework of the plant into new space. Not surprisingly, seasonally indeterminate shoots are especially well developed in well-lighted crown positions of fast growing, pioneer species of pines, birches, cherries, cottonwoods, balsam poplars, aspens, sassafras, and sycamores (Marks, 1975).

Seasonally-determinate long shoots bear only early leaves, whereas indeterminate long shoots bear early leaves along their basal portion and "late" leaves, those formed during the current growing season, along the upper portion. For example, in *Populus*, the late leaves that are produced on leader and terminal shoots after the early leaves develop bear little resemblance to early leaves (Figure 6.3). The different forms are all expressions of the same genotype and are not directly attributable to the external environment. Instead, they are due to the internal environment at the time and place of primordia formation and development. Morphological studies in plant taxonomy, paleoecology, and ecology may be misleading unless this kind of leaf polymorphism is recognized.

PATTERNS OF INTERMITTENT GROWTH

Intermittent (periodic, episodic, rhymthic) growth, rather than continuous growth, is the rule among woody plants. Even trees such as palms and mangroves, noted for continuous growth by producing a leaf or pair of leaves every few months, do not necessarily have a constant rate of shoot extension (Borchert, 1991). Environmental constraints of cold temperature and soil-water stress typically induce meristems or buds to enter a resting or dormant state. Resting buds resume growth within one to two weeks once the favorable conditions resume, whereas dormant buds do not. In temperate areas, the primary control of periodic growth (evidenced by leaf flushing and shoot extension) is favorable temperature; in the tropics it is favorable soil-water. Some trees exhibit repeated cycles of growth and rest unrelated to environmental changes; this intrinsic growth periodicity is often referred to as rhythmic growth (Borchert, 1991).

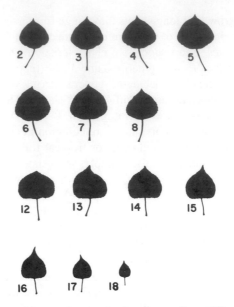

Figure 6.3.   Silhouettes of leaves from a leader shoot of trembling aspen. Leaves 2–7 are typical early leaves; 8 is a transition form to late leaves; and leaves 12–18 are typical late leaves. (After Barnes, 1969. Reproduced by permission of the Bundesforschungsanstalt für Forst- und Holzwirtschaft, Grosshansdorf, Germany.)

For temperate species, Zimmermann and Brown (1971) cite four types of long-shoot extension:

1.   A single flush of terminal growth followed by formation of resting bud, i.e., seasonally determinate long shoots.

2.   Recurrent flushes of terminal growth with terminal bud formation at the end of each flush.

3.   A flush of growth followed by shoot-tip abortion.

4.   A sustained production of leaves including early- and late-leaves up to the time of terminal bud formation, i.e., seasonally indeterminate long shoots.

Type 1.   Many northern species, including eastern white pine, red pine, some oaks, hickories, and buckeyes, make a burst of growth in spring or early summer and then form terminal or end buds (Figures 6.4 and 6.5). Extension growth of the shoots terminates before soil-water stress of midsummer becomes severe. During the remainder of the growing season, a portion of the plant's resources is utilized in developing shoots and foliage inside the protective bud scales of the newly formed terminal buds. The bud contents expand the following year. Therefore, the current year's shoot growth is primarily dependent on environmental conditions of the previous year (when leaf primordia formed). Red and white pines make a single seasonal flush such that groups of branches (false whorls) are separated from one another by long branch-free internodes. The marked nodal occurrence of branches makes aging relatively easy and facilitates the estimation of site quality based on the amount of growth between the nodes (Chapter 13).

Ecological conditions of the native site largely determine when the burst of growth begins. As shown in Figure 6.4, red and eastern white pines exhibit markedly different

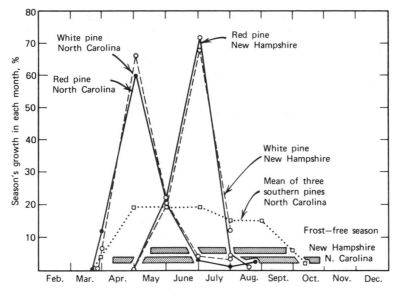

Figure 6.4. Variations in seasonal height growth patterns of red pine and eastern white pine in North Carolina and in New Hampshire, and of three southern pines (loblolly, shortleaf, slash) in North Carolina. The northern pines have preformed shoots and usually have one annual growth flush, whereas the southern pines grow in recurrent flushes. (After Kramer, 1943. © American Society of Plant Physiologists. Used with permission.)

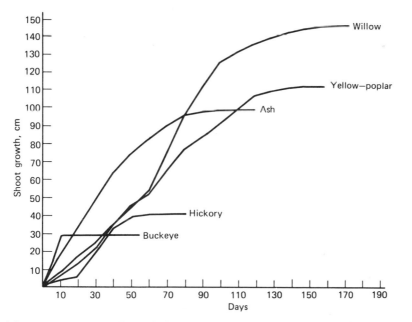

Figure 6.5. Rate and duration of shoot growth among several woody species in the Georgia Piedmont. Measurements of shoot elongation were made biweekly on 9 trees ranging from 8 to 15 years in age: painted buckeye, mockernut hickory, red ash, yellow-poplar (tuliptree), and black willow. (After Zimmermann and Brown, 1971. Reprinted with permission of Springer-Verlag.)

timing of shoot growth in North Carolina as compared to New Hampshire. Except for the oaks, the structure of the crowns seems to be more regular than in the following types.

Type 2.   Trees in regions in which conditions for growth may be periodically favorable, such as the pines of the southeastern United States and Central America, typically exhibit recurrent flushes of growth during the growing season. This type of growth is the most typical one among conifer and deciduous trees in the subtropical and tropical regions. The environmental conditions of the site, especially soil water, determine the number of flushes of growth.

Type 3.   In this common type, the shoot tip aborts following the extension of the pre-formed leaves (beeches) or considerably later in the season after many late leaves have been produced (black willow, birches, mimosa). In many cases, the shoot-tip abortion (and, in the case of birches, shoot tip and terminal bud) is associated with the shortening days of summer. In the following spring, shoot extension precedes from the last fully-formed lateral bud. Zimmermann and Brown (1971) list 16 genera in which shoot-tip abortion occurs. The ecological significance is unclear. However, it may be an effective mechanism to prolong shoot extension as long as favorable conditions exist.

Type 4.   In a few southern species, some shoots make continuous and prolonged shoot growth, including both early and late-leaf development, before setting a terminal bud. Examples of species with southerly distributions include tuliptree, sweetgum, and eastern cottonwood. Within their range, the potential of severe frost damage is low, and selection pressures for shoot-tip abortion may not be strong. They also occupy moist sites so that the pressure for early cessation of growth to avoid drought also may not be great. This type of shoot extension and true terminal bud formation lead to a more symmetrical crown and also to a more pronounced main stem than in trees exhibiting shoot-tip abortion.

Since long shoots are responsible for height growth and crown extension, it is of advantage to use the growing season effectively. The previously-mentioned types of long-shoot extension illustrate various ways trees make full use of favorable and potentially favorable parts of the growing season. While seasonally indeterminate long shoots extend the crown, the early leaves of short shoots and seasonally determinate long shoots contribute photosynthate to satisfy current growth and respiration requirements.

SYLLEPTIC AND PROLEPTIC SHOOTS

The terms **syllepsis** and **prolepsis** identify two types of lateral long-shoot growth that affect crown form. Development of a lateral shoot directly from a growing terminal meristem (or shoot) is referred to as sylleptic growth; a sylleptic shoot is formed. This pattern is common in the tropics but rare in temperate forests, although sassafras, alternate-leaf dogwood, sweetgum, and alders may regularly form sylleptic shoots (Wilson, 1984). In prolepsis, the lateral shoots develop after a rest period and lateral bud formation. In temperate plants, proleptic growth typically takes place following the winter dormant period. However, lateral meristems of some temperate species may resume growth briefly after setting buds if late season weather is warm and moist. Such proleptic shoots are common in fast-growing pines (Rudolph, 1964). In temperate forests, when the *terminal* shoot exhibits such a resumption of growth following bud set, the condition is called **lammas** growth. If lateral meristems develop late in the season (prolepsis), but the terminal shoot does not, in following years several laterals may dominate the terminal and eventually become replacement terminals. A much-forked tree form is the result as often seen in young, open-grown white pines (Wilson, 1984). Late-season shoots apparently occurred about the time of European holidays, Saint's Day (June 24) and in England, Lammas Day (August 1). They are termed in various languages: lammas shoots, Johannistriebe, pousses de la St. Jean, or Sint Jansloten.

ROOTS

Tree roots perform many significant functions. Two important ones are the firm anchoring of the tree in the soil and the absorption of water and nutrients. Anchorage is accomplished by the whole root system, whereas absorption mainly occurs by fine (< 1 mm), nonwoody roots. Other functions of larger roots include storage of carbohydrates and other materials, synthesis of organic compounds, transport of water and nutrients to the crown, secretion of chemical substances, and generation of vegetative shoots (as in aspens, American beech, black locust, and sweetgum). The growth and functioning of roots has received attention from many authors, and the major review papers are those of Röhrig (1966), Lyr and Hoffman (1967), Sutton (1969), Kozlowski (1971b), and Pregitzer and Friend (1996).

Important differences exist in the root systems of angiosperms and conifers. Deciduous trees, evolving later than conifers, have generally developed root systems that are more extensive and efficient (Voigt, 1968; Kozlowski et al., 1991). The roots of the more primitive conifers probably have a greater ability to extract nutrient ions from the soil. Leaves of most deciduous trees decompose rapidly (Ellenberg, 1988, p. 59), and these species may have evolved extensive and finely divided root systems to intercept and absorb the nutrient ions released from the leaves upon decomposition. In general, fine roots of deciduous species are smaller in diameter than those of conifers (Voigt, 1968). Thus nutrient absorption may be more efficient because absorption per unit of surface area increases as root radius decreases (Nye, 1966). The processes of ion uptake and assimilation are discussed in Chapter 19. The relative capacity for root development of several conifer and deciduous trees was studied by Kozlowski and Scholtes (1948). Several of their comparisons are presented in Table 6.1, and demonstrate that seedlings of the deciduous species developed more extensive root systems than those of the loblolly pine in greenhouse and forest environments.

Table 6.1 Comparative Root Development of Seedlings of Deciduous and Coniferous Species.

Species	Age in Months	Number of Roots	Total Root Length meters	Growing Conditions
Black locust	4	7124	325.5	Greenhouse
Loblolly pine	4	419	1.6	Greenhouse
Flowering dogwood	6	2657	51.4	Greenhouse
Loblolly pine	6	767	3.9	Greenhouse
White oak	12	196	2.3	Forest
Loblolly pine	12	148	1.0	Forest

Source: After Kozlowski and Scholtes, 1948. (Reprinted with the permission of the Society of American Foresters.)

Kinds, Forms, and Occurrence

The tree root system is characterized by a framework of relatively large, woody, long-lived roots supporting a mass of small, short-lived, non-woody absorbing roots, many of which are associated with fungi (these mycorrhizal relationships are discussed in Chapters 15 and 19). In many species, seedlings develop tap roots that provide early stability and survival, especially on dry and fire-prone sites. Tap roots are common in pines and in species that have large seeds with much reserve food (oaks, hickories, chestnuts, walnuts). Roots of the main framework are relatively fast growing (up to 1 m per year) and grow at a consistent depth in conformation with the topography for distances up to 20 m or more (Wilson, 1984). They extend rapidly into new soil and have relatively large root tips. As the lateral system becomes widespread, vertical **sinker** roots may develop at intervals and reach into lower soil horizons.

The form and structure of the root system is to a large degree genetically controlled. Nevertheless, site conditions markedly influence the form and pattern of root development. Species adapted to sites characterized by summer droughts may possess deep-penetrating tap roots, as found in pines and upland oaks and hickories. Where the water table is near the surface as in swamps or peat bogs, the root systems are shallow. The uprooting of trees by wind (i.e., **windthrow**) is common on such sites where rooting is shallow. A sample of the great variety of rooting forms that develop in relation to different soil and water table conditions is presented in Figure 6.6. Although many species exhibit deep rooting in sandy soils, their rooting depth may be severely curtailed where impermeable or poorly- aerated layers occur near the surface. Exquisite illustrations of root systems of individual European species were presented by Köstler et al. (1968).

Not illustrated in Figure 6.6 are aerial roots that are formed by some tropical trees, especially strangling roots and stilt and prop roots (Bell, 1991). In eastern North America, yellow birch is noted for its aerial rooting habit. Birch seeds germinate on decaying logs and stumps, and their roots grow downward and into the soil. The roots enlarge and are exposed as the woody debris decays and slumps away. Yellow birch also germinates in the moss on boulders. Descending roots penetrate the soil, and eventually a large tree is perched on the boulder or outcrop of bedrock. In humid environments, such as in the Southern Appalachian Mountains, old yellow birch trees even have the ability to generate aerial roots from the stem cambium. These roots develop downward, either on the outside of the trunk or through the decaying heartwood of the standing tree, and into the soil (Spurr and Barnes, 1980, p. 91).

Fine Root Relations

As in shoots, long and short roots may be distinguished for many species. At varying distances along the main roots, lateral roots are formed that may become long roots or remain short. The root system is thus characterized by several orders of woody, perennial long roots and one or more orders of small or fine non-woody roots (Figure 6.7). Fine roots (< 1 mm) have fine root tips. They branch and rebranch to form many root orders down to a minimum size of about 0.07 mm in northern red oak. Fine roots proliferate to exploit new soil space. Lyford (1980) estimated that a mature northern red oak (40 to 70 years old) would have 90 million fine root tips in its system. An idea of the number of roots in the forest floor comes from Lyford's count of 1,000 fine roots in an average 1-cm cube of the forest floor in a northern red oak–mixed hardwood forest in Massachusetts. These fine roots were estimated to occupy only 3 percent of the cube but have the surface area of more than 6 cm^2, not counting mycorrhizal hyphae or root hairs.

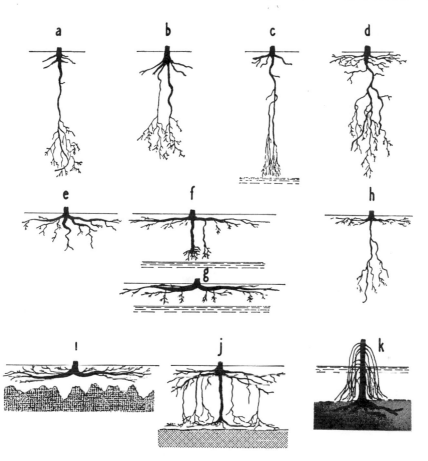

Figure 6.6.   Modification of root systems of forest trees by site. (*a, b*) Taproots with reduced upper laterals: patterns found in coarse sandy soils underlain by fine-textured substrata. (*c*) Taproot with long tassels, a structure induced by extended capillary fringe. (*d*) Superficial laterals and deep network of fibrous roots outlining an interlayer of porous materials. (*e*) Flattened heartroot formed in lacustrine clay over a sand bed. (*f*) Plate-shaped root developed in a soil with a reasonably deep ground water table. (*g*) Plate-shaped root formed in organic soils with shallow ground water table. (*h*) Bimorphic system of platelike crown and taproot, found in leached soils with a surface rich in organic matter. (*i*) Flatroot of angiosperms in strongly leached soils with raw humus at surface and hardpan below. (*j*) Two parallel plate-roots connected by vertical sinkers in a hardpan spodosol. (*k*) Pneumatophores of mangrove trees in tidal lands. (After Wilde, 1958, *Forest Soils: Their Properties and Relation to Silviculture;* Courtesy of the Department of Soil Science, University of Wisconsin-Madison and John Wiley & Sons, Inc.)

In most soils, fine roots are concentrated in the upper 10 cm of soil (organic and mineral). In northern red oak, for example, many roots arising from the lateral root system grow upward and into the humus layers of the forest floor. Here they proliferate into many higher-order, small-diameter, non-woody roots (Lyford, 1980) that absorb water and nutrients. Coutts and Nicoll (1991) found a similar inherent upward movement of Sitka spruce roots and lodgepole pine seedlings. Figure 6.7 illustrates the development of roots upward and downward from a lateral root of northern red oak and the proliferation of short roots with fine tips in the organic matter layers of the forest floor.

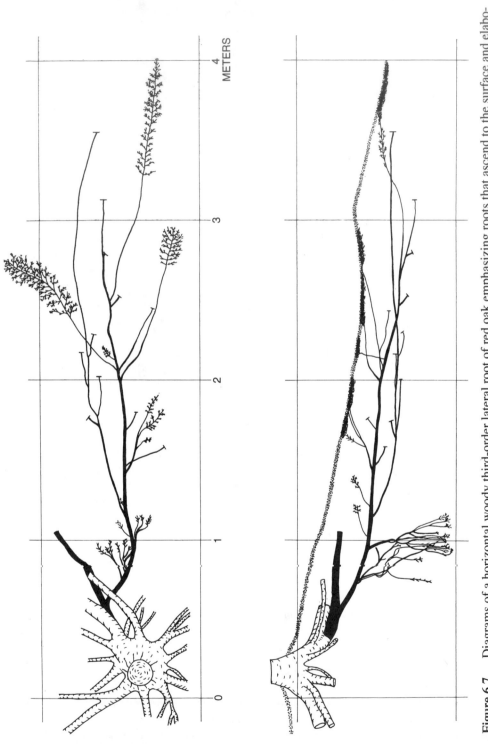

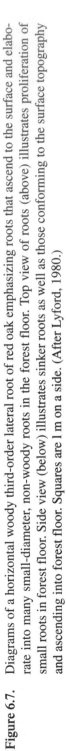

Figure 6.7. Diagrams of a horizontal woody third-order lateral root of red oak emphasizing roots that ascend to the surface and elaborate into many small-diameter, non-woody roots in the forest floor. Top view of roots (above) illustrates proliferation of small roots in forest floor. Side view (below) illustrates sinker roots as well as those conforming to the surface topography and ascending into forest floor. Squares are 1 m on a side. (After Lyford, 1980.)

The majority of roots, especially the small absorbing roots, are located in the upper soil horizons where favorable aeration, nutrients, and soil-water conditions occur. Throughout much of eastern North America and northern and central Europe, considerable precipitation falls during the growing season. Following a rain, the wetting front may saturate the upper 5 to 10 cm or more of soil. This layer is a major source of much of the water required by woody vegetation during the growing season. Decomposing organic matter of the surface layers holds this water effectively and in addition provides nutrients for fine roots. Numerous studies have reported a high concentration of fine roots in the surface layer (Kozlowski, 1971b). For example, in two pine and two oak forests of the Piedmont of North Carolina, from 94 to 97 percent of the roots less than 2.5 mm in diameter were located in the upper 13 cm of soil (Coile, 1937).

The number of fine roots per unit volume varies greatly during the growing season and by soil depth. For example, in a north German beech forest, peak concentration occurred in the spring and in organic matter layers where soil water, aeration, and mineralization of nutrients were most favorable. For Scots pine stands in sandy soils of Finland, Kalela (1957) reported a similar relationship of fine-root distribution. He reported a continual flux of death and replacement of fine roots. His conclusion was that the number of fine roots was closely related to the favorableness of local ecological conditions. Under unfavorable conditions (drought, cold, lack of photosynthate) fine roots died *en masse*, but they were rapidly formed again once conditions became favorable. In some ecosystems, mortality of fine roots may be high in winter due to low temperatures and lack of photosynthate. However, Hendrick and Pregitzer (1992) observed that fine root mortality in a northern hardwood forest was continuous throughout the growing season, with a peak in autumn. Root elongation also was continuous, and the highest rates occurred early in the growing season. The massive and rapid fine-root turnover described by Kalela has been confirmed in studies of carbon balance and nutrient cycling (Chapters 18 and 19).

Horizontal and Vertical Root Development

The horizontal development of perennial long roots is often considerable. In sandy sites, lateral roots of pines, birches, and black locust may extend 10, 20, or even 40 m as they continually occupy new volumes of soil (Lyr and Hoffmann, 1967). Lateral spread is related to the nature of the rooting medium and is more extreme in sandy soils than in clay. For example, in a sandy Wisconsin soil, roots of 20-year-old red pine trees extended about 7 times the average height of the trees, and those of eastern cottonwood for more than 60 m from the source (Kozlowski, 1971b).

Soil water may also be obtained by roots penetrating deeper soil layers; this function may be of survival value for many species in times of drought. For example, prairie-inhabiting species of *Quercus*, *Gleditsia*, *Juglans*, and *Maclura* are able to survive droughts by their inherent deep-rooting ability (Albertson and Weaver, 1945). Although rooting depths of two to three meters are not uncommon for many forest trees, deep-rooting species (rooting depths of three to six meters or more) include eastern redcedar, hackberry, black and honey locust, bur oak, and osage orange (Kozlowski, 1971b). Lyr and Hoffmann (1967) cited various studies reporting rooting up to 10 m for apple, 15 m for *Prosopis* (mesquite), 20 m for *Robinia,* and 30 m for *Tamarix* (tamarisk). In southern Arizona, roots were found 53 m below the surface and were thought to be those of the desert shrub mesquite (Phillips, 1963).

Root growth in depth, however, is curtailed in perennially-saturated soils due to lack of oxygen (anoxia). Nevertheless, roots of some species extend and live for months or years in anoxic conditions between periodic lowering of the water table. One such species is the slash pine, which occurs in shallow, freshwater swamps and depressions of pine lands of the Atlantic and Gulf coastal plains of the southeastern United States. Research by Fisher and Stone (1990a, 1991) indicates that oxygen is transported from the atmosphere through the root tissues to taproots and large sinker roots whose wood has unusually large air contents. Thus the submerged roots are internally aerated, which enables aerobic nutrient uptake (Fisher and Stone 1990b).

Periodicity of Primary Root Growth

Extension growth of roots, like that of shoots, is intermittent. For many trees of temperate latitudes, roots typically begin growth earlier in the spring and continue growth later in the fall than the shoots. However, species vary, and in larch species short-shoot needles are expanded before root growth begins (Lyr and Hoffmann, 1967). Root growth of many trees, particularly angiosperms, exhibits a peak period of activity in the spring and fall when soil water and temperature conditions are especially favorable. Roots typically show reduced activity during summer drought periods, and, with the onset of winter and low soil temperatures, root activity declines markedly. The course of root growth and the marked difference in root and shoot growth of eastern white pine are illustrated in Figure 6.8. Spring and fall peaks of activity are interrupted by a summer rest. Root growth continues long after shoot growth has ceased.

With the onset of winter and low soil temperatures, roots tend to cease growth. However, unlike shoots of temperate-latitude species that remain dormant over the winter period, roots of many species may grow during the winter provided temperature, soil water, and aeration are favorable (Lyford and Wilson, 1966; Lyr and Hoffmann, 1967). Winter root growth has been reported from regions with mild winter temperatures and frost-free

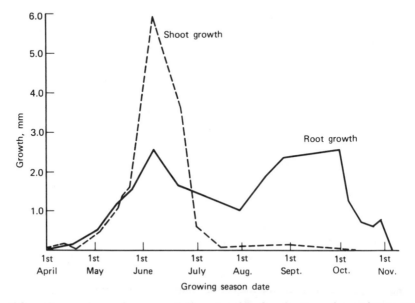

Figure 6.8.   Comparison of average daily extension for shoots and growing roots. Root and shoot growth follow distinct but different seasonal patterns. The growth rate of roots declines in mid-summer and ceases in winter. (Modified from Stevens, 1931.)

soils, such as in the southern United States, coastal British Columbia, the Crimea, and in parts of Europe. For example, Woods (1957) reported considerable growth of longleaf pine roots in the surface layer of sandy soils of the southeastern United States' coastal plain. The roots of scrub oaks and pine were concentrated in the upper 12 cm of soil. Warm temperatures of 16 to 22 °C in the upper 8 cm (6 °C warmer than soil at depths of 20 to 30 cm) apparently stimulated the root development.

Root Grafting

Natural root grafting among individuals of a species is relatively common in woody plants (Graham and Bormann, 1966). It is probably most frequent in pure stands of species (especially pines, spruces, and other conifers) that exhibit a wide-spreading system of roots concentrated in the surface soil layer. Living stumps and girdled trees in forest plantations provide striking evidence for intertree grafts. Grafting is accomplished when a union forms between the cambium layers of two closely associated roots. The phloem and xylem tissues connect so that food, water, growth substances, and pathogens may be transported from one tree to another.

Root interconnections are frequent and effective in red pine stands. Stone (1974) studied red pine plantations in which approximately 30 percent of the stems had been girdled in thinning operations. Many of the girdled trees were still alive up to 18 years after girdling, and completely girdled stems accounted for about one third of the live basal area of the stands 5 to 10 years after thinning. Water transport of the girdled trees still functioned and food was supplied to their roots through phloem connections with intact trees. The work of Stone and others with red pine indicates that up to 90 percent of stems in plantations may be grafted to at least one other individual. The physiological and ecological implications of root grafting are discussed by Graham and Bormann (1966) and Stone (1974).

Roots play an important but different role in the transmission of systemic wilt diseases such as Dutch elm disease and oak wilt (caused by *Ceratosystis ulmi* and *C. fagacearum* respectively) and in root rots caused by *Armillaria mellea*, *Heterobasidon annosum*, and *Phellinus weirii*. Research on root grafting in American elm has shown that nearly all trees within 2 m of one another and about 30 percent of those within 8 m are connected by root grafts (Verrall and Graham, 1935; Himelick and Neely, 1962). This interconnection permits rapid transmission of the fungus from infected to uninfected trees. Spores move passively through the vessel elements of the roots, then germinate and grow through impediments such as pits. In the root rots, however, this vascular transmission is not found. Instead, the fungus (1) grows and decays its way between adjacent trees via root grafts (as in *H. annosum*); (2) spreads over the outer surface of roots, occasionally penetrating the root and causing a new infection (as in *P. weirii*); (3) grows from plant residue, including old dead roots, to the roots of an actively growing plant that happens to come into contact with it (as in *A. mellea*, among others).

Specialized Roots and Buttresses

Site-related root projections are the knees or pneumatophores (roots that function as a respiratory organ) produced by various species that grow in periodically flooded river valleys or subtropical and tropical swamps. Kozlowski (1971b) lists 14 genera for which these structures have been reported. In baldcypress, knee-roots may be formed as a response to the air–water interface or injury. The upperside of the root, exposed to the air, exhibits greatly increased cambial activity compared to the submerged underportion of the root.

Knee-roots in baldcypress are largely site-specific; they are not formed when cypress trees are grown in upland, non-flooded sites unless they are injured. Knee-roots in baldcypress apparently are not important in root aeration but may act to strengthen the roots to give a firmer anchorage in a yielding soil medium (Mattoon, 1915; Kramer et al., 1952; Gill, 1970). However, pneumatophores in some species may contain lenticels and facilitate the exchange of gasses with the roots. Buttresses are also characteristic of species growing in swamps and poorly-drained sites.

Buttresses—flattened triangular plates that develop between the trunk and lateral roots—prop or guy tree crowns under site conditions unsuitable for extensive vertical root growth. They are typical of many tropical rain forest trees but also occur in temperate-latitude trees, especially American elm, and even the Lombardy poplar (Senn, 1923). In baldcypress and water tupelo trees of the southern United States, buttresses develop as a response to the level of flooding. The kind of buttress formed (shallow, cone, bottle, bell) corresponds to a different water-level regime (Kurz and Demaree, 1934). The trunk produces a marked increase in the number or size of individual cells (hypertrophy) at the air-water interface where the trunk is neither completely aerated or completed saturated. This growth increase is caused apparently by the inhibition of growth-regulator translocation at the floodline or by the buildup of ethylene. Thus the diameter at any height up the stem is directly proportional to the frequency of occurrence of that height of flooding.

Mechanical stresses may also affect buttress development. Based on studies of 191 trees of *Triplochiton scleroxylon* in Ghana, Johnson (1972) reported that buttressing was closely associated with site conditions. Small buttresses were found on well-drained upland soils, whereas large buttresses occurred on mid- or lower-slope sites of poorly drained soil. Apparently, buttresses develop in response to mechanical stresses occurring at the juncture of the lateral roots and the stem. Soil conditions that permit deep rooting allow a strong vertical root system to absorb stresses that would otherwise concentrate at the lateral root-stem juncture. If the soil is effectively shallow, due to impermeable layers or a high water table, vertical root development is impeded, and stresses concentrate at the stem base, thus inducing marked buttress growth.

A remarkable adaptation in the Amazonian rain forest are **apogeotropic** roots that grow out of the soil and up tree stems (Sanford, 1987). These roots, occurring in 12 species of 5 families, originate as fine roots in the mineral soil and grow upward on stems as high as 13 m (mean height 1.5 m). Apogeotropic root growth is apparently an adaptation to obtain nutrients from precipitation that flows down along the stem of trees in forests characterized by extremely low soil nutrient availability.

STEMS

The ability to form consecutive layers of structural tissues to the primary stem distinguishes woody species from all other plants. This secondary growth strengthens the stem and increases the transport of food and water between the shoots and the roots. The main tissues of a tree's main stem (bole or trunk) are illustrated in Figure 6.9. The meristematic sheath of cells that surrounds the stem, shoots, and roots, is the **cambium**. It originates the successive layers of secondary growth and is functionally a single cell layer, although a zone of actively dividing cells is typically observed. Evolution of the cambium was a significant event because it provided the structural modifications required to withstand the stresses of aboveground and belowground environments.

Toward the center of the stem, the cambium gives rise to the water- and nutrient-conducting cells, the **xylem**; toward the outside of the stem the cambium generates the photo-

Outer bark

Inner bark

Cambium

Sapwood

Heartwood

Figure 6.9.   Generalized structure of a tree trunk (i.e., a stem) showing the position of major tissues. Tissues include the outer bark (dead and compressed phloem cells), the inner bark (living phloem–food-conducting cells), cambium (sheath of living cells that give rise to phloem and xylem cells), sapwood (outer band of xylem, water-conducting tissue), and heartwood (inner core of xylem–water conducting cells). (Photo courtesy of Champion International Paper Company)

synthate-conducting inner bark or **phloem**. The xylem cells become lignified and form the dead, woody axis of the tree. It is remarkable that the indispensable function of conducting water rapidly over long distances necessarily requires dead cells (Zimmermann, 1971)! Observe in Figure 6.9 the two types of xylem—the central core, or **heartwood**, and the outer portion, or **sapwood**. Dead xylem tissue has important functions of transporting water and maintaining the strong support that gives forest its characteristic vertical stratification and distinguishes it from prairie and tundra. The heartwood is readily decayed in old trees, and, surprisingly, huge, completely hollow trees (for example, basswood, American beech, sycamore) live for decades supported by a thin shell of phloem, cambium, and sapwood.

The phloem cells of the inner bark in Figure 6.9 serve as transport channels for photosynthetic products, hormones, and many other substances within the crown, from the crown to other parts of the tree, and from the roots to the crown. Phloem cells are not lignified and must be regularly differentiated because they function for only a few years before collapsing and becoming part of the outer bark (Figure 6.9). The anatomy and physiology of the xylem, phloem, and bark have received considerable attention, and the reader is directed to Esau (1977) and Bell (1991) for anatomical relations and to Zimmermann and Brown (1971) and Kozlowski and Pallardy (1997a) for physiological considerations.

Main points concerning stem growth and environmental factors are discussed briefly in the next sections.

Xylem Cells and Growth Rings

Diameter growth of trees is primarily due to the periodic formation of layers of xylem cells by the cambium. In conifers, the primary water-conducting cells are the vertically oriented, overlapping **tracheids** that are seen in cross section in Figure 6.10. These thick-walled, tapering cells are 3 to 5 mm long and up to 100 times longer than wide (Kozlowski, 1971b). Conifer wood is relatively homogeneous and composed primarily of tracheids that are arranged in uniform radial rows. Other cells, parenchyma and ray tracheids, are oriented horizontally and function in the lateral transport of water, food, and other substances. Perforations in the walls of the tracheids provide the means for transfer of water and other substances to adjacent cells. Conifer growth rings are distinguished by differences in cell diameter and the cell-wall thickness of the tracheids produced during the early (**earlywood**) and late (**latewood**) parts of the growing season (Figure 6.10).

Xylem cells of angiosperms are more specialized, and their longitudinal components include vessels, tracheids, fibers, and parenchyma. **Vessels** are the primary water-conducting elements although the majority of the xylem cells are fibers. Vessels are composed of single cells whose end walls have disintegrated, thus creating tubes that may be up to several meters long. Therefore, they may rapidly transport water and nutrients to the shoot sys-

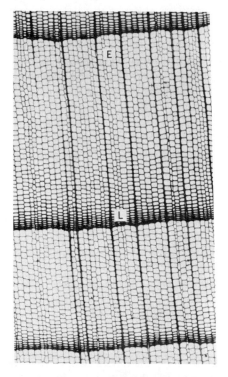

Figure 6.10. Transverse section of wood of ponderosa pine showing a transition in size and shape of earlywood (E) and latewood (L) in successive growth rings. (After Zimmermann and Brown, 1971; Photo by Claud L. Brown. Reprinted with the permission of Springer-Verlag.)

tem of the crown. Vessels vary greatly in size, and, in certain genera (*Fraxinus*, *Carya*, and *Quercus*), vessels that form early in the growing season may have diameters 100 times as great as those formed late in the growing season (Figure 6.11). Growth rings of such **ring-porous** species are easily distinguished because of the concentration of large vessels in the earlywood. Many other deciduous trees, **diffuse-porous** species, have smaller vessels that are more uniformly distributed throughout the growth ring (Figure 6.12). Growth rings of diffuse-porous trees are not as clearly distinguished as those of ring-porous trees, although in some species (beeches) the vessels become progressively smaller and the proportion of fibers and tracheids increase toward the end of the growing season (Figure 6.13).

Periodicity and Control of Secondary Growth

Diameter growth of North Temperate Zone trees is typically characterized by a single annual ring. In addition to variation caused by species, age, and soil conditions, the width of the growth ring depends largely on weather conditions of the current year, especially temperature and soil water. Temperature is more limiting than soil water in subarctic regions and high elevations where cambial growth may continue for only a few weeks (Kozlowski et al., 1991). In contrast, growth may be nearly continuous throughout the year in tropical and subtropical areas. The number and size of xylem cells of a given species is directly related to the size of the crown and the amount of hormones and photosynthate produced by the foliage. Although diameter growth generally continues longer into the growing season

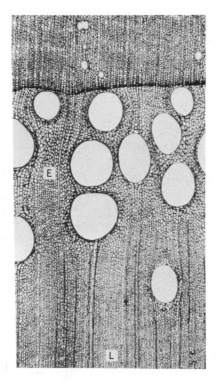

Figure 6.11. Transverse section of wood of a ring-porous black oak showing very large vessels in the first formed earlywood (E) and the preponderance of fibers in the latewood (L). (After Zimmermann and Brown, 1971; Photo by Claud L. Brown. Reprinted with the permission of Springer-Verlag.)

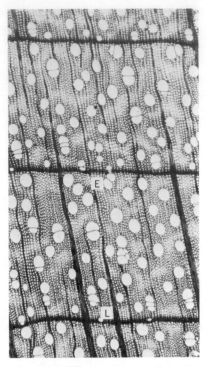

Figure 6.12. Transverse section of a diffuse-porous hardwood, tuliptree, showing fairly uniform distribution of vessels throughout the growth ring and radial flattening of the last formed latewood cells. (After Zimmermann and Brown, 1971; Photo by Claud L. Brown. Reprinted with the permission of Springer-Verlag.)

than shoot extension, most of it occurs in the spring and early summer when soil water conditions are favorable. However, species vary greatly in the duration of the growth period. For 21 tree species on one site in the Georgia Piedmont, the most rapid period of growth varied from 70 to 209 days (Jackson, 1952). In a study of European species, Norway spruce, larch, and ash formed the growth ring rapidly, with little wood laid down after July (Ladefoged, 1952). Species of birch, beech, alder, maple, and oak laid down wood from May until early September, with up to one-third of the ring being formed in the late summer. The width of the growth ring and the proportion of earlywood to latewood cells are closely related to the amount of hormones produced by the foliage and also to the availability of soil water.

The initiation of cambial growth in the spring is closely associated with renewed bud activity and leaf development. Growth-regulating hormones, auxin, and other substances (promoters and inhibitors) from the expanding buds and leaves trigger cambial activity. In conifers, cambial activity begins at the base of actively developing buds and leafy shoots (before shoot extension begins) in the crown and moves rapidly downward throughout the shoots, branches, and the main stem. In diffuse porous hardwoods, the process is similar except that the wave of activity proceeds rather slowly. In contrast, in ring-porous species, cambial activation is almost simultaneous throughout the entire stem. This adaptation is important because many of the large earlywood vessels of the previous year are nonfunctional, and the newly forming vessels are vital in supplying water to the foliage that will soon develop.

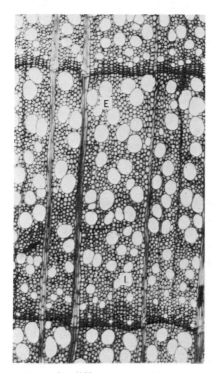

Figure 6.13. Transverse section of a diffuse-porous hardwood, American beech, showing a higher proportion of vessels in the earlywood (E) than in the latewood (L) and a corresponding increase in the proportion of thick-walled fibers in the latewood. (After Zimmermann and Brown, 1971; Photo by Claud L. Brown. Reprinted with the permission of Springer-Verlag.)

CONTROL OF EARLYWOOD AND LATEWOOD FORMATION

Differences in the properties and physiology of formation of earlywood and latewood have attracted interest for over a century. Latewood cells have a higher specific gravity than earlywood cells because of a greater proportion of cell wall substance per unit volume. In conifers the specific gravity of latewood is about two to three times that of earlywood (Kozlowski, 1971b). Because the amount of latewood in structural timber and paper pulp affects the properties of these materials, considerable research has been marshaled to determine the factors controlling early- and latewood formation with the aim of manipulating the amount of latewood cells. Research has demonstrated that the balance of growth hormones controls the kind of cells produced (earlywood or latewood) and that environmental factors, such as soil-water stress, act indirectly to influence earlywood or latewood production through their influence on the balance of hormones. The results of many investigations indicate that any condition that enhances leaf flushing, rapid shoot growth, and continued leaf development results in high levels of auxin production and large diameter cells of the earlywood type (Zimmermann and Brown (1971). Conversely, low temperature, drought, or short photoperiods, which adversely affect shoot extension and leaf development, lower the levels of auxin and bring about the formation of the small-diameter latewood cells.

The study of tree rings, **dendrochronology**, is eminently useful in assessing global climate change because the natural variation of the past climatic record is contained in tree rings. By analyzing both ring width and its maximum wood density, tree ring chronologies from northern Sweden, Finland, and Siberia are used to reconstruct variation in average summer temperatures over the last 6,000 years (Pearce, 1996). Both gradual and abrupt changes are recorded in tree rings of living trees and from buried Scots pine logs up to 6,000 years old. At these high latitudes inside the Arctic Circle near the northern limit of Scots pine, small changes in average summer temperature can greatly affect the number of days when temperatures rise above 5 °C, the threshold for radial growth. The result is dramatic differences in the thickness and density of tree rings. In particular, the thickness and density of latewood cells appear to depend directly on the average mean temperature. Thus an understanding of tree wood structure is at the forefront of distinguishing natural versus human-influenced global warming.

Winter Freezing and Water Transport

In winter, the freezing of water in xylem cells causes problems in water conduction for trees of northern latitudes. Dissolved gasses in the xylem water causes bubbles to form upon freezing; upon thawing, gaps develop in the water columns, thus breaking continuity and forming an air lock. The problem is potentially most serious for the ring-porous species that have very large earlywood vessels. Woody species cope with this situation in three ways (Zimmermann and Brown (1971):

1. **Small-diameter xylem cells.** In species with small conducting cells, the bubbles may be so small that they redissolve when the water thaws, and the vertical water columns rejoin before transpiration begins. Conifers and certain diffuse-porous trees with very small vessels (firs, spruces, alders, willows, trembling aspen, paper birch) can easily survive the extreme freezing and thawing regime of the boreal or alpine forest.

2. **Root pressure.** Enough pressure may be developed by roots of certain diffuse-porous species (birches) that the gaps in small vessels can be filled.

3. **Spring formation of large vessels.** Many ring-porous species with large earlywood vessels conduct water mostly in the last-formed (youngest) growth ring. Possessing very rapid cambial activation, they are able to produce some large earlywood vessels in the spring before the leaves flush out. Upon flushing, these large vessels quickly transport quantities of water as it is required by the developing and transpiring leaves.

The mechanism of ring-porous species is especially interesting because some species of this group are particularly late flushing, for instance, oaks, hickories, walnuts, and ashes. Flushing is delayed until the xylem system is ready to supply the water required by the foliage. Delayed flushing has the disadvantage that the leaves develop very rapidly in the warm temperatures of late spring when transpiration is high. However, the large vessels are highly efficient in rapidly supplying large amounts of water to the crown.

The system is advantageous in that these late-flushing species may have a greater probability of escaping late frost than earlier flushing species. However, by specializing in water conduction by the youngest ring, such species are susceptible to severe damage if significant injury befalls the newly formed rings. Two North American trees, the American chestnut and the American elm, have been dramatically affected by diseases that block their water vessels (see Chapter 16).

The worldwide distribution of tree species is at least partially related to the structure of xylem cells controlling water conduction. Boreal forest and alpine trees are adapted to maintain water conduction, in spite of freezing and thawing, by their small xylem cells. No ring-porous trees have a boreal distribution. Many oaks, hickories, walnuts, and other ring-porous trees are most abundant in the central or southern portions of North America. For example, approximately 80 percent of the world's oak species are found between 0 and 20° north latitude, and only 2 percent occur north of 40° (Axelrod, 1983).

WATER DEFICITS AND TREE GROWTH

Tree growth responds more to water stress than any other perennial factors of the forest site. Therefore, soil water is the key to forest site productivity for many species in many parts of the world. Apical, radial, and reproductive growth of trees, as well as seedling germination and establishment (Chapter 5), and virtually every aspect of plant morphology and physiology (see also Chapters 7, 11, 13, 18, and 19) are highly correlated with environmental soil-water stress. Reviews are available by Zahner (1968), Kramer and Boyer (1995), and Kozlowski and Pallardy (1997a, b).

In temperate climates, soil-water deficits during the middle of the growing season directly affect growth during both the current and succeeding growing seasons. Indirect evidence of the marked effect of soil-water stress on height growth is seen in the low heights that trees attain in dry climates or on dry sites, compared with trees growing on moist sites. Direct measurement of shoot growth of seedling and sapling trees under soil-water stress confirms that growth is closely correlated with water potential within the plant and to soil-water deficits (Chapter 11).

The effects of soil-water stress for a given species depend partly on the species' seasonal pattern of shoot flushing. Some species, such as birches, tuliptree, and loblolly pine are capable of maintaining shoot elongation during the complete growing season, and hence are affected by late season droughts. Many other species, however, including red pine, eastern and western white pine, white ash, sugar maple, American beech, and northern red and white oaks, complete height growth and set buds by midsummer. Although their current year's growth is unaffected by late-season water deficits, their current height growth is affected by late-summer droughts of the preceding year. For example, Zahner and Stage (1966) accounted for 72 percent of the variation in annual shoot growth in 5 stands of young red pine by water deficits of the previous and current growing seasons together. The water deficit of the previous summer (June 15 through October) accounted for as much reduction in annual height growth as the deficit for May 1 to July 15 of the current year. As expected, tree species that exhibit continued shoot flushing (tuliptree) or recurrent flushing (loblolly, pitch, and Monterey pines) show little correlation between total annual height growth and the previous year's rainfall (Zahner, 1968).

Soil-water stress plays an equally important role in the radial growth of trees. It affects the size of the annual ring, the proportion of earlywood and latewood, and various wood properties, particularly wood specific gravity. As we have seen, the transition from large, weak, thin-walled earlywood xylem cells to small, strong, thick-walled latewood cells in conifers may be directly affected by water deficits in the cambium, as well as markedly affected through reduction of auxin levels in the crown (Larson, 1963a, 1964; Zimmermann and Brown, 1971).

Periodic changes (daily and annual) in the radial growth of conifers and angiosperms repeatedly have been shown to be directly correlated with soil-water availability (Figure 6.14). For several deciduous species over a five-year period in Ohio, high air temperatures during periods of midsummer soil-water stress were consistently associated with temporary and sometimes permanent cessation of radial growth in all species (Phipps, 1961). Typically, water stresses are not serious prior to midsummer for trees growing in fully

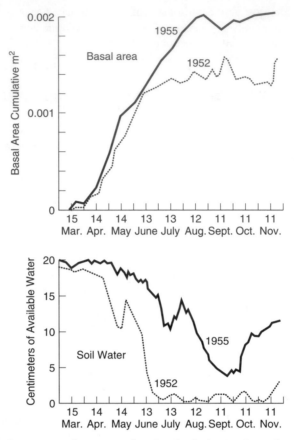

Figure 6.14. Basal area growth per tree for shortleaf pine and trends of available moisture for relatively wet (1955) and dry (1952) growing seasons. Note that the
growth rate slowed in mid-June of 1952 but did not slow until mid-August
of 1955, at the time in each year when available soil water had been depleted to about 5 cm or below. (Redrawn from Boggess, 1956; after Zahner,
1968.)

stocked stands on upland sites in temperate climates. Thereafter, however, it is normal for
water absorption by roots to lag far behind transpiration; the resulting dehydration of tissues in the crowns and stems causes important limitations of growth below the potential
for that time of year (Zahner, 1968). If the water stress is alleviated by late-season rains,
radial growth usually resumes if the mid-season water deficit has not been severe.

Many studies demonstrate that water stresses normally occurring in densely-stocked
stands may be alleviated by silvicultural practices such as thinning or wide spacing of
trees. Radial growth of residual trees is faster and more prolonged in thinned than in unthinned stands (Figure 6.15). Furthermore, heavy thinning of loblolly pine stand may alle-
viate summer soil-water stresses such that residual trees may continue to grow longer
throughout the season (Bassett, 1964).

Drought years leave their record in the growth rings of trees, and the high correlation
of ring width with summer water deficit is widely documented. Up to 90 percent of the
variation in width of annual rings of conifers has been attributed to water stress in semi-

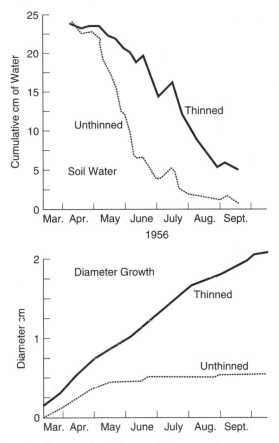

Figure 6.15. Trends of soil water depletion and diameter increment per tree for average dominant loblolly pine during one growing season, thinned plots and unthinned plots. (Redrawn from Zahner and Whitmore, 1960; After Zahner, 1968.)

arid climates (Douglass, 1919; Fritts et al., 1965) and up to 80 percent in humid temperate climates (Zahner and Donnelly, 1967).

Plants resist drought by avoiding and tolerating drought stress (Levitt, 1980b). Primarily, woody plants *avoid* drought stress; they do so by morphological and anatomical adaptations that maintain favorable internal water potentials despite the low potential in the environment to which they are exposed. The major avoidance mechanisms are: (1) the ability of roots to extract large amounts of water from the soil, (2) high root-to-shoot ratio, (3) reduced leaf surface area (including rolling, folding, shedding of leaves, and maintaining a dense pyramidal crown), (4) stomatal control to reduce transpiration—stomata closed most of the day and the ability to close stomata very rapidly in response to stress, (5) thick cuticle to reduce cuticular transpiration, and (6) a high proportion of water-conducting tissue to nonconducting tissue, i.e., veins close together so more water can be delivered per unit area. Reduced transpiration may be of less importance in resisting drought stress than the ability of trees to absorb large quantities of water. The efficient water-absorbing species can keep stomata open for photosynthesis and can transpire and still maintain a favorable internal water potential. Avoidance by stomatal closure alone has the disadvantage that it leads to low photosynthesis and starvation. Drought tolerance is less common than drought avoidance in woody plants. Nevertheless, many species can withstand considerable dehydration before closing their stomata. Besides drought, woody plants are markedly influenced by many other site factors which are considered in Part 3.

Suggested Readings

Barthelemy, D., C. Edelin, and F. Hallé. 1991. Canopy architecture. In A. S. Raghavendra (ed.), *Physiology of Trees*, pp. 1–20. John Wiley, New York.

Borchert, R. 1991. Growth periodicity and dormancy. In *Physiology of Trees,* A. S. Raghavendra (ed.), pp. 221–245. John Wiley, New York.

Hallé, F., R. A. A. Oldeman, and P. B. Tomlinson. 1978. *Tropical Trees and Forests.* Springer-Verlag, Berlin. 441 pp.

Kozlowski, T. T. 1982. Water supply and tree growth. Part I. Water deficits. *For. Abstr.* 43:57–95.

Kozlowski, T. T., S. G. Pallardy, 1997. *Physiology of Woody Plants,* 2nd ed. Academic Press, San Diego, CA, 411 pp. Chapters 2, 3, and 12.

Kramer, P. J., and J. S. Boyer, 1995. Water Relations of Plants and Soils. Academic Press, New York, 495 pp.

Lyford, W. H. 1980. Development of the root system of northern red oak (*Quercus rubra* L.) *Harvard Forest Paper* No. 21:1–30.

Oldeman, R. A. A. 1990. *Forests: Elements of Silvology.* Springer-Verlag, New York. 624 pp. Second Part: Forest components, pp. 27–108.

Tomlinson, P. B. 1983. Tree architecture. *Am. Scientist* 71:141–149.

Wilson, B. F. 1984. *The Growing Tree.* Univ. Mass. Press, Amherst. 138 pp.

Zimmermann, M. H., and C. L. Brown. 1971. *Trees, Structure and Function.* Springer-Verlag, New York. 336 pp.

PART 3

The Physical Environment

FOREST ENVIRONMENT

The volumetric, landscape ecosystems that comprise nested segments of the Earth consist of an atmospheric layer overlying a land–water layer, with the biological component of plants and animals located at the energized surface. The forest environment divides naturally into the physical environment surrounding the aboveground and belowground portions of the trees and associated biota. Although we stress the integrated nature of ecosystem components, it is useful to examine their physical factors as they affect plant distribution and processes. The sum total of these physical factors determines the **forest site**. In addition, the effects of the plants and animals occupying the site locally modify the physical environment. For example, as organic matter decomposes it interacts with mineral soil particles and water to create different soil horizons. These interactions of abiotic and biotic factors emphasize the integrated wholeness of forest ecosystems and the difficulty of treating factors separately, despite the usefulness of such an approach. The chapters in this section fall within the subdivision of ecology known as **autecology,** the study of the organism—in this case the forest tree or other organisms—in relation to environment.

In this section we examine the physical factors that influence the plant where it lives. Thus the forest site is the place where plants live—defined by specific geographic location and by the suite of physiographic, edaphic, hydrologic, and climatic factors that are predominant in sustaining the biota. To a greater or lesser extent these abiotic factors are modified by the plants and animals of the locality. A given physical site factor may directly affect plants and other forest organisms, but more often its action is affected, intensified, or diminished by interactions with biotic factors as well as other physical factors.

149

The term *habitat* is widely used in the literature. It is conventionally used when the major focus is on organisms. For example, "the white-tailed deer's *habitat* is the northern white-cedar swamp" (as if the swamp belonged to the deer) where the vegetation that provides shelter and food seems to define the habitat. However, the cedar's *habitat* is a complex of particular physical factors of air, soil, and drainage. Here, the physical site factors define habitat. Thus, like many others, habitat is a fuzzy term that may take on several meanings. Because ecosystems, not organisms, are our primary focus we use the term, site (or habitat supporting plants), to refer to the physical environment (as modified by biotic effects) of landscape ecosystems.

SITE FACTORS

It seems fairly easy to enumerate the factors making up the environment that are indispensable to the existence and growth of forest organisms. It is exceedingly difficult, however, to understand and evaluate the sum total of the interactions among the environmental factors, and their interactions with biotic factors, that make up the complex we term *site*.

The tree grows with its crown in the atmosphere and its roots in the soil. To the crown come the light, warmth, carbon dioxide, and oxygen; and to the roots come the mineral nutrients and water necessary for photosynthesis and other processes. These are the basic factors that the site must supply. Their availability to the tree, however, depends upon an endless system of changing climate, day length, and soil development—changes that are in part related to the developing vegetation and associated animals.

Organization of Site Factors

Physical and biotic site factors, therefore, may be organized into broad groups, which may be considered separately. In the following diagram we see that the site factors interact to yield the light, heat, water, etc., that are directly available and used by the plant. In this diagram, observe the multiple effects of physical and biotic factors on the plant.

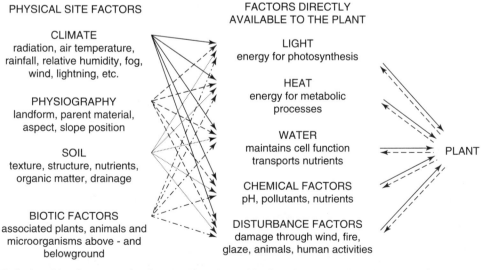

Relationships between site factors directly and indirectly responsible for plant distribution and growth. (Modified from Ellenberg, 1968. Reprinted from *Wege der Geobotanik zum Verständnis der Pflanzendecke* © by Springer-Verlag Berlin, Heidelberg 1968. Reprinted by permission of Springer-Verlag New York, Inc.)

Climatic factors (Chapter 7) are those relating to the atmosphere in which the above- and belowground portions of plants grow. These include solar radiation, air temperature, air humidity, wind, lightning, and the CO_2 content of the air. Climate also determines belowground temperature, moisture, CO_2, and the weathering of nutrients from rock substrate. Precipitation, whether rain or snow, affects plant growth primarily through its effect on soil water. The reception of light and the effects of light on plant growth are considered in Chapter 8, and the influence of temperature is examined in Chapter 9.

Physiographic and soil factors (Chapters 10 and 11) include the topographic and structural features of landscapes plus all the physical, chemical, and biological properties of the soil. The relief and form of the land, the nature of the parent material, the physical and chemical soil properties, and soil microorganisms are particularly important in plant growth and in determining site quality.

Activities of plants and animals, both visible and microscopic, change climate, physiography, and soil—and therefore affect site quality. Large organisms (such as trees, grazing animals, and humans) create the most obvious changes in the microclimate and the soil. Humans also have the potential to alter broad-scale climatic factors by the burning of fossil fuel and production of "greenhouse" gases. Small animals and plants occurring in great numbers (including fungi, bacteria, earthworms, rodents, and many others) can also bring about substantial changes in the site. Because the influence of plants and animals on site is exerted through changes in the climate and soils, a discussion of various biotic factors is integrated into the chapters on climatic and soil factors. In addition, the roles of animals, including humans, on site quality are discussed in Chapter 14.

Fire is also a very important factor affecting forest species and site quality, and fire as a physical force is considered in Chapter 12. The burning of soil organic matter and the heating of the surface horizons of mineral soil result in changes in the physical and chemical properties of the soil and its soil biota.

Once the individual factors affecting forest site have been considered, it becomes necessary to integrate them into a whole if the forest site is to be characterized simply and accurately. Approaches to the determination of forest site quality and the evaluation of ecosystems are considered in Chapter 13.

Interrelationships Among Site Factors

Whereas it is both conventional and convenient to discuss the various factors of the forest site one at a time, a word of caution must preface such a discussion. The presentation of the principal factors one by one may create a dangerous tendency to think of each factor as an independent force affecting the tree or any forest organism. We tend to think and talk in terms of simple direct relationships, and to make such statements as "at a temperature of 55 °C, the plant suffers direct heat injury" or "precipitation is the limiting factor in determining the distribution of this species." Such statements have their value and even truth, but they do tacitly ignore the fact that the plant lives in the total complex of the environment; a change in any one factor of this complex may well bring about a changed requirement of tolerance of the plant for other factors. In other words, these individual environmental factors are not isolated, independent forces operating on the plant, but rather interdependent and interrelated influences that must, in the last analysis, be considered together.

Much of the previous discussion about interrelated factors is embodied in the changing concept of the law of the minimum. As originally enunciated by Liebig in 1855, this law holds that the rate of growth of an organism is controlled by that factor available in the smallest amount. It has been demonstrated repeatedly, however, that such a law does not hold true—but that the rate of growth of a plant can be increased by changes in the supply of more abundant factors that can compensate for the so-called "limiting factor." The law still retains much value and is part and parcel of the thinking of the forest ecologist. In its modern form, though, it is restated to take into account the interactions existing between the various single factors that operate at different scales. We may thus say with authority that "whenever a factor approaches a minimum, its relative effect becomes very great." At the broad, regional scale, low precipitation or a very-cold climate have an overriding affect on the nature of regional vegetation and the range of individual species. However, at the local scale of a forest swamp, severe lack of oxygen in soil may strongly determine the composition, structure, and function of such an ecosystem.

The changing concept of the law of the minimum has arisen from the natural development of ecological research methods. In early years, most ecological theories were based upon observations of trees and other plants under natural growing conditions. For instance, observations that seedlings die under dense forests may lead to conclusions that low light intensity is responsible for the mortality. On reflection, however, the ecologist will realize that many environmental factors are affected by a dense forest cover. Light intensity, of course, is low, but so also is the supply of water in a soil permeated by the roots of the many trees present. Under a dense forest, the temperature regime and wind speed are greatly changed from that in the open or even that under an open forest. The organic matter and the soil microorganisms, too, will be greatly affected. In short, the whole environmental complex is related to forest cover density, and it is misleading, if not downright incorrect, to attribute changes in plant response to any single factor (Chapin et al., 1987).

IMPORTANCE OF SITE IN FOREST ECOLOGY

There is good reason for considering the site before one considers the biota. In a forest ecosystem, the site is more concrete, more stable, and more easily defined than the animals and plants that occupy it. Therefore, the site constitutes a better basis for the description of the forest ecosystem than do the trees, other plants, and animals. The emerging emphasis on earth science as the basis for ecology is seen in the books on **geoecology** by Matthews (1992) and Huggett (1995).

The forest site occupies a given geographical area and is capable of fairly precise definition, especially by physiographic features or landforms. Depending upon chance, upon past history, and upon changing environment conditions within a site type, various types of forest communities may develop on that particular site type (Chapter 1). The superiority of site as a basic unit of ecosystem classification and mapping becomes apparent when consideration is given to the number of quite different communities that may be characterized by a given species or even a given group of species. Ideally, it is the volumetric combination of site and the biota, the landscape ecosystem itself, that is necessary in describing forest landscapes, and more significantly in understanding how forest ecosystems function.

CLIMATE

Climate, defined as the characteristic patterns, means, and extremes of **weather** (local, short-term atmospheric conditions), is extremely important to an understanding of forest ecology. Acting on specific physiological mechanisms of organisms, climate is a major factor determining genetic differentiation and speciation (Chapter 4), species distributions (Chapters 3 and 9), competition (Chapter 15), disturbance regimes (Chapter 16), and growth rates and carbon balance (Chapter 18). Variation in climate with latitude, elevation and proximity to large water bodies and mountain ranges is closely tied to the distribution of vegetation on global, regional, and local scales (Chapter 2). Characteristic temperature and precipitation patterns, as they interact with vegetation, parent materials, and physiographic position, are important in determining soil processes and soil development (Chapter 11). **Microclimate**, or local variation in climate, influences the spatial patterning of local ecosystems and their species composition. Also, it is especially influential in maintaining some species near the limits of their ranges. Long-term changes in climate have continuously altered the spatial distribution and species composition of forested ecosystems globally (Chapter 3). Both natural and anthropogenic changes in climate could potentially alter the distribution of forests and their productivity in the future.

In addressing the effects of climate on terrestrial ecosystems, we first consider climatic controls at the broad scale of incoming radiation, air masses, and atmospheric movements. Next, we examine specific climatic features such as temperature, precipitation, fog, and snow, as well as extreme events of storms, drought, and glaze that markedly affect local plant distribution and growth. Finally, we consider some of the ways that climate has been classified and climatic change is exacerbated by human actions.

Climate is a broad topic, the subject of an entire and constantly evolving field of study. Only highlights pertinent to forested ecosystems are discussed in this chapter. References such as Budyko (1986), Sellers (1965), and Woodward (1987) provide a broader

153

background on climate and the global and local controls that affect weather and plant growth.

CLIMATIC CONTROL OF VEGETATION DISTRIBUTION

Radiation

The Earth's climate arises from the interaction of solar radiation and the atmosphere that surrounds the Earth. From the sun comes, directly or indirectly, the light that makes possible photosynthesis and the heat that warms the air and soil to the point that the life processes of plants and animals occur. Also from the sun comes the energy that drives horizontal and vertical atmospheric movements, evaporation, and precipitation.

The sun is an enormous atomic furnace that constantly emits intense radiation. A very minute fraction of the sun's radiation reaches the Earth. All lighted portions of the Earth receive radiation, but on an annual basis, the equatorial regions receive the most incoming solar energy, whereas the poles receive the least. This discrepancy, combined with a tendency for hot air to rise and cold air to sink, drives most of the air movement on Earth.

Solar radiation (total energy) reaches the outside of the Earth's atmosphere at an average rate of about 1,396 Watts (W) m^{-2}. This value, the so-called solar constant, varies about 1.5 percent with variation in the activity of the sun and an additional 3.5 percent depending on the distance from the sun to the Earth (List, 1958).

The spectrum of solar radiation is divided into three regions: the ultraviolet, the visible, and the infrared (Figure 7.1). Wavelengths are measured in micrometers (μm, 10^{-6} m, one ten-thousandth of a centimeter). The shorter ultraviolet rays, at the left in Figure 7.1, are almost completely absorbed by the atmosphere. Radiation with wavelengths ranging from approximately 0.4 to 0.7 μm is visible to the human eye and is termed **light**. The infrared region, characterized by long waves ranging upward from 0.7 μm, is shown at the left in Figure 7.1. This distribution of energy received from the sun is shown on the curve in Figure 7.1. The intensity of energy rises rapidly through the ultraviolet range and peaks in the visible portion of the spectrum. Fifty percent of total radiation is received in the infrared range. Most of the incoming radiation is shortwave radiation, or radiation having wavelengths shorter than 2 μm. Radiation from the Earth into space is mostly longwave radiation, or radiation in the infrared region.

The division of the light portion of the spectrum into colors is also shown in Figure 7.1. Radiation is absorbed selectively by vegetation, and this energy is used to drive photosynthesis. Most leaves have high absorptivity in the ultraviolet and visible range, low absorptivity in the high energy infrared, and high absorptivity of far-infrared where solar irradiance is very low (Gates, 1968). Plants reflect most green light, thus the characteristic color of green leaves. Some alpine plants have adaptations (such as white hairs) which may serve to protect the plant against excess shortwave radiation while minimizing losses of infrared radiation.

As solar radiation reaches the atmosphere, a portion of it is reflected from clouds and dust particles in the atmosphere (Figure 7.2). Additional radiation is reflected from the Earth's surface. The **albedo**, consisting of approximately 30 percent of the incoming radiation on average, is that proportion of the incoming radiation that is reflected back into space. It is lost from the Earth-atmosphere system.

The remaining 70 percent of solar radiation is absorbed by the atmosphere, Earth's surface, and vegetation. Some of the incoming radiation is absorbed by the atmosphere and some by the Earth's surface, including water and vegetation. Ultimately, for the Earth to maintain a constant average temperature, all of this must radiate back out into space.

Radiation emitted by the Earth is in the infrared portion of the spectrum. Some of the

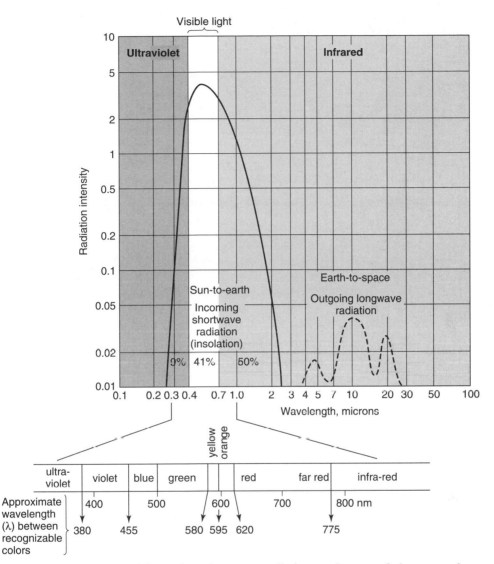

Figure 7.1. Intensity of incoming shortwave radiation at the top of the atmosphere (left); longwave radiation from the Earth to outer space (right) (After Strahler and Strahler, 1989. Reproduced from *Elements of Physical Geography,* 4th ed. by A. N. and A. H. Strahler, © John Wiley & Sons, Inc. Reprinted with the permission of John Wiley & Sons, Inc.); divisions of the visible light region of the spectrum (below). (Hart, 1988. Reprinted from *Light and Plant Growth* with the permission of Chapman and Hall.)

atmospheric gases, especially water vapor, carbon dioxide and various "greenhouse" gases have the property that they absorb relatively little shortwave radiation but absorb a large amount of infrared radiation. As a result, the atmosphere functions much like a greenhouse that allows visible light to enter and that retains infrared radiation or heat. Much of this retained energy is radiated in a downward direction (counterradiation), thus being retained temporarily in the Earth's atmosphere. This counterradiation is important in reducing temperature contrasts throughout the world. The effectiveness of water vapor

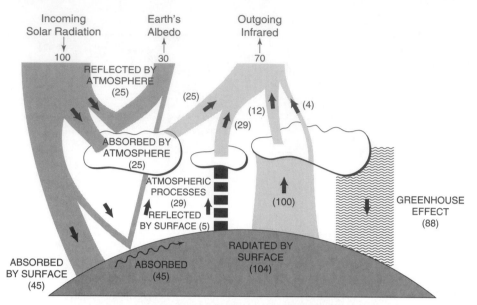

Figure 7.2.   The Earth's energy balance. (From Lilley and Webb, 1990 as modified from Schneider, S. H. 1989. The greenhouse effect: science and policy. Reprinted from *Science* 243:771–781 with permission of the American Association for the Advancement of Science.)

in retaining heat can be seen clearly on cloudy nights, in which a high portion of the surface radiation is captured and radiated downward rather than being lost into space.

The total amount of solar radiation, or **insolation**, reaching an exposed surface depends on latitude, time of day, atmospheric clarity, and altitude. An area of surface at a high latitude receives, over the course of a year, less radiation than is received by a similar area at the equator since the incoming radiation hits the surface at an angle and must pass through a greater amount of atmosphere before reaching that surface. Similarly, more radiation is received at noon, and in the summer, than early or late in the day or during the winter. Plants are highly sensitive to these differences and many use changes in the amount of radiation received, or the changing temporal patterns in its receipt, to control leaf flushing, entrance into dormancy, and other phenological events.

Insolation is less in cloudy areas. Low altitudes receive less radiation at the surface than high altitudes because a greater portion of the shortwave radiation is absorbed or reflected in the atmosphere above. Representative values reported by Abbot (1929) for clear sky conditions with the sun directly overhead give the total solar energy received on Mt. Whitney in California (4,418 m) as 1,222 W m^{-2} as compared to 1,012 W m^{-2} on Mt. Wilson (1,737 m), also in California, and 803 W m^{-2} for Washington, D.C., near sea level.

Air Circulation

The movements of the atmosphere result largely from the imbalance between the amount of radiation received near the equator and the lesser amounts received near the poles. The imbalance sets up vertical and horizontal currents of air that result in warm air flowing toward the poles and currents of cool air flowing toward the equator. The Earth's axis is also tilted relative to its orbit around the sun. This tilt results in the Northern Hemisphere being hit more directly by the sun's radiation during the summer and less so during the winter.

The seasons are, of course, reversed in the Southern Hemisphere. As a result, the various air mass features, discussed next, shift poleward during the summer and equatorward during the winter.

Hot air rises, and cold air sinks. Differentially-intense heating of air near the equator causes it to rise. In doing so, it leaves a relative void at the surface, and cooler air from slightly higher latitudes moves toward the equator to replace it. These areas of rising air are less dense than air in areas where it is descending. We call these areas of rising air **low pressure** systems. The air eventually stops rising and moves laterally toward the poles. As it rises and moves poleward, it cools and begins to sink. The areas of sinking air are **high pressure** areas. Bands of rising and sinking air occur at specific latitudes and result from currents of air that carry warm air poleward and cool air toward the equator. Overall, the Earth's atmosphere has a three-dimensional circulation pattern (Figure 7.3). This general

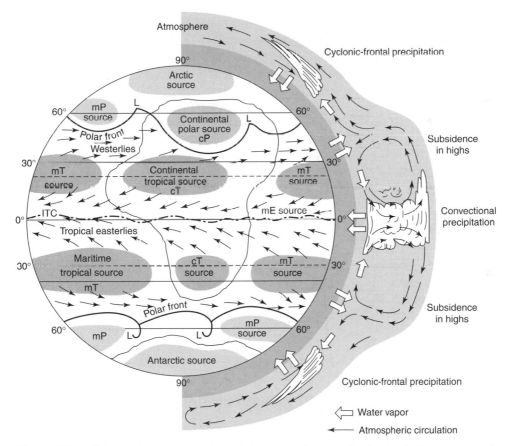

Figure 7.3. Schematic representation of the general atmospheric circulation and zonal wind patterns of the Earth. Arrows show the direction of prevailing surface-level winds. Shaded areas represent high pressure regions where major air masses originate. Arrows on the edge of the globe depict the major vertical circulation patterns in the atmosphere. Average location of the polar fronts and the intertropical convergence zone (ITC) are indicated. (Modified from Strahler and Strahler, 1989. Reproduced from *Elements of Physical Geography,* 4th ed. by A. N., and A. H. Strahler, © 1989 by John Wiley & Sons, Inc. Reprinted with the permission of John Wiley & Sons, Inc.)

pattern of air movement is largely responsible for transferring warm air poleward and cool air toward the equator. This pattern is also closely related to the distribution of major plant formations of the earth.

Air Masses

Air masses are huge bodies of air (subcontinental scale) within which temperature and humidity changes are very gradual on a horizontal plane. Air masses are created principally in the subtropical and polar high-pressure areas where the air is sinking slowly as depicted in Figure 7.3. These air masses tend to pick up temperature and humidity based on the character of the surface over which they form. Air masses that form over land tend to be dry. Those that form over oceans pick up large quantities of moisture. The boundaries between air masses are regions of rapid change in the character of the air known as **fronts**. It is the character of air masses, their movements, and the dynamics of the boundaries between them that result in our weather.

The equatorial area of rising air is characterized by numerous thundershowers and is known as the **intertropical convergence zone**. It can be seen on satellite photographs as a band of clouds in the equatorial region. This equatorial zone is associated with tropical rain forests. In contrast, a zone of sinking air occurs around 30° north and south latitude. The air masses that form over continents in this zone tend to be hot and dry, and these areas are characterized by a lack of clouds and lack of rain. The Sahara Desert is in this zone as is the Sonoran Desert of the North American southwest. The polar regions are also areas of sinking air and high pressure. The air that sinks near the poles is cold and dense. At the ground surface it moves gradually toward the equator. These air masses can be moist or dry depending on the surfaces over which they form.

A zone of rising air, much less well defined than the intertropical convergence zone occurs around 55 to 60° north and south latitude where air moving northward from the subtropics meets air moving southward from the poles. It is also an area of generally rising air and low pressure. The result, is great variability in weather both spatially and temporally. A wide variety of vegetation types occur in this region. All are temperate, but they vary greatly in their moisture characteristics. These include broad-leaved deciduous forests, coniferous forests, prairie, and desert. Much of North America, Europe, and Asia are in this zone. Relatively smaller temperate areas occur in the southern hemisphere because there is less land in the temperate latitudes.

The Earth's daily rotation is also important to the movements of air in the atmosphere. As a result, in the northern hemisphere, any moving air is deflected to the right. South of the equator, it is deflected to the left. Air moving southward between 0 and 30° north latitude is deflected westward forming the northeast trade winds.

The Earth's rotation also affects the movements of air masses and the air within those air masses. As air moves outward from the centers of high pressure areas, it is deflected to the right causing a clockwise outward spiral of air movement to occur near the surface. These high pressure systems are termed **anticyclones**. They are usually associated with clear weather. Air flowing toward low pressure areas is deflected to the right forming a counter-clockwise inward spiral. This airflow pattern is seen in the big **cyclonic storms** of the temperate latitude and in the **hurricanes** of the subtropics.

The temperate zones lie between 30 and 60° north and south latitude. In this area, surface air moves poleward but is deflected to the east. Entire air masses and the fronts between them move from west to east. This eastward movement is characteristic of much of North America and Europe. It results in prevailing wind directions from the west. At high altitudes where there is less surface friction than at the Earth's surface, the winds are deflected almost perpendicular to the original direction of flow forming very large, extremely fast moving currents of air or **jet streams.** These jet streams, characterized by large waves, lie between the large anticyclonic air masses. The mid-latitude cyclonic lows develop and track along them. The jets are typically stronger and lie further toward the equator in winter. They weaken and lie further poleward during the summer. Average summer and winter jet stream trajectories have been shown to correspond closely with the northern and southern limits of boreal forest.

Water

The properties of water and its distribution across the surface of the Earth and in the atmosphere have an enormous influence both on climate and on the biota. Water has a high heat capacity, making it slow to cool and slow to warm. Warm water is an important storage for energy in the environment. By contrast, land and air have much lower heat capacities. Land and air warm and cool extremely rapidly. Without the presence of liquid water, on and near the ground surface and in plants, conditions for life would not exist.

The major storages of water on the Earth are the oceans and the atmosphere. Others include ground water, streams, lakes, and plants. The processes by which water is transferred between these include precipitation, evaporation, transpiration, and sublimation. Forest ecosystems participate directly in these processes, both depending on them and influencing them as illustrated schematically in the hydrologic cycle (Figure 7.4).

Large amounts of water are stored in the atmosphere. Most atmospheric water is in the form of vapor, and is referred to as **humidity**. The amount of water that can be held by the air, or **water-holding capacity,** varies with temperature, with warm air being capable of holding more water than cold air. At 27 °C, air can hold twice as much water as it can at 16 °C. **Relative humidity** is the percentage of water vapor actually present in the air relative to the total water holding capacity. Atmospheric humidity is closely tied to transpiration rates of plants, and spatial humidity patterns may play a role in determining regional vegetation patterns. Unless stomates close, transpiration is higher when the humidity is low. Simultaneously, evaporating water from transpiration increases the relative humidity. Trees tend to have leaf shapes, stomatal distributions, and stomatal characteristics that fit the environments in which they occur. For example, the water conserving needles of conifers may provide an advantage in arid areas, whereas broad-leaved forests are typically associated with humid regions.

Water enters terrestrial ecosystems from the atmosphere as precipitation, which is depicted in the downward "input" arrow in Figure 7.4. If a humid air mass cools too much, it can no longer hold water. Drops sufficiently small to remain suspended in the air form **clouds**. If the cloud is at or near the ground, it is termed **fog**. If the particles grow sufficiently large, they are too heavy to remain in the air, and **precipitation** results. If the particles are frozen, **snow** is formed. If the temperature of the particles is above freezing, **rain** ensues. Rain that freezes as it falls through subfreezing layers becomes **sleet** or **hail**. If liquid precipitation freezes when it hits a surface, it forms **ice** or **glaze**. Much rain received at the surface forms in subfreezing conditions aloft but melts as it falls into warmer air below. Water may also condense directly onto a cool surface near the ground. This water is known as **dew**, if the temperature is above freezing, and **frost, hoar-frost,** or **rime** if the temperature is below freezing.

The cooling that causes precipitation results from air being lifted to higher altitudes. Lifting may result from surface heating of moist air within an air mass, causing the surface air to become buoyant and rise (Figure 7.5a); one air mass moving faster than another and tending to override or push under the air mass in front of it (Figure 7.5b); or air being forced up over an obstacle, such as a mountain range (**orographic lifting**) that is in

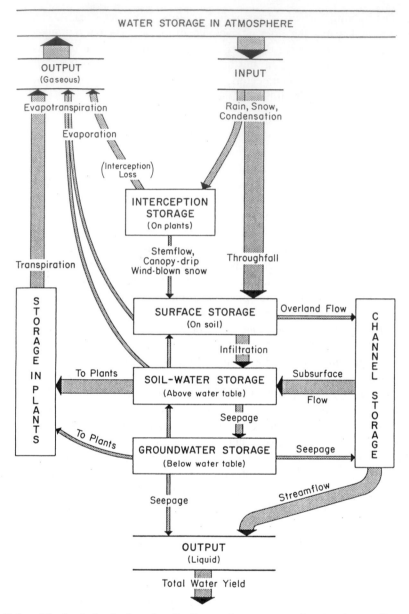

Figure 7.4.   The hydrological cycle of a forested ecosystem showing inputs from the atmosphere and outputs to groundwater, streams, and the air. (After Anderson et al., 1976.)

its path (Figure 7.5c). When surface heating causes moist air to rise, it cools yet remains warmer than the air around it and so continues to rise. When the air has cooled to the extent that it can no longer hold the moisture that it carries, clouds form. With continued rising, rain and sometimes thunderstorms occur. These are frequent throughout the eastern United States during summer periods when the area east of the Rocky Mountains is dominated by air masses that have moved north from the Gulf of Mexico. Fronts form along the edges of air masses. A **cold front** occurs when a cold air mass overtakes a warm air

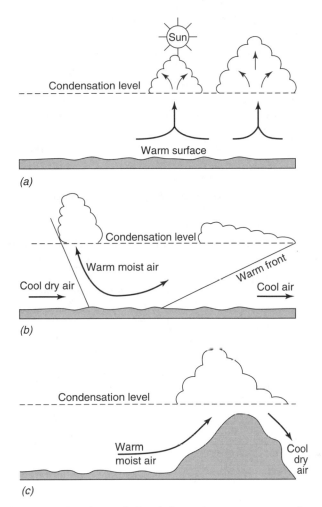

Figure 7.5. Diagrammatic view of cloud formation processes and precipitation. The first panel (*a*) depicts the type of lifting that occurs when moist, unstable air passes over or sits above a warm surface. The surface air is heated, expands, becomes lighter than adjacent air, and rises. As it rises, it cools and eventually reaches the temperature where it is too cold to hold water in vapor form. Condensation, cloud formation, and ultimately precipitation occur. The second panel (*b*) depicts warm and cold fronts that occur where air masses differing in moisture content and temperature meet. The warmer of the two is forced upward. Cooling with increasing altitude causes formation of clouds and precipitation. The third panel (*c*) depicts orographic lifting where warm, moist air is forced upward over a mountain range.

mass and pushes under the trailing edge of the warm air displacing it upward. Thunderstorms and snow storms are characteristic of cold fronts. A **warm front** occurs when a warm air mass overtakes a cooler one and rises up over it. Typical weather associated with warm fronts is cloudiness and drizzle. Cyclonic storms develop as huge eddies along the edges of air masses. These storms are associated with rain and snow, depending on temperature. They typically have both warm and cold fronts extending from them along the frontal boundaries. Shifting air masses and weather associated with the fronts between them characterize the climate of temperate forests.

Once in an ecosystem, water may be stored briefly on the plant surfaces or in the soil (Figure 7.4). From there, it can evaporate back into the atmosphere, or move to other surfaces by dripping from the canopy, running down stems, or being blown by wind. When water reaches the soil, it can evaporate to the atmosphere, infiltrate the soil to be absorbed by roots, or percolate downward to groundwater. Water in excess of that which infiltrates the soil may flow overland toward streams, lakes, or oceans. The processes of water movement in the soil and the acquisition and use of soil water by plants are discussed in Chapter 11.

Water is evaporated back into the atmosphere from surfaces and by transpiration through the stomates of plants as depicted by the upward pointing "output" arrow in Figure 7.4. Evapotranspiration, or the combination of evaporation from surface water, land and plant surfaces, and transpiration over continents accounts for 70 percent of the water returned to the atmosphere. Evaporation from oceans accounts for the remaining 30 percent. As depicted in Figure 7.4 by the widths of the arrows, transpiration accounts for a major portion of evapotranspiration.

Figure 7.4 depicts the hydrologic cycle of terrestrial ecosystems in a very general sense. It is important to note that there is a very large degree of site-to-site variability both regionally and locally in the relative amounts of water entering ecosystems from each source. The causes of precipitation (orographic lifting, frontal lifting, and rising of surface-heated moist air) vary in importance regionally, locally, and temporally. The amounts of water evaporated or transpired vary with atmospheric humidity, soil-water holding characteristics, and availability of soil water.

In mountainous regions, where much of the annual input of precipitation may occur as snow in winter, the melting of the snow pack over the spring and summer months may become a major source of overland flow and the major water input to the soil. In swamps, upward or horizontal inputs of water from groundwater or stream channels may form the dominant source of water for ecosystem processes.

Variation in the hydrologic cycle of two local ecosystems and interconnections between the two are shown diagrammatically in Figure 7.6. Shown are a cypress dome wetland (a swamp of pondcypress trees whose collective crowns often have a domelike shape) and flatwoods (a pine forest with low understory including saw palmeto), two common local ecosystems on the southeastern Coastal Plain. The flatwoods receives most of its water from precipitation; any excesses either run off the surface or percolate downward to the water table. The water table is generally below the rooting zone, but it may be high at some times of year for upward movement from storage to the soil rooting zone, and thus may be important to the plants. Most water is returned to the atmosphere by evapotranspiration since flatwoods soils are typically highly porous and water that fails to evaporate soon after impact percolates downward rapidly. The cypress dome wetland in Figure 7.16 is situated at a lower elevation in the landscape where the ground surface may be below the level of the water table for much of the year. Measured water losses from cypress domes are greatest when the deciduous cypress trees have leaves pres-ent and temperatures are warm. The flow of water through the cypress-dome ecosystem is dependent on the processing of water in the adjacent uplands. In this landscape location, water that percolates downward helps determine the water table height and the level of water in the swamp.

A complete discussion of the hydrologic cycle is beyond the scope of this text. A more complete discussion of the flux of water from the atmosphere to terrestrial ecosystems and back is given by Brooks et al. (1991). In Chapters 11, 18, and 19, we focus on water as a key factor controlling ecological processes.

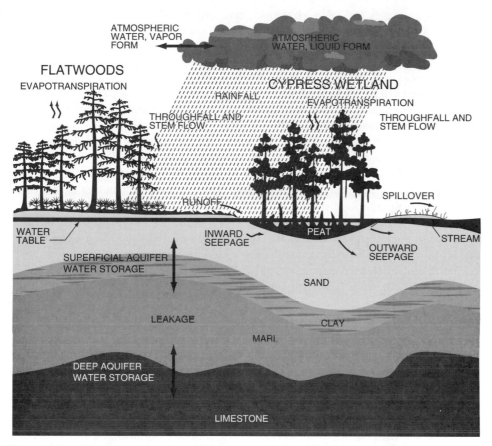

Figure 7.6. Diagrammatic view of the water relationships for two adjacent ecosystems that shows interactions and major water transfers between the ecosystems and the surrounding environment.

Continents and Physiography

The relative positions of the continents, their sizes and shapes, and the locations and orientations of major mountain ranges are important to global and regional weather patterns. Most of the Earth's major land masses are in the northern hemisphere. These include North America, Eurasia, and parts of Africa and South America. By contrast, in the southern hemisphere, there is relatively more open water and virtually no land extending into latitudes south of 40°. As a result, most of the major deserts, temperate forests, and boreal forests are in the northern hemisphere.

Regions dominated by air masses that formed over oceans are said to have **oceanic** or **maritime** climates. Regions dominated by air masses that formed over continents, or which have lost much of their maritime character due to passage over mountains or vast expanses of land, have **continental** climates. Air masses that form over continents tend to be drier than those forming over the oceans and they are likely to be more extreme, hot or cold, in temperature. Dry air masses form in the interiors of North America and Eurasia. At the same time, the northern hemisphere has ample ocean over which air masses can form that are dominated by maritime conditions. A relatively higher portion of air masses in the southern hemisphere are maritime since there are fewer large land masses below the equator.

Mountain ranges and their orientations have major effects on climate. To some extent, high mountain ranges can block or channel air flows. Air is forced to rise over ranges that are perpendicular to zonal air flows, and then it descends beyond the mountains (Figure 7.5*c*). This orographic lifting affects rainfall patterns since the windward side of mountain ranges will have relatively high rainfall and the downwind side, where sinking air is warming, will have low rainfall, forming a **rain shadow**. In western Europe, where no mountain ranges exist parallel to windward coasts, moist air moving in from oceans retains its maritime character for great distances inland.

The major air masses are classified either as tropical, polar, or Arctic depending on the latitudes over which they originate. Most tropical air masses that influence North American weather build over the Gulf of Mexico and tend northward (Figure 7.7). During the summer months, they may penetrate the entire half of the continent east of the Rocky Mountains and up into southern Canada. During the winter they are restricted to a more southern extent. Only the southern half of Florida is almost continuously under the influence of maritime tropical air masses. It is the exceptional brief occurrence of continental polar air in winter that is associated with freezing temperatures in that state.

Northwestern North America is dominated by maritime polar air. These cool, moist air masses originate over the North Pacific Ocean. In the summer, they are largely restricted to the northern Pacific coast from northern California to Alaska and do not penetrate far inland without significant modification. During the winter, they may be forced all

Figure 7.7.   Air masses and their general directions of movement. The air mass types are numbered as follows: 1, Arctic; 2, Polar continental; 3, Polar maritime; and 4, Tropical maritime.

the way across the continent, gradually changing in character as they move. These air masses are responsible for much of the mountain snowfall in the Pacific Northwest. Air masses that form over central and northern Canada are termed continental polar. These are cold, dry, but not extraordinarily frigid air masses. The range of the polar air masses extends as far south as the Gulf coast during the winter and is restricted to the northern United States and Canada during the summer.

Lastly, Arctic air masses form over the northern polar regions, over the pack ice and islands of the extreme north. These are classified as continental in character since the pack ice has temperature exchange characteristics with air that are more similar to land masses than to open water. Arctic air is cold at all times of year.

DESCRIBING CLIMATE

Ecologists have long recognized that there is a general correspondence between the geographic distributions of plants and climate (Chapter 2). Considerable effort has been expended to identify descriptors of climate that best correspond to the observed vegetation patterns.

In the days before satellites, spacecraft, and high flying airplanes, most of our knowledge about weather was obtained from weather stations. Each weather station measures one or more characteristics of the weather—temperature, precipitation, humidity, evaporation, wind, solar radiation, etc. Weather stations are scattered around the world. In the United States, there is typically one weather station capable of measuring maximum and minimum temperature, rainfall, and snowfall in each county. A typical state has about 100 weather stations. A few stations in each region may continuously measure temperature, precipitation, humidity, evaporation, wind, or radiation. Detailed climatic measurements are often available from United States Weather Service offices, major airports (measured by the U.S. Weather Service or Federal Aviation Administration), and agricultural research stations.

In recent years, additional data has become available that is being used more and more frequently. Satellite pictures can be combined with surface observations to track air mass locations and characteristics, wind direction, fronts, and other features. Satellite data provide a way to "see" weather between weather stations. For instance, nighttime surface temperatures can be mapped for clear nights over broad areas thus enabling detection of frost pockets and urban heat islands.

Temperature

The mean annual temperature at any given spot on the Earth's surface is a function of insolation modified by secondary heat transfers arising from terrestrial radiation and air movements. Based on long-term averages, seasonal variations in temperature likewise track incident radiation. Over the course of a year, average monthly temperature trends generally follow a sinusoidal curve with the maximum and minimum temperatures generally lagging the corresponding incident radiation curve by about a month.

In general, temperatures decrease with increasing distance north or south from the equator, and the range of temperature between summer and winter increases. This pattern likely restricts the northern range limits of many plant species.

Heating of the surface layers of air during the day is greatest under conditions where the most radiation is received, that is, in tropical latitudes, at high elevations, and where the air is most free from water vapor, clouds, and atmospheric impurities. Temperatures during the night, on the other hand, depend largely upon the amount of heat absorbed by the underlying land or water during the day and the rate at which this heat is given off as infrared radiation.

There is a lower range in seasonal and daily temperature in maritime climates because temperature is moderated by adjacent water bodies. Continental climates have much greater temperature differences between day and night, and between summer and winter. Florida is surrounded by water on three sides and has far less temperature variation than inland areas to its immediate north. Cold northern air masses must pass over the warm Gulf of Mexico to reach most of the peninsula and are substantially warmed before they do so. Even though Florida lies far to the south of the American Midwest, extreme high temperatures in St. Louis, Missouri regularly exceed those in Florida.

Large lakes can have a similar, though less pronounced, effect on temperatures. Near-shore temperatures on the downwind (usually eastern) side of the Great Lakes show less annual and daily variation than those on the upwind side or further inland. Notably, maritime climates are important for the success of many commercial tree fruit crops such as citrus (which cannot survive extreme cold) in Florida and cherries in along the eastern shore of Lake Michigan in Michigan.

Position relative to major water bodies may influence latitudinal temperature changes. For example, in the American West, from the Pacific Ocean to the crest of the Cascade–Sierra Nevada chain, the prevailing on-shore winds are the dominant influence in the climates, and they greatly diminish temperature variation from north to south (Baker, 1944). On the west side, there is a difference in mean annual temperature of less than one-half degree for each degree of latitude (0.4 °C), or about 0.3 °C per 100 km. This change is only about one-third of that found along the east coast. However, east of the Cascade–Sierra Nevada barrier, continental conditions generally prevail. The overall change here in the interior West is nearly 0.8 °C per degree of latitude, or 0.7 °C per 100 km.

Temperatures also change as altitude changes. In the interior of western North America where rising air is typically not saturated with water vapor, the decrease for each 300 m averages about 2 °C . However, temperature changes with altitude are less for air that is saturated with water vapor. In the maritime climate of northern California, where the air forced upward over the coastal mountains is often saturated with water vapor, the average temperature decreases between 0.5 and 1 °C with each 300 m (1,000-foot) rise in elevation for the coastal ranges and 1.4 to 2 °C for the Sierra Nevada Mountains.

Vegetation often changes similarly with both latitude and altitude, primarily in response to similar temperature regimes. This zonation is particularly striking in parts of western North America where there are dramatic and predictable shifts in vegetative communities with altitude. For instance, Figure 7.8 depicts the lowering of tree line along the mountains that run from Central America to Alaska (North American cordillera). It also shows that, at any given latitude, tree line is lower in the coastal mountains where the summer climate is more maritime and cooler. Grace (1989) suggested that trees cannot grow at combinations of latitude and altitude where the warmest month has a mean temperature below 10 °C. Daubenmire (1978) suggested that timberline occurs at the point where trees are unable to accumulate enough energy to both survive and successfully reproduce.

Growing season length is typically defined as the number of days between the last recorded freezing temperature in the spring and the first recorded freezing temperature in autumn. It is used to represent the length of time a tree may have for aboveground growth during the year. It is pertinent to note that some metabolic processes (e.g., respiration) occur throughout the year. Growing season length decreases from the equator to the poles. At any given latitude, it also decreases from maritime to continental areas. Continental

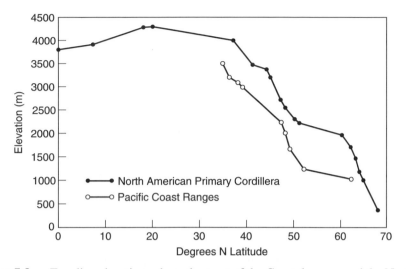

Figure 7.8.   Tree line elevations along the crest of the Coastal ranges and the North American cordillera (the major mountain "spine" running from Central America to Alaska and including the Rocky Mountains and the Alaska Range). (Modified from Daubenmire, 1978. Reprinted from *Plant Geography,* © 1978 by Academic Press, Inc. Reprinted with permission of Academic Press, Inc.)

areas typically have a shorter, hotter growing season than do areas dominated by maritime conditions. Likewise, higher altitudes have a shorter growing season. Growing season length has a major effect on the relative occurrence of conifers versus broad-leaved angiosperms (Chapters 3 and 9).

Growing season **heat sum** is defined as the summation of temperature, above a base temperature, over the course of the growing season. The base temperature for computing a heat sum is typically taken to be 4.4 °C (40 °F) or 10 °C (50 °F) and assumed to be the minimum temperature at which meaningful growth can occur. Heat sums are crudely proportional to the total solar radiation received and have been correlated with initiation of mitosis in vegetative buds (Owens et al., 1977) and with the timing of bud burst and flowering (Lindsey and Newman, 1956; Kaszkurewicz and Fogg, 1967). In Michigan, northern range limits for some deciduous species parallel isopleths (contours) on heat sum maps (Denton, 1985, Denton and Barnes, 1987).

Extremely low winter minimum temperatures have the potential to kill trees, especially those associated with warm climates. Species associated with cold climates typically have the ability to withstand temperatures much lower than those that actually occur within their ranges (Chapter 9). Interactions between cold temperatures and water have also been implicated in setting limits of tree distributions. Specifically, Hocker (1956) found a strong correlation between the northern and western edges of the range of loblolly pine and a combination of cold temperatures occurring simultaneously with low soil water.

Precipitation

On a global basis, forests occur where precipitation exceeds transpiration. In areas where there is less precipitation, grasslands and deserts dominate the landscape. As with air temperature, precipitation varies with latitude and altitude. The distribution of precipitation over the face of the Earth, however, depends primarily upon the interrelationships between air currents, large water bodies, and topography.

The greatest precipitation occurs in the convergence zones at the equator and in the temperate latitudes. The zone near the equator is generally characterized by frequent rain showers and thunderstorms that occur daily in many areas. Not surprisingly, these high precipitation areas are associated with the tropical rain forests. In the temperate latitudes, precipitation is typically associated with frontal systems, cyclonic storms, and unstable air masses. Heavy precipitation also occurs in areas where warm, moist air undergoes orographic lifting. Over 2,500 mm of rainfall per year are normal in such forest regions as the Pacific Northwest of the United States and Canada, Hawaii, New Zealand, and India. Precipitation from orographic lifting may be very localized. Mount Kawaikin (1,590 m) on Kauai, Hawaii may be the rainiest point on Earth. The western side of the Cascade Mountains in Oregon is very wet, portions of it being temperate rain forest, while only a few miles east, on the opposite side of the mountains, there is desert. Similarly, in Hawaii, the windward side of each island is very wet and the downwind side is extremely dry. Orographic lifting and surface cooling create lake-effect snowfalls in Michigan, New York, and Ontario.

Precipitation patterns are particularly affected by air currents and mountain barriers. Precipitation caused by orographic lifting is not closely correlated with elevation above sea level. In very high mountains, the maximum precipitation is commonly reached somewhere below the summit. Air currents become too depleted of moisture to provide as much precipitation at extremely high elevations. Along the Pacific Coast, mean annual precipitation at the lower elevations increases sharply with altitude, at rates varying from 13 to 17 mm per 100 m in the coastal range of Washington to 7 to 8 mm in the Sierra Nevada. The maximum precipitation occurs at the middle elevations—ranging from perhaps 900 m in the Olympics to 1,500 m in northern California and 2,400 m in the southern Sierra Nevada. Above this elevation, precipitation decreases with altitude. East of the coastal mountain barrier in the mountains of the interior, the effect of elevation on rainfall is much more predictable, being approximately 3 to 4 mm per 100 m rise in elevation (range 900 to 1,500 m). These values seem to hold from the east side of the Cascade–Sierra Nevada ranges east through the Rockies, and from Canada south to Mexico.

Precipitation varies from west to east across North America. Moving eastward from the wet coastal ranges along the Pacific coast, the mountain ranges become progressively drier. In the major rain shadow that characterizes the ranges and in the prairie regions east of the Rocky Mountains, precipitation is generally too low to support forests. East of the mountains, rainfall increases gradually as maritime moisture from the Gulf of Mexico becomes a major factor that increases growing season rainfall. Precipitation quantities typically continue to increase toward the east. High precipitation is found in the southeast and throughout the northeast. Conspicuous high precipitation areas occur in the Southern Appalachian Mountains and in the White Mountains of the Northeast.

FOG AND DEW

The same conditions that lead to precipitation may lead to fog formation, because fog is merely a cloud at ground level. However, fog may form under local conditions that are not adequately intense to produce rainfall. **Radiation fog** may form at night when surface radiation cools the surface and causes air immediately above to cool with it. If air is high in moisture, it may cool below its saturation temperature, or dew point, and fog may form. **Advection fog** may form when warm air moves over cold water. Advection fog may occur over the Great Lakes in the summer when warm Gulf air crosses the cold waters of the lake. This fog may be blown onshore where it usually dissipates rapidly. Advection fog is also common during the summer months along the Pacific shoreline. Similar conditions occur when warm air moves over cold ocean waters. Fog also forms when warm moist air is forced up mountain slopes.

The relationship between condensed moisture and trees has been the subject of much speculation. Fog and dew may well be quite important in determining the growth and distribution of forests, but the extent and mechanism of the effect have proved difficult to demonstrate in precise experimentation. In addition to reducing transpiration, there can be no doubt that forests condense appreciable amounts of moisture from fog. Also, appreciable amounts of moisture can be condensed on tree foliage during the night through dew formation. Whatever portion of this condensed moisture falls to the ground through fog drip or dew drip may be added to the soil and thus benefit the forest. Fog may be collected and condensed by tree crowns and dripped to the ground in appreciable quantities when no moisture is caught by rain gauges in the open. For example, Isaac (1946) found that, on a ridge 3 km from the Pacific Ocean in the Oregon fog belt, precipitation was one-quarter greater under trees than in the open because of fog drip, whereas in a valley 8 km inland, precipitation was one-third less under the forest due to interception.

In detailed studies of the "fog-preventing forest zone" on the southeast coast of Hokkaido, Japan, Hori (1953) found that the forest could remove $3,400$ l ha^{-1} hr^{-1} of moisture under standardized conditions of fog density and wind movement. Under similar conditions, grassland captured not more than 10 to 18 percent of this amount. The windward vertical edge of the forest captured as much fog water as a horizontal surface three times as large in area.

In high-altitude conifer forests around the world, much of the total precipitation is received as fog, rime, or hoar-frost. At 1,150 m in the Bavarian Mountains of southern Germany, Baumgartner (1958) reported that 42 percent of the yearly total precipitation of 2,000 mm was received as fog. Numerous experiments have demonstrated, through use of mechanical collectors simulating coniferous needles, that the foliage of conifers in mountainous fog zones is efficient in trapping moisture. Various studies have found that simulated needles collect 1.7 to 4.5 times as much as gauges or rain collectors without simulated foliage (Vogelmann et al., 1968; Schlesinger and Reiners, 1974). The needle-like foliage and twiggy character of the spruce-fir forests serve as effective mechanical collectors of wind-driven cloud droplets. A detailed review of mist precipitation and vegetation was given by Kerfoot (1968).

There is an obvious correlation in many parts of the Earth between the distribution of certain trees and the presence of fog belts. Coincidence between the coastal zone of heavy summer fogs does not prove that a cause-and-effect relationship necessarily exists between fog and the trees' distribution. Perhaps the most spectacular example occurs along the Pacific Coast of North America where Sitka spruce (from Oregon northward) and redwood (in California) characterize the very dense and fast-growing temperate rainforest (Figure 7.9). The summer fog itself may affect the tree climate in many ways. It may well have a direct effect in supplying summer moisture. In addition, however, it also has indirect effects in reducing the hours of summer sunshine, in reducing the summer daytime temperatures (Byers, 1953), and in increasing the supply of carbon dioxide (Wilson, 1948). The coastal zone also differs climatically in many ways from adjacent areas by having heavy winter rains, little cold or hot weather and equable temperatures throughout the year. Thus the relationship between tree distribution, fog, and other ecosystem factors is complex; we cannot conclude that redwood and Sitka spruce must owe their survival to the moisture-giving powers of summer fog along this coast.

Figure 7.9. Redwood forest along California's coastal fog belt with luxuriant understory vegetation. (U.S. Forest Service photo.)

SNOW

In temperate and boreal areas, much of the annual precipitation falls as snow. Snowfall is
distinct from rain. First, frozen water is unavailable to vegetation. On warm days in early
spring when the soil is still frozen, trees with transpiring foliage may experience water
stress. However, melt water may provide a source of soil moisture later in the season.
Snow also serves a protective function and may affect reproduction of some species.
Snow may protect seedlings and foliage from harsh temperatures and winds. Falling seeds
may be blown across a smooth snow surface, thus dispersing seed further than likely to
occur if the same seed fell on bare ground. Holes in the snow adjacent to tree trunks near
timberline may catch wind-blown seed and encourage establishment of seedlings close to
existing trees, resulting in clumping. Persistent snow cover may also prevent establishment. Finally, avalanches may periodically eliminate trees from steep mountainous slopes
and may favor establishment and persistence of clonal species, such as trembling aspen.

Evapotranspiration

Precipitation does not correlate well with the distributions of most species or with the
boundaries of most major plant formations. The reason for this is that evaporation and
transpiration are functions not only of water availability but of temperature. **Potential**

evapotranspiration is the amount of water that could be evaporated from land, water, and plant surfaces if water in the soil was in unlimited supply. Potential evapotranspiration is "potential" because it does not actually occur. An early and well-known estimate was provided by Thornthwaite (1948). It has been widely used in ecological literature because the approach relies on standard weather station data and because of the availability of tables and equations from which it can be easily calculated (Priestly and Taylor, 1972). It is widely recognized as being less applicable than estimates that have a close bearing to the physical properties of water, surfaces, and the atmosphere. Regardless of the method used to calculate it, potential evapotranspiration has proven to be useful in assessing the distribution of forest types. For instance, Patric and Black (1968) found that potential evapotranspiration patterns were more closely related to forest distribution in Alaska than temperature patterns.

Actual evapotranspiration is the amount of water evaporated from water and plant surfaces. It is a function of temperature, air movement, actual humidity and the amount of water that the atmosphere can hold, water availability, and the vertical and horizontal structure of the vegetation. In many areas, such as deserts, more water could evaporate (high potential evapotranspiration) than exists to evaporate (actual evapotranspiration).

Many ecological relationships exist with precipitation and actual or potential evapotranspiration. Several of these will be discussed when we examine climatic classifications, since these relationships form one of the classification factors in most cases. A typical, descriptive tool (Zahner, 1956) uses potential evapotranspiration and transpiration, combined with estimated water storage in the soil to predict the pattern of water availability and stress likely to be experienced by the forest tree (Figure 7.10).

Wind

Wind plays important roles in the life history of trees. For instance, many trees rely on wind for dispersal of pollen and seeds. Trees in dense forests and in areas where wind is inadequate for pollen or seed dispersal typically rely on animals as pollinators or dispersal agents. Wind also circulates air within the canopy providing a continuous source of carbon dioxide for photosynthesis.

Wind sometimes plays an important role in determining the growth form of forest trees. Specifically, wind may alter growth forms on exposed shores and at treeline in mountainous areas. Wind may cause stress-related changes in growth, abrasion of leaves and twigs, and induce transpiration losses. Near tree line, healthy foliage and branches may be found in the lee of ridges, in depressions, behind rocks, or in other protected areas. Branches that are below the snow are typically luxuriant while those which extend above the snow are stunted and deformed. Wind exacerbates the effects of low temperatures and causes transpiration losses in the early spring when the air is warm but the soil is still frozen. Trees near tree line may be unable to restrict water loss, either because the epidermis may be damaged by abrasion during winter winds or because the cuticle does not develop properly in a short, cold growing season (Grace 1989). The deformed trees found near the tree line are called **krummholz**.

EXTREME EVENTS

Climate includes unusual and extreme events as well as those that occur with regularity. These events may be as important, or more important, in controlling some forest processes as are typical conditions. Extreme events may be favorable or unfavorable. For

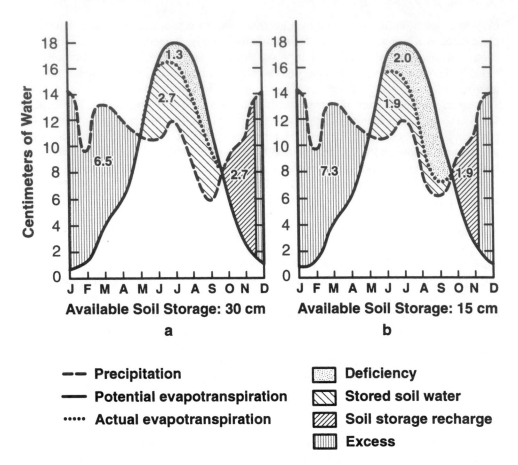

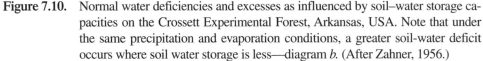

Figure 7.10.   Normal water deficiencies and excesses as influenced by soil–water storage capacities on the Crossett Experimental Forest, Arkansas, USA. Note that under the same precipitation and evaporation conditions, a greater soil-water deficit occurs where soil water storage is less—diagram *b.* (After Zahner, 1956.)

example, favorable combinations of temperature and precipitation, or specific weather patterns may be necessary for reproduction in some species. For instance, pondcypress reproduction in deep water swamps can only occur when the swamp is dry and only when the overall pattern of climate provides appropriate conditions for seedlings to survive and grow sufficiently to have their foliage above water when normal conditions return. The same conditions favorable for reproduction in a deep swamp could represent drought and stress in a shallow swamp in the same area.

Storms

Several types of extreme weather with ecological importance are associated with thunderstorms. Tornadoes are intense cyclonic vortices in which air spirals rapidly around a twisting, vertical axis. The winds are extremely strong, in the 170–500 km hr^{-1} (100 to 300 mph) range. They also have extreme low pressure at their centers. Tornadoes can occur throughout the world, but they are found most commonly in regions that produce strong thunderstorms with regularity. In the United States, Iowa, Kansas, Arkansas, Oklahoma, and Mississippi have the highest frequency of tornadoes per unit area. Tornadoes may also occur in association with hurricanes and air-mass thunderstorms. Tornadoes cause local disturbances because they knock down trees along a relatively narrow path that may extend for kilometers.

Gust fronts are extremely strong winds caused by down-drafts along the leading edges of intense thunderstorms. These gusts are rare, cover small spatial areas, and last for an extremely brief period of time. However, they may cause extensive treefall within the area that they hit. Canham and Loucks (1984) documented the effect of one gust front with winds in excess of 50 m s^{-1} which broke and uprooted tress on an estimated 344,000 ha of forest in Wisconsin.

Hurricanes are extremely intense tropical cyclones that form over warm oceans during the summer. They are known for intense wind, heavy rain, storm surges, and a calm "eye" in the center of the storm. All of these, except the eye, can cause damage to both natural and anthropogenic landscapes. Wind may uproot trees, snap them off, or strip them of their leaves and smaller branches. The intense winds of Hurricane Hugo, which hit South Carolina in 1989, killed or severely damaged trees. Most pines snapped or fell; the old-growth areas in the Francis Marion National Forest were particularly hard hit. Hardwood survival was higher, but many trees were stripped of leaves and smaller branches. After the storm, epicormic branching was the norm. Old, high-volume tree stands generally suffered higher mortality than young, low-volume stands (Putz and Sharitz, 1991; Sheffield and Thompson, 1992). Hurricane Andrew, which hit Florida in 1992, damaged forests across the southern tip of Florida. Acres of dead mangroves felled by the wind were still apparent in the southern Everglades, with little new growth, in 1994. Storm surges also cause damage by direct action of the surf dousing upland areas with salt water. Wind as a disturbance factor is considered in Chapter 16.

Other types of damage significant to forest ecosystems include extreme flooding and even landslides. For example, landslides occurred in Virginia when the remnants of Hurricane Camile (1969) experienced orographic lifting in the Blue Ridge Mountains. On the other hand, the high rainfall associated with these storms may provide an important input of water for some regions. In Florida, water table elevations tend to drop slowly for years but recover rapidly when replenished by a hurricane. This pattern is particularly important for hardwood and cypress swamps that are adapted to high water table conditions.

Lightning strikes are another climatic factor of ecological significance. West-central Florida has the highest lightning frequency in the United States. Lightning may damage trees either by direct hits or by causing fire. Lightning and fire are discussed in Chapter 12.

Drought

Drought is also a recurring climatic disturbance to plants. Drought can be defined as a prolonged and abnormal water deficiency that is relatively extensive in time and space (Porter et al., 1994). By this definition, at least some part of North America is likely to be experiencing drought at any one time. Though not always the case, droughts are usually accompanied by unusually warm conditions.

Drought affects forest trees both directly and indirectly. Directly, it may reduce growth for the duration of the growing season. Also, flowering, seed production, seed germination, and seedling survival may be greatly reduced. Growth of balsam fir in northern Michigan has been shown to be reduced by drought (Drew and Alyanak, 1981). The most damaging effect of summer drought to firs growing on sandy soils was to inhibit bud development when needle primordia for the following year are being formed. The number and size of needles and shoot growth the following year were reduced even though soil water was adequate. Drought in combination with insect attack and disease was implicated as the cause of a severe die-back of hemlocks in Wisconsin (Secrest et al., 1941). Drought may also have indirect effects. The most obvious of these is fire, which is more likely to start, burn hot, and spread quickly under dry conditions. Severe fires that burned in Yellowstone National Park in 1988 were associated with a multiyear drought, high winds, and old-growth lodgepole pine forests that provided ample fuel.

Drought may also be associated with disease. For example, pole blight of western white pine, a disease reducing growth and causing substantial mortality in pole-sized stands in northern Idaho, has been linked to adverse temperature and soil water conditions for growth from 1935 to 1940 (Leaphart and Stage, 1971). Drought conditions on sites with shallow soils led to root deterioration and rootlet mortality, crown decline, reduced growth, and ultimately, the death of many trees. Recurrence of an adverse climatic pattern may increase the probability of disease and would act to favor more drought-resistant species over western white pine on the sites with high soil-water stress.

Glaze

Glaze is rainfall that freezes on contact with the ground, trees, or other surfaces. It has little impact on tree growth unless it becomes so heavy that branches break under its weight. Occurrence of glaze at damaging levels is common at times throughout the East north of Florida. It is particularly common in parts of the Appalachians. Glaze damage is typically increased by winds that occur after ice has formed around branches and twigs. Wind may prevent ice from coating branches if it is present during deposition.

Not all species are equally susceptible to glaze damage. Studies in Virginia (Whitney and Johnson, 1984) and New York (Downs, 1938) have shown that some species are much more likely to be injured than others. Fast growing species such as black cherry, aspen, willow, and basswood are more apt to suffer damage than oaks, sugar maple, white ash, or hickory. Conifers are particularly resistant to damage, despite the presence of foliage. Pines, hemlock, and northern white-cedar are damaged far less than hardwoods. Ironically, severe glaze may create canopy gaps enabling the species most likely to suffer damage, such as black cherry, to grow into the canopy. Gaps created by damage may also favor understory tolerant shrubs, vines, and herbs. If the weight of the ice is not so great as to cause breakage, glaze may protect the plant from freezing conditions. For instance, water is sprayed on citrus trees in Florida to protect the leaves from freezing.

Extreme Cold

Unusually cold temperature can be an extreme event. Whereas most temperate tree distributions appear to be limited by growing season temperature and water availability, unseasonably cold temperatures also may be important. When buds are bursting in the spring unseasonably cold temperatures may hinder reproduction and growth. Cold may also limit the northern extent of tropical and semitropical species, but it seldom seems to have permanent effect on native forest plants. Cold appears to limit the northern spread of several, aggressive, introduced (exotic) species, which are severe problems in southern Florida (especially melaluca, Australian pine, and Brazilian pepper tree). These subtropical species have the capacity to outcompete native species in much of southern Florida but become much less competitive to the north. Repeated winter top kill appears to be one of the factors that limit their expansion.

The timing of cold temperatures may be more important to most native species than extreme minimum temperatures. Freezes that occur after the growing season has commenced may be particularly damaging. Spring and early fall freezes have been associated with the death of oaks in the Appalachians. The freezes are most damaging if the previous growing season was droughty or if some other form of damage, such as insect attack, occurs in the same year (Nichols, 1968). Freezing temperatures during periods of active plant growth have a high potential to preclude reproduction. For instance, failure of the acorn crop was caused by freezing of catkins on low elevation gambel oak at a site in Utah (Neilson and Wullstein, 1980). High elevation oaks, which had not experienced bud burst, were not damaged while lower elevation oaks were injured. The frequency and timing of freezing temperatures relative to physiological development may be one factor controlling the distribution of gambel oak in the southwestern United States. Spring freezes may also be important for plantations of trees in which trees native to one region have been moved to another. For instance, Sitka spruce, which is native to the maritime climate of the Pacific Northwest, has been planted frequently in Scotland where plantations from some seed sources are much more prone to freezing damage than others (Cannell et al., 1985).

MICROCLIMATE

Microclimate refers to very local variations in climate. For instance, the south side of a hill (in the northern hemisphere) is typically hotter than the northern side. Valleys often have morning fog and "pockets" where frost is more likely to occur than on the surrounding hillsides. Local topography may also affect the severity of wind damage, incidence of lightning, extreme minimum temperatures, glaze, and other climatic factors. High winds may damage trees on ridges and on the upwind sides of slopes but leave those on the downwind sides undamaged. Topography can also channel winds through canyons in mountainous areas. Temperature differences, discussed in Chapter 9, are the most pronounced. Microclimate may be sufficiently important in some areas that the north and south slopes of small hills, as well as ridges and valleys, may have almost totally different vegetation.

CLASSIFYING CLIMATE

To understand the interaction between climate and vegetation, and to predict changes that may occur in the future, it becomes useful to classify climate in ways that relate to growth, survival, and reproduction of the biota. A bewildering number of attempts have been made to combine one or more statistical measures to delineate climatic zones that relate to vegetation. Some of the most widely used include the classifications of Köppen, Thornthwaite, Holderidge, and Bodyko. These were designed to classify climates on a global scale. They have also been used for regional classification with varying degrees of success. Multivariate statistics have likewise been used to define regional classifications and those that focus on more local levels. Lastly, patterns of variation, expressed either as monthly summaries or as evaluations of air mass movements and the variability of weather patterns, have been developed and successfully related to patterns of observed vegetative distributions.

Köppen (1931) produced the best known of a series of climatic classifications based on simple summaries of data available from weather stations. The classification scheme of this German scientist has achieved considerable use by geographers. Basically, the climate is classified by three code letters, the first referring to one of five zones of winter temperature, the second to seasonal rainfall pattern, and the third to one of three zones of summer temperature. As with other global schemes, this system is broadly related to major vegetation types, but shows limited correlation with the actual distribution of vegetation regionally. Köppen's classification has received world-wide use, and several recent geographers have found it useful in studies of climatic change (Bailey, 1996).

A number of classifications have been developed that are based on temperature or radiation and some measure of moisture availability. One of the earliest was produced by Thornthwaite (1948). His classification introduced the ratio of precipitation to evapotranspiration as an ecologically-important climatic statistic. He later modified his classification to use an estimate of potential evapotranspiration based on air temperature (Thornthwaite and Mather, 1955). Thornthwaite's classification has received widespread application, and studies based on its concept have been used to describe seasonal variation in precipitation and potential evapotranspiration.

Holderidge's (1947, 1967) classification is based on three factors: biotemperature, potential evapotranspiration, and annual precipitation. Biotemperature is a heat sum based on summing monthly temperatures within the 0 to 30 °C range. The classification identified a series of broad "life zones", such as rain forest, wet forest, dry forest, tundra, and steppe positioned along three axes (Figure 7.11). Within each zone, plant and animal communities and their associated environmental features, such as topography, soils, and precipitation distribution are used to identify ecosystem units, termed associations. The Holderidge life zones are conceptually simple, but like other classifications, they fail to delineate plant formations whose boundaries depend on factors not included in the classification. For instance, North American areas that have tall grass prairie fall within the "Montane Moist Forest of the Warm Temperate Region" (Sawyer and Lindsey, 1964). The discrepancy may be due to absence of annual temperature range as a classification factor (Greller, 1989). Holderidge's life zones have been widely applied in Central and South America. They have also been used in studies of global vegetative changes that might be expected under various scenarios of climatic change (Skole et al., 1993).

Budyko (1974, 1986), a Russian scientist who developed a radiation balance statistic and a dryness index, asserted that he could associate existing vegetative zones with distinct climatic ones. Budyko's dryness index characterizes the dryness (or wetness) of climates as the ratio of potential evapotranspiration to precipitation. His radiation balance is the computed difference between absorbed radiation and net longwave radiation. It is assumed to be proportional to the amount of energy available for plant growth. He used this radiation balance statistic to make a bioclimatological classification of major Russian plant formations. It was used recently (Tchebakova et al., 1993), with minor modification, to create a global climatic classification that appears to have a better correlation with vegetation than global climatic classification models currently in use.

Multivariate classifications of climatic data have proven useful in identifying relatively climatically homogeneous regions. In this context, several authors have used principal component analysis, cluster analysis, and discriminant analysis to: (1) detect major contributors to the climate of a region, (2) group climatic stations into homogeneous regions, and then (3) evaluate the effectiveness of the classification. These models have varied greatly in their performance and the ease with which they can be interpreted. Sowell (1985) produced a climatic classification specifically to see if there were predictable trends in standard weather station data for North American plant formations as defined by Whittaker (1975). Accuracy was high (89 percent), but it was difficult to assign causal climatic conditions to any of the identified formation boundaries.

Denton and Barnes (1988) used a multivariate procedure to classify climates in Michigan by producing separate classifications for growing season temperature, winter temperature, and growing season precipitation variables. These classifications were then combined with physiographic data to create a single classification that was consistent with

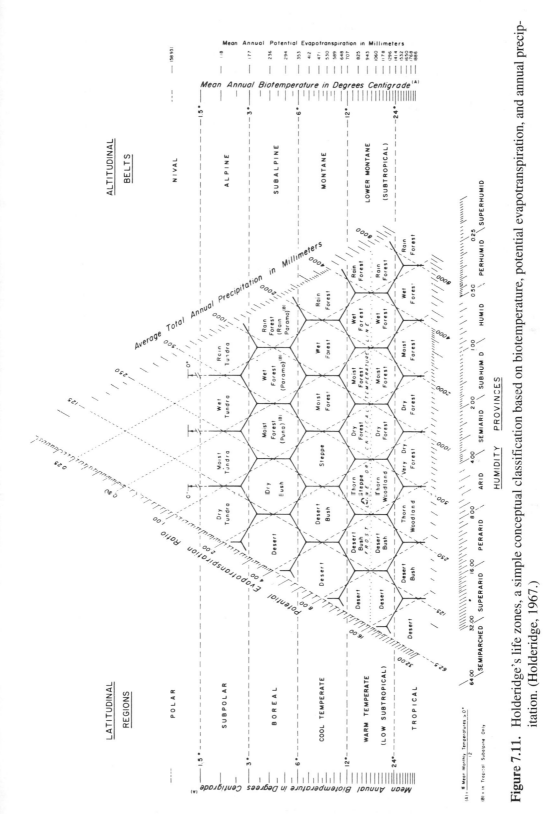

Figure 7.11. Holderidge's life zones, a simple conceptual classification based on biotemperature, potential evapotranspiration, and annual precipitation. (Holderidge, 1967.)

both. A later refinement, which included more specific physiographic and soils data, was used to produce a general hierarchical classification of ecosystem regions and districts in the state (Albert et al., 1986; Chapter 2).

Classifications based on air masses and movement patterns have also proven to correlate well with some major changes in biota. Maps of average jet stream and air mass trajectories yield visual means to assess which air masses are dominating which areas over time. They also provide indications of the paths of cyclonic storms that generally move with the jet streams. Plotted seasonally, they show the latitudinal shifts in position and intensity that occur. Several of the trajectory boundaries correspond closely with major vegetative boundaries. This relationship is particularly true for the boreal forest (Figure 7.12). The typical southern extent of Arctic air masses (the Arctic frontal zone) during the winter corresponds closely with the southern limit of the boreal forest. The southern extent during the summer corresponds closely with the northern limit (Bryson, 1966; Barry, 1967). Denton (1985) noted a strong correlation between the boundaries of the northern hardwood and beech-maple regions in Michigan and the typical southern extent of the continental polar air as mapped by Barry (1967).

CLIMATIC CHANGE

So far, we have discussed climate as though it is constant. It isn't. The Earth has a history of almost continuous climatic change, and some of these changes have been dramatic. Climate was much warmer during the Mesozoic Period, and tropical and subtropical climates were much more widespread than they are today. Glacial periods were associated with generally colder climates. Even since the last deglaciation, which began approximately 18,000 years ago, climate has not been constant (Chapter 3).

A wide variety of natural and anthropogenic factors affect climate. Factors operative on

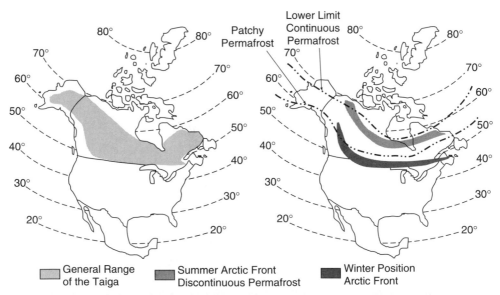

Figure 7.12. Mean locations of the Arctic front in summer and winter and the range of permafrost throughout the boreal forest as compared to the general range of the boreal forest. (After Oechel and Lawrence, 1985. Reprinted from *Physiological Ecology of North American Plant Communities,* © 1985 by Chapman & Hall. Reprinted by permission of Chapman & Hall.)

a geologic time scale include variations in the Earth's orbit and axis of rotation, changes in the Sun's radiation output, continental drift and associated shifts in oceanic currents, shifts in the extent of volcanic activity, impacts of large meteors, and changes in atmospheric composition. Climatic changes operative in mid- to short-term time scales include volcanic eruptions, variations in oceanic currents, and changes in atmospheric composition.

Volcanic eruptions spew millions of tons of ash and sulfur dioxide (SO_2) into the air thereby blocking passage of solar radiation into the lower layers of the atmosphere. The effects of recent large eruptions have been quantified as having marked effect. For instance, the eruption of Mount Pinatubo in the Philippines in 1991 produced observed cooling of 0.3 to 0.4 °C following the eruption (Pearce, 1993). Eruptions of individual volcanoes produce relatively short-term effects because the ash settles out of the atmosphere.

Altered oceanic circulation patterns are major causes of climatic fluctuations since the temperature of surface waters has a direct influence on the temperature and moisture content of air masses. If circulation patterns change such that either major currents within the ocean shift or that there are vertical changes in the temperature profile of the ocean, then there will be direct impacts on the atmosphere and weather. A good example is the 'El Niño,' a weather condition caused by a warm current in the Southern Hemisphere that periodically affects the weather in southern North America. There is also evidence that the Gulf Stream may have remained further south during the ice ages than its current northeastern path. The shifts in oceanic behavior are not fully understood, but given their enormous size and the heat capacity of water, any change is likely to have major effects on climate.

The relative abundance of carbon dioxide and other greenhouse gases (methane and nitrous oxide) is extremely important to the maintenance of the Earth's temperature. Current concerns about climatic change and global warming stem from increases in the quantities of these gases, especially carbon dioxide due to human activities.

Estimating the CO_2 content of the atmosphere is relatively easy, whereas determining the magnitudes and shifts in the various sources and sinks is extremely difficult. For instance, photosynthesis is often limited in the forest by availability of CO_2. Increased CO_2 can result in greater photosynthesis and increased fixing of carbon on land and in the ocean. The oceans and their biota take up a great deal of CO_2 and convert much of the carbon into forms that ultimately sink to become part of the ocean-floor sediments. These processes are not constant, and they shift depending on temperature, upwellings of bottom water, phytoplankton concentrations, and numerous other factors. Overall, atmospheric scientists estimate that carbon dioxide concentrations are increasing in the atmosphere by 2 ppm per year. Concentrations increased from 270–280 ppm in the early 1800s to 360 ppm in 1995. The increase appears to be steady (Figure 7.13).

Based on a synthesis of the available data as reported by Henry et al. (1994), the global pattern is one of increasing temperatures from 1860 until 1940, followed by a drop between 1940 and the 1965–1972 period, and then of rising temperatures since that time. Global records indicate a warming of 0.5 °C since 1800. Warming since 1900 is in the 0.27 to 0.39 °C range.

In North America, some parts of the continent have experienced general warming trends while others have not become warmer. A 10-year moving average of annual mean temperatures at nine sites in the continental United States was computed from United States National Oceanic and Atmospheric Administration data (Figure 7.14). Stations in California, New Mexico, Minnesota, Ohio, Maine, and Utah show an apparent slight warming trend of approximately 2 °C over the past 100 years. However, the southeastern United States, as shown by graphs for Texas and South Carolina, has had a slight cooling trend from 1895 to 1990. This trend is apparently due to a winter shift in the position of the polar jet around 1950 that has allowed cold air masses to penetrate the southeast with greater frequency.

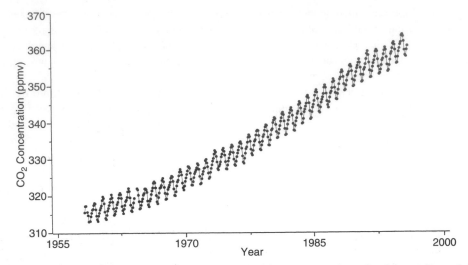

Figure 7.13. Changing atmospheric carbon dioxide concentrations for Mauna Loa site, which is a barren lava field of an active volcano, Hawaii, USA. The Mauna Loa record shows a 14.3 percent increase in the mean annual concentration, from 315.83 parts per million by volume (ppmv) of dry air in 1959 to 360.90 in 1995. (After Keeling and Whorf, 1996.)

Warming is not the only change that could occur. The addition of anthropogenic substances in the atmosphere can potentially block incoming radiation and change the abundance and distribution of clouds in the atmosphere thus changing the Earth's albedo. Precipitation patterns could also shift the balance between precipitation and evaporation. If the Earth becomes warmer, melting of ice may cause sea levels to rise, thus flooding coastal regions.

What are the likely biological consequences of climatic shifts? We don't know, but we can look at the past, and judging from that past, we can be relatively sure that changes could potentially affect the distribution of major plant formations. The boreal forest today is well defined by the northern limit of the Arctic Front in summer and the southern limit in winter (Bryson, 1966). It is consequently marked by a sharp, growing-heat-sum gradient—if temperatures are too low during the growing season, there is tundra. Evidence suggests that the northern limit of the forest shifts with climatic changes. A study done by Bryson et al. (1965) showed that the boreal forest extended further north at times in the past. Between 5,500 and 3,500 years ago, the boreal forest advanced to a point about 280 km north of the present tree line. It then failed to recover after fire. Another advance occurred about 1,000 years ago when the forest again failed to regenerate after fire. Tundra has remained since that time. The southern boundary of the forest may be subject to similar change. Recent investigations have shown that temperatures in the boreal forest have increased significantly since 1892 and that precipitation has decreased (Singh and Powell, 1986). These conditions appear favorable to replacement of boreal forest by grassland in transitional areas between Alberta and northern Minnesota.

The record since the last ice age indicates that species migrated individually since the last glaciation. Plant formations will likely not shift as units. Rather, new plant and animal communities can be expected, and warming temperatures could have effects on various ecosystem processes.

In summary, we don't know how much climate may change due to human alterations of the planet. We do have evidence that some changes are affecting the atmosphere, and

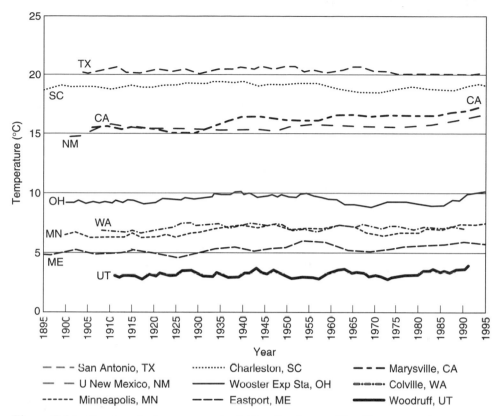

Figure 7.14. One-hundred-year record of climatic change. The 10-year moving average of mean annual temperatures was computed for nine weather stations located in different regional areas of the United States. A variation in trends is revealed with many stations showing slight increases over the last 100 years, whereas a few in the South show none or even somewhat lower temperatures over the most recent 40-year period. Data from the U.S. National Oceanic and Atmospheric Administration.

we have educated guesses of the types of changes that could potentially occur. Because of the strong connection between climate and the geographic distributor of forests, any change in climate is likely to result in a concomitant change in plant distribution.

Suggested Readings

Bryson, R. A. 1966. Air masses, streamlines, and the boreal forest. *Geogr. Bull.* 8: 228–269.

Budyko, M. I. 1986. *The Evolution of the Biosphere.* D. Reidel Pub., Dordrecht. 423 pp.

Canham, C. D., and O. L. Loucks. 1984. Catastrophic windthrow in the presettlement forests of Wisconsin. *Ecology* 65: 803–809.

Daubenmire, R. 1956. Climate as determinant of vegetation distribution in eastern Washington and northern Idaho. *Ecol. Monogr.* 26: 131–154.

Petterssen, S. 1969. *Introduction to Meteorology.* 3rd ed. McGraw–Hill, New York. 333 pp.

Sellers, W. D. 1965. *Physical Climatology.* Univ. of Chicago Press, Chicago. 272 pp.

Woodward, F. I. 1987. *Climate and Plant Distribution.* Cambridge Univ. Press. Cambridge. 174 pp.

LIGHT

The ultimate source of energy for the ecosphere and its biota is light. Light provides the energy used in **photosynthesis** and the signals used in the photoregulation of plant growth and development. Photosynthesis is a process of energy transfer, carried out primarily in the chloroplasts of higher plants as indicated in this simplified equation:

$$CO_2 + H_2O \xrightarrow[\text{chloroplasts}]{\text{light}} (CH_2O)_n + O_2$$

It is a photochemical process whereby packages of light energy, called photons, excite electrons of molecules, thus causing them to change from a lower to higher energy state. Green plant chloroplasts have a reaction center containing two chlorophyll molecules. The chloroplast absorbs a photon of light, and in a trillionth of a second this causes an electron to be transferred out of the reaction center. This reaction leaves the system with enough energy to split water in the cell into oxygen, which is released into the air:

$$2\,H_2O \longrightarrow O_2 + H^+ + 4e^-$$

The hydrogen ions and the electrons released in the cell then bind to carbon dioxide to form sugar molecules, the source of energy used in **respiration**, which is the process responsible for tissue biosynthesis and maintenance (Chapter 18). In this way, light energy is converted to and stored as chemical energy in photosynthetic products. Thus photosynthesis is the initial, basic step in the transfer of energy from light to the food chains of the Ecosphere.

Besides energy, light also provides signals that enable plants to respond and adapt to changes in environment factors. Plants of forest ecosystems, as well as animals, fungi, and microorganisms, have photosensory systems that acquire and process information about light direction, duration, intensity, and spectral quality. In higher plants, photoregulation has been demonstrated for all phases of the plant life cycle, from seed germination and seed formation to the genetic adaptation of populations to their site conditions (Chapters 4 and 5). Preparation for dormancy is initiated by seasonal changes in light quality, and daily changes in light conditions elicit leaf orientation and chloroplast distributions with leaf cells.

In this chapter, we examine the distribution of light at the Earth's surface, the interception of light by plant canopies and its use in photosynthesis, and the effects of light on tree growth and leaf morphology. In Chapter 18, we focus on the ecological aspects of carbon fixation and the allocation of photosynthate to growth, maintenance, storage, and defense, processes that have important implications for the cycling and storage of carbon in forest ecosystems.

DISTRIBUTION OF LIGHT REACHING THE ECOSPHERE

Recall from Chapter 7 that clouds, smoke, ozone, smog, and greenhouse gases may intercept or reflect most if not all of the direct solar radiation. Values obtained for radiation on overcast days commonly range from one-quarter to one-half of those for cloudless days under the same conditions. Most of the radiation under such overcast conditions is scattered radiation—solar radiation that has been scattered by the atmosphere and other interceptors, and then reaches the ground more or less circuitously. As the sun departs from the zenith and the radiation is slanted through the atmosphere, the total amount is even further diminished by absorption and scattering.

A distinction must be made at this point between total incident solar radiation per unit surface area and radiation in the visible range, or **light** (Figure 7.1). Many past ecological studies have been based on measurements that represent the stimulation of the human eye by radiant energy. This measure of radiation is called **illuminance.** Illuminance is commonly expressed in the English system as foot-candles and in the metric system as luxes. Radiometric methods measure energy in Joules (J), or when integrated over time, **Watts** (1 W = 1 J s^{-1}). Today, quantum methods are used to measure the flux density or number of photons (electromagnetic energy) received in the visible region of the spectrum. Photosynthetically active radiation (**PAR**) of the 0.4- to 0.7-micron waveband is measured using quantum or photon flux density (**PFD**) expressed in moles or Einsteins per square meter per unit time. PAR is a measure of the total number of photons within the visible spectrum striking a surface and is appropriate for the investigation of the effects of radiant energy on plants. The term flux is used in physics to denote the rate of flow of a substance expressed as a quantity per unit time. Flux density is the flux through a unit surface area measured either in Watts (radiation) or in moles or Einsteins (photons) per unit area.

Solar radiation reaching the Earth's surface is in the range of 800 to 1,200 W m^{-2} (Chapter 7), which applies to radiation received by a horizontal surface. The slope and aspect of the ground also affect the amount of radiation received. Radiation striking a surface at angles of incidence other than perpendicular to the plane of irradiation is distributed over a greater area (Figure 8.1). Thus the amount of energy received, the photon flux (moles or µE m^{-2} s^{-1}) is less. This relationship is formalized by Lanbert's cosine law: irradiation of a surface varies with the cosine of the angle of incidence (Figure 8.1). In the Northern Hemisphere, south-facing slopes receive more radiation per unit area than north-facing slopes, the highest amounts being received by slopes most nearly facing the sun at its highest elevation during the day. The relative irradiance on a slope of any given percent and aspect may be approximated from a standard formula. The effects of these factors are

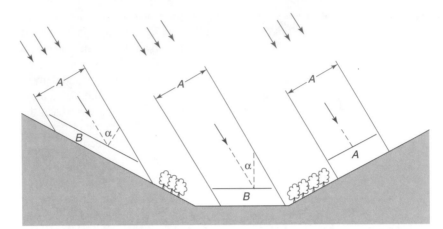

Figure 8.1. Distribution of insolation at different aspects and slopes. Photon flux changes according to the cosine of the angle of incidence ($\propto$). Angle of incidence varies with the orientation of the receiving surface as well as with the position of the light source. Energy received at B = photon flux (moles m^{-2} s^{-1}) of $A\cos\propto$. (After Hart, J. W. 1988. Reprinted from *Light and Plant Growth,* figure 3.3, p. 38. © J. W. Hart 1988. Reprinted with permission of Chapman & Hall.)

illustrated in Figure 8.2 by curves for north-, south-, and east-facing 100 percent (45°) slopes at north latitude 35° 30' on June 21. They show lower maximum irradiance levels on north slopes as opposed to south, a morning maximum for east and an afternoon maximum for west slopes (Figure 8.2). Tables of direct solar radiation for various slopes and latitudes are given by Buffo et al. (1972).

INTERCEPTION OF RADIATION

Plant foliage is well adapted to utilize solar radiation through its absorptance, transmittance, and reflectance. These properties are remarkably similar for many species (Reifsnyder,

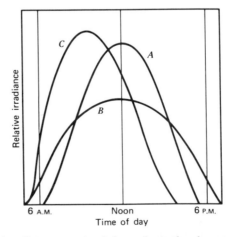

Figure 8.2. Relative irradiance received throughout the day on a 100-percent slope in the southern Appalachians. (*A*), South; (*B*), North; (*C*), East. The west slope curve is a mirror image of that for the east slope; the maximum irradiance occurs in the afternoon. (After Bryam and Jemison, 1943.)

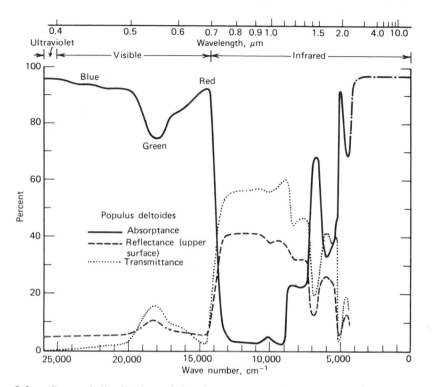

Figure 8.3. Spectral distribution of the absorptance, reflectance, and transmittance of a leaf of eastern cottonwood, *Populus deltoides*. (After Gates, 1968. Reprinted with permission from *Biometeorology,* © 1968 by the Oregon State University Press.)

1967) and are illustrated for a leaf of eastern cottonwood in Figure 8.3. The curves indicate high absorptivity in the ultraviolet and the visible range, strikingly low absorptivity in the near infrared (the 0.7–1.5-μm range, high in energy—see Figure 7.1), and high absorptivity of far-infrared radiation where solar irradiance is very low (Gates, 1968). About 4 to 5 percent of gross solar energy is absorbed and converted to chemical energy. However, only 1 to 2 percent of total solar energy is used in photosynthesis, a remarkably small amount considering the amount of plant biomass produced globally on an annual basis (see Chapter 18).

The effect of the forest itself in intercepting radiation is quite obvious. Only a small percentage of the incident sunlight reaches the floor of a dense forest. Many determinations have been made of the **relative illumination** (RI) within the forest, expressed as a percentage of total solar illumination[1]. Under leafless deciduous trees, the relative illumination may be as high as 50 to 80 percent of full sunlight; under open, even-aged pine stands, 10 to 15 percent represents a common range; under temperate hardwoods in foliage, values from <1 to 5 percent are common. In a dense beech forest in southeastern Michigan, PAR values are as low as 0.2 to 0.4 percent of that in full sunlight. Beneath the tropical rain forest (Carter, 1934; Huber, 1978), relative illumination may be as low as 0.1 to 2 percent. Among the densest temperate conifer forests are those formed by pure stands of Norway spruce. Under high closed spruce canopies in Switzerland, total solar radiation may average only 2.5 percent of that in the open (Vézina, 1961).

---

[1]Ecologists have used relative illumination either to express the relative amount of radiation received in the visible spectral range (= **light irradiance**, measured in Watts m^{-2}), the visual response of radiation (= **illuminance**, measured in luxes or foot candles), or PAR (measured in μE m^{-2} s^{-1}).

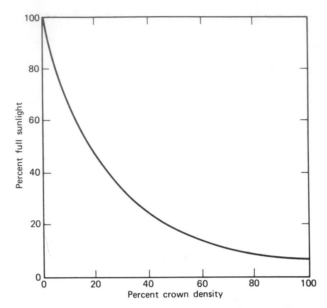

Figure 8.4. Effect of crown density on penetration of light irradiance into conifers in the California Sierra Nevada Mountains during spring snow-melt conditions. (After U.S. Army Corps of Engineers, 1956.)

The general relationship between percent forest cover and percent light transmission, based upon conditions existing under conifers at high elevations in the Sierra Nevada of California, is shown in Figure 8.4. It indicates a marked dropping off of relative illumination as crown density increases from zero to about 36 percent (U.S. Corps of Engineers, 1956). Thereafter, the decrease is moderate. In this particular example, the relative illumination was 18 percent for 50 percent density and only 6 percent for full crown density. Similar, strong inverse relationships between increasing stand basal area[2] and relative illumination are reported for western white pine in Idaho (Wellner, 1948) and shortleaf pine in Georgia (Jackson and Harper, 1955). It is interesting to note that the shape of this generalized curve varies among different types of tree canopies.

Canopy Structure and Leaf Area

The simplified relationship of light availability to crown density shown in Figure 8.4 provides a general basis for illustrating the importance of ecosystem-specific factors of canopy structure and leaf area that influence light absorption and hence forest composition and productivity. The canopy structure in the vertical plane of forest ecosystems directly reflects the aboveground architecture of individual species (Chapter 6) and their ability to tolerate overstory shade. In some wet tropical forests, the crowns of trees and shrubs occupy virtually all vertical space, extending from the tallest overstory tree down to herbaceous plants occupying the forest floor. The widely different canopy architecture of topical trees, in combination with a range of shade tolerance, gives rise to the multi-layered canopy structure illustrated schematically in Figure 8.5*a* (see also Figure 8.13 and 22.3). In other forest ecosystems, canopy structure can be much less complex. In temperate trembling aspen or paper birch forests, for instance, overstory trees form a single-lay-

[2]Basal area is the cross-sectional area of a tree at 1.3 m above the ground. It may be expressed on a per-tree basis or summed for all trees giving an amount per unit area, for example, 80 m^2 ha^{-1}.

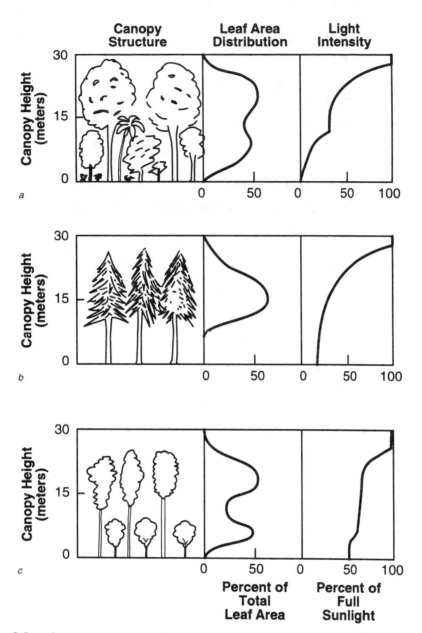

Figure 8.5. Canopy structure, leaf area distribution, and light intensity in a wet tropical (*a*), western conifer (*b*), and a trembling aspen ecosystem (*c*). Notice the relationship between canopy structure and the distribution of leaf area within each forest. Differences in the vertical distribution of leaf area result in dramatically different light profiles within these ecosystems.

ered canopy consisting of sparsely foliated crowns, which allow ample light to reach the well-developed shrub and ground flora (Figure 8.5*c*). The tall-stature coniferous forests of the Pacific Northwest may also have a single canopy layer, but their vertical canopy depth is much different from the aforementioned example. The crowns of conifers in this region extend deeply toward the forest floor (Figure 8.5*b*) and extinguish light to a much greater extent than aspen or birch forests. However, canopy depth varies substantially

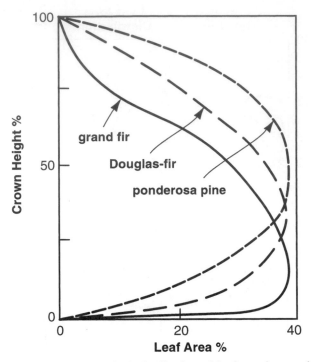

Figure 8.6.   Differences in the vertical distribution of leaf area in grand fir, Douglas-fir, and ponderosa pine. The crown of grand fir, a very shade-tolerant species, has a deeper vertical distribution of leaves, compared to the shade-intolerant ponderosa pine.

among shade intolerant, midtolerant, and tolerant coniferous in the Pacific Northwest. Look at Figure 8.6 and notice that the canopy of grand fir, a very shade-tolerant conifer, is much deeper than that of Douglas-fir (midtolerant) and ponderosa pine (intolerant). Shade tolerance allows grand fir to maintain photosynthetically-active branches at a much lower depth in the canopy.

Differences in canopy structure directly influence the amount of leaf surface area present to capture light energy, and forest ecologists use **leaf area index** (m^2 m^{-2}) to quantify differences in leaf surface area among forest ecosystems. This measure expresses the area of leaves (m^2) displayed over a unit of ground surface (m^{-2}). Consider the multi-layered tropical forest in Figure 8.5a, and notice that the vertical distribution of leaf area causes sunlight to be rapidly extinguished as it passes through the canopy. In contrast, the sparse, single-layered overstory canopy of the aspen forest (Figure 8.5c) extinguishes light to a lesser extent.

Leaf area indexes have been determined for a wide range of temperate forest ecosystems; they are typically greater than 5 m^2 m^{-2} during most (deciduous forest) or all (coniferous forest) of the growing season. In the Pacific Northwest, leaf area indexes can vary from 1 m^2 m^{-2} in sparse western juniper ecosystems to 23 m^2 m^{-2} in coastal ecosystems dominated by western hemlock (Gholz, 1982). These values are based on the upper surface area of leaves, but two-sided leaf areas have been calculated for some forests. In broadleaf trees, total leaf area is twice the one-sided value, whereas total leaf area is 2.5 times the single-sided value for trees with needle-shaped leaves (Waring, 1983).

The leaf area index is a particularly important ecosystem characteristic, because it is a direct measure of the photosynthetically-active surface area which can convert light en-

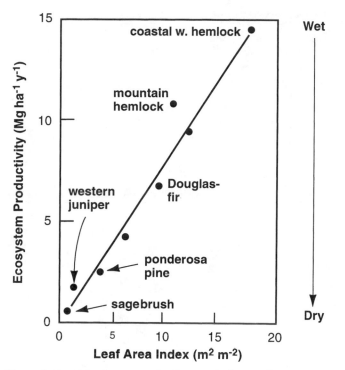

Figure 8.7. The relationship between leaf area index and ecosystem productivity among selected forest ecosystems of the Pacific Northwest. Productivity is expressed in megagrams of carbon per hectare per year.

ergy into plant biomass. It should not be surprising that there is a strong relationship between leaf area index and productivity in many terrestrial ecosystems (Figure 8.7). In the dry climates east of the Cascade Mountains, for example, low amounts of precipitation constrain both leaf area and ecosystem productivity (i.e., Rocky Mountain juniper). Greater amounts of precipitation falling on the Pacific Northwest coast allow forest trees to attain much greater amounts of leaf area (e.g., western hemlock), which result in higher leaf area indexes and greater rates of ecosystem productivity. Realize that climate can place bounds on leaf area by limiting the amount of water available for transpiration, but the canopy architecture and shade tolerance of individual trees control how the leaf area is displayed in vertical space.

Sun Flecks

In many forest ecosystems where soil water is not limiting, in natural stands and plantations, much of the forest floor is in heavy shade most of the time. However, **sunflecks** sweep over the ground as the day progresses, bathing areas momentarily with radiation perhaps half as much as sunlight in the open (Figure 8.8). Canham et al. (1990) studied light transmission beneath closed forest canopies at several temperate sites and one tropical rain forest site. They reported that the proportion of total-growing-season-PAR received as sunflecks ranged from 47 to 68 percent of the total amount of light penetrating through canopy openings. However, sunfleck duration was brief, on average only 5.7 to 7.1 minutes. Under such conditions, it is likely that photosynthesis may occur primarily during the times that a leaf is in the sunfleck, especially if they are of long duration or

Figure 8.8.   Sunlight penetrates the dense redwood forest canopy and sunflecks sweep over the ground, creating an important source of radiation for understory vegetation. (U.S. Forest Service photo.)

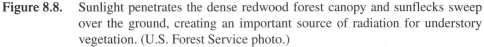

occur in rapid succession. Two species native to deeply shaded rain forests of Australia, the shrub *Alocasia macrorrhiza* and the tree *Toona australis* also can use brief periods of high irradiance as effectively as lower energy levels (Chazdon and Pearcy, 1986a, 1986b). Some tree seedlings, such as sugar maple and white ash that open stomates rapidly and exhibit slow stomatal closure with increasing light, may be able to make use of sunflecks (Davies and Kozlowski, 1974). The efficiency in response to rapidly changing light conditions may explain in part their ability to endure shaded conditions of the forest floor. However, sunflecks represent marked increases of energy, and plants may not always be able to use them. Plants may require several minutes or more to adjust their photosynthetic mechanism to adapt to this new high radiant energy flux (Fitter and Hay, 1987).

In contrast to rain forests or mesic sites where seedling establishment may be favored, sunflecks in drought-prone ecosystems may actually lead to patchy distribution of seedlings, establishment failure, and mortality. Such is apparently the case for red fir seedlings on south slopes in upper montane forests of the central Sierra Nevada Mountains of California (Ustin et al., 1984). Seedling distribution and density were compared on south and north slopes. The high irradiance on open south slopes is apparently inhibitory to red fir seedlings compared with the low-intensity irradiance on more shaded south slope sites and on all north slope sites. On the south aspect, plots with few seedlings had mean daily irradiance 2 times higher than plots with many seedlings, primarily due to 3.5 times more frequent occurrence of sunflecks at irradiance levels above 1,025 µE m^{-2} s^{-1}. Photon flux densities >500 µE m^{-2} s^{-1} exceed the light saturation point for net photosynthesis and substantially increase the energy load (heat) without corresponding increases in carbon gain. The relative absence of seedlings on high-light areas was attributed to temperature and water stresses induced by higher irradiance early in the growing season. Severe effects of high irradiance in high elevations on seedling establishment of Engelmann spruce and the importance of shading has been documented by Nobel and Alexander (1977).

LIGHT QUALITY BENEATH THE FOREST CANOPY

Leaves of the canopy transmit from 10 to 25 percent of the visible radiation they receive. The quality of radiation reaching the understory depends on the optical properties of leaves as well as incoming direct and scattered light penetrating openings in the canopy.

The broader transmission ability of conifers compared to the more selective absorption and transmission by deciduous canopies has been observed by many workers. In comparisons of light beneath sugar maple and red pine canopies, Vézina and Boulter (1966) found far-red radiation to predominate in the sugar maple understory on a clear day, with lesser amounts of green, blue, and red light and ultraviolet radiation (Figure 8.9). Other investigators also report a lower red to far-red ratio (R:FR) in shade light (Smith, 1981; Messier and Bellefleur, 1988). Sunflecks were strikingly evident in the morning and early afternoon in Figure 8.9, and during these periods red light increases sharply. The red pine canopy was much less selective in transmission than sugar maple foliage, and for red pine no significant differences were found among the transmission values of the various wavelengths in the visible spectrum. On cloudy or overcast days, the relative percent of radiation transmitted by forest canopies is higher in general than on clear days, and the sugar maple canopy was not as selective as it was on clear days. Similar relationships for conifer and deciduous species and for cloudy and clear days were reported by Federer and Tanner (1966) in studies of scattered shade light. Under sugar maple, clearly defined minimums were found at 0.47 and 0.67 µm (probably resulting from high absorption by chlorophyll) on clear days.

Although light quality is important, quantitative differences in light intensity during the growing season, even in the forest understory, are primarily responsible for plant response. Light used by understory plants comes from two sources, sunflecks and diffuse shade light filtered through the canopy. Sunflecks are qualitatively similar to light at the top of the forest canopy, but shade light is characterized by a lower red to far red ratio. Tropical shade tolerant species have not been found to respond to changes in light quality (Kwesiga and Grace, 1986; Riddoch et al., 1991).

LIGHT AND GROWTH OF TREES

The most obvious importance of solar radiation to forest trees lies in the dependence of growth and development upon photosynthesis and the dependence of photosynthesis in turn on light. The term **light** is used here, because it is quite well established that it is the solar radiation in the visible bands of the spectrum that affects the photosynthetic process. We use **light irradiance** to express the amount of radiation received per unit area in the visible spectral band.

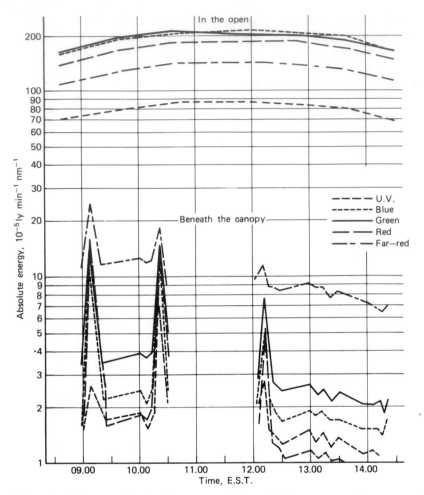

Figure 8.9. Radiation quality in the open and beneath the sugar maple canopy. (After Vézina and Boulter, 1966. Reproduced with permission of National Research Council Canada.)

Because chlorophyll is green, it follows that green foliage reflects a higher percentage of the green wavelengths than the blue-violet or the longer yellow-red wavelengths. Such indeed is the case (Figure 8.3). A corollary is that the blue-violet and the yellow-red wavelengths, being absorbed by the plant instead of being reflected, would exert a relatively greater influence on photosynthesis. This concept is confirmed in a classic study by Hoover (1937) with wheat, in which he found that the effectiveness of solar radiation on photosynthesis was almost entirely confined to the visible spectrum and that the most effective bands were the violet-blue and the orange-red. Similar relationships apparently hold for most plants and forest trees, although Burns (1942) found that, for Norway spruce, eastern white pine, and red pine, the orange-red wavelengths were most effective in stimulating photosynthesis, whereas the shorter violet-blue waves were relatively unimportant, a finding contrary to the earlier study of wheat by Hoover. Linder (1971), working with Scots pine seedlings, found absorption maxima in the red and the blue regions, but the photosynthetic efficiency of blue light was less than that of red light. In general, absorption in both regions is needed for maximum efficiency.

Since the growth rate of trees is closely related to their rate of photosynthesis, and since the rate of photosynthesis can be measured under controlled conditions by measuring the uptake of CO_2 from the air by plant leaves, there have been many studies of the effect of varying amounts of light upon the rate of photosynthesis. At very low irradiance levels, photosynthesis takes place at rates lower than plant respiration, a process that returns CO_2 to the atmosphere. Under such conditions, CO_2 is actually given off by the plant rather than being absorbed by it from the atmosphere, i.e., a net loss of CO_2 occurs. The **light compensation point** is the light intensity at which carbon gain from photosynthesis and carbon loss from respiration are equal. In other words, CO_2 is neither given off nor taken up.

For forest trees under otherwise optimal conditions, the light compensation point appears to occur at from 1 to 2 percent of full sunlight (relative illumination = RI). For example, Grasovsky (1929) found that this point for well-watered eastern white pine occurred at approximately 1.7 percent RI. It thus appears that, in very dense forests, **net photosynthesis** (i.e., photosynthesis greater than dark respiration) occurs only when the leaves are bathed in sunflecks and where plants are specially adapted to trap diffuse light of the forest understory (Huber, 1978). In a test of 14 species, Burns (1923) found that ponderosa pine and Scots pine required light irradiances at a compensation point three times as high as that of eastern white pine. At the other extreme, eastern hemlock, American beech, and sugar maple required the least amount of light to maintain themselves. Similar results were obtained for 12 species in the West, with redwood and Engelmann spruce having the lowest light requirements, and limber and pinyon pines the highest (Bates and Roeser, 1928). For shade and sun plants in general, approximately 0.5 to 1.5 percent RI, respectively, are required at the compensation point under natural conditions (Perry et al., 1969).

Plants may live for a considerable length of time on stored carbohydrates after environmental conditions develop to the point at which respiration exceeds photosynthesis. For example, Grasovsky (1929) found that eastern white pine seedlings remained in good condition throughout the growing season under 0.6 percent RI, even though 1.7 percent RI was seemingly required at the compensation point. Sooner or later, though, such plants are bound to die. In studies with seedlings of several species, Grime (1966) observed the greatest height growth and least mortality in shade for species having large seeds. Mortality of black birch seedlings (0.7 mg dry weight of seed reserve) was four times that of northern red oak (1,969.2 mg dry weight of seed reserve). The occurrence of one- or even two-year old seedlings under dense forest stands does not demonstrate that light under those conditions is sufficient for growth; the seedlings may still be living on stored food material from seed.

As the irradiance is increased above the compensation-point, photosynthesis is increased proportionately. After reviewing the evidence of other workers, as well as carrying out tests of his own, Shirley (1945a) concluded that, in the range from 1 to 15 percent of full sunlight, photosynthesis is directly proportional to irradiance if other factors are favorable. The increase in photosynthesis will continue until other factors combine to bring growth to a halt. At very high irradiances, such factors as high respiration, water deficit causing stomatal closing, and over-accumulation of photosynthate in the leaves may result in decreased photosynthesis.

The dependence on net photosynthesis of leaves of a variety of plants is illustrated in Figure 8.10. Uptake of CO_2 initially increases in proportion to light intensity but continued increase varies from one organism to another. At very high light intensity, photosynthesis continues to increase only slightly or not at all, as seen for European beech (Figure

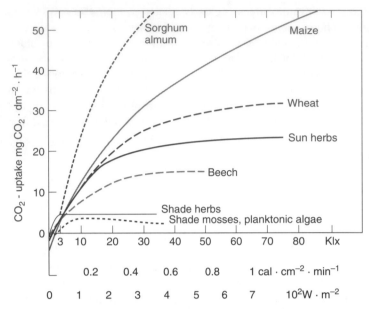

Figure 8.10. Light-dependence of net photosynthesis in various plants at optimal temperature and with a natural supply of CO_2. (After Larcher, 1980. Reprinted from *Physiological Plant Ecology* 2nd ed., © 1980 Springer-Verlag Berlin. Reprinted with the permission of Springer-Verlag.)

8.10). At about 35 to 40 klux for beech the photosynthesis reaction is **light saturated**, and the curve shows no further increase of CO_2 absorption. At this point, the reaction is not light-limited but is affected by the supply of CO_2, PO_4^{3-}, and enzymatic processes. In contrast, for a high light-requiring plant such as maize the saturation curve continues to increase exponentially even at 80 klux.

In very low light intensities, the curves for European beech in Figure 8.11 show that more CO_2 is given off by respiration than is fixed by photosynthesis (rates of net photosynthesis < 0). At somewhat greater intensities the light compensation point is reached. At this **compensation light intensity** (I_K) exactly as much CO_2 is fixed in photosynthesis as is given off in respiration. Note in Figure 8.11 that leaves adapted to shaded portions of a beech tree exhibit a lower light saturation point (I_S in upper diagram). In addition, they respire at a lower rate than those adapted to high light (sun leaves) and reach the light compensation point (I_K in lower diagram) at a lower intensity.

The relative effect of light upon photosynthesis in different species has been studied by a number of investigators. For example, both Kramer and Decker (1944) and Kozlowski (1949) have shown that seedlings of loblolly pine show increased photosynthesis for increased illuminance at all levels up to full sunlight, whereas associated hardwood species reach their maximum photosynthetic rate at 30 percent or less relative illumination. Such a finding explains in part the better relative competitive ability of the hardwoods under open pine stands where the relative illumination is commonly in the neighborhood of 30 percent full sunlight (Figure 8.12).

Other factors affecting the relative ability of different species to compete under given light conditions are the color, shape, presence of waxes and hair, and the arrangement of leaves—factors that regulate the amount of light actually reaching the chlorophyll. Pines, for instance, characteristically have rounded needles in dense clusters. Such an arrangement results in much scattering of light and much mutual shading (Kramer and Clark,

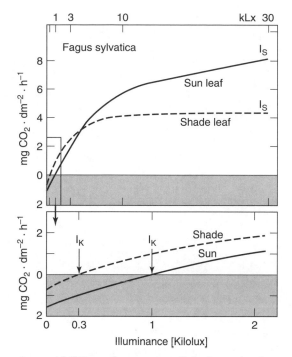

Figure 8.11. Dependence of CO_2 exchange upon light intensity in sun and shade leaves of European beech. Measurements were made at 30 °C. The region of low light (the section enclosed in the box in the upper diagram) is shown in the lower drawing with the abscissa expanded. I_S = light saturation point; I_K = compensation light intensity. (After Larcher, 1995. Reprinted from *Physiological Plant Ecology*, 3rd ed., © 1995 Springer-Verlag Berlin. Reprinted with the permission of Springer-Verlag.)

1947). Many hardwoods associated with pines in the eastern United States, on the other hand, have broad, thin leaves, arranged perpendicular to the direction of incident radiation and spaced by their branching habit so that shading of one leaf by another is at a minimum. It follows that the pine chlorophyll receives a much smaller fraction of the light incident to it than does the dogwood, to cite a specific example. This is at least one of the reasons why pine can maintain a high net photosynthetic rate at high light irradiances, which actually inhibit the growth of dogwood. However, the dogwood can carry on photosynthesis efficiently in low light conditions under a loblolly pine stand.

Actually, conifers and evergreens in general have a relatively low photosynthetic capacity per unit area of foliage throughout their wide range of distribution (Waring, 1991). From the tropics to the boreal forest their maximum photosynthetic capacities are usually half to less than a quarter of that of associated deciduous species. However, in many environments they develop a canopy of leaves dense enough to absorb nearly all the incident radiation in the visible range (Jarvis and Leverenz, 1983). Once a dense, closed-canopy stand of conifers dominate an area, as after fire, co-occurring deciduous species, except for the most shade tolerant ones, are put at a disadvantage.

Branching habit and growth rate indirectly affect light capture. Tree species exhibit multiple layers of foliated shoots in both vertical and horizontal planes of the live crown. Branches of fast-growing shade intolerant species are more widely spaced than those of shade-tolerant species whose shade leaves accommodate to the light climate of the inner

Figure 8.12.   A conspicuous hardwood understory in an open 80-year-old loblolly pine
stand, North Carolina. (After Roth, 1990. Reprinted with permission of
Gebrüder Borntraeger Verlagsbuchhandlung, Stuttgart, Germany.)

crown by anatomical features (see following discussion). Many floodplain species (willows, cottonwoods, silver maple, American elm) have broad and often wide-spreading
crowns that facilitate light capture by foliage of the external and internal crown.

The orientation of the leaf blade, both its angle and azimuth, affects three separate
aspects of solar radiation interception: daily integrated radiation, peak instantaneous irradiance, and diurnal distribution of instantaneous incident irradiance (Ehleringer and Werk,
1986). In many deciduous species, leaf petioles act to orient the leaves to either increase
absorption of radiation or avoid it. Positioning the leaf perpendicular to incoming radiation maximizes the amount of light absorbed and heat received. The horizontally flattened
petioles of aspen species around the world (trembling aspen, bigtooth aspen, European
and Asian aspens) enable the leaf blade to move even in slight breezes. Such leaf movement permits light to bathe leaves more effectively than if they were stationary, especially
in the interior crown. Also, their movement removes the moist boundary air layer surrounding the stomata, thus increasing transpiration (Kirchner et al., 1927).

The failure of seedlings in shade is almost invariably associated with fungal attack
(Grime, 1966). Compared to tree species intolerant of shade, shade-tolerant species are
less susceptible to infection both above and below the compensation point (Grime and
Jeffrey, 1965). The predisposition of shade-intolerant species to fungal attack may be due

to characteristics arising from their adaptation for survival in habitats of high light irradiance. These characteristics are (1) attenuation of the stem and mechanical collapse when shaded, (2) inherently high rates of respiration, and (3) marked rise in respiration with increasing temperature (Grime, 1966; see Chapter 18).

The results of studies detailed above have dealt with the rate of photosynthesis as measured by the uptake of carbon dioxide from the atmosphere. It is also possible, of course, to measure the actual height or weight growth of plants over a period of years under different degrees of irradiance. This approach has been followed by a number of investigators, who, for the most part, have reduced the possible effect of soil moisture by watering the plants grown under all irradiances to keep the soil at a more or less optimum soil-moisture level.

In general, at least 20 percent of full sunlight is required for survival over a period of years, the actual amount varying with the species and with growing conditions. For species that are usually found growing in full sunlight, growth commonly increases with irradiance up to full sunlight. Pearson (1936, 1940) grew ponderosa pine under various shade conditions in Arizona for periods up to 10 years. He found the best growth under full sunlight, although the trees grown at 50 percent showed only slightly less height growth and about one-half the diameter growth. Shirley (1945b) and Logan (1966a) obtained similar results for red pine and jack pine.

Many other species, however, make as much or more height growth under partial light as under full light. Among conifers, white spruce, white pine, and Douglas-fir provide examples (Gustafson, 1943; Shirley, 1945b; Logan, 1966a; Lassoie, 1982). Among hardwoods, this relationship seems to be the rule. Seedling height growth data for six hardwood species in Missouri (McDermott, 1954) showed that five of the six grew tallest in either 50 or 33 percent relative illumination (Table 8.1). For five- to six-year-old hardwood seedlings in Ontario, Canada, Logan (1965, 1966b) reported that the maximum height of white birch, yellow birch, silver maple, and American elm was at 45 percent RI, whereas it was at 25 percent for basswood and sugar maple (both highly shade-tolerant). Few species, conifers or hardwoods, however, make better weight growth under partial light than under full light because of major allocations to the roots (Chapter 18).

Regardless of whether or not height growth is increased or decreased by shaded conditions, there seems little doubt that root development of seedlings is sharply impaired. The lower the irradiance, the greater is this impairment, so that, at low irradiance levels, seedlings of most forest trees have relatively shallow and poorly developed root systems. The data of McDermott (Table 8.1) serve to illustrate this point. American elm may make the greatest height growth at 33 percent RI, but the top is 3.2 times as heavy as the roots under these conditions. The shorter elms that develop under full sunlight have a relatively better-developed root system, the top being only 1.7 times as heavy. Similar results have been reported for many other North American species (Baker, 1945; Shirley, 1945b; Logan, 1965, 1966b) and for European species (Harley, 1939; Brown, 1955; Leibundgut and Heller, 1960). A detailed consideration of carbon allocation to shoots and roots is presented in Chapter 18.

The marked effect of reduced light in reducing root growth is of major importance in explaining the growth and survival of plants in the understory. Under a growing forest, root competition substantially reduces the amount of soil water and nutrients available to the seedlings. The combination of reduced root size due to low light irradiances and reduced soil water due to root competition is frequently fatal to seedlings.

The fact that, under the forest canopy, both light and soil water are reduced makes meaningless any attempt to relate relative-light irradiance to survival and growth of seedlings under uncontrolled conditions. Curves showing the effect of light irradiance

Table 8.1   Seedling Height (HT) and Top/Root Weight Ratio (T/R) of Newly Germinated Seedlings as Influenced by Amount of Sunlight

SPECIES	20% RI		33% RI		50% RI		100% RI	
	HT(cm)	T/R	HT(cm)	T/R	HT(cm)	T/R	HT(cm)	T/R
American elm								
(*U. americana*)								
13 weeks	73	3.1	81	3.2	69	3.0	37	1.7
Winged elm								
(*U. alata*)								
13 weeks	59	5.2	71	3.9	76	4.4	28	1.9
Sycamore (*P. occidentalis*)								
15 weeks	43	4.6	41	3.5	40	4.1	33	2.0
River birch								
(*B. nigra*)								
10 weeks	26	13.4	33	9.3	52	8.3	41	3.6
Red maple								
(*A. rubrum*)								
13 weeks	26	4.1	27	3.7	29	3.5	26	2.0
Alder								
(*A. rugosa*)								
14 weeks	21	3.4	20	3.5	38	4.7	30	2.8

Source: After McDermott, 1954.

upon seedling development often ignore the existence of variations in soil water, nutrients, and other environmental factors which also vary with different canopy densities.

In studies where establishment and growth involve several environmental factors, compensating effects may be expected. In a study of yellow birch regeneration, Tubbs (1969) employed 64 different combinations of three factors—light irradiance, soil water, and seedbed type—to assess germination, survival, and height growth. On well-drained sites good germination occurred under full sunlight on mineral soil, but on heavy and organic soils heavy shade was required. On mineral soil, height growth was best in full sunlight on the well-drained site, whereas shade was necessary for maximum growth on the somewhat poorly drained site. Although partial shade has been found generally to promote better seedling development of yellow birch, full sunlight may be required under certain conditions. The study also demonstrated the remarkable phenotypic variation of the seedlings, all from one mother tree, in germinating and surviving over a wide range of environmental conditions.

LIGHT AND TREE MORPHOLOGY AND ANATOMY

Plants grown under shade develop a structure and appearance different from the same plants grown under full sunlight. These morphological changes are of ecological importance in understanding the capacity of a given species to become adjusted to shaded conditions and the reaction of such a plant when suddenly released, following windstorm or by cutting of the overstory.

Since the leaf is the principal photosynthetic organ of the tree and therefore presumably most affected by changes in radiation and light, it has been the subject of many detailed investigations. The structure of leaves of typical understory plants has been compared with those of typical overstory plants; sun leaves have been compared anatomically with shade leaves of the same plant. The findings are in essential agreement (Büsgen and Münch, 1929; Roth, 1984).

Some species exhibit little plasticity in leaf anatomy with the result that there is little difference between their sun leaves and their shade leaves. Such plants are usually found only in the overstory or only in the understory—their anatomy is suited for their survival under only one set of environmental conditions. They are either obligate sun plants or obligate shade plants.

Most forest trees, however, have the faculty of developing different anatomical structures in leaves grown in the shade than in those developed in the sun (Hanson, 1917; Büsgen and Münch, 1929; Jackson, 1967; Roth 1984). Typically, shade leaves are thinner and less deeply lobed. They have a larger surface per unit weight, a thinner epidermis, less palisade, more intercellular space and spongy parenchyma, less supportive and conductive tissue, and fewer stomata than comparable sun leaves off the same tree (see Figure 8.15). Similar differences in leaf structure are found between species that characteristically grow in shaded conditions and those that grow under fully exposed conditions.

The typical shade leaf is adjusted to carry on photosynthesis when efficiently protected from the detrimental effects of too much light. Furthermore, shade leaves require less light at the compensation point, so they typically survive and show net photosynthesis with very little light.

Despite the assumption that shade leaves develop in response to reduced light, it must be remembered that other factors also are involved. Temperature is obviously different in sun and shade situations. Furthermore, leaves within the crown and in its lower portions are under markedly less water stress on clear days than are the sun leaves at the top of a tree. Thus we distinguish between sun versus shade leaves when discussing differences elicited by light, and contrast xerophytic versus hygroscopic leaves when considering differences in soil-water stress of the exposed upper canopy and the humid understory. Studies in tropical forests emphasize the extremes in adaptation of many different leaf characteristics to light and moisture conditions along a gradient from the top to the bottom of the forest.

Leaf Structure and Stratification in Tropical Forests

Nowhere are there more striking contrasts in leaf morphology and anatomy than in tropical forests. Monographic studies of tree leaf morphology and anatomy in humid tropical rain forests of Venezuelan Guiana (Roth, 1984; Rollet et al., 1990) and the montane cloud forest of Henri Pittier National Park of northern Venezuela (Roth, 1990) illustrate the remarkable adaptations of leaves in a vertical microclimatic gradient of over 40 meters.

Studies in the humid tropical rain forest were conducted in an area of about 156 ha, located south of the Orinoco River at an elevation of 150 to 550 m. The annual precipitation is 2,000 mm or more, and a dry season of varying length occurs between January and April. Trees, shrubs, and vines 10 cm dbh (diameter at breast height) and above were studied in three categories: above 30 m in height, 10 to 29 m, and less than 10 m. Studies were conducted on leaves of 232 species representing 48 families.

Similar studies were conducted in a montane cloud forest of approximately 950 to 1,400 m elevation in northern Venezuela near the Caribbean coastline (Roth, 1990). A cross section of the montane cloud forest in Figure 8.13 illustrates the enormous complexity of stratification and species composition. Twenty species were studied in each of four strata: an upper stratum of trees between 5 and 40 m and three strata from 0 to 5 m, the lowest of which included low herbs and seedlings of tree species between 0 and 1.30 m

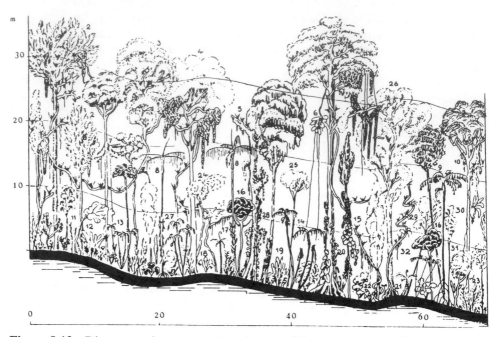

Figure 8.13. Diagrammatic cross section of a part of the montane cloud forest at Rancho Grande, northern Venezuela, near the Portachuelo Pass (950 m). Distances in meters. Numbers refer to species given in Roth (1990). (After Roth, 1990. Reprinted with permission of Gebrüder Borntraeger Verlagsbuchhandlung, Stuttgart, Germany.)

(Roth, 1990). Special emphasis was placed on plants of the lower layers in this study in contrast to the study in the low-elevation humid rainforest. This Rancho Grande area is one of the richest if not the richest in plant species in the world (903 species of 387 genera belonging to 121 families). It has very good physiographic and soil conditions and a most favorable climate in temperature and humidity.

The cloud or mist forest is a montane rain forest characterized by the regular occurrence of orographic mist. Mist affects entire ecosystems by increasing air humidity and simultaneously reducing light intensity. Orographic mists are different from other mists or fog by their regular presence at more or less constant altitudes throughout the year and by their higher moisture content. Due to mist formation the diffuse light in the forest, and especially in the understory and ground-cover layers, has a great influence on the photosynthetic system of plants.

The combined results from the humid and cloud forest provide an excellent overview of leaf response to changes primarily in light intensity and moisture from bottom to top of complex forest systems. Figure 8.14 shows the progressive transformation of the hygromorphic shade leaf into the xeromorphic sun leaf that accompanies change in tree height and position along the microclimatic gradient. Roth (1984, p. 422) contrasts these two types of leaf structure in no less than 35 different characters! Together with this change in leaf structure along the vertical microclimatic gradient is the concomitant change in leaf structure along a temporal gradient of tens to hundreds of years from a young plant of a given species in the understory to an adult in the exposed canopy. The characteristics of the hygromorphic shade leaf of the juvenile disappear little by little while xeromorphic sun leaf features of the adult leaf develop. Leaves of a young plant one meter tall may differ so much from those of the fully grown tree that they appear to belong to two different

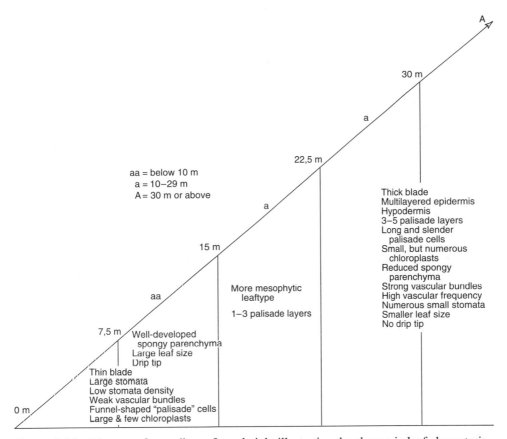

Figure 8.14. Diagram of a gradient of tree height illustrating the change in leaf characteristics from the hygromorphic shade leaf type of the shaded, humid understory to the xeromorphic sun leaf type of the drier, exposed upper canopy. (After Roth, 1984. Reprinted from *Stratificaiton of Tropical Forests as Seen in Leaf Structure* (Scheme #II, p. 452), © 1984 by Dr. W. Junk Publishers, The Hague, Reprinted with kind permission from Kluwer Academic Publishers.)

species. The time and rate of change for the many different characters vary for the different species.

Xerophytes are adapted to a dry climate and are almost always exposed to strong irradiance. Both these factors elicit certain structural responses in plant leaves. In xeromorphic plants, leaf surface is reduced; leaves are often coriaceous, leathery, and hard. Cell walls become thickened and a thicker cuticle develops. Stomatal density is higher and the stomata may become hidden in cavities, holes, or crypts. These and many other features characterize the xeromorphic leaf in response to soil-water stress and wind. Although the xeromorphic leaf type is usually related to the sun leaf type, they should not be confused with one another. The sun type develops independently from the xeromorphic type, according to the degree of irradiance that reaches the leaf. Sun leaves are mainly characterized by longer and narrower palisade cells; spongy parenchyma becomes reduced and intercellular spaces are always smaller. The number of chloroplasts increases considerably, while their size is very much reduced. However the well-developed palisade cells are a response to high light intensity. A cross section of a typical sun leaf of an adult tree of the upper canopy is illustrated in Figure 8.15*a*.

The corollary to xerophytes with sun leaves is hygromorphic plants with shade leaves that maximize trapping of light. These plants live in the humid understory and are

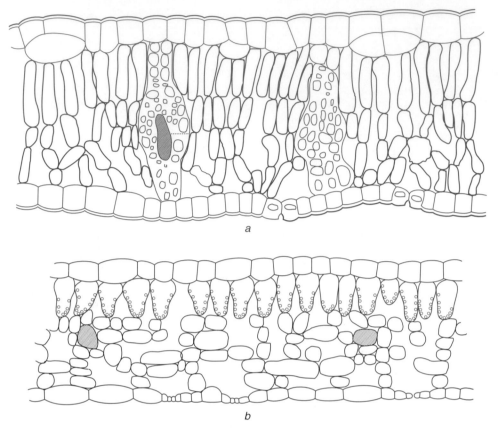

Figure 8.15. Leaf cross sections illustrating xeromorphic sun leaf (*a*) and a hygromorphic shade leaf (*b*), humid tropical rain forest. *a*, *Triplaris surinamensis,* tree of upper canopy (> 30 m); xeromorphic sun type leaf with transcurrent vascular bundles, a water-storing upper epidermis and leaf structure in which the entire mesophyll transforms into palisade parenchyema to adapt to strong insolation. *b*, *Hieronyma laxifolia*; one of the tallest trees of the upper canopy; leaf of 1-m-tall seedling shown here; typical hygromorphic shade-type leaf with funnel-shaped photosynthetic cells, loose spongy parenchyma, and weakly developed vascular bundles; the very large leaves of young plants are extremely well adapted to deep shade and high humidity. (After Roth, 1984. Reprinted from *Stratificaiton of Tropical Forests as Seen in Leaf Structure* (Fig. 44, p. 484), © 1984 by Dr. W. Junk Publishers, The Hague, Reprinted with kind permission from Kluwer Academic Publishers.)

characterized by traits relating to moisture adaptation, such as good aeration by a large intercellular system (Figure 8.15*b*). The humid environment also favors very thin cell walls, a reduced vascular system, lack or absence of hairs, and sparse, large stomata that may be elevated above the surface. In contrast to the sun leaf type, the shade leaf has much reduced palisade parenchyma that are often short and loose. Leaves of the lower strata characteristically have elongated narrow blades with drip tips, whereas broad leaf blades are characteristic of the uppermost layer. A high proportion of red light in the understory accentuates leaf blade elongation. Drip tips are common (rare in upper canopy trees) as they promote drying of the leaf surface; moisture films reduce light collection.

Very well adapted shade leaves even adopt a very peculiar shape of photosynthetic cells resembling that of a funnel (Figure 8.15*b*). In such cells, the large, convex, dark bluish-green chloroplasts are few in number. They lie at the bottom and along the oblique lateral walls of the funnel so that light capture for chloroplasts becomes more favorable and diffuse light penetrating from different directions may be better absorbed. The frequency of funnel-shaped cells is higher in the cloud forest where the presence of mist increases the effect of diffuse light. Not surprisingly, the light compensation point for these plants is very low. For example, Huber (1978) found that the values of the light compensation point of 54 vascular plants of the herb and shrub layer in the cloud forest ranged between 1.5 and 5 $\mu E \, m^{-2} \, sec^{-1}$ (0.07 to 0.23 percent of full sunlight).

The above examples from tropical forests illustrate the enormous variety of genetic adaptation and plasticity in numerous interrelated leaf characters to interconnected gradients of light, moisture, wind, and nutrient availability that are mediated by the biota themselves.

Adventitious Buds and Epicormic Sprouting

Much less work has been done on the effect of light upon bark thickness and the stimulation of adventitious buds. Adventitious buds in the bark may be activated to form **epicormic branches** when the bark is exposed to direct solar radiation. It does not follow, though, that there is a necessary cause-and-effect relationship between light and bud stimulation. The removal of surrounding trees, for example, will affect the environment of the remaining trees in many ways, and carefully controlled experimentation is required to determine which factor or combination of factors actually initiates the change of physiological circumstances that lead to the development of epicormic branches. It is known that the crown exerts a substantial control on sprouting; decapitation of the crown (Books and Tubbs, 1970) and pruning of live branches (Kormanik and Brown, 1969) stimulate sprouting. It is generally agreed that sprouting is triggered by a change in hormones affecting growth within the tree. All that can be deduced from field observations is that epicormic branches form on the exposed boles of many trees and especially those with weak, poorly developed crowns. European foresters have been able to reduce the amount of epicormic branching on oaks by maintaining vigorous crowns and keeping the bole clothed with understory trees such as beech and hornbeam. The occurrence of epicormic branches after thinning is particularly serious with many of the oaks, maples, birches, and other hardwoods where stem quality is of paramount economic importance. Under extreme conditions of change from shade to light, epicormic branches may develop in most forest trees, both conifers and hardwoods.

PHOTOCONTROL OF PLANT RESPONSE

The growth and development of all parts of the plant, including stem elongation, root development, dormancy, germination, flowering, and fruit development are subject to the same causal photocontrol, whereby light is absorbed by a reversible pigment system in the plant. This system is the physiological basis of photoperiodism, which has been repeatedly demonstrated to affect seasonal rhythm, timing, and amount of tree growth. Photoperiodism is the response of the plant to the relative length of the day and night and the changes in this relationship throughout the year.

The term photoperiodism was proposed by Garner and Allard (1920) and Garner, (1923), who carried on extensive tests that demonstrated that both vegetative development and the initiation of reproductive processes in plants could be greatly affected by varying the relative length of day and night. Although many plants were not found to be especially sensitive to day length, some could be characterized as short-day plants because flowering could be induced under suitable conditions by exposure to days shorter than a certain critical length. Some could be termed long-day plants since these responded to days longer than a given critical length or even to continuous illumination (Galston, 1994). Responses to differing day lengths are clearly different for different species.

As we have seen in Chapter 4, photoperiodic requirements are not constant within a species, but vary with the latitude and altitude of the source. Thus genecological differentiation is characteristic of many tree species, having arisen as a result of natural selection for a particular photoperiod associated with the limiting factors of the growing season.

When plants respond to light they do so because light is absorbed by the phytochrome pigment system of the plant (Taiz and Zeiger, 1991; Galston, 1994). The bluish protein phytochrome has been detected in all parts of higher plants and exists in two forms:

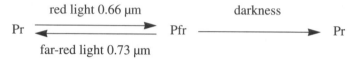

One form, **Pr**, has an absorption peak (0.66μm) in the red region of the spectrum; the other form, **Pfr**, is physiologically active, and has an absorption peak (0.73μm) in the far-red region. Absorption of light at the appropriate wavelength readily converts one form to the other; when the red-absorbing form receives red light, it is changed into the far-red-absorbing form (Pfr), and vice versa. As a result, growth responses elicited by red light are reversed by far-red light. Besides reversibility, the system is characterized by a slow drift in darkness from the far-red absorbing form to the red-absorbing form. The detection of phytochrome, its properties, and effects in morphogenesis of plants are reviewed by Vince-Prue (1975), Larcher (1980), Hart (1988), and Kozlowski and Pallardy (1997a, b).

Growth cessation under short-day conditions and continuous growth under long days may be explained by the phytochrome system. At the end of the daily light period, in which red light predominates, more than 70 percent of the pigment is in the Pfr form. During the dark period that follows, the pigment slowly reverts to the Pr form, and if the dark period is long enough less than 10 percent remains in the Pfr form (Downs, 1962). If the pigment remains in the Pr form for a substantial time during each dark period, woody plants cease growth and enter dormancy, because the amount of biologically active Pfr form is inadequate. Thus it is the length of the night, rather than that of the day, that initiates, through phytochrome, the biochemical reactions controlling growth. Under long days or continuous light, the pigment is in the Pfr form for an appreciable time, and growth is promoted. This physiological system is responsible for the development of a continuous sequence of photoperiodic races in the North Temperate Zone trees through their ranges (Chapter 4).

Unlike growth cessation, leaf flushing in the spring is not appreciably controlled by photoperiod. Once the plant's chilling requirement has been met, flushing is primarily dependent on temperature. The photoperiodic effect on the flowering of trees is not well understood. It has not been possible to establish clearly long-day and short-day flowering types as has been done in herbaceous species. Therefore, most woody plants are regarded as day-neutral.

LIGHT AND ECOSYSTEM CHANGE

Of all site factors affecting tree growth and compositional change over time, light is the one most strongly affected by vegetation. As landscape ecosystems at a given place change over time, and vegetation is one of the most visible and measurable expressions of this change. Vegetation change, often termed succession (Chapter 17), is the result of change in site factors (landform, soil properties, temperature, drainage, etc) and disturbance regimes. Light is also in this process, but fundamentally different from the others. We observe adjacent dry and wet sites, hot and cold sites, fertile and infertile sites—but not dark and light sites, except by night and day! The sun shines the same on adjacent dry and wet sites and infertile and fertile ones.

The vegetation itself modifies the site factors and disturbance regimes over time and, therefore, plays a key role in ecosystem change. Vegetation has relatively little effect on geologic change of landforms. It has a significant effect on soil development over hundreds to thousands of years. Vegetation, especially the dominant trees, however, may quickly (within decades) effect an enormous change in the light available in the forest community—whether the site is wet, infertile, hot, or any combination thereof. Therefore, the occurrence and species composition in the vertical layers of the forest below the dominant overstory (subdominant overstory, understory, shrub, and herbaceous layers) are, in large part, dependent upon light levels mediated by the density and crown architecture of species in the overstory canopy (Figures 8.5 and 8.13).

Over time, a typical pattern is expressed in many ecosystems with a fully-developed overstory canopy. Trees of greater shade tolerance than that of the overstory species tend to establish and grow-up in the shaded understory. Certainly, species composition and density in any ecosystem are strongly influenced by the interaction of many site factors as well as light. However, of all site factors affecting a given ecosystem the plants themselves have the strongest effect on light availability—a key factor in the dynamics of vegetation (see Chapter 17). Temperature is closely associated with light availability in forest ecosystems, and it is our focus in the next chapter.

Suggested Readings

Downs, R. J., and H. A. Borthwick. 1956. Effects of photoperiod on growth of trees. *Bot. Gaz.* 117:310–326.

Ehlcringcr, J. R., and K. S. Wcrk. 1986. Modifications of solar-radiation absorption patterns and implications for carbon gain at the leaf level. In *On the Economy of Plant Form and Function*, T. J. Givnish (ed.). Cambridge Univ. Press, Cambridge.

Fitter, A. H., and R. K. M. Hay. 1987. *Environmental Physiology of Plants*. Academic Press, New York. Chapter 2 (Energy and carbon).

Hart, J. W. 1988. *Light and Plant Growth*. Unwin Hyman, Boston, MA. 204 pp.

Kozlowski, T. T., and S. G. Pallardy. 1997. *Physiology of Woody Plants,* 2nd ed. Academic Press, San Diego, CA.

Kozlowski, T. T., and S. G. Pallardy. 1997. *Growth Control in Woody Plants* Academic Press, San Diego, CA. 641 pp.

Pearcy, R. W., and D. A. Sims. 1994. Photosynthetic acclimation to changing light environments: scaling from the leaf to the whole plant. *In* Caldwell, M. M., and R. W. Pearcy (eds.), *Exploitation of Environmental Heterogeneity by Plants*. Academic Press, New York.

Pearcy, R. W., R. L. Chazdon, L. J. Gross, and K. A. Mott. 1994. Photosynthetic utilization of sunflecks: a temporally patchy resource on a time scale of seconds to minutes. *In* Caldwell, M. M., and R. W. Pearcy (eds.), *Exploitation of Environmental Heterogeneity by Plants*. Academic Press, New York.

Roth, I. 1984. *Stratification of Tropical Forests as Seen in Leaf Structure*. W. Junk, Boston, MA. 522 pp.

CHAPTER 9

TEMPERATURE

Solar radiation is the source of the heat that controls the temperature regime near the surface of the Earth. The great importance of terrestrial radiation and of air movements in affecting the level and distribution of temperature, however, makes it desirable to discuss solar heat and air temperature separately from solar radiation and light.

The mean annual temperature at any given spot on the Earth's surface is basically a function of the incoming solar insolation at that spot modified by secondary heat transfers arising from terrestrial radiation and air movement. Heating of the surface layers of air during the day is most intense under conditions where the greatest amount of infrared radiation is received, i.e., in tropical latitudes, at high elevations, and where the air is free from water vapor, clouds, and impurities. Temperatures during the night, on the other hand, depend largely upon the amount of heat absorbed by terrestrial objects and atmosphere during the day and the rate at which this heat is given off as terrestrial thermal radiation.

In this chapter, we first consider temperatures at the soil surface where the critical phases of plant germination and establishment take place. Temperature within the forest is examined in the next section followed by a consideration of temperature in relation to topographic position. The main part of the chapter is devoted to temperature and plant growth, with emphasis on dormancy and resistance of plants to freezing. Lastly, the distribution of plants in relation to cold hardiness is considered. Detailed treatments of temperature and plant growth, dormancy, and low-temperature relations are available in the works of Fitter and Hay (1987), Sakai and Larcher (1987), and Kozlowski and Pallardy (1997a, b).

TEMPERATURES AT THE SOIL SURFACE

The focal plane of temperature variation is the line of contact between the atmosphere and the ground, or, more accurately, the surface exposed to the sun. Both the surface of the

206

soil and the adjoining surface layer of air heat up considerably under direct sunlight and cool off greatly at night as heat is lost through terrestrial thermal radiation. Direct solar insolation commonly produces a surface temperature at midday of at least 70 °C, even in northern latitudes in the summertime (Vaartaja, 1954; Maguire, 1955).

The exact temperature reached at the soil surface depends upon the rate of absorption of solar energy and the rate at which it is dissipated once absorbed. This rate, in turn, is dependent primarily upon the amount of vegetation and litter cover, and only secondarily upon the color, water content, and other physical factors of the soil itself, if exposed.

The importance of soil color has been demonstrated by Isaac (1938). He found that charcoal-blackened soils will reach a temperature of 73 °C when the air temperature is 38 °C. At this point, comparable gray mineral soil heated up only to 64 °C and yellow mineral soil to 62 °C. These values were obtained on bare soil in the Douglas-fir region of southern Washington.

The rate at which surface materials dissipate heat received was found to be highly important in determining surface soil temperatures by Smith (1951), who worked with eastern white pine in Connecticut. A surface of white pine needle litter in a small clearing heated up to 68 °C on a day when the air temperature reached only 24 °C. In contrast, the surface temperature of bare mineral soil reached 46 °C and that of polytricum moss reached only 39 °C. The specific heat and conductivity of the surface materials are apparently the important physical factors involved. Wind speed, however, is most important of all.

Much of this variation in surface temperatures is due to the water content of materials. Moist material has a much greater capacity to dissipate heat through evaporation of water, while the evaporated water itself tends to reduce the amount of incoming solar energy. Thus Maguire (1955), in California, sprinkled bare mineral soil with the equivalent of 25 mm of rain on May 17 and found that the watered soil was 23, 14, and 9 °C cooler on the succeeding three days than the comparable unwatered soil.

Temperature variation drops off sharply within the soil because both mineral soil and organic layers are poor conductors. Van Wagner (1970) reported heat gradients of 7.2 °C per 0.25 cm of mineral soil and 51.7 °C per 0.25 cm depth of organic matter. Thus a surface temperature reaching 427 °C (800 °F), as in the case of forest fires, would have little effect below 5 cm depth in most forest soils.

Diurnal variation within the soil may disappear within 30 cm of the surface and annual variations within several meters. In fact, Shanks (1956) has argued that soil-temperature data from a 15 cm depth measured at weekly intervals serve as a useful integration of radiation and air temperature conditions during the preceding week. Indeed, temperatures measured deep in the soil, such as may be measured in caves, frequently remain constant throughout the year and may well measure the mean annual temperature of the area (Poulson and White, 1969).

Soil temperature is one of a suite of factors that greatly influences the growth and composition of forest ecosystems. A striking example is that for forest ecosystems of the Chena River floodplain in interior Alaska (Viereck, 1970). The annual soil temperature regime varies markedly for a sequence of ecosystems dominated by willow, balsam poplar, white spruce, and black spruce at successive distances from the river. Near the river's edge the gravely soil is cold in winter and warm in summer. However, this wide annual fluctuation is reduced inland until in permafrost soil under black spruce, soils are cold throughout the summer and waterlogged when not frozen. Many other examples could be given of the importance of soil temperature, illustrating not only differences among landscape ecosystems but its influence in bringing about vegetation changes over time at a specific site.

TEMPERATURES WITHIN THE FOREST

Within the forest, light crown cover and trees without foliage, as in the case of deciduous trees during the leafless season, tend to reduce air movement relative to outside the forest, while allowing solar radiation to penetrate the canopy. Under such conditions, the mean air temperature may be higher within the forest than outside it. For example, Pearson (1914) found that the mean annual air temperature within the open ponderosa pine forest of northern Arizona was 1.5 °C higher than in adjoining parklike openings. He attributed this phenomenon partly to lower radiation losses from the forest and partly to the effect of the forest in deflecting cold air masses moving down from surrounding mountains. In the Copper Basin of Tennessee, Hursch (1948) found that air temperatures were 0.3 to 1.1 °C higher in the deciduous forest than in the open during the winter months, although they ranged from 1.2 to 1.9 °C lower during the summer months.

When trees are in full leaf, the extremes within the forest are generally less than outside, and the diminution of radiation within the forest may result in lower mean annual air temperatures. For example, data collected within and adjacent to an eastern white pine plantation by Spurr (1957) gave a summer air temperature range within the forest of 15.9 °C compared to 21.6 °C in the open (Table 9.1). Throughout the year, at a height of one meter from the ground, the maxima and means were lower and the minima higher within the forest as contrasted to the open station.

Higher up in the forest stand, temperature variations are less than they are near the soil surface. Fowells (1948) measured the temperature profile up to a height of 37 m in a mature Sierra Nevada mixed conifer forest ranging up to 60 m in height. He found that temperature variation decreased with increased distance from the soil surface. At 37 m, minimum temperatures ranged up to 2 °C higher than at 1.4 m while maximum temperatures were similarly lower. At the top of the tree crown itself, however, surface conditions similar to those described for the forest floor exist. Consequently, extreme variation in temperature may be expected and has been described for a beech forest in Ohio by Christy (1952).

TEMPERATURE VARIATION WITH TOPOGRAPHIC POSITION

The effects of local topography upon local temperature have been studied by a number of investigators (Pallman and Frei, 1943; Hough, 1945; Wolfe et al., 1949; Spurr, 1957, among others). All have found that low concave landforms tend to radiate heat rapidly on still cold nights and to accumulate cold air, which flows in from surrounding high land. As a result, such sites will frequently have air temperatures near the ground as much as 8 °C lower than that of the surrounding terrain. The resulting condition is known as an **inversion** in that the temperature increases with height in the layer of air near the ground in contrast to the usual decrease in temperature with height. The concave areas are known as **frost pockets** because the temperatures commonly occurring near the ground result in late spring frosts, early fall frosts, and consequent short growing seasons. The concave area need not be deep. Spurr (1957) found that minimum temperatures in a small depression only 1 m deep were comparable to those in a nearby deep valley 60 m below the general land level.

In contrast, relatively high convex surfaces will tend to drain off cold air as radiation from the ground proceeds, so that night minima remain fairly high. During the day, the low concave landforms tend to accumulate radiant energy and reach high maximum temperatures near the ground, while mounds, ridges, and other convex surfaces tend to remain cooler during the day.

Table 9.1 Mean Weekly Maxima, Minima, and Mean Temperatures (°C) in the Open and Under a Dense 20-year-old Eastern White Pine Plantation[a]

	WINTER	SPRING	SUMMER	FALL
Open				
Maximum	5.1	22.8	29.7	14.2
Minimum	−18.4	−2.2	8.1	−7.3
Range	23.5	25.0	21.6	21.5
Mean	−6.7	10.4	18.9	3.4
Under forest				
Maximum	2.7	19.9	25.6	11.0
Minimum	−16.7	−2.8	9.7	−5.7
Range	19.4	20.2	15.9	16.7
Mean	−7.1	9.8	17.7	2.7

[a]Petersham, MA; by 13-week quarters, 1943–44.

Although variation in temperature due to local topography undoubtedly exerts a strong influence on the distribution of plants, and almost certainly plays a part in inhibiting tree growth in frost pockets, it must be remembered that soil-water drainage is frequently an interacting factor. The same factors—especially the existence of convex surfaces—that make a soil very well drained are apt to insure good drainage of cold air and to indicate a site with lower maximum and higher minimum temperatures than would be indicated by a regional climatic average. Thus a very well-drained site is apt to have a moderated climate suitable to plants of a generally southern distribution. Similarly, a very poorly drained soil is apt to result from a concave land surface that inhibits the drainage of cold air as well as of water. Thus the very poorly drained sites are apt to be characterized by temperature extremes and a short growing season, and might prove suitable for plants of a generally northern distribution. Such examples are common in northern Ohio and southern Michigan where boreal and northern forest species (black spruce, tamarack, northern white-cedar, black ash, yellow birch, leatherleaf, among others), relics from their retreat northward in late-glacial times, still persist in wet depressions (kettles) that are also frost pockets.

Finally, it should be realized that frost pockets can be created by the forming of small clearings in forest stand by windthrow or cutting. The surfaces of the tree crowns of the surrounding forest channel cold air into the clearings as terrestrial radiation chills the surface layers of the air on still, clear nights.

TEMPERATURE AND PLANT GROWTH

Plants regulate their temperature by dissipating part of the energy they absorb and thus preventing injury or death due to excessively high temperature. Of the three major mechanisms, reradiation, transpiration, and convection, reradiation dissipates one-half of the energy that plants absorb (Gates, 1980). Transpiration accounts for an additional heat loss as the leaf is cooled when energy is expended in changing water to water vapor. In addition, convection across the thin air zone that surrounds all surfaces in still air, the **boundary layer,** acts to transfer heat from the leaf to the cooler air. When the leaf is cooler than the air, as is typical at night, heat is transferred to the leaf from the air by convection and conduction. Through the interaction of these and other adaptations described next, the plant maintains a heat balance with its environment. These interactions tend to favor overall plant efficiency, tending to keep leaves warmer than cool air but cooler than warm air.

Plant processes function across a broad range of tissue temperature, generally 0 to 50 °C, as long as living cells and their proteins are stable and enzymatically active. As the seasons change, the foliage becomes conditioned and functions well at temperatures associated with the season. In forest trees, photosynthesis can take place at air temperatures below freezing, down to about −8 °C, although at such temperatures tissues are usually warmed to near or above freezing by solar and terrestrial radiation. Low temperatures depress the rate of photosynthesis (Chapter 18); nevertheless appreciable photosynthesis in conifers may take place in winter. For example, although the optimum temperature range for Douglas-fir is between 10 and 25 °C, net photosynthesis at 0 °C is still 70 percent of the amount occurring at 10 °C (Lassoie, 1982).

As temperature increases, plant activities increase up to an optimum temperature and then decrease until, at very high temperatures, death occurs. The processes influenced most strongly by temperature are (1) the activity of enzymes that catalyze biochemical reactions, especially photosynthesis and respiration, (2) the solubility of carbon dioxide and oxygen in plant cells, (3) transpiration, (4) the ability of roots to absorb water and minerals from the soil, and (5) membrane permeability. Because different growth processes may require different optimum temperatures, one simply cannot characterize the growth or biomass production of a species by a certain optimum temperature. Actually, various phases of the temperature regime—day temperature, night temperature, heat sums, and the difference between day and night temperature (thermoperiod)—all affect growth. Also, optimum growth requirements vary between species and populations within species and vary in a way related to the environmental conditions under which a population has evolved.

The experimental work of Hellmers and associates (1962; 1966a,b; 1970) and Kramer (1957) demonstrate that seedlings of tree species may respond to one or more of the following temperature conditions. Night temperature elicited the greatest growth response of Engelmann spruce (Figure 9.1), whereas redwood responded mainly to day temperature, reaching its maximum growth in the moderate range of 15 to 19 °C. Although redwood and Engelmann spruce grew best with the same day temperature (19 °C), spruce grew better under warmer nights and tolerated warmer days better than did redwood. Redwood grew best with only slight differences in day-night temperature or at constant temperature, apparently reflecting the lower diurnal changes in its native coastal climate. In contrast, Engelmann spruce is apparently adapted to the greater diurnal changes of continental, mountain environments.

Several species (Jeffrey pine, erectcone pine, eastern hemlock) show a marked growth response to total heat received during a day, irrespective of the time of application. For example, Jeffrey pine exhibited the most growth when 300 to 400 degree-hours (hours x temperature above some threshold = heat sum) of heat were received, regardless of the specific different day and night temperatures.

Although some species show a primary response to day temperature, night temperature, or total daily heat, many plants require nights considerably cooler than days for best growth. This differences between day and night temperature is termed **thermoperiod**. Low night temperatures coupled with moderate day temperatures are important in the flowering and fruit set, flavor, and quality of various crop plants and fruit trees (Treshow, 1970).

Seedlings of some forest trees respond strongly to thermoperiod. Maximum shoot growth of loblolly pine seedlings occurred under thermoperiods of 12 °C, with night temperature colder than day temperature (Kramer, 1957). A similar response has been consistently reported for Douglas-fir, although the optimal temperature differential has varied

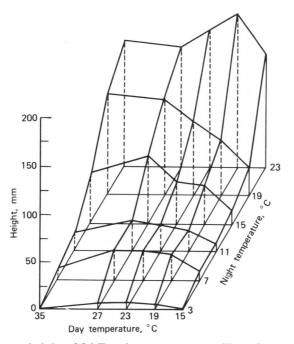

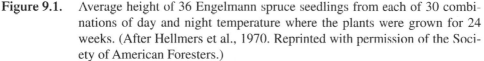

Figure 9.1. Average height of 36 Engelmann spruce seedlings from each of 30 combinations of day and night temperature where the plants were grown for 24 weeks. (After Hellmers et al., 1970. Reprinted with permission of the Society of American Foresters.)

among the studies conducted and the provenances tested (Hellmers and Sundahl, 1959; Lavender and Overton, 1972).

Two effects of thermoperiod were observed for red fir (Hellmers, 1966a). First, for maximum height growth to occur, a warm day must be followed by a cool night, whereas a cool day must have a cold night. Second, maximum height growth was obtained when the thermoperiod was 13 °C. Although maximum growth under a 17 °C day and a 23 °C day was nearly equal, this growth occurred only when the cooler day was followed by a 4 °C night and the warmer day with a 10 °C night, i.e., a 13 °C thermoperiod in both cases.

In contrast to the three species previously described above requiring cold nights and warm days, seedlings of ponderosa pine grow best with warm days and warm nights of 23 °C (Larson, 1967), or with warm nights and cool days (Callaham, 1962). In the latter study, the optimum growth of six provenances occurred with a thermoperiod of 5 °C, with nights warmer than days. Furthermore, pines from diverse parts of the range showed significantly different responses to the temperature regime (Figure 9.2). Seedlings from east of the Rocky Mountains (Figure 9.2c) required a high night temperature for optimum growth. Seedlings from the Southwest grew remarkably fast under cold days and hot nights. Seedling from the west slope of the Sierra Nevada Mountains in California grew well with lower night temperature (14 °C). Root growth of ponderosa pine is more dependent on soil temperature and shoot growth more dependent upon air temperature (Larson, 1967). Optimum root growth occurred in 15 °C air and 23 °C soil. Optimum root growth for northern red oak, basswood, and white ash also occurs at relatively high soil temperatures (Larson, 1970). Low soil temperatures tend to reduce metabolic activity and reduce membrane permeability so that uptake of water and nutrients is limited. Temperature of the crown also affects root growth. As crown temperature increases so does respiration and transpiration; carbohydrates and water are utilized in the crown and become less available for use by the

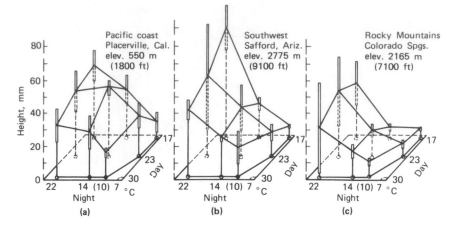

Figure 9.2. Mean height growth of ponderosa pine progenies from three provenances in three regions. Treatment included 16-hr days and nine combinations of three day and three night temperatures. The vertical bars show the range of the mean ± standard errors of the mean. (After Callaham, 1962. Reprinted from *Tree Growth*, T. T. Kozlowski (ed.), © 1962 by The Ronald Press Company. Reprinted with permission of John Wiley & Sons, Inc.)

roots. For example, root growth of silver maple seedlings decreased markedly as temperature of the crown increased from 5 to 30 °C (Richardson, 1956).

The range of response in net photosynthesis of woody species to temperature has been demonstrated by Pisek and associates (Pisek et al., 1969). Species were chosen to represent alpine, montane, and Mediterranean climates of central Europe. The high-elevation and alpine species exhibit lower temperature optima for net photosynthesis than Mediterranean and low-elevation species. The photosynthetic apparatus of the alpine species is more efficient in cold temperatures and less efficient in high temperatures than that of the Mediterranean species.

Studies with various tree species have shown that light interacts with air temperature in controlling growth rates. For example, the optimum temperature for photosynthesis is known to vary with light irradiance. Increasing temperature increases photosynthesis up to a point, and thereafter light irradiance must be raised if photosynthesis is to increase.

With increases of temperature above the optimum, the photosynthetic rate lessens. At about 30 °C, many enzymes tend to become disrupted, and if high temperature persists, enzymes become nonfunctional, halting growth processes. Furthermore, respiration increases greatly until at very high temperatures (perhaps 50 °C) respiration exceeds photosynthesis. Finally, at even higher temperatures, death of cells occurs. A number of studies (Baker, 1929; Lorenz, 1939) seem to indicate that the lethal point occurs at about 55 °C.

The temperature values referred to are those of the plant tissues themselves and not of the outside atmosphere. It stands to reason that air temperatures must be even higher to warm up the plant tissues to the lethal point. Thus prolonged exposure to air temperatures of 50 °C may eventually induce death while the same plants might be able to withstand brief periods of exposure to outside temperatures as high as 66 °C. This time factor has been demonstrated by a number of workers and is of considerable importance in seedling survival in that extremely high temperatures may be reduced by even the lightest shade, such as is the case when cotyledons of the seedlings shade the basal portion of the seedling stem. High temperature stress and heat injury of plants is discussed in detail by Levitt (1980a).

Leaf arrangement and orientation, a response to light irradiance, may act to reduce the amount of solar energy absorbed and hence be of survival value by preventing over-heating of the leaf. For example, when red maple seedlings are subjected to high light intensities, the leaf blades are deflected downward until they hang in a vertical position; when shaded they return to the horizontal (Grime, 1966). Such a mechanism may explain in part why red maple is a versatile species, colonizing both dry, exposed sites and shaded habitats.

Furthermore, seedlings may be able to maintain lower temperatures through the cooling effects of transpiration and thus avoid injury or maintain a better metabolic balance under very high outside temperatures. Shirley (1929) found that, for northern conifer seedlings, the killing temperature was higher in dry air, which favored seedling transpiration, than in moist air, which reduced it. In addition to the mechanisms of reradiation, transpiration, convection, leaf arrangement, and orientation already cited, leaf morphology, coloration, leaf pubescence, and maintenance of protein integrity may also act to enable the plant to function effectively at high temperatures (Levitt, 1980a).

Since temperatures above 50 °C are largely confined to the exposed ground–air boundary in forests, direct heat injury in forest trees is most significant in its effect on small seedlings, which have relatively unprotected live tissues in this critical zone. However, leaves of many mature hardwoods and conifers suffer leaf damage due to water deficiency in cells, particularly along the leaf margin and the tip of conifer needles. For example, widespread leaf damage to a wide variety of woody species occurred in northern California when temperatures suddenly rose above 38 °C in areas that had had an exceptionally cool spring (Treshow, 1970).

Cold Injury to Plants

Unlike lethal high temperatures, lethal cold temperatures occur periodically throughout the entire zone of tree growth in the temperate and boreal regions of the Earth. Thus they affect to a greater or lesser degree the distribution and growth of trees in these zones.

Death of plant tissues, particularly of actively growing plants and succulent tissues, may occur from rapid freezing and formation of ice crystals within the protoplasm. Rapid thawing is very harmful, causing sudden changes that disrupt cell membranes. Slow freezing also may kill many semi-hardy plants at temperatures of −15 to −45 °C at rates of cooling that commonly occur in nature (Weiser, 1970). Water between cells freezes, dehydrating cell contents until a point is reached when only "bound" (unfreezable) water remains in the protoplasm. As temperature decreases further, bound water is pulled away from the protoplasm, initiating protein denaturation and ultimately causing death. In tropical plants, death may occur at above-freezing temperatures ranging from 0 to 10 °C.

Most trees in the temperate and boreal zones, however, become increasingly inactive as the day length shortens and temperatures drop at the end of the growing season. As dormancy sets in, the water content of protoplasm is reduced as the concentration of other contents of cytoplasm increases; water moves out of cells and freezes in extracellular spaces, and many species are able to survive subfreezing temperature without damage. The more hardy the species, the greater the capacity of its cells to tolerate freeze-induced dehydration. Of critical importance is the ability of individuals to initiate and continue the cold-acclimation process to correspond with the progressively decreasing temperatures of their native site in autumn by reducing the water content of their cells. Cold hardiness is a gradual process, and it is timed by the natural selection process to allow plants to survive the gradual lowering of temperatures in autumn and winter and their gradual rise in spring. Before considering frost resistance and the role of temperature in the distribution of woody species, a basic understanding of dormancy in woody plants is appropriate.

Dormancy

Acclimation or hardening are terms used to describe the change of plants from a succulent (tender) to a hardy or dormant condition. The diagram in Figure 9.3 of the annual plant vegetative cycle in temperate climates illustrates the external and internal influences affecting the entrance into dormancy and the resumption of growth. Three stages of dormancy are normally recognized (Vegis, 1964; Perry, 1971; Sakai and Larcher, 1987): predormancy (early rest), true dormancy (winter rest), and postdormancy (afterrest) (Figures 9.3 and 9.4). Short days act as an early warning system, triggering growth cessation and initiating metabolic changes characteristic of predormancy or the first stage of acclimation at temperatures of 10 to 20 °C (Figures 9.3 and 9.4). These changes facilitate further plant response in the second stage of acclimation (also accompanied by metabolic changes) that is triggered by low temperature below about 5 °C and especially subfreezing temperature (Figure 9.4). The third stage of acclimation and attainment of true dormancy is a purely physical process and is induced by low temperatures of −30 to −50 °C. Truly dormant buds and seeds cannot be induced to immediate normal growth by any means. Some species—birches, European beech, and some oaks—do not show true dormancy and hence pass readily from predormancy to post dormancy. In the truly dormant condition, plants in their native ecosystems are not at risk from damage. However, when entering or leaving dormancy abrupt departures from the general temperature decline or rise, as in frost in early fall or late spring, may cause severe injury or death. These and other aspects of cold hardiness and resistance to injury are considered next.

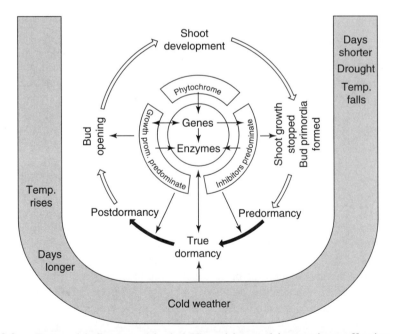

Figure 9.3.   External influences (shaded U) and internal interactions affecting the seasonal alternation of vegetative and floral activity and dormancy in woody plants. Note that the growth inhibitors that influence bud formation in the fall and the growth promotors that initiate bud opening in the spring are activated by the phytochrome system. (After Sakai and Larcher, 1987. Reprinted from *Frost Survival of Plants,* © Springer-Verlag Berlin Heidelberg 1987. Reprinted with permission of Springer-Verlag.)

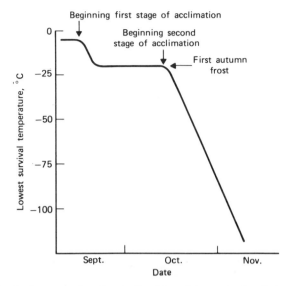

Figure 9.4. A typical seasonal pattern of cold resistance in the living bark of red-osier dogwood (*Cornus stolonifera*) stems in Minnesota. In nature, acclimation in this hardy shrub and in a number of other woody species proceeds in two distinct stages, as shown. The beginning of the second stage of acclimation characteristically coincides with the first autumn frost. (After Weiser, 1970. Reprinted with permission from *Science* 169:1269–1278, © 1970 American Association for the Advancement of Science.)

Frost Hardiness and Cold Resistance

Frost hardiness is strongly related to the timing of the acclimation. Northern races of a given species begin the process earlier than southern races because day-length changes are more rapid in northern latitudes than in southern latitudes. Northern races are more resistant to freezing than southern ones. Racial differences have been reported for eastern cottonwood (Mohn and Pauley, 1969), northern red oak (Flint, 1972), red maple (Perry and Wang, 1960), sugar maple (Kriebel, 1957), sweet gum (Williams and McMillan, 1971), and white ash (Alexander et al., 1984).

Flint (1972) studied the hardening of northern red oak populations grown from seeds collected from 39 localities and grown at a site at the Arnold Arboretum, Weston, Massachusetts. Stem sections were collected six times during the course of a year. The temperatures required to cause frost death in stems, collected 9 October, 14 November, and 12 December 1969, showed a highly significant relationship to the latitude, annual minimum temperature (Figure 9.5), and estimated extreme minimum temperature of their original site. The progression of hardening and dehardening of twigs from four populations at the extremes of the range sampled is shown in Figure 9.6. The earlier hardening of northern sources and their greater hardiness in midwinter (January) are apparent.

Genetic adaptation to low temperatures not only follows the continuously changing gradient of temperature but is related to the low-temperature minimum that occurs in midwinter. Thus species are resistant to freezing injury far below the minimum temperatures typically encountered in their native sites. For example, woody plants of the boreal and northern forest (paper birch, spruces, pines, tamarack, trembling aspen, willows) survive prolonged subzero temperatures. Moreover, if the freezing rate to −30 °C is relatively slow and plants are fully acclimated, these and numerous hardy species are known to withstand experimental freezing to −196 °C, the temperature of liquid nitrogen (Weiser,

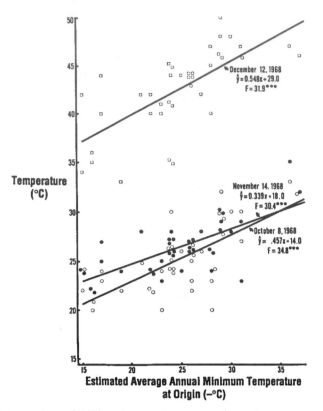

Figure 9.5. Regression of killing temperature on estimated average annual minimum temperature at seed-source origin of northern red oak at three sampling dates: 9 October (○), 14 November (●), and 12 December (□) 1969. (After Flint, 1972. Reprinted with permission of the Ecological Society of America.)

1970; Sakai and Weiser, 1973). Extreme cold resistance is exhibited by the seeds of certain pine and spruce species, which cannot be frozen at any normally occurring temperature when in a dry, dormant condition. In contrast to boreal and alpine species, many trees of the Pacific Coast region and those of the lower Mississippi Valley and southern Coastal Plain (characterized by mild fall and winter climate) are injured at temperatures colder than −20 to −30 °C.

The freezing resistance of plants, as outlined by Levitt (1980a), is a function of both tolerance and avoidance mechanisms (Figure 9.7). With the plant as the focus, an environmental stress, such as cold temperature, may produce injury or death. Direct injury by freezing occurs only as a result of freezing within the protoplasm. Upon freezing, plant tissues actually contract. The frost splitting of trees in winter that occurs with a "crack like that of a gun" can be explained by an asymmetrical contraction, despite the belief that plant tissues expand upon freezing (Levitt, 1980a). According to Levitt, at temperatures occurring in nature, intracellular freezing is always fatal; there is no tolerance to it (Figure 9.7). However, ice normally forms in intercellular spaces and may form masses many times larger than the cells themselves. The primary mechanism of freezing resistance in plants (i.e., of being "hardy") is tolerance of extracellular freezing. Two mechanisms, tolerance of freeze dehydration and avoidance of freeze dehydration (Figure 9.7), may act together to provide such resistance. Details of these mechanisms are provided by Levitt (1980a) and Sakai and Larcher (1987).

Plants avoid freezing by mechanisms that either avoid freezing temperatures altogether or by avoiding ice formation (adaptations 2, 3, 4 in Figure 9.7). Plants may live in

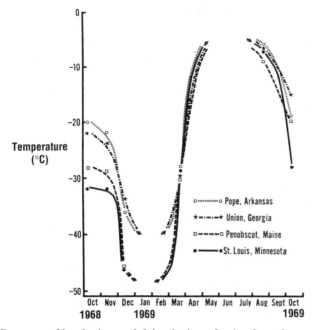

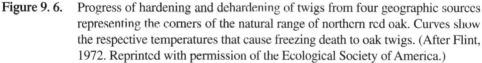

Figure 9. 6. Progress of hardening and dehardening of twigs from four geographic sources representing the corners of the natural range of northern red oak. Curves show the respective temperatures that cause freezing death to oak twigs. (After Flint, 1972. Reprinted with permission of the Ecological Society of America.)

habitats where frost is absent or infrequent. They may have insulating mechanisms such as bud scales and bark that may provide protection up to a point. If the plant cannot avoid freezing, three adaptations may allow it to avoid ice formation and thus freezing injury when cooled below 0 °C. These are: antifreeze, dehydration, and supercooling (undercooling) (Levitt, 1980a). Mechanisms that reduce the amount of water in plant tissues, and

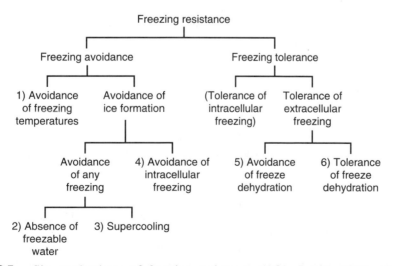

Figure 9.7. Six mechanisms of freezing resistance. (After Levitt, 1980a. Reprinted from Responses of Plants to Environmental Stresses, Vol. 1, © 1980 by Academic Press, Inc. Reprinted with permission of Academic Press, Inc.)

thus decrease the probability of ice formation, include increasing water loss via transpiration and decreasing water uptake by suberization of root surfaces. Increasing the proportion of unfreezable (bound) water and increasing the concentration of solutes (especially sugars and carbohydrates) reduces the likelihood of ice formation. Sugars, by accumulating in the vacuoles, can either prevent ice formation or decrease the amount of ice formed and thus also increase the avoidance of freeze dehydration (5 in Figure 9.7).

Supercooling refers to the reduction in freezing point of water of a plant below that of pure water (Levitt, 1980a). The supercooling point is the lowest subfreezing temperature attained before ice formation. Cytoplasm freezes between −1 and −3 °C and at increasingly lower temperatures depending on the concentration of the solutes it contains. Supercooling on the order of −1 to −3 °C appears to be the mode of freezing resistance in the meristematic tissues of flower buds of woody plants, such as azalea, blueberries, cherries, and plums (Weiser, 1970; George and Burke, 1977). As a rule, supercooling represents a transient, unstable state that may provide protection against brief radiation frosts (Sakai and Larcher, 1987). In very small cells that are hardened slowly, persistent supercooling to very low temperatures is possible (George et al., 1974; Quamme, 1985; Sakai and Larcher, 1987). This deep supercooling of the ray parenchyma cells of xylem of many woody species prevents freezing at temperatures as low as −37 to −47 °C. The distributions of 49 tree species of North America were found by George et al., (1974) and George and Burke (1976) to be related to the degree of deep supercooling. Plants protected by this resistance mechanism are confined to an area where minimum winter temperatures never drop much below −45 °C. Woody plants found in areas where temperatures below −45 °C occur, including the extremely hardy boreal deciduous species of *Salix*, *Betula*, and *Populus*, therefore have no supercooling mechanism to protect them. Plants of these genera are known to have thin and elastic cell walls that permit freeze dehydration of their cells (Sakai and Larcher, 1987). Thus although supercooling is an important resistance mechanism for some tissues and species, most freezing tolerance depends on the capacity of living cytoplasm to tolerate freeze-induced dehydration.

Some plants may avoid freezing by the presence of glaucous waxes on their leaf surfaces. As described in Chapter 4, a cline of waxy glaucousness is evident for leaves of the Tasmanian species *Eucalyptus urnigera* of Mount Wellington. Low elevation leaves are green, lacking wax; leaves of high elevation plants exhibit intense wax formation. Waxy glaucousness not only increases reflectance, thereby decreasing absorption of solar radiation at high elevations, but also increases the water repellency of leaf surfaces. Leaves with a dry surface (waxy) freeze at a leaf temperature of −9 to −10 °C; if wet (green, non-waxy) they freeze at −2 to −4 °C (Thomas and Barber, 1974). Water freezing on the leaf surface of non-waxy leaves leads to internal dehydration. Such freezing always kills the leaves; whereas dry leaves supercool sufficiently to avoid damage at −2 to −4 °C.

THERMOTROPIC MOVEMENTS IN RHODODENDRONS

Leaf movements and orientation not only protect plants from injury in hot and dry ecosystems, as noted previously, but they are important adaptations in response to cold temperatures and freeze–thaw cycles in the genus *Rhododendron* (Nilsen, 1992). The widespread distribution in the northern hemisphere as shrubs and trees in this genus of over 600 species is related to such adaptations. Temperature-sensitive or **thermotrophic** movements, leaf drooping, and curling into a pencil-shape coil (Figure 9.8) have been widely reported by scientists and rhododendron enthusiasts. Harshburger (1899) observed that the petiole takes a sharp bend downward through an angle of about 70 degrees. Overall, rhododendron species that exhibit the most intense movements are the most hardy.

Detailed studies by Nilsen (1992) revealed that leaf curling and drooping are distinct movements with different responses to climatic factors and different ecological signifi-

Figure 9.8.   Drooping and curled leaves of rosebay rhododendron in winter. (After Nilson, 1990; Photo by Rácz and Debreczy, Photographic Archives of the Arnold Arboretum, © 1990 by the President and Fellows of Harvard College.)

cances. Leaf curling occurs in direct response to leaf temperature regardless of light or water availability. Leaf drooping occurs in response to leaf water potential, which is influenced by leaf temperature, light intensity, and soil-water stress. The lower the temperature or the higher the light, the more pendant the leaf becomes.

In the Southern Appalachian Mountains on winter days when air temperature is constant and slightly below 0 °C, leaf angle of rhododendrons is vertical and leaf curling is 100 percent during the daylight hours. If the light intensity increases and leads to an increase in leaf temperature, there is some uncurling. On other days if the air temperature increases from well below 0 °C to well above zero during the morning (a 24 °C diurnal change is possible) there is rapid leaf uncurling and a moderate change in leaf angle toward the horizontal.

The ecological significance of these leaf movements appears to be the avoidance of injury due to freezing and thawing. Leaves of rosebay rhododendron in the Southern Appalachians often freeze in winter. The leaf freezing point is about −8 °C, whereas leaf temperature routinely decreases to −15 °C (Nilsen, 1985, 1987). Daytime winter temperatures may reach 15 °C and initiate rapid thawing. Therefore, rhododendrons that persist in cold northern, alpine, or Arctic sites must have leaf adaptations to tolerate repetitive freezing and thawing. Thermotrophic movements may be effective at reducing the *rate of thaw* following freezing. If frozen leaves were positioned horizontally, irradiance hitting a leaf would warm them rapidly. For example, the leaf temperature can increase 16 °C in a matter of seconds when a sunfleck hits a rhododendron leaf (Bao and Nilsen, 1988). However, if the leaf is pendent and curled, the thaw in response to a sunfleck or irradiance through the leafless canopy is slow.

A second advantage of thermotrophic movement relates to the interaction of cold temperature and high light on photosynthesis. Although the evergreen rhododendron leaves do not photosynthesize in winter, the photosynthetic mechanism may be crippled by high irradiance. The drooping of rhododendron leaves probably serves to protect cell membranes from damage by intense winter radiation when they are cold. Membranes in chlorophyll-rich chloroplasts can be injured under such conditions. The damage may impair photosynthesis as much as 50 percent in the following summer (Bao and Nilsen, 1988). Rhododendrons growing under the leafless tree canopy experience the highest annual radiation of the year during the coldest conditions (Nilsen, 1985). Therefore, curling acts to reduce the quantity of light impinging on leaf surfaces, and injury to the photosynthetic mechanism is avoided. Rosebay and many other rhododendron species grow in dim light of forest understories, and their evergreen leaves function for several years. A fully functioning leaf system is, therefore, critical for their survival; winter injury to the photosynthetic mechanism would substantially affect their net carbon gain and their survival.

Detailed considerations of plant tissue freezing, freezing resistance and avoidance, and freezing injury are given by Levitt (1980a) and Sakai and Larcher (1987). Winter dormancy has been considered in detail by numerous authors (Wareing, 1959, 1969; Romberger, 1963; Vegis, 1964; Perry, 1971; Sakai and Larcher, 1987; Borchert, 1991).

Winter Chilling and Growth Resumption

Buds and seeds of most woody species of the Temperate Zone require a period of **winter chilling** before growth is resumed in the spring. Temperatures near 5 °C are most effective (Perry, 1971). As the postdormancy period proceeds, physiological changes take place enabling growth to resume once minimum temperature requirements are met. Such a chilling requirement prevents death that might otherwise occur by premature resumption of growth triggered by a sudden warm spell in the middle of winter. The nature of dormancy and the chilling requirement vary with the adaptation of a local race of a species to its environment. For example, red maples from New York State attain true dormancy and require a month or more of chilling before resuming active growth (Perry and Wang, 1960). Red maples in south Florida cease active growth, drop their leaves in late autumn, and resume growth in early spring without a chilling requirement. They are unable to withstand severe freezing temperatures, but such events occur infrequently in southern Florida. Growth resumption is a complex process, and factors other than temperature may be involved.

Once the winter-chilling requirement has been satisfied and dormancy is broken, temperature primarily regulates the time of bud burst (leaf flushing), bud-scale initiation, and shoot axis elongation. Vegetative bud dormancy is broken when mitotic activity begins in the buds. It precedes flushing of these buds, usually by several weeks in conifers (Owens et al., 1977). After winter dormancy is broken, temperature sums may be used to predict accurately when a stage of development, such as flowering or flushing, will occur or how rapidly the shoot will elongate. The strong genetic control of this process was discussed in Chapter 4, and Levins (1969) has discussed dormancy and growth resumption as an adaptive strategy.

Seeds of many temperate forest species require a cold and moist period over winter in the forest floor or **stratification** before germinating the following spring. The cold treatment acts to alter the balance between growth inhibitors and promoters to favor spring germination when conditions are ideal for growth. Seeds of other species, including trees that fruit in the spring (elms, cottonwoods, aspens, willows) and many fire-dependent species (including most hard pines), lack a cold requirement. They may germinate quickly following dispersal, once sufficient soil water is available.

NATURAL PLANT DISTRIBUTIONS AND COLD HARDINESS

Low temperature is the single most important factor limiting the distribution of plants (Parker, 1963). However, the natural distribution of woody species is not necessarily related to the cold resistance of adult individuals. Such individuals may survive injury to foliage, buds, and xylem. The northern distribution of species is probably more related to the cold resistance of propagative organs and developmental stages, such as overwintering flower buds and immature fruits, germinating seeds, and young seedlings. Reproductive buds are known to be far less resistant to low temperatures than vegetative buds (Larcher and Bauer, 1981). The failure to reproduce sexually, combined with competition with more hardy species, can markedly affect migration and range extension.

Figure 9.9 illustrates frost resistance of several tissues of Holm oak, a relatively hardy, evergreen Mediterranean oak, whose seeds germinate in autumn and winter as do many species of the Mediterranean area because water availability is favorable (Larcher and Bauer, 1981). Information on flower bud and seedling resistance is therefore particularly important in assessing its range extension. A comparison of anatomical parts and developmental stages of Holm oak (Figure 9.9) shows that germinating seedlings and surface-feeding roots are especially prone to freezing. Although the adult trees are able to thrive where temperatures of −12 °C occur each winter (for example in northern Italy and in parks and gardens of the British Isles and North America), natural regeneration of the species is completely suppressed if winter temperatures regularly drop below −8 °C. Sakai and Larcher (1987) emphasize that for this oak and other Mediterranean sclerophyllous species, absolute temperature minima are seldom useful in determining their northern and altitudinal limits. It is the regular recurrence of an average stress (as measured by mean temperature minima), and not catastrophic drops in temperature, that limit distribution.

Similarly in North America, it is observed repeatedly that species with southerly distributions can survive, reach large size, and even produce fruit in climates far north of

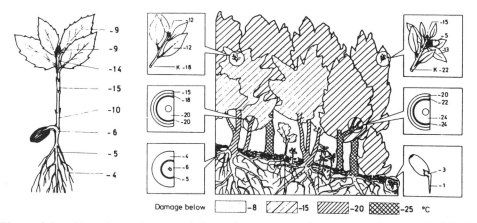

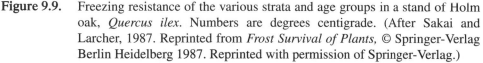

Figure 9.9. Freezing resistance of the various strata and age groups in a stand of Holm oak, *Quercus ilex*. Numbers are degrees centigrade. (After Sakai and Larcher, 1987. Reprinted from *Frost Survival of Plants,* © Springer-Verlag Berlin Heidelberg 1987. Reprinted with permission of Springer-Verlag.)

their natural range in parks, gardens, and lawns where their establishment has been provided by humans. For example, sweet gum is used for landscaping in many northern states
far from the northernmost limits of its natural range. Northern catalpa, whose range is
from northeastern Missouri to southeastern Illinois, has been planted up to 650 km north
of the northern part of this range. A striking example is the osage orange, whose northernmost natural range is in southern Oklahoma and Arkansas (34° N). It has been a widely
planted and successful hedgerow species as far north as southern Michigan (42° 30′ N) or
about 1,000 km north of its northernmost natural range! At more northerly sites these
species fail to regenerate and compete in natural communities.

Deciduousness and Temperature

Deciduousness in angiosperms apparently evolved first in the tropics where leaf
shedding was a selective advantage in a lengthy drought period. In the Temperate Zone, it
is an adaptation to the cold season (Walter, 1973, p. 144). Deciduousness therefore, is in
part an adaptation to physiological drought, whether due to a tropical dry season or a
boreal frigid season.

The shedding of thin deciduous leaves in winter and the protection of meristems
from water losses represent an energy savings compared to maintaining a mass of thick
evergreen leaves over winter (Walter, 1973). However, because little photosynthesis is
possible during the leafless seasons (except in the cortex of thin-barked species) the warm
period must be long enough, warm enough, and favorable in air and soil water and nutrients to enable deciduous species to produce photosynthate sufficient at least for the year's
maintenance, growth, and reproduction. Therefore, deciduous woody species dominate
forest ecosystems and attain the highest productivities in areas of long, warm, and humid
growing seasons where soil water and nutrients are readily available. The most favorable
zone for deciduous species in North America is approximately from 35 to 45° N latitude
and from 70 to 95° W longitude. To the north and at high elevations, deciduous forests
tend to be replaced by conifers as conditions become harsh and extreme. To the west, deciduous forests give way to grassland (Figures 2.8 and 2.10) and xerophytic conifers (junipers, pinyon pines, and ponderosa pine) at low elevations where increasing atmospheric
aridity, high soil-water stress, and frequent fires become limiting factors. To the south,
where long growing seasons and increasingly warm summer and winter temperatures prevail, less snow and ice, increasing fire frequencies, and especially nutrient-poor soils of
the Coastal Plain create conditions more favorable to conifers and evergreen angiosperms.

With increasing latitude and elevation, the ever shorter growing season and cooler
summer temperature provide environments where conifers outcompete deciduous species.
Thin, nutrient-poor soils and prolonged, drying winds of high velocity, are also associated
with these boreal and alpine environments. By inhibiting biotic activity and decomposition of organic matter, cold conditions markedly reduce nutrient availability compared
with comparable soils at lower latitudes and elevations. Low temperatures injure plant tissues directly by sudden drops (early and late frosts) or by extreme lows in winter. They affect plants indirectly by causing: (1) a shorter growing season, (2) cold soils that are unfavorable for root growth and for absorption and conduction of water and nutrients, and (3)
permafrost formation. Therefore, as one proceeds north, those deciduous and conifer
species, and forest shrubs and herbs as well, that are better adapted to extreme conditions
outcompete those adapted to more favorable conditions. Thus competition and mutualisms among species, particularly in the establishment and early growth stages, together
with the direct and indirect effects of adverse site factors, markedly influence the distribution and abundance of species. Temperature and other site factors such as soil water that

significantly affect plant distribution are very strongly affected by physiographic features which are considered in the next chapter.

Suggested Readings

Burke, M. J., L. V. Gusta, H. A. Quamme, C. J. Weiser, and P. H. Li. 1976. Freezing and injury in plants. *Ann. Rev. Plant Physiol.* 27:507–528.

Kozlowski, T. T., P. J. Kramer, and S. G. Pallardy. 1991. *The Physiological Ecology of Woody Plants.* Academic Press, New York. Chapter 5 (Temperature).

Larcher, W., and H. Bauer. 1981. Ecological significance of resistance to low temperature. In O. L. Lange, P. S. Nobel, C. B. Osmond, H. Ziegler (eds.). *Physiological Plant Ecology I, Responses to the Physical Environment*, New Series, Vol. 12A. Springer-Verlag, New York.

Levitt, J. 1980. *Responses of Plants to Environmental Stresses.* Vol. 1. Academic Press, New York. Chapter 5 (The Freezing Process), Chapter 6 (Freezing Injury), Chapter 7 (Freezing Resistance—types, measurement, and changes).

Sakai, A., and W. Larcher. 1987. *Frost Survival of Plants.* Springer-Verlag, New York. 321 pp.

Villiers, T. A. 1972. Seed dormancy. In T. T. Kozlowski (ed.), *Seed Biology.* Academic Press, New York.

Weiser, C. J. 1970. Cold resistance and injury in woody plants. *Science* 169:1269–1278.

Wilson, B. F. 1984. *The Growing Tree.* Univ. Mass. Press, Amherst. 138 pp. (Survival in hard times, pp. 121–127.)

CHAPTER 10

PHYSIOGRAPHY

Many of our most vivid and unforgettable mental images are of the landscape where we grew up or where we live. Mountains with their distinctive configurations and striking seasonal changes inspire a sense of natural grandeur and power. John Muir repeatedly emphasized physiographic diversity in his writings. The first sentence of his book *The Mountains of California* (Muir, 1894) superbly introduces this viewpoint:

Go where you may within the bounds of California, mountains are ever in sight, charming and glorifying every landscape.

Still in the introductory paragraph his fine sense of geomorphologic and ecological awareness stands out:

The Coast Range, rising as a grand green barrier against the ocean, from 2000 to 8000 feet high, is composed of innumerable forest-crowned spurs, ridges, and rolling hill-waves which include a multitude of smaller valleys; some looking out through long, forest-lined vistas to the sea; others, with but few trees, to the Central Valley; while a thousand others yet smaller are embosomed and concealed in mild, round-browned hills, each with its own climate, soil, and productions.

Similarly, the prairie landscape with its sweep of flat or gently rolling land as far as the eye can see has an irresistible and memorable quality whose beauty and harshness is portrayed in many books. The waterscapes of lakes, rivers, and streams and their associated lake shores and riparian landscapes provide another unforgettable mental map to which we can link a diverse array of biotic and human patterns of occurrence and activities in space and time. Around the world, other landscapes of rocky coastlines, sand hills, wetland depressions, tropical lowlands and highlands, tidal flats, arid plateaus, and mountain slopes all reveal a close relationship between form of land or water with the climate above, the parent material and soil below, and the organisms at their energized surfaces. At different spatial scales, from subcontinental plains and mountains to pit and mound microsites in a hemlock–northern hardwood forest, physiographic features or landforms provide our best clue to understanding the structure and dynamics of landscape ecosystems and the interrelationships of organisms in populations and communities. Concepts in the field of ecological geomorphology are considered in the following sections by examining concepts and ecological attributes of physiographic features and by illustrating examples of its pervasive importance for the forest ecologist.

CONCEPTS AND TERMS

The term **physiography** is an abbreviation for physical geography—the surface features of an area (Neufeldt and Guralnik, 1988). Because surface features, such as mountains and outwash plains, have both surface *form* and geologic *parent material* beneath their surface, we include both these elements in the definition. Thus in forest ecology we define physiography as the surface features, their form and substance, of a given regional or local area. The specific physiographic features themselves, the **landforms** (e.g., mountain slopes, plateaus, outwash plains, and river floodplains) are distinguished not only by their form but the parent material that is characteristic of a given landform and in which characteristic soils also develop. For example, crystalline and sedimentary rock typically give rise to landscapes with markedly different landforms and soils. We conceive physiography as a major ecosystem component characterized by many different factors which not only give spatial form and structure to a landscape and its constituent ecosystems but significantly affect ecosystem function.

The term **geomorphology** is commonly used in the literature. It is the science dealing with the nature and origin of topographic features of the earth (Neufeldt and Guralnik, 1988). We use this term when emphasizing the geologic form or shape of an area. However, in addition to form, the ecologist must understand the geologic surface materials (i.e., parent material) that are associated with each feature since they reflect the geomorphic processes by which form and substance were derived. Thus it is possible in the field and on aerial photographs to correlate geomorphology with the parent material beneath, and this integration of form and mineral composition we refer to as physiography. Therefore, in understanding the physiography or ecological geomorphology of an area we study the surficial features, their *form and substance*, and their relationship to one another and the biota they support. For instance, land in a high position influences adjacent low-lying lands and their biota through its effect on climate and hydrology. Land adjacent to a river affects its course, rate of flow, and water quality; and the river affects the land as flood waters erode and deposit sediments, forming levees and floodplain bottoms.

A **landform** is defined as any physiographic feature on the Earth's surface, such as a plain, valley, hill, etc., caused by erosion, sedimentation, or movement (Neufeldt and Guralnik, 1988). Many specific landforms (rocky ridges, alluvial river floodplains, and sandy outwash plains) are formed from distinctive parent material in which soil develops. Thus we use the term landform in the context of specific physiographic features of the earth's surface and the term physiography in the broad context of a major ecosystem component.

As we have seen (Chapter 2), we use the major ecosystem components of climate, physiography, soil, and vegetation to distinguish and map landscape ecosystems at regional and local levels and also to understand ecosystem structure and function. Physiography is exceedingly important because it is the most stable component, i.e., least affected by short- and long-term natural and human disturbances. The form and geologic substance of the earth's surface layer form the relatively permanent framework of landscape ecosystems (Rowe, 1984b). Furthermore, physiography markedly influences ecosystem functioning because it controls the climatic regime at and above the surface; below the surface it controls soil development and acts to regulate soil processes. Surficial shapes and their parent materials modify the fluxes of radiation, soil water, and nutrients thereby regulating plant establishment, distribution, growth, and productivity.

Figure 10.1 illustrates a newly-exposed landform that gives rise to different local climates on its sunny and shaded slopes. Depending on their competitive ability for temperature and soil water, plants initially colonize these different slopes. Animals also respond differentially to the microclimate of the different slopes and to their respective vegetation types that provide shelter and food. Soils develop in the parent material of the landform through interactions of geologic substrate, local climate, biota, and time (Figure 10.1). In-

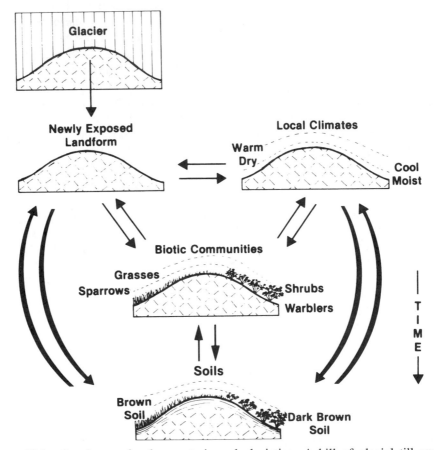

Figure 10.1. Landscape development since deglaciation. A hill of glacial till emerges from the melting ice and "forms" its catena of local climates. These and the parent-material substrate select arriving plants that in turn select adapted animals. Soils develop through interactions of land form, local climate, and biota. Arrows indicate interactions and feedback so that, for example, the properties of developing soils and communities influence the evolution (by erosion and deposition) of the land form. (After Rowe, 1984b.)

teractions of all components through time result in the development of similar ecosystems on similar landforms. Repetitive patterns of landforms give rise to repetitive patterns of vegetation. The size, shape, and relation of one landform to another determines the particular kind of plant and animal communities and races that develop throughout a landscape.

From the foregoing discussion and Figure 10.1, we can see that a map of landforms may be regarded as a kind of soils map as well as a map of local climate. Therefore, it is our best approximation of a map of regional and local vegetation (Rowe, 1969). We have also seen in Chapter 2 how major physiographic features of mountains, oceans, and lakes can affect macroclimate on a continental and subcontinental basis and thereby the major vegetation types. In addition, a landform map reflects the history and patterns of disturbances such as wildfire and flooding. The shape and parent material of specific landforms may either promote or inhibit fire and the incidence and duration of flooding.

CHARACTERISTICS OF PHYSIOGRAPHY AND THEIR SIGNIFICANCE

What are the specific characteristics of physiography that are useful in understanding ecosystem structure and function? We list and briefly describe these characteristics in the hierarchical framework that follows. Keep in mind that many of them may be significant at many different spatial scales, from continental scale to a local wetland depression less than 0.1 ha. Following the list of characteristics we will briefly describe the importance of many of them in understanding ecosystem structure and function.

PHYSIOGRAPHIC CHARACTERISTICS
Physiographic setting (glaciated or unglaciated terrain, coastal or interior landscapes, etc.)
Specific landform (mountain, outwash plain, river terrace, etc.)
　Elevation
　Size
　Form (shape or configuration)
　　Level landforms
　　Sloping landforms
　　　Slope shape or configuration (straight or planar, concave or convex)
　　　Slope position (upper slope, mid slope, lower slope)
　　　Slope aspect (direction a slope faces)
　　　Slope inclination (slope percent or degree of slope)
　Parent material of the feature or landform (kind of substrate, including rock type, particle size, mineralogy)
Position of landform in the landscape (high-level, low-level)
Position of landform in relation to other landforms

Physiographic Setting

An overriding consideration is whether the landscape has been covered by a continental glacier or not because some landforms only occur in glaciated terrain. Many features, such as rivers, lake beds, mountains and hills, occur in both glaciated and unglaciated terrain, and their characteristics accordingly may be different.

Specific Landforms

At scales from continental to local (<1 to 100 ha), the specific kind of landform and its size are of great importance in determining the physical factors of ecosystems and the ranges and patterns of occurrence of organisms and their productivity. Continental mountain ranges, plains, and large river basins are of primary importance. Their properties and effects are appropriately considered in geography and geology texts (Fenneman, 1931, 1938). Many kinds of local landforms are described in the following sections.

Elevation

For a given physiographic region or specific landform, elevation above sea level is especially important when the climate associated with it becomes a limiting factor to plant establishment and growth. Elevation is only a rough indicator of climatic factors affecting plants. Climatic factors at a given elevation are dependent on latitude and other physiographic elements (aspect, slope percent, etc.) that may have an equal or greater importance than elevation. Nevertheless, elevation of a site at a known latitude is often a useful indicator of plant performance when details of specific climatic factors are not known. Elevation is important to determine when comparing the vegetation of similar landforms within a mountain range or comparing vegetation from high and low latitudes. For example, the vegetation of ecosystems along streams is markedly different at low elevations as compared to high elevations whether one is in the Cascade Mountains in Washington state or the Smoky Mountains in Tennessee. Similarly, communities dominated by red spruce at sea level in eastern Canada have many similar species to those found at elevations over 1,500 m in the Southern Appalachians.

Form of Landforms

The two key elements of specific landforms are their shape and their parent material. Shape, in particular, influences the reception and disposition of radiation, soil water, and nutrients and hence the repetitive patterns of biota of an area.

LEVEL TERRAIN

Of the continuum of forms that physiographic features exhibit, a key, initial distinction is whether land is level or sloping. Level or nearly level terrain such as plains, river terraces, old lake beds, prairies, and plateaus are typically water-laid or water-influenced features. However, the flat till (unsorted glacial drift) plains of parts of Indiana and Ohio were deposited by the glacier as it rapidly retreated at a relatively uniform rate. Some plains, now deforested, are remarkably flat and widespread. The Tipton Till Plain in central Indiana is so flat and monotonous that one observer noted: ". . . the traveler may ride upon the railroad train for hours without seeing a greater elevation than a hay stack or pile of saw dust" (Petty and Jackson, 1966).

Level landforms greatly influence the disposition of water because its lateral movement is curtailed. Water from precipitation may move vertically through the soil depending on the nature of the parent material of the feature. Level surfaces having coarse-textured materials are permeable to vertical water movement and tend to become dry, often excessively so. In warm seasons of low precipitation, water drains rapidly through the soil and is absorbed by plant roots or is evaporated from the soil surface. The presence of layers or bands of heavy-textured material or bedrock at or near the surface slows or restricts downward water movement, thereby subtly or markedly affecting plant composition and growth. Lakes and wetlands of marshes, swamps, and bogs are often found in such terrain.

Most of the world's wetlands are found on relatively flat terrain, especially coastal plains, on old lake beds and areas surrounding lakes, and associated with river floodplains where the ground water table intersects the surface.

Flat terrain also limits air drainage, and flat land surrounded by higher land creates natural sinks for accumulation of cold air at night (Chapter 9). Low night-time temperatures in such frost pockets further influence the composition of vegetation and animal populations in flats where air movement is restricted.

SLOPING TERRAIN

Three shapes of sloping land are recognized: straight or planar (surface with zero curvature), convex (curving outward), concave (curving inward). These types are combined in a variety of ways in different landforms. On convex surfaces water, colloidal organic matter, and nutrients in solution tend to move out of the feature. In contrast, concave surfaces tend to accumulate soil water, nutrients, and organic matter. Thus conditions here are more favorable for plant growth and plants attain larger sizes than individuals of the same species upslope (Chapter 13).

Other important slope features are **aspect**, direction a slope faces, and **slope inclination**, typically expressed as a percentage or degrees from the horizontal. Length of slope (in m or km) is important in affecting plant distribution in certain landscapes; for example, a short, steep slope versus a long, gradual slope. In a study of communities of the broad deciduous swamp area of northern Ohio, Sampson (1930) reported that an abrupt elevational change of 18 to 25 cm above the wet swamp forest of American elm, black ash, and silver and red maples gave sufficient local aeration for American beech and sugar maple or even oak and hickory trees to occur. However, a slope of "three or four feet per mile may lead to an elevation of several feet above the lowest depression yet not be sufficiently drained for beech–maple."

Sloping terrain is ecologically different from flat terrain because lateral drainage of water is possible. Water can move internally or overland along a slope so that undrained or poorly drained soils may seldom occur. Mountains exhibit a stunning array of sloping terrain from rugged features of young mountains such as the Rocky Mountains, Sierra Nevada, Uinta, and Cascade Mountains of the western United States to the relatively low, undulating, and long-weathered Appalachian Mountains of the eastern United States.

In glaciated terrain, end and lateral moraines are often high and moderately-to-steeply sloping features of a landscape, whereas ground moraines are flat, gently sloping, or rolling. Ice-disintegration features or **ice-contact terrain** exhibits hilly topography of kames and eskers with relatively abrupt and short slopes interspersed with marked ice-block depressions termed kettles (often containing lakes and swamps) and localized outwash aprons. Ocean and lake beach ridges, terraces, and dunes exhibit locally distributed rolling features of relatively low stature. The recurrence of landforms and their distinctive parent materials in the landscape typically result in predictable patterns of ecosystem types and plant communities, whether along unglaciated coastlines and lakes or in glaciated boreal and northern landscapes.

SLOPE CHARACTERISTICS

Slope position, slope aspect, and its inclination strongly affect microclimate and soil depth, profile development, and the texture and structure of the surface soil. These in turn influence the composition, development, and productivity of the ecosystem.

Position on Slope

Ridge top or upper convex slope surfaces are exposed to intense solar radiation, experience high winds, and are subject to erosion and soil movement. Therefore, they tend to be drier than is the average for the region. At the other extreme, lower slopes with concave surfaces tend to be sheltered from high winds, subject to accumulation of organic matter and soil rather than to erosion, subject to cold-air drainage, and moister than average for the region. Midslopes are generally intermediate in their characteristics. Level surfaces are very stable, their site properties being determined by climate, parent material, soil texture, and drainage conditions.

Relative elevation, i.e., ridge top; upper, mid, lower slope, is frequently one of the most useful criteria for classifying and mapping forest ecosystems. Practically, position on slope has the advantage of being capable of delineation on aerial photographs viewed stereoscopically. Its significance is documented by the many studies that have found position on slope to be the single most useful factor in evaluation of the growth potential of forest trees (Ralston, 1964; Carmean, 1975, 1977).

Aspect

The orientation of the slope with regard to the sun's position, i.e., its aspect, is of great importance. In the north Temperate Zone, the sun is to the south during the warmest part of the day, and south-facing slopes receive more intense sunlight than any other. At any given latitude then, the hottest and driest sites are those that most nearly face the sun's angle during the middle of the summer day. Both steeper and more gradual slopes receive less insolation. The amount of insolation received on a site governs other related factors, including air and soil temperature, precipitation, and soil water, all of which are important for establishment and growth of plants. For example, in a detailed study of eight environmental factors of four southerly aspects in desert foothills of Arizona, Haase (1970) found the sequence of warmest and driest to be S, SSW, SW, and SSE.

North slopes, on the other hand, receive less sunlight and are invariably cooler, moister, and warm up slower in the Northern Hemisphere (these relations are, of course, reversed in the Southern Hemisphere). East and west slopes show similar but less extreme variation. East-facing slopes are exposed to direct sunlight in the cool of the morning and are normally somewhat cooler and moister than west-facing slopes. In mixed upland oak forests of the Appalachian Mountains, northeast aspects are the most productive, being approximately 15 percent more productive than the south and west aspects, which were the least productive (Chapter 13). Figure 10.2 illustrates the complexity of aspect and slope percent in mountainous terrain. The fine-scale variation in just aspect and slope percent (Figures 10.2*a* and 10.2*b*) emphasizes the local slope heterogeneity of mountainous landscapes that results in marked differences in biota and ecosystem function.

Slope Inclination

The slope gradient or the angle of repose of geologic material is usually measured in terms of percentage or in degrees. The steeper the slope, the greater is the surface per hectare or other area measured horizontally. For this reason, good forest sites of moderate slope usually contain more trees and produce greater yields per hectare (measured horizontally as it always is) than do comparable level sites. A steep slope exhibits more rapid movement of water, snow, and soil than a gentle slope, and the danger of erosion, avalanche, and mass soil movement is much greater. In steep terrain of the Pacific Northwest, heavy winter precipitation and young soils act to initiate subtle, slow mass-movement processes of creep, slump, and earth flow. These processes operate simultaneously and sequentially to bring about rapid soil mass movements of debris avalanches and uprooting of trees (Swanson and Swanston, 1977; Swanson et al., 1982a, b; see Chapter 21). Species dependent on disturbances that expose mineral soil and cause canopy openings

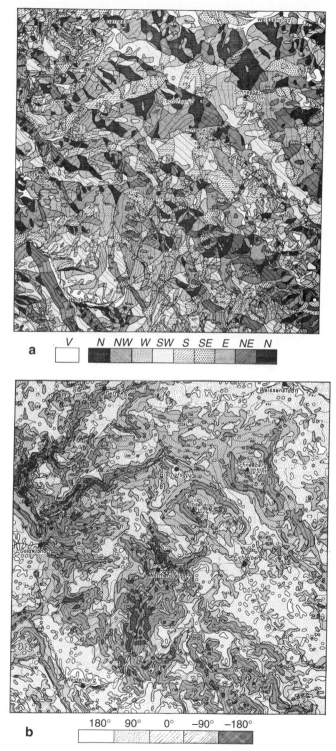

Figure 10.2. Physiographic variation in mountain slopes. *a*, slope aspect in eight 45° categories and level land (V): *b*, slope percentage classes. Spruce Mountains (Fichtelgebirge) of northeastern Bavaria, Germany; study area 347 km². Lower edge of the map is south and is 18.5 km wide. (After Lee and Baumgartner 1966. Reprinted with permission of the Society of American Foresters.)

231

tend to be favored. For example, in southern Wisconsin on steep hillsides, a solid canopy is rarely formed so that light penetration from the side is always possible and species such as black and white oak are favored. Succession from one oak species to another and to more mesic species is most rapid on flat lands where light available to plants of the forest floor is mainly in control of the tree canopy (Curtis, 1959).

Parent Material in Relation to Landform

Parent materials, whether residual or transported, are typically landform specific (Chapter 11 and Figure 11.1). Residual limestone or sandstone rock, upon which different soils develop, support quite different plant communities. Through long-term weathering of their constituent rock, stream action, and geomorphic processes such as avalanche and debris flow, mountains take on different configurations. Wind action initiates different land shapes, sand dunes and loess deposits, that reflect the particle size of the parent material: sand versus silt. The multiple landforms of river floodplains, front, levee, first bottom, backswamp, reflect the nature of their parent material and flooding patterns. Flat (old lake beds, glacial outwash plains) and hilly (moraines, ice-contact terrain) glacial forms have characteristic parent materials that distinguish them and, together with the shape of the feature, help determine species composition, productivity, and successional pathways.

Position of Landform in the Landscape

A specific landform (outwash plain, ravine, ridge) may occur at different positions in the landscape and thereby affect plants and animals differently. Landforms (mountain slopes, valleys, plateaus) that are in the rain shadow of a mountain range exhibit different climate, vegetation and hence a different soil development than similar landforms on the windward side where precipitation is markedly higher for a given elevation. An exposed ridge high on a mountain may support different plants and provide a markedly different growth opportunity than one with the same rock and soil constitution in a lower and protected position. In non-mountainous glacial terrain, an elevated, sandy outwash terrace provides more favorable growing conditions than an equally dry terrace below it in a cold, frost-prone basin. Thus a given landform, due to its position in the landscape, may provide markedly different site conditions for plant establishment and growth.

A related phenomenon is the position of landforms in relation to one another. A forested wetland with its relatively high water table affects land adjacent to it much more than land at a greater distance. At a broad scale, lands adjacent to oceans or large lakes are more strongly affected by the water body than lands at some distance inland. At a local scale, a first-bottom ecosystem adjacent to a river is markedly different in soil and vegetation than an elevated second-bottom ecosystem due to its close proximity to the river and greater likelihood of flooding. Generally, the smaller the ecosystem the more it is influenced by surrounding ecosystems and their processes.

In summary, many different physiographic characteristics act together to influence the biotic composition of landscape ecosystems and their structure and function. These characteristics have been used in many instances to provide the stable basis for the management of forested ecosystems. For example, in glacial landscapes, Angus Hills (1952) pioneered the use of physiography (integrating position on slope, aspect, slope gradient, and parent material) as the basis of a system to determine forest composition and productivity in Ontario, Canada (Chapter 13). In a given macroclimate, the importance of local topography as it determines late-successional forest composition is shown in Figure 10.3. Whereas sugar maple–beech is perceived as the late-successional forest on "normal" microclimate and moist soil, topographic-mediated differences in microclimate (warm ver-

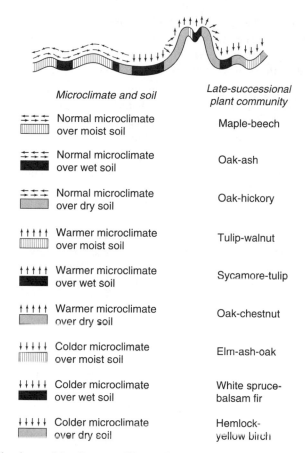

Microclimate and soil	Late-successional plant community
Normal microclimate over moist soil	Maple-beech
Normal microclimate over wet soil	Oak-ash
Normal microclimate over dry soil	Oak-hickory
Warmer microclimate over moist soil	Tulip-walnut
Warmer microclimate over wet soil	Sycamore-tulip
Warmer microclimate over dry soil	Oak-chestnut
Colder microclimate over moist soil	Elm-ash-oak
Colder microclimate over wet soil	White spruce-balsam fir
Colder microclimate over dry soil	Hemlock-yellow birch

Figure 10.3. Physiographic diagram illustrating the effect of local topography (slope position and aspect) in determining late-successional species composition of landscape ecosystems in the Temperate Continental Zone of southern Ontario, Canada. (Modified from Bailey, 1988 and Hills, 1952.)

sus cold) and soil water can bring about marked differences in late-successional communities.

MULTIPLE ROLES OF PHYSIOGRAPHY

Geologic and geomorphic processes create the physiographic features of the landscape, and physiography and its characteristics already described strongly influence ecosystem structure and function. This interrelated suite of characters influences ecosystem processes at regional and local scales and markedly affects soil development, climatic and disturbance regimens, and the distribution of vegetation. The roles of physiography in affecting ecosystem function are several (Swanson et al., 1988):

Quantity of materials received. The kind of landform and its elevation, slope aspect, inclination, and position influence the quantity of solar radiation, air, water, nutrients, and other materials received at the site.

Flux of materials. The same factors mediate the flow of air, water, materials (organic and inorganic particulate matter, dissolved materials), organisms, propagules, and energy within and across ecosystem boundaries by: (1) influencing gravitational gradients, (2) guiding flows of wind and water, and (3) forming barriers to movement.

Disturbance. Landforms and their slope and aspect features influence the spatial pattern, frequency, severity of natural disturbances caused by fire, wind, water, avalanche and soil movement, and human-caused disturbances of erosion, pollutants, and barriers to natural fluxes of water and nutrients.

Spatial position of landforms. The juxtaposition of one landform in relation to another may affect regional and local climate and hydrology, movement of materials, and disturbance regimes.

Through these roles, physiography directly or indirectly affects all aspects of plant life: regeneration, geographic distribution and abundance, growth and productivity, and death. In the following sections, we examine the influence of physiography at multiple spatial scales on landscape ecosystems and their biota. We begin at the subcontinental level and examine physiography to the microtopographic level.

PHYSIOGRAPHIC DIVERSITY, LANDSCAPE ECOSYSTEMS, AND VEGETATION

Throughout the world, distinctive patterns of ecosystems and the physiognomy and composition of vegetation are strongly related to geomorphology and geomorphic processes. Vegetation, in turn, influences the physiographic processes. Examples from mountains, plains, and other settings illustrate this interconnectedness.

Mountainous Physiography

Rugged terrain resulting from rigorous geomorphic processes produces close linkages between environment and organisms. Several examples serve to illustrate the intricacy and critical importance of physiography and mountain landforms.

MOUNTAINOUS TERRAIN OF CALIFORNIA AND THE PACIFIC NORTHWEST
The significance of physiography and specific landforms are conspicuous and complex on the grand scale of mountains throughout the world. A gradient from the coast across the San Joaquin Valley and the Sierra Nevada Mountains in California serves to illustrate the effect of physiographic position of valley and mountain features on precipitation and temperature that markedly pattern the distribution of vegetation from near sea level to over 3,650 m (12,000 ft) (Figure 10.4). East of the dry valley and oak woodland, conifer-dominated forest ecosystems occur on western slopes above about 610 m (2,000 ft). Annual precipitation increases from 50 cm in the oak woodland to 125 cm in the red fir forest. Conifers maintain their dominance even at higher elevations as precipitation declines and where low temperature, very short growing season, and thin soil prevail. On the eastern slopes of the Sierra Nevada range, precipitation drops sharply, and only the most drought-adapted species can survive in this rain shadow. Figure 10.4 illustrates the overall control of a mountain landform on macroclimate and the striking change in vegetation with elevation.

Further north, no better example of physiographic diversity is found than in the Pacific Northwest states of Washington and Oregon. Geomorphic processes from mountain building to local erosion and mountain stream dynamics have created highly varied landscapes and an enormous diversity of regional and local ecosystems (Franklin and Dyrness,

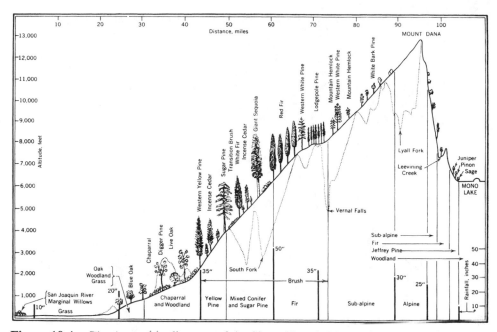

Figure 10.4. Physiographic diagram of the Sierra Nevada Mountains illustrating the distribution of dominant plants in relation to elevation and precipitation. Variations in topography along this generalized gradient are shown by the dotted lines. They define main stream valleys and ridges in this rugged terrain. (After Show and Kotok, 1929. Reprinted with permission as modified in Harlow et al. *Textbook of Dendrology,* © 1996 by McGraw-Hill, Inc.)

1980). The broad physiographic provinces (i.e., regional ecosystems) and the close association of broad vegetation types with them are shown in Figure 10.5. Coastal marine terraces and broad inland valleys (Puget Sound area and Willamette Valley) exhibit the lowest relief and support distinctive vegetation types (Figure 10.5b). At the other extreme are the mountains of the northern Cascade Range and Olympic Mountains of Washington (Figure 10.5a) where relief from ridge crest to adjacent valley floor may be 1,000 to 1,400 m or 20 times the typical height of old-growth forest (Swanson et al., 1990). Again, the vegetation (Figure 10.5b) reflects these major differences in physiography and climate. In the broad regional overview, physiographic provinces delineate major landscapes characterized by differences in climate and geology. The geologic substrates contrast sharply in physical and chemical properties. Gradients in precipitation, temperature, and concomitant changes in vegetation are associated with gradients in elevation, latitude, aspect, and distance from the ocean (Swanson et al., 1990). In addition, other factors that characterize landscape ecosystems and the pattern vegetative types, such as wildfire, snow accumulation and snowmelt, and wind velocity, are all markedly influenced by physiography.

In the Pacific Northwest, the position of a given landform in spatial relation to other landforms is especially noteworthy. For example, the position of marine terraces adjacent to the ocean, the position of the Willamette Valley between the Coast and Cascade ranges of mountains, and the position of the Columbia Basin in the rain shadow of the Cascade Mountains (Figure 10.5a) markedly affect ecosystem processes and biota of their ecosystems. In addition, the parent material of Pacific Northwest landforms is important in determining the great diversity of soils (Franklin and Dyrness, 1980). Finally, regional and local physiographic features, through their influence on temperature and precipitation,

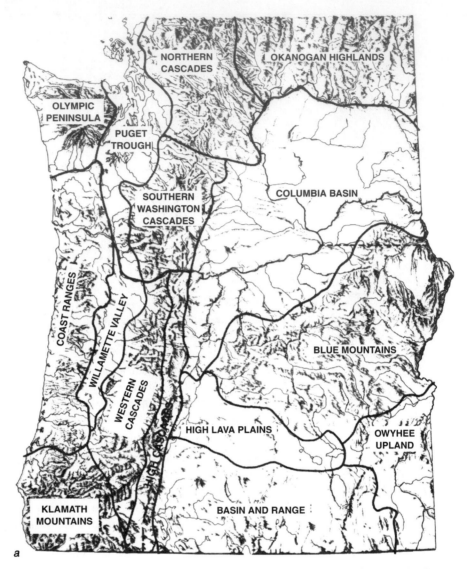

Figure 10.5.   Comparison of physiographic provinces and generalized vegetation areas of Oregon and Washington. *a*, Physiographic–geologic provinces, *b*, Major vegetation areas. (After Franklin and Dyrness, 1980.)

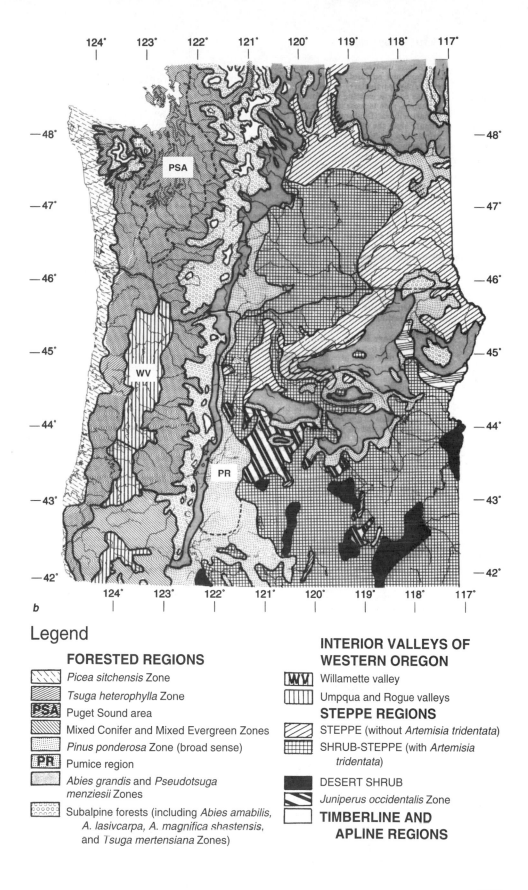

Legend

FORESTED REGIONS

- *Picea sitchensis* Zone
- *Tsuga heterophylla* Zone
- **PSA** Puget Sound area
- Mixed Conifer and Mixed Evergreen Zones
- *Pinus ponderosa* Zone (broad sense)
- **PR** Pumice region
- *Abies grandis* and *Pseudotsuga menziesii* Zones
- Subalpine forests (including *Abies amabilis*, *A. lasivcarpa*, *A. magnifica shastensis*, and *Tsuga mertensiana* Zones)

INTERIOR VALLEYS OF WESTERN OREGON

- **WV** Willamette valley
- Umpqua and Rogue valleys

STEPPE REGIONS

- STEPPE (without *Artemisia tridentata*)
- SHRUB-STEPPE (with *Artemisia tridentata*)
- DESERT SHRUB
- *Juniperus occidentalis* Zone

TIMBERLINE AND APLINE REGIONS

provide the template for the regional and local genetic differentiation of tree species (Chapter 4).

The broad zones of natural vegetation, from coastal Sitka spruce forest and dry steppes to sub-alpine timberlines, correspond closely to the physiographic provinces mapped and described by Franklin and Dyrness (1980; Figure 10.5*b*). On a local scale, forest communities on a 900-m mountainside segment of Mount Rainier in western Washington illustrate changes in dominant species along an elevational gradient (Figure 10.6). The different stand ages and avalanche tracks reflect landform influence on fire history and soil stability. Vegetational zones may occur as sequential belts on mountains slopes, but more often they interfinger (Figure 10.7). Because the site conditions in valleys are markedly different in temperature and soil water than on adjacent slopes of similar elevation, a species or a given community attain its lower elevation limit in valleys and its highest limits on ridges. Zones are relatively broad so that a given species, Douglas-fir for example, may occur in moist sites in the ponderosa pine Zone and on relatively dry ridges higher up in the montane grand fir Zone. The same principle of broad and seemingly tidy vegetation (life) zones (Merriam, 1890) dissected by ridge and valley systems with marked temperature variation (MacDougal, 1900 as cited by Maienschein, 1994) was demonstrated a century ago for the San Francisco Peaks of northern Arizona.

Physiography, which largely controls climate and soil conditions, is the primary ecosystem component responsible for the diversity of these remarkable forests. The diagram in Figure 10.8 illustrates that forest vegetation zones may be sharply contrasted using two factors that are strongly influenced by physiography: precipitation (drought stress) and temperature (temperature growth index). Oak–juniper woodlands and ponderosa pine zones form a group of xerophytic forests that occur in eastern Washington and Oregon and in southwestern Oregon. The Mixed Conifer and Coastal Temperate groups form extensive temperate forests. Within this group are the unique temperate conifer forests of the world (zones: 1, *Tsuga heterophylla*; 11, *Abies grandis*; 12, Mixed evergreen; including Sitka spruce, Douglas-fir, and others) (Franklin et al., 1981). They grow under highly favorable temperature and moisture conditions and reach maximum development in size and longevity. Finally, the subalpine and boreal groups are severely constrained by growing season length and harsh mountain environments.

In this section, we have emphasized the major effects of physiographic features on broad vegetation zones. We should also stress their effects at the local scale. As described in Chapter 4, sharp local differences in microclimate associated with aspect and elevation elicit significant genetic differences in populations of Douglas-fir. In addition, landform features of the riparian zone, the interface between terrestrial and aquatic ecosystems, greatly influence the structure and functioning of riparian and aquatic ecosystems.

PHYSIOGRAPHY AND FORESTS OF THE CENTRAL APPALACHIANS

Throughout the Appalachian Mountains and their different sections (Southern, Central, Northern, Blue Ridge, Ridge and Valley, etc.) the occurrence of forest species and communities are closely associated with the local geologic and topographic conditions (Hack and Goodlett, 1960; Leak, 1982). The classic and comprehensive study of geomorphology and forest ecology by Hack and Goodlett (1960) is outstanding for its detail in demonstrating such interrelationships. They studied a 142 km^2 (55 mi^2) densely-forested area in the drainage of the Little River that is located in the Ridge and Valley Section of the Central Appalachians in northern Virginia (Fenneman, 1938). They demonstrate that the local distribution of species and forest communities in an elevational range from about 600 to 1,300 m is roughly coincident with well defined differences in topographic form and the associated parent materials and disposition of water.

a

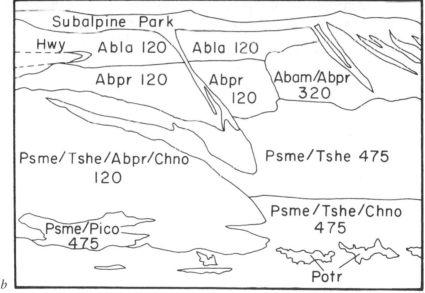

b

Figure 10.6. Photograph, *a*, of the pattern of forest stands on Sunrise Ridge, White River drainage of Mount Rainier, Washington. The east- southeast facing slope extends from 900 to 1,800 m elevation. Diagram, *b*, shows the forest stand types and ages reflecting landform influences and disturbance. Snow avalanche tracts cut through the upper elevation forests. The White River (bottom of photograph and diagram) has experienced major mudflows generated on Mount Rainier and extensive lateral channel migration. Tree species: Abla, *Abies lasiocarpa* (subalpine fir); Abpr, *Abies procera* (noble fir); Psme, *Pseudotsuga menziesii* (Douglas-fir); Tshe, *Tsuga heterophylla* (western hemlock); Chno, *Chamaecyparis nootkatensis* (Alaska-cedar); Pico, *Pinus contorta* (lodgepole pine); Potr, *Populus trichocarpa* (black cottonwood). (After Swanson et al., 1990. Reprinted from *Changing Landscapes: An Ecological Perspective,* © 1990 by Springer-Verlag. Reprinted with permission of Springer-Verlag.)

239

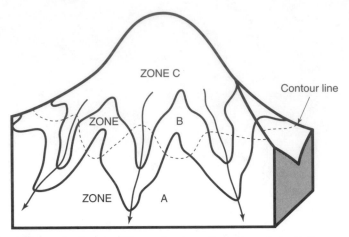

Figure 10.7.   Schematic diagram illustrating interfingering of vegetation zones in mountainous topography. Streams are shown running downslope in valleys. A given vegetation zone is distributed higher in elevation on ridges than in valleys. (After Franklin and Dyrness, 1980.)

The local topography of a typical first-order valley is shown in Figure 10.9*a* (Valley 1). For a north-facing valley (Valley 3), the relationship of forest communities to valley position is shown in Figure 10.9*b*. Three general forest types—dry pine forest, dry-mesic oak forest, and mesic "northern hardwood" forest—are found in characteristic positions in such valleys. Pine or pine–oak forests, dominated by pitch pine, table mountain pine, and several oaks, are characteristically found on noses, ridges, and other slopes that are convex away from the mountain (Figure 10.9*a*). Dry-mesic oak communities generally are found on the straight slopes between nose and valley but sometimes occur in drier hollows. These forests are variable in composition but consist largely of oaks; pines and basswood, sugar maple, and yellow birch are absent or rare. Mesic northern hardwood communities of sugar maple, basswood, yellow and black birches, and northern red oak generally are restricted to moist hollows and other concave slopes where water is concentrated. This mesic community occupies the floodplain of larger valleys, extends as a narrow strip up the floors of the smaller valleys, and in the first-order valleys enlarges into a tadpole-shaped area in the hollows at the valley heads. Trees of the northern hardwood type in the hollow of north-facing Valley 3 (Figure 10.9*b*) are the tallest, largest, and have the greatest volume compared with those of the other two types.

The strong coincidences between the local distribution of species and communities and their position in a valley are modified by: (1) the size of the valleys, a relatively large area being essential for maintenance of the northern hardwood community in a hollow; (2) the orientation of the valleys and side slopes, north and east slopes being commonly forested with northern hardwoods and south and west slopes being typically oak; and (3) the nature and attitude of the bedrock, formations favoring soil-water accumulation and retention being more favorable for the more mesophytic species. Whereas northern hardwoods occupy the hollow of Valley 3 (Figure 10.9*b*), in Valley 1 the oak forest type is found in the hollow as well as the side slopes (Figure 10.9*a*). This occurrence is due to the southwest-facing orientation of Valley 1 that leads to relatively drier soils in this hollow compared to the north-facing hollow of Valley 3 where northern hardwoods prevail. Not surprisingly, due to better soil water and nutrient conditions, trees of the oak type in the hollow of Valley 1 exhibit greater diameter and height growth than those of the same species on the side slope. In addition, the ground-cover species are markedly different in

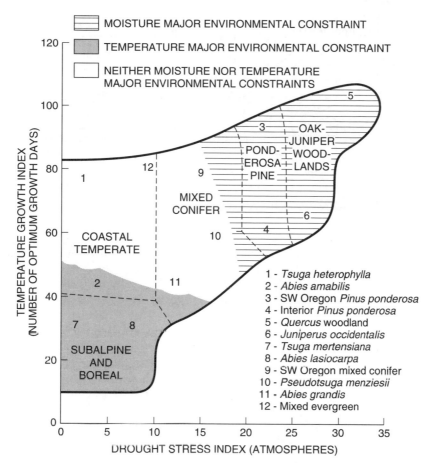

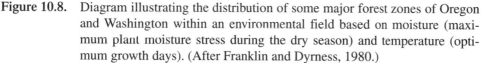

Figure 10.8. Diagram illustrating the distribution of some major forest zones of Oregon and Washington within an environmental field based on moisture (maximum plant moisture stress during the dry season) and temperature (optimum growth days). (After Franklin and Dyrness, 1980.)

the hollow compared with those on side slopes. Thus we recognize three functionally different landscape ecosystem types (nose; side slope; and hollow, footslope, and channel) and two forest cover types (oak forest and northern hardwood forest) in Valley 1.

Detailed studies of vegetation were made in the axes of five additional first-order valleys. Species composition varied from valley to valley and from hollow to hollow. The hollows of three valleys supported northern hardwood forest, whereas the hollows of the others supported oak forest similar to that of Valley 1 in Figure 10.9*a*. The valleys supporting northern hardwoods exhibit topographic features that provide a relatively moist environment. They face either northeast or northwest, are relatively large and are deeply cut into the mountain, have bedrock favoring water retention, and are at a slightly higher elevation.

Hack and Goodlett (1960) stressed the great importance of geomorphic processes in controlling the local distribution of species. Species occurrence is closely related to well-defined differences in topography, which is in turn controlled by geomorphic processes. Cloudbursts and ensuing run-off cause debris avalanches and floods that greatly affect local site conditions. Stream sculpture and debris avalanches produce slopes that are concave-upward, thus favoring mesic northern hardwood species. In contrast, the process of

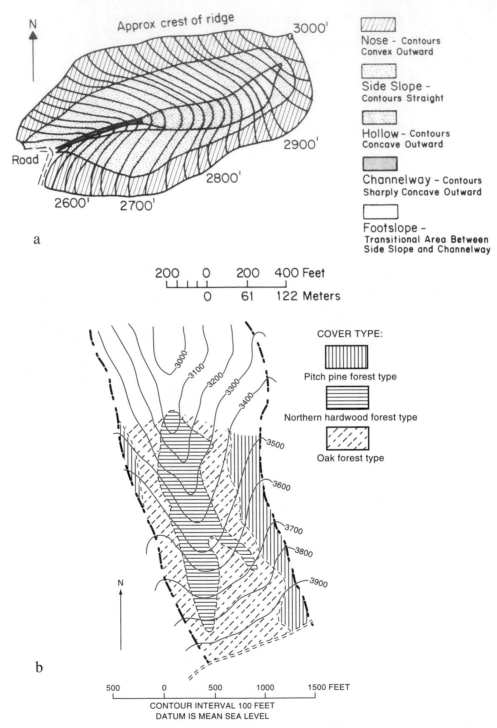

Figure 10.9.   Contour maps of typical first-order valleys, Little River basin, central Appalachian Mountains, northwestern Virginia. *a*: Map of landforms of upper part of Valley 1, west side of Crawford Mountain. Legend gives landform types of first-order valleys. *b*: Map of forest type distribution for Valley 3, northeast of Reddish Knob. (After Hack and Goodlett, 1960.)

soil creep produces convex-upward slopes that typically support pine forests. Vegetation reflects local differences in these processes that determine slope shape, character of surficial materials, and soil water and nutrient disposition. Vegetation also influences geomorphic processes by its density and rooting depth. Generally, the present landscape is the product of a long period of interaction between vegetative and geomorphic processes.

Flatlands

Flat and gently sloping terrain is less diverse physiographically. However, this has distinctive effects on the distribution of plants and biotic communities by its parent material and subtle differences in topography. Flat land may promote fire or preclude it, speed or retard nutrient cycling, and favor early- or late-successional species.

THE GREAT PLAINS

In the Great Plains of North America, lying immediately east of the Front Range of the Rocky Mountains, grasslands occupied most of the smooth topography—the flat or rolling plains and gentle slopes—at the time of settlement (Wells, 1970). A significant feature of the plains, however, is the existence of forests, not only riparian woodlands along rivers, but upland forests. However, they were more or less restricted to rough, dissected, boulder-strewn escarpments. In such sites, the forest was sheltered from wind-driven prairie fires. Trees also occurred on relatively level sites on the leeward sides of lakes and rivers that provided firebreaks. Borchert (1950) described how the distribution of treeless grasslands was guided by the "master hand of climate." However, evidence also supports the argument that the "master hand of physiography," through its influence on climate and fire, is the key (Wells, 1970). Across the entire width of the Central Plains the abrupt topographic segregation of woodland and grassland prevails. The rougher and the more dissected the topography, the greater the occurrence and spread of forest at the expense of grassland (Wells, 1965).

Besides the dry climate (itself a result of regional physiography), the vast flat or rolling smoothness and continuity of the relatively undissected sedimentary deposits play a major role in supporting the grassy flora and setting the stage for fire. Wind-swept grass fires, lightning and human ignited, spreading across a flat or rolling plain would continue indefinitely until quenched by rain or stopped by an abrupt break in topography. Sparsely-grassed escarpments harbored fire-sensitive trees that would encroach on grassland in the fire-free periods. Recurrent fires, however, would readily destroy tree seedlings and saplings, and particularly those of conifers that lack the ability to sprout. Thus the mosaic of landscape ecosystems of the Plains is due not simply to climate but to the regional flatness and continuity of the physiography; the fuel-producing, annual dieback of grasses; and the *sine qua non*, fire (Wells, 1970).

SOUTHWESTERN PLATEAUS AND PINE FORESTS

Plateaus and gentle mountain slopes in the Southwest, at higher elevations with somewhat moister climate than the plains, favored the spread of fire through the grassy ground cover and the dominance of ponderosa pine forests (Cooper, 1960). As late as the 1850s, an almost continuous band of ponderosa pine forest, nearly 480 km long and 40 to 65 km wide spread across northern Arizona. Lt. Edward Beale, leader of a famous camel expedition through northern Arizona, observed:

> We came to a glorious forest of lofty pines, through which we have travelled ten
> miles. The country was beautifully undulating, and although we usually associate

> *the idea of barrenness with the pine regions, it was not so in this instance; every foot being covered with the finest grass, and beautiful broad grassy vales extending in every direction. The forest was perfectly open and unencumbered with brush wood, so that travelling was excellent (Beale, 1858)*

The parklike natural conditions were maintained by light surface fires that were set by lightning or Native Americans. The fires burned throughout the pine forest at regular intervals of 3 to 10 years. They acted as natural thinning agents, maintained the patchy character of pine forests, restricted invasion by other tree species, and reduced surplus fuel. Fire exclusion and grazing have dramatically changed the character of the southwestern pine forests (see Chapters 14 and 16).

PINE SAVANNAS OF THE WESTERN GREAT LAKES REGION

Flat and gently rolling areas of the western Great Lakes region and adjacent Canada, coupled with hot spring–summer periods and droughty, sand soils, favor frequent fires and tend to perpetuate fire-dependent species and communities. In parts of Minnesota, northern Wisconsin and Michigan, and southern Ontario, flat outwash or rolling plains, nutrient-poor, sand soils, and flammable vegetation led to frequent fires that maintained communities of nearly pure jack pine or jack pine and oak (Curtis, 1959; McAndrews 1966; Vogl, 1964; 1970). Unique to the upper Great Lakes region (Heinselman, 1981a), and termed jack pine savanna or barrens, these lands were estimated to occupy over 400,000 ha in Michigan (Voss, 1972, p. 62) and over a million hectares in Wisconsin (Curtis, 1959). The account of a reporter accompanying an expedition by botanists William J. Beal and Liberty Hyde Baily across the northern lower peninsula of Michigan in June, 1888 carried the title: A BARREN WASTE. MOSQUITOES THE LARGEST AND MOST PROMISING PRODUCT YET FOUND (Voss and Crow, 1976). The landscape is vividly described:

> *These plains are clothed with a scant vegetation, the most conspicuous and common characteristic plant being the jack, or scrub pine. . . . Fires often sweep over the plains, destroying all vegetation. Young pines soon spring up in these burnt areas, . . . These groves of young trees are often miles in extent, and are so dense that the traveler is completely hidden from view at the distance of a few paces. . . . The sward-like glades are found to be clothed with stunted huckleberries or blueberries, miserable growths of sweet fern, and a few other pinched and starved plants which can endure the heat and dryness of the sands. . . . On the whole the plains are exceedingly uninviting in aspect.*

Although typically considered a single community type, i.e., pine barrens, many ecosystem types occur in these landscapes due to local topographic and soil differences (Zou et al., 1992; Barnes, 1993). For some rare animals and plants, such as the Kirtland's Warbler, Allegheny Plum, and Hill's thistle, these systems are not so uninviting.

TILL PLAINS OF THE MIDWEST

The extensive till plains of Indiana and Ohio are relatively flat or gently rolling and crossed only by a number of low moraines, ridges, and shallow drainage ways. They form the central core of the Beech–Maple Forest Region of Braun (1950; Figure 2.8). The extensive, flat topography, coupled with heavy-textured soils with moderate to poor drainage favor overwhelming dominance of American beech and sugar maple to the virtual exclusion of other tree species and to fire. Both beech and sugar maple are highly tolerant of shaded understory conditions, and both cast an enormous amount of shade such that virtually no other tree seedlings can exist for long periods of time. In addition, great amounts of moist leaf litter cover the forest floor, denying fire a flammable fuel, or following rapid decomposition, insufficient fuel to carry a fire. Therefore, through the indirect and direct effects and interactions of all factors, fire is virtually excluded.

SOUTHEASTERN AND SOUTHERN COASTAL PLAIN

Along the southeastern and southern coast of the United States, from New Jersey to Texas, stretches a relatively flat plain, supporting a patterned mosaic of landforms. These include flat plains and terraces, embayed rivers, and sandy ridge and swale terrain near the Atlantic coast that merges into rolling hills adjacent to the Piedmont Plateau. Early in the 20th century, ecologists such as Roland Harper (Harper, 1906, 1911, 1914; Delcourt, 1978) and J. W. Harshberger (1903) were geoecology pioneers who recognized the significance of physiography in controlling plant occurrence. Christensen (1988a) and Myers and Ewel (1990) provide detailed descriptions of Coastal Plain ecosystems and the complex vegetation relationships with physiography, soil, hydrology, and fire.

In mountainous areas, vegetation zones and community types may be widespread and change gradually. However, in only several hundred meters horizontal distance with a 10 m elevation rise in the Coastal Plain, community physiognomy may vary from grassland and savanna to shrubland, to needle- and broad-leaved sclerophylous woodland, and to rich mesophytic forest (Christensen, 1988a). Also, the most diverse assemblage of freshwater wetlands in North America occur here (Ewel, 1990). Such remarkable vegetative differences are due primarily to physiography and fire and the ecosystem processes they influence. Water table, drainage patterns, and hydroperiod (length of time soils are saturated during a year); soil water and nutrient cycling; and fire frequency and severity are essentially controlled by regional and local physiography features.

Fire-dependent communities prevail throughout much of the Coastal Plain (Chapter 16), and the occurrence of rare local ecosystems with late-successional mesic communities is the result of favorable physiographic and soil conditions together with their position in the landscape. For example, Figure 10.10 illustrates the position of four distinctive ecosystem types in western Louisiana that were reconstructed using the American Land Office Survey of 1821 (Delcourt and Delcourt, 1974). The mesic magnolia–holly–beech upland hardwood community occurred on thick loess deposits of the upland, the magnolia–beech–holly bottomland hardwoods in ravine and river lowlands, and the tupelo-gum–cypress community in swamp land adjacent to the Mississippi River (Figure 10.10). Where the loess cap thinned out toward the north-eastern part of the parish a mixed oak–pine–beech ecosystem was found. The dominance of southern magnolia, American beech, and their mesophytic associates in the Coastal Plain was largely due to their protection by firebreaks from natural and human-set fires.

FLOODPLAINS

Throughout the Ecosphere, large rivers exhibit similar fluvial (river) processes and patterns of physiography: specific landforms are identifiable, and complex patterns of ecosystems and microsites occur within landforms. Rivers and their landforms are linear features so that similar ecosystems with characteristic riverine vegetation may extend for tens or hundreds of kilometers along a major river, at least in times before rivers were tamed by dams, channels, and human-made levees. The geomorphic processes of flooding, transport and deposition of sediments, and the erosive and abrasive forces of ice and water movement impress a certain uniformity of physiography on the landscape, whether a lowland river or a mountain stream. Regardless of size, stream floodplains or **riparian** zones are functionally three-dimensional zones of direct interaction (i.e., in flooding, bank cutting, and sedimentation) between terrestrial and aquatic ecosystems (Gregory et al., 1991).

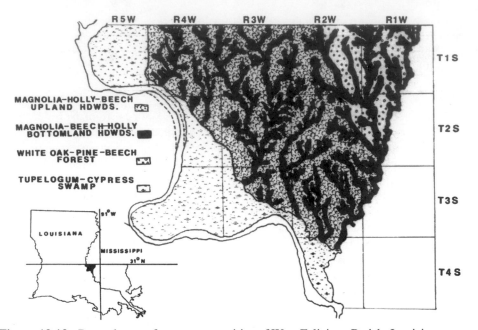

Figure 10.10. Presettlement forest communities of West Feliciana Parish, Louisiana, as reconstructed from 1821 land survey records. The dashed line represents the position of the Mississippi River in 1821. (After Delcourt and Delcourt, 1974. Reprinted with permission of the Ecological Society of America.)

Riparian characteristics and processes have received detailed attention (Leopold et al., 1964; Swanson et al., 1982b; Brinson, 1990; Gregory et al., 1991; Malanson, 1993), and special consideration is given to riverine ecosystems of mountain landscapes in Chapter 21.

The pattern of zonation of landforms, ecosystems, and their vegetation is generally similar for large lowland rivers where lateral channel migration is the dominant process of land form development on valley floors. However, differences arise due to geographic location, size, and velocity of the stream, kinds of materials transported, and the physiography and parent material of adjacent terrain (Brinson, 1990). The river constantly meanders in the floodplain—cutting away the outer bank, and depositing sediments and forming point bars and new land downstream on the opposite, inner, river bank (Figure 10.11). Sediments during floods build the level of new land above ordinary high-water stages. Flood waters deposit coarser sediments near the river channel, forming relatively high, well-drained ridges or natural levees. "Gallery" or riverside forests are typical of this land-form where the edge effect of the river enables trees to lean over the bank and develop large crowns. The riverside species not only must tolerate periodic inundation but drought when the coarse sediments drain rapidly and water levels drop. Silts and clays are deposited behind the levee in a low, poorly-drained zone or backswamp of the first bottom. A series of bottoms or terraces typically lie behind the first bottom and are progressively less frequently flooded.

Additional physiographic variation and vegetation patterning occurs as entire channels are abandoned and oxbow lakes form and fill in with sediment and vegetation. Oxbow depressions and lakes (sloughs) support plant species with the greatest adaptation to inundation and anaerobic soils. Most floodplains exhibit ridge and swale topography where channels, point bars, levees, and backswamps were cut off and abandoned by the

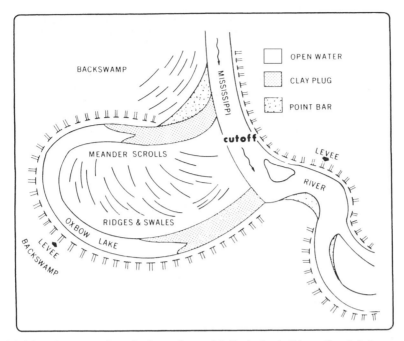

Figure 10.11. Diagram of typical section of Mississippi River floodplain near False River, Louisiana. Physiographic features include: (1) natural levees adjacent to the channel; (2) ridge-and-swale topography (meander scrolls) associated with lands on and between former point bar/levee deposits as the channel migrated laterally and downslope; (3) oxbow lake where former channel has been cut off; (4) backswamps of lower topography behind the levees, and (5) point bars on inside curve of channel where deposition is rapid. (After Brinson, 1990. Reproduced from *Forested Wetlands* with kind permission of Elsevier Science-NL, Amsterdam, The Netherlands.)

meandering stream (Figure 10.11). As a result, an enormous number of landforms and microsites are created due to fluvial processes. Many different tree, shrub, vine, and herbaceous species assort themselves throughout this complex pattern of sites based on their physiologic and genetic ability to tolerate different hydroperiods and substrates.

A generalized cross-sectional diagram of southeastern river bottomlands (Wharton et al., 1982) illustrates the pattern of land form zonation and corresponding species occurrence (Figure 10.12). Although rivers vary in the number and kinds of landforms and the names associated with these features, we can identify in Figure 10.12 typical landforms of riverine ecosystems: (1) the river channel margin where a point bar with increasing elevation by deposition becomes new land and then the "front," (2) the ridge or natural levee where coarse sediments are deposited, and (3) back of the levee a series of bottoms or terraces. The first bottom is characterized by marked local relief of ridges, swales (areas between ridges), and oxbows or sloughs. The range in elevation between ridge tops and slough bottoms may be as little as 2 m. Finer sediments are deposited behind the levee, forming low, broad areas of poorly-drained, slackwater clay or silty clay soils. Higher land of the second bottom is also characterized by local relief, but, because it is flooded less frequently and for a shorter time, a different set of species occurs on its better-drained sites.

Patterns of species and community occurrence mirrors the pattern of topography, soil, and flooding frequency and duration. Trees reach huge sizes and communities attain

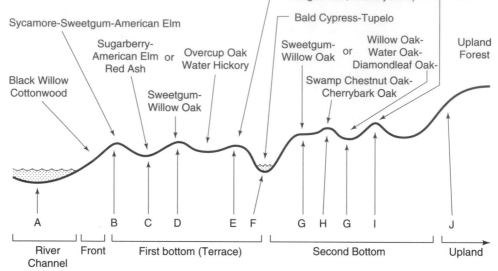

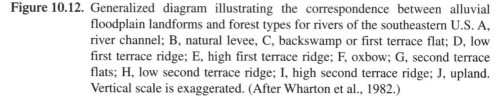

Figure 10.12. Generalized diagram illustrating the correspondence between alluvial floodplain landforms and forest types for rivers of the southeastern U.S. A, river channel; B, natural levee, C, backswamp or first terrace flat; D, low first terrace ridge; E, high first terrace ridge; F, oxbow; G, second terrace flats; H, low second terrace ridge; I, high second terrace ridge; J, upland. Vertical scale is exaggerated. (After Wharton et al., 1982.)

extremely high productivity in many bottomland ecosystems. Sweetgum (Figure 10.13) reaches large sizes (up to 4 m in diameter) on fertile, first-bottom sites. Sycamores on the Wabash River bottomlands in Indiana once attained diameters over 4.6 m. The physiological tolerances of species to flooding related to topographic positions has been intensively studied (Gill, 1970; Hook, 1984), and the effects of flooding are reviewed by Gill (1970). Associated with glacial meltwater channels and floodplains throughout eastern North America are swamp ecosystems of diverse kinds. Those of the Coastal Plain are described by Penfound (1952), Monk (1966), Christensen (1988a), and Ewel (1990).

A great many studies emphasize floodplain landforms and vegetation for rivers from all regions of the United States, including northern Alaska (Bliss and Cantlon, 1957), Oregon (Hawk and Zobel, 1974), North Dakota (Johnson et al., 1976), Texas (Chambless and Nixon, 1975), New Jersey (Wistendahl, 1958; Frye and Quinn, 1979), Virginia (Osterkamp and Hupp, 1984; Hupp and Osterkamp, 1985), and Indiana (Lindsey et al., 1961). An excellent review of North American rivers is presented by Brinson (1990), and those of other continents are described in a book on forested wetlands of the world (Goodall et al., 1990).

PHYSIOGRAPHY AND FIREBREAKS

Firebreaks consist of rivers and streams, lakes, and rough topography. Whereas fire spreads rapidly over flat land, rough topography disrupts the continuity and flammability of fuels, creates erratic wind movements, retards air flow, and generally slows and stops advancing fires (Grimm, 1984; see Chapter 16). In the Great Plains, physiographic fea-

Figure 10.13. Giant sweetgum trees in old-growth bottomland hardwood forest near Holly Bluff, Mississippi. (U.S. Forest Service Photo.)

tures such as escarpments and rivers created natural firebreaks and enabled trees to survive in an otherwise flat landscape that was an effective firepath. Farther east, in south-central Minnesota, firebreaks, more than any other factor, enabled tall, dense deciduous forests to exist on prairie soils (Daubenmire, 1936; Grimm, 1984). Early French explorers discovered these Big Woods (Bois Grand, Bois Fort) ecosystems that were dominated by fire-sensitive species such as American elm, basswood, sugar maple, and hop-hornbeam.

In northern and boreal forests, physiographic features also affect fire movement and, hence, forest composition. In the Boundary Waters Canoe Area of northern Minnesota, Heinselman (1973) reported that the location, size, shape, and compass alignment of physiographic features (lakes, streams and wetlands, bedrock ridges, valleys and troughs) were related to historic fire patterns and distribution of vegetation because they were either firebreaks or firepaths.

In the Coastal Plain of the southern United States, the highly understory-intolerant and fire-dependent longleaf pine dominated presettlement forests throughout much of the area (Marks and Harcombe, 1981a,b; Christensen, 1988a; Myers, 1990). In 1911, the widely traveled Roland Harper observed that all longleaf pine forests he had seen in 200 counties in 7 states all bore the marks of frequent fires. However, he observed that firebreaks enabled mesophytic communities to form. For example, in central Florida late-successional communities ("hammocks") lacking longleaf pine and dominated by evergreen and deciduous angiosperms occurred on fire-protected islands and peninsulas. Similar vegetation in the Altahama Grit region of eastern Georgia (Harper, 1906, p. 35, 98) invariably occurred along creeks and rivers offering protection from fire.

Because of the pervasive presence of fire on uplands, mesophytic species such as American beech and southern magnolia were relegated to the most protected sites. Ecosystems on high bluffs east of the Apalachicola River in the panhandle of Florida supported these species. A variety of mesophytic species occurred here, including two rare species that are endemic to this bluff region—Florida yew and Florida torreya. This mesic forest persisted largely because of its protected position from fire and rich, moist soil. According to Harper (1914), "there is probably no area of equal extent on the mainland of Florida that is better protected from fire."

MICROLANDFORMS AND MICROTOPOGRAPHY

In addition to its pervasive and striking effects at mega-, macro-, and meso-scales (Figure 2.2), physiography also plays a significant role at the micro-scale. Microtopographic features, at scales less than 10 m^2 or even 1 m^2, influence regeneration of plants, succession, and local soil development. For example, the uprooting of large trees and the formation of **pit** and **mound** microlandforms create multiple microsites where plants may establish in an otherwise unfriendly environment. Although such uprooting events occur at a microscale, collectively, they affect soil–site heterogeneity and plant regeneration and dynamics over millions of hectares throughout forested landscapes. Microtopography is also created by the fallen trees lying on the forest floor and by the elevated bases of living and dead trees in wetlands. The form and substance (organic matter) of these biotic "landforms" provide favorable microsites for plant establishment and growth.

Tree Uprooting and Pit and Mound Microtopography

Probably the most widespread example of microphysiographic effect is that of microlandforms, and microsites on these landforms, created by the uprooting of trees. The felling of trees by wind occurs continually in forests. Trees are uprooted annually by cyclonic storms and in longer cycles by hurricanes or tornadoes. During uprooting, a tree falls with much of its large roots intact, thereby ripping up the soil. A depression or pit is typically formed at the former position of the main mass of structural roots (Figure 10.14). The upthrown root and soil mass, the root plate, typically forms a treethrow mound adjacent to the pit. The fallen tree decays over time and itself forms another kind of microtopographic feature on the forest floor that is itself colonized by woody plants.

Pits and mounds usually occur in pairs, and their size is determined by soil-site conditions, tree size and rooting habit, and the quantity of sediment that returns into the pit via soil slumping and erosion. Because of the complexity of the tree uprooting process and the time elapsed since uprooting, the size of pits and mounds varies greatly (Schaetzl

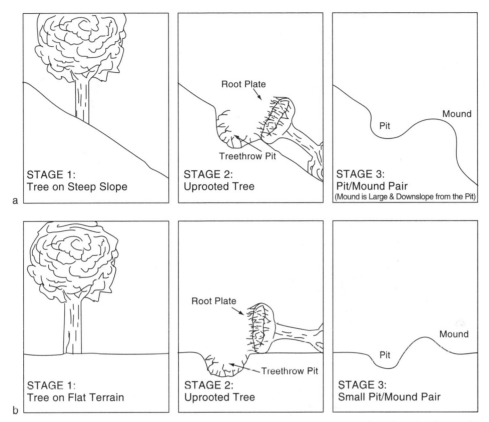

Figure 10.14. Diagrammatic illustration of the uprooting process, showing the formation of a pit and mound pair by soil slump off the root plate, and the resulting mixed horizons within the mound. *a,* steep slope; *b,* flat terrain (After Schaetzl and Follmer, 1990. Reproduced from *Geomorphology* with kind permission of Elsevier Science-NL, Amsterdam, The Netherlands.)

et al., 1989a). Figure 10.15a illustrates the size, shape, and spatial location of prominent pits and mounds in a landscape of sandy, glacial drift in New Brunswick (Lyford and MacLean, 1966). Throughout this site 36 percent of the area was occupied by mounds and 12 percent by pits; 1,255 mounds and 1,455 pits per hectare were identified. Approximately 10 to 50 percent of the forest floor in temperate forests may be covered by pit and mound topography (Schaetzl et al., 1989a). In contrast, only 0.09 percent coverage was reported in a mature Panamanian tropical forest (Putz, 1983). Rapid leveling of microtopography, in only 5 to 10 years, is apparently due in part to torrential rainstorms acting on soil unprotected by thick litter layers (Putz, 1983).

The age of treethrow microtopography is difficult to determine; estimates vary greatly but are usually less than 500 years. Longevity may range from 10 to over 2,000 years; the documented maximum of 2,420 years was determined by radiocarbon dating (Schaetzl and Follmer, 1990). Thus the possible lifespan of these features may exceed 500 to 1,000 years in many environments.

Tree uprooting and pit and mound topography are not randomly distributed. They are strongly associated with ecosystems having parent material and soil that favor shallow rooting and systems with severe windstorms. Ecosystems with wet mineral or organic soils (high water tables), rocky soils, or soils developing root-restricting horizons typically have a high incidence of uprooting. The abundance of pit and mound microland-

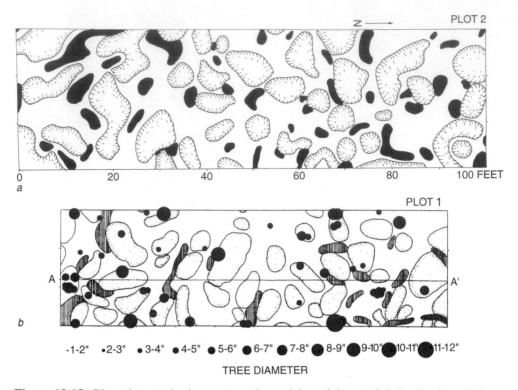

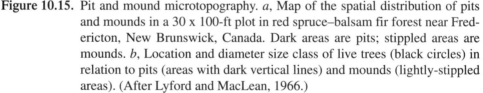

TREE DIAMETER

Figure 10.15. Pit and mound microtopography. *a*, Map of the spatial distribution of pits and mounds in a 30 x 100-ft plot in red spruce–balsam fir forest near Fredericton, New Brunswick, Canada. Dark areas are pits; stippled areas are mounds. *b*, Location and diameter size class of live trees (black circles) in relation to pits (areas with dark vertical lines) and mounds (lightly-stippled areas). (After Lyford and MacLean, 1966.)

forms is closely related to landforms positioned low or high in the landscape, i.e., low, wetland areas with high water tables and high positions with steep slopes or exposed to strong winds.

Pit and mound topography creates heterogeneous soil and microclimatic conditions in ecosystems predisposed to tree uprooting. At the base of uprooted mature trees an average of 12 to 16 m^2 of soil to a depth of one meter or more may be disturbed (Lutz, 1940; Brewer and Merritt, 1978; Putz, 1983; Peterson et al., 1990). Considerable soil mixing occurs as lower soil horizons are brought to the surface. Then slumping and erosion of sediments from the root plate expose mineral soil and produce irregular, discontinuous, or interfingered horizons within the soil mound (Schaetzl et al., 1990). Soil development is affected, and in many areas is accelerated beneath pits more than on mounds (Stone, 1975). On mountain slopes, trees usually fall downslope, and uprooting produces net downslope transfer of sediment and can trigger debris avalanches and debris flows (Swanson et al., 1982a; Schaetzl et al., 1990).

The heterogeneity of microsites in pits and on mound surfaces provides a variety of environments for colonization by woody species. As illustrated in Figure 10.15*b*, establishment of trees is almost exclusively on mounds, especially on lower slopes where soil water and nutrients are most favorable. Although pit centers may be colonized soon after treefall (Peterson, et al., 1990), trees fail to establish and develop there over time due to wet conditions with occasional standing water (and ice) and deep accumulations of leaf litter (Schaetzl et al., 1989*b*). Mound surfaces are clear of competing roots, clear of litter, and for a time free of overhead shade. Mounds shed leaf litter and provide mineral-soil seedbeds that are ideal for species (e.g. hemlock and birch) having tiny, wind-dispersed seeds. Seedlings of such species fail to establish in sites with thick accumulations of organic matter such as the pit bottoms and the adjacent undisturbed forest floor.

In addition to favorable soil conditions, the gap in the canopy created by the uprooting process often provides sufficient light for establishment and growth of even shade-intolerant species. In the mature tropical forest of Panama, Putz (1983) reported that although seedlings and saplings of pioneer tree species are found elsewhere in treefall gaps, they are concentrated on the soil disturbed by uprooted trees. Thus the uprooting process and pit and mound microlandforms not only play a major role in the community composition of different ecosystems but also markedly influence microscale ecosystem changes in light, topography, and soil conditions that, in turn, mediate vegetational change (see Chapters 16 and 17). Excellent reviews of the extensive literature on uprooting and microtopography are provided by Schaetzl et al. (1989a, b; 1990) and Schaetzl and Follmer (1990).

Microtopography and Regeneration in Hardwood Swamps

Microtopography and its effects in forest ecosystems is not associated only with treefall pits and mounds. As noted above, fallen and rotting tree trunks and other coarse woody debris on the forest floor provide moist and competition-free establishment sites for seedlings of certain species. In swamps of the southeastern Coastal Plain such "nurse-logs" often bear distinctive vegetation (Schlesinger, 1978; Huenneke and Sharitz, 1986). In addition, many other kinds of microtopography provide sites for regeneration in these waterlogged, periodically flooded landforms, and generally harsh environments. For example, in a riverine swamp in Florida, Titus (1990) identified 19 types of microsites in 4 different categories: (1) frequently submerged soil sites such as swamp bottoms, (2) raised-soil sites (elevated soil, soil near a stump, cypress knee, or shrub base, etc.), (3) dead wood sites (fallen log or branch), and (4) live wood sites (trunk, root, cypress knee, palm tree base). Distribution patterns of 25 woody species corresponded to the duration of inundation by flood waters and microsite substrate type. Although the elevational range in the study area was only 1.5 m, microsite elevation was the factor most strongly correlated with seedling distribution. Microsite substrate type also was closely associated with species presence. The highest densities of seedlings were on raised-soil sites. Generally, in wetland sites around the world where high water table and low oxygen adversely affect seed germination, seedling establishment, and growth of woody plants, the role of microtopography is enormous.

Suggested Readings

Grimm, R. C. 1984. Fire and other factors controlling the Big Woods vegetation of Minnesota in the mid-nineteenth century. *Ecol. Monogr.* 54:291–311.

Hack, J. T., and J. C. Goodlett. 1960. Geomorphology and forest ecology of a mountain region in the central Appalachians. Geol. Survey Prof. Paper 347. 66 pp + map.

Heinselman, M. L. 1981. Fire and succession in the conifer forests of northern North America. In E. C. West, H. H. Shugart, and D. B. Botkin (eds.), *Forest Succession, Concepts and Application.* Springer-Verlag, New York.

Huggett, R. J. 1995. *Geoecology, An Evolutionary Approach.* Routledge, New York. 320 pp.

Osterkamp, W. R., and E. R. Hupp. 1984. Geomorphic and vegetative characteristics along three northern Virginia streams. *Bull. Geol. Soc. Am.* 95:1093–1101.

Rowe, J. S. 1988. Landscape ecology: the study of terrain ecosystems. In M. R. Moss (ed.), Symp. Proc. *Landscape Ecology and Management*. Polysci. Publ. Inc., Montreal.

Swanson, F. J., R. L. Frederiksen, and F. M. McCorison. 1982a. Material transfer in a western Oregon forested watershed. In R. L. Edmonds (ed.), *Analysis of Coniferous Forest Ecosystems in the Western United States*. Hutchinson Ross, Stroudsburg, PA.

Swanson, F. J., T. K. Kratz, N. Caine, and R. G. Woodmansee. 1988. Landform effects on ecological processes and features. *BioScience* 38:92–98.

Swanson, F. J., F. J. Franklin, and J. R. Sedell. 1990. Landscape patterns, disturbance, and management in the Pacific Northwest, USA. In I. S. Zonneveld and R. T. T. Forman (eds.), *Changing Landscapes: An Ecological Perspective*. Springer-Verlag, New York.

Wells, P. V. 1970. Postglacial vegetational history of the Great Plains. *Science* 167:1574–1582.

CHAPTER *11*

SOIL

Forest trees, like all other terrestrial vegetation, require five primary resources for growth and development: radiant energy, carbon dioxide (CO_2), water (H_2O), mineral nutrients, and a porous medium for physical support. Although plants obtain energy from solar radiation and CO_2 from the atmosphere, the remaining resources are provided by soil. Consequently, soil forms the "foundation" of forest ecosystems in more ways than one. As you will see, soil is critical to the cycling of nutrients (Chapter 19), a process that influences the growth of individual trees and the functioning of entire ecosystems. In this chapter, we provide an overview of the physical, chemical, and biological properties of soil that regulate the availability of soil resources to plants, particularly forest trees. We begin with a discussion of the soil forming process and explain how climate, geology, and biota influence the geographic distribution of forest soils.

There are many definitions of soil as it pertains to the growth of terrestrial plants. For our purpose, we define soil as a porous medium consisting of minerals, organic matter, water, and gases. The combined influences of climate, topography, biota, and time differentiate geologic materials into soil. As such, soils are as diverse as the climates in which they occur, the landforms on which they develop, and the plant life that grows upon and within them. One would expect soils supporting tropical rain forests to differ markedly in their physical, chemical, and biological properties from those beneath forests in temperate or boreal climates. In the pages that follow, we review the processes that give rise to such differences and focus on how they influence plant growth and ecosystem function.

PARENT MATERIAL

The Earth's surface is blanketed by a wide array of geologic materials, differing in their chemical composition and degree of consolidation. The relatively-unweathered geologic

255

material from which a particular soil has developed is called **parent material**. It constitutes the basic substrate for soil formation and exerts a substantial influence on many soil properties. Parent materials, as you will read, are typically associated with characteristic kinds of landforms. **Weathering** is an important component of soil formation, because physical abrasion and chemical dissolution differentiate freshly-exposed geologic material (i.e., parent material) into soil. Living organisms play an integral role in this process too, wherein organic acids produced by plant roots and soil microorganisms solubilize minerals, allowing their constituents to be leached and deposited at depth. Additionally, the hydrolysis of CO_2 resulting from root and microbial respiration produces acidity, which further contributes to the dissolution of minerals and the weathering process.

Parent materials are broadly classified as **consolidated** and **unconsolidated** (Figure 11.1). Consolidated parent materials include igneous, sedimentary, and metamorphic rock. A description of them can be found in most introductory geology texts, and Fairbridge (1972) provides a particularly detailed discussion of their formation and chemical composition. Soil developing in consolidated geologic substrate is said to be formed in residual parent material (Figure 11.1). Forest soils derived from residual parent materials occur throughout North America, primarily in areas that have not been influenced by glaciation, moving water, or oceanic uplift. Forests of the Piedmont Plateau and Appalachian uplands in the eastern United States occur on these materials, as do forests of the Sierra Nevada, Cascade, and Rocky Mountains in the western United States.

Rates of soil formation on residual parent material composed of hard minerals can be quite slow, and, in some situations, deep soils may never develop because erosion rates exceed those of soil formation. Such a situation commonly occurs in mountainous regions where steeply-sloping topography and exposed rock combine to form relatively thin soils. Nevertheless, not all consolidated parent materials weather at slow rates nor give rise to thin soils. The relative resistance of rock to physical and chemical weathering is as follows (Birkeland, 1974):

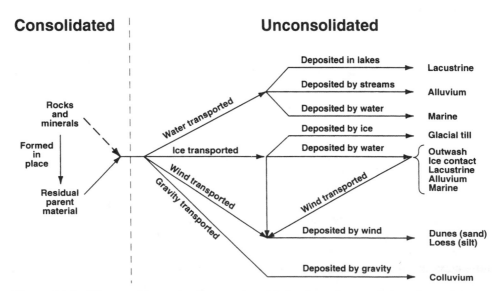

Figure 11.1. Diagram illustrating how various kinds of parent material are formed, transported, and deposited. Modified from Brady, 1990. Reprinted with permission of the Macmillan Publishing Co., Inc. as modified from *The Nature and Properties of Soils,* 11th ed., by Nyle C. Brady. Copyright ©1996 by Macmillan Publishing Co., Inc.

quartzite, chert > granite, basalt > sandstone, siltstone > dolomite, limestone

In general, rocks composed of insoluble minerals (i.e., quartzite, SiO_2) are relatively more resistant to weathering than those containing soluble minerals (i.e., calcite, $CaCO_3$ or dolomite, $CaCO_3 \cdot MgCO_3$), which rapidly break down in warm, humid climates. Residual parent materials also yield very different chemical constituents as they are abraded and dissolved during the weathering process. For example, the average SiO_2 and Al_2O_3 contents of igneous and sedimentary rock are similar; however, their calcium (Ca) and sulfur (S) content can dramatically differ (Table 11.1).

There can be considerable differences in the chemical constituents among soils derived from residual parent material, with implications for tree growth. In the Ozark Mountains of Arkansas, for example, different forest ecosystems develop on two kinds of residual parent material: black and white oak are overstory dominants on soil derived from chert (SiO_2), whereas eastern redcedar and northern red oak are dominant on limestone-derived ($CaCO_3$) soil (Read, 1952). Although both of these parent materials are derived from sedimentary rock, differences in their chemical composition and weathering rate markedly influence the distribution of forest trees.

Unconsolidated parent materials are mineral particles that have been transported by water, ice, wind, or gravity (Figure 11.1). They are chemically similar to the rock from which they originate, but are distinguished from residual parent material by being moved from their point of geologic origin. The agent of transport has a substantial influence on the physical, chemical, and biological properties of soils formed in unconsolidated parent materials; differences that often influence forest composition and ecosystem productivity. In general, sediments deposited by water and wind have a narrow particle size distribution, whereas those deposited by ice contain fragments that range in size from microscopic clay particles to boulders several meters in diameter. A complete discussion of transported parent materials and the soils that they give rise to can be found in Buol et al. (1980).

One example of how unconsolidated parent materials influence forest composition and ecosystem productivity comes from the glaciated portions of eastern North America. In this region, the Wisconsinan Glaciation (14,000 years before present) left behind a landscape consisting almost exclusively of unconsolidated parent materials. Stratified materials were deposited by glacial melt waters (outwash), semi-stratified materials were deposited near the margins of stagnant ice (ice contact), and unstratified materials were deposited directly by glacial ice (till). These parent materials differ markedly in their particle size distribution and their ability to supply water and nutrients for plant growth. In the northern portion of Michigan's Lower Peninsula, for example, dry oak-dominated (northern pin, black and white oak) ecosystems consistently occur on sandy glacial outwash (72 percent coarse and medium sand). Mesic northern hardwood ecosystems occur

Table 11.1 Generalized Chemical Composition of Igneous, Metamorphic and Sedimentary Rock

	SiO_2	Al_2O_3	Fe_2O_3	FeO	MgO	CaO	K_2O	P_2O_5	S
					%				
Igneous	69.9	15.2	2.0	2.0	2.2	4.0	3.3	0.20	0.04
Metamorphic	58.2	15.5	2.9	4.8	3.8	6.0	2.6	0.30	0.03
Sedimentary	49.9	13.0	3.0	2.8	3.1	11.7	2.0	0.16	0.18

Source: Modified from Ronov and Yaroshevsky, 1972.

on till-derived soils with lower sand contents (55 percent coarse and medium sand; Host et al., 1988). In addition to differences in species composition, aboveground productivity varies by a factor of three between the dry oak ecosystems (1.3 Mg ha^{-1} y^{-1}) and the mesic northern hardwoods (3.4 Mg ha^{-1} y^{-1}). Similar relationships also have been observed among geology, soil, forest composition, and ecosystem productivity in the glaciated portions of Wisconsin (Pastor et al., 1984).

SOIL FORMATION

As parent material weathers and is occupied by plants and animals, it differentiates into more or less distinct horizontal zones, giving rise to a **soil profile**. The type of soil profile that develops depends upon the interaction of (1) climate, (2) parent material, (3) plants and animals occupying the soil, (4) relief of the land, and (5) the amount of time that has elapsed. Soil formation is in part a chemical process resulting from the weathering of geologic material exposed to air and water, and in part a biological process resulting from the activities of organisms growing on and in soil.

Soil Profile Development

Forests comprise the natural vegetation in many of the moister parts of the earth regions where precipitation supplies more water than can be evaporated or transpired over the normal year. Under such conditions two factors dominate the soil-forming process: 1) precipitation in excess of evaporation and transpiration moves downward through the soil removing soluble minerals, and 2) tree roots remove both water and nutrients from the soil, transpiring most of the former and eventually returning most of the latter to the soil surface as leaves, twigs, fruits, cones, seeds, and fine roots. In temperate regions, the typical forest soil can be differentiated into five zones or **horizons**, so identified by the soil forming process occurring within them (Figure 11.2). A soil horizon is differentiated from the over- and underlying layers by attributes that can be easily identified in the field.

The accumulation of organic matter at the soil surface, or **O horizon**, is an attribute unique to forest soils. Material contained within this horizon consists almost entirely of leaves, twigs, flowers, fruits, cones, and seeds that have been deposited on the soil surface. The O horizon lies above the mineral soil and is distinguished from it by a high organic matter content (> 20 percent organic matter if soil has no clay; >30 percent if soil is more than 50 percent clay). It can be divided into three subordinate horizons, each reflecting different stages of decomposition: 1) the Oi horizon contains relatively "fresh" organic matter whose origin is easily recognized, 2) the Oe horizon is composed of partially decomposed plant parts, and 3) the Oa horizon reflects the latter stages of decomposition and consists of well-decomposed organic matter of unrecognizable origin (i.e., humus).

The **A horizon** marks the surface of the mineral soil and is characterized by: 1) the leaching or **eluviation** of many soluble minerals that migrate in the downward flow of water, and 2) the accumulation of organic matter originating from the overlying O horizon. In many forest soils, the A horizon is dark in color and is well structured, owing to its relatively high organic matter content (4 to 12 percent). The majority of aggregates contained within this horizon are of crumb (< 1 to 5 mm dia.) or granular (<1 to 10 mm dia.) size (see soil structure that follows). The A horizon also is characterized by a large number of fine roots which actively forage for nutrients released during organic matter decomposition. Because of the shallow distribution of fine roots and their decomposition products, most forest soils are characterized by a thin A horizon. In contrast, A horizons of grassland ecosystems support deeply rooted grasses that incorporate organic matter to depths of 50 cm.

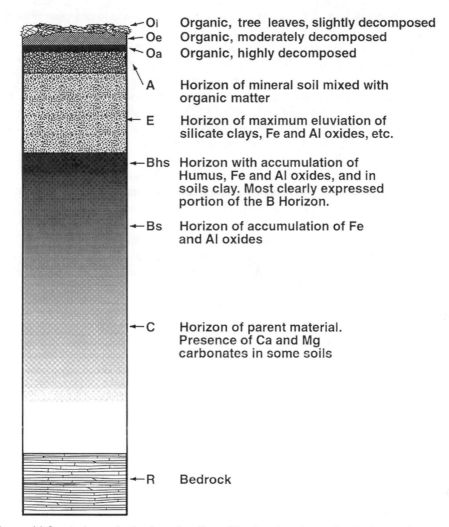

Oi — Organic, tree leaves, slightly decomposed

Oe — Organic, moderately decomposed

Oa — Organic, highly decomposed

A — Horizon of mineral soil mixed with organic matter

E — Horizon of maximum eluviation of silicate clays, Fe and Al oxides, etc.

Bhs — Horizon with accumulation of Humus, Fe and Al oxides, and in soils clay. Most clearly expressed portion of the B Horizon.

Bs — Horizon of accumulation of Fe and Al oxides

C — Horizon of parent material. Presence of Ca and Mg carbonates in some soils

R — Bedrock

Figure 11.2. A theoretical mineral soil profile showing the major horizons that may be represented. Reprinted with permission of the Macmillan Publishing Co., Inc. from *The Nature and Properties of Soils,* 11th ed., by Nyle C. Brady. Copyright ©1996 by Macmillan Publishing Co., Inc.

In humid climates, the downward flux of water can remove humus, silicate clays, Fe oxides, and Al oxides from the surface soil, leaving behind light-colored, resistant minerals, especially quartz (SiO_2). The layer so formed is an **E horizon**, and it is distinguished from the overlying A horizon by its light color. In cool, humid climates, E horizons often develop under coniferous forest growing on course-textured (sand-to-sandy loam) parent material. The acidity of this horizon is typically higher than the over- or underlying horizons due to the loss of base cations and presence of inorganic acids. Root densities within the E horizon are low because few plant nutrients reside within this highly leached horizon.

Materials leached from either the A or E horizon migrate downward and are deposited at depth to form the B horizon. The B horizon results from the process of **illuviation** (i.e., accumulation) and is distinguished from other soil horizons by this important soil-forming process. Materials that accumulate in the B horizon of temperate forest soils include silicate clays, humus, Fe and Al oxides—materials that can have a great impact on the physical, chemical and biological properties of the B horizon. In arid and semi-arid regions, $CaCO_3$, $CaSO_4$, and other salts can accumulate in the B horizon.

The **C horizon** is the unconsolidated parent material underlying the A, E, and B horizons and is outside the influence of the processes giving rise to the horizons above it. The A, E, and B horizons can be derived from the same material contained within the C horizon. However, in areas where geologic activity has deposited a relatively thin layer of mineral material over a previously existing deposit, the A, E, or B horizon may be derived from a different parent material. This situation is common in the glaciated portions of North America where advancing ice has overridden preexisting deposits of glacial drift. In regions where soil forms in residual parent material, the C horizon is replaced by an **R**, the horizon that denotes the presence of the underlying consolidated rock.

Note that the A, B, and C horizons refer to zones that have been leached, enriched, and unaffected by soil forming processes, respectively. It does not necessarily follow that the upper mineral horizon is always the A horizon, or that all horizons are present in every soil. Following sheet erosion, the B or C horizon may be exposed on the surface. Likewise, very "young" or unweathered soils, like those forming on sand dunes, may lack a B horizon and consist only of an A and C horizon. In these landscape positions, clay, humus, or Al- and Fe-oxides often have not accumulated to any great extent in the B horizon.

PHYSICAL PROPERTIES OF SOIL

As plant roots grow within the soil, they anchor the aboveground portion of the plant and supply it with water and nutrients. Plant roots also require oxygen for respiration, the supply of which is controlled by the rate at which oxygen diffuses through water and other gases in soil. The physical properties of the solid, liquid, and gaseous phases of soil have a substantial influence on the supply of water, nutrients, and oxygen for metabolism, and the availability of physical space to anchor aboveground plant structures. Providing physical support for aboveground tissues is of particular importance, because plants must properly orient themselves to capture the sun's energy for use in photosynthesis. In shallow and poorly-drained soils, windthrow is common because physical space is limited by the occurrence of shallow bedrock in the former and anoxic (without O_2) conditions in the latter. In the pages that follow, we discuss the physical properties of soil that are of particular relevance for the growth of forest plants and the functioning of forest ecosystems.

Soil Texture

Soils are composed of mineral particles with a wide array of sizes. These particles, or soil separates, are grouped into three size classes: sand, silt, and clay. Sand particles range in size from 2.00 to 0.02 mm in diameter, silt ranges from 0.02 to 0.002 mm in diameter, and the clay fraction is less than 0.002 mm. **Soil texture** refers to relative proportion of sand, silt, and clay-sized particles contained in a particular soil. This physical property plays an integral role in regulating the availability of water and mineral ions for plant uptake, and the rate at which gases (O_2 and CO_2) are exchanged between soil and the overlying atmosphere.

In addition to grouping mineral particles by size, soil separates in different size classes can be distinguished by their physical and chemical properties. For example, sand- and silt-sized particles are chemically identical to the rock from which they originate and are called primary minerals. They are round or irregular in shape, composed primarily of quartz (SiO_2) or other silicate minerals like orthoclase ($KAlSi_3O_8$) or plagioclase ($[Ca, Na][Al, Si] AlSi_2O_8$). As a consequence of their size and shape, sand particles have relatively low surface areas ($1-2$ m^2 g^{-1}) with large pores between individual particles. (One gram of soil would form a mound in the palm of your hand equal to the size of a United States quarter coin). These attributes provide sandy soils with good aeration, but limit the amount of water available for plant use. Because silt-sized particles have higher surface areas (45 m^2 g^{-1}), they hold relatively larger quantities of water compared to a soil consisting primarily of sand.

The clay fraction of soil is comprised of **secondary minerals**, which result from the physical and chemical weathering of primary minerals. In temperate soils, clays consist primarily of aluminosilicate minerals that differ markedly from sand and silt in their shape and mineralogy. These minerals form plate-like structures or **micelles** that are referred to as phyllosilicate (leaf-like silicate) clays. Due to their shape, they have a high surface area (80 to 800 m^2 g^{-1}), enabling them to hold relatively large quantities of water for plant use. Phyllosilicate minerals differ widely in their physical and chemical properties from Al and Fe oxides, which compose the clay fraction of soils in the humid tropics. Differences in clay minerals found in temperate and tropical soils will be discussed in more detail later in this chapter (see Chemical Properties of Soil).

Soils are grouped into textural classes based on their sand, silt, and clay content. For example, a soil consisting of equal proportions of sand, silt, and clay is classified as a loam (Figure 11.3). For a soil to be classified as a clay, 60 percent of its particles must be

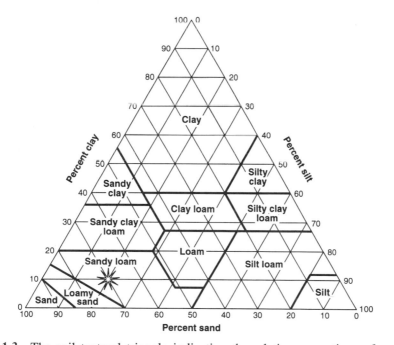

Figure 11.3. The soil textural triangle indicating the relative proportions of sand, silt, and clay composing each textural class. Note that some textural classes encompass a range of particle size distributions (i.e., the clay class ranges from 60 to 100 percent clay), whereas other textural classes are narrowly defined (i.e., sand class ranges from 90 to 100 percent sand). The particular soil (70 percent sand, 20 percent silt, and 10 percent clay) represented by an asterisk in the sandy loam class can be used to illustrate the use of this diagram. Begin by finding the lines of equal value for 70 percent sand and 20 percent silt. The intersection of those lines with the 10 percent clay lines locates the position of the asterisk, and the textural class of the soil.

less than 0.002 mm in diameter. In contrast, sands are soils in which more than 90 percent of the mineral particles range from 0.02 to 2.00 mm in diameter. The textural classes illustrated in Figure 11.3 have been delineated with specific reference to plant growth. As you will read later in this chapter, water availability, nutrient supply, and aeration all are substantially influenced by this all-encompassing soil property.

Soil Structure

Primary soil particles (i.e., sand, silt, and clay) are arranged into secondary structures called **aggregates** or **peds**, which result from the combined activities of plants and soil microorganisms. Plant roots enmesh and compress mineral particles and bind primary particles to one another. Further pressure can be exerted during wetting and drying cycles, because clays expand as they hydrate and contract as they dry. Organic compounds (polysaccharides) excreted by plant roots, and produced during the microbial degradation of plant litter, function as cementing agents binding to the surface of one or more soil particles. In combination, these processes make soil aggregates highly-stable structures that often remain intact even when immersed in water.

Aggregates range from single-grain structure, in which soil particles are totally unattached, to massive structure, in which all soil particles adhere to one another in large clods. Beach sand is an example of the former, whereas the latter often occurs in poorly drained soils with a high clay content. The degree of aggregation most conducive to plant growth lies somewhere between these extremes. Within this mid range, aggregates vary in size from granules several millimeters in diameter to blocks, prisms, or columns several centimeters in size. Granular aggregates commonly occur in the surface horizons of many forest soils, whereas blocky aggregates often occur in subsurface horizons. Plant roots generally occupy the spaces between aggregates, rather than growing through or within them. These spaces, or macropores, have a substantial influence on the rate at which water and gases move into and through the rooting zone. As a result, a well-aggregated soil will typically hold more water and will have better aeration than an unstructured soil of the same texture.

Soil Color

Color provides insight into many physical and chemical properties of soil, particularly organic matter content and drainage (or aeration). The surface of mineral soils are generally dark in color, reflecting organic matter additions from leaf and root litter. A deep surface soil dark in color (e.g., black or dark brown) usually contains relatively greater amounts of organic matter than a thin, light-colored surface soil. The subsoil of sandy forest soils in northerly climates can contain accumulations of organic matter and iron that have been leached from the surface soil. This subsurface accumulation is easily identified by its dark chocolate-brown color.

In addition to providing insight into organic matter content, soil color can provide qualitative information regarding soil drainage. Most soils contain large amounts of iron (Fe) which, in an oxidized state (Fe^{3+}), is bright orange or red. This condition occurs when soils are well aerated and O_2 rapidly diffuses into the soil profile. However, during prolonged periods of water saturation, O_2 in soil can be depleted if the demand by plant roots and soil microorganisms exceeds the diffusion rate of O_2 in water. In these situations, Fe is reduced (Fe^{2+}) producing compounds blue-gray to gray in color. Zones of mottling, a patchwork of yellow-orange and blue-gray colors, indicate the presence of both oxidized and reduced conditions. This situation occurs where the level of the water table fluctuates within the soil profile, and color can be used to determine if it lies within the rooting zone of plants. High water tables in poorly-drained landscape positions greatly restrict the rooting depth of trees, making them prone to windthrow. Using soil color, one can easily identify soils with poor drainage that often restrict the growth of many forest trees.

Soil Water

Water availability controls the global, regional, and local distribution of vegetation on the Earth. For example, forests occur in regions where the annual amount of water supplied by precipitation exceeds that which is lost through evaporation. Although broad-scale patterns of precipitation control the *total* amount of water entering forest ecosystems, it is the interaction of water molecules with soil particles that largely influence the amount of water that can be used by an individual plant for growth.

Water flows along a continuum extending from the atmosphere, through the plant, and into soil. The force driving water movement along this continuum is transpiration; water is literally "pulled" from soil and through plants by this process. Transpiration at the leaf surface creates tensions that are translated down water columns extending through plants and into soil. Soils can be "dry" to plants but can still contain substantial amounts of water. This relationship may occur if the forces holding water in soil exceeds the force (i.e., tension) created by the transpiration of water at the leaf surface. Understanding the dynamics of water along the atmosphere–plant–soil continuum has clear relevance for the study of plant growth and ecosystem function, because these dynamics directly control the amount of water available for plant use.

PHYSICAL PROPERTIES OF WATER

The physical properties of water greatly influence its availability to plants. Water (H-O-H) molecules have a net positive charge to one side of the molecule and a net negative charge to the other. Mineral particles in soil also have charged surfaces that attract water molecules. The attraction of positively and negatively charged bodies is termed **adhesion**, a force in soil that greatly influences the amount of water available for plant use. Adsorbed water molecules, those attracted to charged surfaces in soil, are linked to others through hydrogen bonding, a chemical bond linking the oxygen atom (−) of one water molecule to a hydrogen atom (+) of another. Hydrogen bonding gives rise to the cohesive force, or **cohesion**, joining water molecules into chains or polymers that extend away from the surface of mineral particles.

Due to the strong attraction of water molecules to charged surfaces in soil, adsorbed water molecules are closely packed and exist in an energy state less than that of pure water. Although water in direct contact with mineral particles is strongly held to their surface by adhesion, that force diminishes as the distance to the solid surface increases, much like the attraction of a magnet for iron diminishes as the distance between them increases. When soils are saturated, some water molecules are only weakly attracted to the surface of mineral particles, because they lie at relatively large distances from any charged surface. Water draining from saturated soil does so because the Earth's gravitational pull exceeds the adhesive and cohesive forces holding a portion of soil water. As soil continues to dry, either through plant uptake or evaporation, the forces holding the remaining water molecules steadily increase. Adhesive and cohesive forces holding water in soil are tensions that must be overcome if plants are to extract water from soil. Because the forces holding water in soil can exceed those imposed by plants, only a proportion of the water in soil is available for plant use.

In addition to being attracted to charged surfaces, water molecules also are attracted to ions with net positive (cations) or negative (anions) charges. Salts, like NaCl, dissolve in water because the attraction of water molecules for cations (Na^+) and anions (Cl^-) is much greater than the attraction between them. The strong attraction of water for positively- or negatively-charged ions causes water molecules to lose energy as they hydrate either type of ion. Because water molecules lose energy as they associate with cations and anions, water in soil has a lower energy status than pure water.

The semi-permeable membrane surrounding plant cells (i.e., plasmalemma) functions as a "molecular sieve", allowing water to transverse while excluding larger, hydrated ions. The movement of water molecules across any semi-permeable membrane in response to differences in ion concentration (i.e., the energy status of water inside versus outside the cell) is **osmosis**. Water molecules moving across a semi-permeable membrane exerts a force known as osmotic pressure. Osmosis, and the energy it produces, is of particular relevance to plant growth because it influences the movement of water into and out of plant cells. For example, plants under salt stress suffer from a lack of available water, because dissolved ions in soil water lower its energy status to a point where it is less than that in the plant cell. Water flows from the plant into soil solution by osmosis. This can cause the water content of plant cells to decrease to such a low level that physiological processes are impaired and the plant is no longer able to maintain turgor.

SOIL WATER POTENTIAL

The movement of water in soil, its uptake by plant roots, and its loss to the atmosphere from the leaf surface are all energy-related phenomenon. In soil, adhesion, cohesion, the presence of dissolved ions, and the Earth's gravitational pull are the primary forces influencing the energy status of soil water, and hence the movement of soil water and the proportion of it available for plant use. Forces in soil acting upon water molecules can be pressures (gravity) or tensions (adhesion and cohesion), both of which are measured in megapascals (MPa; values less than 0 MPa are tensions and those greater than 0 MPa are pressures).

Soil water potential refers to the energy status of soil water; it also can be thought of as the effective concentration of water in soil. By definition, the potential of pure, liquid water at 20 °C and at standard atmospheric pressure is 0 megapascals (MPa). Pure water is used as a standard reference point from which we measure the influence of adhesion, cohesion, dissolved ions, and gravity on the energy status of soil water.

Adhesion and cohesion, the forces holding water molecules to charged surfaces and to one another, give rise to the **matric potential** of soil water. Because adhesion and cohesion lower the energy status of soil water (i.e., relative to pure water) matric potentials are less than 0 MPa. As such, matric potentials represent tensions holding water molecules to one another and to the surfaces of charged particles in soil. The presence of dissolved ions, which also lower the energy status of soil water relative to pure water, give rise to the **osmotic potential** of soil water. **Gravitational potential** results from the downward force of gravity and its ability to extract water from soil. Gravitational potentials are greater than zero (i.e., pressures), because the downward pull of gravity extracts water from soil. In combination, matric, osmotic, and gravitational potentials give rise to the **total soil water potential**, which represents the summed energy status of soil water. In most well-drained soils, matric potential is the most important factor regulating the supply of water to the root surface.

Because adsorption and dissolved ions lower the free energy status of soil water, plants must expend energy to remove water from soil. As such, the water potential (or energy status) of plants must be lower than that of soil, if water is to flow from soil into plant roots. Transpiration at the leaf surface drives the flow of water along the energy-related path from soil to the atmosphere. The **atmospheric water potential**, which greatly influences the transpiration rate of plants, is largely determined by relative humidity and air temperature. The concentration of water in the atmosphere is much less than the concentration of water in either plants or soils, and as a consequence, atmospheric water potentials are more negative (i.e., at a lower energy status) than those of plants or soil. It is not unusual for atmospheric water potentials to attain values of -100 MPa, values 10 to 100 times more negative than those in soil (Bidwell, 1974). Because water flows from a region of high potential (i.e., high energy state) to one of low potential (i.e., low energy state), water moves from soil into plant roots, through the vascular system of the plant, and into the atmosphere. Large negative atmospheric water potentials drive the process of transpiration and the flow of water from soil, through plants, and into the atmosphere.

Although large negative water potentials at the leaf surface are translated downward to the root surface, plants are generally unable to extract soil water held by potentials less than -1.5 MPa (i.e., a tension of 1.5 MPa; Figure 11.4). Under these conditions plants wilt and are unable to regain turgor even following the addition of water. At a potential of -1.5 MPa, soil water has attained the **permanent wilting point**, which defines the lower limit of plant available water (Figure 11.4). **Field capacity** represents the upper limit of plant available water and is the amount remaining in soil after it has freely drained due to the downward pull of gravity (a potential of -0.01 MPa). The quantity of water bounded by field capacity and the permanent wilting point represents the **available water content** (mL cm^{-3}) of soil. It differs substantially from the **saturation water content**, which is the total amount of water that can be stored within soil pores (Figure 11.4). The reader is referred to Kramer (1983) for a complete treatment of plant water relations and soil water dynamics.

Soil texture substantially influences the available water content of soil. Figure 11.5 illustrates the relationship between water content (mL of water per cm^3 of soil) and soil water potential for a clay, loam, and sand. At field capacity (i.e., potential of -0.01 MPa), the clay soil holds approximately 0.6 mL of water per cm^3 of soil, almost 2.5 times more water than the sand (calculated from Figure 11.5). The adhesive properties of water, in combination with the large surface area of clay-sized particles, allow the clay to hold more water at any given potential than the sand. Notice that silt loam in Figure 11.5 contains the greatest quantities of plant available water. It does so because of the favorable distribution of macro- and micropore spaces. The available water content of sand, calculated from Figure 11.5 is 0.15 mL cm^{-3}, approximately 50% of that held by the silt loam (0.36 mL cm^{-3}) and clay (0.33 mL cm^{-3}).

CHEMICAL PROPERTIES OF SOIL

Plant life is constructed from a surprisingly small suite of elements, whether we consider the majestic redwoods of northern California or the single-cell alga growing on the soil surface. In Table 11.2, we summarize the chemical building blocks from which plant life is constructed and their source within terrestrial ecosystems. Each is required for plant growth and development, albeit in different quantities. **Macro-nutrients** are those ele-

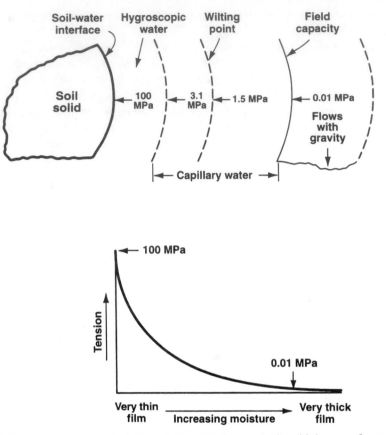

Figure 11.4. Diagrams showing the relationship between the thickness of water films and the tension with which water is held by soil particles. Tensions are presented in megapascals in the upper illustration. The thickness of water films in relationship to matric potential is presented in the lower figure. Reprinted with permission of the Macmillan Publishing Co., Inc. from *The Nature and Properties of Soils,* 8th ed., by Nyle C. Brady. Copyright ©1974 by Macmillan Publishing Co., Inc.

ments required in relatively large amounts. They are commonly found as constituents of nucleic acids, proteins, carbohydrates, lipids, and chlorophyll (Table 11.3). **Micro-nutrients** (Fe, Mn, Bo, Mo, Cu, Zn, Cl, and Co), as their name implies, are required in relatively small amounts and occur as co-factors in enzymatic reactions. Although micro-nutrients are required in small amounts, they are nonetheless important in the biochemical functioning of plants and entire ecosystems; all are supplied to plants by chemical processes in soil. Further discussion regarding the biochemical and physiological function of plant nutrients can be found in Salisbury and Ross (1992).

Although plants assimilate carbon and oxygen from the atmosphere, the majority of macro- and micro-nutrients are supplied by ion exchange reactions, mineral weathering, or organic matter decomposition—processes all occurring within soil! The supply of nutrients often limits the growth of individual plants and entire ecosystems. Nitrogen (N) availability, for example, is known to limit the growth of many boreal and temperate forests (Flanagan and Van Cleve, 1983; Pastor et al., 1984), whereas phosphorus (P) has

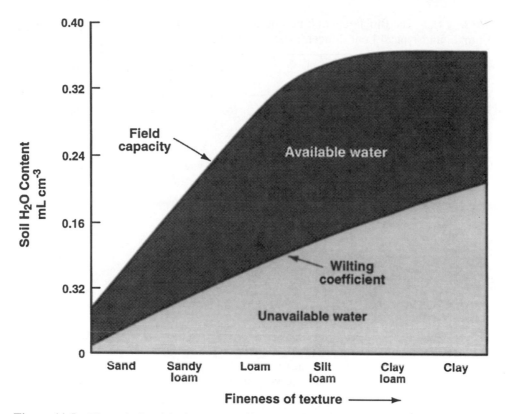

Figure 11.5. The relationship between soil texture and the available water content of soil. Note that field capacity increases from the sand to the silt loam and then levels off as the proportion of clay increases. Because the permanent wilting point increases linearly as a function of soil texture, the largest amount of plant available water occurs in soil with a silt loam texture. Reprinted with permission of the Macmillan Publishing Co., Inc. from *The Nature and Properties of Soils,* 11th ed., by Nyle C. Brady. Copyright ©1996 by Macmillan Publishing Co., Inc.

Table 11.2 Macro-nutrient Elements Required by Plants and their Source within Terrestrial Ecosystems

Element	Symbol	Source
Carbon	C	Atmosphere
Hydrogen	H	Water
Oxygen	O	Atmosphere, water
Nitrogen	N	Organic matter, atmosphere
Phosphorus	P	Mineral soil, organic matter
Potassium	K	Mineral soil, organic matter
Sulfur	S	Mineral soil, organic matter, atmosphere
Magnesium	Mg	Mineral soil
Calcium	Ca	Mineral soil

Source: After Brady (1990).

Table 11.3 The Biochemical Function of Plant Macro-nutrients, their Form of Uptake, and Typical Leaf Concentrations in Plants

Element	Biochemical Function(s)	Form Assimilated	Leaf Concentration
Carbon (C) Hydrogen (H) Oxygen (O)	Form the basic building blocks of all biologically-active compounds	CO_2, H_2O	90-98%
Nitrogen (N)	Nucleic acids, amino acids, proteins, chlorophyll, anthrocyanins, alkaloids	NH_4^+, NO_3^-	1-4%
Phosphorus (P)	Nucleic acids, nucleitides, sugar phosphates, phospholipids	$H_2PO_4^-$	0.1-0.4%
Potassium (K)	Enzyme co-factor, osmotic regulation, cell ion balance	K^+	1%
Calcium (Ca)	Pectin synthesis and cell wall formation, metabolism/ formation of nucleus and mitochondria, enzyme activator	Ca^{2+}	0.8%
Sulfur (S)	Amino acids, proteins, sulfolipids	SO_4^{2-}	0.2%
Magnesium (Mg)	Chlorophyll, enzyme co-factor	Mg^{2+}	0.2%

Source: After Salisbury and Ross (1992).

been observed to constrain forest growth in the humid tropics (Vitousek, 1984; Vitousek and Sanford, 1986). In the following discussion, we explore some of the chemical processes in soil that regulate the supply of nutrients for plant growth in both temperate and tropical forest soils. These processes, in combination with plant uptake and litter decomposition, control the cycling of nutrients within forest ecosystems (Chapter 19).

Clay Mineralogy

In studying Table 11.3, note that many macro-nutrients exist as cations, the ionic form assimilated by plant roots. The mineralogy of clay particles, and the ion exchange reactions mediated by their negatively-charged surfaces, substantially influence the supply of cations for plant growth. Ion exchange reactions in soil also are an important mechanism influencing the retention and loss of nutrients from forest ecosystems, especially following disturbances like harvesting, large-scale windthrow, or fire.

Clay minerals form during the weathering process and, to some extent, chemically reflect the primary minerals from which they originate. Phyllosilicate minerals dominate the clay fraction of temperate soil and originate from a wide array of primary minerals including feldspar, orthoclase, and hornblende. Although phyllosilicate clays may differ in mineralogy, they are all formed from the same chemical building blocks. The primary

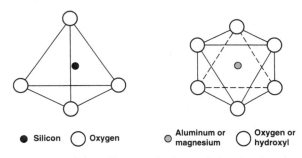

Figure 11.6. The structure of the silica tetrahedra and the alumina octrahedra that form the basic building blocks of phyllosilicate minerals.

structures of these minerals are the silica tetrahedra (SiO_4) and the alumina-magnesia octahedra (AlO_6 or MgO_6; Figure 11.6). By sharing O atoms at their corners, these subunits can link to form tetrahedral ($[AlO_6]_n$) or octahedral sheets ($[MgO_6]_n$), secondary structures that confer many unique properties to phyllosilicate minerals.

Fine-grained mica, vermiculite, chlorite, montmorillonite, and kaolinite are common phyllosilicate minerals in temperate soil, and some of their physical and chemical properties are summarized in Table 11.4. The weathering process, which gives rise to clay minerals, substantially influences on the extent to which phyllosilicate clays attract and bind cation nutrients. As a consequence, unweathered soils and highly weathered soils differ markedly in clay mineralogy and their ability to supply nutrients for plant growth.

Isomorphic substitution conveys a permanent net negative charge on phyllosilicate minerals enabling them to attract cations and supply them for plant growth. This process occurs as one type of clay mineral (i.e., montmorillonite) weathers and gives rise to another (i.e., kaolinite). During the formation of phyllosilicate minerals, atoms of the same size, but of lower charge, replace either Si in tetrahedral sheets or Al in the octahedral sheets. The extent to which a particular clay is substituted is influenced by the chemical

Table 11.4 Physical and Chemical Properties of Some Common Phyllosilicate Minerals

Property	Fine-Grained Mica	Montmorillonite	Kaolinite	Chlorite
Ratio of Octahedral to Tetrahedral Sheets	2:1	2:1	1:1	2:1:1
Size (μm)	0.1-5.0	0.01-1.0	0.5-5.0	0.1-2.0
Shape	Plates Flakes	Flakes	Hexagonal Crystals	Variable
External Surface ($m^2\,g^{-1}$)	50-100	70-120	10-30	70-100
Net Negative Charge ($cmol\,kg^{-1}$)[a]	100-180	80-120	2-5	15-40

[a]centimoles of negative charge per kilogram of dry soil

Source: After Brady (1990).

constituents of the parent material and the degree to which it has weathered. This property is a permanent attribute of phyllosilicate minerals and is relatively unaffected by changes in soil acidity. As a consequence of a low degree of isomorphic substitution, kaolinite, a highly weathered clay mineral, has a low net negative charge that is balanced by a small number of cations adsorped into its crystalline surface (Table 11.4). Montmorillonite, a clay mineral less weathered than kaolinite, is highly substituted in its octahedral sheets and has a relatively large net negative charge. The aforementioned examples illustrate how weathering directly controls the ability of phyllosilicate clays to adsorb and exchange cations with soil solution, a factor that has a profound effect on the supply of nutrients for plant growth and the ability of ecosystems to retain nutrients against leaching.

Following long periods of intense weathering (i.e., 100,000 to 1,000,000 years), Si is lost from the structure of phyllosilicate minerals leaving behind oxides of Fe and Al, which constitute the clay fraction of some soils in the humid tropics. These oxides differ markedly in their chemical characteristics from the phyllosilicate minerals from which they originate. In contrast to the ordered crystalline structure of phyllosilicate minerals, oxides of Fe and Al are amorphous (without form) and exist in a less-ordered semi-crystalline state (Schwertmann and Taylor, 1977; Sposito, 1989). More importantly, these minerals exhibit a pH-dependent variable charge, in contrast to the stable, negative charge of phyllosilicate minerals. This property results from the large number of hydroxyl groups ($-OH$) contained within these minerals; the formulae for goethite [$FeO_3(OH)_3$] and gibbsite [$Al(OH)_6$] illustrate this point.

Soil acidity in highly-weathered tropical soils plays an important role in the functioning of clay minerals and their ability to supply plants with nutrients. In acidic soils, the hydroxyl groups of the Al and Fe oxides are protonated, conveying a net positive charge and the ability to adsorb and exchange anions. However, these minerals lose protons (H^+) in relatively alkaline soil (e.g., $AlO(OH)_5^- + H^+$), thus producing a net negative charge and the ability to adsorb and exchange cations. As a consequence, soils dominated by Al and Fe oxides have a pH-dependent point of zero charge at which neither cations nor anions are adsorbed; both cations and anions are susceptible to loss through leaching. Land management practices which alter soil pH clearly have the potential to alter the ability of tropical soils to adsorb and retain plant nutrients. For further elaboration on the dynamics of variable-charge soils and nutrient mobility within them, we refer readers to Sollins et al. (1988) and Uehara and Gillman (1981).

Cation Exchange and the Supply of Nutrients

The **cation exchange capacity** of soil is a general measure of plant nutrient availability and represents the total amount of cations (centimoles of positive charge) that can be adsorbed by a kilogram of soil. Cations in soil solution exist in an equilibrium with those adsorbed to cation exchange sites on clay micelles. When cations in soil solution are assimilated by plants, those adsorbed to clay particles are released and a new equilibrium is established. As such, adsorbed cations can be thought of as a "reservoir" of plant nutrient in soil. Because cations vary in size and charge, they are adsorbed and exchanged in a predictable manner reflecting their affinity for exchange sites on clay micelles. In general, cations with a small hydrated radius and a large positive charge are most strongly held. We have listed the cations commonly found in soil solution in order of decreasing affinity:

$$Al^{3+} > H^+ > Ca^{2+} > Mg^{2+} > K^+ = NH_4^+ > Na^+$$

Weakly adsorbed cations like K^+ and NH_4^+ are more available for plant growth; they also are more susceptible to leaching than others higher in the order.

Table 11.5   The Relationship Among Soil Texture, Clay Mineralogy, and the Cation Exchange Capacity in Surface Forest Soils of the Eastern United States

Vegetation/ Physiography	Texture	Clay (%)	Mineralogy	Cation Exchange Capacity (cmol kg^{-1})	Organic Matter (%)
I. Northeastern US					
jack pine/ outwash plain	sand	2	mixed	13	6.4
northern pin oak– black oak–white pine/outwash plain	sand	4	mixed	11	6.0
sugar maple– basswood/moraine	loamy sand	6	mixed	18	12.0
II. Southeastern US					
loblolly pine/ coastal plain	silt loam	7	montmorillonitic	11	3.9
loblolly pine/ coastal plain	silt loam	20	montmorillonitic	20	3.7
loblolly pine/ coastal plain	silty clay	37	montmorillonitic	28	2.3
water oak–willow– oak/central prairie region	silty clay	39	montmorillonitic	30	7.0
loblolly pine/ upper coastal plain	loamy sand	5	kaolinitic	4	0.9
loblolly pine/ upper coastal plain	sandy loam	11	kaolinitic	7	1.3
loblolly pine/ upper coastal plain	clay loam	28	kaolinitic	26	2.3

We summarize the cation exchange capacity for forest soils of different texture and mineralogy to illustrate the combined influence of weathering and parent material on supply of cations for plant growth (Table 11.5). Note that cation exchange capacity increases as the percentage of clay rises. Also note that soils dominated by kaolinite, a weakly-substituted and highly-weathered phyllosilicate mineral, generally have lower cation exchange capacities than those soils containing montmorillonite. The soils of mixed mineralogy contain a mixture of phyllosilicate clays and have relatively high cation exchange capacities even though clay contents are relatively low. As you will read later in this section, the high organic matter contents of these soils greatly contribute to their ability to supply plants with cation nutrients.

The proportion of cation exchange sites occupied by Ca^{2+}, Mg^{2+}, K^+, and Na^+ is the **percent base saturation** of soil. These cations are not technically bases, because they do not directly neutralize H^+ in soil solution. Nonetheless, they reduce soil acidity when they are adsorbed in place of H^+, a topic that we will further elaborate in the following section. Percent base saturation is a chemical property of particular relevance to plant growth, because it is both a general measure of cation nutrient availability and soil buffering capacity. In general, soils that have a high proportion of exchangeable bases have a high capacity to both supply plants with base cations and buffer acidic inputs.

Soil Acidity

Soil pH is commonly used to quantify acidity and, by definition, is the negative log of the hydrogen ion concentration in soil solution ($pH = -\log [H^+]$). In combination with adsorption and exchange reactions, soil acidity substantially influences the supply of nutrients for plant growth. It does so by controlling the solubility of soil minerals. Figure 11.7 illustrates the availability of plant nutrients along a pH gradient ranging from very acidic soils to those with high pH. Note that the availability of most nutrients is greatest at neutral pH values (Figure 11.7). By influencing mineral solubility, soil pH also affects the weathering rate of parent material, the formation of clay minerals, and soil development. Thus, the weathering process and soil acidification often go hand in hand. The activity of soil microorganisms also is influenced by soil acidity as we see in Figure 11.7.

There are several sources of H^+ that contribute to the lowering of soil pH, the weathering of parent material, and the removal of base cations. Perhaps the most important source of H^+ in soil results from the respiration of plant roots and soil microorganisms. Carbon dioxide produced during respiration dissolves in soil solution and forms carbonic acid (H_2CO_3). The prolonged exposure of soil minerals to this relatively-weak acid results in their solubilization and the removal of base cations. Much stronger organic acids (e.g., fulvic and humic acids) are produced as by products of the microbial decomposition of plant tissues. Plant roots also exude organic acids that similarly act upon soil minerals. Over extended time periods (i.e., 10,000 to 1,000,000 years), these sources of acidity facilitate the weathering of soil minerals and lower pH.

The industrial activities of humans also have the potential to influence soil chemistry. Oxides of nitrogen (NO_x) and sulfur (SO_2), released during the burning of fossil fuels, can further oxidize in the atmosphere to produce nitric (HNO_3) and sulfuric (H_2SO_4) acids. The addition of these acids in precipitation has raised concerns in eastern North America and central Europe. Because the constituents of "acidic deposition" are relatively strong acids, they have the potential to act upon soil minerals in the same manner as the acids produced by the metabolism of plants and soil microorganisms. Some soils in eastern North America, particularly those with coarse textures where bedrock is shallow, are sensitive to acidic deposition because of their low cation exchange capacity, and low base saturation.

It is important to consider soil acidity, mineral weathering, and the removal of base cations simultaneously, because they are co-occurring processes in soil. As mentioned earlier, percent base saturation represents the ability of soil to buffer the input of H^+ from chemical and biological sources. Base cations buffer the soil reaction when they weather from soil minerals and replace H^+ adsorbed to exchange sites. The H^+ so released initially enters soil solution; however, it is easily leached from the soil resulting in a decrease in acidity. Clearly, over long time periods, the ability of soil minerals to relinquish base cations to weathering can be exhausted. In such a situation, pH declines to the point where the soil reaction is dominated by Al. At a very low soil pH (< 4.0), Al exists as Al^{3+} and,

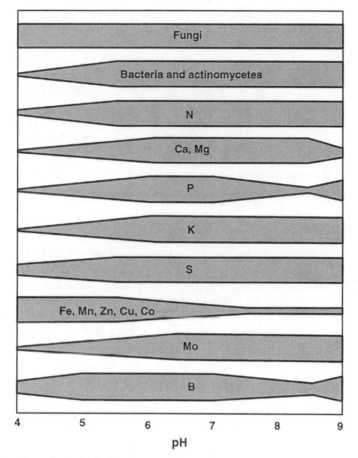

Figure 11.7. The relationship between soil pH and the availability of plant macro- and micro-nutrients is shown. The influence of soil pH on microbial activity within soil also is illustrated.

along with H^+, occupies the majority of cation exchange sites. In soil solution, Al^{3+} can react with water in the following manner:

$$\text{Adsorbed-}Al^{3+} \rightleftharpoons Al^{3+} + H_2O \rightleftharpoons AlOH^{2+} + H^+$$

In studying this equation, note that the reaction of H^+ with bases in soil solution will shift the equilibrium to the right, producing a "new" H^+ during formation of $AlOH^{2+}$. This mechanism constitutes a buffering system that maintains acidic soils at a low pH. Furthermore, Al^{3+} released in the reaction is toxic to plants and can greatly restrict root growth (Runge and Rode, 1991). High concentrations of Al^{3+} in soil solution are known to reduce root elongation, kill root meristems, and interrupt the functioning of the plasmalemma in above-ground tissues (Foy et al., 1978).

Most forest soils range from extremely acid (pH 4.0) to slightly acid (pH 6.5). Where a particular forest soil lies along this gradient is substantially influenced by organic matter additions (e.g., leaves, roots, twigs, reproductive structures) from overstory trees and the acids produced during microbial decomposition. The general trend is for conifers such as pines, spruces, hemlock, and Douglas-fir to increase surface soil acidity (i.e., decrease pH) to a greater extent than hardwoods or northern white-cedar.

An example of how individual trees influence soil acidity is provided by tuliptree and eastern hemlock in eastern Kentucky (Boettcher and Kalisz, 1990). Although these trees co-occur on the same soil parent material, the soil pH under tuliptree (pH 4.7) is consistently greater than that beneath eastern hemlock (pH 4.0). In addition, quantities of the exchangeable bases Ca^{2+} and K^+ also are lower beneath eastern hemlock. In eastern Washington, organic matter additions from western hemlock (pH 4.0) lower surface soil pH to a much greater extent than western red cedar (pH 5.9) when both species occur on the same soil parent material (Alban, 1969). These examples illustrate that differences in the leaf litter chemistry of forest trees can have a substantial influence on the chemical properties of forest soils.

Soil Organic Matter

Although organic matter composes a relatively small fraction of most forest soils (e.g., < 1 percent to 15 percent), it has a profound effect on a wide array of physical, chemical, and biological properties. As noted earlier, soil organic matter (i.e., plant litter) contributes to aggregate formation, which in turn influences the amount of soil water available to plants. It also functions as a "storehouse" of plant nutrients, supplying most of the nitrogen used in the annual growth of forest ecosystems. Finally, soil organic matter is the substrate used for the growth and maintenance of microbial populations in soil. It is through the metabolic activities of these organisms that nitrogen and other plant nutrients are released from soil organic matter. Because the nutrients so released can be re-assimilated by plants, soil organic matter represents an important "weigh station" in the cycling and storage of nutrients within forest ecosystems (see Chapter 19).

The organic matter entering the soil originates from above- and belowground sources of plant litter. Aboveground sources consist of leaves, reproductive structures, twigs and tree stems, whereas roots (fine and coarse) are the primary belowground source of litter. In most forests, belowground litter from fine roots equals or exceeds aboveground litter production (i.e., leaves, seeds, flowers). In general, plant litter contains approximately 15 to 60 percent cellulose, 10 to 30 percent hemicellulose, 5 to 30 percent lignin, and 2 to 5 percent protein (Paul and Clark, 1996). In soil, these compounds are metabolized by microorganisms, producing energy, CO_2, H_2O, and humus as end products. Humus, which composes the Oa horizon, is a complex and chemically-resistant material that gives surface soils their dark color and unique chemical properties. Due to its advanced state of decay, humus does not physically or chemically resemble the plant material from which it originated. Humus also is chemically resistant to further microbial degradation and can remain in soil for periods of 100 to 3,000 years (Paul and Clark, 1996). The surface of humus can have a net negative charge, resulting from the dissociation of H^+ from hydroxyl ($-OH$), carboxylic ($-COOH$), or phenolic ($C_6H_{11}-OH$) groups. At high pH values, the cation exchange capacity of humus (150 to 300 cmol kg^{-1}) can exceed that of many silicate clays. As such, cation or anion exchange reactions mediated by humus represent an important mechanism influencing nutrient availability in soil. In some soils, approximately 50 percent of the total cation exchange capacity of soil can be attributed to humus. The relatively high cation exchange capacities of the sandy soils of mixed mineralogy in Table 11.4 result from high organic matter contents.

The organic matter content of soil reflects a balance between the addition of organic matter from plant production and its loss during microbial decomposition. Because forest harvesting can alter both rates of litter input and loss through decomposition, it also has the potential to alter the quantity of organic matter and associated plant nutrients stored in soil. In Chapter 19, we further consider the impact of forest harvesting on soil organic matter dynamics and the cycling of plant nutrients.

The organic matter content of soil exerts an important influence on the available water content of soil. Soil organic matter holds relatively large quantities of water at field capacity, but its permanent wilting point also is proportionally high, providing only small quantities of plant available water. However, organic matter content is the primary factor influencing soil aggregate formation. In turn, soil aggregation influences the proportion of micro- and macropore space, which directly controls the water-holding characteristics of soil. Consequently, soil organic matter exerts its main influence on the water-holding characteristics of soil through its influence on soil structure. Well-aggregated, fine-textured soils with ample organic matter contents (5 to 10 percent) generally hold large quantities of available water, making them good substrates for plant growth.

SOIL CLASSIFICATION

A taxonomic system of soil classification, referred to as the Soil Taxonomy, is widely used throughout North America. It was developed by the soil survey staff of the United States Department of Agriculture to classify soils in regard to their potential for agricultural management (Soil Survey Staff, 1975). This system is based on measurable morphological characteristics present within a particular soil profile. The primary advantage of this approach is that the soil profile itself, rather than the soil-forming process, is classified. Soil taxonomy is modeled after the plant taxonomic system, with categories ranging form order (broad grouping) to series (narrowest category).

The main soil orders supporting forests in North America are Entisols, Inceptisols, Spodosols, Alfisols, Histosols and Ultisols. **Entisols** (recent soils) are mineral soil without, or with only the beginnings, of horizon development. They often occur on talis slopes, flood plains, sand dunes, and where bedrock lies close to the land surface. **Inceptisols** (from Latin inceptum) are more weathered than Entisols, and contain a weakly developed B horizon. These soils have a wide geographic distribution, and, with the exception of arid climates, can be found in most regions in North America. Inceptisols are common forest soils in the Pacific Northwest, Rocky Mountains, and the eastern United States. They often occur in well-drained, upland landscape positions, but also can be found along river corridors.

Alfisols (from the chemical abbreviation for aluminum and iron) typically form in cool to hot humid areas and are common under deciduous forests in the eastern United States. They are characterized by gray to brown surface horizons, medium to high base saturation, and the accumulation of silicate clay in the B horizon. Alfisols appear to be more strongly weathered than Inceptisols, but are less weathered than Spodosols.

In cold and temperate climates, the process of leaching can give rise to the formation of **Spodosols** (from Greek *spodos*, wood ash). These soils are characterized by the presence of a strongly-developed E horizon and the accumulation of humus and oxides of Fe and Al in the B horizon. Spodosols often form beneath boreal forests occurring on coursetextured parent materials. In general, these soils are best developed beneath spruce and fir, species whose litter generally acidify the surface soil, and are common in the northeastern United States, northern Lake States region, and the Pacific Northwest. It should be noted that Spodosols also can form beneath coniferous forest in warm climates. In Florida, Spodosols are often encountered in landscape positions in which coniferous forests are seasonally flooded.

Histosols form in poorly-drained landscape positions and are characterized by a high organic matter content ($\geq$ 20 percent). These soils can form anywhere the land surface is continually saturated with water, and thus occur in all climates and have a global distribution. Forest vegetation occurring on these organically-derived soils include: black spruce bogs in the northern Lake States, black ash–red maple swamps in the northeastern United States, and the pocosin and cypress swamps of the southeastern United States. Large expanses of forested Histosols also can be found in Scandinavia, Siberia, and Canada.

Ultisols (form the Latin word *ultimus*) are a common forest soil on old land surfaces in warm, humid climates, such as those of the southeastern United States. These soils are characterized by an accumulation of silicate clay in the B horizon. However, Ultisols are distinguished from Alfisols by a low base saturation—the result of more intense weathering. These soils are widely distributed in the eastern US, extending southward from Maryland to Florida and westward from the east coast to the Mississippi River Valley. They also occur in portions of the Pacific Northwest and eastern California, and can be found on old land surfaces in Australia, Africa, India, southern China, and southern Brazil.

Oxisols (from the French *oxide* and the Latin word for soil, *solum*) support forest vegetation in the tropical and sub-tropical regions of Central and South America, southeast Asia, and Africa. The subsoil of these highly-weathered soils contains an accumulation of kaolinite and oxides of Al and Fe. The old land surfaces on which these soil occur, in combination with the intense weathering of humid tropical climates, can give rise to profiles exceeding 15 meters in depth!

Forest and agricultural ecosystems differ in ways that limit the use of the soil taxonomy to classify forested landscapes. In the western United States, for example, a wide range of forest habitat types can be found on the same taxonomic unit of soil (Neiman, 1988), making it difficult to use soil classification to predict the occurrence of forest vegetation. The primary reason for such a disparity is that forest vegetation reflects a myriad of interacting factors like physiography, harvesting frequency and intensity, and prior land uses that are not considered by soil taxonomy. Nevertheless, soil factors that reflect moisture and nutrient regimes, like texture, aggregation, and coarse-fragment content, are often related to the occurrence of some forest habitat types (Neiman, 1988). Because the relationship between forest communities and the soil developing beneath them is multifactorial and dynamic, it is likely that any single-factor classification, like the soil taxonomy, will be of limited use in predicting the distribution and growth of forest ecosystems. In Chapter 13, we further discuss the limitation of single-factor systems for classifying forested landscapes.

LANDFORM, SOIL, AND FOREST VEGETATION: LANDSCAPE RELATIONSHIPS

Changes in topography, vegetation, and parent material can give rise to marked differences in soil formation, even within a relatively small geographic area. In Michigan's Upper Peninsula, for example, forest ecosystems dominated by white pine, sugar maple, and red oak can occur in well-drained, upland landscape positions where a thin blanket of glacial drift overlies bedrock (Fig. 11.8). Soils forming in these landscape positions are Inceptisols, characterized by a shallow profile and the minimal development of a B horizon. The deeper deposits of glacial drift downslope give rise to Spodosols, which occur beneath a canopy of sugar maple and basswood. In lower slope positions also dominated

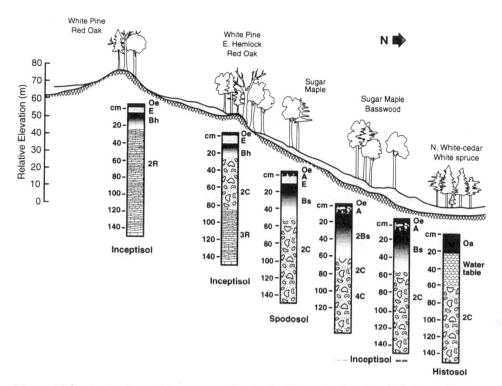

Figure 11.8. A physiographic cross section in the Upper Peninsula of Michigan illustrating the influence of topography and forest vegetation on soil profile development. (After Pregitzer et al., 1983. Reprinted with the permission of the Soil Science Society of America, Segoe, Wisconsin, USA.)

by sugar maple, Inceptisols form in relatively recent deposits of colluvium. The formation of an Inceptisol in this landscape position is related to the relatively short duration over which the parent material has weathered. Swamp forest dominated by black ash and northern white-cedar occur in the poorly-drained bottom slope positions where the accumulation of organic matter has lead to the formation of Histosols (Fig. 11.8). The patterns of soil formation described above are repeatable features in the landscape, occurring in other locations with a similar set of soil-forming factors (i.e., parent material, vegetation, time).

In the southeastern United States, topography exerts a similar influence on the soil developing beneath the flatwood vegetation of the lower coastal plain (Fig. 11.9). In this region, relatively small elevational differences differentiate well-drained from poorly-drained landscape positions. This land surface is relatively old compared to the relatively recent (ca. 10,000 years before present) deposition of glacial materials in Michigan's Upper Peninsula. As a consequence, the parent material giving rise to this soil has been effected by climate, biota, and topography for a longer duration.

In uplands dominated by longleaf pine, relatively-dry conditions give rise to Inceptisols—poorly-developed soils with minimal horizon formation. The warm, humid climate of this region, in combination sandy parent material and coniferous vegetation, give rise to Spodosols in somewhat poorly-drained landscape positions. Histosols also form in very poorly drained landscape position in the southeast United States, similar to the landscape distribution of these soils in other regions.

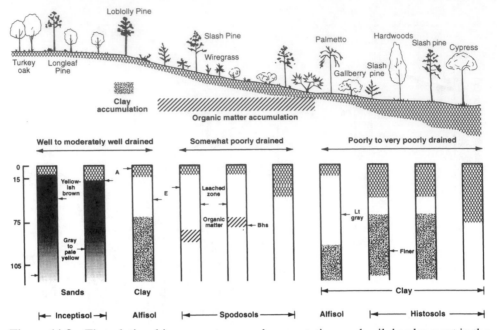

Figure 11.9. The relationship among topography, vegetation, and soil development in the coastal plain flatwoods of the southeastern US. Organic matter accumulation in the surface horizon dramatically increases as one moves from the well-drained upslope positions to the poorly-drained bottom slope positions. In this region of the United States, relatively small changes in elevation elicit large changes in soil drainage, profile development, and overstory composition. (Adapted from Pritchett and Smith, 1970. Reprinted with permission from *Tree Growth and Forest Soils,* © 1970 by Oregon State University Press.)

Suggested Readings

Binkley, D. 1995. The influence of tree species on forest soils: processes and patterns. pp. 1–33. In Mead, D.J., and I.S. Cornforth. (eds.). *Proceedings of the Trees and Soils Workshop*, Agronomy Society of New Zealand Special Publication No. 10. Lincoln Univ. Press, Canterbury, New Zealand.

Brady, N.C., and R.R. Weil. 1996. *Nature and Properties of Soil*, 11th ed. Prentice–Hall, New Jersey. 740 pp.

Paul, E.A., and F.E. Clark. 1996. *Soil Microbiology and Biochemistry,* 2nd ed. Academic Press. New York. 340 pp.

Sanchez, P.A. 1976. *Properties and Management of Tropical Soils.* John Wiley, New York. 618 pp.

CHAPTER *12*

FIRE

Fire has affected a substantial portion of the forests of the world at one time or another. In North America, virtually all of the upland forests in the South, the Lake States and adjacent Canada, the West, and most of those in the Northeast, Appalachian Mountain Region, and Central States have been burned more or less frequently. In the boreal forests of Alaska and the Canadian North, fire has been a powerful natural factor affecting vegetation and wildlife. Wetlands, such as swamps (Cypert, 1973; Ewel, 1990), bogs, marshes, and even moist tropical forests (Goldammer, 1990b) also have been burned, although less frequently, and their vegetation markedly affected. In addition, fire was extensively used by Native Americans prior to European settlement. Natural fire has patterned the forests and savannas of the Mediterranean Region (Naveh, 1974; Gill, 1977), South America (Coutinho, 1990; Soares, 1990), much of Africa (Komarek, 1972; Phillips, 1974; Booysen and Tainton, 1984), and of Europe, Australia (Gill et al., 1981, 1990; Pyne, 1991), and Asia (Goldammer and Penafiel, 1990; Stott et al., 1990). In addition, humans have introduced fire into the wet tropics during deforestation for economic development (Kaufman, et al., 1990; Levine, 1991). Books by Kozlowski and Ahlgren (1974), Pyne (1982, 1991), Wein and MacLean (1983), Goldammer (1990a), Levine (1991), Johnson (1992), Agee (1993), Whelan (1995), and Bond and van Wilgen (1996) provide an entry into the voluminous literature on fire.

Fire has always been a natural and extremely important process shaping the evolution of species and the functioning of ecosystems in which they reside. Fire in prehistory and the remarkable interactions of fire with humans and wildlife is considered by Schüle (1990). Its frequency, intensity, extent, and time of occurrence are characteristic of different regional and local landscape ecosystems. It is a principal influence on plant traits and life cycles as well as ecosystem processes: carbon, nutrient, and water cycling; biomass accumulation; succession; and diversity. Specifically, fire plays many major roles in land-

279

scape ecosystems around the world (Wright and Heinselman, 1973; Swanson, 1981). It influences:

geomorphic and hydrologic processes of hillslopes and stream channels;

physical and chemical properties of soil;

nutrient loss;

biomass accumulation;

genetic adaptations of plants;

plant composition and diversity, mortality, regeneration, growth, and succession;

wildlife habitat and wildlife population dynamics;

presence and abundance of forest insects, parasites, and fungi.

In this chapter we consider fire as a physical site factor, examining its effects on forest species and forest site quality. The role of fire as a disturbance factor in forest ecosystems and their communities is considered in Chapter 16.

FIRE AND THE FOREST TREE

Causes

Evidence of natural fires in the form of fossil charcoal, termed **fusain**, has been found in the Carboniferous coal deposits of 400 million years ago and in the Tertiary deposits of brown coal (Harris, 1958; Komarek, 1973). Lightning was the prime cause of these fires before the advent of humans. Meteorites were an ignition source, and falling igneous rock from volcanic eruptions undoubtedly caused local fires then as it does today. Spontaneous combustion (Viosca, 1931) and sparks from falling quartzite rocks (Henniker-Gotley, 1936) are rare but documented possible causes of fires.

Lightning is estimated to cause about 50,000 wildland fires worldwide each year (Taylor, 1974). This number is less than 1 percent of the estimated 182 million cloud-to-ground discharges occurring annually in the forests and grasslands of the world. Actually, between 70 and 100 lightning flashes are estimated to occur every second worldwide (Pearce, 1997), but not all strike the ground. About 10,000 lightning-caused fires occur in the United States each year, and about 80 percent of these are in the Rocky Mountain and Pacific Coast states. Here a single lightning storm may start many small fires when fuel and climatic conditions are conducive to ignition by lightning.

Throughout much of the modern world, however, humans have been the most significant cause of fires. In pre-Colonial America, Native Americans set many fires. Europeans have followed suit. Because of their high intensity and frequency, such fires, often associated with logging and land clearing, have markedly changed the character of forests and affected site quality.

Fire Regime

Studies of fire history have led to the definition of **fire regime** as the kind(s) of fire and the prominent immediate effects of fire that characterize an area. A fire regime is typically characterized by the following features: type, frequency, intensity, severity, size, and timing (season of burning). However, three factors—fire type, frequency, and intensity—are the most important. The **type** of fire, according to the level at which they burn, are

ground, **surface**, and **crown** fires. **Frequency** refers to the recurrence of fire in a given area over time and is expressed in a number of ways (Agee, 1993). The return interval, the average number of years between successive fires, may be expressed for a given point (for example, a single fire-scarred tree or small group of trees) or for an area. Frequency is also expressed as fire rotation or fire cycle, the length of time required to burn over an area equal to that under consideration. Agee (1993) describes and contrasts these approaches and their computation.

Fire **intensity** refers to the length of the flame or amount of energy generated, whereas **severity** expresses the *effect* of fire on the soil or the vegetation (seed bank, mortality of plants). For example, an intense spring fire, when the soil is moist, may rank low in severity, whereas a low intensity summer fire during drought conditions may be severe in its effects on soil properties. Intensity is typically estimated in kilowatts per meter length of fire front ($kW\,m^{-1}$) and in flame length (m) along the fire front (Johnson, 1992; Agee, 1993). The **timing** of fire may have differential effects on vegetation and soil. For example, in the southeastern coastal plain of the United States, the proper proportion of winter fires to fire-free years is critical to the perpetuation of pure longleaf pine forest because pine regeneration is prevented by annual burning and summer fires (Marks and Harcombe, 1981b).

Three primary fire regimes are recognized—based on whether fires are nonlethal to the dominant aboveground vegetation or result in stand replacement: nonlethal understory fires (including frequent or infrequent surface fires), stand-replacing fires (short or long-frequency crown fires), and variable or mixed fires. The latter regime applies to combinations of understory and stand-replacing fires that may occur in two ways: (1) a *variable* fire regime of frequent, low-intensity surface fires typically followed by a long return time, stand-replacing fire, or (2) a *mixed* fire regime of individual fires alternating between nonlethal understory burning and stand-replacing fires, creating a fine-scale pattern of young and older trees. These types are illustrated by examples in Chapter 16

FIRE TYPES, FREQUENCY, AND SEVERITY

From the viewpoint of physical behavior, Van Wagner (1983) identifies five main types of fire in the northern forest, including two kinds of surface and crown fires:

ground fires, smoldering in deep organic layers (less than $10\ kW m^{-1}$)

surface backfires, burning against the wind (100 to $800\ kW m^{-1}$)

surface headfires, burning with the wind (200 to $15,000\ kW m^{-1}$)

crown fires, advancing as a single front (8,000 to $40,000\ kW m^{-1}$)

crown fires, high-intensity spotting fires (up to $150,000\ kW m^{-1}$)

The most common type, the surface fire, burns over the forest floor, consuming litter, killing aboveground parts of herbaceous plants and shrubs, and typically scorching the bases and crowns of trees. The greater the fuel accumulated on the surface, the greater the mortality of shrubs and trees. These fires are very sensitive to wind speed. Surface headfires may attain quite high intensities in brush in leafless hardwood stands and in open forest where trees are sparse or crowns are high above ground. In addition to the intensity of the fire, the amount of tree mortality depends on the species, the age of the tree, and rooting habit. Young pines may succumb to a surface fire, whereas older individuals of the same species survive due to thicker bark protecting the cambium from heat damage and the higher elevation of the crown above the flames. A shallow rooting habit, whether due to the inherent nature of the species or the site conditions (wetland or bedrock), increases the susceptibility to fire injury compared to that of the deep-rooting habit typical of up-land species such as oaks and hickories.

Surface fires tend to kill young trees of all species (often, however, just the above-ground portion) and most of the trees of less fire-resistant species of all sizes. However, pole-size to mature trees of fire-resistant species survive light surface fires in varying proportions. Survival in a surface fire for most fire-resistant tree species is not typically dictated by damage to the stem cambium but by their susceptibility to root injury and to scorching of the crown by hot gases rising above the flames. For example, Van Wagner (1970) reported that a light surface fire will leave mature crowns of red pine undamaged, whereas a hot surface fire will kill a red pine stand just as surely as a crown fire 10 times as intense. Observations of fire-damaged red and white pines suggest that if more than 75 percent of the crown is killed, the tree will die.

Fires sweeping the forest floor may generate ground fires, which burn in thick accumulations of organic matter, often peat that overlies mineral soil. When they burn below the surface, they are flameless, and they kill most plants with roots growing in organic matter. Ground fires burn slowly and generate very high temperatures. In moist organic matter, heat from the fire dries out material adjacent to the burning zone and perpetuates a zone of combustible fuel. Ground fires tend to be persistent and serve as reignition sources for surface fires.

Surface fires, fueled by accumulations of organic matter and whipped by winds, may scorch and ignite crowns of trees, thus generating a crown fire. Traveling from one crown to another in dense even-aged stands, most trees are killed in its path. However, even in intense crown fires, unburned strips may be left due to powerful, downward air currents (Simard et al., 1983). Where the forest is patchy and broken, consisting of small groups of trees such as in much of the dry-climate, ponderosa pine forests of the West, some groups may carry a crown fire, but an extensive and devastating crown fire is highly unlikely. Conifers are most susceptible to crown fires because of the high flammability of their foliage and the greater likelihood of their occurrence in pure stands than broad-leaved species. Sparks and burning debris may start new surface fires often far away from the site of the crown fire.

Fire is irregular in frequency, intensity, severity, and burning pattern. These characteristics are primarily controlled by climate, fuel accumulation and flammability, soil water, and especially topography. Fire frequency was probably greatest in presettlement grasslands where burns every two to three years were common (Wells, 1970). In the dry ponderosa pine forests of the West, the average interval between fires can vary from 2 years to about 18 years (Weaver, 1974; Dieterich and Swetnam, 1984; Savage and Swetnam, 1990). Averages vary because of the method of determination (Agee, 1993) and ecosystem type; pine forests of drier low-elevation ecosystems burn more frequently than those of moist slopes and higher elevations.

Arno (1976) reported fire frequency over an elevational gradient (1,150 to 2,600 m) for three watersheds in the Bitterroot Mountains of western Montana for the period 1735 to 1900. Low elevation forests of ponderosa pine and Douglas-fir burned at about 9-year intervals; the minimum and maximum intervals between burns was from 2 to 20 years. Intervals between burns increased with increasing elevation; at high elevations the average interval was about 35 years (range 2 to 78 years).

Intense surface fires may damage the cambium and leave a scar. The fire scar record of individual trees indicates the actual fire frequency of a single place (Spurr, 1954). One ponderosa pine in Arno's study was scarred by 21 fires from 1659 to 1915, an average interval of 13 years (Figure 12.1). Nearby trees were scarred by additional fires, demonstrating that fires do not always burn hot enough to cause injury, particularly when they occur so frequently that only light accumulations of fuel have built up.

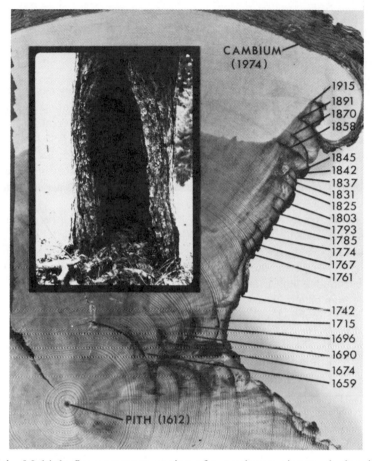

Figure 12.1. Multiple fire-scar cross section of a ponderosa pine trunk showing 21 fire scars from 1659 through 1915. (Arno, 1976. U.S. Forest Service photo.)

Similarly, Dieterich and Swetnam (1984) studied a small, suppressed ponderosa pine in Arizona containing 42 fire scars over the 178-yr period from 1722 to 1900. A detailed historical fire chronology of the 97-ha area showed that whereas the mean fire interval for the tree was 4 years, the interval for the area was 2 years, i.e., not every fire was recorded by this tree. Therefore, records of fire frequency may reliably indicate the frequency of severe fires but may underestimate the frequency of all fires. The light to moderate surface fires, although often undetected, are nevertheless important in regulating seedling composition and distribution and reducing the concentration of fuels on the forest floor.

In contrast to frequent fires in the dry, low-elevation, submontane ponderosa pine and mixed conifer forests, fires occur infrequently in high-elevation subalpine ecosystems of the Pacific Northwest and northern Rocky Mountains (Agee, 1993). Dominant trees are firs (typically subalpine fir), mountain hemlock, and whitebark pine. Because of local variation in ecosystem conditions and fuel accumulation, fire regimes are heterogeneous in frequency. For example, fire-return intervals varied from over 1,500 years for ecosystems dominated by mountain hemlock to differing intervals for whitebark pine ecosystems: from 50 to 300 years (Arno, 1980) to a low of 29 years (Morgan and Bunting, 1990).

In the Sierra Nevada Mountains of California, a frequency of nine years between fires was found at one locality in the giant sequoia–mixed conifer forest during the period 1705 to 1873 (Kilgore, 1973). This is comparable to the overall eight-year frequency of fire reported for ponderosa pine forests (Weaver, 1951). As is typical of many other western forests, the moister east and north slopes do not burn as readily as drier south and west slopes. However, when the mesic slopes burn, they may burn more intensely than those that burn more frequently. A similar relationship develops along elevational gradients from warm and dry low elevations to the cool and moist slopes of high altitudes.

Studies of redwood in California also illustrate that fire-return intervals are ecosystem-related and not specific for a given species. The fire-return interval for moist coastal redwood sites at the northern part of the range may be as long as 500–600 years (Viers, 1980). However, on drier, interior and more southerly sites, intervals of 33–50 years (Viers, 1980) and 20–29 years are reported (Jacobs et al., 1985; Finney and Martin, 1989). Such variation in return interval by specific ecosystem conditions is also reported for subalpine forests.

In the conifer forest of the Boundary Waters Canoe Area of northern Minnesota, detectable fires were relatively frequent (Heinselman, 1973, 1981a). Fires burned at approximately 4-year intervals during the presettlement period 1727 to 1886. Settlement activities increased fire frequency, and the interval dropped to two years. Major fires, burning over large areas, occurred at longer intervals, 21 to 28 years.

The emerging pattern from studies of fire history is one of cyclic occurrence that is determined by the multiple and interrelated physical and biotic features of site-specific ecosystems. Fire regimes of regional and local ecosystems are affected by climate, physiography, vegetation type (affecting fuel accumulation and its flammability), and human activity. Thus fires of different intensities have occurred—from frequent and light surface fires that reduce fuel accumulations and pattern seedling and sapling distribution, to rare, catastrophic crown fires that regenerate entire stands. This ecosystem-dependent range of fire frequency, intensity, and severity was instrumental in the evolution of the life span and species characteristics and, more generally, the kind of vegetation present in each region (Chapter 16). In Figure 12.2, we see the remarkable degree to which fire frequency and severity has patterned regional forest composition throughout North America. In mountainous areas surface fires are common and their frequency is related to climatic factors of temperature and precipitation associated with elevation. However, these surface fires can lead to moderate and severe stand-replacing fires given the right climatic and fuel conditions.

Fire Adaptations and Key Characters

Certain key characteristics and adaptive traits of forest trees and shrubs appear to have developed in response to fire. Many individual characteristics are typically cited, for example, thick insulative bark and sprouting ability. Probably however, few of these are genetic adaptations that are exclusively fire determined. Seasonally hot, dry sites favor fire and support species that not only are perceived to have fire adaptations but also are physiologically adapted to live in such severe environments. Selective forces of the total environment affect the plant's evolutionary response. This point is confirmed by the discovery that many species of tropical rain forests persist in human modified ecosystems now characterized by frequent fire. Such "fortuitous adaptations" allow species to survive fires or quickly establish following fire regardless of evolutionary derivation (Kauffman and Uhl, 1990). These adaptations include thick bark, vegetative sprouting, and seeds stored in the soil. Therefore, our emphasis in this section is on those traits of woody species that are suited to survive and reproduce in ecosystems characterized by fire, i.e., fire-dependent ecosystems.

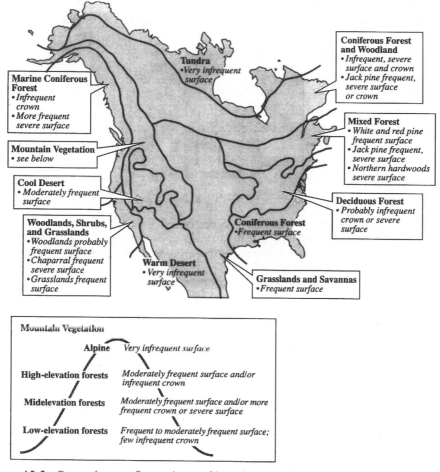

Figure 12.2. Presettlement fire regimes of broad vegetation types. Based on broad ecoregions of Bailey (1995). (After Vale, 1982. Reprinted with permission of the Association of American Geographers.)

The characteristics probably most directly elicited by fire are closed cones (cone serotiny, i.e., delayed opening of cones); thick fire-resistant bark; buried or protected buds and subsequent sprouting; early flowering, fruiting, and dispersal; and the "grass stage" in longleaf and other pines. In addition, the relative life span of many forest species has been determined in part by the frequency of devastating fires (Gill et al., 1981). As a result, moderately long-lived, fire-dependent species such as Douglas-fir and eastern white pine, are able to compete with long-lived, late-successional associates (western hemlock and western red cedar; sugar maple and American beech) on mesic sites that burn infrequently but within their life span.

The key characteristics of forest species that enable them to persist in fire-prone ecosystems may be grouped in four general categories to emphasize major life history features related to fire and site. Thus species exhibit traits that we perceive act (1) to **avoid** fire damage, (2) to **recover** following fire damage, (3) to **colonize** sites after fire, and (4) to **promote** or **facilitate** fire. Especially fire-resistant or susceptible species typically exhibit the extreme expression of these characteristics. Examples of each category are cited below:

Avoiding Fire Damage

Thick, insulated bark (many pines and oaks, western larch, giant sequoia, redwood);

Buried buds protected by soil (aspens, sweetgum, sumacs, many shrubs);

"Grass stage"—longleaf pine of the United States (see Chapter 16), several pines of Mexico and the Caribbean region (Mirov, 1967, p. 417), and chir and Merkus pines of Asia;

Deep rooting—tap root in young plants (upland oaks and hickories);

Rapid juvenile growth—crown grows above the surface fire zone and heat-resistant bark is formed (pines);

Basal crook—dormant buds on the lowermost stem are protected from fire by a crook of the stem that brings the buds in contact with mineral soil (pitch, shortleaf, pond, and other hard pines);

Branch habit and self-pruning ability—rapid self-pruning of branches decreases the likelihood of crown fire, whereas low or drooping branching habit and poor self-pruning ability increase the likelihood of crown fire (larches, pines, and Douglas-fir self-prune well in closed stand conditions, whereas true firs, hemlocks, and spruces retain their branches and often exhibit drooping branches);

Stand habit—open-grown stands decrease the probability of crown fire and also afford less fuel (western larch, ponderosa pine, longleaf pine);

Fire-resistant live foliage—hardwoods are much less flammable than conifers; among conifers, larches have less flammable foliage than pines, Douglas-fir, spruces, and true firs;

Rapid foliage decomposition—retards fuel accumulation and reduces the opportunity for fire ignition and spread (ashes, elms, sugar maple, basswood).

The differential resistance of tree species of the northern Rocky Mountains to fire damage and mortality was shown by Flint (1930). He determined their relative resistance using characteristics such as bark thickness of old trees, rooting habit, resin in old bark, branch and stand habit, relative flammability of foliage, and abundance of lichens on stems. Western larch, ponderosa pine, and Douglas-fir are very resistant, whereas western hemlock and subalpine fir are low to very low in resistance. Although species vary greatly in resistance, severe fires largely erase differences in resistance (Wellner, 1970).

Recovering from Fire Damage

Sprouting—new shoots arise from living stems or roots after fire damage:

Stems:

Rhizomes (many shrubs and herbaceous plants)

Root collar or crown (oaks, paper birch, black cherry, redwood, chaparral species)—basal sprouting is rare in conifers, but in hard pines (shortleaf, pitch, Monterey, etc.) sprouts may occasionally arise from dormant buds formed in the axils of primary needles at the base of the stem (Stone and Stone, 1954);

Lignotubers (burls)—relatively large stem swellings at or below ground incorporate many clusters of dormant buds, common in redwood and nearly all eucalypt species and Mediterranean shrubs (Gill et al., 1981; James, 1984);

Bole—dormant buds along the bole initiate new shoots after the crown is scorched or killed (pitch pine, redwood, big-cone Douglas-fir, eucalypt species).

Roots:

Important in some trees and shrubs (aspens, rock elm, sweet gum, sassafras, sumacs, Acacia species).

Deep Rooting—tap root or sinker roots provide food reserves for rapid regeneration of new shoots (upland oaks and hickories).

Colonizing Burned-Over Areas

Early flowering and seed production—enables a species to reproduce itself sexually on a site where there are short intervals between fires (jack pine, lodgepole pine, pitch pine, and *Xanthorrhoea* species of Australia (Gill et al., 1981);

Light, wind-borne seeds—facilitates widespread dissemination of seeds, some of which may reach burned sites (species of *Betula, Populus, Salix, Tsuga, Larix*);

On-plant seed storage and fire-triggered dehiscence of fruits and cones—closed cones that contain viable seeds persisting on branches are typical of many species of the pine family (including jack, lodgepole, pitch, pond, sand, table mountain, knobcone, Monterey pines) and cypress species of arid sites of southern California (Vogl et al., 1988). Black spruce exhibits semiserotinous cones; the cone scales open upon drying and close when wet. Seed dispersal is therefore periodic, sometimes occurring over a two-year period.

Shrubs of the Australian genus Banksia retain follicles on their branches and fail to open unless they have been in contact with flames (Gill et al., 1981). Many other taxa of the Australian flora also have this wood-fruit habit. Also, some eucalypt species of southeastern Australia retain their fruit four or more years with seed release following fire;

No dormancy—certain species of fire-prone ecosystems have seeds that do not enter true dormancy. They germinate readily at any favorable time following postfire rains (jack, lodgepole, longleaf, and pitch pines);

Heat-induced germination—hard-coated seeds of certain species of *Arctostaphylos*, *Ceanothus*, and *Rhus* tend to lie dormant in the soil (in redstem ceanothus up to 150 years (Mutch, 1976). Germination is favored by fire, which cracks the seed coat and generates the heat needed to stimulate germination.

Promoting Fire Occurrence

Traits that increase the likelihood of fire and thereby favor regeneration of the species over others:

Flammable foliage and bark—needles of pines, many conifers, and some angiosperms are highly flammable, decompose slowly, and form a ready fuel source for surface fires (Mount, 1964; Mutch; 1970, Gill, 1977). The bark of certain species, such as paper birch and some eucalyptus species, is highly flammable.

Retention of foliage near the ground—promotes crown fires (understory conifers, especially firs, northern white-cedar and western redcedar; juvenile oaks).

Short stature—brings foliage close to the ground where a surface fire may spread to the crown (young or slow growing trees with flammable foliage, jack and lodgepole pines).

> **Retention of dead, lower branches**—serves as "ladder" fuel promoting fire spread to the crown (narrowleaf cottonwood, jack pine, some oaks).

Strategies of Species Persistence

Using one or more of these and other adaptations, species persist in fire-prone ecosystems, regenerating themselves either by seed or by vegetative means. Rowe (1983) distinguished five kinds of "strategies," based on an understanding of fire adaptations, life-cycle, and mode of persistence of plants in ecosystems of northern and boreal forests. Disseminule-based plants—those propagating primarily by seeds—are termed **invaders**, **evaders**, and **avoiders**. Vegetatively-based plants—those propagating primarily by horizontal and vertical extension of vegetative parts—are the **resisters** and **endurers**. Although these groups were conceived for boreal conditions they have wide application elsewhere, as Agee (1993) has shown for the species of western North America.

Invaders are the early arrivers that are successful due to copious production of short-lived, wind-disseminated seeds. Once established, they flower and fruit profusely or spread vegetatively. They are typically the shade-intolerant pioneering, "fireweed" plants that establish after fire regardless of its cycle length and intensity. Many herbs exemplify this group; woody species include paper birch, aspens, and willows.

Evaders store seeds in the canopy, humus, or mineral soil. Their strategy is placement of seeds to evade high temperatures, followed by rapid seed germination and establishment. The plants themselves may be killed by fire, but their seeds are protected. They include short-lived ephemerals as well as intolerant to shade-tolerant species that persist into later successional stages (Rowe, 1983). Woody species of the boreal forest include jack and lodgepole pines and black spruce.

Avoiders arrive late in succession and prosper where fire cycles are long. They essentially lack direct adaptations to fire and are often said to occupy unburned areas. They are the mesophytic and tolerant members of ecosystems relatively undisturbed by fire. Included in the group are many herbs and shade-requiring mosses and lichens; balsam fir and white spruce are woody examples of northern and boreal forests. Sugar maple, basswood, beech, and southern magnolia are deciduous examples of eastern and southern forests.

Resisters are the few intolerant species whose adult stages can survive *low-severity* fires. They continue growing *vegetatively* in spite of fire, whereas evaders are likely to be killed by fire. Boreal examples include jack and lodgepole pines and cottongrass, a circumpolar sedge, whose dense tussock form resists fire. Giant sequoia; ponderosa, red, shortleaf, loblolly, and longleaf pines; western larch; and many thick-barked oaks are also resisters. This group is not mutually exclusive with evaders.

Endurers are composed of the large and diverse group of species, shade-tolerant and -intolerant, that resprout following fire. They regenerate from roots, root collars, rhizomes, and other belowground organs. Their persistence and abundance is strongly related to the vertical positioning of the perennating parts in the insulating humus and mineral soil. Aspens, birches, and many shrubs, including ericaceous bog species, are woody representatives of the group.

These groups of species are closely related to the site-specific ecosystem where they occur and its particular fire cycle and fire behavior. Ecosystems characterized by high severity fires favor invaders, evaders, and endurers; those with low intensity surface fires favor resisters as well as evaders and endurers. Species may belong to more than one group. For example, jack and lodgepole pines may be invaders, evaders, and resisters because of their multiple adaptations in fire-prone ecosystems.

CLOSED-CONE PINES

Where fire is a dominant ecosystem force, many life history characteristics appear strongly related to its frequency and severity. Such is the case with the varied group of pines bearing serotinous cones: jack and lodgepole pines of northern and western North America; pitch pine and pond pine of the Coastal Plain of the eastern United States (Lutz, 1934; Ledig and Fryer, 1972; Givnish, 1981; Ewel, 1990); table mountain pine of the Southern Appalachians (Zobel, 1969); and knobcone, Bishop, and Monterey pines of California (Vogl, 1973). Closed cones may persist on the tree several decades, up to 75 years in lodgepole pine (Clements, 1910) and still bear viable seeds. In lodgepole pine, millions of seeds per hectare may be stored in serotinous cones.

Generally, populations of a given species with a very high proportion of cone serotiny typically occur in the most frequently burned ecosystems (Givnish, 1981). For example, serotiny is common in Coastal Plain populations of pitch pine but rare in the Appalachian Mountains where it occurs with serotinous populations of table mountain pine that grow on the more fire-prone sites.

Both jack and lodgepole pines may bear cones that open readily (nonserotinous) or remain closed under normal climatic conditions. The cone scales are bonded by resin that melts from fire-generated heat above 45 °C; the scales open as they dry, and the seeds are disseminated. Cones in the tree crown may be exposed to temperatures of 900 °C for 30 seconds, and seeds still have high viability. The range of closed cones per tree in jack pine is from 0 to 100 percent; the average for the species is about 78 percent (Schoenike, 1976). Open-cone trees are found in the more southerly portion of the range, whereas closed-cone trees predominate in the boreal and western range (Figure 12.3). In the southern range, jack pine occurs in mixed stands with oaks and red pine, and fires are more often of the lighter, surface type than in the boreal forest. In the boreal forest, jack pine is regenerated by crown fires that open the cones and prepare the seedbed, thereby perpetuating the predominance of the serotinous habit.

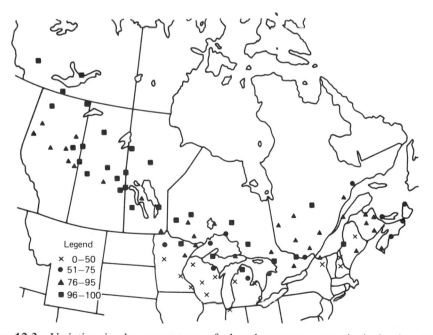

Figure 12.3. Variation in the percentage of closed cones per tree in jack pine (After Schoenike, 1976.)

Similarly, in lodgepole pine, stands originating after a severe burn typically produce trees with a high percentage of closed cones (Muir and Lotan, 1985). In Yellowstone Park in northwestern Wyoming, Tinker et al. (1994) found proportionally more serotinous cones on xeric sites than mesic sites and a significant negative correlation of closed cones with elevation. These factors suggest that high fire frequency and severity increase the proportion of cone serotiny.

The importance of natural selection by fire in maintaining the serotinous habit was emphasized by Givnish (1981). He studied pitch pine in a local portion of its distribution on the Coastal Plain of New Jersey, the Pine Barrens. In areas of the barrens that topographically favor frequent fire, the Pine Plains, 99 percent of the individuals are serotinous. Cone serotiny decreases with distance in all directions from the Plains; infrequently burned lowland sites averaged 84 percent serotinous individuals. Lutz (1934) estimated that the Plains appeared to burn every 6 to 8 years, on average, whereas the surrounding Barrens burned every 16 to 28 years. Givnish and Lutz stressed the flatness and physiographic location of the Plains as key factors affecting the incidence of fires and thus very high cone serotiny. Fires from many directions and distances could burn the shrubby Plains vegetation, whereas firebreaks of swamps and streams protected adjacent areas from frequent fires. Many authors believe that frequent fires are responsible for the stunted form of Plains vegetation. Furthermore, frequent fires and stunted vegetation tend to reinforce one another because the commonly occurring ground fires are easily converted into devastating crown fires, thereby stunting the plants (Lutz, 1934; Andresen, 1959).

FIRE AND THE FOREST SITE

In the early days, the tendency of writers was to consider fire as a destructive agent with few or no beneficial aspects. The development of prescribed or controlled burning as a tool in silviculture, fire hazard reduction, and fire management, however, has caused a reevaluation of the effect of fire on the site. This reevaluation has led to the realization that the effect of fire on forest ecosystems is complex and may often be entirely beneficial. The many important functions of fire and its beneficial effects in forest ecosystems are considered in Chapter 16. In this section, we examine fire effects on site conditions, which affect the regeneration and growth of plants. We may distinguish between the *indirect* effects of fire on site quality through its effects on vegetation, litter accumulation, and soil organisms, and its *direct* effects on soil properties and microclimate.

Indirect Effects

The indirect effects of fire depend upon changes in the vegetation. These are discussed in detail in Chapters 16 and 17. Since an intense fire will kill most or all of the plant above the soil surface, the succeeding vegetation tends to be made up of light-seeded species that can move in from outside the burned area, species with perennial root systems capable of sending up new sprouts, and species with dormant seeds stimulated by heat. Many legumes and species of chaparral (*Ceanothus* spp.; Hanes, 1988) fall in these categories, and the abundance of these and other nitrogen-fixing plants is often increased by burning. In such a case, although previously accumulated nitrogen is volatilized, there may shortly be a new increase in available nitrogen, and the overall site quality may be temporarily improved. In addition, burning of organic matter leads temporarily to a high pH, thereby increasing microbial activity that often gives rise to increased mineralization and nitrogen availability.

However, in many parts of the world, recurrent fires favor the development of a shrubby vegetation composed of sprouting species with characteristically tough foliage, which is low in nutritive value and slow to decompose. Heather in northern Europe, blueberry and other Ericaceous species and bracken in many countries, scrub oaks around the Northern Hemisphere, and many chaparral types (the broad-sclerophyll scrub or brushland vegetation) of California (Hanes, 1988) and the Southwest—all are plants that become dominant after heavy and repeated fires.

Indirectly, fire may substantially change the heat regime of soils by removal of forest tree cover and organic matter of the forest floor. In the South, Komarek (1971) noted that soils of openings are colder than those under forest, whereas in the north the reverse is true. In the boreal forest, frost penetrates deepest under the tree canopies, and in permafrost areas (ground that is permanently frozen) soils thaw least at the surface (Rowe and Scotter, 1973).

A large part of the boreal forest in Alaska lies in the zone of discontinuous permafrost (Van Cleve et al., 1986). In winter, the canopy intercepts snow that would otherwise provide insulation, and in summer organic matter, built up under forest cover, provides insulation from heat accumulation. Fire, by removing tree cover and surface organic matter, as well as blackening the surface, therefore contributes both to increased heat flow into the soil in summer (black surface and less organic cover) and decreased heat outflow in winter (increased snow cover). As a result, permafrost is melted at increasing depths below the surface. Viereck (1982) reported that thaw depth increased steadily for nine years following fire. Such belowground changes significantly affect surface hydrology, vegetation composition, and site productivity.

Fire regulates dry matter accumulation, thereby controlling the severity of burning. This affects the density and composition of forest vegetation, which influence site quality. Abnormally long intervals between fires, such as those caused by prolonged fire exclusion, usually lead to high concentrations of organic matter. The intense and destructive fires, which follow sooner or later, may preclude or delay the reestablishment of normal vegetation on the site or change the kind of vegetation present. The indirect effects of change of vegetation or lack of vegetation, however, are much less than the direct effects of severe burning.

Soil flora and fauna significantly influence site quality by decomposing organic matter, fixing nitrogen, and providing aeration. The effects of burning on soil organisms are highly variable depending on the intensity of the fire, depth below the surface, time elapsed following burning, and the nature of the soil and vegetation of the site (I. Ahlgren, 1974; Rundle, 1981). Changes in populations of microorganisms and fungi are most evident in the upper 2 to 5 cm. Decreases are typically observed immediately following intense fires. No postfire decreases are normally found on light burns in soil below the surface 5 to 8 cm, and at 6 months to 1 year after burning. Increases in microorganisms are typically found after fire, often multifold, apparently due to the sudden availability of organic substrate available in the soil, increases in soil pH, and other soil chemical changes associated with burning. Reinoculation may occur quickly from windblown spores or invasion from subsurface layers. Soil water is very important; moist soil and rainfall following fire favor reinoculation and increased microbial populations.

Little is known about the response of soil fauna to fire. Populations of earthworms, beetles, spiders, mites, collembola, centipedes, and millipedes are typically reduced by burning but increase thereafter (I. Ahlgren, 1974). Earthworms tend to be affected more by postfire loss of soil water than by the heat generated in burning. Ants are less affected than other fauna because their behavior enables them to survive in lower soil layers. Furthermore, they are adapted to xeric conditions of postfire topsoil. Although many ants are often destroyed by fire, their social organization and rapid colonizing ability enable them to reestablish populations rapidly after burning. In general, soil fauna are more severely affected in forest sites than in grasslands. The effect of fire is greater in the forest because forest fires may be hotter than grassland fires due to a greater accumulation of fuel. In addition, there is a change from cool and moist forest floor to xeric postfire conditions, there are greater temperature fluctuations, and the fauna may lack food.

Direct Effects

The direct effects of fire on site quality arise from two principal sources: the burning of organic matter above and on the mineral soil, and the heating of the surface layers of the soil. The burning of organic matter results in the release of carbon dioxide, nitrogenous gases, and ash to the atmosphere and the deposit of minerals in the form of ash. Wood and litter ash is more soluble than the organic matter from which it was formed. Thus the effect of fire is to increase the amount of available minerals, at least temporarily, to lessen the soil acidity and increase base saturation, to decrease total soil nitrogen, and to change the soil water and temperature conditions of the site.

Two examples are indicative of many studies on fires effects. In the ponderosa pine region of Arizona, burning was found to increase the soluble nutrients as a result of the ashing of the surface layer of unincorporated organic matter (Fuller et al., 1955). This caused an increase in pH, available phosphorus, exchangeable bases, and total soluble salts, and a decrease in organic matter and nitrogen to a depth of 20 to 30 cm. Microbial activity, particularly of bacteria, increased as a result of burning. On the negative side, the surface was compacted by rains following the removal of litter, resulting in a decrease in the rate of water penetration. In another study in eastern Washington (Tarrant, 1956a), however, burning was not found to be detrimental and perhaps even slightly beneficial in its effects on permeability and associated physical properties of the soil.

In the Douglas-fir region, Tarrant (1956b) investigated the effects of slash burning on physical soil properties. Light burning increased the percolation rate of water within the surface 8 cm of soil, but severe burning confined to less than 5 percent of intentionally burned, logged-over sites did seriously impede water drainage about 70 percent.

Unfortunately, the same characteristics of small particle size and high solubility that render minerals in ash readily available to plants, also render the ash susceptible to leaching and erosion by rainwater. If the ash is washed down into the soil so that the roots can absorb the nutrients dissolved from it, site quality is usually improved, at least temporarily. If, on the other hand, the ash is leached down below the tree roots (or is washed off the surface) then site quality is lowered. In general, the former may occur on level soils of sandy to loamy texture, whereas the latter is apt to happen on very coarse sands or clayey soils, particularly those with considerable slope.

Tropical savanna soils are notably low in nutrients, so nutrient losses could have a critical effect on site productivity. However, Kellman et al. (1985) found that burning does not significantly reduce fertility of the topsoil in Belize savannas. On both savanna (Kellman et al., 1987) and more fertile soils (Matson et al., 1987) of tropical forest sites, nutrients increase in surface soils for a short time before returning to normal; plant uptake by recovering understory vegetation following fire is very important for retaining nutrients. The varied effects of fire on soil-site conditions of tropical ecosystems are discussed by Koonce and González-Cabán (1990) and Coutinho (1990).

The loss of total nitrogen through volatilization is widely recognized and is related to the intensity of the fire (Rundel, 1981). Knight (1966), working in the coastal Douglas-fir region, found no nitrogen loss in soils heated to 200 °C, a 25 percent loss at 300 °C, and a 64 percent loss at 700 °C. Nitrogen loss is also proportional to the amount of fuel consumed, and considerable nitrogen may be lost during intense fires. For example, the severe Entiat fire in a second-growth, mixed-conifer stand in north-central Washington resulted in a loss of about 97 percent nitrogen originally in the forest floor and a loss of two-thirds of the nitrogen of the A horizon (Grier, 1975). Although replacement of the nitrogen by precipitation alone would require about 900 years, nitrogen reaccumulation will result much sooner due to N_2 fixation by symbionts associated with snowbrush (*Ceanothus velutinus;* see Chapter 19). Generally, despite major volatilization of nitrogen in fires, the available forms of nitrogen are commonly higher on burned than unburned sites (Rundel, 1981).

Much of the nitrogen lost through burning of litter and vegetation is not in a form available to the plant. For example, the organic matter layer in coniferous forests, especially in boreal regions, contains a large amount of nitrogen, but only a minute part of it is in a form that the vegetation can use (Viro, 1974). To be made available for uptake, it must first be mineralized and converted into ammonia (Chapter 19). The ability of the succeeding vegetation and soil bacteria to replace the available nitrogen lost in burning is an important factor determining the effect of fire on site quality. The higher pH, which is due to release of mineral bases in the soluble ash, can provide a more favorable soil environment for the free-living, N_2-fixing bacteria and thus can initiate an increase in available nitrogen, albeit at a slow rate. Although the loss of nitrogen is widely cited as a deleterious effect of fire, the significance of the loss for the new regeneration and the overall nutrition of specific ecosystems is not well known. For boreal forests dominated by Norway spruce, Viro (1974) concludes: "burning unquestionably results in great losses of total nitrogen from the site, but simultaneously results in an increase in mineralized nitrogen. The former is practically unimportant, the latter of great consequence."

The actual heating of the mineral soil is of relatively less importance than the action of fire on the organic matter. The heat of the fire does not penetrate far into the soil. Even under a hot fire in logging slash, temperatures seldom exceed 90 °C at 3 cm down in the mineral soil. Light surface fires only heat the top centimeter of the mineral soil to near the boiling point.

In this heated zone, soil aggregates may be broken down, first by the heat and later by the direct striking action of raindrops, resulting in loss of soil structure and in lowered infiltration capacity of the surface soil. In extreme cases, which are rare, clay soils may be baked hard, and soil organisms will be killed. Burning of logging debris can provide such an extreme example. In many cases, burning heavy accumulations of slash in piles causes intensely hot fires and alters the physical structure of the soil. The underlying soil may be baked and its structure markedly altered. For example, burning piles of coniferous slash in northwestern Montana resulted in lower tree densities and markedly slower growth of conifers on burned slash pile sites compared to adjacent unburned areas (Vogl and Ryder, 1969). The growth depression was attributed to impaired physical soil properties, especially decreased water infiltration. In general, spreading the slash and "broadcast" burning is the more desirable technique because it approximates naturally occurring wildfire in spreading burning effects over a greater surface area (DeByle, 1976).

Although the effects of direct heating the mineral soil are many and varied, in general their sum total does not alter the site quality to any marked extent for any substantial period of time. Except in the rare burn that creates extreme heat within the mineral soil, the effects of fire on site quality are best interpreted in the light of its effect upon the soil through the destruction of organic matter and erosion.

Detrimental and Beneficial Effects

There are some situations where fire is obviously catastrophic. These include cases where the soil is composed almost entirely of organic matter, and those where the destruction of organic matter exposes highly erodible soils to heavy rain.

The burning of a peat bog after drainage or a series of dry seasons literally results in the complete destruction of the soil and the return to swamp conditions. Thousands of years of peat accumulation are necessary to replace the lost organic soil, and such areas can be virtually eliminated from our productive sites for forest growth. In the United States, many bogs in the Atlantic Coastal Plain and in the recently glaciated parts of the Northeast and the Lake States have been destroyed as forest sites by fire.

Similarly, the burning of organic matter lying directly on top of rock will eliminate the soil. In glaciated portions of Canada and the northeastern United States, and in many mountain regions, thin accumulation of humus provides the only nutrition for forest trees. Fire in such cases often burns down to bedrock with disastrous results (Kelsall et al., 1977).

Still another bad but natural situation occurs where highly erodible soil of steep slopes is exposed by fire burning the organic protection (Swanson, 1981; Rundel, 1981). Fires can cause serious erosion in the northern Rocky Mountains where erosive soils occur on steep slopes; the fires tend to be catastrophic, and multiple burns frequently occur. Furthermore, the classic example is in southern California, where chaparral species and the organic matter from them protect the granitic or clayey soils lying at approximately the angle of repose in steep mountains. Swanson (1981) estimates that over 70 percent of the long-term sediment yield from steep chaparral slopes is the result of fire effects.

Even in the absence of fire, erosion is high. In the San Gabriel Mountains near Los Angeles, studies of debris movement on steep slopes covered by old chaparral revealed that each year an average of 8,000 kg of debris per hectare moved down the slopes to the stream channels (Anderson et al., 1959; Krammes, 1965). Chaparral is highly susceptible to burning, especially when it reaches about 30 years of age, because of the dense, highly flammable fuel accumulations, shrubby growth habit, and seasonal dry, windy periods (Biswell, 1974; Keeley and Keeley, 1988). Following wildfire, Biswell (1974) reports that debris movement may increase dramatically, and on south-facing slopes may reach 10 times that of the already high prefire rate (Figure 12.4). In this instance, most of the movement was during the dry period. However, if, after a fire, the heavy rainstorms characteristic of this semiarid region strike before revegetation of the burn, whole slopes may wash downhill. Such massive erosion not only lowers soil fertility of the soil, but frequently wreaks havoc on the valleys below.

The burning of litter and organic matter in the soil may be significant in causing reduced infiltration, increased surface runoff, and erosion in many areas of the western United States where water-repellent soils have been reported (DeBano et al., 1967; Meeuwig, 1971; Dyrness, 1976). For a number of years California scientists were puzzled by the sight of "dusty tracks in the mud" in freshly burned areas after fall and winter rains. Now it is known that a variety of soils can become resistant to wetting. These are soils in which the particles repel water; droplets do not readily penetrate and infiltrate but "ball up" and remain on the soil surface for variable periods of time. This phenomenon is widespread in sandy soils throughout much of the wildland areas of western North America

Figure 12.4. Dry-creep erosion may be severe immediately after intense wildfire in old-growth chaparral on slopes above the angle of repose. (Photo courtesy of the Pacific Southwest Experiment Station, U.S. Forest Service.)

supporting chaparral or coniferous vegetation, as well as many other parts of the world (Foggin and DeBano, 1971).

Fire plays an important role in the formation of water-repellent soils. In unburned areas, litter decomposition produces nonwettable or hydrophobic organic molecules that coat surface soil particles, creating a weak water-repellent layer in the upper soil profile between the litter layer and the mineral soil (Figure 12.5; DeBano, 1969). During a fire, litter is consumed, and the hydrophobic substances are volatilized and diffuse downward into the soil and condense on cooler soil particles (Figure 12.5). After summer and fall wildfires, water repellency is high, and rain falling on the soil surface infiltrates readily until impeded by the shallow nonwettable layer. After the wetting front encounters the repellent layer, infiltration is slowed, surface runoff begins, and erosion may readily occur.

Just as there are cases where fire is catastrophic in its effects on site, so there are cases where it is clearly beneficial. Such is the case for sites in the far north where dampness and cold retard the decomposition of organic matter, giving rise to thick mats of highly acid organic matter or "raw humus." In Norway, Sweden, and Finland, considerable success has been achieved by burning such sites to raise the site quality by improving nutrition, soil water, and temperature conditions (Viro, 1974).

Postfire temperature conditions are often changed drastically by fire. Temperature extremes are typically greater on burned sites than on unburned sites. Average maximum soil surface temperatures on burned sites may be from 3 to 16 °C higher than on comparable unburned areas (C. Ahlgren, 1974). Increased soil temperatures hasten spring development of roots and shoots on burned areas, speed decomposition, and promote the activity of soil organisms. Extremely high temperatures, due to the blackened soil surface and the presence of charcoal, may cause seedling mortality and delay forest regeneration. On fire

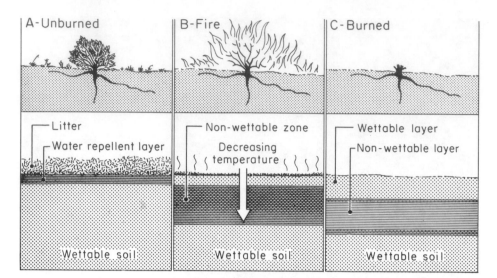

Figure 12.5. Soil non-wettability before, during, and after fire. A, before fire, the non-wettable substances accumulate in the litter layer and mineral soil immediately beneath it. B, fire burns vegetation and litter layer, causing non-wettable substances to move downward along temperature gradients. C, after a fire, non-wettable substances are located below and parallel to the soil surface on the burned area. (After DeBano et al., 1967.)

sites, however, fire-dependent species are typically adapted to tolerate extreme conditions. In boreal climates, where thick layers of raw humus tend to develop under spruce and fir vegetation, site quality is markedly improved by burning because it enhances the postfire thermal regime.

In most situations, however, the effect of fire on site quality is relatively less pronounced. Repeated burning—once haphazard and now more or less controlled—in the sand plains of the southern pine region has apparently had no major detrimental effect on site quality and, in fact, has been shown locally to be beneficial to the soil (Metz et al., 1961). In loess soils under even-aged shortleaf and loblolly pines on flat terrain in Arkansas, nine successive annual burns had little effect on the nutrient content or structure of the topsoil (Moehring et al., 1966). Studies of soil biota after 20 years of annual prescribed burning on a very fine sandy loam in the coastal plain indicated that burning had no effect on the total number of fungi per gram of soil, although it did reduce their total number through a decrease in weight of the organic horizon (Jorgensen and Hodges, 1970). The number of bacteria in the organic layer was reduced by annual burning but not in mineral soil. For the organisms studied, there was little indication that prescribed burning adversely affected metabolic processes.

Controlled burning in a Douglas-fir–larch forest on sandy loam soils in northwestern Montana generally reduced accumulated fuels without nutrient loss, runoff, or erosion (Stark, 1977). The "biological life of the soil," the years that a particular soil is capable chemically of supporting trees, was quantified using a formula based on nutrient losses through burning, erosion, harvesting, and the like. The estimate of 55,000 years showed that burning could be conducted on this soil a very long time with no problem of soil fertility. The biological life of the soil was so long that major catastrophic climatic, erosional, or glacial events are likely to occur and change natural processes over such a long time span.

In the sand plains of the Lake States, organic matter provides the major source of nutrients for plant nutrition. Burning the sand plains, therefore, may be undesirable for growing trembling aspen, a species of relatively high nutrient requirements (Stoeckler, 1948, 1960). In contrast, 10 years' experience with light prescribed burning on sandy soils of Minnesota indicates that site productivity will not be altered for red pine, a species with low nutrient requirements that occurs naturally on fire-prone ecosystems (Alban, 1977). There seems no reason to fear site deterioration in burning the more level pinelands of the Western states. When dealing with clay soils and steep slopes, however, fire may do no harm if gentle rains impound the ash, and revegetation anchors the soil.

Suggested Readings

Agee, J. K. 1993. *Fire Ecology of Pacific Northwest Forests*. Island Press. Washington, D.C. Chapters 1, 4, 5, and 6.

Arno, S. F. 1980. Forest fire history of the northern Rockies. *J. For.* 78:460–465.

Goldammer, J. G. 1990. *Fire in the Tropical Biota*. Ecological Studies 84. Springer–Verlag, New York. 497 pp.

Johnson, E. A. 1992. *Fire and Vegetation Dynamics*. Cambridge Univ. Press. 129 pp. Chapters 4 and 6.

Kauffman, J. B., and C. Uhl. 1990. Interactions of anthropogenic activities, fire, and rain forests in the Amazon Basin. Chapter 8. *In* J. G. Goldammer (ed.), *Fire in the Tropical Biota*. (Ecological Studies 84). Springer–Verlag, New York.

Rundel, P. W. 1981. Fire as an ecological factor. *In* O. L. Lange, P. S. Nobel, C. B. Osmond, and H. Ziegler (eds.) *Physiological Plant Ecology I., Responses to the Physical Environment*, New Series, Vol. 12A. Springer–Verlag, New York.

Wein, R. W., and D. A. MacLean. 1983. *The Role of Fire in Northern Circumpolar Ecosystems*. John Wiley, New York. 322 pp.

Whelan, R. J. 1995. *The Ecology of Fire*. Cambridge Univ. Press, Cambridge. 346 pp.

Wright, H. E., Jr., and M. L. Heinselman. 1973. The ecological role of fire in natural conifer forests of western and northern North America: introduction. *Quaternary Res.* 3:319–328.

CHAPTER 13

SITE QUALITY AND ECOSYSTEM EVALUATION

In the preceding chapters, the forest tree and the individual environmental factors that affect it were considered separately, yet within a landscape ecosystem framework (Chapters 1 and 2). In this chapter the focus is on the "site" of the forest ecosystem—the constellation of factors that not only influence tree and forest productivity but affect the entire range of management and conservation applications. The forest ecosystem may be divided into (1) geographic position in space and the physical factors associated with these places (**site**), and (2) the forest trees and associated plants and animals (**biota**) that occupy these places and that are supported by site factors (Chapter 1, Figure 1.3). The factors of climate, physiography, soil, and disturbances, such as fire and wind, are predominant in determining the site conditions that affect the occurrence and diversity of organisms, their productivity, and their changes in space and time. In addition, the biota themselves modify, more or less, the direct effects of the physical factors. For example, the forest canopy trees affect understory light and temperature and soil development. Also, uprooted or fallen trees in uplands and wetlands (Chapter 10) create microlandforms that affect forest composition and dynamics. The forest site is thus the sum total of all the factors affecting the distribution and growth of forests or other vegetation (see site diagram, p. 150).

For centuries, hunter–gatherers, farmers, foresters, and land managers have confronted the problem of evaluating the quality of forest sites for a variety of purposes. How good is the land? What's it good for? How does one measure this complex interaction of physical factors that have been modified by historic and biotic influences, including human disturbances? Simple and complex methods have been tried with varying degrees of success. It is fitting, therefore, that we end this section on the forest environment with a discussion of approaches to site quality evaluation because of the strong influence of site factors on management activities.

Because forests dominate the lands under consideration, forest scientists took the lead in devising evaluation procedures. Forest managers were primarily concerned with the forest tree segment of the ecosystem and typically with the potential land quality for fiber production. Single-factor methods of evaluation tend to reflect these objectives. However, the emphasis has shifted from the stand (bioecosystem concept) to the land (geoecosystem concept), i.e., the ecosystem. We can no longer think only of site quality evaluation in terms of tree or stand production. We seek an understanding of landscape ecosystems so as to assure their sustainability at local and regional scales (Rowe, 1992b; Franklin, 1997).

Foresters and other land managers today seek to evaluate sites and ecosystems for many reasons: recreation, aesthetic values, water and wildlife conservation and management, ecosystem and biological diversity, and maintenance of ecosystem processes in addition to, or to the exclusion of, fiber production. Thus broader methods and detailed ecosystem approaches, involving multiple factors and mapping of entire landscape ecosystems, have captured center stage. Such methods provide both the basis for biomass production and the ecological framework for ecosystem conservation, management, and restoration at multiple scales.

In this chapter, you will encounter a broad array of approaches in evaluating sites and ecosystems. These include consideration of effects of single site factors and conventional use of tree height to estimate site quality and complex methods of ecosystem classification and mapping. For over a century, this process has produced a wealth of insights concerning plant growth and occurrence. Thus elucidation of these important site–plant relationships is also a major objective of this chapter as well as examining site-evaluation approaches.

To estimate forest site quality in terms of forest productivity, the forest scientist is faced with the problem of integrating many physical site factors. The site factors are not only interdependent but are also dependent in part upon the forest, which is itself a major site-forming factor. Because of these interactions, the simple regression technique of estimating site quality from an evaluation of a few important site factors, important as it is in practical forest ecology, can only be approximate.

Nevertheless, the estimation of forest productivity is of the utmost importance in both forest ecology and in ecosystem management. This productivity, or actual site quality, may be measured directly for a few forests where accurate long-term records of stand development and growth have been maintained. Generally, however, it can only be estimated indirectly by one or more of the alternatives that are considered in this chapter:

Vegetation of the forest
Tree height (site index method)

Ground vegetation (indicator species and species groups)

Overstory and ground-cover vegetation in combination

Factors of the physical environment
Climate

Physiography

Soil survey and soil-site methods

Multiple-factor and multiple-scale approaches (using some or all of the above factors, disturbance regime, and forest land-use history)

The multiple-factor methods have broader applicability than just for productivity estimation and have evolved along with the emphasis on multiple forest values. Significant reviews of traditional forest site quality estimation have been published by Coile (1952), Rennie (1962), Ralston (1964), Jones (1969), and Carmean (1970a, 1977, 1996), and comprehensive reviews of site quality evaluation, its history, methods, and application, by Carmean (1975) and Hagglund (1981) also are available.

DIRECT MEASUREMENT OF FOREST PRODUCTIVITY

Actual forest productivity is generally measured in terms of the gross volume of bole wood per acre or hectare per year over the normal rotation. This gross mean annual increment (m.a.i.) may be computed from long-term permanent sample plot data. For instance, on a pumice soil site on the North Island of New Zealand, Douglas-fir has been computed to yield a gross m.a.i. of 31 m^3 ha^{-1} yr^{-1} (Spurr, 1963). Yields of 36 m^3 ha^{-1} yr^{-1} may be expected from Monterey pine on similar sites (Spurr, 1963). In the United States, gross mean annual increments range upward to perhaps 15 m^3 ha^{-1} yr^{-1} on the best sites. Average productivity in the temperate forests of North America and Europe is approximately 5 m^3 ha^{-1} yr^{-1}. The growth is presented in gross values—the total amount of wood put on by all trees within a given unit of time without deduction for natural mortality, removal by humans, or decrease in wood volumes by rot. By consistently using such gross values, comparable increment measurements of aboveground production can be obtained.

Unfortunately, actual gross productivity data like these are scarce. Furthermore, actual yield is conditioned not only by site factors but also by genetic factors (of species and race), by age or rotation, by the history of the stand, and by stand density. Nevertheless, actual growth represents the proven productivity of a site and therefore may be taken as the closest available approximation of potential productivity.

Theoretically, a stand of a given species of a given age on a given site will produce the same amount of wood per year at various densities of stocking as long as the site is fully occupied. As long as the trees in the unthinned stands retain good crown development and vigor, they will fully occupy the site. Similarly, if the trees in thinned stands or even in open stands fully occupy the soil to the extent of being able to fully utilize available soil water and nutrients, they will normally fully occupy the site even if excess crown space is available in the stand. These being the cases, thinned and unthinned stands, otherwise comparable, should give the same total increment. To make such a comparison correct, however, gross growth should be computed by adding in mortality within a growth period, whether from cutting activity by humans or from natural causes.

In recent years, forest ecologists have been attempting to estimate forest productivity in terms of all components of the forest ecosystem rather than the growth of the tree boles alone. Researchers around the world have been sampling not only the stems but also the branches, leaves, and roots; the organic matter in the forest floor; and even the animals inhabiting the forest, to provide a more exact appraisal of the entire forest ecosystem. For practical forest site evaluation purposes, however, the wood content of the boles or main stems of the forest trees remains the best-known measure of forest productivity.

TREE HEIGHT AS A MEASURE OF SITE

The height of free-grown trees, individuals that have grown from the time of their establishment without suppression by an overstory canopy of a given species and of a given age, is more closely related to the capacity of a given site to produce wood of that species than any other one measure. Furthermore, the height of free-grown trees is less influenced by stand density than other measures of tree dimensions and may thus be used as an *index* of site quality in even-aged stands of varying density and silvicultural history. Use of height is especially appropriate for understory intolerant and midtolerant species compared to tolerant species whose individuals may be suppressed in the understory for variable periods of time.

The height of the dominant portion of a forest stand at a specified standard age is commonly termed **site index** although tree height is but one of many indices of site quality used in forest ecological and silvicultural investigations. The conventional application of site index is illustrated in Figure 13.1. A set of height/age curves are developed for a geographic area, in this case for the Coastal Plain and Piedmont physiographic regions of the southeastern United States. Using this set of curves one can determine the site index for any given stand of loblolly pine in the area by determining the mean total height and age of a representative sample of dominant "site trees." The example in Figure 13.1 shows that the site index is 50 for a site supporting dominant pines that are 55 ft tall at 60 years, whereas it is 95 for a site supporting pines that are 85 ft at 40 years. Site index is simply a number that is used to indicate the relative productivity of a stand of a given species at one place in comparison to stands of the same species on other sites. The site index number doesn't indicate why a given site is poor, medium, or good for growth and yield of the species currently growing there. Although the same site index may be determined for a species on two different areas, they may not necessarily be the same ecosystem type or have the same volume production. Thus the broad geographic application of the method may have notable limitations.

In the United States, site index has long been defined as the average height that the dominant portion of an even-aged stand will have at a specified age. This standard age is generally 50 years in the eastern United States, and 100 years for the longer-lived species

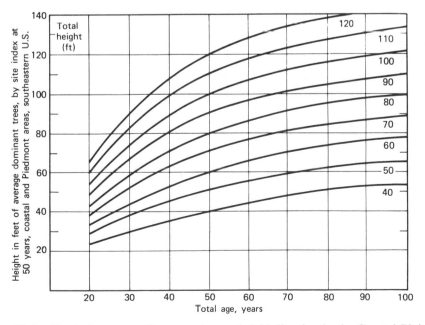

Figure 13.1. Site-index curves for second-growth loblolly pine in the Coastal Plain and Piedmont areas of the southeastern United States. (After Hampf, 1965.)

of the West. Occasionally, other standard ages are specified for a particular species or region, for example, 25 years for pulpwood rotations in the South.

Because of the problems of determining and measuring dominant or codominant trees, there is a tendency to restrict more carefully the trees that should be measured for site determination. A preferable practice is to measure the height of an objectively determined sample of the larger trees in the stand. In Britain and some other countries, the concept of **top height** is widely used. This term refers to the arithmetic mean height of the 250 largest-diameter trees per hectare. However, because it is seldom feasible to measure so many trees per hectare for height, and since heavy thinning practices frequently reduce the stand to fewer than 250 trees per hectare, a smaller sample of the very largest trees is used. Thus a mean height based on the largest 100 trees per hectare (sometimes defined as mean **predominant height**) is used. Top heights based upon the 60 largest-diameter trees per hectare, the 8 to 12 largest trees, or even upon the 1 largest tree per hectare are also used.

Although height is perhaps the single tree measure best related to the site productivity of a given species, it does not necessarily follow that it is completely unrelated to other factors, nor that a perfect correlation may be obtained between stand top height and site productivity. In particular, stand density, particularly extremes of stand density, may influence height growth. Under such circumstances, site-index curves should be developed separately for different stand density classes, as has been done for lodgepole pine in the Rocky Mountains (Alexander, 1966; Alexander et al., 1967). Moreover, although the same site index may be determined for a given species on two different sites, the stand density may be different, and hence the productivity of the species on the two sites is not necessarily the same (Assman, 1970; Curtis, 1972). For example, for Douglas-fir in northern Idaho and northwestern Montana, the density of trees is greater in even-aged pure stands than in uneven-aged mixed-species stands of the same dominant height (Sterba and Monserud, 1993). Although the site index for Douglas-fir for these two sites is the same, the productivity is different.

Genetic factors may well control height growth to a great extent on a regional ecosystem basis, especially over a heterogeneous area in climate and landform (Monserud and Rehfeldt, 1990). Within a local area, the strength of genetic control of height growth of trees within a species tends to be low, and site factors, such as soil water and nutrients, are major determinants of height growth. Nevertheless, even in a given stand, significant differences in individual genotypes may occur such that they must be accounted for in the sampling procedure.

Naturally occurring aspen clones on sandy soils in northern lower Michigan have been found to differ greatly in height on the same site (Zahner and Crawford, 1965). Some clones are more than twice as tall as adjacent clones of the same age. Such variation is much more likely in species developing natural multi-stemmed clones than in species where each stem is a different genotype. In clones, competition is primarily between stems of the same genetic constitution; the slow-growing clones are more likely to survive than in stands where competing individuals are of different genotypes. Sampling the largest dominant trees (but perhaps only one clone) might result in overestimating site quality. Therefore, the mean total height from at least five different clones, rather than five "site trees" (that might be measured from a single genetically superior clone) is recommended.

Finally, the condition of the site at the time the stand is established, as well as competition from other vegetation in early years, may affect height growth markedly. Both naturally reseeding and planted pines will usually show different growth trends and amounts on old fields as contrasted with cutover sites.

Site-Index Curves

The usual method of determining site index on the basis of tree height depends upon the use of a height-over-age growth curve to estimate the height at a standard age. Most such curves for North American species in the past were developed from a series of regression curves, based upon a single guiding curve and harmonized to have the same form and trend (Figure 13.1).

The weaknesses of this approach are by now well known (Spurr, 1952b). First and foremost, the technique is sound only if the average site quality is the same for each age class. If, however, as is often the case, younger stands are found on generally better sites (perhaps because of early logging on these sites) while the remaining old growth stands are concentrated on the poorer sites, the average curve will be warped upwards at younger ages and downward at older ages. The reverse situation can also occur.

A second major weakness of the conventional technique is the assumption that the shape of the height-growth curve is the same for all sites. Although this generalization gives good results in some instances, it does not hold for all because of differences in climate, landform, or soil conditions. For instance, if the depth of soil is limited by physical reasons, growth of a tree may be normal up to the point where the depth of the soil becomes a limiting factor (curve B; Figure 13.2). In another ecosystem and soil, the same species may grow slowly until roots reach an underlying enriched horizon or a deep lying water supply, after which growth will be accelerated (curve C). The shape of these two growth curves may differ markedly from the normal growth curve on a normal soil (curve A).

A corollary of this problem is the assumption in the standard technique that site differences are apparent at early ages. The process of harmonizing site-index curves assumes that, if a site produces a higher tree at age 50 or 100, that tree will be higher at all preceding and all subsequent ages. The assumption is in contrast to the fact that many plantations and even-aged natural stands on marginal sites may grow normally in youth

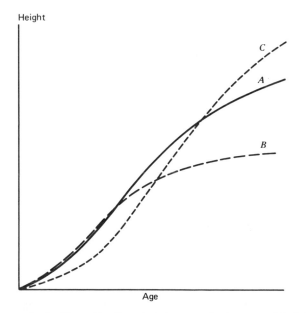

Figure 13.2. Theoretical effect of soil on height growth. *A*, normal height growth in homogeneous soil. *B*, height growth on good but shallow soil. *C*, height growth on soil poor at the surface but with a fertile horizon beneath.

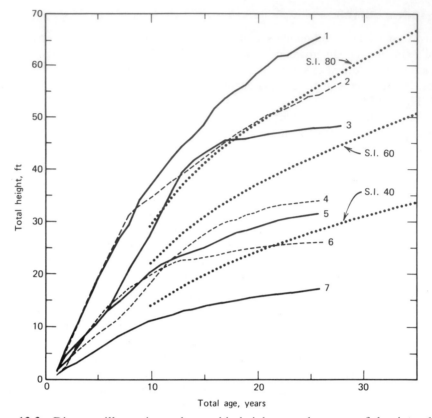

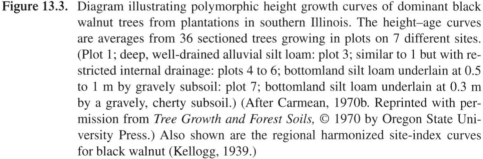

Figure 13.3. Diagram illustrating polymorphic height growth curves of dominant black walnut trees from plantations in southern Illinois. The height–age curves are averages from 36 sectioned trees growing in plots on 7 different sites. (Plot 1; deep, well-drained alluvial silt loam: plot 3; similar to 1 but with restricted internal drainage: plots 4 to 6; bottomland silt loam underlain at 0.5 to 1 m by gravely subsoil: plot 7; bottomland silt loam underlain at 0.3 m by a gravely, cherty subsoil.) (After Carmean, 1970b. Reprinted with permission from *Tree Growth and Forest Soils,* © 1970 by Oregon State University Press.) Also shown are the regional harmonized site-index curves for black walnut (Kellogg, 1939.)

and only in middle life exhibit sharply decreased growth. Planted black walnut trees on seven contrasting sites in southern Illinois (Figure 13.3) show rapid early growth, even on the poorer sites, but may slow abruptly after 10 years (Carmean, 1970b). The polymorphic patterns are closely related to soil conditions. Trees on plot 1 are growing on a deep well-drained alluvial silt loam, whereas those on plots 4 to 7 are growing on a bottomland silt loam soil underlain at 1 m or less by a gravely subsoil. Curves based on the premise of harmonization with a standard average curve cannot show such plant–soil relationships.

Height-growth patterns are known not only to vary in different geographic parts of the range of a species but also in local areas of contrasting soil and topography, i.e., different local ecosystem types. Height-growth patterns of oak, for example, not only vary between different soil texture groups but also vary with aspect and slope within soil groups. Therefore, it is not surprising that polymorphic site-index curves have been repeatedly demonstrated to characterize the variable height-growth patterns of forest trees better than the simple monomorphic pattern portrayed by regional harmonized curves (Carmean, 1970b, 1975).

Sometimes relatively small differences exist in curve shape over a broad area. For example, polymorphic site index curves were developed for jack pine throughout northern Ontario encompassing several different glacial landforms and soils (Carmean, 1996). Only slight differences were found in average height-growth patterns, and only one set of curves were developed for this broad range. However, polymorphic patterns of height growth were evident between extreme sites. Pines on the poorest sites (8 m at 50 yr) exhibited a relatively linear growth pattern, whereas those on the best sites (22 to 24 m at 50 yr) were more curvilinear (rapid growth rate for the first 50 years and a gradually decreasing rate to 100 yr). Black spruce exhibited a similar polymorphic pattern; the most linear pattern was on the poorest sites where spruce grew on nutrient-poor, organic soils. Polymorphic site-index curves have been prepared for many species (Carmean, 1975, 1996), although not usually for local ecosystems. However, polymorphic curves for specifically mapped ecosystem types were developed for Norway spruce and other species for intensive forest management in southwestern Germany (Barnes, 1984).

Considering the weaknesses of the standard site-index curve techniques, height-growth curves should be based upon actual measured growth of trees for specific sites or ecosystem types and not upon the harmonized method of averaging together height and age values from plots for the entire or regional range of sites upon which the species is found. Such a conclusion is not new. Most present-day European growth curves are based upon actual measured growth rather than upon temporary plots.

In studying height aspects of forest growth for correlation with site characteristics, then, our basic objective should be to work with actual recorded tree growth and to collect our data in such a way that separate growth curves can be developed for different soil-site conditions should evidence of differing curve shapes become apparent. There are three general sources of growth information that meet these requirements. The first and the best are the records of long-term permanent sample plots. Where enough plots have been established and measured for a considerable period of years, growth curves should be developed from the data these plots provide.

In the absence of sufficient permanent plot data, **stem analysis** may be employed. By sectioning trees from top to bottom, their course of growth can be reconstructed. This technique, too, has its pitfalls, but they are less serious and more easily overcome than those of the temporary plot techniques. The appropriate regression methods for estimating site index based on stem analyses are discussed by Curtis et al. (1974a,b).

A third source of information is the internode distance on the boles of trees that put out distinct annual whorls. On Douglas-fir and many of the pines, the whorl pattern shows clearly the height of the tree each year in the past back to an early age. Careful measurements of such trees can be used to produce accurate height-growth curves (Beck, 1971; Carmean, 1996). For the middle part of the height-over-age growth curve, height growth is relatively constant, and the average annual distance between whorls may be assumed to hold true for a future short period. The use of mean annual height increment over the middle period of height growth, the **growth-intercept method**, as a measure of site quality is old and well established. For loblolly, shortleaf, and slash pines (but not for longleaf), the 5-year intercept above breast height has proved better correlated with site quality than total height (Wakeley and Marrero, 1958). The technique has been adapted to red pine, Douglas-fir, southern pines and other species on which the annual whorls are apparent (Carmean, 1975).

Comparisons Between Species

Just as trees of differing genetic character within a species may be expected to show variation in height response to a given forest site, different species growing on the same site will show the same variation but to a much greater degree. The height-over-age site-index curve may be quite different for two different species on the same site. Nevertheless, the site index as predicted from the measurement of the height of one species may be used to give an estimation of the site index of the other. Carmean (1975, 1996) lists such studies, including examples from the southern pines, eastern hardwoods, northern and boreal trees, and western conifers; examples for northern Idaho conifers are given by Steele and Cooper (1986).

Advantages and Limitations

Of the various methods for determining site quality, site index is the one most often employed in North American forests. Site index can provide a convenient and reliable estimate of forest productivity. It is also a useful guide in selecting the most productive tree species for specific sites. However, because of the various problems cited previously, it must be applied with great caution. Overall, the development of site index curves for specific ecosystem types or groups of ecosystems with similar site conditions and genetic potential is useful.

However, site index is not an appropriate guide for determining silvicultural practices for regeneration and care of stands. Therefore, site-quality studies and ecosystem classification using vegetation and physical site factors have attracted universal attention. They not only provide a basis for estimating productivity, but they also provide the ecological basis for determining silvicultural practices for various management goals. In the following sections, we will discuss these other approaches used to estimate site quality—not only for understanding wood production potential, but in understanding the larger ecological significance of site quality in managing ecosystems for wildlife, water, recreation, and wilderness as well as timber.

VEGETATION AS AN INDICATOR OF SITE QUALITY

The presence, relative abundance or coverage, and relative size of the various species in the forest reflect the nature of the local forest ecosystem of which they are a part and thus may serve as indicators of site quality. The correlation may or may not be apparent because the vegetation also reflects the effects of happenstance and factors that are unrelated to site quality: plant competition and mutualism, herbivory, low light conditions in the understory, past events in the history of the vegetation such as drought, insect outbreaks, and human disturbance. Nevertheless, site characteristics are sufficiently reflected in the vegetation to make it a successful index of site quality in many instances. The plants themselves are used as the measure of site: they are **phytometers**. They integrate the effects of many climatic and soil factors that are difficult to measure directly. However, they are not precise phytometers because of the factors already noted. Furthermore, their indicator value changes with changes in regional macroclimate and physiography.

Tree species are useful indicators. They are long-lived, relatively unaffected by stand density, and easily identified in all seasons of the year. Some species have such a narrow ecological amplitude that their occurrence is indicative of a particular site. Demanding hardwoods such as black walnut, white ash, and tuliptree reach their best development only on moist, well-drained, protected sites rich in soil nutrients and characterized by a well-developed forest floor. Most trees, however, have a wide ecological amplitude; they may occur and prosper on a wide variety of sites. Their presence is thus of little indicator value. Their relative abundance and their relative size, however, may be. In the same eastern hardwood forests, the greater the proportion of northern red oak to black and white oak in the forest, the better will be the soil water conditions and the general site quality. The sizes of free-grown dominant oaks of these three species in even-aged stands of uniform density may be used as an index of site quality.

Plant species of the shrub, herb layer and moss–lichen layers (termed groundflora or ground cover) of the forest understory—although they are more apt to be influenced by stand density, past history, and the composition of the forest than the tree species—have in many cases a more restricted ecological tolerance and may therefore be more useful than tree species as plant indicators. This relationship is particularly true in the circumpolar boreal forest, where the dominant tree species—the spruces, firs, pines, and birches—are few in number and widespread in their distribution on various sites, thus having relatively poor indicator value. Under such circumstances, site classification methods based upon indicator species of the groundflora have been markedly successful. They are most easily applied in regions where variation in altitude, precipitation, and soil conditions are great and where humans have not markedly altered the original vegetation.

The classic example is Cajander's system of site types for Finland (1926), designed to distinguish quality classes in a forest characterized by spruce, pine, and birch in both pure and mixed stands. Height-growth curves of pine on four such site types are presented in Figure 13.4. These may be taken as indicative of the system, which is still useful for the site conditions where it was developed (Lahti, 1995). The Cladina type is characterized by a lichen understory (*Cladonia alpestria* in particular) and occurs on dry sandy heaths of the poorest site quality. On the Calluna type, mosses and lichens are generally present, but heather (*Calluna vulgaris*) is typically the predominant species. The Vaccinium type is typified by the lingonberry, *Vaccinium vitis-idaea*, and occurs on moderately dry sandy

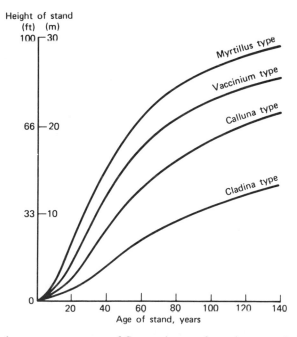

Figure 13.4. Height-over-age curves of Scots pine on four site types in Finland characterized by different indicator species.

ground and on glacial ridges. Mosses and lichens are of less importance and dwarf-shrubs of greater importance than in the preceding types. Finally, the Myrtillis type, named for the predominant *Vaccinium myrtillis,* is developed on a richer soil supporting a late-successional vegetation of spruce but frequently converted to pine by fires, felling, and silvicultural control. Lichens are unimportant and herbs more important and richer in numbers than in the other types cited.

At a more complex level, all of the resources of vegetation description may be used in the segregation of forest site classes and different kinds of landscape ecosystems. It should be remembered, though, that site variation typically takes the form of a gradient rather than of distinct and mutually exclusive site classes. The latter are found only when a distinct break in site factors occurs. For example, a topographic corresponding soil and microclimatic break may occur between an upland and adjacent wetland swamp or between a sandstone-derived residual soil and a limestone-derived residual soil. Otherwise, site changes tend to be gradual. One may express the changes in terms of an ecological gradient or segment the gradient into classes or types that are useful in ecological comparisons and in land management. Thus Rowe (1956), working with the mixed boreal forest in Manitoba and Saskatchewan, identifies five classes along a soil water gradient from dry to wet. He assigns understory vegetation to each class depending on their soil water requirement and thereby forms groups of species that indicate distinctive positions along the soil water continuum.

The plant-indicator concept has been adapted successfully to many temperate forests of North America. Increasingly, overstory dominants and physical ecosystem factors are used in conjunction with ground-cover vegetation to estimate site quality and classify and map forest ecosystems.

Species Groups of Groundflora

It is possible under certain conditions to characterize a site in terms of a very few species. However, the key indicator plants may or may not be present at a given locality because of chance, past forest history, or present competitive conditions. It is, for instance, subjectively difficult to classify a site or ecosystem type as an Oxalis-Cornus type if it contains no *Oxalis* and no *Cornus.* Yet the sum total of the vegetation and the growth characteristics of the ecosystem may well fit the area into this category.

The disadvantages of the foregoing approach may be overcome by using many ground-cover species and by identifying groups of species having similar environmental requirements. Species groups have been determined by a number of different approaches (Daubenmire, 1952; Rowe, 1956; Sebald, 1964; Mueller-Dombois and Ellenberg, 1974; Barnes et al., 1982; Ellenberg, 1988; Klinka et al., 1989). Typically, many species are used, and the indicator value of each species is determined subjectively by intensive field study. The concept of ecological species groups is attributed to Duvigneaud (1946), but plant ecologists have long recognized the occurrence of plants with similar distributions in relation to environment. The particular method employed depends on the nature of site and vegetation conditions of the application area and the purpose for which the method is to be used.

Two different approaches are used. In the first, each species is assigned an indicator value (an integer or a class name, e.g., low nitrogen, high nitrogen) for individual site factors. For example, in the western part of Central Europe, Ellenberg (1974) listed the indicator value (in classes 1–9) of approximately 200 vascular plants for each of the gradients of light, temperature, continentality, soil water, pH, and soil nitrogen. In the second approach, groups of species that reflect a similar combination of site factors are constituted.

In virtually all approaches, the "importance" or abundance of the groundflora species present on a site is quantitatively rated in percent **coverage**. Whereas ecologists use the number of stems or the basal area of tree species to determine their relative importance on a site, crown or foliage coverage is typically used for the ground-cover species. Coverage for a given species is defined as the proportion of a given area that is covered by a vertical projection of the foliage onto the ground surface of all individuals of the species. This process is tedious, time consuming, and subjective! Therefore, ecologists estimate coverage on a scale from <1 to 100 percent using cover classes; a 6-class-, 10-class-, or 12-class-scale may be used depending on user purpose, site conditions, and number and rarity of the species present (Mueller-Dombois and Ellenberg, 1974; Spies and Barnes, 1985b; Klinka et al., 1989).

INDICATOR PLANTS OF COASTAL BRITISH COLUMBIA

For decades V. J. Krajina and his students have pioneered studies of plant–environment relationships, characterizing over 3,000 species in British Columbia. As part of this endeavor, Klinka et al. (1989) characterized the indicator value of 416 species of coastal British Columbia by 4 site attributes: climate, soil water, soil nitrogen, and ground surface material. The gradient of each attribute was divided into classes (6 for climate, 6 for soil water, 3 for soil nitrogen, and 5 for ground-surface material), and each indicator species for an attribute placed in a class, thereby forming 20 indicator species groups. For the soil-water attribute, each group represents a relatively wide and overlapping segment of the soil water gradient. For soil water, indicator values were assigned to 337 species in the 6 classes: (1) excessively dry to very dry (17 species), (2) very dry to moderately dry (50 species), (3) moderately dry to fresh (74 species), (4) fresh to very moist (107 species), (5) very moist to wet (59 species), (6) wet to very wet (39 species). Each indicator species can be assigned to one of the groups in one or more of the four attributes. A given site or ecosystem is then characterized by the presence–absence and abundance (determined by coverage) of different indicator species in groups of each of the four attributes.

ECOLOGICAL SPECIES GROUPS

The use of ecological species groups for forest ecosystem classification, mapping, and site quality evaluation was implemented first in the southwestern German state of Baden-Württemberg (Sebald, 1964; Dieterich, 1970; Barnes, 1984; see Table 13.1 and following discussion). Plants that repeatedly occur together in areas with similar combinations of soil water, nutrients, light, and other factors are perceived to have similar requirements or tolerances and are therefore grouped together. Groups consist of several to many herbs, shrubs, and less commonly mosses and lichens of the groundflora. Each group is named for the most characteristic species. The approach also has been applied in distinguishing and mapping landscape ecosystem types in old-growth forests in Michigan (Pregitzer and Barnes, 1984; Spies and Barnes, 1985b; Simpson et al., 1990) and in highly disturbed oak-hardwood forests of southern Michigan (Archambault et al., 1990) and Wisconsin (Hix, 1988). Methods used to develop and evaluate the groups, once established, are given by Spies (1983), Spies and Barnes, (1985b), and Archambault et al. (1990). An example of 2 contrasting groups of the 16 groups used to map forest ecosystems in upper Michigan (Spies and Barnes, 1985b) illustrates their indicator value:

> *Clintonia borealis* group (6 species): Most common on moist, very infertile soils supporting conifers. Dry to wet. Very infertile to infertile. Shade tolerant.
>
> *Caulophyllum thalictroides* group (5 species): Characteristic of very fertile, moist to very moist soils. Moist to very moist. Fertile to very fertile. Shade tolerant.

As noted here, the groups constituted represent the integrated effects of multiple factor gradients, i.e., a factor complex encountered in the field. The co-occurrence of plants in the field is due to many interacting physical and biotic factors, including mutualism and competition with other plants. Thus the absolute indicator value of a plant for each separate factor may be difficult to determine.

Once constituted, the groups are used to help distinguish and map landscape ecosystems in the field by their presence or absence and by the relative coverage of plants in each group (see spectral format in Table 13.1, p. 323). They are never used alone, but always with attributes of physiography, soil, microclimate, and the composition and vigor of overstory trees (see below).

These indicator approaches may be useful locally, but how widely one can extrapolate the perceived indicator value of an individual species or species group? Can ecological species groups, once constituted for a given area, serve unchanged in different regional landscape ecosystems (i.e., areas with different macroclimate, gross physiography, and soil)? Does a given species always indicate the same relative level (high fertility, low soil water) of a given site factor in different regions? We can't give definitive answers to these questions, but it is clear that indicator species and groups should be used with great care. Applications of this approach in Germany and Michigan show that the usefulness of a given species as a plant indicator changes from one regional landscape ecosystem to another. As macroclimate, kinds and patterns of landforms, genetics of the species, and species competition and mutualism patterns change from one region to another, the relative indicator value of a species also changes. The usefulness of certain species increases (especially as new species assert dominance) and that of other species declines. For example, in mapping landscape ecosystems in four macroclimatically different areas of northern Lower Michigan and Upper Michigan four different sets of ecological species groups were required. Thus a regional landscape classification (e.g., Albert et al., 1986; Figure 2.10) is a useful framework to determine how widely to extrapolate a set of ecological species groups.

Plant Associations and Habitat Types in the Western United States

Because of widespread deforestation in many parts of the world (Perlin, 1989), it is rarely possible to use the native or presettlement forest trees in conjunction with the groundflora to evaluate or classify sites. However, in the northern Rocky Mountains relatively pristine old-growth forests facilitated the development of a system employing both trees and understory plants as indicators. Rexford Daubenmire (1952; R. and J. Daubenmire, 1968) used groups of understory species, termed **subordinate unions**, in combination with late-successional overstory species (**dominant unions**) to distinguish forest or plant associations. The collective area of a given plant association, termed the **habitat type** (literally, the type of late-successional vegetation on a particular site or habitat), may indicate similar environmental conditions, and therefore its use in practical land management is examined next.

OPERATIONAL SITE CLASSIFICATION BASED ON VEGETATION

A vegetational approach to land classification, based on the habitat-type method (Pfister et al., 1977; Pfister and Arno, 1980), has been applied in forested lands in federal ownership over millions of hectares of the Rocky Mountains, the Intermountain Region, parts of the Southwest and California (Wellner, 1989), and in the Pacific Northwest (Hall, 1989). It is typically applied to parts of states (northern Idaho, western Montana) or individual National Forests, i.e., without a regional ecological framework.

The rationale of the habitat type approach is that the *natural potential climax* (poly-climax theory of Tansley, 1935; R. and J. Daubenmire, 1968; Pfister and Arno, 1980) overstory and undergrowth vegetation integrate and express the environmental complex for a specified geographic area. A habitat type represents, collectively, all parts of the landscape that support, or have the potential of supporting, the same primary climax vegetation (Alexander, 1986). Climax community is defined as: "The culminating stage in plant (forest) succession for a given habitat, that develops and perpetuates itself in the absence of disturbance, natural or otherwise." (Cooper et al., 1991, p. 134). Note that the method identifies areas of similar climax vegetation, not necessarily areas of the same ecosystem type. The climax vegetation upon which the classification is based is called a "plant association."

The habitat type and other vegetationally-based classification systems have been widely used in forest management on public lands in timber, wildlife, range, and watershed management (Layser, 1974; Ferguson et al., 1989). They have been used more specifically in: growth and yield evaluation (Monserud, 1984; Stage, 1989), recreational use studies (Helgath, 1975), forest protection (Arno, 1976), fire effects (Fisher, 1989) and management (Arno and Fisher, 1989), and natural area preservation (Schmidt and Dufour, 1975; Wellner, 1989). Habitat-type and community-type workers have made a major contribution to inventorying, describing, and classifying forested lands and providing a framework for management.

In this approach, habitat types (h.t.) are based on potential climax vegetation of the overstory and understory. Late-successional overstory dominants (termed **series**) that occur along an elevational gradient from grassland to alpine tundra are determined. Figure 13.5 shows the sequence of these late-successional tree series that are typical of the northern Rocky Mountains. Within each series, undergrowth species (shrubs and herbs) are used to identify many different plant associations. Habitat types are named for the potential climax community type or plant association, for example the *Pseudotsuga, menziesii/Calamagrostis rubescens* h.t. In this type, the series level is denoted by the first part of the name identifying the late-successional tree species, usually the most shade tolerant tree adapted to the site. The second part of the name refers to a dominant or indicator undergrowth species of the plant association.

The category of series acts to order habitat types along an elevational (climatic) gradient. For example, ponderosa pine occupies areas that are warmer and drier than areas where Douglas-fir or Engelmann spruce and subalpine fir become dominant (Figure 13.5). A third level, termed **phase**, is used to designate major within-type variation in understory vegetation associated with geographic, topographic, or edaphic features (Young-blood and Mauk, 1985). For example, the ubiquitous *Pseudotsuga menziesii/Calamagrostis rubescens* h.t. in Montana has four phases which range geographically from northwestern to southwestern Montana and are found at elevations from 823 to 2,377 m (Pfister et al., 1977).

Wellner (1989) summarizes the history of the habitat type approach from 1952 to 1987 and documents 127 classifications that were conducted primarily on U. S. Forest Service lands. Classifications of plant associations and keys to the habitat types they represent are typically developed for forest lands of a state or part of a state (e.g., Montana, Pfister et al., 1977; northern Idaho, Cooper et al., 1991), a national forest or parts thereof (Alexander et al., 1986), or a community type (trembling aspen, Mueggler and Campbell, 1986). Alexander (1985) lists and briefly describes 909 habitat types, community types, and plant communities for the Rocky Mountains alone. The use of the designations: habitat type, community type, and plant community by different workers illustrates primarily one's perception of whether the vegetation encountered is climax or not.

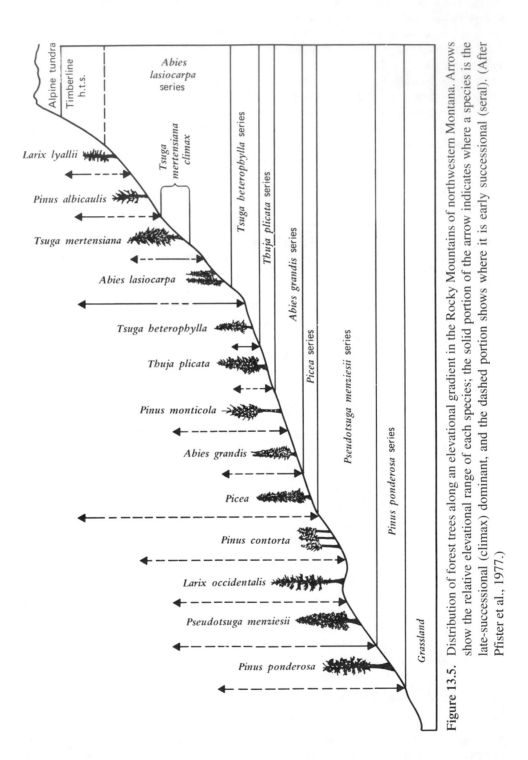

Figure 13.5. Distribution of forest trees along an elevational gradient in the Rocky Mountains of northwestern Montana. Arrows show the relative elevational range of each species; the solid portion of the arrow indicates where a species is the late-successional (climax) dominant, and the dashed portion shows where it is early successional (seral). (After Pfister et al., 1977.)

Regardless of the name used, this approach is a modern version of a very old tradition of vegetation classification (Daubenmire, 1989)—a taxonomy of types by which sites can be classified. In the early days of forestry, attention was narrowly focused on forest cover types and their delineation. Foresters were stand-centered rather than land-centered, so a taxonomy of stands based on vegetation was very useful. The phytosociological classification developed by Cajander (1926) and transplanted by his students to Canada, served very well. The approach is taxonomic, beginning with a classification of vegetation developed from plot samples, and then applying the typology to other sites. This approach is also the essence of the habitat-type classification (Pfister and Arno, 1980). It remains useful especially where the forest land management problems to be addressed concern particular cover types that are to be tended, logged, and site-regenerated with the same species (Rowe, 1984b). Using a taxonomic key to habitat types or communities, and supplemented by selected topographic and soil properties, each forest stand or location can be classified and assigned an appropriate prescription for multiple-use management.

The method is applied for the goal of providing a framework for extensive management of large areas. In central Idaho, 64 habitat types and 55 phases are described for 6.7 million hectares (Steele et al., 1981). Recognizing the need for greater resolution of types, a reworking of approximately 4.9 million hectares of northern Idaho, including the area of the Daubenmires' (1968) original study, yielded a five-fold increase in taxonomic units. Because a habitat type is a relatively large areal unit—all land areas potentially capable of producing a similar plant association at climax—considerable variation may occur in physiography and soils within a given type. Hanks et al. (1983), working in northern Arizona, report:

Two land areas with obvious differences in measurable environmental factors may fall within the same habitat type, if they are equivalent with respect to plant requirements. For example, a habitat type can occur on two sites with different soils and climatic regimes when greater moisture holding capacity of soils on one site compensates for a drier climate. Because of compensating factors a vegetation type may occur on different physical environments.

Daubenmire (1976) also cites an example from the northern Rocky Mountains: "...the *Pseudotsuga/Physocarpus* forest occurs on steep northfacing slopes at its lowest altitudinal limits, moves onto zonal soils at intermediate elevations, then onto the shallow soils of steep southfacing slopes at it highest limits" (Daubenmire and Daubenmire, 1968). Thus a given habitat type may not necessarily identify a landscape ecosystem type that is homogeneous in physical environment.

With increasing disturbances (clearcutting and stand thinning, livestock grazing, and wild and prescribed fire) and disruption of the old-growth and near-natural forests by forest management, site classification increasingly depends on a secondary successional plant community that may have little floristic similarity to its climax community (Neiman, 1988). Neiman reports that "even highly trained plant ecologists find this to be a speculative and frustrating task." Thus attention is turned increasingly to studies of successional relationships of habitat types related to management (Arno et al, 1985; Fisher and Bradley, 1987; Steele and Geier-Hayes, 1995), fire effects (Fisher, 1989; Bradley et al., 1992), and the use of abiotic factors such as soil to identify ecosystem units. However, little relationship has been found between taxonomic soil units and habitat types (Daubenmire and Daubenmire, 1968; Neiman, 1988). This result is not surprising because separate disciplines have different goals, assumptions, and methods. As Daubenmire (1970) recognized, the soil-site properties that play key roles in patterning vegetation are not among those emphasized in soil classification.

Neiman's (1988) study over a broad area of northern Idaho revealed no statistically or ecologically significant correlations between habitat types and taxonomic soil units. However, physical soil characteristics of soil aggregate size and cobbles were useful in classifying specific pairs or groups of habitat types. He concluded that further delineation of habitat types, based on soil variation, would permit greater accuracy in predicting site capabilities and response to disturbance.

Applications and Limitations of Vegetation

Vegetation is universally recognized as a component of major importance in site evaluation and classification: it integrates the effects of many interacting factors, and key species may indicate specific site conditions (Coile, 1938; Rowe, 1969; Daubenmire, 1976). Nevertheless, like any single ecosystem component, vegetation has attributes that markedly affect its use. The limitations of vegetation in forest land management are discussed by Rowe (1984b). In summary, we emphasize that vegetation: 1) is strongly controlled by macro- and microclimate so that its use within regional ecosystem hierarchies (Albert et al., 1986; Bailey, 1996) is well advised; 2) is highly sensitive to disturbance and understanding historic disturbance regimes is important. Existing vegetation, as well as potential natural vegetation, need to be understood for land management applications (Eshelman et al., 1989); 3) is floristically complex and may require the identification of the entire complement of vascular plants and often mosses and lichens; 4) is dynamic; successional patterns should be understood; 5) varies in occurrence, abundance, coverage, and biomass; plants may be lacking due to chance or historic causes. Reliable sampling methods are required to assess plant indicator value; and 6) varies greatly in vertical layering; this attribute is of major significance in animal ecology and wildlife management.

ENVIRONMENTAL FACTORS AS A MEASURE OF SITE

Many situations exist where the forest vegetation, whether the trees or the understory plants, are less useful than site factors in providing an index of forest productivity. These include agricultural or other non-forested lands to be reforested; areas recently subjected to severe fire, logging, heavy grazing, or other disturbances; and areas to be converted from one forest type to another. In these and similar cases, potential site productivity must often be estimated from an analysis of the physical environment rather than from the vegetation.

Physical factors may be used singly or in combination. Included are all the climatic, physiographic, and edaphic factors discussed in the previous chapters. To be useful as an index, however, any factor should be capable of simple and inexpensive measurement and should furthermore be highly correlated to forest productivity. Many site factors are disqualified because of lack of information concerning them or because of the difficulty or expense in obtaining this information. Others are not used because of their lack of sensitivity as a measure of site quality. In general, the most useful environmental factors are those that are in short supply with regard to the demand by forest trees, so that small changes in the supply of the factor will result in measurable changes in tree growth. It follows that different environmental factors will prove useful in different regional and local forest ecosystems.

Climatic factors are generally useful in providing a rough index of productivity among adjacent forest regions or among altitudinal zones within a geographic region. They are much more responsible for genetic differences over a species' geographic and altitudinal range than are soil factors. In particular, temperature and precipitation data may be used to compare forest growth in various geographic regions or altitudinal zones, assuming similar soil conditions, or at least assuming soil conditions that in themselves are closely related to the climate.

Within a given climatic region, growth will vary greatly, depending upon physiographic and soil conditions. Soil factors, therefore, are apt to be particularly useful in local studies of forest site quality.

Since the sum total of climatic, physiographic, soil, and biotic factors define the site portion of the forest ecosystem, the more factors used in an index of forest productivity, the better will be the correlation. However, increasing the number of factors usually means increasing the cost and the time needed to complete the analysis. Thus accuracy and reliability must be weighed against the cost. In efforts to construct environmental site classifications applicable to a wide area, multiple factors should be used at both regional and local scales.

Climatic Factors

Forest climate is obviously related to tree growth because the crowns and boles live in the air and are affected by it. Macroclimatic factors have been appropriately used to distinguish differences among major forest regions (Schlenker, 1960; Rowe, 1972; Findlay, 1976; Ecoregions Working Group, 1989). The applications are limited though, because long-term climatic data from forest sites are either nonexistent or difficult to obtain. Data interpretation for the various interrelated variables is problematical, and furthermore, vegetation itself is perceived as a useful and more easily measured integrator of the complex of climatic factors. In addition, most site quality evaluations for management decisions take place within a regional ecosystem unit (Chapter 2) where the average climate may not vary widely. However, local climate may vary significantly from place to place within the unit.

Despite the obvious importance of local climate, however, it has seldom been used in site evaluation. For one reason, other variables such as land-use history and forest management history have obvious as well as masking effects on forest growth. For another reason, local climate is often strongly related to local topography and soil, and an ecosystem classification based upon topography and soil will also carry with it an implied classification with respect to local climate. The same factors that make a soil very well drained are apt to insure good cold air drainage. Similarly, a very poorly drained soil is apt to result from a topography that inhibits the drainage of cold air as well as of soil water. Thus the very poorly drained sites are apt to be characterized by temperature extremes and a short growing season.

Physiographic Land Classification

Because physiography directly or indirectly affects so many other key ecosystem factors it is highly useful at both regional and local scales to provide a framework for ecosystem management. As emphasized in Chapter 10, specific landforms, their shape, parent material, and location in relation to one another, exert a powerful influence over the reception of solar energy as well as run-off and retention of precipitation (Rowe, 1984b). In the mid-South, Smalley (1991) has developed and applied a taxonomic system based on landforms for about 12 million hectares of the Cumberland Plateau and Highland Rim physiographic provinces of Virginia, Tennessee, Kentucky, and Georgia. Following the system of land stratification and system of Wertz and Arnold (1972, 1975), Smalley used five hierarchical levels (Province, Region, Subregion, Landtype Association, Landtype) to classify land-scapes. Landtypes, the most detailed level (scale 1:10,000 to 1:60,000), are described by nine factors of geography, geology, soil, and the most commonly occurring woody species. Figure 13.6 illustrates the importance of slope position and aspect in distinguishing landtypes characteristic of highly dissected terrain in the Cumberland Mountains of Kentucky, Tennessee, and Virginia (Smalley, 1984).

Factors encouraging the use of this approach are lack of soil survey information and a long history of indiscriminate cutting, burning, grazing, and clearing for agriculture that has drastically affected the natural vegetation-site relationships. Consequently, existing forests are a varied mixture of stand conditions, productivity is far below its potential, and too few suitable stands are available for a direct measure of site potential. In rugged terrain, land-forms may have as many, or even more, recognizable relationships with tree growth than do soil series or the existing tree vegetation. Once adopted, landtypes become the basic unit of land management rather than existing forest stands whose boundaries are typically artifacts of past land use. And this broad landform base facilitates subdivision, as needed, into finer ecosystem types based on local topography, soil, and ground-cover vegetation.

Physiographic and Soil Factors: Soil-Site Studies

The challenge of relating physiography and soil to site quality has attracted many investigators. Of principal concern has been the determination of site quality for areas that have highly variable physiography (specific landforms, topographic factors; see Chapter 10), soil, and stand conditions and for areas that are either unstocked, stocked with unwanted species, or stocked with trees unsuited for site index measurements. Such methods have received great emphasis. Carmean (1975, 1977, 1996) discusses the applications and limitations of this method and cites over 180 soil-site studies for the United States and Canada; Burger (1972) cites many additional ones for Canada.

Depending upon the nature of the specific site, many individual soil and physiographic factors may serve as useful indices of forest productivity. They may be correlated with the site index of the desired forest species. Specifically, **soil-site** studies involve measuring or scoring many soil and site variables, termed independent variables (e.g., soil depth, texture, and drainage class; slope position; aspect) and relating these through multiple regression analyses to tree height or site index. By combining these and other soil and topographic factors, useful formulas have been evolved by which site index can be estimated approximately. For example, Zahner (1958), restricting his regressions to soil groups within a limited geographical region, related the site index of two southern pines to the thickness of the surface soil, the percentage of clay in the subsoil, the percentage of sand in the subsoil, and the slope percentage.

The equations derived from soil-site studies are used for developing site-prediction tables and graphs for estimating site index in the field. In successful soil-site studies, the combination of independent variables may explain 65 to 85 percent of the variation in tree height or site index observed in the field plots (Carmean, 1975). Although many variables may be used to develop precise equations, some of the variables may be difficult or tedious to measure in the field. Therefore, somewhat less precise equations are often developed using the variables that are most easily identified and used in the field.

Before going further, however, two warnings are in order. First, the problem is one of correlation and not necessarily of cause and effect. Too many investigators have read causal relationships into regressions based upon soil and topographic factors, which are not justified. Such correlations may merely reflect a causal relationship attributable to another and unmeasured soil characteristic. Second, frequently the dependent variable is site

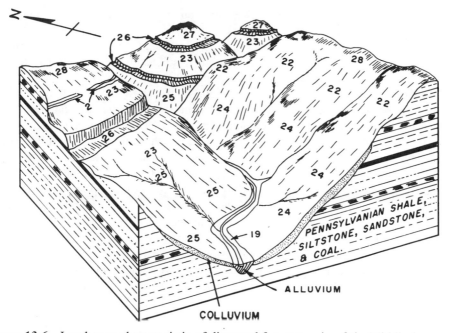

Figure 13.6. Landtypes characteristic of dissected forest terrain of the Middlesboro syncline and Wartburg Basin-Jellico Mountains of Kentucky, Tennessee, and Virginia. Legend: 2, shallow soils and sandstone outcrops; 19, mountain footslopes, fans, terraces, and stream bottoms; 22, upper mountain slopes–north aspect; 23, upper mountain slopes–south aspect; 24, colluvial slopes, benches, and coves–north aspect; 25, colluvial slopes, benches, and coves–south aspect; 26, surface mines; 27, narrow shale ridges, points, and convex upper slopes; 28, broad shale ridges and convex upper slopes. (After Smalley, 1984.)

index as read from harmonized site-index curves. As previously pointed out, such values are suspect by the very nature of the method used to construct the site-index curves. Thus caution is advised in using soil-site equations.

Generalizing from the many efforts to relate soil properties to forest site, the growth potential of forest trees is chiefly affected by the amount of soil occupied by tree roots and by the availability of soil water and nutrients in this limited space. Of prime importance, then, is the **effective depth** of the soil or surface soil depth—the depth of the portion of the soil that is either occupied or capable of being occupied by roots of the tree. This effective depth may obviously be limited by the occurrence of bedrock near the surface (Green and Grigal, 1979, 1980). The position of the water table during the growing season likewise sharply limits root penetration. Less obviously but of equal significance, a coarse, dry stratum may prove an effective barrier to root penetration just as may a highly compact and impervious stratum, such as a highly developed hardpan.

Consequently, many measures of effective soil depth have proved significant in correlating soil factors with site quality (Coile, 1952; Carmean, 1975). The soil factors most frequently found important are the depth of the A horizon above a compact subsoil, the depth to the least permeable layer (usually the B horizon), the depth to mottling (indicative of the mean depth to restricted drainage), and thickness of the soil mantle over bedrock. All these measures quantify the effective rooting depth of trees. They are important when soils are shallow but are relatively unimportant for deep soils where downward root development is unimpeded.

Next in importance are soil profile characteristics that affect soil water, soil drainage, and soil aeration. The physical nature of the profile—soil texture and structure of the least permeable horizon (again usually the B)—is of principal importance here.

Actually, however, the topographic position of the site (slope position) and other landform factors are often closely related to microclimate and to the physical properties of the soil that govern soil–water and aeration relationships. This influence of local climate near the ground, plus belowground climate induced by landform control of soil water, together with landform influence on nutrient regime (via parent materials and soil water availability), assures the importance of landform as the chief correlate of the patterns of both soil and vegetation and hence of site productivity. Moreover, topographic site can be quickly recognized and evaluated using aerial photographs and topographic maps, without the necessity of soil measurement. For example, an index of forest site quality, based only on aspect (6 classes), slope percentage (5 classes), and slope position (5 classes), was developed for use in the Ridge and Valley Physiographic Province of the central and southern Appalachian Mountains (Meiners et al., 1984). By subjectively ranking field sites from 1 to 5(6) for each of the three variables an index is derived (3=lowest quality; 16=highest quality) that provides a rapid way of evaluating relative site quality where available soil water is a major limiting factor. This physiographically-based index is highly correlated with oak site index in the Allegheny Mountains of West Virginia and Maryland and with mean annual increment of trees on steep slopes in the Ridge and Valley Province of Virginia (Ross et al., 1982).

Many useful site relationships have been developed based upon the relative topographic elevation, aspect, slope position, and degree of slope. Studies of oak site quality in the Appalachian Mountains and the Appalachian Plateau (Trimble and Weitzman, 1956; Doolittle, 1957; Carmean, 1967; McNab, 1987) have found that relative position between ridge top and cove, aspect, and degree of slope are all closely related to site quality. Carmean (1967) found that the equations based solely on topographic features explained more than 75 percent of the variation in total height of black oak in southeastern Ohio. These close relations between topography and site quality occur because topography is closely related to important soil features such as A horizon depth, subsoil texture, stone content, organic matter content, and nutrient availability. The relationships of aspect, slope steepness, and site index are illustrated in Figure 13.7.

The general relationship of site quality to aspect for mixed upland oak forests in the Appalachian Mountains resembles a cosine curve (Figure 13.8; Lloyd and Lemmon, 1970). In hilly terrain, the importance of topography and its close relationship with microclimate must be stressed. Northeast aspects and lower slopes usually have cool, moist microclimates and thus are better sites; southwest aspects and upper slopes and ridges have dry and warm microclimates and hence are usually the poorer sites.

Although height growth and site quality are strongly related to landform and topographic variables, their predictive value has been inconsistent among studies. Part of this inconsistency may derive from different methods of scoring or measuring the variables. Therefore, McNab (1987, 1991, 1993) developed indexes for quantifying key landform features of slope position and local landform shape. Tests of this approach at four sites in the Blue Ridge Physiographic Province showed that the site index of tuliptree and landform index were significantly correlated (McNab, 1989). At least for this site-sensitive species, the results suggest that microsites may be a more important source of spatial site variation in the southern Appalachians than previously recognized. Such areas represent local ecosystem types where the beneficial effects of leaf litter accumulation and decom-

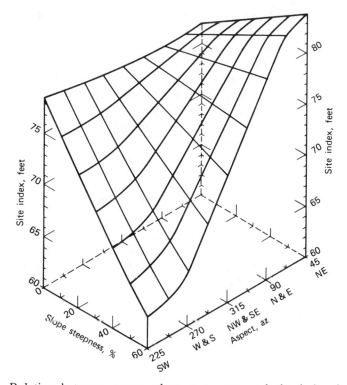

Figure 13.7. Relation between aspect, slope steepness, and site index for black oak growing on medium-textured, well-drained soils. Site index increases from southwest-facing slopes to northeast-facing slopes. These increases are very pronounced for steep slopes, but site index increases related to aspect are relatively minor on gentle slopes. For southwest-facing slopes, site index decreases drastically with increased slope steepness, whereas it increases slightly on northeast slopes as slopes become steeper. (After Carmean, 1967. Reprinted from *Soil Science Society of America Proceedings,* Vol. 31, p. 808, 1967 by permission of the Soil Science Society of America.)

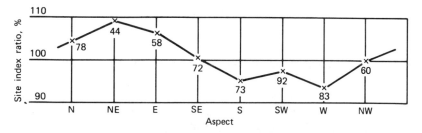

Figure 13.8. Productivity curve with a site-index ratio over aspect; based on 560 soil-site index plots on 27 soil series in mixed upland oak forests of the Appalachian Mountains. The site-index ratio is the ratio of the plot site index to the average site index of all plots of that soil series. (After Lloyd and Lemmon, 1970. Reprinted with permission from *Tree Growth and Forest Soils,* © 1970 by Oregon State University Press.)

position (Stone, 1977; Welbourn et al., 1981), combined with gravitational movement of subsurface water on mountain slopes (Hewlett and Hibbert, 1963; Dwyer and Merriam, 1981), create especially favorable nutrient and soil water conditions.

Minor variations in topography may be highly important in very flat locations, reflecting the effective depth of the soil over poorly aerated lower horizons. For instance, Beaufait (1956), working with southern river bottomland oak sites, found that topographic features varying only centimeters in elevation were correlated with marked differences in the silt and clay content of the soil, with flooding conditions, and with soil aeration, and thus were highly related to site quality.

Soil Survey

The soil survey approach, using soil series and phases, was developed for application in agriculture, not forest ecosystem management (Chapter 11). Although this approach provides a taxonomic classification of forest soils and maps of these taxonomic classes, it has generally proved unsatisfactory for the precise estimation of forest site quality (Grigal, 1984). Grigal stresses that land management does not deal with taxonomic units, but with mapping units. There is a clear and basic difference between soil taxonomic units and soil mapping units. Typically, variation in forest productivity, as estimated by site index, within a given soil-taxonomic unit is too great to be acceptable. On soils of the Rustin series, loblolly pine site index ranged from 59 to 105 (Covell and McClurkin, 1967). Excessive site variation within soil-taxonomic units also has been reported for numerous species in eastern hardwood forests (Carmean, 1970a, 1975). It is widely recognized (Rowe, 1962; Jones, 1969; Carmean, 1970a, 1975; Grigal, 1984; Neiman, 1988) that soil series alone are too heterogeneous to serve as a basis of site evaluation. They can prove satisfactory if they are refined to incorporate specific soil, landform, and local topographic factors that are closely related to forest productivity (Richards and Stone, 1964; Carmean, 1967, 1970a; Grigal, 1984).

MULTIPLE-FACTOR METHODS OF SITE AND ECOSYSTEM CLASSIFICATION

In the previous sections we have considered single-factor approaches to the evaluation of site quality, such as site index and vegetation indicators, physiography, and soil. However, these represent only individual components of the ecosystem, whereas the site quality is the sum total of factors affecting the land's ability to produce forests. The more factors taken into account, the better is the estimate of site productivity, and the greater our ability to understand ecosystem processes as the basis for ecosystem management and its many perspectives. Intensive multifactor methods have been employed with success for over 60 years in Europe, and similar, more extensive methods have proliferated widely in Canada and more recently in the United States.

A word about terminology is appropriate here. Although we conceive landscape ecosystems as volumetric, layered segments of the Earth's surface with inseparable physical and biotic components, these parts are typically treated separately. A common feature of multifactor methods in practice is the consideration of a given ecosystem in two parts: the site and its biotic cover, i.e., the physical ecosystem factors of a specific geographic place and the plants and animals occurring there. One or more physical factors may be used to determine and characterize this space, which may be termed: land type, site type, site unit, geosystem, physiographic site, or forest site. For the biotic part, terms such as vegetation type or cover type are used. The names for ecosystem units (i.e., the volumetric land units integrating site and biota) are also various, from the broadest (ecodomain) down to the smallest (ecosystem type). Due to the importance of the geographic site, terms such as site unit, site type, and total site are sometimes used to indicate landscape ecosystems because *regardless of present vegetation*, these spatial units have the same or similar properties and potential vegetation.

In the following sections, examples are presented of different approaches beginning with the most intensive landscape ecosystem methods. Despite the names of the various methods and their terminologies, their common focus is on geographic area, or *site*: the stage where each unique complex of climate-biota-soil-landform carries on its dynamic and ever-changing performance, directed and invigorated by solar energy (Rowe, 1992b).

Ecosystem Classification and Mapping in Baden-Württemberg

The first operational and systematically applied multiscale, multifactor system of ecosystem classification and mapping was developed in the southwestern German state of Baden-Württemberg (Barnes, 1984; Barnes, 1996). The area is extremely diverse in climate and geology, including the Rhine River basin, adjacent Black Forest, a limestone plateau (Swabian Alb), and glaciated terrain stretching nearly to the Swiss Alps. Much of the presettlement forest of beech and oak was replaced by monocultures of Norway spruce, often several generations old. Thus to diversify and improve multiple-use land management, an ecological approach integrating geology, physiography, climate, soil, vegetation, and forest history (effects of extensive grazing and agriculture) was developed to classify and map forest ecosystems. The flow diagram in Figure 13.9 illustrates that this detailed and systematic approach is initiated by classification and mapping at multiple scales. Studies of the growth and biomass production of major tree species are used to evaluate differences in the major ecosystems (termed "site units"). Then, detailed silvicultural recommendations are made for ecosystems at regional and local scales. This system, applied for over 50 years, is used in long-term planning and on-the-ground forest/wildlife management, with increasing emphasis on natural regeneration of native species (Mühlhäusser and Müller, 1995), and for conservation of unique ecosytems as forest preserves (Dieterich et al., 1970; Arbeitsgruppe Biozönosenkunde, 1995).

The far-sighted, interdisciplinary team who developed this muiltiscale and multifactor approach emphasized integrating ecosystem components at a time in the 1940s when disciplinary fields of geology, climatology, soil science, and vegetation science were pursued for their own sake. This team realized that the practical application of these disciplines in forestry required their combination in a flexible way to match landscape characteristics. Thus they pioneered the simultaneous integration of ecosystem components in classification and field mapping (Schlenker, 1964; Mühlhäusser et al., 1983).

The synthesis of factors at the regional level leads to a division of the state into major landscapes (termed growth areas), which are in turn subdivided into minor landscapes (growth districts) (Figure 13.9). This regional framework acts to limit sweeping generalizations and prescriptions that were often made for species over wide areas having vastly different environments and histories. On the local level (within districts), individual ecosystem types (site units) are identified, mapped at a scale of 1:10,000, and described. Ecosystem types are distinguished and mapped in the field by local differences in physiography; microclimate; soil factors of texture, structure, pH, depth, and water and nutrient status; and overstory and ground-cover vegetation. Each type is defined to include individual sites, which, although not identical, have similar physical site characteristics, silvicultural and management potential, disturbance regime, incidence of disease and insect attack, and growth rates and tree biomass.

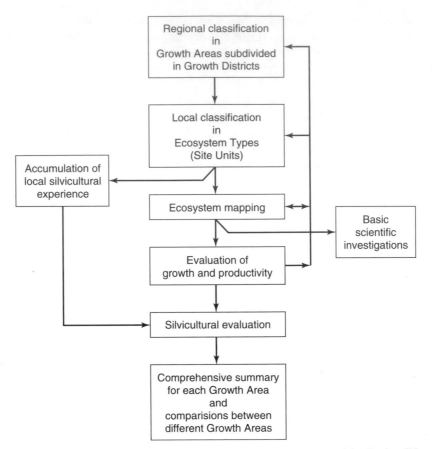

Figure 13.9. Model of the multiscale and multifactor system used in Baden-Württemberg, southwestern Germany to classify, map, and evaluate regional and local landscape ecosystems.

The German team pioneered the use of **ecological species groups** to distinguish and characterize local ecosystem types (Schlenker, 1964; Sebald, 1964; Dieterich, 1970). In Table 13.1, a representative sample of the 24 ecological species groups and 30 ecosystem types of the Upper Neckar Growth District (Sebald, 1964) provides a simple illustration of the use of species groups along soil water gradients to differentiate the ecosystems. Each group is composed of several plant species, which, because of similar environmental requirements or tolerances, indicate certain site-factor complexes. Some species groups have a wide ecological amplitude, like the *Milium effusum* group, whereas others have a narrow amplitude, like the *Arnucus silvester* group (Table 13.1).

An ecosystem is characterized by the presence or absence of groups or the relative abundance of the species in the respective groups. The gradual trend of differences when units are arranged along two soil water gradients (from moderately fresh to moderately dry, units 1 to 10, and from fresh to wet, units 11 to 15) is clearly seen. Units at opposite ends of the respective gradients are easily distinguished by the species groups. However, adjacent site units may be similar in their species groups and would be differentiated in the field by soil and topographic features. For example, units 1 and 2 have similar vegetation, but 1 is a podzolized loamy sand on level terrain, whereas 2 is a podzolized sand on a moderately steep slope of south to southwest aspect.

Table 13.1 Representative Ecological Species Groups and Ecosystem Types (Site Units) Along Moisture Gradients, Upper Neckar Growth District, Baden-Württemberg, Southwestern Germany[1]

Ecological Species Group	Ecosystem Type (Site Unit) Mod. Fresh ⟶ Mod. Dry										Fresh ⟶ Wet				
	1	2	3	4	5	6	7	8	9	10	11	12	13	14	15
Vaccinium myrtillus group	●	●	●	●	·						●	·			
Leucobryum glaucum group	●	●	·												
Bazzania trilobata group	●	·									·				
Deschampsia flexuosa group	●	·	●	●	·	·	·				●	·	·		
Pirola secunda group				·	·	·	●	·	·	·		·			
Milium effusum group	·	·	●	·	●	●	·	·	·	●	·	●	●	●	●
Elymus europaeus group						●	●		·					●	
Aruncus silvester group														·	●
Ajuga reptans group					·	·	●	·			·	●	●	·	·
Stachys silvaticus group					·	·						·	·	●	●
Chrysanthemum corymb. group							·	●	·	·					
Carex glauca group							●	●	●	·					
Molinia coerulea group											·		·		

Source: After Sebald, 1964.

[1]Thirteen of Sebald's 24 groups are shown. Major differences among the groups is indicated by space between the sets of groups. Two sets of site units (ecosystem types) are ordered along gradients from moderately fresh to moderately dry (units 1 to 10) and from fresh to wet and somewhat poorly drained (units 11 to 15).

Key: ● species of the group abundant
      ● species of the group moderately abundant
      · species of the group rare

In field mapping, ecological species groups are used simultaneously with physiographic and soil characteristics to delineate ecosystem boundaries. The indicator value of each group is reliable only within the rooting zone of the species in the group. As seen in Table 13.1, certain units are well defined by vegetation and could be mapped by vegetation alone. However, the combined technique, using soil and physiography as well, is always faster and more reliable.

Applications of Multifactor Methods in the United States and Canada

ECOSYSTEM CLASSIFICATION AND MAPPING IN MICHIGAN

In Michigan, a modified version of the Baden-Württemberg approach was applied at both regional and local levels. The regional landscape ecosystems of Michigan were classified and mapped at three hierarchical levels: regions, districts, and subdistricts. The map is illustrated in Figure 2.11 (p. 37; Albert et al., 1986). The climatic foundations of the method are discussed in Chapter 7. This approach has been extended at the two upper levels throughout Wisconsin and Minnesota (Albert, 1995).

At the local level, landscape ecosystem types have been classified and mapped for both old-growth forests (Barnes et al., 1982; Pregitzer and Barnes, 1984; Spies and Barnes, 1985a; Simpson et al., 1990) and highly disturbed landscapes (Archambault et al., 1990; Zou et al., 1992; Pearsall, 1995). Characteristic of this approach is the physiographic diagram illustrating the correspondence of landform, soil, and vegetation for each ecosystem type. A physiographic diagram illustrating part of the McCormick Experimental Forest in upper Michigan is presented in Figure 13.10 together with a portion of the ecosystem map of this area (Figure 13.11). The map illustrates the location, size and shape of the ecosystems, and the pattern of their occurrence over the landscape. This map is introduced to illustrate the fine scale of ecosystem occurrence in formerly glaciated landscapes. Ecosystem types, relatively homogeneous in site conditions and groundflora

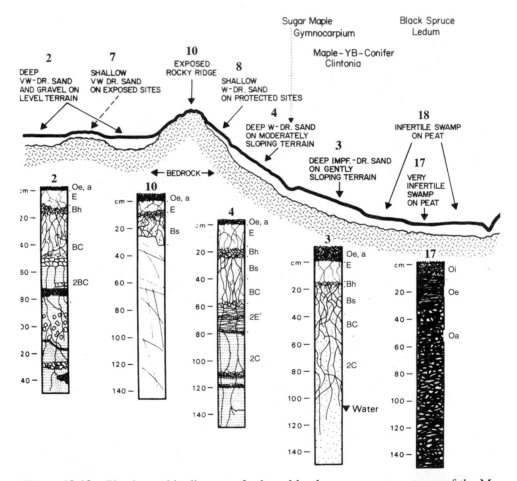

Figure 13.10. Physiographic diagram of selected landscape ecosystem types of the McCormick Experimental Forest, Ottawa National Forest, in upper Michigan. Ecosystem types (identified by number) are characterized here by late-successional overstory species and characteristic ecological species groups, soil profile, and their physiographic position in the along the landscape gradient. (Modified from Barnes et al., 1982. Reprinted with the permission of the Society of American Foresters.)

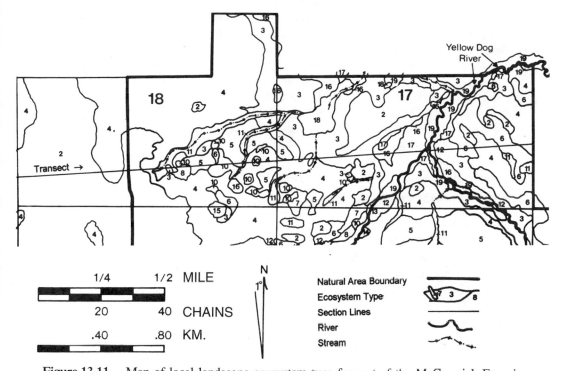

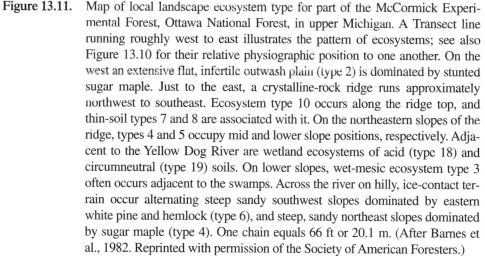

Figure 13.11. Map of local landscape ecosystem type for part of the McCormick Experimental Forest, Ottawa National Forest, in upper Michigan. A Transect line running roughly west to east illustrates the pattern of ecosystems; see also Figure 13.10 for their relative physiographic position to one another. On the west an extensive flat, infertile outwash plain (type 2) is dominated by stunted sugar maple. Just to the east, a crystalline-rock ridge runs approximately northwest to southeast. Ecosystem type 10 occurs along the ridge top, and thin-soil types 7 and 8 are associated with it. On the northeastern slopes of the ridge, types 4 and 5 occupy mid and lower slope positions, respectively. Adjacent to the Yellow Dog River are wetland ecosystems of acid (type 18) and circumneutral (type 19) soils. On lower slopes, wet-mesic ecosystem type 3 often occurs adjacent to the swamps. Across the river on hilly, ice-contact terrain occur alternating steep sandy southwest slopes dominated by eastern white pine and hemlock (type 6), and steep, sandy northeast slopes dominated by sugar maple (type 4). One chain equals 66 ft or 20.1 m. (After Barnes et al., 1982. Reprinted with permission of the Society of American Foresters.)

in this old-growth forest, recur in intricate but *predictable* patterns. The transect line (Figure 13.11) running roughly west to east illustrates some of these patterns as seen in the physiographic diagram in Figure 13.10. Rocky and fire-prone ridges support white pine and northern red oak; outwash plains and slopes support northern-hardwood forests, and wetlands of various kinds are found along the Yellow Dog River. A typical topographic transect from rocky ridge to stream in this area was presented in Figure 11.8. For management purposes, the major ecosystem types, occupying over 95 percent of the area, can be grouped into four ecosystem groups: sandy outwash and ice-contact terrain (types 2, 4, 6), fertile lower slopes and valleys (5, 11), rocky sites with shallow soils (7, 8, 9, 10), and wetlands and wet-mesic adjacent slopes (3, 16, 17, 18, 19).

Resolution of ecosystems to the fine scale shown in Figure 13.11 is currently perceived as too expensive except where management is very intensive. Such is the case in Baden-Württemberg and other German states, and elsewhere in Europe and Scandinavia, where long traditions of intensive land use are much different than those in North America. However, this fine level of detail is useful in understanding the ecosystem diversity of landscapes to which the diversity of plants and animals (biodiversity) is closely related (Lapin and Barnes, 1995; Pearsall, 1995). Increasing interest in biodiversity as a part of ecosystem management and the intensification of land use in North America tends to increase the usefulness of the fine-scale resolution of landscape ecosystems. However, the fine scale of ecosystems is only one part of the system as developed in southwestern Germany and applied in Michigan. The ecosystem approach is applicable at several scales to meet appropriate management objectives.

ECOSYSTEM CLASSIFICATION IN THE SOUTHEASTERN UNITED STATES

In the southeastern United States, landscape ecosystem classification is recognized as the first step toward ecosystem management (Jones and Lloyd, 1993). The multifactor approach has been applied in the Blue Ridge Province of the southern Appalachian Mountains (McNab, 1991) and in South Carolina (Jones, 1991). Using the physiographic map of South Carolina (Meyers et al., 1986) as the regional hierarchical framework, landscape ecosystems are distinguished and mapped using a combination of landform, soil, and vegetation (Jones, 1991). The physiographic diagram of the model for the Interior Plateau Subregion of the Midlands Plateau Region of the Piedmont Province (Figure 13.12) illustrates five ecosystems named by their landscape position in relation to soil water availabil-

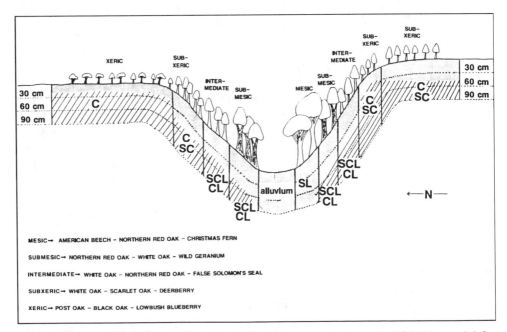

Figure 13.12.   Physiographic diagram and landscape ecosystem classification model for the Interior Plateau Subregion of the Midlands Plateau Region of the Piedmont Province, South Carolina. Soil designations: C, clay; CL, clay loam; SC, sandy clay; SL, sandy loam; SCL, Sandy clay loam. (After Jones, 1991.)

ity (xeric to mesic) and the vegetative associations occurring on these sites. Maps of these ecosystem types provide the basis for predicting potential habitat for endangered species, wetlands delineation, and ecological restoration projects in addition to the traditional forestry interpretations of productivity and silvicultural management.

ECOLOGICAL LAND CLASSIFICATION IN CANADA

The most active and widespread use of multiple-scale and multiple-factor methods has been in Canada. Ecosystem classification in Canada, as in Europe, is characterized by many different systems and complex terminologies. However, it is strikingly different due to its broad scale of application. Because of the need to describe and classify Canada's immense land resources for management and conservation, a multifactor approach using physiography, climate, soil, and vegetation, and relying heavily on remote sensing (air-photo and landsat imagery), computer-based ordinations, and GIS (Geographic Information System) methods is widely applied. Ecological land survey programs have permitted completion of mapping and establishment of GIS data systems for the entire nation at several levels of ecological land classification. These levels include Ecozones (15 units across Canada; Wicken, 1986), Ecoprovinces (45 units), Ecoregions (178 units), and Ecodistricts (5400 units) (Rubec, 1992). The enormous diversity of Canada's forests has, in part, dictated the many different approaches and terminologies that characterize ecosystem classification and mapping. The diversity and detail of nine of these approaches, representing eight provinces and northern Canada, are presented in a single issue of the Forestry Chronicle (Canada Inst. Forestry, 1992).

Hills' Physiographic Approach

Much of the early attention regarding site quality in Canada was given to ground vegetation due to the influence of Cajander's method and that of the Zürich-Montpellier school of plant sociology (Lemieux, 1965; Burger, 1972). The phytosociological phase began to decline in the 1950s and 1960s with the pioneering work of Angus Hills in Ontario who stressed physiography in evaluating site quality. He and his students developed a *total site* or ecosystem approach integrating climate, soil, and vegetation on a landform basis. The idea that landform influenced not only local climate, drainage, and soil formation but the distribution of plants and animals was new and radical at the time (Rowe, 1992a). The total site system, developed by Hills and his associates (Hills, 1952, 1960; Hills and Pierpoint, 1960; Burger, 1972; Hills, 1977; Burger, 1993) in Ontario, was undertaken originally to meet the need for accurate, descriptive resource maps for land-use decisions and in estimating forest productivity. At a time when vegetative associations had been the major focus, his multiscale/multifactor approach and emphasis on physiography was timely and needed. Hills' conceptual model is illustrated in Figure 13.13.

To provide the regional ecosystem framework, Hills divided Ontario hierarchically into 13 regional landscape ecosystems, termed **site regions** (Burger, 1993). To gain additional homogeneity, a site region was subdivided into site districts on the basis of relief and type of bedrock or parent materials. Physiographic features are the basic frame of reference because "they remain most easily recognizable in a world of constant change" (Hills, 1952). Thus **physiographic site types** and **forest types** (characterized by both overstory and groundflora) are combined to form the total site types (Figure 13.13).

Other Canadian Approaches

Following Hills' work, forest ecosystem classifications in Ontario have been developed intensively since the early 1980s for a broad geographic zone across north-central Ontario (Jones et al., 1983; Sims et al., 1989; Sims and Uhlig, 1992). A major consideration of

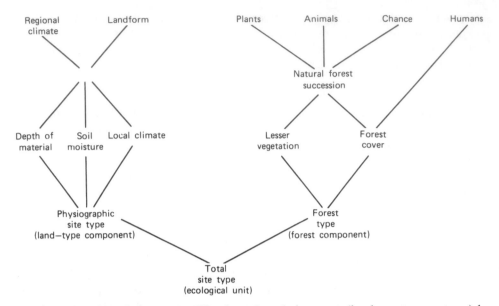

Figure 13.13. Model of the classification of total site types (landscape ecosystems) by Hills' method in Ontario, Canada. (After Hills and Pierpoint, 1960.)

these classifications is that they are purposive. They were first constructed for a specific and predefined range of uses and are primarily stand-level planning tools. They are designed to help foresters and other resource users in developing management strategies and alternatives, including prescribed burning guidelines, estimating susceptibility of a site to erosion, predicting productivity of white and red pines, and rating habitat for the white-tailed deer. Examples of application are given by Sims and Uhlig (1992).

Vegetation types are the primary focus in many systems and are especially useful where human disturbances are minimal, for example in Newfoundland and Labrador (Meades and Roberts, 1992) and British Columbia (MacKinnon et al., 1992). In Newfoundland, the detailed floristic research of Damman (1964) facilitated forest site classification. In British Columbia, the phytosociological approach has remained strong with Krajina and his students developing a more inclusive biogeoclimatic approach that also considers climate, soil, and topography (MacKinnon et al., 1992). In the 1990s, the emphasis is on computer-based multivariate mathematical techniques and GIS technology to distill the most important indicators of forest land potential.

In summary, landscape ecosystem classification and mapping, using multiscale and multifactor landform-based approaches, provide the hierarchically nested units that are required for extensive and intensive ecosystem management. Ecosystem management emphasizes the sustainability of ecosystems instead of sharply focusing on the productivity of individual or competing resources. Today, it requires, above all, spatially mapped ecosystem units at the levels appropriate to the purposes at hand. German methods illustrate that ecosystem classification and mapping can meet the most intensive (and changing) management and conservation practices even in areas of highly variable landforms, soil, and climate, and for sites that have a long history of disturbances. With the increasing refinement of remote sensing, geographic information systems, and computer-based techniques, systematic ecosystem classification and mapping provide the basic framework for ecosystem management and land planning around the world.

Suggested Readings

Carmean, W. H. 1975. Forest site quality evaluation in the United Sates. *Adv. Agronomy* 27:209–269.

Forestry Chronicle. 1992. Forest site classification issue. Vol. 68(1):21–120.

Grigal, D. F. 1984. Shortcomings of soil surveys for forest management. *In* J. G. Bockheim (ed.), Symp. Proc. *Forest land classification: Experience, problems, perspectives.* NCR-102 North Central For. Soils Com., Soc. Am. For., USDA For. Serv., and USDA Soil Cons. Serv.

Haggulund, B. 1981. Evaluation of forest site productivity. *For. Abstr.* 42:515–527.

Host, G. E., and K. S. Pregitzer. 1992. Geomorphic influences on ground-flora and overstory composition in upland forests of northwestern lower Michigan. *Can. J. For. Res.* 22:1547–1555.

Pfister, R. D., and S. F. Arno. 1980. Classifying forest habitats based on potential climax vegetation. *For. Sci.* 26:52–70.

Rowe, J. S. 1984. Forestland classification: limitations of the use of vegetation. *In* J. G. Bockheim (ed.), Symp. Proc. *Forest Land Classification: Experience, Problems, Perspectives.* NCR-102 North Central For. Soils Com., Soc. Am. For., USDA For. Serv., and USDA Soil Cons. Serv.

Rowe, J. S. 1991. Forests as landscape ecosystems: implications for their regionalization and classification. *In* D. L. Mengel and D. T. Tew (eds.), Symp. Proc. *Ecological land classification: applications to identify the productive potential of southern forests.* USDA For. Serv. Gen. Tech. Report SE-68. Southeastern For. Exp. Sta., Asheville, NC.

Rowe, J. S., and J. W. Sheard. 1981. Ecological land classification: a survey approach. *Env. Manage.* 5·451–464.

Sims, R. A., and P. Uhlig. 1992. The current status of forest site classification in Ontario. *For. Chron.* 68:64–77.

PART 4

Forest Communities

Forest ecosystems are characterized by a layered structure of functioning parts—Earth features of physiography and soil, the layered atmosphere above, and at the energized interface connecting these two, the forest biota. Forest communities of trees and associated plants and animals form a key structural component of forest ecosystems—one part of the whole.

Plants are an extremely important driving force in forest ecosystems—and they are integral parts of all ecosystem processes. They evolved in a structured environment, and developed in response to Earth beneath and light and gasses above. Their individual forms and their community physiognomy, therefore, reflect this vertical stratification. Plants of the forest community, individually and collectively, are instrumental in a myriad of changes that occur in the soil and atmosphere of forest ecosystems.

Animals are an indispensable part of forest ecosystems, and they are considered in Chapter 14. Animals, small and large, affect all aspects of the life history of plants from reproduction to death, and they even promote "life-after-death" in clonal plants. Therefore, the role of animals in forest ecosystems is emphasized in this chapter. Because animals attack and consume plants at all stages in their life cycle, we consider in detail the plant dilemma: to defend or grow. In addition, responses of animals to fire and adaptations to it are examined. Finally, the enormous effects of large animals, including humans, on forest ecosystems receive special treatment.

The forest community is considered in Chapter 15 as an integral part of the forest ecosystem. In simplest form, community refers to the aggregate or collection of organisms in a particular place at a given time. Traditional views of forest community are examined, but our emphasis is placed on the grounding of communities in specific environments and their role as a notable part of landscape ecosystems. Interactions among organisms—mutualisms and competition—are emphasized. Finally, because of its importance in community dynamics, the concept of understory tolerance of forest trees is examined in detail. These considerations of forest biota and their communities prepare the way for the chapters in Part 5 on forest ecosystem dynamics.

CHAPTER *14*

ANIMALS

A host of animals of all sizes forms an indispensable part of forest ecosystems. As biotic factors they markedly influence forest community composition and ecosystem processes. Animals, in turn, are strongly affected both by the physical environment and by the plants with which they associate.

Plants provide shelter and food for animals. The foods produced by green plants are the foundation of plant–animal relationships of ecosystems; they initiate all food chains. Each food chain consists of a food plant, the animals that eat it (**herbivores**: **browsers** and **grazers** or **phytophages**), the animal predators and parasites that feed on the phytophages, and the scavengers that eat animal remains and excrement. The plant–animal cycle is completed by decomposers (both animals and plants) that degrade and mineralize plant litter and animal residues.

In this brief treatment, we emphasize the reciprocal adaptations and interactions of animals and plants, including: plant defense; the role of animals in regulating plant life history and production; wildlife habitat and fire; and the effects of large animals, including humans, on forest ecosystems.

Our focus in this chapter is on understanding the role of animals, including humans, in forest ecosystems. Much attention in the natural resource literature is given to the destructive attributes of forest animals, for example insect epidemics and destruction of forest regeneration by native herbivores and livestock. This focus derives from society's interest in wood and wildlife outputs from forests for human use. In adopting an ecosystem perspective, we are encouraged to see that mutualisms and beneficial aspects of native animals may far outweigh their perceived destructiveness. We cannot overemphasize the significant, but often little-appreciated, contributions of native forest animals in the evolution of plants and their indispensable roles in ecosystem processes. The interaction of animals and plants is probably nowhere better illustrated than the area of plant defense, which is considered next.

PLANT DEFENSE

Many examples of mutual adaptations of woody species and animals are observed. We accept many as logical deductions without rigorous demonstration of cause and effect. The animal–plant adaptation, however, also must be consistent with other selective pressures of the physical environment, including fire, drought, and low nutrient availability. Woody plants exhibit a great array of defense mechanisms that evolved under the selection pressure of herbivores and seed predators in site-specific ecosystems. Woody plants are subject to attack at all stages in their life cycle, and all their parts are subject to attack. Above all, however, the particular site conditions of the ecosystem that supports a plant determine the nature of plant defense (Mattson and Haack, 1987; Herms and Mattson, 1992).

The physical defenses employed by woody plants include texture and composition of the plant surface such as leaf toughness and presence of trichomes (as in oaks; Hardin, 1979) and presence of specialized organs or tissues such as thorns or resin ducts. Many tree species, including black locust, honey locust, osage orange, hawthorns, junipers, and some hard pines, as well as many woody shrubs and vines (roses, smilaxes, blackberries, raspberries) have prickles, spines, thorns, or sharp needle-leaves that deter browsing. In tree species, these structures are particularly concentrated in the juvenile phase when foliage and stem feeding by rodents, rabbits, and other herbivores is most likely.

Chemical defenses include the presence of hormone-like substances that affect the development of insects: unsuitable pH or osmotic pressure, absence of nutrients required by the pest, or accumulations of chemicals of secondary metabolism such as alkaloids, terpenes, phenolics, and cyanogenic glycosides (Levin, 1976; Herms and Mattson, 1992). In all higher plants, an enormous number and diversity of secondary chemicals are known; thousands—4,000 alkaloids alone—have been described. Similar to physical defenses, some secondary chemicals have their highest concentration during early stages of leaf expansion and seedling growth (Herms and Mattson, 1992) as well as in adventitious shoots produced following severe browsing. Of course, there are no absolute barriers to herbivory (Mattson et al., 1988; Herms and Mattson, 1992). Nevertheless, secondary chemicals that deter colonization and feeding are primary reasons why plants escape the great majority of herbivores of their ecosystems. However, in addition to allocating carbon to a defense system, plants must also allocate carbon to grow—hence the plant dilemma.

The Plant Dilemma: To Grow or Defend

In unfavorable sites, plants allocate photosynthate to structural and chemical traits favoring the retention and efficient use of scarce resources of light, moisture, and nutrients and defenses to herbivory. In favorable sites, plants allocate the photosynthate to growth and only relatively small amounts to defense—thereby maximizing their acquisition of growth-requiring resources of light, moisture, and nutrients. They tolerate and compensate for herbivores by vigorous growth that keeps them competitive. Simply stated, plants must outgrow herbivory or defend against it, not both (Herms and Mattson, 1992). This relationship applies to defense against both vertebrate and insect herbivores.

Within its particular site-specific ecosystem, the plant balances growth (using processes of cell division and enlargement) with defense mechanisms (chemical and morphological changes) that limit herbivory. The general relationship of the allocation of photosynthate to growth and secondary chemicals is illustrated in Figure 14.1. The allocation to defense is high in harsh sites where resource availability is low and where the effects of herbivores, pathogens, and physical conditions on growth may be severely limiting. The mix of genotypes that survive

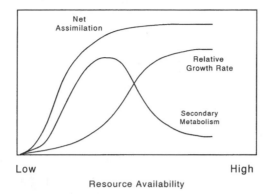

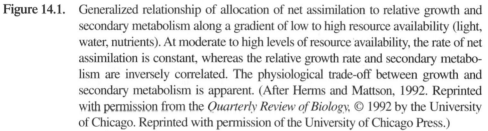

Resource Availability

Figure 14.1. Generalized relationship of allocation of net assimilation to relative growth and secondary metabolism along a gradient of low to high resource availability (light, water, nutrients). At moderate to high levels of resource availability, the rate of net assimilation is constant, whereas the relative growth rate and secondary metabolism are inversely correlated. The physiological trade-off between growth and secondary metabolism is apparent. (After Herms and Mattson, 1992. Reprinted with permission from the *Quarterly Review of Biology,* © 1992 by the University of Chicago. Reprinted with permission of the University of Chicago Press.)

in a given site depends on pressures exerted by competition, mutualisms, and herbivory. The relative importance of these is determined by the site factors of the plant's ecosystem.

In Figure 14.2, a conceptual model illustrates the effects of competition and herbivory on the evolution of genotypes along an environmental gradient (Herms and Mattson, 1992). In diagram *a* of Figure 14.2, genotypes A and B represent populations occurring in contrasting sites where the evolutionary importance of herbivory (H_1 and H_2) and competition (C_1 and C_2) are markedly different. Thus the amount of photosynthate allocated to primary growth is high for A and low for B. In diagram *b*, the importance of competition (C_3) and herbivory (H_3) are equal. The population of genotype A is under pressure from herbivores to evolve increased defense, whereas the population of genotype B is under pressure from competition to evolve a faster growth rate at the expense of secondary metabolism. The two populations may converge on genotype C, which will be maintained by selection exerted by opposing pressures of competition and herbivory. Herbivory and competition will favor a certain subset of carbon allocation patterns (X and Y axes) for each given environment.

Insects

Of all plant–animal mutualisms, insects are probably the most highly co-evolved with plants. Plant defense strategies range from complete tolerance, whereby the plant can fully compensate for any and all injury, to strategies of chemical and physical defense that provide complete or nearly complete immunity (Mattson et al., 1988). However, even "immune" plants may be susceptible when under mass attack or during periods of stress (Mattson and Haack, 1987). Most plants evolve some form of partial tolerance whereby they regulate damage to an acceptable level by chemical and morphological defenses. The extent and nature of the defense system depends on the type of tissue on which a herbivore feeds (for example, meristematic tissue or heartwood of trees), the timing of the attack (early versus late in the growing season), and the size of the individual herbivore or the social-feeding unit (single insect or mass attack). Based on these relationships Mattson et al.

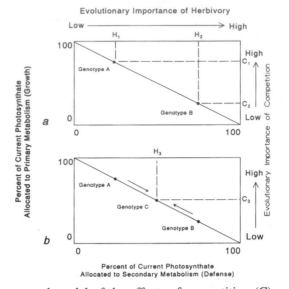

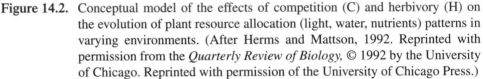

Figure 14.2. Conceptual model of the effects of competition (C) and herbivory (H) on the evolution of plant resource allocation (light, water, nutrients) patterns in varying environments. (After Herms and Mattson, 1992. Reprinted with permission from the *Quarterly Review of Biology,* © 1992 by the University of Chicago. Reprinted with permission of the University of Chicago Press.)

(1988) ranked 13 insect feeding guilds according to their potential injury to woody plants. The least damaging are the leaf and twig gall-formers and insects consuming late season or prior year's leaves. Intermediate are the phloem and sapwood borers and root feeders. Most severe are the phloem and cambium borers on main stem and roots. The latter guilds, bark beetles for example, are especially damaging because they can destroy the plant's essential conductive tissue at a rate exceeding its replacement by the plant.

EXAMPLES OF INJURY AND PLANT DEFENSE
In conifers, and in particular the Pinaceae, the physical (rate and duration of flow, quantity produced, viscosity, exudation pressure) and chemical properties of oleoresin exuded from resin ducts in needles, shoots, and bark significantly deter foliage feeders and bark beetles that mine and construct egg galleries in the inner bark tissues. At the same time, however, insects use resin vapors to find their hosts. They are often attracted to damaged trees where the resin vapor is highly concentrated. Vigorous, standing trees are typically more resistant than lightning damaged or freshly cut trees (Hanover, 1975). For example, western larch trees are virtually immune to bark beetle attack while standing but are immediately attacked after being felled (Furniss, 1972). In moisture-stressed pines, abnormal abscission of needle fascicles and exudation of resin were shown to attract insects (Heikkenen et al., 1986).

In contrast to gall-forming, mining, chewing, and sap-sucking insects, plants require an iron-clad defense against cambium-feeding bark beetles, because their damage, the girdling of the conducting phloem and xylem, is devastating (Mattson and Haack, 1987). The gregarious behavior of many of these insects makes them especially dangerous. If one insect becomes established it rapidly attracts others through pheromonal communication. However, a healthy tree is a non-existing resource at endemic population levels, and it is only when trees are subjected to stress that they become available to beetles (Larsson,

1989). In young and vigorously growing pines, oleoresin pressure is high and beetles entering the bark contact the resin directly. They are either physically repelled ("pitched out") or physiologically impaired by chemical properties of the resin, thus preventing them from breeding and reproducing. In healthy ponderosa pine, the terpenes myrcene and limonene kill western pine beetles feeding on its needles or bark (Smith, 1966). Bark beetles exhibit various degrees of tolerance to the toxicity of resins. In general, beetles as a group are often host specific and are more tolerant to resins of their own host species than those of other pines.

The severity of bark beetle attack is greatest when the resin defenses are low. This may be caused by seasonal variation in resin level by the natural aging process, or by major stresses on the plant such as extreme competition, drought, pollution, logging damage, disease, and defoliation.

In certain hardwood trees, resins may also deter feeding of insects as in cottonwoods (Curtis and Lersten, 1974). Young leaves of creosote bush, a desert shrub, have two to three times the resin content of older leaves and are accordingly more resistant to defoliation by various insects (Rhoades, 1976). Other chemicals are also important. Juglone, produced by walnuts and shagbark hickory, deters feeding by some bark beetles but not others. Tannins, when in high concentration, tend to deter insect feeding. Oaks concentrate tannins around the embryo at the apical end of oak seeds, thereby confining weevil activity to the less protected basal end (Steele et al., 1993). Tannin also deters insect feeding on oak leaves. However, during leaf and shoot development in the spring, when tannins are absent or scarce, oaks may be infested with insects. Nevertheless, the late and rapid flushing of preformed leaves and shoots of oaks minimizes the time insects may feed and reproduce using these tissues.

This rapid leaf-flushing trait of trees of north temperate forests has apparently elicited reciprocal adaptations of insects and their hosts with respect to time of flushing. For example, flushing time may differ by as much as three weeks between different trembling aspen clones (Barnes, 1969). Populations of tortricid caterpillars predominantly infest leaves of early flushing clones (Witter and Waisanen, 1978). A similar relationship was also reported for larvae of tortrix moths on oaks in Russia (Sukachev and Dylis, 1964) and Europe (DuMerle, 1988). In sugar maple, early flushing buds suffered greater damage by pear thrips than trees with late budburst (Kolb and Teulon, 1991).

Site conditions may affect the intensity and nature of insect attack of plants and genetic differences may be elicited in response to herbivory. For example, pinyon pines that colonized cinder fields in Sunset Crater of northern Arizona following an eruption about 800 years ago live in a highly stressful environment. Moisture is limiting, and nutrient status is low. The pines suffer unusually high levels of chronic and severe herbivory by many insects (Whitham and Mopper, 1985). Attack of a shoot and cone-boring moth (*Dioryctria albovitella*) is so severe that tree architecture is significantly altered, growth is reduced, and female reproductive function eliminated (Whitham and Mopper, 1985; Mopper et al., 1991a). However, numerous trees are comparatively resistant to herbivory and were found to produce twice as much wound-resistant resin as susceptible trees (Mopper et al., 1991b). These trees were found to be genetically different from susceptible trees even when growing side by side. In contrast to severe herbivory in Sunset Crater, pinyons growing on less severe sandy loam soils adjacent to the cinder fields are rarely attacked and exhibit normal reproduction and growth. The trees on sandy-loam soil produced the least amount of resin and yet exhibit the lowest levels of herbivory. Thus when the incidence of herbivory is low, defense is not as critical, and photosynthate is used primarily for growth rather than for defense.

Research on two different insect herbivores of pinyon pines at Sunset Crater demonstrated that they not only reduced growth directly by consuming aboveground tissue but also indirectly by reducing the amount of ectomycorrhizal fungi associated with roots of susceptible individuals (Gehring and Whitham, 1991; Del Vecchio et al., 1993). Pines of ages 60–150 years that were susceptible to shoot moth attack had 33 percent fewer ectomycorrhizae than resistant trees. Thirty-year-old pines susceptible to a sap-sucking scale had 28 percent fewer ectomycorrhizae than resistant trees. In both cases, removing insects from susceptible trees revealed that mycorrhizal levels rebounded to a level comparable to that of the resistant trees, demonstrating in each case that herbivory reduced the amount of photosynthate allocated to maintain mycorrhizae.

BETWEEN-PLANT AND WITHIN-PLANT HETEROGENEITY

In the example above, genotypes resistant and susceptible to herbivory occur side by side on the same site. An even more extreme example is that of black pineleaf scale insects (*Nuculaspis californica*). Localized populations are selected for increasing adaptation to individual host trees of ponderosa pine (Edmunds and Alstad, 1978). When infestations occur scale-free pines are observed standing for years beside trees infested with many scales. Scale-free trees tend not to become infested although their branches may intertwine with those of infested trees, and larvae from infested trees may crawl upon their branches. When twigs are reciprocally grafted from adjacent infested and uninfested trees, the grafted shoots retain their characteristics.

For a given tree, developmental changes in resistance and susceptibility are important in determining the distribution of herbivores. In certain tree species, juvenile and adult phases exhibit marked differences in insect resistance that reflect differences in tissue chemistry. In narrowleaf cottonwood, the performance of two species of insects transferred onto different-aged ramets of the same naturally-occurring clones showed opposing and significant changes in host resistance (Kearsley and Whitham, 1989). The gall-forming aphid *Pemphigus betae* was 70 times as common on mature ramets as on juvenile ramets. Aphid survivorship on mature ramets was 50 percent higher than on juvenile ramets. However, the leaf-feeding beetle *Chrysomela confluens* exhibited the opposite distribution, with densities 400 times as high on juvenile as on mature ramets. The pattern was again adaptive as larvae transferred to mature hosts had 50 percent lower survival. Because of differences in the plant maturation process (Chapter 6), an individual genotype may exhibit a pattern of different developmental stages in its branch structure which, in turn, may affect the resistance to herbivores. Even early and late leaves on the same indeterminate shoot (Chapter 6) of yellow birch have been demonstrated to have significantly different levels of phenolic contents and toughness (Schultz et al., 1982).

NUTRITION

Food quality greatly affects insect feeders. For plants growing in sites of limited moisture and nutrients, low nutritional quality may be detrimental to insect performance and may provide a substantial barrier to herbivory (Herms and Mattson, 1992). On the other hand, plant stress-herbivore theory (plant stress benefits insects) predicts that low moisture or other limiting factors may increase the concentrations of sugar and other usable substances in foliage or create a more favorable balance of nutrients (Mattson and Addy, 1975; White, 1978). Evidence for both situations comes from experimental studies of sawflies on pinyon pines at Sunset Crater, Arizona (Mopper and Whitham, 1992; see pp. 337–338). This seeming paradox, whereby abiotic stresses may decrease or increase performance, may be explained by the context in which stress is conceived, i.e., the difference between *long-term* sustained and *temporary* simultaneous plant stress. Mopper and Whitham (1992) contrast *sustained plant stress*, such as chemically-poor soil conditions or prolonged drought prevailing while insects are both active and inactive, with *simultaneous plant stress*, such as low precipitation while the insect is feeding or ovipositing. In

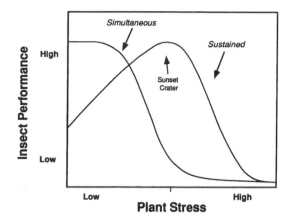

Figure 14.3. Hypothetical relationship between insect performance and simultaneous and sustained plant stress. Conditions at Sunset Crater, Arizona, are indicated. (After Mopper and Whitham, 1992. Reprinted with permission of the Ecological Society of America.)

Figure 14.3, their model predicts that under simultaneous stress, insect performance will be high when plant stress is low while insects are feeding. Performance drops rapidly as stress increases, and at the highest level the plant is an inadequate food source. In contrast, under sustained stress, insects in sites of low chronic stress are relatively successful (Figure 14.3). Insect performance increases as stress becomes intermediate, but then declines when high stress makes the plant an inadequate food source. At the Sunset Crater site, the positive relationship between environmental stress and high insect densities is well documented (Mopper et al., 1990; Mopper and Whitham, 1992). Nevertheless, much of their experimental results refuted the plant stress theory and demonstrated that plant stress can be detrimental to insect performance.

Generally, insects adapted to feeding on nutrient poor plants should be fundamentally different in morphology, physiology, and behavior than those primarily adapted to nutrient-rich plants. These features are described by Mattson and Scriber (1987). Although herbivory may select for traits that further lower the nutritional quality of host plants, insects may also increase their consumption to compensate for low-quality nutrition. However, if insect mortality rates increase due to increased exposure to parasites and predators or reproduction is reduced, the plant may nevertheless benefit (Herms and Mattson, 1992). Plant–animal biochemical systems, having evolved for millions of years, are closely interdependent, and the balance between their systems is usually finely tuned. Thus insects respond readily to slight changes in plant chemistry that make their diet more nutritional, less toxic, or allow easier access to the plant tissues, as in the case of bark beetles.

In summary, woody plants exhibit an incredible array of defenses against insect herbivores including those associated with different individual genotypes and the spatial and temporal heterogeneity within a given individual. Finally, the likelihood of somatic mutations in modular shoots of long-lived plants has led to the idea that a single long-lived plant or clone may represent a mosaic of genotypes that might effectively prevent herbivores from evolving metabolic pathways to overcome plant defenses (Whitham and Slobodchikoff, 1981).

PLANT HYBRID ZONES AS RESERVOIRS FOR INSECTS

In a series of elegant studies, Whitham and associates have demonstrated the importance of zones of natural hybridization as centers of insect abundance and diversity (Whitham, 1989; Kearsley and Whitham, 1989; Floate et al., 1993; Whitham et al., 1994). For example, 85 to

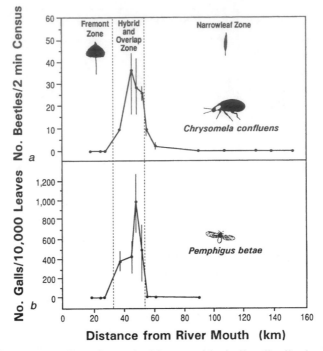

Figure 14.4. Occurrence of two insect herbivores with similar distributions in the hybrid zone between Fremont and narrowleaf cottonwood along the Weber River, northern Utah. *a,* free-feeding beetle, *Chrysomela confluens,* values are 3-yr means (± 1 SE); *b,* galling aphid, *Pemphigus betae.* (After Floate et al., 1993. Reprinted with permission of the Ecological Society of America.)

100 percent of the *Pemphigus betae* gall aphid population occurs on less than 3 percent of its host population in a 13-km zone in Weber Canyon, Utah, where hybridization and back-crossing occur between Fremont cottonwood and narrowleaf cottonwood (Whitham, 1989). Figure 14.4*b* illustrates the importance of this restricted zone for the aphid population where it is more viable on hybrids than on the parent species. Essentially, hybridization has altered the well-engineered defense of the pure species. The concentration of aphids on such a small segment of the host population suggests that susceptible hybrid plants not only act as insect reservoirs (or "sinks") in ecological time, but may also have prevented aphids from adapting to the more numerous resistant hosts in evolutionary time.

The free-feeding beetle, *Chrysomela confluens,* also occurs in the same restricted hybrid sink zone in Weber Canyon (Figure 14.4*a*). The hybrid zone is a superior beetle habitat because (1) the early leaf flush of narrowleaf cottonwood and the hybrid provides the first source of abundant food for beetles in spring, and (2) staggered leaf phenologies in the hybrid zone allow beetles to shift onto newly flushed Fremont cottonwoods as foliage of the sympatric hybrid and narrowleaf trees declines in quality. For this beetle, the hybrid zone is a phenological reservoir or sink that increases beetle fecundity and leads to chronically high herbivory year after year.

Mammals

In defense against browsing mammals, plants also employ chemicals in stems and leaves, in addition to anatomical armaments of prickles, spines, and thorns. The pattern of woody plant food selection by boreal mammals that feed on a variety of evergreen and deciduous species, growth phases, and plant parts, is primarily a consequence of avoidance of secondary chemicals (Bryant, 1981; Bryant et al., 1985; Huntly, 1991). These chemicals strongly influence the palatability and food value of winter-dormant woody plant tissues. In circum-boreal deciduous forests, birches and willows are heavily browsed by voles and snowshoe and mountain hares. Therefore, overwinter survival requires antiherbivore defenses that reduce the probability of herbivory during the critical juvenile period. Secondary chemicals, especially papyriferic acid, have been demonstrated to regulate mammal browsing on birch taxa in winter. Small droplets containing resin on the surface of birch twigs and young stems are the site of the chemical defense. The palatability of birch seedlings (one-year old) and saplings (seven to eight years old) to the mountain hare is strongly and negatively correlated with the number of resin droplets. Hares are known to be discriminating feeders. They can select palatable plants among different birch species growing in seedling containers spaced at a 5 cm interval (Rousi et al., 1989) and even select among seedlings of the same family within tightly tied bundles of twigs (Rousi et al., 1991). Substantial genetic variation in such resistance in birches by the terpenes and phenolics has been detected among species, geographic races, and families (Rousi, 1990; Rousi et al., 1989, 1991).

Generally, birches are especially low quality browse species. For example, when a previously multispecies site on Alaska's Kenai Peninsula became white birch-dominated after a forest fire, the moose population starved (Oldemeyer et al., 1977). The palatability of birches can be made even poorer through genetic selection. Although the defense theory already described assumes a trade-off between growth and defense, the birches haven't yet discovered the theory! For the birch taxa tested in Finnish studies (Rousi et al., 1991, 1996) there were no indications of this trade-off; on the contrary, the tallest seedlings were also the most resistant. Therefore, suggests Rousi (1990), fast-growing and resistant birches can be selected, cloned, and outplanted in Nordic countries where they are economically important species.

A common response of various woody plants to severe browsing is the production of sucker shoots in the juvenile phase (Chapter 6) that contain chemicals in their bark and leaves that deter browsing. The juvenile phase of all boreal plants, even of fast-growing plants that allocate most resources to replace browsed tissues by compensatory growth, has been strongly selected for defense against vertebrates in winter (Bryant et al., 1985). For example, juvenile sprouts of trembling aspen contain a chemical that deters beavers from browsing young saplings and changes their foraging habits (Basey et al., 1988, 1990). When juvenile saplings are common, for example in areas of high beaver activity over a long period, beavers select large trees that likely have low concentrations of a specific phenolic compound. However, where juvenile trees are uncommon, as in newly occupied areas, beavers select against the largest trees to maximize their net energy intake from smaller, *non-juvenile* trees as predicted by optimum foraging models.

ROLES OF ANIMALS IN PLANT LIFE HISTORY

Both invertebrate and vertebrate animals affect ecosystem processes, such as nutrient and water cycling, and the regeneration and succession of forest trees by dispersing pollen and seeds, feeding on live plant tissues, decomposing dead organic matter, burrowing in the soil, and damaging and killing trees. In many instances, they regulate forest composition and growth. In the following sections, we examine some of the key roles of animals in affecting the plant from birth to death.

Pollination

Because pollination is a critical process in the life cycle of plants, the role of animals as important pollinators of woody plants cannot be overstated. Wind pollination is dominant in temperate and boreal forests, and coniferous trees are exclusively wind pollinated. Animal pollination is widespread among woody species in tropical regions but less common in temperate forests. Pollination is accomplished by insects (bees, wasps, flies, beetles, butterflies, and moths), birds (especially hummingbirds in the new World), and bats (Baker et al., 1983). Only rarely do mammals other than bats pollinate trees and shrubs; a small primate, the bush baby, is known to pollinate the African baobab tree (Coe and Isaac, 1965). The primary attractants of animal pollinators are nectar; fragrance; flower color, shape, and size; and in the case of birds, insects visiting the blossoms. A comprehensive account of animal–plant interactions in pollination ecology is given by Faegri and van der Pijl (1979).

In North America, 40 families and 69 genera of woody species are dominantly or wholly insect pollinated (Haack, 1994). Understory species are primarily insect pollinated, whereas most upper tree canopy species are mainly wind pollinated. Major insect pollinated groups include all species of the Ericaceae Fabaceae, and Rosaceae, as well as species of *Acer, Aesculus, Catalpa, Cornus, Magnolia, Liriodendron, Nyssa, Salix, Sassafras, Tilia,* and *Zanthoxylum.* In the case of tuliptree (*Liriodendron*), insect pollination is inefficient (Boyce and Kaeiser, 1961), and only about 10 percent of the seeds may be viable. Nevertheless, enough germinable seeds are produced per tree that natural regeneration is not limited.

Seed Dispersal

Animals are highly significant seed dispersal agents and thus instrumental in maintaining and spreading woody plant populations by connecting site and plant. Most of the enormous dispersal literature is presented by biologists in the context of animal–plant mutualisms, coadaptations, and coevolution (Janzen, 1983). However, dispersal is part of plant regeneration, a process that links organisms (animal and plant) and site, the stage on which the success of dispersal is played out. The prevailing paradigm in fruit and seed dispersal, particularly for tropical regions, has been one of "specialization" versus "generalization" (Howe, 1993). Some trees produce nutritious fruits adapted for use by a small group of specialized frugivores that provide reliable seed dissemination. Other tree species are expected to offer superabundant fruits of lower nutritional reward, relying instead on common opportunistic frugivores that are individually less reliable, but collectively disperse seeds effectively. Most animal-dispersed temperate woody species fall between these extremes. However, some of the stone pines (discussed below) have a single predominant bird disperser (see section below on Birds). Howe (1993) reviews the specialist–generalist paradigm for tropical forests and suggests alternative ecological frameworks based on studies of "keystone" mutualists and density-dependent seed or seedling mortality as it affects vegetation dynamics. In either case, regional and local ecosystem studies involving site factors as well as animal–plant mutualisms provide a useful framework.

Animal agents of seed dispersal are not only important for many angiosperms, especially in equable climates, but may be dominant for some gymnosperms as well. Vertebrate animals (birds, mammals, fish, and reptiles) are the primary dispersal agents; insects and mammals transport fungus and moss spores; and earthworms are thought to disperse seeds of some orchids. Regardless of the vertebrate dispersal agents, three key features must characterize the animal–plant–site dispersal system to guarantee establishment and persistence of the plant species. First, animals must be attracted to the fruit or seed by sight, smell, or taste. Second, the attractant must be timed to coincide with the maturity of the seed; premature ingestion destroys the developing seed. Finally, enough viable seeds must "escape" the dispersal agent and be deposited in a microsite where the chances of successful establishment are high. Some escape mechanisms include burying of seeds; regurgitation of seeds; hard, smooth, seed coatings that assure undamaged passage through digestive tracts; and darkly or inconspicuously colored fruits coupled with brightly colored accessory parts. In the latter case, the animal is attracted to the fruit by a red or orange aril, peduncle, or bract, but if the dark-colored fruit is dropped, it is not readily found (Janzen, 1969). Regarding dispersal of seeds via vertebrate guts, Janzen (1983) observed that virtually all traits of seeds and fruits have probably been modified in response to selection for protection of the seed in passing through the guts of dispersing animals.

FISH AND REPTILES

Fish eat pulpy seeds of various woody species growing along tropical rivers, but the fate of these seeds is unknown. Although the seeds of many tropical trees are dispersed by water, the fish may provide upstream transport that is not normally possible.

Reptiles, turtles and tortoises, alligators, and lizards have a keen sense of smell and eat fruits after they have dropped off trees or when borne close to the ground. The fruits of *Celtis iguana* are eaten by climbing iguanas. Most modern reptiles, however, are not vegetarians. Fruits eaten by reptiles typically have a smell and they may be brightly colored.

BIRDS

Birds are primary dispersal agents, and many adaptive interrelationships between plants and birds have evolved. The two main dispersal methods are disgorging fruits or seeds carried in the mouth and excreting seeds contained in fruits that have been eaten. Rarely do birds carry tree fruits on the outside of their body. One exception is the sticky fruits of the dwarf mistletoe which are carried on the bodies of birds.

Birds destroy many seeds when they eat and digest them. However, many wood pigeons, thrushes, nutcrackers, crows, and waxwings can disgorge whole fruits and seeds, which they carry in their beak. More important are the birds that store part of their food but neglect to recover it. Two groups are notable. The first are birds that disperse and cache acorns and other nuts above ground, usually in tree cavities and bark crevices. The acorn woodpecker is an excellent case in point (Stacey and Koenig, 1984). The acorn woodpecker of California is noted for imbedding thousands of acorns, almonds, and hickory nuts in the bark of standing trees (Figure 14.5). Squirrels and other rodents may then carry them to a germination site, completing the dispersal. The acorn woodpecker harvests acorns and caches them individually in small holes it drills in communal storage trees known as "granaries" (Pavlik et al., 1991). A granary is developed over time by many generations of woodpeckers. A tree (pines are preferred in California; cottonwoods are utilized New Mexico) may be riddled with thousands of storage holes from its trunk to its upper limbs. The prospects for regeneration from aboveground caches, however, are poor because the nuts may be eaten or dry out and lose viability.

The second group, birds that routinely cache nuts and seeds in the ground, is represented by jays and nutcrackers. Placement in the ground markedly increases the probability of successful establishment. Seeds of many woody angiosperms (oaks, beeches, hickories, chestnuts, hazelnuts) are dispersed in this manner. Johnson and Webb (1989) list 15 species whose nuts are reported to be dispersed by the blue jay. In the eastern deciduous forest, the blue jay appears to be the only animal that disperses nuts more than several hundred meters and caches them in the ground (Johnson and Webb, 1989). Flights by blue jays of up to 1.9 km in Virginia and up to 4 km in Wisconsin are reported as conservative

Figure 14.5. The California acorn woodpecker (*Melanerpes formicivorus*) has imbedded many acorns of the California black oak (right) in this Jeffrey pine tree. (Near Julian, California; Photos by Terry Bowyer.)

dispersal distances (Darley-Hill and Johnson, 1981; Johnson and Adkisson, 1985). From a single Wisconsin woodlot, jays were estimated to disperse 150,000 viable beechnuts in only 27 days (Johnson and Adkisson, 1985). The critical role of the blue jay in facilitating the holocene migration of heavy-seeded species was discussed in Chapter 3.

One of the most striking mutualisms of worldwide significance is that between wingless-seeded white pines of semiarid (pinyon pines) and subalpine (stone pines) environments in North America, Asia, and Europe and nutcrackers, jays, and woodpeckers (Lanner, 1981; Lanner, 1990). The birds disperse millions of pine seeds over long distances and bury them in multi-seed caches where microsite conditions for germination and establishment are favorable. Mammals also disperse these seeds, but bird dissemination is usually much more important. At irregular intervals these "bird pines" provide their avian dispersers with an abundant and highly nutritious food source for survival and reproduction. Germination and establishment of unrecovered seeds appears to be the only way many of these species, limber and whitebark pines for example, become systematically established (Lanner, 1980, 1982). Dispersal is more than simply a bird-pine interaction; insects may indirectly affect the dispersal process by both birds and mammals (Christensen and Whitham, 1991; 1993). Furthermore, the specific site conditions, whether xeric southwestern woodlands or subalpine mountain sites, affect the life histories of both plant and animals and their mutualistic association.

In the southwestern United States, pinyon jays and Clark's nutcrackers disseminate pine seeds and bury them 2 to 3 cm deep in loose soil at communal caching areas. South-facing slopes that are usually free of snow by late winter (February) are preferred (Vander Wall and Balda, 1977). Availability of cached seeds provides the energy required to initiate breeding which primarily occurs in late winter (Ligon, 1978). Seeds stored from mid-August to January makes possible the efficient retrieval of seeds when nesting commences. Spring breeding is chancy, except in those springs following a major pinyon seed crop that occurs at approximately six-year intervals. Jays and nutcrackers can carry many seeds in their sublingual pouch. An average of 55 (maximum 95) is reported for the Clark's nutcracker (Vander Wall and Balda, 1977). When pinyon seeds are abundant, entire flocks of 200 to 300 pinyon jays gather and store them within their home range day after day over a period of months. Using conservative figures of 30 seeds per trip and 4 trips per day, Ligon (1978) estimated a flock of 250 jays would store 30,000 seeds per day (120 seeds per day per bird) and approximately 4.5 million seeds over a five-month period. For the smaller-seeded whitebark pine, the estimated dispersal quantity for a single Clark's nutcracker was 1225 seeds per day for 80 days or 98,000 seeds per individual per year (Hutchins and Lanner, 1982). Energetic studies show that nutcrackers cache pinyon seeds containing two to three times the necessary energy to survive the fall and winter months. Since jays and nutcrackers can cache seeds up to 22 km from pinyon stands their role of disseminating pinyon seeds is enormous.

The integrated bird–plant–site dispersal system largely determines tree morphology, successional status, population age structure, and tree spacing of limber pine and the stone pine group, including whitebark of western North America, Swiss stone pine of Europe, and Korean, Japanese, and Siberian stone pines of Asia (Lanner, 1990). These species are found at the highest forest elevations and at timberline in rocky, cold, snowy, windy, and moist conditions. Tree and cone morphology are attributed to selection by nutcracker species (Lanner, 1980). Cones are displayed on steeply upswept limbs and rigidly attached where they are highly visible. Seed-filled cones are retained in the crown, and being effectively indehiscent, seeds are visible but cannot fall out even when shaken or rotated. These characteristics exquisitely adapt the stone pines to the efficient foraging needs of nutcrackers. Furthermore, the population structure of pine stands is in part a result of the nutcracker filling its sublingual pouch with seeds from only one or a few trees before caching. The multiple seedlings of whitebark pine in a single cache are much more closely genetically related (half-siblings or full-siblings) than individuals from distant clumps (Furnier et al., 1987). Thus the local population structure and possibly the mating system of the pines are strongly related to avian behavior in seed harvest and caching.

Among the dominantly wind-dispersed conifers the soft pines already described illustrate one remarkable kind of site–plant–bird mutualistic system. In particular kinds of mountain landscapes, avian visitors are "enticed" by attractive displays of cones retaining thin-shelled, wingless seeds. *Juniperus* is the other exceptional group of bird-dispersed conifers. In this case, entire berrylike cones are effectively displayed, thereby enticing birds to consume whole cones that contain the seeds. Figure 14.6 illustrates the predominance of long-distance bird dispersal for eastern redcedar. As in the case of the "bird pines," the site supporting eastern redcedar is again a key part of the system. Pre-European settlement landscapes of barrens and rock outcrops (cedar glades) of the mid-south and Appalachian Mountains were open, dry, and relatively fire-free. The early-successional redcedar produces large crops of small cones at an early age. Widespread bird dispersal to these few safe sites, coupled with rapid germination and physiological adaptation to xeric conditions, would enable the species to persist in areas overwhelmingly dominated by deciduous forests. Paradoxically, these same attributes also make the species today a successful colonizer of old fields and abandoned pastures, a relatively recent and abundant habitat in eastern North America.

In contrast to *Juniperus*, seeds of the closely related genus *Cupressus* are encased in round, heavy, woody cones that travel only as far as they roll down slope or rodents carry them away. The difference in dispersal systems may help to explain the fact that junipers are widely distributed in North America, whereas species of *Cupressus* are limited to a few isolated groves (Grant, 1963).

The amount of fruit and seeds a tree produces may be adapted to the effectiveness of its dispersal agents. In the above case we have considered species that periodically produce large

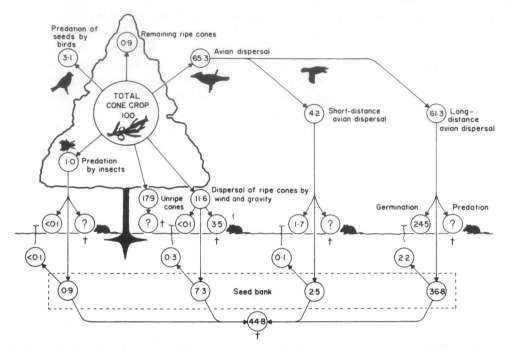

Figure 14.6. Descriptive model of eastern redcedar cone-crop dispersal from June through May of the following year. Numbers in circles are percentages of the total cone crop and are means of four sample trees in southwest Virginia. (After Holthuijzen et al., 1987. Reprinted with permission of National Research Council Canada.)

cone and seed crops. A tree producing a limited number of accessible fruits should rely on an efficient and dependable obligate fruit-feeding dispersal agent, i.e., a specialist. Howe (1977) describes such a mutualistic relationship of the Costa Rican rain forest tree *Casearia corymbosa* whose seeds are reliably dispersed by only one bird species, although 21 other avian visitors feed on its fruit. The masked tityra (*Tityra semifasciata*) is an effective seed dispersal agent because it is a common and regular visitor throughout the fruiting season, has high feeding rates, removes seeds from the vicinity of the parent tree before processing them, and regurgitates viable seeds. In contrast, two parrot species remain in the tree feeding on arils bearing the seeds that then drop to the ground under the tree. Here seedling mortality is virtually complete.

A comprehensive treatment of dispersal ecology is presented by van der Pijl (1972), and a monumental early reference is available in the work of Ridley (1930). The anatomical mechanisms of seed dispersal are described by Fahn and Werker (1972), and dispersal in relation to seed predation is reviewed by Janzen (1971, 1983). Symposium proceedings on frugivore and seed dispersal studies are also available (Estrada and Fleming, 1986; Fleming and Estrada, 1993).

MAMMALS

Mammals, including rodents, ungulates, bats, and humans, are also important dispersal agents. In temperate forests, rodents are the primary agents. Tropical forests are rich in species because they have fruits adapted for mammal dispersal the year round; many fruits are eaten by both mammals and birds.

Some animals, especially rodents, destroy the seeds of oaks, pines, and various other groups. However, many seeds are cached and forgotten so that germination and establishment are accomplished. Acorns and hickory nuts are dispersed up and down slope by squirrels for distances up to 50 m from the parent tree. Many hickory species rely on gaps in the oak forest or the forest edge for their eventual development into the overstory. Wind and gravity provide negligible dispersal away from the parent tree except on steep slopes where fruits might roll down hill.

Squirrels prefer acorns of the white oak group to those of the red oak group (Short, 1976). Tannins in the acorns of both groups (three to four times higher in the red oak group) deter rodent feeding so that if other more palatable foods are available, acorns may be cached rather than eaten on the spot. Also, oaks are known to concentrate tannin in the apical part of the seed where the embryo is located (Steele et al., 1993). Squirrels and bird acorn consumers were observed to eat only a portion of the seed from the basal part (cap end). The apical part is less palatable, thus increasing the probability of embryo survival of seeds discarded by the consumers.

Acorns of the white oak group germinate soon after falling and diminish in palatability after sprouting (Smith and Follmer, 1972). Acorns of the red oak group lie dormant over winter and germinate the following spring. They are available but of low palatability during this time. Compared to other foods available to squirrels (legume and other seeds, dried and fleshy fruits), acorns have relatively low levels of protein and phosphorus (Short, 1976). They supply adequate energy but do not satisfy the metabolic requirements of squirrels for nitrogen and probably phosphorus. Therefore, a finely-balanced adaptation system between squirrels and oaks has evolved chemically, physiologically, and morphologically accounting at least in part to the persistence of both groups. The relatively large size of the nuts of many temperate oaks and hickories provides a valuable energy source for mammals and birds but size is probably also site related. Many of these species grow on dry or dry-mesic sites where energy is required for initial development of a tap root to penetrate a surface mat of organic matter and reach subsurface moisture during the growing season.

The fruit characteristics of plants are therefore adapted closely to site conditions where they establish and their dispersal agents. As dispersal agents, mammals differ from birds in that they have a keener sense of smell and have teeth that masticate seeds. They are also typically larger, rarely lead an arboreal life, and are mostly color-blind night feeders (van der Pijl, 1972). The corresponding characteristics of fruits eaten by mammals are a favorable smell; a hard skin; a more evident protection of the seed itself against mechanical destruction (often a stone-like covering of the seeds as in all drupes), often assisted or replaced by the presence of toxic or bitter substances in the seed; nonessentiality of color; and in many cases large size, typically causing them to drop to the ground.

Ungulates, including tropical and savanna ruminants and elephants, consume a wide variety of vegetation. African ruminants of savannas (springbok, gemsbok, eland) consume considerable amounts of tree fruits, particularly legumes. Many acacias provide leathery, nutritious pods containing extremely hard and smooth seeds, which evade or resist strong molars. Extinct horses, gomphotheres (mastodon-like proboscidians), glyptodonts, ground sloths and other Pleistocene megafauna may have also been important in seed dispersal of certain plants in Central American lowland forests (Janzen and Martin, 1982). Even some temperate deciduous species with sweet-fleshed large fruit, for example Kentucky coffee tree, osage orange, honey locust, pawpaw, persimmon, and ginkgo, may have been dispersed by megamammals and as a result may have had denser populations and much wider ranges in the past. In fact, seed dispersal by animal guts may date back at least as far as dinosaurs if the sweet fleshy covering of seeds of Chinese ginkgo trees and African podocarp trees are as old as we think they are (Janzen, 1983).

Bats are important dispersal agents of woody species, primarily in tropical Asia and Africa (van der Pijl, 1957). Fruit bats are nocturnal, color blind, and have a keen sense of smell. Bats eat fruits that are of drab color, have a musky odor, and are often large and exposed outside the foliage. Bats typically consume the juice after intense chewing of the fruit, and the remnants are regurgitated. They transport seeds within about 200 m of the fruit source.

Monkeys and apes are mostly destructive, eating everything edible, ripe or unripe, with apparently a limited dispersal role. However, the macaques, an Old World monkey species, are apparently responsible for evolution of the Asian dogwood, characterized by large, compound, and red-seeded fruits (Eyde, 1985). Big-bracted dogwoods, with small simple fruits and tart or bitter taste, formerly extended round the Northern Hemisphere until monkeys came in contact with them about five million years ago. The monkeys' appetite for soft, sweet flesh; good color vision for the large, conspicuous fruits with red seeds; and dispersal habits to spit or void seeds apparently led to the new species.

Humans are significant dispersal agents, spreading plants, insects, and pathogens into diverse areas outside their native ecosystems. Many significant horticultural and forest introductions are important parts of our lives. At the same time introduced pests have decimated species such as the American chestnut, American elm, and western white pine. The significance of human-introduced invasive species is considered below.

Germination and Establishment

Animals influence germination of temperate tree species by caching seeds in the forest floor where dormancy requirements (if any) may be satisfied. In addition, many seeds are adapted to pass through the digestive tracts of birds and mammals unharmed. In fact, digestive juices may weaken the seed coat, thus favoring the absorption of water and increasing eventual germination.

Animals aid in plant establishment and plant growth, especially on nutrient-poor sites, by providing inoculum for mycorrhizae that help the plant obtain water and nutrients (Maser et al., 1978; Maser and Maser, 1988). Fruiting bodies of mycorrhizal fungi, sporocarps, are the link between the mycorrhizal fungi and many forest animals, including deer mice, voles, chipmunks, squirrels, and other rodents that depend on hypogeous fungi for food. When these animals eat sporocarps, they consume fungal tissue that contains nutrients, viable fungal spores, nitrogen-fixing bacteria, yeast, and water. Fungal sporocarps are digested and the undigested material, containing these components, is excreted as fecal pellets in the forest and in disturbed areas. Under favorable conditions, roots of seedlings or established trees in contact with the pellets may be inoculated with the mycorrhizal fungus when spores germinate.

The digging activities of various animals turn up mineral soil that provides a seedbed suitable for establishment. Sukachev and Dylis (1964) reported about an area where two to three percent of the surface was covered with molehills; the germination of oak and maple seeds here was twice as high as in undisturbed soil. Rooting by wild boars may also remove thick moss or other vegetation that prevents seeds from germinating and establishing. However, rooting animals may destroy young seedlings and cause widespread damage. Rooting by wild boars in Russia may cause the replacement of hardwoods by Norway spruce.

Pocket gophers may eliminate or stimulate the regeneration of trembling aspen in the western United States. In northern Arizona, belowground herbivory by pocket gophers on the roots of aspen prevented its invasion of mountain meadows (Cantor and Whitham, 1989). Rock outcrops, largely inaccessible to pocket gophers, serve as refugia for the aspens. In 32 aspen-meadow associations, the distributions of aspens and pocket gophers were nonoverlapping 93 percent of the time. However, in another western area, the mountain plateaus of central and northern Utah, gophers may favor aspen regeneration. Where clonal aspen stands dominate montane landscapes vigorous young aspen suckers and understory saplings are often observed in aging clones. These occurrences are presumably due to gopher or other belowground herbivory. In these areas, limestone-derived soils support gopher burrowing and nest building. In contrast, no such aspen understory development occurs on sandy Michigan sites where sand soils apparently limit extensive animal burrowing.

Decomposition, Mineral Cycling, and Soil Improvement

Soil animals play a significant role in organic matter decomposition. Soil animals aid in the breakdown of organic matter by:

1. Physically disintegrating tissues and increasing the surface area available for bacterial and fungal action,
2. Selectively decomposing material such as sugar, cellulose, and even lignin,
3. Transforming plant residues into humus,
4. Mixing decomposed organic matter into the upper layer of soil,
5. Forming complete aggregates between organic matter and the mineral fractions of soil (Edwards, 1974).

The physical environment, forest composition, and the amount and kind of humus (mull or mor) strongly affect the number and diversity of micro- and meso-fauna, i.e., earthworms, springtails, mites. The sequential development of vegetation on a site strongly influences the corresponding sequence of soil invertebrates. Together these changes constitute the continuous change of belowground biota of an ecosystem.

Severe insect defoliation of the overstory usually results in more light, warmth, and moisture reaching the forest floor; all these factors markedly increase the decomposition of organic matter and hence the rate of nutrient cycling. Insect defoliation also may regulate nutrient cycling by increasing the rate of litter fall, influencing the rate of nutrient leaching from foliage, stimulating the redistribution of nutrients within plants, increasing light penetration through foliage, and stimulating the activity of decomposer organisms (Mattson and Addy, 1975).

The physical and chemical properties of soils are improved by soil-dwelling invertebrates and burrowing mammals. Organic matter is mixed with mineral soil, and aeration, moisture-holding capacity, and nutrient content are improved. Soil structure is aided by various soil fauna, including insects, myriopods, woodlice, earthworms, moles, and other rodents. Soil invertebrates ingest soil particles and mix them with finely ground and digested organic matter. This material, when excreted, forms small but durable aggregates that improve soil structure. In addition soil animals, through their excreta and carcasses, may return considerable amounts of nutrients to the soil.

Small animals move through the soil profile and leave their feces at different depths; earthworms carry fragments of litter with them. They deepen vertical distribution of humus in the surface mineral soil, thus improving aeration, porosity, and moisture-holding capacity. Earthworms also help neutralize acid soil by secretions of calcium carbonate from calciferous glands.

Soil water is profoundly influenced by animals. First, the action of soil-dwelling invertebrates and burrowing vertebrates increases the percolation of water in the soil. Second, the organic matter distributed by soil animals increases the water-holding capacity of the soil. Third, severe defoliation by canopy insects limits interception and evapotranspiration and thereby increases the incident precipitation available to the understory and ground-cover layers. Finally, the drainage system of lands adjacent to streams and small lakes may be significantly changed by dam building activities of beavers (Hammerson, 1994).

Damage and Death

Animals eat plants, so it is not surprising that much of the literature reports animal damage to forests. Animals browse, chew, gnaw, pierce, strip, debark, girdle, fell, and trample woody plants. Nearly every animal group inhabiting forest land has been cited for some kind of damage (Crouch, 1976). However, the incalculable positive influence of animals on ecosystem structure and function cannot be overlooked.

In presettlement ecosystems, animals were natural thinning agents whose populations fluctuated in relation to climate, food supply, and other animal predators and parasites. Major concern by managers over animal damage arose in North America in the early part of the 20th Century when humans began regenerating forest lands. The high incidence of damage to plantations and naturally regenerated stands typically followed deforestation and European settlement, which had disrupted the natural site–plant–animal interrelationships of forest ecosystems. Timber management devoted to artificial regeneration (planting) must often contend therefore with human-caused problems such as lack of predators and high populations of herbivores. For example, deer browsing is probably the most common mammal injury to forest regeneration in Europe and North America. In many areas of eastern North America, deer populations exploded following heavy forest cutting in the 19th or early 20th Century. Figure 14.7 illustrates the rapid rise of deer populations in Pennsylvania. Under such conditions, stands become parklike, a distinct browse line is maintained, and forest regeneration fails (Tilghman, 1989). The lush regeneration within areas from which deer have been excluded (exclosures) contrast vividly with severely browsed areas outside the exclosure (Figure 14.8). In Central Europe, high deer populations for centuries have severely limited forest regeneration and stand development; most natural regeneration and plantings must be fenced from deer for several years during and after establishment.

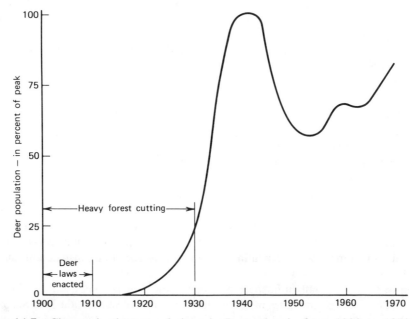

Figure 14.7. Change in deer populations in Pennsylvania from 1900 to 1970. (After Marquis, 1975.)

Figure 14.8. Deer browsing has completely eliminated tree regeneration outside this fenced deer exclosure, Allegheny Plateau, Pennsylvania. (Marquis, 1974; U.S. Forest Service Photo.)

From seed and seedling to mature tree, no forest tree is exempt from animal damage. Cones, fruits, and seeds are destroyed by various insects that may limit regeneration completely in non-mast years. Squirrels cut conifer cones and cache them by the hundreds, storing them for two years or more. Seeds that reach maturity may be widely consumed by birds and small mammals (mice, chipmunks, voles, shrews, and squirrels). In a western Oregon clearcut, birds and mammals caused a 63 percent loss of Douglas-fir seeds, whereas little loss (16 percent) was sustained by the smaller-seeded western hemlock and that for western redcedar was negligible (Gashwiler, 1967). In a two-year study of the fate of seeds of the relatively large-seeded ponderosa pine, only 4 percent of all seeds reaching maturity were available for germination as shown in Table 14.1.

Table 14.1 Effect of Animal Predation on Seed Availability Ponderosa Pine in Western Montana

Fate of Seeds (per 100 reaching maturity)	Number of Seeds	Percent
Seeds used by animals before dispersal	66	66
Seeds dispersed	34	
Seeds used by animals after dispersal	30	88
Seeds available for germination	4	

Source: Schmidt and Shearer, 1971.

In excellent seed years, however, millions of seeds may be produced with the result that more than enough survive even heavy losses to germinate and live.

After establishment, seedlings of all woody species are subject to clipping and bark removal by hares (Dimock, 1970; Crouch, 1976) and stem and root girdling by voles, mice, shrews, mountain beaver, and pocket gophers (Crouch, 1976; Teipner et al., 1983), as well as trampling and browsing by deer and other ungulates. In nurseries and plantations in particular, seedlings are subject to root girdling by white grubs (larvae of scarabacid beetles) causing death or markedly reduced growth (Sutton and Stone, 1974). Sapling and pole-sized trees are subject to browsing by deer (Graham et al., 1963; Crouch, 1969; Marquis, 1974, 1975), moose (Snyder and Janke, 1976; Pastor et al., 1988; Thompson and Mallik, 1989), elk (Gaffney, 1941; Kay and Wagner, 1993), girdling by beavers (Crawford et al., 1976; Barnes and Dibble, 1988; Naiman et al., 1988) and porcupines (Krefting et al., 1962), and defoliation by a variety of insects (Mattson et al., 1991). They are even subject to damage by sapsuckers that make distinctive drill holes in the bark, killing or damaging a wide variety of orchard, shade, and forest trees (Rushmore, 1969; Erdmann and Oberg, 1974). Mature and overmature trees are subject to severe attack by a variety of insect feeders, the primary focus of the following section.

PRODUCTION AND REGENERATION

The typical insect consumption of from 5 to 30 percent of annual foliage usually does not impair tree growth or total annual plant production (Franklin, 1970; Mattson and Addy, 1975). Under epidemic infestation conditions, however, insect grazers may consume 100 percent of the foliage. In North America, about 85 species of free-feeding and leaf-mining folivorous insects periodically cause serious and widespread defoliation of forest trees (Mattson et al., 1991). These expansive outbreak folivores cause individual epidemics exceeding 1,000 contiguous hectares. To this can be added 20 more local outbreak folivores. Insect outbreaks are most likely in very old stands having low net primary production. Spruce budworm outbreaks in overmature balsam fir and spruce stands are a good example (Mattson and Addy, 1975). For example the eastern and western spruce budworm annually affect 5.1 million hectares of commercial forests in the United States (Haack and Byler, 1993). To this may be added the pine-bark beetle cycles; the southern pine beetle alone affects 3.8 million acres of commercial forests. Insects usually do not unilaterally cause the regeneration of senescent forests. They typically act as important disturbance factors (Chapter 16) in conjunction with fungi, fire (Chapter 12), climatic stress (White, 1969, 1974), and windthrow to recycle aging forests, which then may be replaced by fast-growing, productive young stands.

In summary, a complete life-history cycle of reciprocal relations exists between site factors and forest trees and animals—beginning with pollination, seed dispersal, and establishment, through stand development and thinning to the death of old trees and stands, and in clonal plants—life after death.

WILDLIFE HABITAT AND FIRE

All temperate forest and grassland ecosystems are influenced more or less by fire, thus markedly affecting the evolution and behavior of wildlife. Scientists who have reviewed the significance of fires in each part of North America have stressed the importance of fires in providing a mosaic of habitats necessary for diverse wildlife species (Wright and Heinselman, 1973). The effects of fire on birds and mammals have been reviewed by Bendell (1974); prehistory interactions of fire with humans and wildlife are considered by Schüle (1990).

Adaptations to Fire-Dependent Ecosystems

Within temperate forests, no clear difference in animal species diversity has been found between the more combustible conifer forests and the less combustible hardwood forests (Bendell, 1974). Due to the patchiness of the forests after burning and the close proximity or mixture of hardwoods and conifers, the distinctness of habitats is blurred. Animal species are not typically exclusive to one type or the other. For example, relatively few bird species specialize in conifer trees for breeding, feeding, and living (Udvardy, 1969).

The habitat conditions created by fire include: an abundance of vegetative growth on or near the ground; growth of shrubs and trees in open stands with thick branches and twigs; large fruits and seeds that may be retained on the plant; and slowly rotting litter (Knight and Loucks, 1969; Bendell, 1974). These features tend to favor animals that browse and graze and are of relatively large size. Functional adaptations of animals to habitats dominated by fire are many (Komarek, 1962; Handley, 1969): the ability to fly or run quickly and for long distances, the ability to burrow and live underground, effective camouflage and ability to press flat to avoid detection in open areas, and the ability to store food.

Many animals have evolved to exploit periodically-burned grasslands and forests, and in turn some may promote fire and otherwise help perpetuate their habitat. Squirrels typically feed at the base of trees, and the dry cone scales accumulated there provide fine, resinous fuel to ignite a fire following a lightning discharge (Rowe, 1970). Similarly, the placement of nests of grass and fine woody material by birds and mammals on or inside large trees may enhance their flammability (Bendell, 1974). Animals also disperse seeds of favored forage species in new burns (Ahlgren, 1960; West, 1968; Maser, 1989).

Kinds and Abundance of Animals

Fire-dependent ecosystems are remarkably stable in the kinds of birds and mammals found before and after burning. Bendell (1974), in summarizing the findings of many authors (Table 14.2), found that most birds and small mammals stayed in an area after fire. Over 80 percent of the birds and small mammals present following fire were present before fire. Only a few species disappeared, and only a few new species moved in. Most bird populations either showed no change or increased. The greatest increase was shown by

Table 14.2   Change in Species of Breeding Birds and Mammals After Burning.

Foraging Zone	Before Burn	After Burn	Gained[b] %[c]		Lost[b] %[c]		No Change[b] %[c]	
No. of species of birds[a]								
Grassland and shrub	48	62	38	(18)	8	(4)	92	(44)
Tree trunk	25	26	20	(5)	16	(4)	84	(21)
Tree	63	58	10	(6)	17	(11)	82	(52)
Totals	136	146	21	(29)	14	(19)	86	(117)
No. of species of mammals[a]								
Grassland and shrub	42	45	17	(7)	10	(4)	90	(38)
Forest	16	14	13	(2)	25	(4)	75	(12)
Totals	58	59	16	(9)	14	(8)	86	(50)

[a]Source: Modified from Bendell, 1974, p. 105. Reprinted from *Fire and Ecosystems,* © 1974 by Academic Press, Inc. Reprinted with permission of Academic Press, Inc.
[b]Numbers of species are in parentheses.
[c]Based on number of species before burn.

ground-foraging birds; the majority of tree dwellers did not change. The abundance and density of small mammals showed little change in grassland and shrub zones or in the forest following fire compared to prefire conditions. The major factors causing the persistence of birds and mammals in fire-dependent ecosystems are their adaptations to fire itself, their ability to tolerate the wide fluctuation of prefire to postfire conditions, and the fact that due to the erratic burning pattern some prefire habitat is almost always left interspersed with varying degrees of burned habitat. Furthermore, many plants of fire-dependent ecosystems are clonal (grasses and many shrubs especially) and resprout quickly following fire, or like the closed-cone pines regenerate immediately after fire. Thus a relatively stable food source is maintained.

The above findings substantiate many observations that animals of fire-prone ecosystems have a high tolerance for surviving fire; burning does not cause much immediate loss of life (Bendell, 1974). Some animals are killed but most avoid or escape fire. Few are killed by direct heat; most mortality is apparently due to suffocation. Furthermore, wildlife can tolerate wide fluctuations in the physical site conditions following fire. Thus the stability of species composition and abundance indicates that wildlife of fire-dominated ecosystems are broadly adaptable.

Factors Affecting Animal Responses to Fire

Food supply and the pattern and structure of the vegetation are probably the most important factors controlling the kinds and abundance of animals in fire-prone ecosystems. The quantity of food of the right kind is probably more important than food quality. Protein content of browse plants is apparently maintained at a relatively constant level throughout fire cycles, whereas mineral nutrients may vary more widely.

The mosaic of forest, shrub, and open land created by the marked variation in the intensity, frequency, shape, and extent of fire affects not only site factors (wind, temperature, light, snow cover) but also interspecific competition, predators, parasites, and diseases, all of which influence the response of animals to burning. Birds and mammals typically benefit from reduced incidence of parasites for several years after fire. The size of the burn, amount of forest edge, and interspersion of openings with different kinds of cover types positively influence the response of specific animals such as moose (Buckley, 1959) and ruffed grouse (Gullion, 1972).

The pattern of cover greatly influences predator–prey relationships. In an open burn, small mammals may be exposed to new predators. On the other hand, ruffed grouse may be preyed upon by birds concealed in thick clumps of unburned conifers (Rusch and Keith, 1971; Gullion, 1972) and by mammal predators hiding in slash and debris on the ground. Therefore, burning is of value in removing both kinds of cover. The ruffed grouse is strongly dependent on periodic fire that removes conifers and produces a mix of cover types: dense clonal stands of trembling or bigtooth aspen less than 10 years old for breeding and broods, 10- to 25-year-old aspen clones for food (male flower buds borne on thick, stubby branches where birds can rest), nesting, and wintering (Gullion, 1972). The relationship of forest structure and composition to ruffed grouse life history and survival is illustrated in Figure 14.9. Breeding densities and longevity of males are greatest in young stands of the aspens. Stands of aspens and conifers over 50 years of age are not highly productive habitat for grouse.

Like the ruffed grouse of the northern forest, the bobwhite quail of southern pinelands is favored by burning of its habitat. Stoddard (1931) pioneered the use of prescribed fire for improvement of quail habitat in pine forests against the prevailing ideas of the time. Studies by Ellis et al. (1969) confirm Stoddard's early work; both fire and cultivation of patches of crops markedly increase quail populations. Many other game

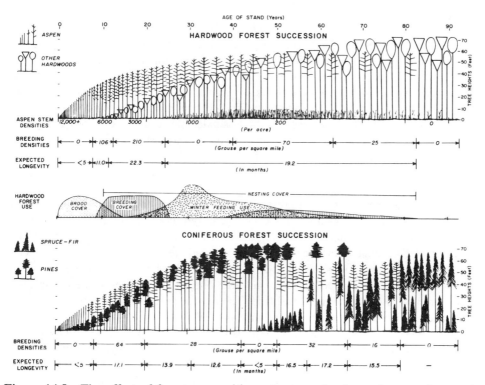

Figure 14.9. The effect of forest composition, structure, density, and succession on the reproduction and survival of ruffed grouse. (After Gullion, 1972.)

birds are favored by fire: wild turkey, ring-necked pheasant, various grouse species (sharptailed, prairie chicken, ruffed grouse, blue grouse, willow ptarmigan), and some waterfowl. Exclusion of wildfire in the sandy, scrub oak plains of the northeast was probably instrumental in the decline and extinction of the heath hen (Thompson and Smith, 1970). Recent large wildfires in Michigan have brought a resurgence of the rare and endangered Kirtland's warbler, the bird of fire (Probst and Weinrich, 1993; see Chapter 21).

Large mammals that respond favorably following fire include moose, white- and black-tailed deer, elk, cougar, coyote, black bear, beaver, and hare (Bendell, 1974). On the other hand, various animals of mature forests may be displaced or eliminated by burning: caribou, marten, grizzly bear, wolverine, and fisher. In summary, the response of animals to fire is highly variable and related to both the kinds and behavior of the animals in addition to the complex of plant and site changes that we have stressed.

INFLUENCE OF LARGE ANIMALS ON FOREST ECOSYSTEMS

The activities of humans, deer, cattle, sheep, and other large animals result in substantial changes in forest ecosystems. Humans have had the most far-reaching and pervasive influence through a variety of direct and indirect activities. One of these, introduction of hoofed grazing and browsing animals, has a great potential in effecting change. Around the world, severe grazing by livestock has probably been more important than any other factor in reducing the productive capacity of uncultivated land.

Livestock Grazing

The effect of herbivores on forest ecosystems and site quality is both indirect and direct. Grazing and browsing animals change the vegetation through their selective feeding habits and the differential ability of plants to survive and prosper. The changed vegetation, often fewer grasses and more woody plants, in turn results in changed litter and soil biotic activity and thus in changed site conditions. The change may be for the better or worse, but it is most apt to be the latter because the less palatable species are woody and generally those with lower nutrient content. The litter of such plants will generally decompose slowly and will inhibit soil biotic activity. Thus changes in vegetation result in changes in soil organic matter, soil organisms, soil chemistry, nutrient availability, and therefore plant productivity.

Heavy, single-species grazing pressure can cause entire changes in the structure of the plant community. For example, in the 19th century following European settlement of the intermountain American West, overgrazing from livestock, particularly cattle and sheep, depleted the bunch-grass vegetation and changed fire regimes (Wagner, 1969). Cottam (1947) queried: Is Utah Sahara Bound? As a result of reduced competition from grasses, woody species such as big sage, junipers, bitterbrush, and serviceberry increased greatly and turned grassland into brushland. The reverse is reported in Africa where elephants and other browsing animals may turn woodlands and brush types into grasslands (Wagner, 1969). In the 20th century, widespread increases of the mule deer placed heavy pressure on the shrubs, causing them to disappear slowly. In some areas, the vegetation is returning to the original bunchgrass type.

The direct effects of heavy grazing on the site result largely from the action of animal hoofs in compacting the surface soil and in breaking up the ground cover. The pounding of animal hoofs results in a breaking down of the soil aggregates that give a crumb structure to the surface soil. The pore space of the surface soil is greatly reduced, often to the point of seriously reducing the supply of air in the soil, and of preventing the infiltration of rain water thereby causing surface runoff. As a result, soils of heavily grazed areas are apt to be poorly aerated, have a lessened infiltration capacity, and become subject to sheet erosion. Biotic activity is minimized in highly compacted grazed soils. Site quality is often lowered substantially.

In a study of paired grazed and ungrazed woodlots in southern Wisconsin, Steinbrenner (1951) found that, in general, the highly compacted soils of the heavily grazed woodlots had a lower initial moisture content in the spring and dried out faster in the summer and late fall, evidently because of the lowered soil permeability and increased run-off. The grazed soils were so compacted that water permeabilities averaged about one-tenth those of comparable ungrazed woodlots.

Grazing has greatly affected forest regeneration and site quality in heavily grazed areas in farm woodlands of the East, Central States, and the South, and throughout most of the open woodland and ponderosa pine forests of the American West. Patric and Helvey (1986) reviewed the modern evidence for effects of livestock grazing on soil and water in the Eastern United States. They emphasized that livestock grazing *in moderation* can have negligible effects on forest soil and water. They recognized the problems of heavy grazing, but concluded that there is no evidence that woodland grazing in the East, as typically practiced, has substantial adverse effects on water quality or on flooding in streams draining grazed woodlands. However, the effects on regeneration may still be substantial.

In the West, severe overgrazing by livestock in the 19th Century affected forest composition, fire regimes, and site conditions. For ponderosa pine forests of northern Arizona, Cooper (1960) concluded that when European settlers brought cattle and fire-fighting crews to the Southwest, the forest began to change. Grazing altered both the composition and quantity of ground cover vegetation. Removal of herbaceous competition and exposure of mineral soil by livestock helped prepare the ground for dense thickets of pine reproduction and simultaneously reduced the likelihood of fire. With few fires to thin dense stands, they soon grew into almost impenetrable sapling thickets.

Intensive cattle grazing in riparian areas of southcentral New Mexico has severely affected populations of the acorn woodpecker (Ligon and Stacey, 1995). The drastic population decline and likely extinction is correlated with the loss of nearly all large granary trees of narrowleaf cottonwood and lack of middle-age trees to take their place. The lack of storage sites is attributed to a period of intensive cattle grazing during which regeneration of young cottonwoods was suppressed.

In forest ecosystems of the Blue Mountains of northeastern Oregon and southwestern Washington, long-term effects of grazing by livestock and big game produced a complex web of interconncted local and distant effects (Irwin et al., 1994; Langston, 1995). Long-term heavy use by domestic livestock and elk changed ecosystem processes, and mixed results have followed. These may include improvement or reduction of forest regeneration and wood-fiber production, drainage of wetlands, elimination of meadows, and reduction in productivity of cattle and elk due to vegetation changes.

In Australia, grazing by livestock occurs over 4.5 million km^2 or 60 percent of the land surface (Wilson, 1990). These lands are home to approximately 50 million sheep and 15 million cattle and are also grazed by 20 million kangaroos, 0.5 million feral goats, and perhaps 100 million wild rabbits! Wilson describes grazing effects on natural ecosystems and emphasizes the progress made over 40 years in reducing overgrazing and erosion in some regions through research and education. Viewing grazing as ncither alien or temporary, he emphasizes that restoration will occur with appropriate management in conjunction with continued use, rather than by grazing exclusion.

Native herbivores such as deer and elk are generally not present in numbers sufficient to compact the soil excessively but have markedly changed the forest composition through the differential browsing of seedlings in many localities in the East, Appalachians, western pine and aspen forests, and elsewhere. Failure of hardwood forests of Pcnnsylvania to regenerate (Marquis, 1975; Tilghman, 1989), lack of sapling regeneration of eastern hemlock in the western Great Lakes region (Frelich and Lorimer, 1985; Mladenoff and Stearns, 1993), and regeneration failure of trembling aspen clones in the Intermountain West (Kay, 1997) is attributed in part to high deer populations. Where deer have been introduced into forests composed predominately of palatable species previously unbrowsed, they may virtually eliminate all vegetation within reach. This has occurred in many indigenous forest areas of New Zealand.

Human Component of Ecosystems

Landscape ecosystems contain humans and other organisms from bacteria to elk, and are supported by the matrix of air-water-soil. Considerations of humans in ecosystems range from that of Marsh (1864) to the more recent works of many authors (Thomas, 1956; Perlin, 1989; Turner et al., 1990; Rowe, 1990; McDonnell and Pickett, 1993; Kay, 1994; Whitney, 1994; and Langston, 1995). Humans are an exceptionally powerful factor in the ecosphere and have profoundly changed regional and local ecosystems in both dramatic and subtle ways. Perlin (1989) provides an exceptionally powerful historical record of human destruction of forests in the development of civilizations from the Bronze Age world in Mesopotamia to 19th Century America. Not the least among human actions in North America have been European agricultural clearing, logging, introduction of livestock grazing, long-term changes in dominants in forest communities, forest fragmentation, climatic change, introduction of forest "pests," and changes in fire regimes. The piercing story of human interventions in the complexity of ecosystems is compellingly described for the Blue Mountains of Oregon by Langston (1995).

Far from being unaware of human effects on ecosystem composition, structure, and function many individuals have sought to understand the effects and correct abuses from different perspectives. For example, Hornstein's (1958) classic book, *Wald und Mensch*, a detailed history of human changes to forests of southwestern Germany, was instrumental in developing an intensive multiscale and multifactor site classification and mapping system (Chapter 13) to provide an ecological framework for ecosystem management and restoration. In North America, however, we probably underestimate the role of Native Americans in changing the biota. For example, studies in the Intermountain West led Kay (1994) to conclude that the modern concept of wilderness as areas without human influence is a myth. By limiting ungulate numbers and purposefully modifying the vegetation with fire, Native Americans changed the composition of entire plant and animal communities.

Confining ourselves to contemporary direct site effects, human activity is responsible for considerable site deterioration through soil compaction. Trucks, tractors, and other heavy equipment used in logging result in substantial soil compaction (Moehring and Rawls, 1970; Dickerson, 1976). In the Atlantic coastal plain, soil compaction on skid roads was found to reduce soil infiltration rate and pore space by 84 and 34 percent, respectively, and to increase bulk density 33 percent (Hatchell et al., 1970). Forty years may be required for infiltration to recover on severely compacted logging roads (Perry, 1964).

Compaction from logging traffic is much more pronounced on wet than dry soils and more severe on clayey than on sandy soils (Moehring and Rawls, 1970). Wet weather logging can cause soil compaction that may markedly reduce growth rates of established seedlings and significantly reduce seedling establishment on skid trails (Youngberg, 1959; Perry, 1964; Hatchell et al., 1970). Hatchell and Ralston (1971) estimated that 18 years may be required for severely disturbed soils to attain normal tree densities.

The human foot itself is an effective compacting agent. On the Mall in Washington, D.C., for example, heavily-trampled soils were found to have a bulk density and particle density similar to that of asphalt and concrete (Patterson, 1976). Generally, the problem of soil compaction in forest parks and other forested recreational areas has been recognized for over 50 years (Meinecke, 1928) and is reaching serious proportions. Death of large and famous trees has been attributed to compaction, and decreased growth rate is frequently apparent. Because of compaction, it has been necessary to fence out tourists from the immediate neighborhood of famous trees, and to move public campgrounds out of old-growth areas as in the redwood and giant sequoia localities of California. The increasingly intensive use of the forests for camping and other recreation is giving added importance to the dangers of site deterioration directly from humans and their vehicles.

HUMAN-INTRODUCED EXOTICS

The distribution of biota of the ecosphere has been restricted by oceans and other natural barriers for millions of years. However, during the last 100 years, human activities, especially international travel and trade, have broken these barriers. Introduced species or "exotic" species are invading new continents at an increasing rate. Human-introduced exotic insects and pathogens have dramatically altered forest ecosystem diversity, function, and productivity. More than 400 exotic insects and 20 exotic fungal pathogens attack woody trees and shrubs in North America (Haack and Byler, 1993; Mattson et al., 1994, Niemelä and Mattson, 1996). Some of the most important introduced forest insects, pathogens, and plants are presented in Table 14.3 (Liebhold et al., 1995). Several have had devastating and lasting effects, including the chestnut blight fungus, white pine blister rust, the Dutch elm disease fungus and its bark beetle, and the gypsy moth. In addition, of the hundreds of

Table 14.3 Selected List of Important Forest Insects and Diseases Introduced into North America that have Severely Disrupted Native Ecosystems and Decimated or Displaced Native Species.

Agent	Latin Name	Origin	Hosts	Tissue Attacked
Insects				
Gypsy moth	*Lymnatria dispar*	Europe, Asia	Hardwoods	Foliage
Winter moth	*Operopthera brumata*	Europe	Hardwoods	Foliage
European pine sawfly	*Neodiprion sertifer*	Europe	Pines	Foliage
Balsam woolly adelgid	*Adelges piceae*	Europe	True firs	Phloem
Hemlock woolly adelgid	*Adelges tsugae*	Asia	Hemlocks	Foliage
Diseases				
Chestnut blight	*Cryphonectria parasitica*	Asia	Chestnuts	Cambium, phloem
White pine blister rust	*Cronartium ribicola*	Europe, Asia	White pines	Needles, stems
Beech bark disease	*Nectria coccinea var. faginata*	Europe	Beech	Bark, cambium
Dutch elm disease	*Ophiostoma ulmi*	Europe	Elms	Xylem, phloem
Plants				
Banana poka	*Passiflora mollissina*	S. America		
Australian pines	*Casuarina spp.*	Australia		
Brazilian pepper tree	*Schinus terebinthifolius*	S. America		
Faya tree	*Myrica faya*	mid Atlantic islands		
Melaleuca	*Melaleuca quinquenervia*	Australia		

Source: After Liebhold et al., 1995. Reprinted with permission of the Society of American Foresters.

intentionally introduced exotic woody plants many have become serious problems; for example, the Japanese honeysuckle in the southeastern United States, the Australian melaleuca tree in Florida, and the Asian kudzu vine in the South. In Hawaii, nearly 900 plant "pest" species have become established and have significantly altered the native vegetation (Cuddihy and Stone, 1990; Liebhold et al., 1995).

The invasion process is defined by arrival, establishment, and spread. Traits that enhance establishment of exotics are: high reproductive rate, wide host preference, tolerance of climatic extremes, efficient mate location, and high genetic or phenotypic plasticity (Liebhold et al., 1995; Reichard and Hamilton, 1997). Ever-increasing urban forests are often first sites for invasions. Ecologically, open and disturbed forest sites favor establishment and spread of introduced organisms. Fire exclusion, also due to human activities, has also put many native species at a disadvantage in competing with mesic native and alien invaders alike. The threat to forest ecosystems by exotics is considered in detail by Liebhold et al. (1995) along with case histories of major pests, their impact, and management recommendations. These introductions represent a major environmental and socioeconomic problem that is likely to escalate in the future.

Suggested Readings

Bendell, J. F. 1974. Effects of fire on birds and mammals. In T. T. Kozlowski and C. E. Ahlgren (eds.), *Fire and Ecosystems*. Academic Press, New York.

Herms, D. A., and W. J. Mattson. 1992. The dilemma of plants: to grow or defend. *Quart. Rev. Biol.* 67:283–335.

Janzen, D. H. 1971. Seed predation by animals. *Ann. Rev. Ecol. Syst.* 2:465–492.

Kay, C. E. 1994. Aboriginal overkill, the role of Native Americans in structuring western ecosystems. *Human Nature* 5:359–398.

Langston, N. 1995. *Forest Dreams, Forest Nightmares, The Paradox of Old Growth in the Inland West*. Univ. Wash. Press, Seattle, WA. 368 pp.

Liebhold, A. M., W. L. MacDonald, D. Bergdahl, and V. C. Mastro. 1995. Invasion by exotic forest pests: a threat to forest ecosystems. *For. Sci. Monogr.* 30. 49 pp.

McDonnell, M. J., and S. T. A. Pickett. 1993. *Humans as Components of Ecosystems*. Springer–Verlag, New York. 364 pp.

Mattson, W. J., and N. D. Addy. 1975. Phytophagous insects as regulators of forest primary production. *Science* 190:515–522.

Perlin, J. 1989. *A Forest Journey, The Role of Wood in the Development of Civilization*. Harvard Univ. Press, Cambridge, MA. 445 pp.

Regal, P. J. 1977. Ecology and evolution of flowering plant dominance. *Science* 196:622–629.

Whitham, T. G., and S. Mopper. 1985. Chronic herbivory: impacts on architecture and sex expression of pinyon pine. *Science* 228:1089–1091.

Whitney, G. G. 1994. *From Coastal Wilderness to Fruited Plain*. Cambridge Univ. Press, Cambridge, MA. 451 pp.

CHAPTER *15*

FOREST COMMUNITIES

In Chapter 1, we emphasized that the focus of forest ecology is, or should be, on two first-order objects-of-interest, the organism and the ecosystem. Ecosystems and organisms are conceptually alike in their three-dimensional integrity; each has an inside and an outside, and thus we can study their physiology and ecology. However, communities and populations are categorically different. Communities (aggregates of organisms) are contained in ecosystems as one compositional component, but they are not ecosystems. Animal and plant organisms and their fluctuating populations make up the biotic component of forest ecosystems. Animals as ecosystem components were considered in Chapter 14. In this chapter, we consider aggregates of plants as ecosystem components. These collections of plants form the vegetative cover of forest ecosystems—from the tree dominants to the groundflora layer of shrubs, herbs, bryophytes, lichens, fungi, and algae. In particular, we will examine the community concept, communities as parts of landscape ecosystems, historical views of community, mutualistic and competitive relationships among members of plant communities, and the vertical structure of forests especially related to the ability of plants to survive and grow in the forest understory. These considerations lead to a discussion of disturbance in Chapter 16 and ecosystem change and succession in Chapter 17.

COMMUNITY CONCEPT

The concept of **community**, simply stated, refers to the collection or assemblage of organisms within a particular area at a given time. If only plants are considered it is a plant community; if all organisms are considered, it is a biotic community or **biome**. Community is a generic term of convenience that is used to designate sociological units of any degree of extent and complexity (Cain and Castro, 1959). Thus the aggregate of plants over a broad area may be considered a community (the plant formation or the plant component of a biome) as well as a local collection of plants associated with a specific site. Beyond the co-occurrence of plants that occupy space and have a spatial boundary, opinions vary as to the "nature" of the community.

The early concepts of plant communities mainly emphasized co-occurrence or mixture (Clements, 1905, p. 316) of species, physiognomy (life form), its organism-like properties, and its classification as an idealized "type." In their detailed study of the community concept, Shrader-Frechette and McCoy (1993) observed that, by the middle of the 20th century, the community concept was a hodge-podge of ideas. In the second half of the century, the dominant concept was focused on communities "as units of interacting species and habitats." Note the addition of the non-community element, "habitat." Also at this time communities were defined as: distinct ecological units, ecological units of every degree, and units of dynamic stability. Whittaker (1975, p. 359) defined plant communities as living systems. Over 40 years ago Cragg (1953) observed that the meaning of community was limitless, ranging "from a piece of shorthand denoting an assemblage of organisms to something endowed with the attributes of organization which, in the absence of factual support, rivals the daydreams of the alchemists." McIntosh (1993) observes that definition and delineation of the community have been the despair of many ecologists. Why is the concept of a plant community so contentious?

The key to confusion about the nature of "community" is that communities are not ecosystems, ecological units, or living systems. Plant communities are aggregations of organisms that happen to occupy a common segment of Earth space. They are composed of individuals and species whose ecological amplitudes, mutualistic relationships, and competitive abilities allow them to coexist (Rowe, 1984a). Communities are inseparable parts of landscape ecosystems and not systems themselves as are organisms and ecosystems. As emphasized in Chapter 1, plants do not stand on their own; they evolve and exist in the context of ecological systems that confer those properties called life.

A community is a taxonomic construct, an entity or assemblage defined by the spatial association of organisms—neighbors in the same space. As such, they can be classified or arranged in many ways depending on purpose. In this regard, the scale at which vegetation is viewed is an important consideration (Rowe, 1966; Vasileich, 1968). When viewed at the broad scale of subcontinents and regions, no two units within a subcontinent appear alike (see Figures 2.4, 2.5, 2.7). However, if a large and complex area is divided again and again, the scale of differences between units is reduced and the possibility of finding similar communities in limited space is increased. For example, the recurring ecosystems in local areas illustrated in upper Michigan (Figures 2.12 and 13.11) are similar in vegetative cover of tree dominants and groundflora. In brief, those whose sample areas that are large will inevitably discover uniqueness; those whose sample areas that are small, numerous, and in the same vicinity are more apt to discover that essential of classification, *acceptable* alikeness (Rowe, 1966).

The problem of community definition is traceable to the development of concepts with primary focus on the superficial covering of the landscape, i.e., vegetation, rather than the landscape ecosystem. Plants and animals evolve together with site-specific environments (Chapter 4) as integral parts of ecosystems. The perceived individualistic nature of plant communities, therefore, is not solely a plant property but one of ecological systems, past and present, in which plants evolve and are a part. Community composition, physiognomy, vertical structure, and population fluctuation are not so much the result of direct plant to plant interactions as they are of the effects of site factors on plants and their reciprocal interactions. Although we may regard plant life forms, patterns of species distribution, and competitive abilities as "plant" properties, they have been in large part determined through interactions with the sites that support them. Let's examine this interrelationship by using a widely occurring circumboreal plant community.

It is common knowledge that certain patterns of forest composition characterize extensive areas of the Earth. A regional forest ecosystem dominated by spruce and fir trees, with associated pine and larch, extends around much of the Earth's boreal zone. Despite its heterogeneity, the forest vegetation has a characteristic physiognomy and composition that makes it immediately recognizable as a "spruce–fir" community, or regional forest type. Furthermore, many of the smaller plants and many of the animals found in one spruce–fir forest will be found in other spruce–fir forests in different geographic locations. True, the individual species of spruce, fir, other plants, and animals will vary from continent to continent and within a continent. Regional and local climate differ from place to place, and because site conditions are not always the same, the tree, shrub, and herb associates of spruce and fir may be different from place to place, the relative proportions of spruce and fir may differ as well, and the races of these species will vary from place to place (Chapter 4). Obviously, no two communities or ecosystems are exactly alike. Nevertheless, it is clear that a common denominator exists, i.e., an acceptable likeness, so that when we speak of the boreal spruce-fir forest we convey an immediately recognizable concept and mental image to others.

However, this mental image or concept, although named by tree species, is not simply vegetative cover, but regional ecosystems with harsh climate and cold soil, supporting complex biotic communities dominated in the overstory layer by spruce and fir trees. Because we conventionally name both regional and local ecosystems by dominant trees (oak–hickory forest, beech–maple forest), we may loose sight of the key focus—the landscape ecosystem—of which the community is only one part. In addition, being labeled (i.e., assigning it to a class) may suggest a uniformity that is never realized on the ground. Because sites and communities are inseparable it is important to *ground* communities when we examine their composition, layered structure, and their spatial distribution. Several examples are introduced in the next section.

Grounding Communities

At broad and local scales we conceive the landscape as ecosystems; communities are notable and integral parts of them. Thus it is useful to examine community occurrence, composition, and structure in relation to their supporting ecosystems. Although these broad vegetative units may seem discretely bounded and particulate, ecosystems and their communities form more or less continuous patterns or gradients in nature, typically following closely those of changing site conditions.

FLORIDA KEYS
Regional ecosystems and their communities, although distinct from others of equal rank, are actually quite heterogeneous. If we scale down to the local level we see this variation and the close match of site and the vegetation it sustains. In the Florida Keys (Figure 15.1), the plant communities of dry tropical forest, in areas undisturbed for at least 50 years, are shown as parts of the landscape ecosystems (here termed ecological site units) as mapped units along the physiographic diagram (Ross et al., 1992). The community structure and composition reflect the complex of factors and factor gradients, including physiographic position, hydrology, soil depth, salinity, frequency of tidal inundation, fire regime, and hurricane effects.

The map of ecosystem units (top diagram in Figure 15.1) emphasizes the pattern of communities related to the physical configuration of the island. This view shows spatially

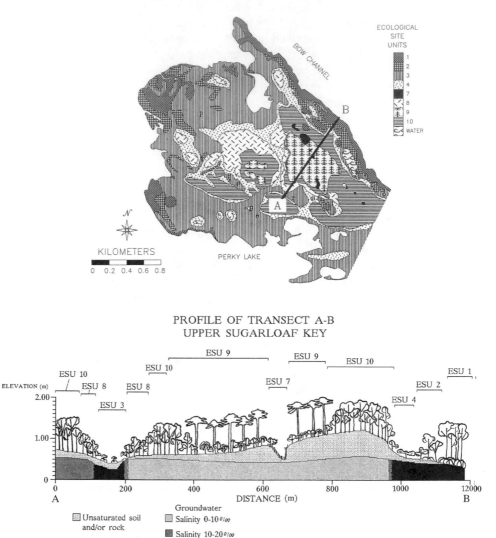

Figure 15.1   Ecological site units (landscape ecosystem types), geomorphology, and ground water characteristics for Upper Sugarloaf Key, Florida. Upper diagram is a map of the ecological site units (ESU), and the lower diagram is a cross section of part of the island along the transect line A–B shown on the map. (After Ross et al., 1992.)

patterned ecosystems as if they were all discrete units. However, we know that, on the ground, the concrete systems and their communities vary from place to place; they are not homogeneous. Nevertheless, recurrent examples of each ecosystem are similar enough in their multiple factors and their spatial occurrence in the island landscape to be grouped together as a unit. Also, each ecosystem and its community would grade more or less gradually or abruptly into ones surrounding it. The nature of this transition zone or **ecotone** would depend on site, disturbance, and community factors. The relative degree of continuity/distinctness of ecosystem types and their community types is portrayed in the physiographic cross section in Figure 15.1. Notice that there are some relatively abrupt transitions between ecosystems (for example between ESU 1 and 2, 7 and 9, 3 and 8 on the physiographic diagram) and also gradual changes as well (between ESU 9 and 10, 8 and 10). The communities differ markedly in species composition and structure. The number of plant species in the respective communities varies greatly—from the relatively species-poor mangrove units (unit 1, peaty mangrove forest; unit 2, peaty mangrove woodland; unit 3, dwarf mangrove mud flats) with only 5 species to the pine rockland forest (unit 9) with 35 species.

INTERIOR ALASKA

Going now to the far north, communities from the taiga of interior Alaska are illustrated in Figure 15.2 (Viereck et al., 1984). In this case, a vegetation classification was first developed. Then the authors undertook development of an ecosystem-type classification by including physiography, soil type, and other environmental factors that were associated with the vegetation. In Figure 15.2, a selected group of communities near Fairbanks illustrates the diverse array of tree-dominated communities (stands) from river floodplain and relatively warm southern slopes to the coldest north slopes at tree line. Notice how closely the communities are related to the physiographic (landforms of river valley and upland flats

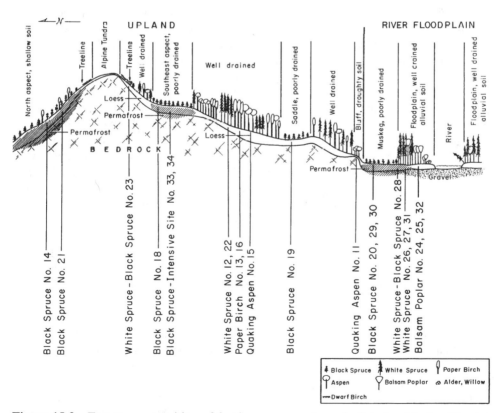

Figure 15.2 Forest communities of landscape ecosystems in interior Alaska. Physiographic diagram of the topography, landforms, vegetation, and parent material in the Fairbanks, Alaska area with locations of sampled stands. (After Viereck et al., 1984.)

and slopes, aspect, elevation) and edaphic conditions. Both early-successional and late-successional stands are included in the stands sampled. Deciduous angiosperms (balsam poplar, quaking aspen, paper birch) that form early-successional stands following fire occupy the warmest sites. The aspen and birch communities (for example, stands 15, 13, and 16 shown in Figure 15.2) eventually succeed to white spruce on the well drained, loess–soil ecosystem type. Many kinds of black spruce-dominated ecosystems, both lowland and upland, are evident. Again, notice that in this landscape continuum some communities are relatively abruptly separated due to interactions with their supporting site features (for example, balsam poplar on alluvial soil and the adjacent white spruce on permafrost; the black spruce muskeg versus the droughty bluff with quaking aspen). Relatively continuous change in site and community is also observed, especially in the transition from forest to treeline and alpine tundra.

SOUTHERN ILLINOIS

In a totally different geologic and macroclimatic region in midcontinent, we find again the patterning of local communities closely related to the interactions of geologic substrate, slope position, soil, microclimate, and disturbance regime. In the Shawnee Hills of southern Illinois, local landscape ecosystems and their communities may be distinguished on a physiographic cross section of ridge and valley terrain (Fig. 15.3; Fralish et al., 1991). In addition to showing their spatial pattern, this example illustrates the dynamic nature of ecosystems and their communities.

In Figure 15.3*a*, the dominant forest cover types in presettlement time (prior to 1800) are shown on the physiographic diagram. In contrast, "old-growth" forest cover types of 1988, that developed during human settlement characterized by fire suppression, are shown in the same physiographic setting in Figure 15.3*b*. Geologic substrate, landform, and local topography of this area change relatively little on a time scale of hundreds to several thousands of years and form the basis for distinguishing plant communities that reflect these site conditions. Using these features together with our *a priori* ecological understanding of their interactions with soil and vegetation, we can distinguish and map the ecosystem types of presettlement time along the continua shown in Figures 15.3: (1) shallow soil over bedrock, (2) rocky upper slope, (3) loess-covered ridge, (4) upper north slope, (5) lower north slope, (6) alluvial floodplain, (7) terraces, and (8) mid-south slope.

In presettlement time, site conditions, fire regimes, and vegetation interacted to produce the distinctive communities along the physiographic gradient shown in Figure 15.3*a*. Both gradual and abrupt changes between ecosystems and their cover types are observed. As expected, eastern redcedar and post oak dominate the driest sites, mixtures of white and black oak dominate the fire-prone ridge and south slope, and river birch (Bn) and American sycamore (Po) dominate the alluvial floodplain. Due to a relatively high fire frequency, white–black oak dominated three ecosystems: loess-covered ridge, upper north slope, and lower north slope.

Examination of the old-growth forest cover types that exist today on the area (Figure 15.3*b*) demonstrates that, for the most part, they are the same as those in presettlement time. Landform and soil are relatively unchanged, and the spatial position of ecosystem types has not changed. However, due to lack of fire, more mesophytic species—red oak and sugar maple—have replaced the white–black oak overstory of the upper north slope and lower north slope ecosystem types, respectively (Figure 15.3*b*). Also, red oak has replaced black oak in the loess-covered ridge ecosystem. Changes in these forest cover types illustrate the dynamic nature of vegetation in relation to human-caused disturbance, in this case the lack of fire (Fralish et al., 1991). However, tree communities of other ecosystems are virtually unchanged. Thus the composition of forest communities now and

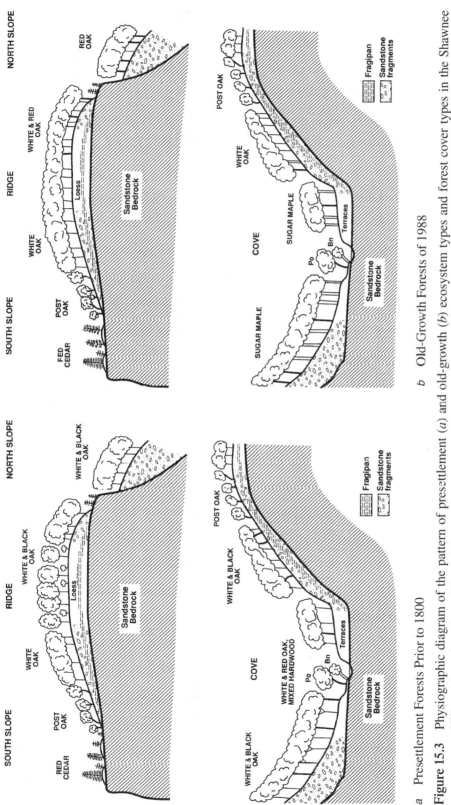

a   Presettlement Forests Prior to 1800

b   Old-Growth Forests of 1988

Figure 15.3   Physiographic diagram of the pattern of presettlement (*a*) and old-growth (*b*) ecosystem types and forest cover types in the Shawnee Hills region of southern Illinois. Ecosystem types (cover types) include: (1) shallow soil over bedrock (eastern redcedar), (2) rocky upper slope (post oak and white oak), (3) loess-covered ridge (white and black oaks), (4) upper north slope (white oak), (5) lower south slope (white oak), (6) stream terraces (white and northern red oaks, sugar maple, basswood), (7) alluvial floodplain (sycamore-Po and river birch-Bn), and (8) mid-south slope (white and black oaks). Presettlement communities before 1800; old-growth communities as of 1988. (After Fralish et al., 1991. © 1988, American Midland Naturalist. Reprinted with permission of the American Midland Naturalist.)

in the past is highly dependent on the interaction of specific site conditions and disturbance regimes that influence the competitive ability of the tree species of the Shawnee
Hills landscape.

In summary, plant communities are visually distinctive features of the landscape. Understanding their composition, stand structure, dynamics, and spatial distribution for any
landscape requires a solid understanding of the site conditions, disturbance regimes, history, and processes of the landscape ecosystems of which they are an inherent part.

VIEW FROM THE PAST: COMMUNITY CONCEPTS

The modern paradigm of the community as one component of landscape ecosystems differs from past and related present concepts of the community because of its first-order
focus on the ecosystem. Traditionally, the primary focus of plant ecologists is on plant
species and communities, their distribution, composition, and classification, and secondarily their relationship to site factors. Over the years, ecologists have established a hierarchy of communities and assigned special names to distinguish them. Because many of
the terms and underlying concepts are entrenched in the literature and used today, their
consideration is instructive.

The early emphasis of early plant geographers, such as Humboldt (see Botting, 1973;
McIntyre, 1985), Schimper (1898), Gradmann (1898), and Warming (1909) was on the
geographical distribution of plants (i.e., phytogeography) and classification. For example,
Warming (1909), regarded by the noted American ecologist Henry Cowles (1909) as the
father of modern ecology, developed his classification based on water relations: water
plants and land plants, with the latter subdivided further into 12 primary groups. A keen
field ecologist and botanist, Cowles was dissatisfied with Warming's classification and pioneered a physiographic classification of plant communities of the Chicago area (Cowles,
1901). In this classic work, Cowles recognized that physiography best accounted for the
distribution of species and communities because of its influence on climate, soil, and geomorphic change (erosion, deposition)—key factors determining local plant occurrence.
Although his stated focus was a community classification, his emphasis was essentially on
ecosystem units and their geomorphic change to which characteristic plants and communities were dynamically related. Cowles physiographic approach to plant ecology was
succeeded by groups in North America and Europe emphasizing community types, floristics, and vegetational continua.

Schools and Terminology

Two schools dominated the North American field of community ecology in the first half
or two-thirds of the 20th century. In the first part, Frederic E. Clements (1905, 1916;
Weaver and Clements, 1929) developed a school of dynamic ecology based on plant succession. It was followed by the Curtis–Whittaker group emphasizing the abstract vegetative continuum (Curtis, 1959; Whittaker, 1962, 1975; Fralish et al., 1993). Clements'
broad understanding of the historical development of the time, his ecological insights (especially regarding physiography and site–plant relations), and emphasis on succession
and vegetation dynamics overshadowed his unacceptable theories of vegetation as an organism, rigid classification, and monoclimax (see Chapter 17). Like many of his predecessors, Clements' major unit of vegetation was the plant **formation**—the most comprehensive kind of plant community. It is characterized by a given physiognomic form that
recurs on similar sites (i.e., grassland, temperate deciduous forest, tropical rain forest).

Historically, each formation was seen as composed of various other distinctive community-types, termed **associations**. Thus the Deciduous Forest Formation may be interpreted as composed of several different associations; beech–maple, oak–hickory, among others. Examples of this broad community classification are several. Braun (1950) mapped and described such associations as "forest regions" (Figure 2.7). Küchler (1964) published a non-hierarchical classification of the "potential" vegetation units of the conterminous United States. In Canada, Rowe (1972) has described the forest regions of Canada, and Strong et al. (1990) have summarized the vegetation zones of Canada. A worldwide classification of plant formations was published by Mueller-Dombois and Ellenberg (1974; Chapter 22).

Europeans and Americans followed markedly different courses in the use of the term, association (Mueller-Dombois and Ellenberg, 1974). Partly because the term, association, was strongly identified with its rigid European definition and usage and partly because in Clementsian usage it related only to "climax" communities, the term, association, went into disfavor among American ecologists and "community" was read in its place. Even the "correctness" of community is questioned today, and alternatives such as plant assemblage and patch are used. A hierarchical system with the narrow definition of association as its core is preferred by many European ecologists involved in describing and classifying vegetation. Other ecologists not so involved also may use the term, association, but in a much broader sense without organismal overtones. The term, community, is not restrictive by itself, undoubtedly accounting for its wide usage. Its particular meaning in a given study depends on the context in which it is used and the modifiers applied to it—oak–pine community, early-successional community.

In the Clementsian system and European schools of phytosociology, finer subdivisions of the association are recognized and named. The subdivision of the community most widely used by American ecologists is the "layer." Thus the spruce–fir formation of northeastern United States and eastern Canada includes the red spruce–balsam fir community, which may in turn be subdivided into an overstory or tree layer, an understory or shrub layer, and a ground-cover layer.

The term **forest type** refers to a forest community defined only by composition of the overstory. Since the community is or should be defined by the total plant complement, its name often takes into account characteristic ground-cover plants. For instance, in northern Idaho and eastern Washington the ponderosa pine type is subdivided into six associations according to whether the undergrowth is characterized by grasses on stony, coarse-textured soils (*Agropyron spicatum, Festuca idahoensis, Stipa comata*) or shrubs on heavier-textured, more fertile soils (*Purshia tridentata, Symphoricarpos albus*, and *Physocarpus malvaceus*; Daubenmire and Daubenmire, 1968).

CONCEPTS OF CLEMENTS AND GLEASON

The prevailing ideas of the plant association as a natural unit for study and classification were deeply embedded in the practice of plant ecology of the early 20th century. Clementsian ecology was a very tidy and orderly system—far more orderly than nature itself (Egler, 1968). At that time species' populations were thought to be more or less uniform: thus it was easy to believe in uniformity of plant formations and communities. This belief is unacceptable today and came under attack early in the 20th century. The most extreme view was that of Clements who likened the formation and association to a complex organism that arises, grows, matures, reproduces, and dies. The final, stable, self-maintaining and self-reproducing state of vegetational development was the **climax** (climatic maximum). According to Clements, only one true or climatic climax (formation or association) occurred in a climatic region, but common sense forced him to recognize other units such as pre-climax, post-climax, and dis-climax.

The less extreme view held by many plant ecologists, including Cooper (1913, 1923) and Nichols (1923), and European phytosociologists, was that the association was not an organism, but a series of separate similar units, variable in size but repeated in numerous examples, and comparable to a species. These views were challenged by Henry A. Gleason who developed the "individualistic concept" of ecology and the plant association (Gleason, 1917, 1926, 1939). He objected to the organismal concept of vegetation and the classification of vegetation into rigid "pigeon-holes" of seemingly uniform types. He pointed out that communities are not uniform but are varying throughout both space and time due to chance and environmental effects. At a time when Clementsian ecologists and European phytosociologists emphasized uniformity of communities, Gleason emphasized their variation. He described the vegetation unit as a temporary and fluctuating phenomenon, dependent, in its origin, its structure, and its disappearance, on the selective action of the physical environment and on the nature of the surrounding vegetation (Gleason, 1939).

Gleason was a first-rate field botanist, and he advocated a floristic, individualistic approach to vegetation (Nicolson, 1990). Contrary to Cowles and Clements, he emphasized species composition as the most fundamental feature of plant communities and that plants themselves are, in many cases, the controlling agents in the environment (Gleason, 1910, p. 35). Gleason used "individualistic" in two senses: (1) for individuals that make up a species, and (2) for species as individualistic components of a community (Egler, 1968). He emphasized that the individualistic concept is revealed in communities as we see them—the juxtaposition of individuals (the aggregate or mixture of species) resulting from two major factors: migration and environmental selection. In noting that every species is a law unto itself, he goes on to say that its distribution in space depends upon its individual peculiarities of migration and environmental requirements. Gleason (1936) noted that two factors determine the true nature of the association: "the physical environment, which decides what kinds of plants may exist in it, and the living plants themselves, which tend to control and to modify the physical environment. Individuals of a species grow wherever they find favorable conditions, disappear from areas where the environment is no longer endurable, and occur in company with any other species of similar environmental requirements." Noticeably lacking in Gleason's writings is consideration of mutualisms that are pervasive in communities, suggesting that individuals and species aren't such laws unto themselves as one might think.

Of various examples of community variation, Gleason (1939) observed: "At the Biological Station of the University of Michigan, the aspen association, with a single continuous community some six miles long, exhibits demonstrable variation from one end to the other, with no visible reason." Recently, we mapped and described both the vegetative cover types (i.e., plant communities) and the landscape ecosystem types of this 4,000-ha Biological Station (Pearsall, 1995). Along a five-mile stretch of the main road through the midst of the "aspen community," we determined the number of mapped bigtooth aspen communities (only those immediately adjacent to the road) and the ecosystem types in which they occur. Although one may perceive, as Gleason did, that there is a single broad aspen community, we mapped 6 different aspen-dominated communities and 17 different ecosystem types. The occurrence of six communities demonstrates the variable nature of communities that Gleason recognized across this aspen-dominated landscape, which was subjected to the major disturbances of pine logging and post-logging fires. However, a major visible reason for the variation is the marked difference in site conditions (including physiography, soil texture, drainage, and nutrient availability) that occur along this small strip. These communities are, therefore, not solely coincidences of chance establishment but are due to variations in physical environment as well as the nature of the vegetation of the surrounding area, chance establishment events, and the previous forest history of the area.

The main strength of Gleason's argument for individuality of communities is at the broad regional scale where differences are most evident. Within a region he asserted that similar environments result in similar floras that could be classified together. Recognizing communities as landscape features, he noted that abrupt and gradual changes in environmental factors were many times the cause of like changes among different communities. Gleason's floristic approach was timely in stressing variation among communities and the importance of chance events in the plant colonization of sites, and the influence of vegetation on the site. However, a fundamental weakness of his approach, and others before and after, was the primary focus on species and community and not the ecosystem of which the community is a part.

Like organisms, communities don't stand on their own. From a landscape ecosystem perspective they aren't necessarily individualistic in that they are influenced in many ways by other communities of their ecosystem neighbors that surround them. Communities of dry-mesic ecosystems protected by swamp-ecosystem firebreaks (Chapter 10) have quite different community composition and structure compared to those on unprotected sites. Also, communities of mesic sites may be burned and markedly changed by fires spreading from adjacent flammable communities. Communities of upper-slope ecosystems influence those on the adjacent lower slopes through movement of wind-blown leaf litter and by water and nutrients flowing downslope. Many such exchanges, from one community to another through landform and soil-based linkages, provide a new perspective on the individualistic concept of the community.

PHYTOSOCIOLOGY IN EUROPE

About the same time Clementsian ideas were taking root, a different kind of community (analogous to the species in plant taxonomy), the "association," was defined in a technical sense by ecologists of the Zürich-Montpellier school of phytosociology and adopted at the International Botanical Congress of 1910: *An association is a plant community of definite floristic composition, presenting a uniform physiognomy, and growing in uniform habitat conditions.* The association is the basic unit of a hierarchical system of vegetation classification, the primary goal of the school of phytosociology identified with Braun-Blanquet (1921, 1964). Braun's approach was essentially established by 1921 when he urged a floristic instead of an ecological classification system (van der Maarel, 1975). Braun's main idea was that plant communities are conceived as types of vegetation, recognized by their floristic composition. According to Braun, the full species composition of communities better express their relationships to one another and to their environments than any other characteristic (Westhoff and van der Maarel, 1973). In this system, the basic unit of classification, the association, has a type specimen complete with author, date, and description analogous to that of a plant species: for example, "*Abieti-Fagetum* Oberdorfer 38;" and the "*Galio-Carpinetum* (Buck-Feuct 37) Oberdorfer 57 em. Th. Müll. 66." Braun never adopted an organismal approach (van der Maarel, 1975), and there is little justification for the naming of communities as if they were species.

Despite shortcomings, the floristic approach has made important contributions to the description and classification of vegetation—hundreds of thousands of communities have been described (Ellenberg, 1988). It is applied widely in eastern and central Europe (Oberdorfer, 1990) in one form or another and used by many plant sociologists throughout the world. Excellent descriptions and perspectives of the Braun-Blanquet approach and the Zürich-Montpellier School of phytosociology are available (Poore, 1955; Becking,

1957; Whittaker, 1962; Shimwell, 1971; Westhoff and van der Maarel, 1973; Mueller-Dombois and Ellenberg, 1974; van der Maarel, 1975).

Continuum Concept

A second major school of plant ecology in the United States, the **continuum concept** of vegetation, includes the continuum approach of John T. Curtis and associates and the **gradient analysis** approach of Robert Whittaker (Curtis and McIntosh, 1951; Curtis, 1959; Whittaker, 1962, 1967; Cottam and McIntosh, 1966; McIntosh, 1967, 1968, 1993). The continuum approach extended Gleason's floristic-individualistic concept of the community, which by the 1940s had been overshadowed by the Clementsian and European association concept. The concept of vegetation continuum was conceived and developed in reaction to the concept of the association as a relatively discrete unit by which vegetation was classified into types i.e., **community types**.

McIntosh (1968), spokesman for continuum adherents, defined the community-type approach as: "the idea that species occur in association with others, that such aggregations have adaptations holding them together as units, and that vegetation is comprised largely of such units with relative limited areas, boundaries, or ecotones separating them." In contrast, the continuum was defined by Curtis (1959) as "An adjectival noun referring to the situation where the stands of a community or large vegetational unit are not segregated into discrete and objectively discernible units but rather form a continuously varying series." Daubenmire (1966) dissented, asserting that "There is no denying that vegetation presents a continuous variable by virtue of ecotones; the argument hinges on the existence or absence of plateau-like areas being of sufficient similarity to warrant being designated as a class." His work in undisturbed forests of northern Idaho had shown that "typal communities" or "noda" that formed the basis of his habitat type approach (Chapter 13, Figure 13.5) could be readily identified in old-growth forests but that disturbance by logging weakened the discontinuities.

There is no question, even by continuum critics, that vegetation or floras are continua (Daubenmire, 1966). This is true for a number of reasons: (1) environmental conditions change in space at varying rates affecting the establishment and, through competition, the composition of the vegetation; (2) genecological variation of tree species is typically clinal, and in addition a given genotype often has wide phenotypic plasticity to tolerate a number of environments (Chapter 4); (3) historical and chance events may reinforce points 1 and 2 so that no communities are exactly alike; and (4) a continuum of successional change in time is superimposed upon changes in composition in space. Furthermore, depending on methods of field sampling or analysis, discontinuity or relative continuity may be demonstrated for a tract of vegetation in the field.

Continuum adherents were opposed to classification, but agreed (McIntosh, 1967) that "if classification is urged simply as a desirable convenience for mapping or information storage (Daubenmire, 1966) for practical ends and specific purposes, the classification being arbitrary and directed to these specific ends (Rowe, 1960), there is no contest." Concerning the perceived dichotomy between classification and ordination,[1] Lambert and Dale (1964) pointed out a common misconception at the time—that classification was only properly applicable to "discontinuous" data, while ordination techniques were more appropriate to continuous vegetation. In contrast, they emphasized strongly that there is no *a priori* reason why the use of either method should be restricted in this way: classified

---

[1]An ordination is an arrangement of species, communities, or environments in sequence along axes with their respective properties determining their relative position to one another.

units can be ordinated, and ordinated units can be classified. Which method to adopt is entirely a matter for the user, irrespective of any subjective concept of the "real" nature of vegetation.

Although the controversy was cast as a dichotomy of two extremes, one would think that a middle ground would soon prevail— that studies of communities in nature would reveal that in many areas gradually changing species aggregations would be found and in other areas relatively abrupt changes and acceptably similar compositional types would be revealed (just as Gleason had observed). However, this was not the case because the crux of the continuum argument was the arrangement of communities in *abstract space*, not necessarily in nature. Recognizing this situation, Lambert and Dale (1964) asserted that continuum and continuity were confused because they were applied both to the actual continuity on the ground and to the continuity in abstract models. Both Curtis (1959) and McIntosh (1967, 1993) recognized that discontinuities or abrupt changes of vegetation type existed and were familiar to all field ecologists. But these were "irrelevant" and not at odds with the abstract continuum.

In retrospect, the continuum school succeeded in changing the viewpoint of many plant ecologists, encouraged countless compositional studies of vegetation, and initiated the introduction of increasingly sophisticated multivariate analyses to classify and ordinate species and communities. However, the approach also generated controversy and confusion (Dansereau, 1968; Austin, 1985) followed by loss of interest in generalizations about plant communities. Noy-Meir and van der Maarel (1987) observed that this may reflect the feeling that the processes in plant communities are probably too intricate and complex to expect any general pattern to be observable at the community level. The basic lesson, however, is that the plant community is an unrewarding subject of study. The plant assemblage is incomplete, and like climate, physiography, or soil, is only one part of the landscape ecosystem. Such an assemblage is not a functional system and therefore has no processes. In its abstraction of organisms from the physical space they occupy, the concept of community frees us from the *necessity* of physiological and ecological thought (Rowe, 1969).

Vegetation is not a thing-in-itself, to be studied in isolation. Plant communities are, among other things, assemblages of carbon-fixing, water-using organisms that function within and as parts of landscape ecosystems from which they derive their resources for growth. It is these communities as ecosystem parts that applied ecologists (foresters, land mangers, naturalists, environmentalists) seek to manage, conserve, and restore.

COMMUNITY AS A LANDSCAPE PROPERTY

The concept of the community as a landscape ecosystem property, as already described, is reinforced by Austin and Smith (1989) who reformulate the continuum concept. They note that the two ideas of community (i.e., community-type and continuum) are based on different frames of reference: (1) community as a landscape ecosystem property, and (2) continuum as an environmental concept referring to an abstract space. For example, Figure 15.4 shows a landscape with four species present. Species associations that may be recognized due to their frequency of occurrence along a transect from low to high elevation are A, AB, B, C, and D (seen along x-axis in Figure 15.4). The combinations BC and CD are regarded as transitions or ecotones. The "communities" composed of coexisting species are a result of the site-specific conditions and spatial pattern of this landscape. If we examine the elevational gradient, however, we find a continuum of species regularly replacing one another in the sequence A, B, C, and D with increasing elevation (right side

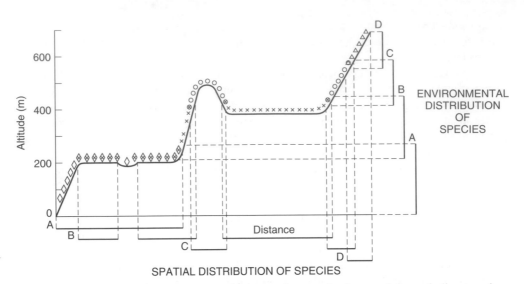

Figure 15.4.   Pattern of co-occurrence of four species on a landscape along an indirect environmental gradient of elevation. Note the distribution of species along the x-axis (Distance) illustrating distinctive species associations of A, AB, B, C, and D that occur at characteristic positions on the diagram. Then observe the continuous variation of composition (regular replacement of A, B, C, and D) along the elevational gradient on the right side of diagram. Elevation is an indirect gradient because its effect on growth is through the site factors of temperature, water, and nutrients. (After Austin and Smith, 1989. © 1989 Kluwer Academic Publishers. Reprinted by permission of Kluwer Academic Publishers.)

of diagram, Figure 15.4). Thus what we see as relatively distinct species associations horizontally along the x-axis of distance, can be viewed as a continuum along the abstract elevational gradient.

If the transect in Figure 15.4 were taken through an adjacent area where the first bench or terrace were 30 m lower, then the combination AB would become rare. At 170 m, only species A would be present. Similarly, if the second bench was 30 m higher, the combination BC would become common. On such a transect the "communities" recognized would be A, B, BC, C, and D with ecotones AB and CD. Generalizing from this example, communities are a function of the whole landscape and specific landscape ecosystems. Abrupt changes or gradual transitions may occur depending on the pattern of recurring ecosystems in a given landscape. Co-occurring groups of species can be recognized for any particular area with a recurrent pattern of landscape ecosystems. Labeling or classifying these communities is useful for communication, management, and research, but extrapolation of these communities to other regions will be accurate only if the regions have similar patterns of site factors (Austin and Smith, 1989).

The continuum in relation to elevation will be valid and applicable within similar transects of the same regional ecosystem, but not in a different region where the growth-influencing variables of climate (temperature, water) change in relation to elevation. Austin and Smith (1989) conclude that: (1) the community is a landscape property, i.e., the concept of a community of co-occurring species can only be relevant to a particular landscape and its pattern of combinations of site factors, and (2) the continuum concept applies to the abstract environmental space, not necessarily to any geographic distance on the ground or to any indirect environmental gradient (such as elevation). Details of this new model of the continuum concept are beyond the scope of this treatment, but are considered by Austin and Smith (1989) and Austin (1990).

EXAMPLES OF SPATIAL VARIATION IN FOREST COMMUNITIES

Discrete Forest Communities

Forests are not particulate by nature. They are probably most accurately considered in terms of gradient patterns, their segmented geographic appearance being due to local steepenings in rates of gradient change, at least in the absence of disturbance (Rowe, 1960). Thus whenever one type of forest community abuts another distinctly different type of forest community, it will be found that this abrupt change is related to an abrupt change in site conditions or to a completely different vegetational history of the two communities. The existence of discrete communities, therefore, is evidence of the existence of discrete differences in growing conditions, either now or in the past.

The boundary between two communities is usually a belt rather than a sharp line. It is a belt or zone, though, which may vary markedly in width. In the forest-grassland transition, there will always be an outer belt of forest that will be modified by the adjacent open areas, and an inner belt of grassland that will be modified by the adjacent forest. As mentioned above, the transition zone between two communities is termed an **ecotone**. It usually embodies some of the ecological features of the two communities, but often has a specific site characteristic of its own.

Many persisting, abrupt site differences can give rise to sharp forest-type boundaries. Among these are: (1) a sharp boundary between two geological formations giving rise to a sharp boundary between two soil types that differ markedly in water and nutrients they provide to vegetation; (2) a sharp boundary in landforms (for example, kettle-kame topography in glaciated landscapes, see Figure 2.12) where abrupt differences in soil drainage conditions occur, such as between a poorly-drained swamp and a well-drained upland; (3) a sharp boundary in topographic position affecting local climate, such as a knife-edged ridge separating a north from a south slope or an air dam impeding cold air drainage so that frost pocket conditions exist below the level of the dammed air; and (4) a sharp boundary in the structure of the vegetation affecting the local climate and soil conditions, such as a forest edge facing grassland, a shrub community impinging upon open rock surfaces, or a logging boundary where a clearcut patch abuts standing forest. Among the historical events that may give rise to sharp boundaries between plant communities are fires, tornadoes and other windstorms, salt spray from the sea, fumes from smelters, logging, and agricultural development of land.

The sharpest community boundaries often occur when abrupt physiographic changes occur that are also associated with soil drainage and microclimate—a situation described in the following section. Then, in worldwide context, we describe ecotones between forest and adjacent low-stature vegetation that have long attracted the attention of ecologists, particularly the forest–grassland ecotone and the alpine timberline.

COASTAL CALIFORNIA: GIANT AND PIGMY FORESTS

Soil scientist Hans Jenny (1961, 1980) recognized the need for an ecosystem concept embodying the joint development of soil, vegetation, and animal life with their interrelated feedbacks. Jenny and others (1980, 1969) described a remarkable example of diverse

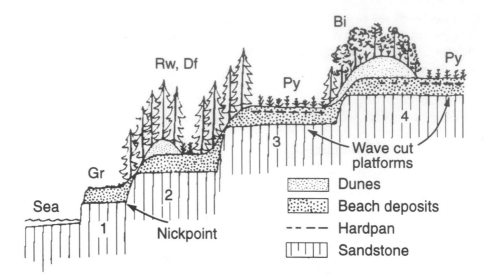

Figure 15.5. Physiographic arrangement of four marine terraces (1, 2, 3, 4) in the Fort Bragg, California area, with a young dune on the second and a very old dune on the fourth terrace. Gr = grassland, Rw, Df = redwood–Douglas-fir forest; Bi = Bishop pine forest; Py = pygmy forest. Horizontal distance is 4.8 km, vertical distance is 152 m above sea level. (After Jenny et al., 1969.)

ecosystems and communities related to landforms along the Mendocino coast of north-central California. During Pleistocene times, rising seas cut terraces into the prevailing sandstone rock, and retreating seas covered them with beach sands, gravels, and clays. Tectonic forces elevated the terraces, and nearly level features occur at elevations of approximately 30, 53, 91, 130, and 198 m above sea level (Figure 15.5). The hillsides and steep canyon slopes are dominated by redwood and Douglas-fir, whereas the three upper and oldest terraces exhibit distinctive hardpan soils and support a sparse, dwarf or pygmy forest of slender, stunted cypresses and dwarfed pines (Bishop pine and Mendocino White Plains lodgepole pine). As much as 25 percent of this ground is bare or covered with lichens. The soil is characterized by extreme acidity, spodosol development (see Chapter 11), and a thick hardpan layer of iron or clay. In late fall, a perched water table forms as water accumulates above the hardpan layer; the entire surface may be flooded. In late spring, the surface water table disappears by evapotranspiration and seepage, and in summer the hardpan dries out, hardens, and imparts extreme xeric conditions. The distinctive vegetation is probably the result of many soil physical and chemical factors associated with soil development in the parent material of the flat terrace surface.

In contrast, soil development has not proceeded as far on the lower two terraces. The lowermost terrace supports grass and pine forests, and the second terrace supports a pattern of different communities from redwood, Douglas-fir, and western hemlock to dwarf pines, depending on soil development. The communities of the terraces are therefore strongly dependent on physiography—the flat terrace form and its parent material on which soil development occurs.

Besides terrace and hillside landforms, sand dunes are present on all terraces. Depending on age and soil development, they exhibit three relatively distinct plant communities. On the lowest terrace, *recent dunes* are still moving inland and are stabilized by lupines and lodgepole and Bishop pines. Second-terrace dunes are thousands of years old and third-terrace dunes even older. Soils of these *young dunes* have developed favorable water and nutrient conditions to support magnificent forests of redwood, Douglas-fir, and grand fir. The *very old dunes* of the fourth and fifth terraces have highly acid and poor nutrient relations but no hardpan formation. Here the vegetation is dominated by Bishop pines with a dense understory of ericaceous species plus wax myrtle and chinquapin. Thus three kinds of dune ecosystem types and their communities are evident due to their age and soil development caused, in part, by the colonizing and developing vegetation itself.

FOREST-GRASSLAND ECOTONE

Abrupt changes between forest and grassland in the tropical and temperate zones may or may not be associated with abrupt changes in site (Beard, 1953). Once the forest edge has been established, such as by fire or land clearing, site conditions (both climatic and edaphic) within the forest may differ so substantially from those in the grassland as to perpetuate the forest border.

Often, grassland originates in forested country as a result of a fire that destroys the forest and creates an environment at the ground more suitable for the development of grasses than for the reestablishment of the forest. Once established, the grassland persists because of the inability of the adjacent forest trees to invade the site—whether due to the recurrent incidence of fires (Wells, 1965, 1970; Rowe, 1966; Veblen and Lorenz, 1988); to the failure of the tree seeds to penetrate the sod and reach a medium suitable for germination; to excessive root competition for soil water provided by the grasses; or to the absence of mycorrhizae (Langford and Buell, 1969). The alpine meadows of the western American mountains, the fingers of prairie extending up into the Black Hills of South Dakota, and the extension of the Prairie Peninsula east into Indiana and Ohio are all examples of fire-caused grasslands that have persisted for hundreds of years under climates suitable for tree growth.

This does not imply that the forest–grassland border is ever static. Invasion of one type by the other does occur. Grasslands within forested regions are being invaded by forest throughout the world, partly because of the improvement in modern fire-suppression techniques and partly because of a change in climate.

ALPINE TREE-LINES

The elevational limit of forests in mountain ranges often provides another spectacular forest edge. As with all ecological phenomena, the timberline is a result of the interaction of the trees and the site over a long period of time. Actually, ecologists distinguish three kinds of altitudinal tree-lines (Huggett, 1995). The forest-line or timberline is the upper limit of tall, erect trees growing at normal forest densities. Above this is the tree-line, the limit at which individuals recognized as trees (> 2 m) grow. Higher still lies the tree-species line, the point up to which tree species will grow, but in deformed habit (Krummholz or "crooked wood"). When two or three of these lines coincide, the discontinuity between trees and low-growing vegetation is very sharp. However, tree-lines are not always sharp (Armand, 1992).

Many causes have been ascribed to timberline formation (Daubenmire, 1954; Wardle, 1985; Huggett, 1995). Within the zone of stunted and recumbent trees, wind, snow blown by the wind, snow pack, and other factors produce an exposed and rigorous climatic zone near the ground through which trees cannot grow. One of the principal factors determining timberline location (as contrasted with climatic factors that cause dwarfing and recumbent growth) is heat deficiency during the growing season at high altitudes, which limits growth and winter-hardening of shoots. Wardle found that summer tempera-ture data bore out this conclusion for timberlines in New Zealand (Wardle, 1985) and in Colorado (Wardle, 1968) where timberlines are among the highest in the world despite desiccating wind and low winter temperatures. He also found that wind kept Colorado sites blown free of snow so that there was a lack of protection for seedlings during the winter and an absence of melt water in the spring to moisten the rocky, coarse-grained soils. In the Austrian Alps, Aulitzky (1967) reported that only the highest tree-line was governed by the 10 °C line of July temperature. In most situations, growth of trees was lower than this level due to other unfavorable factors: wind, snow depth, and snow dura-tion. The causes of tree-lines are much debated, but most ecologists believe that climate is their root cause (Huggett, 1995).

Merging Forest Communities

Whereas distinct plant communities, separated by transitional belts, reflect abrupt changes in site, major disturbance, and land history, gradual changes in site or vegetational history result in similar gradual changes in the composition of the forest. Gradual changes are characteristic of forests of a generally similar history over a geographical stretch of many kilometers (such as north to south or east to west). They also occur over gradual changes in elevation, aspect, soil water, and soil fertility. Furthermore, climate strongly affects for-est composition (Chapter 7), and the relative continuity of macroclimate in space and time favors gradual change.

Continua characterize the composition and structure within forests in the absence of an abrupt change in site or vegetational history. The great deciduous hardwood forest of the eastern United States stretching from the Gulf of Mexico to Canada is, at this broad scale, a great continuum within which distinctive regional ecosystems and communities can be recognized (Figure 2.8).

EASTERN DECIDUOUS FOREST—SOUTHERN APPALACHIANS
The forest of the eastern United States—characterized by deciduous hardwood species, but containing evergreen hardwoods and conifers in parts of it (Barnes, 1991)—shows many gradual changes at the broad regional ecosystem scale and at the local scale as well. The old eroded slopes, valleys, and ridges of the Southern Appalachian Mountains and Cumberland Plateau to the west in Kentucky and Tennessee provide regional ecological gradients of forest composition. The greatest complexity and size of individual trees is found in the Southern Appalachian region.

In the Great Smoky Mountains, Whittaker (1956) studied gradients in composition with altitude and with "moisture gradients" within altitudinal belts. Moisture gradient re-ferred to the complex gradient from valley bottoms to dry slopes without any assumption as to its causation. Along such a gradient, the most numerous trees varied from mesic species in the valley bottoms to xeric species on the driest and most exposed portions of the slope. For instance, between 750 and 1,050 m in elevation, hemlock was the most nu-merous in the bottoms, with silver bell, red maple, chestnut oak, scarlet oak, pitch pine, and table-mountain pine each entering the transect and becoming more numerous toward the drier end of the gradient. Similarly, altitudinal gradients were constructed for a given site moisture class. On mesic sites, hemlock and red maple were most numerous at low el-evations (600 to 900 m) with silver bell, yellow birch, sugar maple, and basswood reach-ing maximum abundance in the 900- to 1,220-m zone, and yellow buckeye, mountain maple, and beech being most common at higher elevations.

Combining the two general gradients, Whittaker synthesized a general vegetation pattern of the Great Smoky Mountains (Figure 15.6). He found the broad forest pattern to

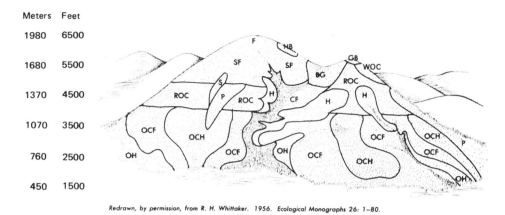

Redrawn, by permission, from R. H. Whittaker. 1956. *Ecological Monographs 26:* 1–80.

Figure 15.6. Topographic distribution of vegetation types on an idealized west-facing mountain and valley in the Great Smoky Mountains. Vegetation types: BG, beech gap; CF, cove forest; F, Fraser fir forest; GB, grassy bald; H, hemlock forest; HB, heath bald; OCF, chestnut oak–chestnut forest; OCH, chestnut oak–chestnut heath; OH, oak-hickory forest; P, pine forest and pine heath; ROC, red oak–chestnut forest; S, spruce forest; SF, spruce–fir forest; WOC, white oak–chestnut forest. (After Whittaker, 1956. Reprinted with permission of the Ecological Society of America.)

be one of continuous gradation of stands along these generalized gradients, yet with certain relatively discontinuous types. As described by Whittaker (1962):

> *The whole pattern was conceived to be a complex continuum of populations, with the relatively discontinuous types confined to 'extreme' environments and forming a minor part of the whole. Allowing for discontinuities produced by disturbance and environmental discontinuity, the vegetation pattern could be regarded as a complex mixture of continuity and relative discontinuity.*

As the above quotation reveals, Whittaker recognized the existence of abrupt discontinuities between cove forests and beech stands on south-facing slopes, between grassy balds and forests, and between heath balds and spruce–fir forests.

In contrast to Whittaker's sampling approach along an elevational gradient, Hack and Goodlett (1960), working in the central Appalachians (Chapter 10), emphasized local physiographic features. The species assemblages were generally coincident with landform units and often changed abruptly with changes in the form of the slope. Thus results can often vary by the scale and detail at which the study is conceived and by the sampling methods employed.

NEW ENGLAND

In central New England, forests in different successional stages on old agricultural fields occupy much of the forest landscape. Following widespread land clearing, farming, and old-field abandonment, revegetation has occurred on soils of glacial origin ranging from poorly to excessively drained. As a result, the composition of the forest community varies in response to two major ecological gradients: one gradient, in time, covering the development of old-field succession, and one, in space, covering the range of site conditions from wet to dry. In this region of moderate elevations and rolling topography, neither absolute elevation above sea level nor aspect is as important in affecting the composition of the forest as the two gradients named.

Table 15.1 Occurrence of Species as Major Components in Transitional Middle-Aged Stands in the Harvard Forest, Central Massachusetts.

Site	Northern Red Oak	Red Maple	Paper Birch	White Oak	White Ash
	Frequency (percent)				
Somewhat excessively drained	67	0	0	33	0
Well drained	95	61	14	13	3
Somewhat poorly drained	81	81	9	6	19
Poorly drained	42	100	8	0	8
Very poorly drained	20	100	0	0	0

In the Harvard Forest in central Massachusetts, Spurr (1956b) classified all existing stands according to both relative position on the successional-development gradient (pioneer, transitional, or late-successional) and relative position on the soil–water gradient (from somewhat excessively drained to very poorly drained). The frequency of occurrences of the individual tree species was found to vary consistently with the two gradients. Table 15.1 summarizes the change in the transitional type associated with changes in the soil-water gradient. Two species were found to be practically omnipresent. Northern red oak and red maple were prominent in all successional stages, and one or the other was prominent on all sites. Both species exhibited a marked relationship to soil water; northern red oak being most frequent on the well drained and red maple on the very poorly drained sites. Other species proved more specific in their site associations. White oak was most frequent on very well drained sites, paper birch on well drained sites, and white ash on somewhat poorly drained sites. Everything considered, all the forest communities found in the Harvard Forest seem to represent a continuous gradation series correlated with successional stage and soil water.

COMPETITION AND NICHE DIFFERENTIATION

As we well know from field experience and the previous considerations, communities are not composed of successive, mutually exclusive sets of species. It has been shown repeatedly that individual species have different genetically-based physiological requirements or tolerances, making effective competitors able to exist in many different sites. A given species may be highly competitive in one community in a given site and hence predominate there. Although it may also exist in adjacent communities having different site conditions, other species may be more competitive and predominate in these ecosystems. Most forest species in a given climatic region probably have their optimal development under similar site conditions of favorable water and nutrients. Hence it is competition and mutualism in space and time, coupled with adaptation of species to particular site conditions, that elicits differentiation of what we may recognize and map as more or less distinct communities.

For example, in a large part of central Europe many forest species reach their optimal development under similar site conditions (Figures 15.7 and 15.8). However, because of the great competitive ability of European beech and two oak species, the other species

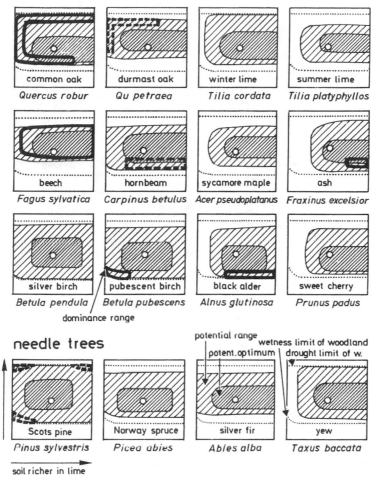

Figure 15.7. Ecograms showing the potential and optimum range of important tree species of Central Europe in the submontane belt of a temperate suboceanic climate as related to soil water and nutrient gradients. Broad-hatch = "physiological amplitude" or potential range of tolerance, narrow-hatch = "physiological optimum" range or potential optimum, area with thick black border = range where the species achieves a natural dominance under natural competition, broken border = species is co-dominant with others or in the case of *Pinus* this co-dominance applies only in the south and east of Central Europe. In each of the ecograms the ordinate represents the degree of wetness of the site. The abscissa covers the range from extremely acid to very basic soils. Above the upper dotted line it is too dry for tree growth; below the lower line it is too wet. The small circle in the center of each ecogram indicates average conditions. (After Ellenberg, 1988. Reprinted from *Vegetation Ecology of Central Europe,* © 1988 by Cambridge University Press. Reprinted with permission of Cambridge University Press.)

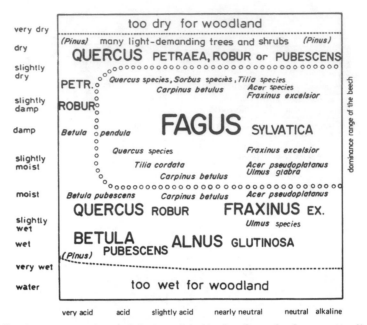

Figure 15.8. An ecogram showing the tree species that form the forests on soils of varying moisture and acidity in the submontane belt of Central Europe in a temperate suboceanic climate. The larger the print the greater the degree to which the species occurs in the tree layer of the potential natural vegetation, as would be expected to result from competition undisturbed by humans. In brackets = only in some parts of Central Europe. (After Ellenberg, 1988. Reprinted from *Vegetation Ecology of Central Europe,* © 1988 by Cambridge University Press. Reprinted with permission of Cambridge University Press.)

occur only in restricted portions of their potential range. Two significant factors, soil water and nutrients (more or less correlated with acidity), are used as the coordinates in the "ecograms" of Figures 15.7 and 15.8. Under these conditions the growth of trees is only limited absolutely by waterlogged soils and by extremely shallow soils over rock. Both limits are shown by dotted lines in the diagrams. The **physiological optimum range** of each species, i.e., the conditions of maximum growth with which they could achieve where competing species were excluded, is hatched more densely (Figure 15.7) than their potential tolerance range, or **physiological amplitude**. Although the optimum ranges of the species generally coincide, under natural competition, however, the ones most likely to succeed under prevailing disturbance regimes are those with a long life span, moderate growth rate, and ability to both tolerate and produce deep shade. As seen in Figures 15.7 and 15.8, beech can exclude all or almost all other trees from its optimal range. Birches can dominate on wet soils only when these are very acid. The proportion of oaks becomes greater the poorer and drier *or* the wetter the soil, given that they can still flourish there. In this humid, submontane climate, deciduous trees flourish on almost all soils and essentially exclude conifers. Only toward the limits of existence of forests, where greater disturbance is likely, does the frugality of Scots pine give it a chance to become dominant either in very dry sites, whether calcareous or acid, or at the edge of raised, oligotrophic bogs. Thus for plants, microsite is an important component of niche.

Although plants may have similar ecological requirements and tolerances, they occur together in recognizable communities not because they react in the same way, but because they compete successfully in various ways due to differences in their genetic makeup and physiological responses. Different tree, shrub, and herb species may exist in a given arid, cold, or wet environment, and seemingly have similar requirements and tolerances, but they occupy different niches (Chapter 4). That is, they occupy different microsites, or they utilize a given site differently through their morphological–physiological responses and life history traits such as time of germination; amount and timing of shoot and root growth; depth of rooting; differential allocation of photosynthate to leaves, stems, and roots; and adaptive mechanisms for obtaining water and retaining it. Thus species are co-adapted in their physical and temporal environment: they are niche differentiated. The complex interactions that, in part, bring about niche differentiation and ecosystem-specific species distributions are considered in the following sections.

INTERACTIONS AMONG ORGANISMS

Plant interactions with animals and other plants are ubiquitous in nature—occurring continuously and in countless ways. The need of ecologists to understand the diversity of interactions and their significance has led to their definition and, for convenience, their comparison using the "interactions grid" illustrated in Figure 15.9. In the grid, interactions are grouped by net effect (plus, minus, or zero) that each species has on its associate(s). By far the most important of these in shaping the way we have viewed organisms and evolution are the antagonistic (minus) interactions of competition and predation. In contrast, interactions with no net effect on one or both partners—commensalism, amensalism, or neutralism—are rarely studied. The remaining type, mutualism, benefits both species, and is sometimes termed symbiosis, cooperation, or facilitation. Although mutualisms tend to be neglected relative to their importance in nature, an enormous amount of information has now accumulated (Bronstein, 1994). Although competition is an important process, a narrow view of a fiercely competitive nature is a biased one. The environment of evolving life is necessarily constituted as well by the persuasive and underlying influences of symbioses, cooperation, and mutualisms (Rowe, 1990, p. 76). Ecological and evolutionary considerations of mutualisms are considered by Boucher (1985).

Mutualisms in Forest Ecosystems

The importance of mutualisms throughout the entire life-cycle of woody plants and in ecosystem processes cannot be underestimated (Janzen, 1985). The main types include: (1) nutritional, including the breakdown and supply of nutrients; (2) protection, either from extreme site conditions or from enemies; (3) transport, by dispersal of pollen,

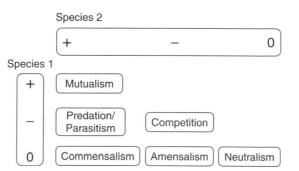

Figure 15.9.   The grid of interactions between species. Interactions are grouped according to their net effect for each of the two interacting species. (After Bronstein, 1994. Reprinted with permission from *Quarterly Review of Biology,* © 1994 by the University of Chicago. Reprinted with permission of the Chicago University Press.)

spores, or seeds; and (4) supply of energy by plants to animal associates through photosynthesis.

SYMBIOTIC MUTUALISMS—MYCORRHIZAE

Probably the most important symbiotic mutualism for plants is the association of roots with fungal hyphae. The resulting structure, known as a **mycorrhizae**, can increase the availability of water and nutrients to the plant and can protect roots from fungal and bacteria pathogens in the soil. To date, approximately 6,500 species of angiosperms have been studied for the occurrence of mycorrhizae, and the roots of 70 percent are consistently associated with these fungi. Another 12 percent are facultatively mycorrhizal, sometimes forming mycorrhizae and sometimes not. Of the coniferous trees studied to date, all are consistently infected by mycorrhizal fungi. Clearly, the association between roots and mycorrhizal fungi is the rule rather than the exception for plants.

The most common mycorrhizal fungi can be placed into two groups: ectomycorrhizal and vesicular-arbuscular mycorrhizal (VAM) fungi. Ectomycorrhizal fungi are characterized by an extensive hyphal development that proliferates between, but does not penetrate, root cells. These fungi give the plant root a characteristic swollen appearance that can easily be identified in the field. Figure 15.10 illustrates the ectomycorrhizae of loblolly pine, Douglas-fir, and western hemlock. Basidiomycotina and Ascomycotina subdivisions contain several thousand species with the ability to form ectomycorrhizae.

Vesicular-arbuscular mycorrhizal fungi differ from ectomycorrhizal fungi in their mode of infection—VAM hyphae actually penetrate and proliferate within root cells. This type of infection does not modify the external appearance of the root, and consequently VAM mycorrhizae can only be identified through microscopic observation of root cells. There are several hundred species of VAM-forming fungi, all of which belong to the family Endogonaceae, subdivision Zygomycotina. Many forest trees form ectomycorrhizae, some form VAM, and a few are capable of forming both.

Perhaps the greatest benefit the plant derives from its symbiotic association with mycorrhizal fungi is an increased supply of water and nutrients. In fact, the successful establishment of forest trees on drought-prone or nutrient-poor soils may depend on the presence of mycorrhizal fungi. Mycorrhizae tend not to form on the roots of trees growing in nutrient-rich soil, even when the soil is inoculated with the fungus. Pines typically grow on nutrient-poor sites, and they have an obligate requirement for ectomycorrhizae; they do not normally grow without them. In many areas of the world, including the former treeless areas of the United States (Hatch, 1937), ectomycorrhizal trees and their associated fungi do not occur naturally (Marx, 1975). In these areas, afforestation attempts were either total or near failures until ectomycorrhizal infection occurred.

One example of the importance of mycorrhizae to plants comes from the high elevations of the Klamath Mountains of southern Oregon where the establishment of white fir following harvesting depends on the presence of mycorrhizal fungi. In this region, clearcut harvesting can change productive white fir forests into open grasslands. Despite numerous attempts since the 1960s, some clearcut areas have not been successfully reforested with conifers.

Comparison of soil beneath clearcut and intact white fir forests indicated substantial changes in soil structure and microbial community composition, but little change in organic matter content or nutrient status. By adding small amounts of soil (ca 150 mL) collected from the root zone of a healthy conifer plantation to each planting hole, ecologists doubled the growth and increased the survival of conifer seedlings nearly 50 percent in the first year after planting (Amaranthus and Perry, 1987). After three years, seedlings that

Figure 15.10. Ectomycorrhizae of loblolly pine, Douglas-fir, and western hemlock. (*a*) Five morphologically different forms of ectomycorrhizae of loblolly pine. (*b*) Smooth-mantled Douglas-fir ectomycorrhiza formed with *Lactarium sanguifluus*. (*c*) Douglas-fir ectomycorrhiza formed with *Corticium bicolor* with a dense covering of mycelium and strands of aggregated hyphae. (*d*) Ectomycorrhiza of western hemlock formed with *Byssoporia terrestris*. (U.S. Forest Service Photos; *a*, Courtesy of Donald H. Marx, *b*, *c*, *d*, Courtesy of James M. Trappe.)

received soil from the healthy conifer plantation remained alive, whereas those without soil additions experienced 100 percent mortality. The small amount of soil added to the planting hole of each seedling did not substantially increase nutrient availability to the seedling. Rather, it provided an inoculum of ectomycorrhizal fungi and other beneficial soil microorganisms that allowed the seedlings to more effectively forage for water and nutrients in the clearcut areas. Similar observations have been made in the southcentral and southeastern United States, where the establishment of pines on mine spoils is contingent on the successful inoculation of roots with ectomycorrhizal fungi. These extreme examples emphasize the importance of mycorrhizae in the nutrition of forest trees. Mycorrhizae are important in nutrient cycling, and we further discuss the physiological mechanisms by which nutrients flow from mycorrhizae to host plants in Chapter 19.

Mycorrhizae are also beneficial in that they protect the host tree from root rot pathogens by their antibiotic exudates (Marx, 1969 et seq.; Marx, 1973). The fungal sheath physically bars invading parasites, and the antibiotics deter root pathogens and the attack of root aphids (Zak, 1965). Other attributes of mycorrhizae include extending the lifetime of small roots and contributing to soil aggregate formation that in turn can increase soil-water storage.

NONSYMBIOTIC MUTUALISMS

Nonsymbiotic mutualisms occur between two species that are not physically connected. Primary examples are pollination, seed dispersal, protection, and decomposition of organic matter. Pollination of flowers by animals is critical to successful sexual reproduction in many woody plant families (Chapter 14). Furthermore, as also emphasized in Chapter 14, dispersal of seeds and fungal spores by animals is also exceedingly important. Protection of woody plants from predators, parasites, and diseases is provided by ants and other insects (Boucher et al., 1982).

Ants are extremely important in the tropics and to a lesser extent in temperate forests (Huxley and Cutler, 1991). Such protective mutualisms range from symbiotic obligate systems in which the plants house and feed the ants to mutualisms in which plants house ants but provide no food. Plants are also "fed" by ants that inhabit the plant and bring debris into their nests, which decomposes and is available for uptake by the plant. In almost all instances, aggressive and predatory ant behavior serves to reduce damage to the plant. In one study (Fonseca, 1994), the Amazonian rain forest tree *Tachigali myrmecophilia* suffered only half the leaf loss compared to trees without ants by keeping colonies of a stinging ant in hollows in their leaves. The ants mounted 24-hour patrols, attacking all creatures that disturbed the tree, whether other insects or mammals. The ants derive energy only indirectly from the tree. The ants' main source of energy is honeydew produced by sucking insects (coccids), which are keep inside ant nests where they feed on substances in the tree phloem. Finally, an indispensable mutualism in all forests is the decomposition of organic matter on the forest floor and in the soil by many kinds of animals, fungi, and bacteria thus ultimately making available nutrients and CO_2 to plants.

Competition

Competition is understood by plant ecologists to have several meanings. We use the definition suggested by Grubb (1985): the relationship between any two species not symbiotic with each other and capable of occupying the same landscape unit, considered over the whole life cycle. This definition has a landscape context and encourages us to think of competition at all stages of the plant life cycle. The term "interference," the sum of the processes by which the yield of one plant is reduced as a result of another plant being present, was used by Harper (1961) in the context of experimental studies of species mixtures. Thus competition is a process whose outcome is affected by a plant's susceptibility to both above- and belowground interference and to its inherent requirements or tolerances of light, temperature, water, nutrients, and other site resources.

Changes in the structure and composition of the forest community result from the constant demand of each individual tree for more crown and root space and from the eventual death of even the most dominant individuals. The increasing size of the main canopy trees results in competition for growing space, with a few individuals gaining space, a few more holding their own, and an increasing majority losing space and eventually succumbing. The death of the dominants, due to lightning, fire, wind, insects, and diseases releases from the main canopy a portion of the site for occupancy by a growing and developing understory.

Competition between the trees of the same species does not affect the composition of the forest ecosystem and therefore has no direct effect on forest succession. Competition among individuals of different species, however, results in a change from a forest of one composition to that of another. The rate of species replacement slows as succession proceeds, and ultimately a group of species having complementary ecological roles characterizes the late successional forest. Competition takes place in even- and uneven-aged stands and in the main canopy of the forest and in the understory.

COMPOSITION, STAND STRUCTURE, AND DENSITY

In species that regenerate en masse at about the same time, as following fire or flooding, an even-aged stand structure is formed. Almost all pines, Douglas-fir, black spruce, cottonwoods, eucalypts, and many other species form natural even-aged stands. As the name implies, the individuals are nearly the same age and the stems are often of a similar height and diameter. In contrast, various long-lived hardwoods and conifers form stands in which great diversity exists in tree age and size. Such uneven-aged stands are typical of naturally-regenerated hemlock–northern hardwood forests (comprising hemlock, sugar maple, beech, yellow birch, and basswood, among others) and of some spruce–fir forests of Central Europe. Here the trees of a given species on a single hectare may range in age 300 years or more. In many forests, a mosaic of small even-aged groups, when considered over a large area, forms the uneven-aged forest. This was probably the characteristic structure of presettlement upland oak forests of the East and ponderosa pine forests of the West.

In even-aged stands, competition for light, water, and nutrients depends largely on the number of stems per unit area. In time, crowns of the trees come together and crown class differentiation becomes pronounced (Figure 15.11). Due to a combination of environmental and genetic factors, some trees develop rapidly and exhibit large, well-formed crowns. Other trees grow more slowly and their crowns become more or less restricted; in time some trees are gradually overtopped and suppressed. Root competition is often severe in the even-aged stand, although it is much less observable than crown competition. In even-aged pine stands, for example, the root system of a tree may compete with several hundred trees, whereas its crown competes only with a few adjacent trees.

A simple classification of crown classes illustrates the results of intense competition in even-aged stands (Figure 15.11). The silviculturist uses the crown classes as a basis for judging the vigor of the stand and for conducting thinnings and other cultural operations. The major crown classes are:

Dominant: trees with crowns extending above the general level of the canopy and receiving full light from above and partly from the sides; trees larger and more vigorous than the average stems in the stand; crowns well developed.

Codominant: trees with crowns forming the general level of the canopy or somewhat below; receiving full light from above but only moderate or little amounts from

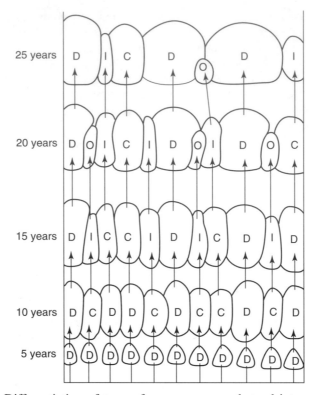

Figure 15.11.   Differentiation of trees of a pure even-aged stand into crown classes. D, dominant; C, codominant; I, intermediate; O, oppressed (suppressed). (After Smith *et al.*, 1997. Reprinted from *The Practice of Silviculture* by D. M. Smith, B. C. Larson, M. J. Kelty, and P. K. S. Ashton, © 1997 by John Wiley & Sons, Inc. Reprinted with permission of John Wiley & Sons, Inc.)

the sides; trees usually with medium-sized crowns that are more or less crowded on the sides.

Intermediate: trees shorter than the preceding classes but with crowns extending into the canopy formed by the dominants and codominants; trees receiving little direct light from above and virtually none from the sides; usually with small crowns that are considerably crowded on the sides.

Oppressed or Suppressed: trees with their crowns entirely below the general canopy level; trees receiving no direct light from above or from the sides.

To study the competition within a species population in an even-aged stand or among different species in an uneven-aged stand, we can examine the age distribution of species over time. For example, in an old-growth hardwood forest in New Hampshire the curves in Figure 15.12 illustrate the sharply falling shape of the curve for the species. Sugar maple and beech exhibit the sharply declining population of species that establish many seedlings periodically in the shaded forest floor (persistent seedlings, Chapter 5). Many survive and eventually some replace the overstory veterans. In contrast, red spruce exhibits a lack of young seedlings, and over time red spruce may diminish in abundance. Without disturbance it is not competitive in the younger age classes with beech and maple

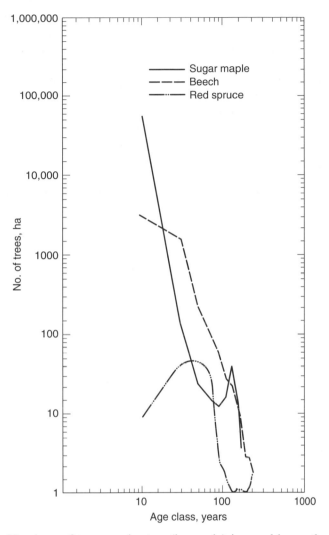

Figure 15.12. Numbers of trees per hectare (log scale) in an old-growth northern hard-wood forest over midpoints of 20-yr age classes (log scale) for sugar maple, beech, and red spruce. (After Leak, 1975. Reprinted with permission of the Ecological Society of America.)

on this site. Such curves, developed at one point in time, do not tell us how the populations actually change over time. However, if the major past disturbances of the forest are known, the general trend of the populations over time can be determined.

Vertical Structure

One result of competition among plant species of the forest is the development of a vertical structure of the vegetation. The multistoried forest has an upper layer of overstory trees, one or more subdominant layers (more or less distinct) composed of younger stems of the overstory trees and mature trees of other species that do not reach the overstory, layers of high and low shrubs, layers of high and low herbs, and a ground layer of mosses, liverworts, and lichens. The species of each layer are genetically adapted or modified to make the best use of the space, light, and microclimatic resources of their respective vertical positions, i.e., they are effective competitors under those specific site conditions.

In the tree layer alone, considerable variation occurs depending on species composition and site conditions. Generally, the more favorable the site conditions (especially water) the greater the number of layers. In North America, the complex structure of the Southern Appalachian Mountain cove forests is noteworthy. In addition to the species-rich overstory, there may be several understory tree layers and below these is often a dense layer of tall ericaceous shrubs, primarily the rosebay rhododendron. In the tropical rain forest, even a more pronounced vertical structure may exist, including huge emergent trees whose crowns are fully exposed to sunlight because they extend 5 to 10 m above the general canopy level. In such multistoried forests, (for example, see Figure 8.10) each layer has a different microclimate and usually its distinct assemblage of insects and other animals.

In contrast, many species, particularly pioneer species that colonize areas following major disturbances, especially fire, typically exhibit initially a single tree layer. In fact, nearly pure natural overstory monocultures may be formed by species such as lodgepole and jack pines, aspens, and the southern pines following catastrophic fire. However, the image of a plantation-like expanse of even-aged and nearly same-sized stems is misleading. In the dense stands that arise following a stand-regenerating crown fire, a patchy distribution of trees is often the case over tens or hundreds of hectares. Although the young and middle-aged stands may be dense and may lack a well-developed understory, eventually gaps form and regeneration begins to develop. The classic example is lodgepole pine forests of the central and northern Rocky Mountains that may exhibit even-aged stand structure over large areas (Clements, 1910; Mason, 1915; Stahelin, 1943).

However, evidence of considerable variation in stand structure is accumulating (Parker and Parker, 1994). This structural complexity may be associated with physiography and soil conditions, variability in fire regimes, and pest infestation patterns. For example, Parker and Parker (1994) found two distinct kinds of stand structures in 120- to 140-year old lodgepole pine in central Colorado. Four stands had closed canopies, relatively high basal areas and stem densities, and low sapling and seedling densities of pine. In contrast, three other stands exhibited relatively open canopies, lower tree densities, and significantly higher densities of pine saplings and seedlings. The structure of such closed stands exemplifies dense regeneration developing rapidly following catastrophic crown fire. In the open stands, nearly continuous recruitment of pine, a very understory intolerant species had occurred over the past 120 to 140 years. Although vertical structure has received much less attention than species composition in forest community studies, a given old-growth forest ecosystem can usually be distinguished by its characteristic structure as well as by its species composition.

The overstory canopy intercepts much of the incoming irradiance, and given an unbroken overstory layer, species of the lower layers are physiologically and structurally adapted to use the continuously decreasing amount of light available as one approaches the forest floor. However, since the forest overstory is never completely unbroken over large stretches, various subdominant trees, shrubs, and herbs that require relatively high light levels to survive utilize the well-lighted microsites and extend their crowns into openings. In addition, overstory trees differ in the density of their crowns so that considerably more light may penetrate one overstory than another. For example, the shrub layer is particularly well developed in oak forests of eastern North America compared to beech–sugar maple forests. The oak crowns are less dense and the canopy is more open than that of beech–sugar maple forests. The increased light, together with more frequent fires in oak forests, favor the growth, sprouting, and clonal spread of understory shrubs.

Stand Density

The number or mass of trees occupying a site has important implications for the trees themselves, the site, and for the silviculturist responsible for controlling forest reproduction, growth, and composition. Density is typically measured in terms of numbers of trees or basal area per unit area.

Except at very low densities, mortality is caused by competition among trees and between trees and associated vegetation for light, water, and nutrients. Mortality is greatest in the seedling stage when the number of seedlings per unit area is highest. Curtis (1959) reported a mortality of 99 percent of sugar maple seedlings over a 2-year period. Hett (1971) followed up this work and found that the mortality rate is relatively independent of age in the early years but declines as the seedlings mature. As plants die, the remaining individuals become larger; the smaller plants are continually eliminated from the population. Therefore, in stands of plant species, there is a strong relationship between plant size and the density of the stand; larger size (or weight of biomass) of the stems are associated with fewer stems per unit area and vice versa.

Various measures of stand density or stand density indices (Spurr, 1952) have been developed. Reineke (1933) showed that plotting the logarithm of number of trees per acre against the logarithm of average diameter of fully stocked, even-aged forest stands typically resulted in a straight-line relationship. This negatively sloping line, expressed by: $\log N = -1.605 \log D + k$ (where N = number of trees per acre; D = diameter of tree of average basal area; k = a constant varying with species) in most cases could be used to define the limits of maximum stocking. This relationship is mathematically the same (Avery and Burkhart, 1994) as the so-called $-3/2$ law of self-thinning developed by plant ecologists. In this case, the slope of the line of logarithm of mean volume or weight of surviving plants versus logarithm of plants per unit areas is approximately $-3/2$ (Drew and Flewelling, 1977; White, 1985; Silvertown and Doust, 1993).

This relationship is illustrated in Figure 15.13 for tree density and volume for five stands of loblolly pine over a 50-year period which have different initial densities (Peet and Christensen, 1987). As stand development proceeds, competition increases and mortality occurs such that stand density decreases and mean tree volume increases (lines directed up in Figure 15.13). When net biomass production equals loss by mortality, the slope of the thinning curve will be -1, and if the rate of biomass loss exceeds the rate of net production, the slope will be greater (less negative) than -1. During the establishment phase of the pines, competition is low, and the slope is much less (more negative) than $-3/2$ (straight line in Figure 15.13). However, as the canopy closes and resources become limiting, stand mortality rate converges on a $-3/2$ slope, with the rate of convergence clearly dependent on initial density. As expected, at high densities thinning begins very early, whereas at low densities convergence is delayed. In Figure 15.13, the mean weight–density relationship for each stand follows a curved path, initially directed straight up, then bending to conform briefly to the $-3/2$ limit, then bending even farther away. Whereas no one stand clearly defines the $-3/2$ thinning limit, all stands together do define a line with a slope of nearly $-3/2$, which individual stand trajectories approach but never cross.

The $-3/2$ relationship, defining the maximum stocking level, is reported to be valid for stands of any species, of any age on any site. The relationship holds for many species in pure and mixed stands in north-temperate forests of the western United States (Long, 1980). Peet and Christensen (1987) reported marked differences for species in mixed-age hardwood stands, but all species together showed moderate conformity. Although controversy has developed and exceptions reported (Lonsdale, 1990; Osawa and Allen, 1993), the view consistent with past and present evidence is that, through time, self-thinning ap-

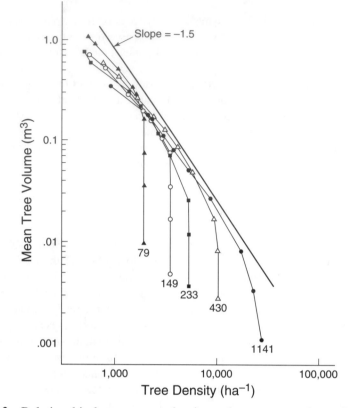

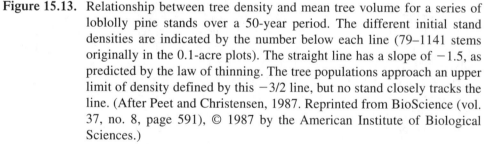

Figure 15.13. Relationship between tree density and mean tree volume for a series of loblolly pine stands over a 50-year period. The different initial stand densities are indicated by the number below each line (79–1141 stems originally in the 0.1-acre plots). The straight line has a slope of −1.5, as predicted by the law of thinning. The tree populations approach an upper limit of density defined by this −3/2 line, but no stand closely tracks the line. (After Peet and Christensen, 1987. Reprinted from BioScience (vol. 37, no. 8, page 591), © 1987 by the American Institute of Biological Sciences.)

proaches a slope of −3/2 although intercept value may vary widely, depending on species and site conditions (Silvertown and Doust, 1993). In forestry, this relationship has been applied to stand-density management of Douglas–fir plantations (Drew and Flewelling, 1979).

For the silviculturist, control of stand density is extremely important. Diameter growth is strongly influenced by density and decreases with increasing density. Thus if large trees are desired, density must be controlled by wide initial spacing via planting or by thinning to an appropriate number of trees (or amount of basal area) per hectare to maximize diameter growth, yet still fully utilize site resources. However, for some uses, wood quality is adversely affected by low density (trees become branchy and the wood knotty; growth rings may be too wide) so that a balance must be reached for a given species on a given site through appropriate density control, depending on the end product desired.

Within a wide range of densities, tree height of free-growing trees is unaffected by stand density. This relationship, together with the fact that different site qualities strongly affect height growth, enables tree height to be used as an effective indicator of site quality (Chapter 13). Finally, as brought out in Chapter 13, as long as the site is fully occupied (trees making full use of available resources), the species will produce the same amount of wood per year over a relatively wide range of densities. Whether there are many small trees or fewer large trees, a similar wood volume is produced. However, as site quality improves (more favorable water supply and higher fertility) a greater number of trees can be grown per unit area and a greater volume of wood can be obtained than on the poor site. Understanding the site quality is therefore of great importance in knowing how to regulate stand density. Detailed considerations of stand density from the standpoint of growth and yield and silvicultural management of forests are available in measurement and silvicultural texts (Oliver and Larson, 1990; Avery and Burkhart, 1994; Smith, et al., 1997).

COMPETITION AND OVERSTORY COMPOSITION

The trees in the main canopy of the forest tend to increase yearly in height, bole size, length of each growing branch, and number of leaves. If the tree is to remain alive and vigorous, it must grow. To grow, it must increase its growing space—its utilization of the site.

In any fully stocked stand, then, it follows that competition between growing individuals in the overstory will result in the elimination of some trees, particularly of those species genetically and physiologically less suited for survival under the particular site condition that may exist. The result is a gradual change, or occasionally an abrupt change, in the composition of the overstory and thus in the forest community. Examples are many.

The internal changes in a developing temperate forest are illustrated in Table 15.2, which summarizes the growth over 60 years of a middle-aged, mixed hardwood stand

Table 15.2 Changing Composition Over 60 Years in a Northern Hardwood Forest on Sugar Island, St. Mary's River, Upper Michigan.

Species	1933	1944	1955	1971	1978	1989	1993
			Number of trees per hectare				
Sugar maple	803	741	652	502	474	403	381
Red maple	314	210	173	158	151	133	133
Paper birch	89	40	22	7	7	7	5
Bigtooth aspen	49	7	7	7	7	5	5
Red oak	44	44	44	37	37	35	35
Yellow birch	35	17	15	10	10	5	2
All trees[1]	1364	1077	927	734	697	598	571
			Basal area in square meters per hectare				
Sugar maple	9.4	11.6	12.7	13.2	13.3	12.4	11.8
Red maple	7.1	7.5	7.9	9.0	9.7	10.1	10.7
Paper birch	2.8	1.6	1.3	0.5	0.5	0.6	0.4
Bigtooth aspen	1.4	0.4	0.5	0.8	0.9	0.8	0.9
Red oak	2.0	2.8	3.6	5.0	5.4	6.2	6.7
Yellow birch	1.0	0.7	0.7	0.6	0.6	0.2	0.1
All trees[1]	24.0	25.0	27.3	29.7	31.0	31.0	31.2

[1]Includes some individuals of species not listed above.

growing on an island in the St. Mary's River between Michigan and Ontario. During this period, the relatively short-lived aspens and birches have practically disappeared from the stand. Sugar maple has also suffered steady mortality; growth slowly increased until 1978 but has declined over the last 15 years. Red maple declined greatly in numbers, but the growth of the residual trees has been such that the basal area of this species has shown a steady increase. Sugar and red maples now make up 72 percent of the basal area. The few red oaks have survived well; they grow relatively fast and make up a substantial proportion of the basal area for their few stems. The better competitive ability of sugar maple, red maple, and red oak at the expense of aspen and birches has brought about major changes in stand composition. The stand basal area has increased slowly and appears to be leveling off in total and for some species.

While permanent plot data record the changing composition of the overstory with time, they do not chronicle the cause of the death of the trees in the interval between measurements. We generally assume that most of these trees simply are suppressed; that is, their leaf area and feeding root area are reduced to the point that the tree dies. The plant is unable to maintain a positive carbon balance under competition for site-specific resources of light, water, and nutrients.

In actual fact, death may be due to many causes. Even in the case of so-called natural mortality, the tree probably dies during some period of extreme stress, such as in a late summer, hot, dry spell or in a severe unseasonable frost, or in an extremely cold winter. In other cases, the tree becomes weakened to the point that it falls prey to some insect or disease. Usually, several factors are interrelated and together account for death (Mueller-Dombois, 1987).

COMPETITION IN THE UNDERSTORY

Whereas competition in the overstory involves relatively few individuals and is spread out over many years, competition near the forest floor involves not only trees but many shrubs and herbaceous plants, often in very large numbers and changing in relationships year by year. Many plants are adapted to spending their entire life cycle in the understory. Included are some herbs that carry on a major portion of their annual growth before the trees reach full leaf. Included also are other herbs, many shrubs, and small trees that are tolerant of understory site conditions. Some trees can both thrive in the understory and also occupy the overstory itself when the opportunity occurs.

An estimate of the sheer numbers of potentially competing understory stems and plants is given by Wendel (1987) in north-central West Virginia oak forests. Sections of the forest floor from each of four different sites were placed in a greenhouse over winter and the plants allowed to develop during the following growing season in the open. The number of tree seedlings and sprouts (9 species) averaged 981,000 ha^{-1}, woody shrub stems (8 species) averaged 243,000 ha^{-1}, and herbs (25 species) averaged 220,000 plants ha^{-1}. Obviously, most of the newly germinating seedlings would not have survived in the forest understory. Nevertheless the potential for competition in the forest understory is enormous in these ecosystems.

In assessing competition within the understory, the problem of the successful plant attracts little attention. It is merely carrying on its growth under existing site conditions without incident. As Decker (1959) points out, it is with the failing and dying plant, the plant intolerant of understory conditions, that we must be concerned. The unsuccessful competitor loses vigor and dies, but the causes of death may be various. Starvation resulting from inadequate light and consequent inadequate photosynthesis is certainly a contributory factor. It leads to a death spiral as the weakened root system is unable to obtain adequate water to support a vigorous cambium and crown. Inadequate soil water is another. Insects and diseases, too, play their role, particularly when the plant has reached a weakened condition. Thus the site factors relating to understory survival are discussed in the following section.

UNDERSTORY TOLERANCE

The problem of survival in the understory is central to an understanding of forest succession. Those forest trees capable both of surviving as understory plants and responding to release to reach overstory size will inevitably form a major portion of the dynamic forest community. A forest tree that can survive and prosper under a forest canopy is said to be **understory tolerant** (or simply **tolerant**), whereas one that can thrive only in the main canopy or in the open is classified as **understory intolerant** (or **intolerant**). Obviously there are not just two extremes but a continuum of tolerance from one extreme to the other.

This use of the term, tolerance, to refer to the relative capacity of a forest plant to survive and thrive in the understory is a restricted application of the general biological meaning of the term, which refers to the general capacity of a plant to be genetically adapted and physiologically compatible with unfavorable conditions. Thus a salt-tolerant plant is one that is adapted to grow in soil with a high salt concentration, and a drought-tolerant plant is one that can grow at low water levels that would be fatal to most other plants. Pioneer species may be intolerant in the context of the understory environment but "tolerant" or most competitive with high irradiance, extreme temperature, drought, and low nutrient status of some exposed, open sites. In this chapter, and elsewhere in the book, however, the unmodified use of the term, tolerance, refers to a plant's vigor in the forest understory due to its genetic and physiological adaptations to this multifactor environment. Understory tolerance is the more acceptable term.

In regard to the term, tolerance, we note that despite its widespread use, tolerance is an inadequate concept (Grubb, 1985). The main reason is that many plants do positively require for maximum growth rate the conditions with which they are associated in the field and do not merely tolerate them. Some plants associated with shade do not merely tolerate it, they require it because they perform poorly in full sunlight. Similarly, some plants may not only be tolerant of flooding but require it to outcompete associates in establishment and for optimum growth. For example, species such as red ash, eastern cottonwood, and sweetgum (Broadfoot and Williston, 1973), and water tupelo (McKevlin et al., 1995) require flooding for efficient growth.

In many cases, relative shade tolerance may be the major factor accounting for a plant's performance in the understory. Although all successful understory species tolerate or require shade, they may have other adaptations to the understory environment such as to soil water stress, temperature, browsing, and disease. The term, shade tolerance, as a substitute for understory tolerance tends to be widely used in the literature, but should be applied specifically to situations where light is clearly the dominant factor.

Relative understory tolerance can be recognized and rated in general terms with reasonable success. If the term is used only in the sense of the ability of a plant to survive and prosper in the understory, then its nature can be investigated in light of physiology and ecology of understory plants and the understory environment (see the following discussion). Often many other characteristics of tolerants and intolerants are cited, and with their inclusion a broader understanding of the nature of tolerant and intolerant species is necessary.

Characteristics of Understory Tolerant and Intolerant Species

Since, by definition and context, a tolerant tree is one that grows and thrives under a forest canopy, it may be recognized by various values that relate to its vigor under such conditions. In contrast, an intolerant tree is frequently characterized by opposite extremes of the same traits. In addition, other characteristics have been cited as criteria or indicators for the determination of tolerance (Toumey, 1947; Baker, 1950).

Tolerance of Suppression in the Understory. Tolerant trees live for many years as understory plants under dense forest canopies. Individuals of many species may germinate and survive a few years under such conditions, but only tolerant trees will persist and continue to grow for decades or centuries.

Response to Release. The vigor of tolerant trees is demonstrated by the fact that, following removal of the overstory, they have the ability of responding to release by initiating immediate and substantial growth. A tolerant tree may survive for years in the understory while putting on as many as 12 to 24 rings per cm of radius, and yet begin growing rapidly soon after release.

Crown and Bole Development. The lower branches on tolerant trees are foliated longer and to a greater extent than those on intolerant trees. Consequently, tolerant trees will have deeper and denser crowns (Graham, 1954). It follows that tolerant trees will prune naturally at a slower rate, will maintain a greater number of leaf layers, and will maintain healthy and vigorous leaves deeper into the crown. The boles of tolerant trees tend to be more tapered than those of intolerant trees because of the greater depth of crown (Larson, 1963b).

Shoot growth and branching pattern of a species are adapted to provide light appropriate to the light requirements of its foliage. For example, the relatively fast branch growth and spreading form of American elm, sycamore, and silver maple allow a well-lighted crown that is required for leaves of these species. In contrast, sugar maple and beech leaves tolerate considerable shade and thrive even in the interior of a crown characterized by a more compact form and slower branch growth than that of more shade-intolerant species.

Stand Structure. Tolerant trees persist over long periods of time in natural mixed stands, tend to be successful in mixture with other species of equal size, and consequently form denser stands with more stems per unit area than do comparable intolerant trees.

Growth and Reproductive Characteristics. Tolerant trees inherently grow more slowly than intolerant ones. This trait is best expressed in moderate   to   high   light conditions; in dense shade the slow growing tolerant may be perceived to grow faster than the dying intolerant. Tolerant deciduous angiosperms (beeches, maples) have seasonally determinate shoots, whereas intolerants (willows, aspens, cottonwoods, birches) have seasonally indeterminate shoots (Marks, 1975). Tolerant trees typically mature later, flower later and more irregularly, and live longer than intolerant trees.

Although the understory tolerance of a species is defined on the basis of its ability to survive and prosper in the understory, it is not this attribute that causes the many other distinctive and important differences between tolerant and intolerant species. Populations of the respective extreme types have evolved under different selection pressures of environment and plant competition. Hence they belong to two complex and markedly different adaptation systems. Tolerant species establish, grow, and have become adapted to conditions markedly different from those of intolerants. Although it is often instructive to compare the two extremes, we are dealing with species of varying degrees of intermediacy along a cline from very intolerant to extremely tolerant.

Intolerant species are typically pioneers that may colonize a wide variety of sites. They are successful because of two major types of adaptations. First, their capacity for rapid establishment on disturbed sites, fast growth in the open, early seed production, and widespread seed dispersal have enabled them to perpetuate themselves wherever fire, windstorm, flooding, cultivation, or other disturbances have eliminated or reduced the existing vegetation. In some climatic regions, disturbance alone may account for the perpetuation of intolerant species since they are replaced in time on most sites by more tolerant species.

However, intolerants have a second strategy—adaptation to extreme site conditions. They are typically adapted to some type of xeric (hot or cold) or infertile site as well as to the climatically extreme, initial conditions of the disturbed site. Thus they may form relatively permanent communities on extreme sites where more tolerant species are at a competitive disadvantage. For example, willows in annually flooded bottomlands, jack pine on nutrient-poor sands of the Lake States, table-mountain pine on the driest and least fertile soils of the Appalachian Mountains (Zobel, 1969), and sand live oak on sands of the southern Coastal Plain are all examples of intolerants occupying extreme sites of diverse kinds.

In contrast, tolerant species occupy and are adapted to more moist, sheltered, and fertile sites (mesic conditions). They replace intolerant species and perpetuate themselves through adaptations favoring survival and growth in a shaded understory, establishment in relatively undisturbed litter layers, and a long life span. Once they attain canopy dominance their presence tends to reinforce the shaded, moist, humid, and relatively disturbance-free environment favorable to their own regeneration (Chapter 5).

Tolerance Ratings of Tree Species

Tolerance is not constant for a given species; it varies with genetically different individuals and races, with different regional climates, with different local site conditions, with different plant associates, and with different vegetative conditions (seedlings versus sprouts), and especially with age. A given species may be perceived to be more tolerant in one part of its range than another, one site in comparison to another, and in one forest type than in another.

For example, eastern white pine is perceived to be more tolerant in Minnesota, where it is considered to be midtolerant, than in New England, where it is relatively intolerant. In central New England, it is more tolerant on dry sandy soils than on moister sandy loams—at least it occurs more consistently under the less dense stands on the former sites. Like most species, it is more tolerant in the juvenile phase than it is as it grows older. Sprouts with well-developed root systems, such as occur in oaks, hickories, and tuliptree, appear more tolerant than seedlings of the same species, at least in part, because sprouts are better able to absorb water and nutrients from the well-established root system of the parent.

Understory tolerance, then, is not only a relative matter, but the relative ranking of a species with regard to tolerance will depend in part upon the regional ecosystem, specific site, age, and its associates. Tolerance ratings must be interpreted with great care and with these points in mind.

At the same time, the approximate tolerance rating of the more important and characteristic forest tree species should be known and understood by ecologists and practicing silviculturists as a general frame of reference. Thus such a frame of reference, an estimate of the relative understory tolerance of selected tree species, assuming representative site conditions for each species, is presented in Table 15.3. Obviously the practicing forest ecologist must learn to understand the relative tolerance and competitive ability of trees under field conditions and not from a textbook table. Nevertheless, a "standard condition" or general starting point and an awareness of the nature of tolerance are appropriate at the outset.

Examples of Understory Tolerance in Forest Ecosystems

Before examining the physiological nature of tolerance, we illustrate various aspects of tolerance by photographs taken in four forest ecosystems. Figure 15.14 depicts a young, second-growth stand of intolerant tuliptree in a cove site of the Southern Appalachian mountains. Sufficient light and water reach the understory to favor development of a conspicuous and diverse community of midtolerant and tolerant species, such as oaks, maples, and beech, as well as a rich shrub and herbaceous flora.

The stand of midtolerant oaks of the Missouri Ozark Region (Figure 15.15) illustrates the dominant oak overstory and an open, partially shaded understory of oaks and associated vegetation. The shade cast by the overstory and the low soil water during the growing season combine to curtail growth of the understory. This is in contrast to the luxuriant understory of the tuliptree forest in the mesic cove site of Figure 15.14.

The dense shade from the overstory of sugar maples in a northern Michigan stand (Figure 15.16) favors the very-tolerant sugar maple seedlings over all other species. In deep shade, they survive but grow slowly; in small openings where more light is available (background of Figure 15.16) they respond with accelerated growth and, barring disturbance, will perpetuate the dominance of sugar maple.

The western white pines of the northern Idaho stand in Figure 15.17 originally colonized the site following a fire. Now, two centuries later they form a dense stand, and their crowns intercept most the incoming light. However, they will be gradually replaced by the very tolerant species seen growing in the understory—western hemlock, grand fir, and western redcedar—in the absence of disturbances such as fire, windstorm, and logging.

Nature of Understory Tolerance

Although the relative tolerance of a given species growing in a given site-specific ecosystem can be estimated with a reasonable degree of accuracy, the explanation of the nature of tolerance is much more difficult. This problem has intrigued ecologists for many years and has been the subject of much controversy and semantic debates.

The nature of tolerance may be examined from the broad, species-adaptation level, and as we have seen, the so-called tolerant and intolerant species belong to two markedly different adaptational systems. Usually, however, the specific trait of survival in the understory is investigated by studying the site factors and the physiological processes involved.

ENVIRONMENTAL FACTORS RELATING TO UNDERSTORY TOLERANCE

The most obvious ecological feature of the understory environment is low light irradiance. In fact, some people to this day associate the capacity of a plant to survive in the understory solely with the capacity of a plant to survive under low light irradiances, equating the general concept of understory tolerance with the specific concept of shade tolerance.

As data have accumulated from studies of the effect of light on tree growth under controlled or semicontrolled forest conditions, however, it has become evident that the light irradiances under most forest canopies are sufficient to permit most forest trees to carry on photosynthesis at rates higher than that required to balance respiration losses of

Table 15.3 Relative Understory Tolerance of Selected North American Forest Trees[1].

Eastern North America		Western North America
Gymnosperms	**Angiosperms**	**Western North America**

	Very tolerant	
Eastern hemlock	Flowering dogwood	Western hemlock
Balsam fir	Hophornbeam	Pacific silver fir
	American beech	Pacific yew
	Sugar maple	
	Tolerant	
White spruce	Basswood	Spruces
Black spruce	Red maple	Western redcedar
Red spruce		White fir
		Grand fir
		Alpine fir
		Redwood
	Mid-tolerant[2]	
Eastern white pine	Yellow birch	Western white pine
Slash pine	Silver maple	Sugar pine
	White oak	Douglas-fir
	Northern red oak	Noble fir
	Hickories	
	White ash	
	Elms	
	Intolerant[2]	
Red pine	Black cherry	Ponderosa pine
Shortleaf pine	Tuliptree	Junipers
Loblolly pine	Sweet gum	Red alder
Eastern redcedar	Sycamore	Madrone
	Black walnut	
	Black oak	
	Red ash	
	Scarlet oak	
	Sassafras	
	Very intolerant	
Jack pine	Paper birch	Lodgepole pine
Longleaf pine	Aspens	Whitebark pine
Virginia pine	Black locust	Digger pine
Tamarack	Eastern cottonwood	Western larch
	Pin cherry	Cottonwoods

[1]Based on representative site conditions for the respective species.

Survival in the understory is related to light irradiance, moisture stress, and other factors. As a general guide to the light irradiance component, we estimate the range of *minimum* percentage of full sunlight for a species to survive in the understory at each of the five arbitrary levels of tolerance.

Very tolerant species may occur when light irradiance is as low as 1 to 3 percent of full sunlight; *Tolerant* species typically require 3 to 10 percent of full sunlight; *Intermediate* species 10 to 25 percent; *Intolerant* species, 25 to 50 percent; *Very intolerant* species, at least 50 percent. For example, an intolerant species competing in the understory is unlikely to survive with less than about 25 percent of full sunlight (unless other compensating factors are favorable).

[2]Many *mid-tolerant* and *intolerant* species are observed to be more tolerant in the juvenile phase than in the adult phase. Notable examples include black cherry, white ash, elms.

Figure 15.14.   Young second-growth stand of tuliptree in a cove of the Appalachian Mountains, North Carolina. (U.S. Forest Service photo.)

the whole plant. Under moderate canopy cover, such as those created by pine and oak forests, ample light reaches the forest floor to provide energy for photosynthesis by many forest tree seedlings. Even under dense canopy cover, such as those formed by spruce–fir and tropical rain forest species, light flecks sweeping through the forest and light at the edges of small gaps permit occasional plants to survive and grow in the understory. Only in those forests where light on the forest floor is less than about 2 percent full sunlight is light obviously a single limiting factor in understory survival.

Light, though, is not the only environmental factor greatly modified by the forest canopy. Under a dense forest canopy, almost all the factors of the climate and soil differ from those characterizing similar physiographic sites in the open. Soil water is foremost among the affected environmental factors.

The importance of soil water in regulating understory establishment and growth was demonstrated spectacularly in Germany in 1904 by Fricke, who cut the roots of competing understory trees by trenching around small, poorly-developed Scots pine seedlings growing under a stand of the same species. These seedlings responded with vigorous growth, indicating that they had been inhibited principally by a shortage of soil water created by the competing roots of the overstory trees rather than by low light. Trenched plots were used with similar results under eastern white pine in New Hampshire (Toumey and Kienholz, 1931), loblolly pine in North Carolina (Korstian and Coile, 1938), and others.

Figure 15.15. Uncut, old-growth oak stand in the Ozark Mountains of southeastern Missouri. (U.S. Forest Service photo.)

Generally speaking, trenching a small plot of a few square meters or so in size under a pine stand so as to remove root competition by severing entering roots will result in a great increase in available soil water and the consequent appearance of luxurious vegetation.

Usually, either trenching or watering understory plants will substantially increase their height growth. However, tying overstory tops back to allow greater amounts of light to reach their leaves will have little effect. It would be a mistake, however, to attribute tolerance solely to soil water just as it would to attribute it to light alone. Obviously, both soil water and light as well as nutrients (Chapter 19) are involved in understory survival, and other factors such as temperature, humidity, and CO_2 content undoubtedly play a role as they become limiting.

A series of studies on the ecological nature of tolerance carried out with loblolly pine and associated hardwoods in North Carolina has done much to explain the cause of death of loblolly pine and other intolerant seedlings under loblolly pine overstories. In contrast to tolerant seedlings, which are able to survive and even grow under pine canopies, the relatively intolerant loblolly pine seedling seems to photosynthesize more than enough to counterbalance respiration losses but not enough to permit its root system to expand and

Figure 15.16. Sugar maple stand in the Upper Peninsula of Michigan. Unbrowsed sugar maple seedling shown on left and browsed seedling on right. (U.S. Forest Service photo.)

reach the deeper soil strata. Over the years, therefore, the loblolly pine seedling develops a somewhat etiolated top without a compensatory root system of sufficient extent and depth. Sooner or later, these seedlings will die during a period of unusually severe soil water stress under hot, dry midsummer conditions. In contrast, the midtolerant oaks and other hardwoods under similar conditions will develop extensive root systems that are deep enough to permit them to survive the droughts.

Ability to survive under the moderate light intensities and severe soil-water shortages characteristic of pine forests, then, appears to be dependent upon a plant's carrying on sufficient photosyhthesis to develop a sufficient root system that will survive midsummer drought in soils kept at low soil-water levels by competing roots of overstory trees and understory plants. Under other situations, either light or root competition may be relatively more important. Under dense Sitka spruce and western hemlock in the Pacific Northwest rain forest, light is at extremely low levels, whereas the site is almost always wet or at least moist. Here, light is obviously the more important factor. Under open oak woodland types or under ponderosa pine in the drier part of its western range, light under the forest is usually above any critical level, whereas soil water is always in short supply. Here, soil water is the more important factor. For example, Nance et al. (1990) demonstrated that certain proteins were produced in drought-stressed ponderosa pine and enabled the pines to survive in full sun. In contrast, stressed seedlings shaded to 10 percent full sun failed to produce the proteins, exhibited reduced drought tolerance, and did not recover upon termination of drought. Always, however, it is the ecosystem-specific interaction of light, water, temperature, and other environmental factors as well that together determine understory survival and growth.

Figure 15.17. The old-growth western white pines will eventually be replaced, barring disturbance, by very tolerant conifers that have become established in the understory; northern Idaho. (U.S. Forest Service photo.)

PHYSIOLOGICAL PROCESSES RELATING TO TOLERANCE

That some understory plants can survive and grow under the site conditions of the understory, while others cannot, indicates a basic genetic difference between species. Such differences are expressed in their physiological response to these environment conditions. The surviving plants must exhibit some superiority over failing plants such as: (1) maintaining a greater photosynthesizing leaf area in the understory, (2) photosynthesizing more efficiently per unit leaf area in the understory, (3) maintaining lower rates of respiration per unit leaf area and per whole plant in the understory, (4) producing more plant tissue per unit of H_2O lost via transpiration, or (5) absorbing water more efficiently.

Present evidence from controlled laboratory experiments of shade tolerance indicates that plant characteristics leading to failure of a genotype or species in one environment may be an indirect consequence of adaptations necessary for survival in another (Grime, 1965a). Thus the adaptation of photosynthetic and respiration mechanisms of intolerant species for full productivity under full sunlight is achieved at the cost of lowered efficiency under shade conditions. Although the rates of photosynthesis have been found closely associated with plant performance in shade (Logan and Krotkov, 1969, Logan, 1970), the differences in rates of respiration between tolerant and intolerant species are probably the most important determinants of success or failure in forest shade (Grime, 1965b; Loach, 1967). In such situations, the plant may spend many more hours below than above the light compensation point. Intolerant species have high rates of photosynthesis, but they are offset to some extent by high rates of respiration (Chapter 18). These species are highly productive in open sites but are less adapted to shaded conditions where relatively high respiration rates lessen carbon gain (Chapter 19).

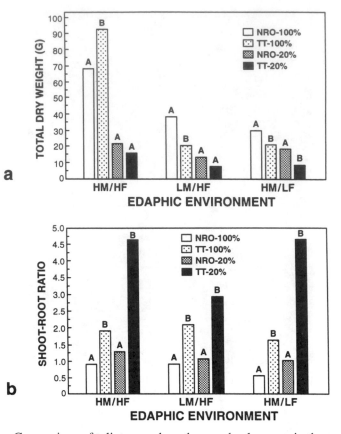

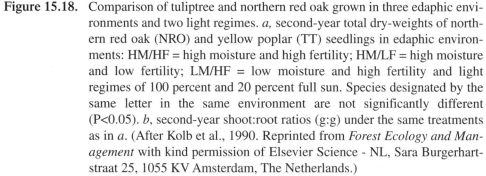

Figure 15.18. Comparison of tuliptree and northern red oak grown in three edaphic environments and two light regimes. *a,* second-year total dry-weights of northern red oak (NRO) and yellow poplar (TT) seedlings in edaphic environments: HM/HF = high moisture and high fertility; HM/LF = high moisture and low fertility; LM/HF = low moisture and high fertility and light regimes of 100 percent and 20 percent full sun. Species designated by the same letter in the same environment are not significantly different (P<0.05). *b,* second-year shoot:root ratios (g:g) under the same treatments as in *a.* (After Kolb et al., 1990. Reprinted from *Forest Ecology and Management* with kind permission of Elsevier Science - NL, Sara Burgerhartstraat 25, 1055 KV Amsterdam, The Netherlands.)

Tolerant species are more competitive in shaded environments through selection for low respiration rates (Went, 1957); they also tend to have lower photosynthetic rates and hence grow slowly in all environments. In addition, shade-tolerant species such as beech and sugar maple not only have the ability to open their stomates in low light but to open them rapidly to take advantage of light flecks over short periods for photosynthesis, even though their absolute rates of photosynthesis may be low (Woods and Turner, 1971; Davies and Kozlowski, 1974).

A comparison of rates of photosynthesis and respiration for sun and shade leaves of several tolerant and intolerant hardwood and conifer species of northern temperate forests showed that intolerants made more efficient use of high light (Loach, 1967). However, they suffer the greatest proportional reduction in photosynthesis when grown in shade.

Significantly, the respiration rates of tolerant species are consistently less than those of intolerants. Thus selection for a high rate of photosynthesis at high light irradiances and high growth rate in full sunlight may inevitably limit the plant in the shade. Two additional experiments bear out this relationship.

Experimental studies with tuliptree and northern red oak illustrate the responses of plants representing markedly different site adaptations and growth strategies. From field experience we characterize tuliptree as a pioneer opportunist that requires full sun, high nutrient status (Mitchell and Chandler, 1939), and moist sites to thrive and dominate the forest overstory. In contrast, red oak is suited to less optimal conditions of light (midtolerant), nutrients (Mitchell and Chandler, 1939), and soil water. With a seed 300 times heavier than tuliptree and the ability to develop a deep tap root, it should be less sensitive to reduced levels of light, nutrients, and soil water. Early growth comparisons between these species with interacting levels of these site factors bear out this expectation (Kolb et al., 1990). The two species were grown from seed in containers for two seasons in six environments consisting of three soil conditions (high moisture/high fertility, high moisture/ low fertility, low moisture/low fertility) and two light intensities (100 and 20 percent full sun). The total dry-weight of tuiliptree was 36 percent greater than that of red oak under combined high levels of light, moisture, and nutrients (Figure 15.18a). In contrast, the total dry-weight of red oak was 38 to 126 percent greater than that of tuliptree where at least one of the site factors was at a suboptimal level. The shoot:root ratio of tuliptree was significant greater than that of red oak in all environments (Figure 15.18b)—a greater weight in leaves and stems of tuliptree coupled with a marked reduction in roots compared to red oak. In summary, the tuliptree illustrates opportunistic use of plentiful resources by higher relative growth rate (Loach, 1970) and high photosynthetic rate. The greater tolerance of red oak to low resource levels is due to several characters including its ability to regulate water content during drought (Bahari et al., 1985), its relatively low respiration in shade (Loach, 1967), and its ability to allocate photosynthate to tap and lateral roots in droughty or nutrient-poor environments.

Northern red oak was also included in a study of growth and CO_2 exchange of several northern hardwood species in controlled environments. This study confirmed the relative photosynthesis and respiration rates of understory tolerant and intolerant species (Table 15.4; Walters et al., 1993). Relative to the tolerant sugar maple, the intolerant paper birch

Table 15.4 Comparison of Photosynthesis and Respiration for Tolerant and Intolerant Species.

Species	Photosynthesis $(nmol\ CO_2\ g^{-1}\ s^{-1})$		Respiration $(nmol\ CO_2\ g^{-1}\ s^{-1})$		
	Leaf		Leaf	Stem	Root
Paper birch	high	225	22.9	23.2	37.5
	low	102	25.3	22.3	42.1
Northern red oak	high	94	11.1	10.4	12.7
Sugar maple	high	69	14.8	10.0	20.5

[1]Source: After Walters et al., 1993. (Reprinted from Oecologia, © Springer-Verlag Berlin Heidelberg 1993. Reprinted with permission of Springer-Verlag.)

Photosynthesis is net leaf CO_2 assimilation expressed on a leaf mass (nmol CO_2 g^{-1} s^{-1}) base of seedlings grown in high and low light conditions in growth chambers. Respiration is in mass-based values (nmol CO_2 g^{-1} s^{-1}) as overall seedling means.

had high relative growth rates (RGR). Fire-dependent and midtolerant northern red oak that may grow in both open and moderately shaded environments was intermediate in RGR, remarkably low in respiration, and had the highest ratio of root weight to total plant weight. Leaf, stem, and root respiration of paper birch were not significantly different for high and low light levels, but these were significantly higher than for sugar maple or red oak.

Generally, physiological studies under controlled conditions support field observations of relative understory tolerance and continue to provide insights to the complex interactions involved. The relationship of photosynthesis and total plant respiration in determining the carbon balance of individual plants and whole ecosystems is considered in Chapter 18.

Suggested Readings

Austin, M. P., and T. M. Smith. 1989. A new model for the continuum concept. *Vegetatio* 83:35–47.

Boucher, D. H., S. James, and K. H. Keeler. 1982. The ecology of mutualism. *Ann Rev. Ecol. Syst.* 13:315–347.

Bronstein, J. L. 1994. Our current understanding of mutualism. *Quat. Rev. Biol.* 69:31–51.

Dansereau, P. (ed.), 1968. The continuum concept of vegetation: responses. *Bot. Rev.* 34:253–332.

Gleason, H. A. 1926. The individualistic concept of the plant association. *Bull. Torrey Bot. Club.* 53:7–26.

Grubb, P. J. 1985. Plant populations and vegetation in relation to habitat, disturbance and competition: problems of generalization. In J. White (ed.), *The Population Structure of Vegetation* (Handbook of Vegetation Science, Part III), W. Junk, Dordrecht.

McIntosh, R. P. 1993. The continuum continued: John T. Curtis' influence on ecology. In J. S. Fralish, R. P. McIntosh, and O. L. Loucks (eds.), *Fifty Years of Wisconsin Plant Ecology*. Univ. Wisc. Press, Madison.

Nicolson, M. 1990. Henry Allan Gleason and the individualistic hypothesis: the structure of a botanist's career. *Bot. Rev.* 56:91–161.

Rowe, J. S. 1969. Plant community as a landscape feature. In K. N. H. Greenidge (ed.), *Essays in Plant Geography and Ecology.* Nova Scotia Museum, Halifax, N.S.

Shrader-Frechette, K. S., and E. D. McCoy. 1993. *Method in Ecology; Strategies for Conservation.* Cambridge Univ. Press, Cambridge. 328 pp.

PART 5

FOREST ECOSYSTEM DYNAMICS

Forest ecosystems, like all other types of terrestrial and aquatic ecosystems, are dynamic entities that change through time and over space. In the previous sections of this book, we have discussed how climate, physiography, and soil influence the geographic distribution of forest ecosystems at local, regional, and global scales. That is, we have considered how the physical environment, in combination with life-history traits of forest trees and associated plants and animals, influence the composition and structure of forest ecosystems from place to place at a given time. In this section, we consider the extent to which disturbance (Chapter 16), an ecosystem process initiating change through time (termed succession, Chapter 17), alters the flow of carbon (Chapter 18) and plant nutrients (Chapter 19) within forest ecosystems. The diversity of forest organisms and ecosystems is considered in Chapter 20, and in Chapter 21, we discuss landscape ecology and how a landscape-ecosystem approach can be used for management and conservation.

Forest ecosystems experience a wide array of disturbances that alter the biotic community and the physical environment in many ways. Fire, flooding, insects, windstorms, and harvesting can remove one complement of plants and animals, and over time, replace it with another. In Chapter 16, we treat the disturbance as an ecosystem process, recognizing that events which destroy or damage forest organisms are an integral component of the ecosystem. Major disturbances that alter both the forest community and physical environment are discussed in detail. The treatment of disturbance leads naturally to the consideration of succession in forest ecosystems. In Chapter 17, we discuss the concepts, causes, mechanisms, and models of succession and use examples from different forests to illustrate how the composition and structure of forest ecosystems change following a particular type of disturbance.

Disturbance and subsequent changes in the forest community fundamentally alter the pattern in which carbon and plant nutrients are cycled and stored within forest ecosystems. In Chapter 18, we discuss the physiological processes controlling the capture, storage, and loss of carbon (C) by forest trees, a set of processes that directly control forests growth. We then build upon these principles to understand the capture, storage, and loss of C by entire forest ecosystems, and the extent to which disturbance alters this ecosystem process. Changes in the growth of forest ecosystems following a disturbance also change the demand for nutrients, which are needed by the plant to construct new tissue. The cycling of nutrients is considered in Chapter 19, in which we trace the flow of growth-limiting nutrients into, within, and out of forest ecosystems. We place particular emphasis on the extent to which both natural and human disturbances alter the pattern in which nutrients are cycled and stored within forests.

The diversity of plants and animals is a key attribute characterizing regional and local ecosystems around the Earth. Furthermore, the diversity of landscape ecosystems determines biodiversity itself. Therefore, in Chapter 20, both biodiversity and ecosystem diversity are considered at regional and local scales. The value of biodiversity and its measurement is treated in detail, as is the maintenance and conservation of ecosystems and their biotic diversity.

In Chapter 21, landscape ecology—the study of spatial patterns and heterogeneity of landscape features, such as ecosystems, communities, and patches—concludes this part on ecosystem dynamics. A diverse sequence of landscape ecology studies is presented, including examples of forest-stream linkages in the Pacific Northwest, hurricane-patterned landscapes in New England, fire-structured areas of Yellowstone Park, and the geomorphic patterning of Michigan landscapes. In addition, ecosystem management at the landscape level is considered using examples of innovative harvesting practices in old-growth forest of the Pacific Northwest, remote sensing techniques, and the ecosystem approach to conserving endangered species.

CHAPTER 16

DISTURBANCE

Disturbances are landscape ecosystem processes that significantly affect ecosystem composition, structure, and function. Disturbances destroy forest organisms and change site conditions. Their direct effects on parent material, soil, and hydrology cause ecosystem change of which alterations in species composition are one part. Disturbances selectively destroy biota and initiate changes in mutualistic and competitive relations among organisms, often directing the course and rate of vegetation change. Natural disturbances, from plate tectonics and volcanism to local avalanches, floods, and fires, shaped landscapes and the biota that evolved with them.

Particular kinds of natural disturbance characterize different kinds of landscape ecosystems—avalanche in mountains, fire in dry plains, hurricane along sea coasts, flood in river valleys. Disturbance acts, in part, to determine genetic differentiation of plants of these systems as well as causing the more visible vegetative changes in ecosystem structure and composition. Natural disturbances of all kinds were prevalent in North American ecosystems before the first migrants from eastern Asia set foot on the continent. Thereafter, however, the cultural landscape developed as Native Americans significantly disturbed environments close at hand in many ways. Opinions vary as to how undisturbed the forests were that greeted the European settlers of North America. Early writings and literature tend to emphasize the "forest primeval" as undisturbed old growth and the Clementsian "climax" forest (Whitney, 1994). However records from various sources document the existence of both catastrophic and localized disturbance patterns of wind and fire associated with regional ecosystems and landforms. Denevan (1992) cites evidence of widespread disturbance associated with Native American presence in the early sixteenth century. Whitney (1994) observes: "If the primeval forest did not consist of stagnant stands of immense trees stretching with little change over vast areas (Cline and Spurr, 1942), neither was it an amalgamation of pioneer species recovering from one form of disturbance or another."

Superimposed on natural ecosystem disturbances are those caused by humans. Throughout the world, human alteration of original landscape ecosystems has been enormous (Chapter 14). Wherever humans occupy a region they significantly affect the soil, microclimate, and biota through the one distinctive human possession, according to Heizer (1955), which we call culture. Trees have been the principal fuel and building material of almost every society for over 5,000 years from the Bronze Age to the middle of the nineteenth century. In documenting the role of wood in the development of civilization, Perlin (1989) describes deforestation and its consequences over enormous scales of space and time. In an equally illuminating way for eastern North America, Whitney (1994) has described disturbances of all kinds, natural and human, and their consequences from 1500 AD to the present. For overall perspective and details of disturbance that are beyond the scope of our treatment, these and other books (Pickett and White, 1985; Oliver and Larson, 1990; Coutts and Grace, 1995) are excellent sources. In this chapter, we consider the concept of disturbance and primarily the kinds of disturbances—fire, wind, insects and pathogens, logging, land clearing, biotic changes—that bring about ecosystem change known as secondary succession. This introduction prepares the way for examination of succession as an ecological process in Chapter 17.

The forest ecologist and silviculturist are primarily concerned with ecosystem change that is initiated by disturbance. Their focus is often regeneration of forest species and changes over time in composition, stand structure, and biomass, i.e., forest dynamics or **secondary succession**. Such change following disturbance of the existing forest is in contrast to **primary succession**, which occurs on previously unvegetated sites such as water or bare land or rock. Throughout any forest region, disturbances of one sort or another are constantly altering the course of forest succession, and initiating or re-starting secondary succession. It is not surprising, therefore, that disturbance ecology is one of the most exciting fields of forest ecology.

DISTURBANCE AS AN ECOSYSTEM PROCESS

The word "disturbance" refers to the breaking up of settled order or orderly working (i.e., any departure from normal; Neufeldt and Guralnik, 1988). Given that many ecologists would not agree that there is any necessary "order" in Nature, we interpret disturbance to mean *any relatively discrete event in time that disrupts ecosystems, their composition, structure, and function*. Such a definition includes disruption of ecosystem components such as landforms (landslide and volcanic activity), soil (flood, erosion, debris flow), vegetation (many factors), and any of the processes that link these components. Definitions of disturbance typically focus on organisms, such as "a force that kills at least one canopy tree" (Runkle, 1985), "the mechanisms which limit the plant biomass by causing its partial or total destruction" (Grubb, 1988), or events that make growing space and other resources available to survivors or new colonists (Glenn-Lewin and van der Maarel, 1992). Context is important, and because organisms are often our primary focus, we tend to conceive disturbance as death and destruction. Even more fundamental, however, is that disturbances (plate tectonics, volcanism, glaciation, fire, windstorm, flood), directly or indirectly, are natural events that occur in all ecosystems.

When flood, fire, or wind kill forest organisms, the sequence of processes occurring in the field, or "order" at the time, is disturbed. Following such disturbances we can study recovery mechanisms and the organisms that replace those destroyed. However, we can't do this without understanding the underlying ecosystem conditions that predispose ecosystems to disturbance or the processes that mediate the changes following disturbance. Therefore, it is of first-order importance to conceive disturbance in the context of ecosystems. In doing so, we find that certain disturbances are characteristic of physiographic features (Chapter 10)—flooding in riverine ecosystems, avalanche on steep mountain slopes, and fire on flat, dry or rolling plains. Notice that we can express disturbances by rates of occurrence per unit of time (flooding and avalanche frequency, fire return interval), just as we can characterize other ecosystem processes by rates of change. For example, estimates of rates of fire return for a variety of forest types are given by Heinselman (1981b), Oliver and Larson (1987), and Whitney (1994).

From an ecosystem perspective several insights are worth considering in the sections that follow.

Scale of Disturbance. The kinds and patterns of disturbance can be characterized at different scales. At the broad regional level some parts of North America are more susceptible to hurricanes (Gulf and Atlantic coasts) than other areas, and the same is true of tornadoes and glaze storms (mid-continent) and lightning fires (western mountain regions). At this scale, macroclimate and physiography strongly affect the kind of disturbances typical for a region. At the local level, specific ecosystem characteristics (physiography, soil, vegetation) and position of an ecosystem in relation to other ecosystems strongly determine the kind, frequency, and severity of disturbance (Chapter 10). With the ability to map regional and local ecosystems, we have the means to characterize disturbance by ecosystem type as well as by plant communities.

Sites Created and Changed. Although we tend to focus on the dramatic effect of disturbance on individual plants and communities, disturbance has major, multiple effects on forest sites that are instrumental in affecting regeneration and forest succession following disturbance. Certain disturbances strongly affect the physical site by creating new landforms, rearranging parent material, or otherwise changing soil properties. These include mountain building, volcanism, earthquake, and glaciation at very broad scales and land movement (avalanche, debris flow, mud slide, soil creep), flood water, wind, and fire at lesser scales. Wind creates new landforms and substrates such as sand dunes and loess deposits; flood water also creates new landforms by its cutting river banks and forming new land (point bars, levees, bottoms). By uprooting trees wind rearranges soil, creates pit and mound microtopography, and thereby creates a diversity of microsites (Chapter 10). Destruction of vegetation changes the light and temperature regimes of the site. Thus these disturbances not only affect biota directly but have far reaching indirect effects on regeneration and forest succession through both catastrophic and subtle site changes.

Differential Disturbance Effects. Disturbance factors have differential effects on site and biota. Those of fire, wind, and flooding will be highlighted below. The markedly different effects on site and vegetation mean that processes of regeneration and succession are different as well. Wind primarily damages the large, old, overstory trees; young stands and the understory of old stands are rarely affected. In contrast, fire always affects the understory—whether catastrophic fire that completely destroys overstory and understory or light surface fire that severely damages only the understory. In contrast to wind and fire, flooding has more significant direct effects on landform and soil, thereby indirectly affecting the biota. Regenerating understory plants are much more vulnerable to flooding than overstory trees. Fire is more likely than wind to maintain and regenerate early-successional communities (via a new seedling or sprout stand), whereas wind disturbance has a greater potential to hasten succession to more tolerant species if they are well established in the understory. Flooding maintains those plants tolerating or requiring its effects.

Exclusion of Disturbance. Infrequent and severe disturbances to an ecosystem may lead to nearly irreversible change. However, if the disturbance is sufficiently regular with temporary effects, then ecosystem components may adapt and eventually require disturbance to maintain a normal, resilient system (Wareing, 1989). In this sense, the effects of such factors then can hardly be considered disturbances. Suspension or exclusion of such disturbances as fire and flooding then constitutes the ecosystem disturbance with new and unexpected consequences. For example, fire is an integral feature of many kinds of ecosystems, from southern pine forests to subarctic black spruce–jack pine forests. When fires are suppressed, fire exclusion becomes the disturbance, or disruption, to those systems with accompanying changes in ecosystem structure, composition, and function. For example, fire-dependent species, such as most oaks and pine, fail to regenerate, and more fire-sensitive invasive species are likely to take advantage of the changed site conditions.

Interactive Multiple Disturbances. Although we tend to think in terms of single disturbance factors (fire, wind, water) or events, ecosystems experience multiple disturbances. Effects on site or vegetation over the long-term cannot always be assigned to a specific type of disturbance, especially when disturbances follow one another. For example, frost, drought, and insect attack may cause severe tree mortality as a result of multiple and interacting stresses. Fire, windfall, fungi, and bark beetles often interact to determine forest development (see Figure 16.10). Prolonged drought typically precedes the most severe wildfires, as in the massive western fires of 1910 and those in the Yellowstone Plateau of 1988. Disturbance factors are often interrelated: the fire-scorched tree is linked to successful attack by bark beetles; drought provokes insect outbreaks (Mattson and Haack, 1987). Widespread windfall often leads to insect attack and fuel accumulations that promote wildfire. Mud slides follow fire in steep-slope, chaparral-dominated ecosystems of California. Whereas coastal mangrove ecosystems around the world experience hurricanes, sea-level surges, flood, and freezes, boreal forest ecosystems experience a markedly different kind, severity, and frequency of disturbances. Not surprisingly, therefore, the pattern and nature of multiple disturbances is characteristic of regional ecosystems with their distinctive climates (Mediterranean, northern continental, coastal/marine), physiographic features (mountains, plains), and soil/hydrology (excessively drained, poorly drained).

Biological and Physical Legacies. Most natural forest communities don't start "from scratch" following a major disturbance because of **biological legacies** (Franklin, 1997). Natural catastrophes of fire, flood, windstorm, avalanche, or outbreaks of insects and disease kill trees and other organisms, but they typically leave behind much of the carbon in the form of snags (standing dead trees) and down logs. Large legacies of living organisms, including trees, are also left behind because most disturbances are patchy. **Physical legacies** left after disturbance include bare or rearranged mineral soil (following fire, flooding, landslide) and new microtopography as the result of tree uprooting by wind, soil movement via debris flows, and the cutting and depositing actions of flood waters. Studies of regeneration following natural disturbances have emphasized the importance of biological legacies for the rapid reestablishment of forests that have high levels of diversity of composition, function, and stand structure. In contrast, old agricultural fields and large clearcuts offer minimal levels of legacies.

Human-Caused Disturbances. Human disturbances may be markedly more devastating to site or biota than those that occurred under natural conditions before human occupation of North America (about 12,000 years ago). For example, land clearing and cultivation drastically change both site and biota. The combination of overstory logging and multiple fires that destroyed forest understories of many areas of the upper Great Lakes, following settlement by Europeans drastically changed conditions for regeneration and hence the structure and composition of forests. Modern commercial clearcutting and timber extraction, followed by planting, generate forests strikingly simplified in structure, composition, and different in function from naturally-regenerated ecosystems (Franklin, 1995). Recovery of herbaceous layers in Appalachian cove forests following clearcutting may require centuries (Duffy and Meier, 1992).

Modern burning of biomass on a global scale is of concern (Goldammer, 1990a; Levine 1991a). Forest biomass accounts for 1,540 teragrams (Tg) of the total 8,680 Tg worldwide (1 teragram = 1 million tons = 10^9 kg of dry material) and contributes substantially (18 percent) to the world's annual production of CO_2 (Levine, 1991b). Fire and flood exclusion over the last 100 years has dramatically changed forest ecosystems. To this could be added urbanization and fragmentation of landscapes by roads, fields, and power lines and the introduction of invasive exotic species. Their effects tend to confound our perceptions of what the structure, composition, and processes of presettlement forests might have been.

Source of Disturbance

Let us next examine the source of disturbance. Disturbance must be initiated from outside the plant community, noted Grubb (1988). Lightning strikes a tree; flood waters kill plants intolerant of low O_2; gypsy moth or spruce budworm arrive en masse. With an organism or community focus, disturbance originates from the outside. However, with an ecosystem focus, the frequency, extent, severity, and consequences of disturbance from external factors are closely related to the site conditions where the plants are growing: the lightning-struck tree is on a ridgetop, not in the protected valley; the flooded plants are on the first bottom not the elevated second bottom. Thus an ecosystem context is paramount in understanding disturbance effects.

Examples are innumerable. Site and biota go hand in hand in determining the frequency and severity of fire: dry, nutrient-poor sites tend to favor conifers with flammable foliage, whereas moist, fertile sites favor species with rapidly decomposing leaves thus reducing fuel accumulation. Windthrow is markedly prevalent on sites with high water tables or hardpan development where rooting depth is limited but where trees may grow tall and are susceptible to wind bursts. The intimate relationship of water table and transpiring tree crowns is revealed when American elms in swamp ecosystems in northern Lower Michigan die in large numbers from disease. The water table rises due to lessened transpiration; this rise puts the root systems of other species at risk. Basswood and red maple trees are next to die and initiate further raising of the water table. At this point the northern white-cedar roots are overwhelmed as well, and only the most tolerant species of high water table, the black ash, remains together with aquatic vegetation that has invaded.

In summary, disturbance is typically ecosystem-specific, and its kind, source, frequency, severity, and consequences are best understood in this context. In the sections that follow, we examine several major disturbances that affect the structure, composition, and processes of forest ecosystems.

MAJOR DISTURBANCES IN FOREST ECOSYSTEMS

The most obvious disturbances to forest ecosystems are those that alter sites or partially or completely destroy the forest community by killing either the trees in the overstory or the trees and other plants in the understory.

Catastrophic and Local Land Movement

Catastrophic events such as volcanism, earthquakes, and landslides create new ecosystems and markedly influence species occurrence and distribution. Volcanic activity on several continents provides new substrate for forest development, including the Cascades of the Pacific Northwest, the Andes in Chile (Veblen, 1987), the Hawaiian Islands, and the Changbai mountains of northeastern China (Barnes et al., 1992). The eruption of Mt. St. Helens provides recent evidence of the massive effect on existing vegetation and the rapidity of succession on newly formed substrates as a result of biological legacies of diverse kinds (Franklin et al., 1985; Franklin et al., 1988; Franklin, 1990). Landslides in mountain terrain provide new substrates for plant occupancy. For example, in the southern Chilean lake district, the 1960 earthquake triggered volcanic activity and thousands of landslides which provided excellent regeneration sites for the southern beeches (*Nothofagus* spp.) (Veblen, 1987). Such large-scale disturbances are frequent enough to influence forest composition and structure of the beeches over the long term. Also, landslides and rock avalanches, probably initiated by prehistoric earthquakes, changed landscapes and affected forests in mountains of the Puget Sound area of Washington State about 1,000 years ago (Jacoby et al., 1992; Schuster et al., 1992). In the Appalachian Mountains the debris avalanche, either a catastrophic or awesome local event, is a major disruptive and land-shaping force (Hack and Goodlett, 1960; Eschner and Patric, 1982; Jacobson et al., 1989). A debris avalanche, a rapid downhill flow of soil, rock, and vegetation, is almost always preceded by heavy rains. The significance of debris avalanches in the Pacific Northwest is considered in Chapter 21.

Fire

Fire is the dominant fact of forest history (Spurr, 1964). The great majority of forest ecosystems of the world—excepting only the perpetually wet rain forest, such as that of southeastern Alaska, the coast of northwestern Europe, and the wettest belts of the tropics—have been burned over at more or less frequent intervals for many thousands of years. Even under present-day conditions, marked by a great awareness of forest fires and forest fragmentation by intervening tracts of farmland and settlements and roads and trails, fire continues to be a major disturbance factor in much of the North American forest. However, from about 1900 to 1940 organized fire protection activities were mounted and have become increasingly effective in reducing the number and size of fires.

The condition was quite different up to the 20th Century. Primitive people throughout the world, and most civilized people as well, had until recently no compunction about burning the forest, and no desire or intent or ability to put out existing fires, whether lit by humans or lightning. In fact, throughout the world, fires have been set deliberately for thousands of years to clear the underbrush, improve grazing, drive game, combat insects, without thought, or just for the hell of it. As more and more historical research is conducted into the ecological history of fire, the more it is realized that frequent burning has been the rule for the vast majority of the forests of the world as far back as we have any evidence (Chapter 12).

Within forest regions, fires have been primarily responsible for the development of heathlands and moors of western Europe and the British Isles, for many of the savannas within the tropical forest belts, for upland meadows within the forests of the American mountains, and in general for the persistence of grassland areas on upland sites within forest regions. Around the world, the dominance of pine and oak forests of virtually all upland species and in virtually all regions is due to fire. So are the vast areas of Douglas-fir in the Rocky Mountains and Pacific Northwest and of eucalyptus in Australia. Even the extensive areas of spruce in the boreal forest of North America and Eurasia are structured to a great extent by past fires (Bloomberg, 1950; Sirén, 1955; Viereck, 1973).

The foregoing statements are sweeping and perhaps overstated. Nonetheless, they reflect the feeling of many forest ecologists who, wherever they have studied and worked, have come increasingly to realize the great importance of forest fires and succession following fires in framing the local forest composition and structure (Cooper, 1961; Agee, 1993; Whelan, 1995).

In the context of the established and growing forest community, fire is clearly a disturbance that disrupts the development of the existing stand. However, from the viewpoint of the forest ecosystem, fire is a natural factor whose effects have long been incorporated in species' adaptations and ecosystem dynamics (Chapter 12). The importance of fire in forest ecosystems is highlighted by examining the similar roles it plays in many different fire-dependent systems.

ROLE OF FIRE IN FOREST ECOSYSTEMS

Around the world fire has played important and similar roles in fire-dependent conifer and hardwood forests in presettlement times. Many ecosystems are regularly recycled by fire; existence for many forest species literally begins and ends with fire. The following major functions and processes are regulated by fire:

Regeneration and Reproduction. Catastrophic fires kill existing stands and set the process of regeneration in motion for the next forest. Fire, as a selective force, also elicits the following reproductive characteristics of forest trees:

1. Asexual reproduction, primarily sprouting, occurs in all fire-dependent angiosperms and in some conifers (Chapter 12).

2. Sexual reproduction by light, wind-blown seeds is favored by fire, and self-fertility in pines may have evolved as a result of fire. Although a small percent of self-fertilized seeds are viable in many pines, red pine is highly self-fertile. The resulting seeds are viable, and the seedlings show no growth depression due to selfing, unlike most other tree species (Fowler, 1965a,b). Thus a single isolated red pine that survives a severe fire may self-fertilize and perpetuate itself by establishing a colony of seedlings.

Preparation of Seedbeds and Dry-Matter Accumulation. Fire reduces the amount of litter, sometimes bares mineral soil, and greatly enhances seedling establishment. Seeds are often partially buried in the ash layer, thus favoring germination. Fire regulates dry-matter accumulation, thus influencing the severity of burning.

Reduction of Competing Vegetation. Severe fires eliminate trees, shrubs, and herbs, thereby favoring establishment of a new stand. Light surface fires thereafter reduce encroaching vegetation and reduce competition for light, soil water, and nutrients. This effect is typical in many mixed conifer–hardwood stands in the South in which understory hardwood regeneration is reduced by periodic burning. Similarly, in oak forests, periodic surface fires kill seedlings of understory-tolerant species that continuously establish on the forest floor and grow into the understory.

Nutrition. Throughout the existence of the stand, recurrent surface fires reduce organic matter to basic components of water, CO_2, and mineral nutrients. Nutrients that are otherwise unavailable to the forest vegetation become available (Chapter 19). This nutrient supply is particularly important in conifer forests and wherever site conditions are unfavorable for decomposition. Although nitrogen is lost in burning, surface soil conditions after burning often favor nitrogen-fixing bacteria and soil organisms.

Thinning. In many conifer-dominated ecosystems, dense pure stands of seedlings that become established after fire, are thinned by periodic surface fires. The larger, faster growing seedlings with thicker bark are favored, and competition for soil water and nutrients is temporarily reduced.

Sanitation. Fire creates the dense, even-aged stand conditions that are conducive to disease and insect outbreaks. Such epidemics generate fuel concentrations leading to intense fires. Fire terminates the outbreaks (such as those of bark beetles and the spruce budworm), destroying living as well as dead trees. It creates conditions for the establishment of a new even-aged stand that is resistant for a time to disease and insect attack. Sooner or later the maturing stand is again susceptible to epidemics, and thus a self-perpetuating cycle is established. Fires also eliminate plant parasites such as mistletoes on ponderosa pine, lodgepole pine, and black spruce.

Succession. Depending on site conditions, fire tends to retain fire-dependent species on an area as long as fire frequency and intensity are balanced with the species' fire adaptations. Ponderosa pine, the southern pines, red and white pines, upland oaks, and many other species may be regenerated generation after generation, although they are not the most understory tolerant or late-successional species of their respective regions. However, if the fire frequency or intensity increases, a given species is likely to be replaced by a more fire-dependent species. For example, in the Rocky Mountains, Engelmann spruce is replaced by Douglas-fir, Douglas-fir is replaced by ponderosa pine, and ponderosa pine is replaced by grasses. On the other hand, if fire frequency decreases (by fire exclusion), a less fire-dependent type would prevail in each elevational zone: Engelmann spruce would replace Douglas-fir, Douglas-fir would replace ponderosa pine, and ponderosa pine savanna or forest would replace grassland. Thus a mosaic of fire-dependent communities of different age classes was typical of many North American upland presettlement forests. This pattern resulted, in part, from the varying frequency and severity of fire and its erratic burning pattern, which in turn was controlled by heterogeneous soil-site conditions and fuel build-up on the different ecosystems characteristic of a given region (Williamson and Black, 1981).

Very understory-tolerant species of both conifers and hardwoods tend to be susceptible to fire (sugar, red maple, beech, hemlocks, firs, northern white-cedar and western redcedar). They typically occupy protected or moist sites that are least susceptible to burning. In intervals between significant fires, they colonize adjacent areas, establishing and eventually replacing fire-dependent species in the absence of fires intense enough to kill them. However, one must be careful not to generalize too broadly. For example, although the eastern hemlock is susceptible to fire due to shallow roots, it may regenerate vigorously following burns, which provide highly suitable seedbeds for its tiny, wind-dispersed seeds.

Wildlife habitat. Fire universally created a mosaic of habitats and niches for wildlife of many kinds (Chapter 14). Species diversity, as with plants, tends to increase following fire until crown closure occurs; then it declines.

From the many examples that may be cited to demonstrate these roles and significance of fire, a selection representing important North American trees and regions will illustrate fire's importance in forest ecosystems.

Pines in New England and the Lake States

In the northeastern United States and adjacent Canada, the occurrence of the two- and three-needled pines (red pine, jack pine, and pitch pine) as well as of even-aged pure stands of white pine (but not of individual white pine in mixed forests) is largely controlled by the past occurrence of forest fires. In colonial days, many of the fires resulted from burning by European settlers in land-clearing operations, but the evidence is ample that fires were commonly set by Native Americans for many hundreds of years before the settlers arrived (Cline and Spurr, 1942; Day, 1953; Curtis, 1959; Little, 1974; Denevan, 1992). Because of the heavy precipitation commonly associated with summer thunderstorms, lightning apparently has played a lessor role in causing fires here than in the West.

Ecological studies of a few relict old-growth pine stands have all shown that fire played an important part in their formation. In both northwestern Pennsylvania (Lutz, 1930; Hough and Forbes, 1943) and southwestern New Hampshire (Cline and Spurr, 1942), more or less pure even-aged stands of old-growth white pine originated from past forest fires. In contrast, nearby mixed types with occasional dominant white pine were relatively free from evidence of past burns. In northwestern Minnesota, extensive even-aged stands of old-growth red pine clearly date from a series of forest fires, many of which antedate the advent of the settlers (Spurr, 1954; Frissell, 1973).

Jack pine in the Lake States and Canada and pitch pine on sandy soil near the mid-Atlantic coast (Little, 1974) are virtually completely fire-controlled (Chapter 12). Jack pine, for instance, grows in nearly pure stands on dry sandy soils, forming a highly flammable vegetational type. At intervals of a few decades, the jack pine stands are burned during hot dry periods, with the fire characteristically crowning and killing, in large patches, all the vegetation. The next generation of trees arises from four sources: (1) in the case of jack pine, from seeds stored over many years' accumulation of serotinous cones in the tree crown; (2) in the case of red pine, by seeds from residual veteran seed trees with bark of sufficient thickness and clear bole of sufficient length to permit them to survive the fires without crowning out; (3) in the case of the hardwoods such as bigtooth and trembling aspen, northern red oak, and red maple, by sprouts arising either from roots or the lower part of the stem not killed by the fire; (4) in the case of pioneer hardwoods, by the dissemination of seeds into the area from afar either by wind for aspens and birches, or by birds for the cherries; and (5) in the case of pin cherry and black cherry, from seeds stored in the seed bank of the forest floor (Marks, 1974).

Western Pines and Trembling Aspen

The remarks made concerning the northeastern United States pines can be applied with little modification to virtually all other pine species growing in the United States. In the interior of the western forest, lodgepole pine plays an analogous part to that of its close relative, jack pine, in the Lake States. It regenerates on recent burns, largely from seeds stored in serotinous cones of trees killed by the fire, to form dense, even-aged, post-fire pioneer stands (Figure 16.1). In 1988, fires burned 570,000 ha of the Yellowstone Plateau in northwestern Wyoming, including approximately 240,000 ha of forested land in Yellowstone National Park (Schullery, 1989). Despite massive suppression efforts, the fires were essentially unstoppable until precipitation and cool temperatures arrived on September 11 (Ellis et al., 1994). As expected in such a fire-dominated system, regeneration of lodgepole pine has blanketed the area in succeeding years in patterns related to site conditions, stand density, and fire severity.

In the wake of the catastrophic fires of 1988, questions were raised as to whether the Park's policy of complete fire suppression from 1872 to 1972 led to abnormal fuel conditions and therefore to abnormal fire spread, behavior, and damage. Results of research suggest that the 1988 fires should not be viewed as an abnormal event for this area (Romme and Despain, 1989). In terms of total area burned, rate of spread, and severity, the 1988 fires were similar to those that occurred around 1700. Although fire suppression may have had some influence on size and behavior of the 1988 fires, they were primarily due to climatic conditions and the normal successional dynamics and fuel build-up following the last major fires approximately 280 years ago.

Given several hundreds of years free from forest fire, the lodgepole pine gradually will be replaced by tolerant Engelmann spruce and subalpine fir. In Alberta, white and black spruce also play a part as late-successional species. Douglas-fir and other more tolerant western conifers become prominent toward the Pacific Northwest. In subalpine and high foothills in Alberta, Horton (1956) estimates that from 225 to 375 years exclusion from fire is required for succession to take place from pine to spruce and fir. On the drier southern slopes, succession takes much longer, if indeed it ever takes place. In this regard, the multiple disturbances that maintain lodgepole pine and affect its stand development are discussed below in connection with Figure 16.10.

At lower elevations and in warmer, drier portions of the western forest, pure stands of ponderosa pine are commonly a product of a long and complex fire history (Figure 16.2; Cooper, 1960; Fischer and Bradley, 1987). In the cooler and moister portions of its range, ponderosa pine comes in as a pioneer following fire and is gradually replaced by more tolerant conifers such as Douglas-fir, incense cedar, and white fir. In the warmer and drier portions of the ponderosa pine range, it may be a late-successional species. Fire regimes and stand development of ponderosa pine are discussed for a variety of regional and local ecosystems by Fischer and Bradley (1987), Agee (1993), and Arno et al.(1995).

Trembling aspen is one of the few hardwood species, along with oaks, able to compete with conifers in the fire-dominated western mountains. It competes as a pioneer on the higher, cooler, and wetter sites, coming in as wind-disseminated seed in the northern Rocky Mountains and in western Canada. In the Great Basin and the central and southern Rocky Mountains, aspen root suckers predominate, and aspen is maintained as a late-successional type by fire in certain localities (Baker, 1918; Mueggler, 1985). In contrast to the eastern and northern parts of its range, aspen clones of the central and southern Rockies may become very large, up to 43 ha in extent and contain thousands of genetically identical stems (Kemperman and Barnes, 1976; Chapter 5, Figure 5.14). The presence of many large clones suggests that individual clones may have been perpetuated for thousands of years following rare events of seedling establishment by a combination of fire and little browsing (Barnes, 1975; Kay, 1997). However, many of these central and southern Rocky Mountain clones are now deteriorating, and, due to fire suppression and excessive browsing, they are not regenerating (Schier, 1975; Kay, 1997). In the moister northern Rocky Mountains, aspen seedlings may establish readily following fire in appropriate microsites. For example, following the major fires in Yellowstone Park, thousands of aspen seedlings were observed over a broad range of vegetation types (Renkin et al., 1994); in 1989, seedling density ranged from 0.6 to 1,014 m^{-2}. Nevertheless, high elk

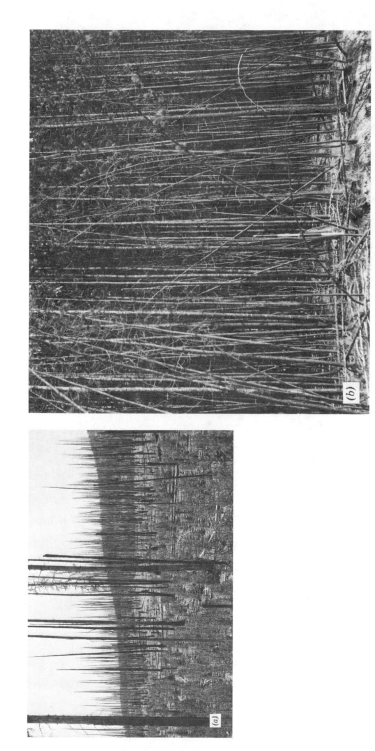

Figure 16.1. Lodgepole pine regeneration following fire in Oregon. *a*) Lodgepole pine revegetates a large burn in central Oregon. Reseeding the area to grass after the burn delayed the establishment of pine, *b*) Under ideal conditions of seedbed and seed supply, dense lodgepole pine thickets develop following fire. This 65-year-old stand of about 25,000 stems per hectare in eastern Oregon averages only 8 m high and 5 cm in diameter at breast height. (U.S. Forest Service photos.)

Figure 16.2. Fire and ponderosa pine. *a*) A surface fire burns grass and litter of this stand in central Idaho. *b*) Fires in presettlement time maintained open, grassy, parklike stands of ponderosa pine such as this one in western Montana. (U.S. Forest Service photos.)

populations pose a threat to their continued regeneration and controlled grazing is advised (Kay, 1997).

Southern Pines

Nowhere is the dependence of the pine forest upon recurring fires more evident than in the southern pine belt of the southeastern United States. Here, earlier travelers wrote of the open character of the "piney woods" due to burning by native Americans. Ever since settlement, local farmers have periodically burned the woods to keep down the "rough" and bring about fresh postfire revegetation suitable for grazing by domestic stock. After many years of futile attempts at complete fire exclusion by foresters, the practice of "prescribe" or "controlled" burning has become well accepted in recent decades to reduce the hazard of a crown fire and to maintain the pine type without reversion to the hardwoods that otherwise would replace it through natural succession.

Of the four most common southern pines, longleaf pine is most clearly dependent upon recurring fires for its perpetuation (Chapman, 1932). Longleaf pine once dominated much of the presettlement Coastal Plain (Christensen, 1988a) but is now regarded as the second most threatened forest type in the United States (Noss et al., 1995). The presettlement longleaf pine forests occurred on three relatively distinct kinds of sites—dry sandhills, intermediate areas, and wetlands (shrub-bogs, savannas, flatwoods)—each with different fire effects and successional pathways (Marks and Harcombe, 1981b). Generally, for the perpetuation of pure longleaf pine forests, the proper ratio of winter fires to fire-free years is critical because pine regeneration is prevented by annual burning and summer fires. Furthermore, longleaf pine stands must be burned in the "grass stage" to control the brown spot disease (*Schirria acicola*) and burned repeatedly after a forest canopy is formed to prevent the establishment of understory hardwoods that invade the site and gradually replace the pine in the absence of fire.

Longleaf pine regenerates itself best in full sunlight, as seen in Figure 16.3, on mineral seedbeds created by forest fires that destroy all but scattered longleaf seed trees. It is one of the most understory intolerant and fire-resistant of all pines. Many of the seedlings in Figure 16.3 appear like dense bunches of grass, and this stage is known as the **grass stage**. Longleaf pine seedlings, unlike those of all other North American pines, remain in the grass stage for many years (typically 6 years; range, 3 to 12 years) while a deep tap root develops. In the grass stage, the bud is protected from fire by a surrounding sheath of dense needles. If the needles are destroyed by fire they are replaced by a new set from stored carbohydrates and nutrients in the large tap root. Eventually, rapid stem elongation begins, quickly elevating the crown above the level of periodically occurring surface fires.

The other common southern pines—loblolly pine, slash pine, and shortleaf pine, as well as sand pine, Virginia pine, pitch pine, and other species of more or less local occurrence—are all pioneer species that become established after destructive forest fires and give way to understory tolerant hardwood mixtures in the long-continued absence of fires (Little and Moore, 1949; Wahlenberg, 1949; Campbell, 1955; Myers, 1990). The clearly established dependence of these species upon fire has given rise to extensive research and practical use of fires as a silvicultural tool that will tend to hold natural succession in the pioneer pine stage.

Douglas-Fir in the Pacific Northwest

In the Douglas-fir region of northern California, western Oregon and Washington, and southern British Columbia, the characteristic summer drought results in highly flammable conditions during the hottest period of the year with the result that extensive forest fires

Figure 16.3.   A dense stand of natural longleaf pine seedlings in the Coastal Plain of South Carolina following a prescribed burn to prepare the seedbed. (U.S. Forest Service photo.)

have not only been characteristic of forests of the 20th century but also of presettlement times. On the west side of the Cascades and over much of the Coastal Range, Douglas-fir is a pioneer species on burns, provided that adjacent Douglas-fir stands survive undamaged to furnish a source of seed. The relatively pure Douglas-fir type, therefore, may represent the first stage of a postfire secondary succession. More tolerant conifers such as western hemlock, Sitka spruce, and western redcedar invade the Douglas-fir forest to form an understory and eventually may achieve dominance after 500 or more years with the decadence and death of the dominant Douglas-fir (Figure 16.4; Franklin et al., 1981). Fire-history scenarios that produce different forest age structures of old-growth forests may vary from stand replacing fires recurring every 200 to 400+ years to a regime of multiple patchy fires of varying severity recurring every 20 to 100 years (Morrison and Swanson, 1990). Fire-disturbance scenarios and stand development of Douglas-fir throughout its Pacific Northwest range are discussed by Agee (1993).

Giant Sequoia
The native giant sequoia–mixed conifer forests of the Sierra Nevada of California are fire-dependent (Kilgore, 1973; Swetnam, 1993; Agee, 1993). Fire plays a major role in sequoia forests by preparing a seedbed of soft, friable soil on which the small seeds of sequoia fall and are lightly buried thus favoring germination (Hartesveldt and Harvey, 1967). Water is a limiting factor in seedling establishment (Rundel, 1972), and partially-burned organic matter may contain more available water than unburned litter (Stark, 1968).

Periodic presettlement surface fires kept the forests open and parklike (Figure 16.5a; Biswell, 1961). The fire regimes favored pioneer species and eliminated the small understory tolerant trees, particularly white fir, which continually invaded the understory. Upon settlement, burning by Native Americans was gradually eliminated, and fire suppression

Figure 16.4. The dominant, old-growth Douglas-fir individual (left) in this western Washington stand will eventually be replaced in the absence of fire by more tolerant firs and western hemlocks in the understory. (U.S. Forest Service photo.)

became increasingly efficient. Surface fuels built up, and white fir and other species grew up and often formed thickets under the sequoias (Figure 16.5*b*). A ladderlike vertical sequence of fuel was thus formed from ground level to low-hanging fir branches to the top of understory crowns, 10 to 30 m above the surface (Kilgore and Sando, 1975). Therefore, fire starting in surface fuels was likely to pass through understory crowns and torch-out in sequoia crowns. Because giant sequoia, unlike redwood, does not sprout, once the crown is killed the tree will die. Prescribed burning at 5- to 8-year intervals is therefore required to reduce surface fuels, kill small understory trees and the lower crowns of larger understory trees, and prevent continued encroachment of a tolerant understory.

FIRE HISTORY AND BEHAVIOR

Detailed studies of fire history in fire-dependent ecosystems increase our awareness of fire as a dominant force over virtually all of the northern and western landscape. Three examples illustrate fire history, behavior, and effects.

Northern Lake States

Heinselman's (1973) classic study in the Boundary Waters Canoe Area in northern Minnesota documents fire occurrence and effects over nearly 400 years. Fire has been a major force in determining the composition and structure of the presettlement vegetation for nearly 10,000 years. By the time logging reached the area, about 1895, recurrent fires had kept nearly three-fourths of the region in recent burns and stands of small-diameter trees. Thus little of the 215,000 ha tract, now reserved as a unit of the Wilderness Preservation System, was subjected to the timber cutting of the time—"high grading" (cut the best and leave the rest) the pines. Effective fire control began about 1911.

Figure 16.5. 1890–1970: Eighty years of fire exclusion in the giant sequoia–mixed conifer forests of Yosemite national Park (Confederate Group, Mariposa Grove). *a*) 1890: A parklike stand as the result of periodic fires. (Kilgore, 1972; Historical documentation by Mary and Bill Hood. Photo by George Reichel. Courtesy of Mrs. Dorothy Whitener.) *b*) 1970: The parklike stand was invaded by thickets of white fir. By 1970 firs obscured all but the fire-scarred sequoia on the left. Such thickets provide ladderlike fuels that could support a crown fire fatal even to mature sequoias. (National Park Service photo by Dan Taylor.)

A fire chronology was developed based on the age of existing stands, tree-ring counts from sections or cores from fire-scarred trees, and historical records. Nearly all of the forest burned one to several times in the period 1595 to 1972. Much of the burned area is accounted for by just a few major fires. In the settlement period, fires of 1875 and 1894 accounted for 80 percent of the total area burned; and in the presettlement period just 5 brief fire periods accounted for 84 percent of the total.

Heinselman found that the major fire periods coincided with prolonged summer droughts of subcontinental extent. Climate combined with fuel accumulations (associated with dry matter accumulations, spruce budworm outbreaks, and blowdowns) to bring optimum burning conditions at rather long intervals. The fires of 1863 and 1864, caused in part by the major droughts of those years, burned 44 percent of the area (Figure 16.6). Fire in 1864 also burned over 400,000 ha in Wisconsin. Clements (1910), working at Estes Park, Colorado, concluded that the burns of that year were the most extensive in the Rocky Mountains.

Heinselman estimates the natural fire rotation, the average number of years required to burn and reproduce new forest generations over the entire area, to be about 100 years. The natural rotation for different ecosystems would vary; those dominated by aspen–birch forests burning at intervals of 50 years or less, and many red and white pine stands lived for 150 to 350 years before being regenerated. A typical sequence for red and white pine ecosystems is occasional light surface fires that reduce the organic layer and understory competition, followed by a severe fire that kills the overstory and provides openings for regeneration. However, if a second major fire comes only a few years after a previous burn, many conifers could be eliminated, being too young to withstand the heat and needle scorch and too young to bear cones. Regeneration then comes from hardwood sprouts, produced by trembling aspen, paper birch, red maple, and northern red oak.

Fires over the entire area produce a mosaic of stands of different composition and ages depending on the complex of site and fire factors. Thus fire generates diversity in stand and species composition. The entire biota are adapted to these ecosystems and community patterns. Moose, beaver, black bear, snowshoe hare, woodland caribou, small mammals and birds are all adapted and dependent on the mosaic of different habitat conditions created by fire. Fire exclusion is now restructuring the entire regional ecosystem, gradually eliminating the niches of many wildlife species as the forest composition

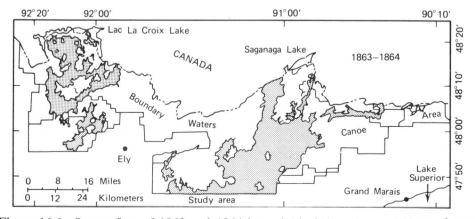

Figure 16.6. Severe fires of 1863 and 1864 burned (shaded area) over 1800 km² (700 mi²) of the Boundary Waters Canoe Area, northern Minnesota. (After Heinselman, 1973. Reprinted by permission of the Quaternary Research Center, University of Washington, Seattle.)

changes. It will require monitored lightning fire and prescribed fires to restore the natural vegetation mosaic for wildlife in this human-dominated wilderness.

Boreal Forest and Taiga

In the boreal forest and taiga of North America, a vegetation mosaic leading to plant and animal diversity is primarily the result of wildfires burning over diverse sites (Lutz, 1956; Slaughter et al., 1971; Rowe and Scotter, 1973; Viereck, 1973, 1983; Bonan and Shugart, 1989). Seven of the 10 major tree species are pioneers and have adaptations for rapid invasion of burned areas (Chapter 12). Only balsam fir and subalpine fir are less well adapted for regenerating immediately after fire; their cones disintegrate at maturity, and seeds are not retained in the crown. The understory and ground layers are characterized by clonal species whose underground stems and roots are undamaged by fire and regenerate rapidly. Fire maintains a patchy pattern of vegetation that assists in maintaining diverse wildlife populations.

North of the boreal forest is the taiga, a widespread area of open, slow-growing spruces interspersed with occasional well-developed forest stands and treeless bogs (Viereck, 1973; Van Cleve et al., 1986). The pattern of the vegetation is closely related to fire history and permafrost. Permafrost is permanently frozen ground in which water is often incorporated as lenses of pure ice. The presence of permafrost itself cools the ground by perching the water table, creating a deep ice layer for long periods of time on undisturbed sites (Bonan and Shugart, 1989). Thick layers of moss in forest stands act as insulators during summer months, limiting the thaw of soils to depths of one meter or less. Permafrost is widespread, and in many areas permafrost and vegetation are in delicate balance. Fire, by burning organic matter and warming the surface, can increase the annual depth of thaw of permafrost soils. This can significantly affect the presence, as well as the composition of the vegetation. Where ice lenses occur, increasing the depth of thaw following fire may cause subsidence of the areas over the ice, creating a polygonal mound-and-ditch pattern. This pattern may occur in paper birch stands at least 40 to 50 years after fire. With succession to black spruce these sites may become stabilized, or small ponds may develop and alternate in the cycle with black spruce.

Northern Rocky Mountains

Fire has burned in every ecosystem and virtually every square meter of the coniferous forests and summer-dry mountainous forests of northern Idaho, western Montana, eastern Washington, and adjacent portions of Canada (Wellner, 1970; Habeck and Mutch, 1973; Arno, 1976). Prior to 1940 and fire protection, fire was second only to precipitation as the major factor shaping the character of forests. Catastrophic fires, killing complete stands or nearly so, were common due to the extremely dry summer climate and the build-up of fuel in the densely forested areas. Forest insects, and at times diseases, contributed tremendous quantities of fuel by killing trees in dense stands over extensive areas. Following prolonged drought, the series of 1910 fires alone burned between 1 and 2 million hectares.

This region is characterized by multiple burns, which usually remain without tree cover for long periods (Larsen, 1925). Fire breeds fire, and many single burns created such hazardous fuels that they reburned. Parts of the 1910 burn reburned in either 1917 or 1919, or both, and even again in subsequent years. Snags were particularly important as a source of lightning ignition. Wind-blown burning materials from tall snags could start new fire far in advance of the burning front. Fire prevention since about 1940 has largely prevented multiple burns and increased the area of forest cover.

Nevertheless, unstoppable, catastrophic fires certainly do occur in regional ecosystems that are characterized by the climatic and vegetational features conducive to fire. One of these with a long fire history (Romme, 1982) is the Yellowstone Plateau that includes the Yellowstone National Park mentioned above (see also Chapter 21). Most of the Yellowstone Plateau is above 2,500 m in elevation, and the dominantly forested area is characterized by relatively even-aged forests of lodgepole pine between 100 and 150 years old. In 1988, multiple fires of the Yellowstone area burned in a summer that was one of the driest in Rocky Mountain history. The dry summer led to extremely dry conditions of fuels that had accumulated for decades. In addition, over twice the number of lightning storms were recorded in 1988 than in the average year (Wuerthner, 1988). A 1990 study of lodgepole pine regeneration in moderately burned, severely burned, and unburned stands revealed that lodgepole pine density ranged from 0.06 to 184 m^{-2} (Ellis et al., 1994). Highest seedling densities were in moderately burned areas. Evidence suggested that in the severely burned areas fire consumed the bulk of a stand's seed supply, resulting in relatively low post-fire seedling densities. However, the varying or patchy pattern of burn severity means that seeds from lightly or moderately burned areas supply much of the seed source for adjacent severely burned areas.

In other parts of the Rocky Mountains, fire is a key factor in forest dynamics and has been examined in detail by Peet (1981) and Veblen (Veblen and Lorenz, 1986,1991; Veblen et al., 1991a), and Knight (1994).

FIRE EXCLUSION

The foregoing sections illustrate the widespread occurrence of past fire and their inseparability from forest ecosystems. Seen in the time scale of centuries, fire is an integral part of the evolution and perpetuation of forest species as parts of landscape ecosystems rather than a "disturbance." However, increasingly efficient fire suppression efforts of this century, if continued indefinitely, constitute a disturbance that will greatly change systems (Smith et al., 1997) in ways rarely or never before experienced. In fact, it is already happening.

Before settlement by Europeans fires were common, set by lightning and Native Americans. During settlement they were a common sight, and increased in frequency in many places. Unless they threatened human life, livestock, or buildings, they were little regarded except as a local nuisance. The rise in public opinion against fire, fueled by promiscuous burning and the large destructive fires of the late 1800s, led to the development of rigid suppression policies. Fire was not considered a natural phenomenon but a disastrous threat to the forest and to human life and habitation. In the eyes of public opinion, fire killed trees, promoted erosion and floods, destroyed the habitat of animals, and often burned nests and killed the animals themselves. As Roland Harper (1962) observed, fire damage could be seen immediately and thus encouraged anti-fire "propaganda," whereas its beneficial effects might be years in materializing.

Today, because of increasingly efficient fire control, due largely to sophisticated communication, fire prediction methods, rapid attack helicopters and water tankers, it is increasingly possible to extinguish fires in an early stage of development. Although complete suppression is not possible, a significant reduction in the area burned is being achieved. Exclusion of fire leads to a marked build-up of fuel, decline in animal and plant species, and an increase in incidence of wind damage and insect and disease attack as the proportion of mature forest increases. Ironically, the prolonged buildup of fuel may eventually lead to more catastrophic crown fires and be more destructive to the site than typically occurred in the original forest ecosystem. In addition, the mosaic of successional forest types is altered in that late-successional species increase in abundance, leading to a decline in wildlife habitat and often in both plant and animal diversity (Chapters 14 and 20). Furthermore, fire exclusion allows thin-barked, slow-growing species to grow into size classes that are fire-resistant (Harmon, 1984). Thus restoration of the original fire-interval alone will not restore the original forest composition and structure.

Reaction to fire exclusion came gradually. Research, especially in the South and West, led to the conclusion that fire was a natural factor—that prescribed burning could be beneficial to plants and animals. Therefore, in many parts of North America, attention turned to letting selected natural fires burn and to conducting prescribed burning. Fire control programs, formerly confined to fire prevention and suppression, evolved into fire management programs.

Based on the work of pioneer fire scientists, such as Roland M. Harper, H. H. Chapman, Herbert L. Stoddard, Harold H. Biswell, and Harold Weaver, the U.S. Park Service and the U. S. Forest Service have instituted fire management programs with the goal that fires play a more natural role in forest ecosystems. Similarly, The Nature Conservancy has an active program of prescribed fire to restore and maintain many of its "last great places." Personnel of many national parks and monuments now let selected natural fires burn and conduct prescribed burning to maintain the natural diversity of plant and animals that have evolved as a result of past fires. Fire management and prescribed burning have been considered from many standpoints and in relation to many different forest ecosystems (Slaughter et al., 1971; Wright and Heinselman, 1973; Kayll, 1974; Kilgore, 1975, 1976a,b,c; Mutch, 1976; Martin et al., 1977; Knight, 1994; Arno et al., 1995).

Wind

The consequences of wind disturbance in forest ecosystems are enormous—affecting their structure, composition, and function at regional and local scales. Like fire, both the frequency and intensity of wind have significant and different effects as wind encounters forested landscapes. The basics of wind as a climatic factor was considered in Chapter 7; here we examine its ecological effects in forest ecosystems.

CATASTROPHIC AND LOCAL EFFECTS

From an ecosystem perspective, certain regions of North America and the world are characterized by distinctive wind patterns. Hurricanes and their catastrophic disturbances occur predictably along the Gulf and Atlantic coasts of North America. Tornadoes are most prevalent in the central part of the United States with "tornado alleys" extending from Texas to Nebraska and Kansas east across central Indiana. Minimum rotation periods of 1,200 to 1,300 years in central Indiana to maximum values greater than 10,000 years for the upper Great Lakes region and the Northeast are reported by Whitney (1994). Sustained winds of hurricanes reach velocities over 200 km hr^{-1} (1989 hurricane Hugo, 216 km hr^{-1}; 1992 hurricane Andrew, 242 km hr^{-1}). Tornadoes have the highest wind velocities, although their damage is more localized than hurricanes. Wind speeds of $>430 \text{ km hr}^{-1}$ were reported for the 1993 Kane tornado in northwestern Pennsylvania (Peterson and Pickett, 1995).

Whereas hurricanes may cause forest destruction over huge swaths across a landscape (50-km wide for Andrew; Pimm et al., 1994), local wind gusts (90–150 km/h), associated with cyclonic storms that break off or uproot a single tree or groups of trees, may affect only small areas (20–100 m^2) in the forest. These gaps in the overstory canopy or exposed patches on the ground occur throughout large areas in North America. Their accumulated ecological significance may be equal to or greater than that caused by catastrophic winds (Bormann and Likens, 1979). However, it is wrong to think only in the dichotomy of extremes. Potentially-destructive winds occur in a broad continuum of velocity. They encounter, at different times of the year, regional and local ecosystems with a great variety of site conditions and vegetation. Thus wind disturbance is variable as well. However, despite the many gradients, the nature and amount of wind damage is relatively predictable by regional ecosystem, specific landform-soil drainage conditions, site exposure, and tree species and height.

PRINCIPLES OF WIND DAMAGE

Near the surface of the Earth, wind is turbulent due to the nature of the ground over which the air is moving. The rougher the surface the more turbulent is the air. Forests form a particularly rough surface, and wind over them is much more turbulent than over farmland or meadow. The turbulence is organized into coherent **gusts** that move widely across the forest (Figure 16.7). As wind moves across the forest it becomes unstable, and each gust consists of a rapid increase in wind speed together with a downward movement of air into the canopy (Figure 16.7, 5). Such gusts are the main cause of damage in forests, with the strongest gusts exerting a force on trees up to 10 times larger than that due to the mean wind velocity (Quine et al., 1995).

Changes in vegetation composition or sudden changes in its height induce additional turbulence and wind acceleration. Forest edges, roads, and other open areas cause upward deflection of the wind and tend to increase damage. The wind force on trees close to a forest edge is also substantially greater than that on trees inside the forest. Therefore, trees suddenly exposed by the formation of a new edge become more vulnerable. Such an example from the Rocky Mountains is illustrated in Figure 16.8; here the cutting edges deflect wind currents, resulting in increased velocities where the deflected currents join others. The practical problems of wind behavior in mountainous forested country have been summarized by Gratkowski (1956) and Alexander (1964), who found that much wind damage is due to local acceleration of wind by either topography or forest borders. Such acceleration can take place where wind velocity increases: (1) on ridge tops, upper slopes,

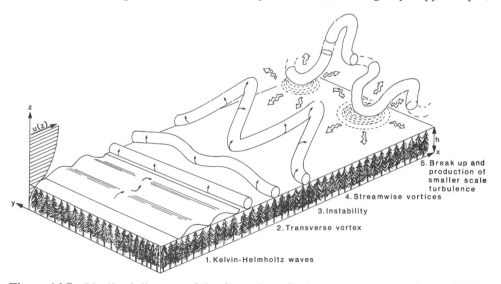

Figure 16.7. Idealized diagram of the formation of coherent gusts over a forest. 1) The rapid change of wind speed (u) at the top of the canopy (z=h) is unstable and leads to the emergence of Kelvin-Helmholtz waves. 2) The waves become transformed into across-wind vortices. 3) These vortices are unstable and begin to distort. 4) The distortion produces coherent gusts aligned in the direction of the wind. The gusts propagate across the forest and if strong enough, lead to wind damage. 5) Eventually the gusts become distorted and break up. (After Finnigan and Brunet, 1995. Reproduced by permission of the British Forestry Commission.)

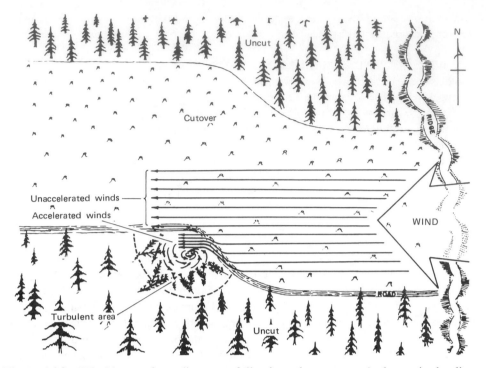

Figure 16.8. Windthrow of standing trees following a harvest cut. A change in the direction of the cutting boundary that was parallel to the prevailing winds acts to funnel wind into standing timber, causing a pocket of blowdown. Routt National Forest, northwestern Colorado. (After Alexander, 1964. Reprinted with the permission of the Society of American Foresters.)

and the shoulders of a mountain, (2) on gradual and smooth lee slopes during severe windstorms, (3) in gaps and saddles of main ridges, (4) in narrow valleys or V-shaped openings in the forest that constrict the wind channel, and (5) where forest cutting edges deflect wind currents as in Figure 16.8.

Windthrow (uprooting of trees) is most apt to occur where the concentration of air currents causes high wind velocities at a particular spot. It occurs primarily in shallow-rooted species on shallow soils and those with impeded drainage. As we learned in Chapter 11, soil is of vital importance in tree stability. The greater the mass of soil adhering to the root system of a tree the more wind-firm it is. This mass depends on its volume, measured by the depth and spread of the central core of roots, and bulk density of the soil. Lateral root spread is not typically limited by soil conditions, but they strongly determine rooting depth. The main causes of restricted rooting depth are inadequate oxygen supply (poorly drained soils) and hardpan layers (bedrock, ortstein, etc.). Tall trees with large crowns are especially susceptible. Furthermore, conifers are more vulnerable than deciduous trees in winter, when deciduous trees are leafless. Thus it is possible to identify high-risk species—black spruce, white pine, and Norway spruce, for example—and high-risk sites.

BROAD-SCALE DISTURBANCE BY HURRICANES
Occasional severe storms, particularly hurricanes and tornadoes, may destroy the over-story canopy on hundreds or thousands of hectares initiating secondary forest succession over large areas.

Southern Atlantic Coast

The September 21, 1989 hurricane Hugo struck the Atlantic Coast and caused widespread damage over 1.8 million hectares—over half of the counties in South Carolina (Sheffield and Thompson, 1992). Conifers, primarily loblolly pine, were much more severely affected than hardwoods. Conifer growing stock (trees larger than 13 cm) was reduced by 21 percent, whereas only 6 percent of hardwoods were destroyed.

In south Florida, Hurricane Andrew struck on August 24, 1992 and severely affected terrestrial and aquatic ecosystems over a swath 50 km wide by 100 km long (Pimm et al., 1994; Loope et al., 1994). In pinelands, from 25 to 40 percent of overstory trees were snapped at heights of 1 to 6 m or uprooted; 2 to 3 times as many pines had broken boles as were uprooted. In contrast, the understory was relatively unaffected. Tall, large-diameter hardwood trees of hammock ecosystems were extensively damaged (20 to 30 percent downed or large branches broken off). Cypress forests (dominated by bald cypress) showed only modest damage. Mangroves suffered catastrophic damage, reaching 80 to 95 percent destruction in the most vulnerable sites by trunk snapping and uprooting (Smith et al., 1994). Interestingly, small patches of living mangroves, surrounded by dead individuals were discovered to be saplings that had regenerated in gaps caused by lightning-induced fires. The small gaps will provide nuclei for recolonization of destroyed forests since the three Florida mangroves reproduce viable propagules on stems less than one meter in height. The consequences of hurricane Andrew must be viewed in light of the multiple disturbances characteristic of southern Florida—fires, floods, freezes, and sea-level fluctuations as well as hurricanes. A primary concern is the alteration of the natural vegetation by the spread of introduced plant species that was already rapid before Hurricane Andrew. The hurricane may accelerate the process by dispersing propagules and opening forest canopies (Loope et al., 1994). Of particular concern is the spread, by wind-dispersed propagules, of invasive trees: the paperbark tree, the most invasive plant in south Florida, the Brazillian pepper tree, and Australian pine.

New England—1938 Hurricane

Forests of the northern Atlantic coast and New England region are also strongly affected by occasional hurricane-force winds as well as disturbances by fire, ice storm, and pathogens. The great 1938 hurricane in New England created an awareness of the past occurrence of severe windstorms in this region at intervals from a few to more than a hundred years. Seven major hurricanes along different tracks from 1620 to 1950 have inflicted light to severe damage (Chapter 21, Figure 21.9). The 1938 hurricane virtually leveled over 240,000 ha of central New England's forests along a 100-km swath from central Connecticut and Massachusetts to the north-west corner of Vermont (Whitney, 1994; Foster and Boose, 1992). Foster (1988b) concluded that damage to species and stands from this catastrophic windstorm occurred quite predictably and specifically; damage was not everywhere catastrophic—quite the contrary. Post-hurricane damage was a mosaic of differentially-damaged stands controlled by the physiography, wind direction, soil type, and nature of the pre-hurricane vegetation, especially species composition, tree height, and density (Spurr, 1956a; Foster, 1988a,b; Foster and Boose, 1992). A case study of disturbance across the New England landscape is presented in Chapter 21.

WAVE-REGENERATED FIR SPECIES

Wave-regeneration of subalpine fir species is another example of broad-scale, wind-induced disturbance. In these situations, persistent wind (rather than wind velocity per se), together with other factors, is instrumental in long-term maintenance of fir regeneration and other ecosystem processes. In high-elevation balsam fir forests of northeastern United States (Sprugel, 1976; Sprugel and Bormann, 1981) and fir forests of central Japan (Sato and Iwasa, 1993), the fir canopy is broken by numerous crescent-shaped strips of dead trees. Each strip is a band of standing dead trees, with mature forest on one side and young, vigorously regenerating forest on the other side (Figure 16.9). The "wave" of death and regeneration moves slowly through the forest; trees die at the wave's leading edge and are replaced by seedlings. The waves move in the general direction of the prevailing wind (Figure 16.9) at a constant rate at speeds of 0.37 to 3.3 m per year. Tree mortality of the dieback zone is presumably due to the prevailing wind, which may cause winter desiccation, summer cooling, increased rime deposition, and breakage of branches and roots. The shallow, rocky soils of these ecosystems also contribute to death of the fir trees. Strong winds induce tree sway and cause root breakage on sharp rock surfaces, thereby reducing water conducting tissue and encouraging root disease fungi (Harrington, 1986). After death of the canopy trees, seedlings that have established under them are released and eventually become canopy trees in 60–80 years (Figure 16.9). The wave-regenerated forest is an unusually orderly one due to omnipresent and highly predictable disturbance (Sprugel and Bormann, 1981). The prevailing wind acts to increase the spatial predictability and organization of the disturbance cycle in wave fashion in these highly stressed, high-elevation forests.

Floodwater and Ice Storm

Few natural forces have been as powerful and important in creating and patterning landscapes around the world as glacial ice and associated meltwater (Chapters 3, 10). In addition, floodwaters in historic time have shaped and rearranged physiographic features of streams of all sizes (Chapter 10). Through inundation, siltation (deposition of fine soil particles), and influence on local hydrology, floodwaters control the regeneration of riverine species and, therefore, strongly influence the distinctive pattern of forest communities associated with riverine landforms. The timing, duration, and temperature of flood waters has elicited genetic differentiation of river valley species; their many different physiological adaptations to lack of oxygen for root metabolism enable particular species and races to dominate the diverse riverine ecosystems (Gill, 1970; McKnight et al., 1981; Hook;, 1984).

   In contrast to wind, which rapidly decimates the overstory with stunning visual effect, flood waters slowly and pervasively change the forest site by bank cutting and soil deposition. The flux of sedimentary deposits is so great that topographic sites available for establishment of vegetation are constantly created and destroyed (Malanson, 1993). Flooding also controls seedling regeneration by continuously and unobtrusively eliminating upland invaders, favoring physiologically-adapted plants, and providing nutrients and water. In these ways, flooding significantly determines the composition and dynamics of riverine vegetation. Flood-control measures and human development of floodplains have markedly changed the ecosystem processes and hence the natural successional patterns associated with flooding regimes. Today, the occurrence of many upland species in bottomland sites where flooding is excluded emphasizes the enormous selective effect flood waters have on seedling establishment. In addition to the effects of flooding on understory vegetation, damage to overstory and understory trees alike by ice floes and woody debris can be considerable (Lindsey et al., 1961).

   The great effect of erosion following heavy rainfall or various natural and human disturbances should also be mentioned for its effects in changing site conditions and thereby affecting forest composition and growth. Finally, ice or glaze storms (Chapter 7), directly affect the overstory by severe branch or stem breakage and indirectly by providing entry for pathogens or weakening trees so they are more subject to insect attack.

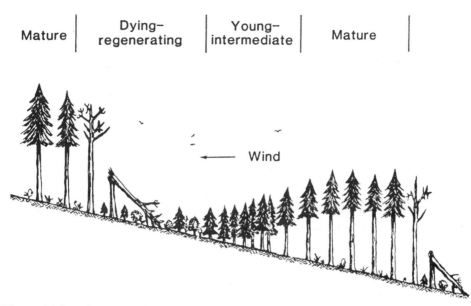

| Mature | Dying–regenerating | Young–intermediate | Mature | |

← Wind

Figure 16.9. Diagrammatic cross section through a regeneration wave. Fir regeneration is initiated in and along the edges of the mature stand and released as the trees die. (After Sprugel and Bormann, 1981. Reprinted with permission from *Science* 211:390–393, © 1981 by the American Association for the Advancement of Science.)

Insects and Disease

Insects and disease-causing pathogens and mistletoes are major regulators of forest ecosystems (Haack and Byler, 1993; Castello et al., 1995). Millions of hectares each year in North America are affected by mortality or growth loss (Chapter 14). Insects and pathogens, through catastrophic epidemics or less obvious chronic effects, create widespread disturbances. In turn, these disturbances reciprocally affect insect populations in many ways (Schowalter, 1985).

Insects and pathogens influence succession by killing or weakening trees singly or in patches. Root fungi, bark beetles, and inner-bark borers are the primary tree killers. They tend to attack one particular species or genus of trees, and strongly control the rate and direction of succession (Franklin et al., 1987). The small gaps resulting from death of overstory trees from pathogens or insects often favors more understory-tolerant species already established in the understory. Also, bark beetle attacks of epidemic proportions may occur. In some Colorado subalpine forests, the effects of spruce beetle outbreaks appear to be as great as those of fire (Veblen et al., 1991b). For example, extensive blowdown of subalpine forest in 1939 led to an epidemic of spruce beetles in the 1940s that devastated 290,000 ha of the White River National Forest alone (Hinds et al., 1965). Insects and disease often interact with other disturbances, such as fire and wind. Figure 16.10 illustrates how interactions of fire, wind, disease, and insects act sequentially to determine the course of stand development.

Logging

As in many other parts of the world, forests are among the most potent factors in the economic life of a developing country; and so it was in the United States. Whitney (1994) describes the pattern of lumbering in the eastern United States in three phases: the highly selective white pine era, diversification of logging with an emphasis on secondary species,

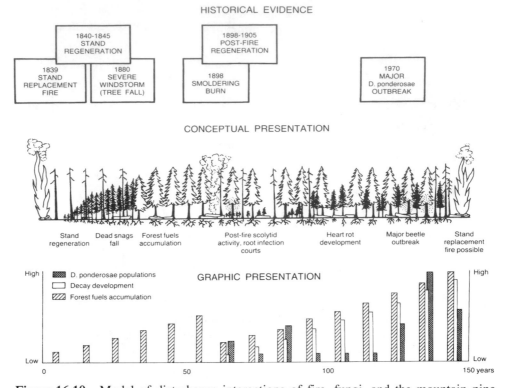

HISTORICAL EVIDENCE

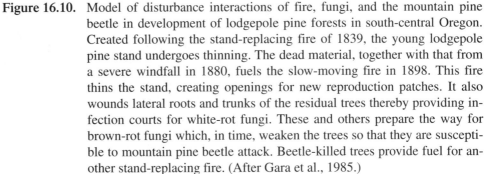

GRAPHIC PRESENTATION

Figure 16.10. Model of disturbance interactions of fire, fungi, and the mountain pine beetle in development of lodgepole pine forests in south-central Oregon. Created following the stand-replacing fire of 1839, the young lodgepole pine stand undergoes thinning. The dead material, together with that from a severe windfall in 1880, fuels the slow-moving fire in 1898. This fire thins the stand, creating openings for new reproduction patches. It also wounds lateral roots and trunks of the residual trees thereby providing in-fection courts for white-rot fungi. These and others prepare the way for brown-rot fungi which, in time, weaken the trees so that they are suscepti-ble to mountain pine beetle attack. Beetle-killed trees provide fuel for an-other stand-replacing fire. (After Gara et al., 1985.)

and "mop-up" operations with the portable mill. For the most part major changes in forest composition followed logging. Post-logging fires, fueled by woody debris left after log-ging, also contributed significantly to change in forest composition (Whitney, 1994).

With the exception of large-scale clearcutting, logging can be similar to windthrow in that the cutting of commercially valuable trees tends to remove the overstory and re-lease the understory. Many understory hardwoods cut in operations connected with log-ging will resprout to form vigorous and fast-growing stems that compete for overstory space in the developing new stand.

As a general rule, the intensity and pattern of the cut will affect the competitive abil-ity of the new crop. Light, partial cuts (thinning and improvement cutting) will favor toler-ant species, particularly those already established in the understory, and thus will tend to push forest succession forward rather than initiate an earlier stage in the process (Figure

Figure 16.11. Partial cutting in this eastern white pine stand in Maine has favored advanced regeneration of tolerant red spruce. (U.S. Forest Service photo.)

16.11). Moderately heavy partial cuts will favor midtolerant species as will the cutting of small groups (group selection cutting) of trees to create gaps in the forest of approximately the same size that would be made by the death of one or several large, old-growth trees. Clearcutting, or the cutting of large gaps with a diameter of at least twice the height of the stand, will favor the invasion of pioneer species, particularly if mineral soil is exposed by the harvest operation. Thus the silviculturist, by regulating the intensity and pattern of cutting, can greatly influence subsequent forest composition and the rate of succession.

The effect of partial cutting versus clearcutting on forest succession is well illustrated by experience in the Douglas-fir region of the west side of the Cascades in Washington and Oregon. Much of the cutting in old-growth stands occurs in more or less even-aged stands composed predominantly of Douglas-fir and dating from past forest fires 150 to 500 years ago (Franklin et al., 1981). Clearcutting in such stands at the time of, or immediately after, a seed year results in the reestablishment of even-aged Douglas-fir. Patches of Douglas-fir ranging from 1 to 15 ha in size will normally come back to Douglas-fir within a few years, with seed blowing in from surrounding uncut forest. On the lower and wetter sites in the Coast Range, red alder comes in prolifically as a pioneer species on clearcut sites. On any site, the new forest is composed of pioneer species that range from intolerant to midtolerant in their ability to compete in the understory in that region.

Early experience with partial cutting in old-growth Douglas-fir forests (Isaac, 1956) led to the belief that the end product of partial cutting is a deteriorating partial overstory with a vigorous understory composed of conifers more tolerant and less commercially valuable than Douglas-fir. This economic situation, coupled with the uncertainty of adequate stocking of Douglas-fir through natural regeneration, led to large-area clearcutting and planting of Douglas-fir as a conventional forestry practice for several decades. However, public concern over large-scale clearcutting and timber extraction practices have encouraged the management of forest ecosystems in ways that less seriously interrupt natural ecosystem processes (Chapter 21). Furthermore, the emerging view today is that Douglas-fir can be regenerated and grown under a partial overstory, such as might be created with a variable retention harvest system or shelterwood (Chapter 21, Figures 21.23, 21.24, and 21.25).

Another effect of logging on forest succession results from the differential removal of one species and the leaving of another—thus changing the composition of the forest. The effect of logging a favored species on forest composition is exemplified in the mixed-wood forests of Maine where logging has been more or less continuous for more than 120 years (Whitney, 1994). Early logging was concentrated primarily on large white pines suitable for masts for wooden sailing ships and house construction. The next stage of logging followed the building of sulfite and groundwood pulp mills and the large-scale cutting of red spruce to supply them. Since the red spruce grew in association with tolerant balsam fir, sugar maple, and beech, the logging of the spruce created a mixed forest of these species. However, the proportion of spruce was drastically reduced. Balsam fir then became more utilized as the supply of red spruce decreased. As clearcutting followed partial cutting, the water table rose precluding forest in many of the wetter flats, and pioneer species such as trembling aspen and paper birch invaded the drier sites. Thus successive waves of logging, each concentrating upon different species, have greatly modified the composition of forest of central and northern Maine.

Land Clearing

Forests occur in moist regions and are therefore eminently suitable for the raising of agricultural crops. Since the great majority of the world's population lives in forested regions, it is inevitable that much of the world's forests have been cut and the land cleared for agriculture. The massive forest clearance in the eastern United States and its effects leading to forest fragmentation are described by Whitney (1994). In the tropics, the use of forest land for agriculture is often transitory, with fields being allowed to revert to forest again after a few years of cropping. Even in the Temperate Zone, much land used for farming in the past has been found unsuitable for continued cropping, or has been supplanted by bringing better lands into production. This land has been planted to forest or allowed to revert naturally to forest. Secondary forest succession following land abandonment, therefore, is an important process, taking place over many hundreds of thousands of hectares in well-settled forest areas. Old-field succession in the eastern United States has been studied in detail and is described in Chapter 17.

BIOTIC COMPOSITION CHANGES

Although secondary successions are considered to originate primarily from disturbances such as fire, wind storms, logging, and land clearing, a second major kind of disturbance—the addition or subtraction of species, whether plant or animal—will inevitably change the succession of the forest. Because these changes in the flora or fauna of the forest may result in considerable disturbance to existing ecosystems, they may well be considered as initiating secondary successions in a very real sense.

Elimination of Species

When a species is eliminated or greatly reduced in abundance, its place in an ecosystem is taken by other organisms. An outstanding example of this is the virtual elimination of the American chestnut in the eastern hardwood forest by the blight caused by *Endothia parasitica*. This disease, introduced from Asia about 1904, killed most of the mature chestnuts in New England within 20 years, and had completed its work in the southern end of the Appalachians by the 1940s. Seldom if ever before in historical times has a major forest tree been so nearly eradicated. American chestnut was a major dominant of many dry and dry-mesic ecosystems in the Southern Appalachian Mountains and in deciduous forests of eastern North America (Braun, 1950). Figure 16.12 illustrates the huge size of chestnut trees that grew in old-growth forests of the Southern Appalachians. Although we still observe sprouts from root systems (Griffin, 1992), few trees survive long enough to reach the overstory.

Succession following the elimination of the chestnut has often resulted in the simple replacement of that species by its former associates (Illick, 1914, 1921; Korstian and Stickel, 1927; Augenbaugh, 1935; Keever, 1953; Woods, 1953; Nelson, 1955; Woods and Shanks, 1959; Day and Monk, 1974; Stephenson, 1974). Chief among the succeeding trees in the southern Appalachians are the oaks (especially chestnut, red, and white oaks), along with various hickories and, on the better sites, tuliptree. Thus the eradication of chestnut in this region has resulted in the replacement of the former oak-chestnut type with an oak or oak-hickory type. Similar changes, but involving some different species, previously occurred in the middle Atlantic states and southern New England (Good, 1968).

In the Allegheny Mountains of western Pennsylvania, logging and fire following the death of chestnut trees created open sites for invasion by early and midsuccessional species (black cherry, black birch, black oak, sassafras, and black gum) together with species present on the site such as sugar maple, red maple, white ash, and beech (Mackey and Sivec, 1973). Without fire or other disturbances in the former chestnut forest, succession gradually proceeds to more tolerant and mesophytic species such as hemlock, sugar maple, and beech.

More recently the Dutch elm disease, caused by the fungus *Ophlostoma ulmi* (introduced into Ohio from Europe about 1930) and the phloem necrosis disease had, in about 50 years, virtually eliminated mature American elm trees from swamp and river floodplain forests of the eastern and midwestern United States. Studies of succession following the elimination of overstory elms in midwest forests indicate that elms are replaced by a number of different species, depending on regional and local site conditions. In Illinois woodlands, sugar maple is the species most likely to increase in dominance except where soils are too poorly drained (Boggess, 1964; Boggess and Bailey, 1964; Boggess and Geis, 1966). In southeastern Iowa, hackberry and box elder were the most frequent trees replacing elm (McBride, 1973).

In southeastern Michigan, American elm was formerly a late-successional dominant in deciduous swamp forests together with red maple, yellow birch, and black ash. These three species now dominate the swamps, but in different proportions depending on site conditions (Barnes, 1976). American elm has not been eliminated from deciduous

Figure 16.12.   American chestnut trees in an old-growth forest of the Great Smoky
 Mountains, North Carolina. On favorable sites American chestnut trees
 could grow over 2 meters in diameter and 36 meters in height. (Photo
 courtesy of the American History Society.)

swamps but makes up about 10 to 15 percent of the understory. Old fields and other open
upland areas are much more important sites for regeneration of elm than swamps. Indications are that, unlike the American chestnut, American elm will be perpetuated for generations by seeds from young elm trees; however, the average life span will be drastically reduced. New resistant clones hold promise for horticultural planting (Townsend et al.,
1995).

In the spruce–fir forest of eastern Canada and adjacent northeastern United States, periodic epidemics of spruce budworm have resulted in the killing of overmature and mature balsam fir over large areas together with lesser amounts of black and white spruces. The budworm apparently has played a major role in the mixed softwood forests in holding down the proportion of the very-tolerant balsam fir as compared to that of the somewhat less tolerant spruces.

In the animal portion of forest ecosystems, the virtual elimination of many predators—particularly the wolf, bear, cougar, and lynx—from most of the American forest has played a role in increasing the number of deer, rabbits, and other herbivores. This in turn results in greater browsing of the understory, including tree regeneration. Such disturbance precludes the regeneration of black cherry in areas of Pennsylvania (Marquis, 1974, 1975, 1981; Tilghman, 1989) and hemlock in many places in the northern United States (Frelich and Lorimer, 1985; Peterson and Pickett, 1995).

Addition of Species

The species occupying a given site are not necessarily those best adapted to compete and grow on that site, but merely the best of those that have access to that site at the time of its availability. Invasion of the site by better competitors often results in substantial changes in forest succession. It makes no difference whether this invasion occurs as a result of natural immigration or through introduction by humans. The result is ecologically the same.

The chestnut blight fungus cited in the previous section is an example of an accidental introduction that has greatly modified the forest. Other organisms may be trees, other higher plants, fungi, bacteria, and animals of all levels (Chapter 14). Important deliberate forest-tree introductions in Europe include: Douglas-fir, eastern white pine, northern red oak, and black locust; in the northeastern and north-central United States: Scots pine, Norway spruce, Norway maple, black pine, and European larch ; and in temperate zones of the Southern Hemisphere: Monterey pine, patula pine, and slash pine. Coconut and *Casuarina* spp. are representative of many littoral species that have been widely disseminated throughout the Tropics, partly through human and partly through other forms of transport. These and many others have become vigorous and often dominant species in the local flora.

INTRODUCTIONS TO NEW ZEALAND

A spectacular example of the rapid successional changes in both the flora and fauna of a region that can occur when better-adapted plants and animals are introduced is provided by the natural history of New Zealand. In this south temperate land, only Southern Hemisphere conifers (primarily podocarps) and hardwoods (primarily *Nothofagus* spp.) had access to the land prior to the coming of the Maoris a thousand years or so ago and of the Caucasians in the past century or two. Birds—many of them without functional wings—constituted virtually all the higher animal life. The climate, however, is temperate, moist, and mild, ideal for many Northern Hemisphere plants and animals. Following extensive introduction by humans in the last hundred years, the introduced flora and fauna are rapidly and effectively replacing their native counterparts (Figures 16.13 and 16.14). Monterey pine, lodgepole pine, and Douglas-fir among the trees; gorse, blue grass, and ragwort among the smaller plants; and the European red deer, sheep, rainbow trout, Australian possum, and Himalayan thar among animals are but a few examples of vigorous organisms better suited to the site than those isolated there by the accidents of geological history. The forest ecosystems and the successional trends within them will inevitably be more and more influenced by these new and vigorous plants and animals.

Figure 16.13. The original old-growth forest of New Zealand dominated by the podocarp Rimu, with characteristic understory of hardwoods and tree ferns. Under impact of logging and grazing, this forest is being replaced by North American conifers, both planted and naturally seeded. (New Zealand Forest Service photo by J. H. Johns, A.R.P.S.)

Increased Animal Impact

It is not necessary that a species be either added or eliminated in order for its disturbance to affect forest succession. Change in the abundance of a species may have a considerable effect in itself. This is illustrated by the great changes brought about by the increase in number of herbivores in the forest through reduction of predators, burning of the woods to increase browse, and other human activities. The herbivores may be domestic or wild.

Among domestic stock, goats are by far the most destructive of forest regeneration, followed by pigs, sheep, and cattle, in approximately that order. The free-running forest-pig industry reached its peak in the Midwest in the first half of the 19th century, and may have resulted in the decline in the abundance of beech and white oak throughout the Midwest and Northeast (Whitney, 1994). Long-continued overgrazing by livestock will result in the elimination of palatable species from the ground up to the browse line, compaction of forest soil, and eventual conversion of the forest to an open scrub of unpalatable species or to grassland (Chapter 14). This has been the history of much of the forest in the Mediterranean regions of Europe and Africa (Dodd, 1994), virtually all of Asia Minor and the countries to its east, and of large areas in Russia. Much of the scrub and grassland in the drier and warmer forest regions of the world owes its origin to the long-continued

Figure 16.14. Monterey pine forest in New Zealand. These 47-year-old trees in a plantation form a taller and higher-volume forest than the old-growth Rimu, several hundred years of age. (New Zealand Forest Service photo by J. H. Johns, A.R.P.S.)

overgrazing by goats, sheep, and cattle. Many of these sites could support high forest if this disturbance were controlled. Furthermore, grazing by livestock may also disrupt disturbance regimes, especially fire (Weaver, 1959; Cooper, 1960; Savage and Swetnam, 1990). Overall, the long-term disturbance effects of animals, both large herbivores and small mammals as well, are enormous.

CLIMATIC CHANGE

A third source of disturbance initiating secondary succession in forest ecosystems is climatic change. The change may be short and extreme, such as a severe drought or cold spell that kills part of the existing forest, thus initiating a secondary succession with a different complement of potential competitors for space than existed previously. Or it may be a long-term climatic change of lesser intensity, but one that changes the relative competitive ability of the species present. The first type is obvious and well understood. The second is increasingly realized as playing an important role in changing forest succession. For example, substantial tree invasion into heather meadows in the Cascade and Olympic Mountains has been associated with regional warming in the 20th century (Agee, 1993).

Also recent colonization of burns by white birch and balsam poplar into previously tree-less tundra areas suggests that deciduous trees with efficient long distance dispersal abilities will become more abundant with global warming (Landhäusser and Wein, 1993). Warmer and drier conditions could make wildfire more common, hence accelerating the process. Evidence for long-term climatic change is presented in Chapter 3, and the likelihood and significance of marked climatic change in the next 50 to 100 years is considered in Chapter 7.

Suggested Readings

Agee, J. K. 1993. *Fire Ecology of Pacific Northwest Forests*. Island Press, Washington D.C. 493 pp. Chapters 5 and 7–12.

Cooper, C. F. 1960. Changes in vegetation, structure, and growth of southwestern pine forests since white settlement. *Ecol. Monogr.* 30:129–164.

Cooper, C. F. 1961. The ecology of fires. *Sci. Amer.* 204:150–160.

Heinselman, M. L. 1973. Fire in the virgin forests of the Boundary Waters Canoe Area, Minnesota. *Quaternary Res.* 3:329–382.

Perlin, J. 1989. *A Forest Journey, The Role of Wood in the Development of Civilization*. Harvard Univ. Press, Cambridge, MA 445 pp.

Pickett, S. T. A., and P. S. White. 1985. *Natural Disturbance and Patch Dynamics*. Academic Press, New York. 472 pp.

Pimm, S. L., G. E. Davis, L. Loope, C. T. Roman, T. J. Smith III, and J. T. Tilmant. 1994. Hurricane Andrew. *BioScience* 44:224–229.

Rowe, J S., and G. W. Scotter. 1973. Fire in the boreal forest. *Quaternary Res.* 3:444–464.

Sousa, W. P. 1984. The role of disturbance in natural communities. *Ann. Rev. Ecol. Syst.* 15:353–391.

Sprugel, D. G., and F. H. Bormann. 1981. Natural disturbance and the steady state in high-altitude balsam fir forests. *Science* 211:390–393.

White, P. S. 1979. Pattern, process, and natural disturbance in vegetation. *Bot. Rev.* 45:229–299.

Whitney, G. G. 1994. *From Coastal Wilderness to Fruited Plain*. Cambridge Univ. Press, Cambridge. 451 pp.

CHAPTER 17

FOREST SUCCESSION

The bases of dynamic change in forest ecosystems have been detailed in previous chapters. All components of ecosystems—climate, landform, soil, biota—change over time. Change is gradual or rapid, often depending on the particular disturbances that characterize regions and local sites (Chapter 16). The most visible expression of ecosystem development and change is that of vegetation, especially the forest trees, which, for example, following fire or windstorm, may change dramatically over a short time. From early times onward, people observed that forest types succeeded one another. This is simply good English usage of the word "succession"—the act of succeeding or coming after another in order or sequence (Neufeldt and Guralnik, 1988).

Forest succession, often characterized as change in species composition or the replacement of the biota of a site by one of a different nature, is just one manifestation of ecosystem change. Many studies have documented ecosystem development and change over time. Soil development has been documented (Olson, 1958; Miles, 1985; Matthews, 1992), and changes involving biomass accumulation and nutrient cycling are described in Chapters 18 and 19. In this chapter, compositional change of species is the primary focus. Forest succession is what happens in one place over an extended period of time measured in 10s to 1,000s of years, i. e., what happens with space fixed and time changing (Rowe, 1961b). It is characterized by a sequential change in relative structure, kind, and relative abundance of the dominant species. Ecosystem change and forest succession never stop, but our primary focus is on intervals between disturbances and within a time period of the same order of magnitude as the life span of the longest-lived organism in the successional sequence (tens to thousands of years).

As part of ecosystem change, forest succession progresses in nearly infinite ways and is driven by many different factors along with simultaneously occurring processes. However, it can only have meaning in the context of a particular geographic framework (i.e., regional and local ecosystem). The heart of the matter is that *location-specific ecosystems* undergo development, change, and succession. The dynamics of each landscape ecosystem are displayed more or less visibly according to the resilience and inertia of component parts when disturbed. Although the visibly changing vegetation draws our attention, attempts to explain and predict its successional path—in abstraction from the geographic matrix that gives it meaning—is unlikely to be rewarding.

In contrast to the landscape ecosystem context, the focus of definitions of plant succession has been the plant species or community. Definitions include those of Grime (1977): "a progressive alteration in the structure and species composition of the vegetation" and Finegan (1984): "the directional change with time of the species composition and vegetation physiognomy of a single site." Because changes in plant composition are more easily seen or recorded over relatively short time periods than are changes in other ecosystem components (climate, physiography, soil), it is not surprising that plants are usually our focus rather than the systems of which they are a part. Obviously, forest succession deals with change in biota, but how we conceive its context (plant or ecosystem, part or whole) is critical in understanding the patterns and processes of succession over the stunning heterogeneity and complexity of regional and local ecosystems. Troll (1963b) was the first to emphasize and explicitly describe succession as a landscape process involving all ecosystem components.

Adopting vegetation as the primary focus undoubtedly contributed to Clements' analogy of community as an organism and the development of an elaborate and often criticized system. Literally thousands of papers have been published, introducing additional terminology and controversy. Miles (1987) wrote: "The confusion in the literature, and the absence of an acceptable unifying theory, seem to spring largely from an unachievable search for an ecological Shangri-La." However, the perceptive review by Pickett et al. (1987) clears up much confusion, and the landscape ecosystem perspective provides the spatial framework for understanding how succession works. Actually, all successions have these key characteristics that affect their course and rate: (1) a sequence of concomitant environmental and vegetational change that, for convenience and practical applications, may be characterized by stages; (2) disturbance regimes; and (3) mechanisms (processes) occurring at different times in the sequence. Above all, the possible successional sequences of interacting organisms, mechanisms, and disturbances are all controlled or mediated by the regional and local landscape ecosystems where succession takes place.

In this chapter, we examine successional concepts, the evolution of the concept of forest succession, causes of ecosystem and forest change, and the mechanisms involved. The focus is on vegetational change, but as far as possible it is placed in an ecosystem context. Examples of primary and secondary succession are presented to illustrate succession in diverse settings. For details beyond the scope of our treatment the following books and reviews provide diverse viewpoints: Clements (1916, 1936); Tansley (1929, 1935); Drury and Nisbet (1973); Connell and Slatyer (1977); Golley (1977); Miles (1979, 1987); White (1979); West et al. (1981); Finegan (1984); Pickett et al. (1987); Glenn-Lewin et al. (1992); Matthews (1992); Peet (1992); and Worster (1994).

BASIC CONCEPTS

Primary and Secondary Succession

Primary succession is ecosystem change that occurs on previously unvegetated terrain and proceeds in the absence of a catastrophic disturbance. For example, primary succession on recently deglaciated terrain in Alaska eventually supports spruce forest after first being colonized by communities of willow or alder. Succession that follows a disturbance to an existing forest, disrupting ecosystem processes and destroying existing biota, is termed **secondary succession**. Often, the distinction between primary and secondary succession is more arbitrary rather than real. It is followed here as a matter of convenience in organizing material on dynamic changes in forest ecosystems.

Biological Legacies

Biological legacies that follow disturbances such as fire, flood, and windstorm are a key feature that distinguish the course of ecosystem change in contrast to succession on bare sites, such as recently deglaciated terrain or a cleared and abandoned agricultural field. Biological legacies are: (1) living organisms that survive a disturbance, particularly a catastrophic, stand-regenerating disturbance, (2) organic debris, and (3) biotically-derived patterns in soils and understories (Franklin, 1995). The living legacies include intact plants and animals, rhizomes, and dormant spores and seeds. Important biotically-derived structures include standing dead trees (snags) and fallen logs, large soil aggregates, and dense mats of fungal hyphae. These diverse biological legacies, derived from predisturbance ecosystems, have important influences on the paths and rates of forest recovery and ecosystem function (see Chapter 21).

Successional Pathways, Mechanisms, and Models

Pickett et al. (1987) define the terms successional pathway, mechanism, and model. A plant **successional pathway** is the temporal pattern of vegetation change. It typically shows change in community types with time and may describe the decrease of particular species populations. Figure 17.1 is an excellent example of successional pathways because the initial site conditions are distinguished together with the physiographic processes that affect ecosystem change. A **successional mechanism** is an interaction or process that contributes to successional change. Examples include general ecological processes such as dispersal, competition, and establishment and physiological processes such as biomass allocation (Chapter 18) and nutrient uptake (Chapter 19) that affect plant form and reproductive ability. A **successional model** is a conceptual map that explains a successional pathway by identifying and specifying the relationship among the mechanisms and the various stages of the pathway.

Autogenic and Allogenic Succession

Traditionally, successional change has been characterized as either **autogenic** (by endogenous factors) or **allogenic** (by exogenous factors) depending whether successive changes are brought about by the action of the plants themselves on the site or by external factors (Tansley, 1926). Primary succession, in the classical view, is an autogenic process (plant driven) and secondary succession is allogenic (driven by periodic disturbances). However, the dichotomy between autogenic and allogenic factors is artificial: the causal factors don't fit strictly into one category or the other (White, 1979; Matthews, 1992). In fact, Tansley (1929, p. 678) recognized this when he proposed the concept, stating that it must never be forgotten that actual successions commonly show a mixture of the two classes of factors.

Today, the ecosystem is or should be the focus, not the plant, because change is elicited by a combination of factors both within the ecosystem and external to it—not the dichotomy of either one or the other (see discussion later in the chapter and Figure 17.12). For example, although allogenic factors (physical processes and disturbances such as fire and windstorm) may be initiated outside a given ecosystem, their effects are often

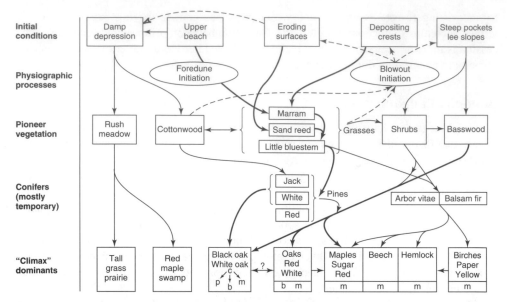

Figure 17.1. Alternative successional pathways from different initial sites on Lake Michigan sand dunes. Beaches, foredunes, and blowout dunes provide diverse sites which undergo different successions. Center of diagram gives oversimplified outline of "normal" succession, from dune-builders to jack or white pine to black oak–white oak with several ground-cover types: choke-cherry–poison ivy (c), "prairie" (p), blueberry–huckleberry (b), or mesophytic herbs (m), depending on topography, water table, and biotic and fire history. Damp depressions (left) and protected lee slopes and pockets (right) may lead to richer forests including basswood, northern red oak, and many mesophytic trees and herbs. Except for physiographic processes, no mechanisms are shown. (After Olson, 1958. Reprinted from *Botanical Gazette,* © 1958 by the University of Chicago. Reprinted by permission of the Chicago University Press.)

strongly dependent upon the site conditions (e.g., topography, aspect, soil depth, drainage) and the associated vegetation within a given ecosystem. The aspect, topography, soil properties, and vegetation of the ecosystem, and its position in relation to fire breaks, strongly affect fire frequency and severity. Windstorm effects are significantly related to soil depth, drainage, rooting ability, and the height of tree species of the affected ecosystem. Furthermore, autogenic factors of litter accumulation and plant-induced changes in microclimate and soil water and nutrients all affect the outcome of allogenic disturbances. In turn, autogenesis is almost always influenced by allogenic factors.

How is Succession Determined?

In either primary or secondary succession, one should track ecosystem and vegetation change at one place over time thus determining an *actual* chronological sequence or successional pathway for a given ecosystem. Such monitoring is rarely possible due to the long life-span of forest trees. Therefore, much of what we know about succession comes from short-term observations at different places. The typical approach used to characterize long-term succession is to *infer* successional trends from stands of different ages, all of which are assumed to have developed on the same site and under the same disturbance conditions. This approach has come to be known as a **chronosequence** (Major, 1951) or **space-for-time substitution** (Pickett, 1988). However, the conventional chronosequence approach is problematical in determining the actual successional trend with space fixed and time changing (Matthews, 1992). An example is the assumption that the forest surrounding a bog has developed through all the stages we observe—from open water and sphagnum mat to the forest itself. The *actual* chronosequence of ecosystem and forest change is one that is site-specific over time.

EVOLUTION OF THE CONCEPT OF FOREST SUCCESSION

The dynamic nature of forests has been recognized at least from the time of the earliest observers who put their thoughts into writing. The plant-based (bioecosystem) concept of forest succession (Spurr, 1952a) dates back well into the beginnings of ecology and forest science and appears in the words of early Roman natural historians. Of the New England forest of the early 1780s, Thomas Pownall (1949, p. 24) wrote:

> *The individual trees of those woods grow up, have their youth, their old age, and a period to their life, and die as we men do: You will see many a Sapling growing up, many an old Tree tottering to its Fall, and many fallen and rotting away, while they are succeeded by others of their Kind, just as the Race of Man is: By this Succession of Vegetation the Wilderness is kept cloathed with Woods just as the human Species keeps the Earth peopled by its continuing Succession of Generations.*

In 1792, Jeremy Belknap recognized the transitory nature of forest trees "on the same soil." The term "succession" was used in a letter by John Adlum (1806; Spurr, 1952a) noting the change of southern pine timber to other species "on the same soil." In Europe, beginning with Hundeshagen in 1830, observed changes in forest composition were the subject of specific articles by professional foresters and botanists. Hundeshagen pointed out instances of spruce replacing beech and other hardwoods in Switzerland and Germany, and of spruce and other species taking the place of birch, aspen, and Scots pine.

The first detailed North American report of composition changes was apparently that of Dawson in 1847, dealing primarily with the Maritime Provinces of eastern Canada. He recognized the effects of windthrow and fire in the forests found by the original European settlers, and distinguished between successional trends in small clearings following cutting, following a single fire, following repeated fires, and as a result of agricultural use of land.

As early as 1860, Henry David Thoreau recognized that pine stands on upland soils in central New England were succeeded after logging by even-aged hardwood stands that today constitute the principal forest type of the region. He named this trend **forest succession** and published an essay on "The Succession of Forest Trees." His late natural history writings, published long after his death (Thoreau, 1993), are filled with ecological-successional insights: pines are "pioneers" to oaks and that "lusty oaken carrots" are a "remarkable special provision for the succession of forest trees." In articles published in 1875 and 1888, Douglas discussed at some length the concepts of forest succession and pioneer species. He presented an explanation of how it is that short-lived, light-seeded pioneer species formed the first forest types on burned-over pine land.

The concept of forest succession as tree or community replacement, then, dates back well into forest and ecological science. It evolved slowly, but was well established by the beginning of the 20th century, when Cowles, Clements, and other American ecologists systematized its study.

Formal Ecological Theory

A general theory of plant succession, and indeed the foundations of plant ecology as a study of vegetation dynamics, were initiated by Henry C. Cowles (1899, 1901). He analyzed succession on sand dunes of Lake Michigan, beginning with uncolonized sand and ending with a mature forest. Although dunes were formed and destroyed by wind, he found that "vegetation profoundly modifies the topography." In 1911, Cowles noted that the original plants in any habitat give way "in a somewhat definite fashion to those that come after." Furthermore, he recognized that changes in vegetation caused by climatic, topographic, and biotic changes occurred concurrently, at different rates, and at times in different directions.

It remained for a contemporary, Frederic E. Clements, to fabricate an elaborate philosophical structure of plant succession (1916, 1949) that attempted to systematize and formalize all eventualities of plant community change. Specific examples of forest succession were documented by William S. Cooper, with his studies of Isle Royale in Michigan (1913) and the colonization by plants following glacial retreat in Glacier Bay, Alaska (1923, 1931, 1939).

Clements, in particular, developed a complex nomenclature to describe plant succession in a meticulous, orderly systematization of nature that has both facilitated and greatly complicated the efforts of his successors. Some of his terms have taken a permanent place in the vocabulary of ecologists (vegetation dynamics, biome); others have persisted but with broadened and changed meanings (climax), while many have been dropped from general usage (cyriodoche, consocies).

HOW DOES SUCCESSION WORK?

Succession involves many different factors and processes operating in a geographic framework at multiple scales, and these are summarized in Table 17.1. As an overview of how succession works, it is first useful to examine the key characteristics of ecosystem change and what we might conclude about succession from them. Key characteristics that interact to determine vegetation change with space fixed and time changing include:

The great diversity of site conditions of the Earth's regional and local ecosystems provide the landscape framework that determine the organisms involved and the course and rate of succession.

An array of disturbance factors and processes associated with particular regional and local ecosystems provide substrates for primary and secondary succession and periodically affect the course and rate of succession.

Organisms colonizing sites possessing an enormous range of life-history traits (genetic adaptations of structural architecture, longevity, and physiological processes) and regeneration strategies that are intimately associated with site conditions and disturbance regimes.

Reciprocal changes in site conditions and organisms with one affecting the other and vice versa.

Competition and mutual interactions among organisms.

Common sense suggests that with these characteristics we can expect that: (1) no two successional pathways will be the same; (2) different successional pathways will be characteristic of diverse regional and local ecosystems; (3) a great number of different and si-

Table 17.1   Causes of Succession: Hierarchical Summary of Factors and
Processes

Landscape Ecosystem Characteristics
   Regional climate, physiography, vegetation patterns
      Local climate, physiography, landform, soil, species distribution

   Site availability
         Disturbance (kind, periodicity, size, severity, dispersion; disturbance history
            of target site and adjacent ecosystems)

   Differential species dispersal and establishment
         Landscape position in relation to seed source availability
         Reproductive mode of species (sexual, asexual) and fecundity
            Dispersal (dispersal agents, landscape configuration)
            Propagule pool for establishment (time since land disturbance,
               land use treatment)

            Differential species performance (in regeneration and post-establishment
               development)
            Genetic adaptations and plasticity (germination and establishment
               traits, understory tolerance, assimilation and growth rates, biomass
               allocation, plant defense)
            Competition and mutualism (presence, kind, amount)
            Environmental stress (climatic cycles, site history, prior occupants)
            Allelopathy (soil chemistry, neighboring species)

---

multaneously operating mechanisms that control the course and rate of succession may be
recognized; (4) more or less distinguishable stages of succession and pattern in succession
will be perceived in ecosystems not severely affected by humans; (5) earlier successional
species modify the environment in ways that may limit their regeneration and that may or
may not act to favor or inhibit the regeneration and growth of species dominating later in
the sequence; and (6) although rates of change vary, no end point of the process occurs.
These outcomes seem reasonable expectations, assuming one can observe succession for a
range of different ecosystems and over a time scale of centuries—at least 200 to 300
years—or based on the return time of major disturbances for the given ecosystem.

   This overview of key characteristics now can be fleshed out in a story of how succession works—simplified, of course, yet highlighting its significant features: (1) following
disturbance, creating either (a) new land (unvegetated substrate for primary succession) or
(b) a disrupted forest and a more or less vegetated site (with secondary succession following), substrate is available for recolonization or continued growth by organisms; (2)
species from surrounding areas, and in the case of the disrupted forest either from surrounding areas or the disturbed area itself occupy the disturbed site in many patterns and
temporal sequences. In the disrupted forest undergoing secondary succession, new
colonists may be dispersed into the site or are already present in the understory, the soil
seed bank, or the vegetative propagule bank (Figures 5.2 and 5.13); (3) the invading plants
either die or become established (processes of regeneration described in Chapter 5) and
develop on the site; (4) plant interactions of competition and mutualism (Chapter 15)
strongly affect plant establishment and development over time; (5) while simultaneously,
the plants are changing the site conditions (light, temperature, soil properties and

processes) which, in turn, may or may not affect them over time in their continuing interactions with other plants and animals; (6) multiple disturbances that are ecosystem-specific in kind, frequency, and severity modify all the above processes with differential effects through time; and (7) over time, some species persist, thrive, and become dominant
in their time, whereas others drop out or diminish in abundance in the short- or long-run.
The nature and rate of change depends on the spatial and temporal scales under consideration, the site conditions of the regional and local ecosystems, species' life history traits
and regeneration strategies, disturbance regimes, chance effects, and the other processes
identified in Table 17.1. If it is known what species are present, as well as what ones are in
the vicinity of a particular land area, and if something of the ecology of the species in the
context of the various land-vegetation patterns that exist is known, then it is possible to
make an educated guess as to what is *likely* to happen in the future (Rowe, 1961b).

Clementsian Succession

It is instructive to compare the scenario just described with the yardstick of classical
Clementsian succession that provided the first model of the vegetative approach, a legacy
for our current understanding and the basis for opposing ideas. Formulated at the turn of
the twentieth century (Clements, 1905), and based on a thorough knowledge of European
and North America work, Clements (1916) described the "essential processes" in the development of a climax formation as follows (bold type added for emphasis):

> *Every sere must be initiated, and its life-forms and species selected. It must progress
> from one stage to another, and finally must terminate in the highest stage possible
> under the climatic conditions present. Thus, succession is readily analyzed into ini-
> tiation, selection, continuation, and termination. A complete analysis, however, re-
> solves these into the basic processes of which all but the first are function of vegeta-
> tion, namely, (1) **nudation**, (2) **migration**, (3) **ecesis**, (4) **competition**, (5) **reaction**,
> (6) **stabilization**. These may be successive or interacting. They are successive in ini-
> tial stages, and they interact in most complex fashion in all later ones.*

Here and elsewhere, Clements (1905; Weaver and Clements, 1929) laid out the rules of
succession in no uncertain terms. The basic processes were clear cut, and stage follows
stage (especially in primary succession) in a systematic stepwise, directional process. Figure 17.2 illustrates how this process might work with ecesis (germination, establishment,
and growth) immediately after plants reach the site (migration). The effects of dominant
plants on environment (reaction) changes the environment to favor or "facilitate" the dominance of the next community that can compete better at the site. The process was orderly,
predictable, and therefore *deterministic*. In stabilization, the end point of a given succession is reached as the **climax** formation or association, the "adult organism," is attained.
The climax dominants modify their environment such that they perpetuate themselves and
are the community of maximum stability (i.e., resisting change) and self-perpetuation.
Furthermore, Clements assumed that all primary successions in a climatic region eventually converged to the same climatic climax from multiple starting points. The term **monoclimax** is therefore applied to this conceptual model. A critical flaw was Clements' analogy of community and organism (Chapter 15) that led to the concept of the climax as an
"organic entity ." In addition, in emphasizing the influence of plants upon the habitat (i.e.,
biotic reaction in Figure 17.2), Clements saw changes caused by internal factors as the
driving force in succession (i.e., autogenic succession) leading to successive stages of
plant communities. Next, therefore, it is important to consider the stages of succession as
a basis for examining insights, problems, and misconceptions about orderly succession.

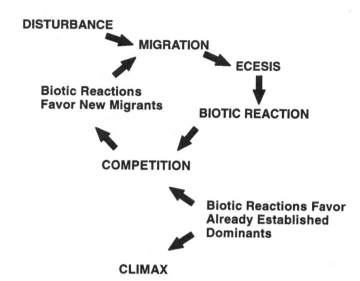

Figure 17.2. Diagrammatic representation of the classical (Clementsian) model of auto-genic plant succession. *Disturbance* brings about nudation followed by ar-rival of propagules (*migration*), that germinate and grow (*ecesis*) and mod-ify the site (*biotic reaction*) in *competition* with one another. Ecesis, reaction, and competition often occur simultaneously. Competition and re-action continue in cycles (left) that favor new immigrants and hence new communities; ultimately (lower right) the reactions lead to the climax or end stage. (From Christensen, 1988b.)

Stages of Succession

The simplest approach to understanding plant succession is to postulate an unvegetated substrate or disturbed forest and then to deduce the successive plant communities that will occupy this site under the assumptions that: (1) the regional climate will remain un-changed, and (2) catastrophic disturbances such as windstorm, fire, or epidemic will not occur. In view of the hundreds, even thousands of years involved in forest successions, these assumptions are unrealistic. Their adoption, however, does provide one view of suc-cession as a more or less orderly and predictable sequence, which depends upon the char-acter of the physical site and associated biota and also provides the framework for us to examine some problems of this perspective.

The recognition of stages, too, is a matter of convenience rather than of their actual occurrence. Stages are, at best, wave-like replacements of species with similar ecologies and, at worst, arbitrary divisions in a continuum (Matthews, 1992). Actually, plant com-munities vary in their rate of change, depending on site and species availability, as new species invade the site and existing species either reproduce or disappear through failure to reproduce. Nevertheless, the arbitrary classification of this continuum into stages, char-acterized by the dominance or presence of certain species and certain life forms of plants, is useful and a convenience worth maintaining.

PRIMARY SUCCESSION

Initial unvegetated sites range from pure mineral material (rock, soil, or detritus) to water; mixtures of soil and water (i.e., moist, well-drained mineral soil) are the most favorable for plant colonization and growth. Thus a continuous range in site exists. Nevertheless, it is convenient to select points along this range at which to postulate succession. Primary plant succession beginning with dry rock material (either as rock or as mineral soil) is termed a **xerarch** succession; that beginning with water is termed a **hydrarch** succession; while that beginning with moist but aerated soil is a **mesarch** succession.

The major stages of primary succession are generally consistent and worthy of study. However, one should remember that many types of primary successions exist (Matthews, 1992), and that both the specific successions and the vegetational stages within each are arbitrarily chosen. In Table 17.2, a series of 10 stages are given for each representative type of primary succession. Some stages are omitted under conditions where the next successional life form (tree, shrub, herb) is capable of directly colonizing an earlier vegetational type. In the Clementsian paradigm, the series of stages at a given site is termed a **sere**.

In relating the possible stages to real-world succession, several misconceptions should be dealt with immediately. First, succession does not necessarily begin with stage 1 and proceed through each successive stage in a unidirectional sequence. Second, stages do not typically proceed separately one after another in relay fashion; considerable overlap usually occurs rather than discrete jumps. Third, there is no set time period for each stage to begin and end; depending on the site, one stage may occupy the site for a long time and seemingly terminate succession until site conditions change. As with taxonomic classes, a false sense of uniformity and rigidity is conveyed by stages, despite their convenience.

Observe that the stages shown in Table 17.2 are not mutually exclusive. The various life forms and developmental stages may be characteristic of more than one stage, and indeed, many persist through many stages. Some mosses, for instance, may invade a site early in succession and persist through to the later vegetational stages characterized by tolerant trees. Tree seedlings often establish themselves at an early stage along with annuals and grasses or beneath shrubs. This is particularly apparent in secondary succession when tree seed sources surround areas burned by fire or cleared by agriculture. Thus all plants of each stage do not necessarily appear and die out abruptly at an appointed time; rather they may overlap in various sequences according to the life-history traits of each species. For example, many clonal shrubs and alder, rhododendron, and mountain laurel thickets persist for 20 to 40 years or more, preventing the establishment of tree seedlings within them (Niering and Goodwin, 1974; Damman, 1975; Vose et al., 1994).

The actual composition of the different stages will be dependent upon those species that have access to the site in question, either by virtue of their proximity or by the capacity of their seed to reach the site by various avenues of dissemination (Clements, 1905;

Table 17.2 Stages of Primary Succession

Stage	Xerarch	Mesarch	Hydrarch
1	Dry rock or soil	Moist rock or soil	Water
2	Crustose lichens	(usually omitted)	Submerged water plants
3	Foliose lichens and mosses	(usually omitted)	Floating or partly floating plants
4	Mosses and annuals	Mostly annuals	Emergents
5	Perennial forbs and grasses	Perennial forbs and grasses	Sedges, sphagnum and mat plants
6	Mixed herbaceous	Mixed herbaceous	Mixed herbaceous
7	Shrubs	Shrubs	Shrubs
8	Intolerant trees	Intolerant trees	Intolerant trees
9	Midtolerant trees	Midtolerant trees	Midtolerant trees
10	Tolerant trees	Tolerant trees	Tolerant trees

Abrams et al., 1985; McCune and Allen, 1985; McClanahan, 1986; Matthews, 1992; Fastie, 1995). The actual plant communities on any given site, of course, depend upon the available plants as well as upon the site. These may be plants already on the site, as buried seeds and propagules (roots, rhizomes) or may be transported by wind, water, or animals from adjacent or distant ecosystems. For example, Clements (1905, p. 239) recognized that "dormant disseminules" were present early in succession "as well as those constantly coming into it."

In the diagrams of Figure 17.3, Egler (1954) illustrated the dichotomy of two models, "relay floristics" and "initial floristic composition" (IFC), that contrast the invasion strategies of species in old-field succession. Relay floristics (Figure 17.3a) refers to the successive appearance and disappearance of stages of species at a site through time. This

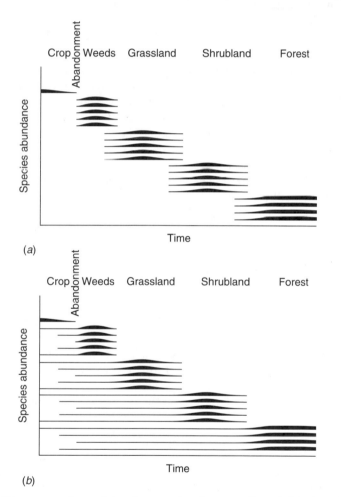

Figure 17.3. Diagrammatic models of succession from studies of abandoned agricultural fields (old-field succession). *a*, relay floristics—invasion of successive groups of species occurs in sequential phases; *b*, initial floristic composition (IFC)—all species establish at or shortly after initiation of succession but still dominate sequentially in stages. (Reprinted from Egler, F. E. 1954. Vegetation science concepts I. Initial floristic composition, a factor in old-field vegetation development. *Vegetatio* 4:414, © 1954 by Uitgeverij Dr. W. Juink, Den Haag. Reprinted with kind permission from Kluwer Academic Publishers.)

model demonstrates that, by modifying the environment, one group of plants favors or *facilitates* the establishment of the next group in the sequence. Although this rigid scheme is attributed by some to Clements, he stated in 1905 that all the migrants into a new, denuded, or greatly modified habitat are sorted by the establishment process into three groups: (1) those that are unable to germinate or grow and soon die, (2) those that grow normally under the conditions present, and (3) those that pass through one or more of the earlier stages in a dormant state to appear at a later stage of the succession. Matthews (1992, p.294) observes that the relay floristics model, as shown here, does not realistically represent the lack of discreteness of the species groups in primary succession.

The IFC model (Figure 17.3*b*), in contrast to the relay floristics extreme, shows that all species are present at or shortly after the initiation of succession and then occur as separate stages over time, as in the relay floristics model. The IFC model demonstrates the idea that succession is a function of individual species' life history traits rather than facilitation by earlier arriving species. Egler's IFC model shows succession with all species present at the start and without additional species invasions—a questionable assumption. For example, although shrubs and small trees (cherries, juniper), some pines and red maple are early colonizers in old fields, oaks and hickories are significantly later colonizers (Finegan, 1984). Finegan marshals other evidence to demonstrate situations where succession does and does not resemble the IFC model. Matthews (1992) examined this model in studies of primary succession on recently deglaciated terrain in Alaska and cites examples of it in glacier foreland. He concludes that there is generally an appreciable lag between colonization by pioneer species and the subsequent colonization of successive loose groupings of later colonizers. The differential patterns of establishment clearly limit the applicability of the IFC model, and more sequences resemble the relay floristic model but lack the discreteness shown in Egler's diagram (Figure 17.3*a*).

SECONDARY SUCCESSION

Ecologists working in various forest regions have produced markedly similar conceptual models of post-disturbance forest succession with four relatively distinct phases or stages. Three examples are shown in Table 17.3. In 1947, Watt described European beech forests as having upgrading and downgrading periods that he divided into four phases, of which the **gap phase** has become the most widely known. Following disturbance most forest ecosystems regenerate rapidly. In temperate and tropical forests alike we recognize two basic stages, known variously as upgrading, building, or aggrading versus downgrading, senescing, degenerating, or stand decline. Most well-known in North America are the stand development model of Oliver (1981; Table 17.3) and the population approach of Peet and Christensen (1980a,b; Table 17.3). The former was developed in New England

Table 17.3 Stages of Secondary Succession

Stand Development Model (Oliver, 1981)	Population Model (Peet and Christensen, 1980a,b)	Beechwood Model (Watt, 1947)
Stand Initiation	Establishment	Upgrading < Gap → Bare
Stem Exclusion	Thinning	
Understory Reinitiation	Transition	Downgrading < Oxalis → Rubus
Old Growth	Steady-State	

and the Pacific Northwest and is illustrated in Figure 17.4a. The latter applies particularly to the Piedmont Plateau in the southeastern United States where abandoned agricultural land reverted to pine-mixed hardwood forest. However, both apply to forests in many other regions as well (Oliver, 1981; Peet and Christensen, 1987). These two models apply particularly to forests: (1) following major disturbance, (2) with single or several age classes, and (3) with stems regenerating during a relatively short period following disturbance. In contrast, Watt's beechwood example is typical for forests on mesic sites, but could apply in many forest regions where small gaps are characteristic.

In the **stand initiation** or **establishment stage**, plants regenerate the area and may develop from (1) newly dispersed or buried seeds (e.g., species A in Figure 17.4a), (2) sprouts from pre-existing stems and roots, and (3) **advanced regeneration** (species B, C, D)—saplings and seedlings of the forest understory that accelerate growth when released by disturbance. Depending on the kind and severity of disturbance, the regeneration strategies of the species, weather, herbivory, and many other factors, stand initiation lasts over a period of less than 5 to over 100 years (Oliver and Larson, 1990). During this period invasion continues until resources become limiting.

In the **stem exclusion** or **thinning stage**, virtually all growing space is utilized, crown closure occurs, overtopped seedlings die, and tree regeneration is limited or ceases (Figure 17.4a). Herbs and shrubs, however, are typically present so the ground cover is not always as bare as depicted in Figure 17.4a. In addition, severe thinning of the initial cohort takes place (Peet and Christensen, 1987; Figure 15.13). Figure 17.4a illustrates that the pioneer species A has dropped out and other species (B, C, D) assort themselves vertically depending on growth rate and biomass allocation (Chapter 18). Many different combinations of species are possible, and examples of development in single- and multiple-cohort are described in detail by Oliver and Larson (1990).

In the **understory reinitiation** or **transition stage** (Figure 17.4a), the overstory begins to thin out or "break up" thereby increasing the light reaching the understory; gaps may begin to form. Tree regeneration begins, and individuals of advanced regeneration may live for a few years or decades and then die. Longevity of this regeneration in the understory varies from 3 or 4 years in cherry bark oak to over 100 years for Pacific silver fir (heights less than 1 m; Oliver and Larson, 1990).

The **old-growth** or **steady-state stage** (i.e., the late-successional forest) follows inexorably as overstory trees age and increase in height and biomass (Figure 17.4a). Invasion and regeneration continue, and advanced regeneration increases in size as site conditions of the understory improve. Senescence, chronic insects and pathogens, and minor and major disturbances cause tree crowns to disintegrate or entire trees to die and disintegrate singly and in groups. Mueller-Dombois (1991) emphasizes that in some Pacific island ecosystems this stage is not just a matter of scattered groups but occurs as entire stand-level dieback which is associated with harsh sites and low-diversity, even-structured stands. In the old-field Piedmont forests, hardwoods dominate over the declining loblolly pines. Relatively even-aged patches, coincident with previous gaps, now undergo a miniature version of the three previous stages. The old-growth stage is defined not only by tree age and size but by biotic properties of living and dead trees (standing and fallen) and ecological processes markedly different from the stand initiation stage (Franklin et al., 1981; Harmon et al., 1986; Oliver and Larson, 1990).

Obviously, there are many possible variations on this four-stage, stand-development theme, depending on the ecosystem-specific site, vegetation, and disturbance characteristics. One alternative is illustrated in Figure 17.4b where pioneer species A dominates in the first two stages. Then, in the midst of the understory reinitiation stage, a major disturbance occurs causing accelerated succession of species B and C. Thereafter, a B–C old-

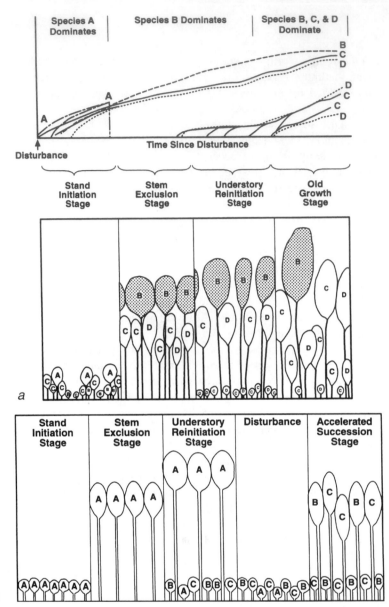

Figure 17.4. Diagram illustrating stages of stand development following major disturbance. *a*, All trees forming the forest are already present in the stand or invade
soon after disturbance. However, the dominant overstory tree species changes
as stem number decreases and vertical stratification of the species progresses.
Height attained and duration of each stand varies with species, site conditions, and disturbances. Barring intervening disturbances, the "old growth"
stage may be reached in less than 200 to over 500 years. (After Oliver, 1981.)
b, Alternative diagram of disturbance-mediated accelerated succession in a
pioneer forest community. Species A is a pioneer tree, whereas B and C are
later successional species. Disturbances may include logging, windthrow, ice
storm, fire, and insect/disease epidemic. The old-growth stage of species B
and C is not shown. (After Abrams and Scott, 1989.) (Part *a* is reprinted from
Oliver, C. D. 1981. Forest development in North America following major
disturbances. *Forest Ecology and Management* 3(3):156, © 1981 with kind
permission of Elsevier Science-NL, Sara Burgerhartstrast 25, 1055 KV Amsterdam, The Netherlands. Part *b* after Abrams and Scott, 1989. Reprinted with
the permission of the Society of American Foresters.)

growth stage would be characterized either by areas of no disturbance, or by minor, or major disturbance which, in turn, might lead to replacement of B by C, C by B, or B and C by A. Logging or windthrow of the overstory when advanced regeneration is present (Chapter 16, Figure 16.11) is a generic example of the Figure 17.4*b* alternative, and specific examples are given by Abrams and Scott (1989).

In summary, it must be emphasized for both primary and secondary succession that complete, stage-by-stage successions rarely, if ever, occur in nature. Disturbances disrupt the gradual internal changes of the ecosystem and may set back, accelerate, or permanently change the course of succession. However, succession is typically ecosystem-dependent, and a more or less predictable sequence of vegetational stages over time can be expected in most ecosystems. Often, disturbances only temporarily alter the sequence. However, disturbances such as erosion, deposition, fire, and windstorm may restart succession or permanently change the site, thereby initiating an entirely new successional sequence. For example, on well-drained sand soils, nutrients are retained primarily in the accumulated soil organic matter. A severe fire that destroys the stand and the organic matter may actually change soil conditions so drastically that a new succession is initiated (Damman, 1975).

Even a vegetational stage itself can so permanently change the site that there is a concomitant change in succession. In Newfoundland, for example, Damman (1975) reported a relatively predictable succession following fire on most sites. However, once *Kalmia* heath became firmly established after fire, it initiated soil changes leading to thin, iron-pan formation, water logging, and peat bog formation. This prevented the return of forest vegetation and created a new successional sequence. Although succession does not always follow the expected pattern, changes in site conditions may strongly indicate the course of the new pattern.

Successional Causes, Mechanisms, and Models

Some of the key characteristics, causes, mechanisms, and models of succession are discussed in this section. They arose as the role of disturbance in forest ecosystems was increasingly appreciated, as succession was examined in a variety of different ecosystems, and in reaction to the classical model.

KEY CHARACTERISTICS AND REGENERATION STRATEGIES

Succession is a function of the interaction of different species with markedly different life histories and regeneration strategies that are determined genetically. Grubb (1977) concludes that there seem to be almost limitless possibilities for differences between species in their requirements for regeneration, giving opportunity for replacement of the plants of one generation by those of the next. Although we attribute these traits and strategies to the species themselves, they are intimately related to the site conditions that elicited them (Chapter 4). The paramount characteristics in forest trees include: understory tolerance, seed production and seed size, dispersal, growth rate and longevity, tree crown architecture, resistance to insects and pathogens, biomass production, allocation of photosynthate (Chapter 18), and nutrient requirements (Chapters 18 and 19). Tilman (1985, 1988) advanced the resource-ratio hypothesis of succession on limiting conditions of nutrient availability and light. He hypothesized that the competitive ability of species is regulated over time by the changing ratio between the two resources. However, the resource-ratio hypothesis has not been found to be generally applicable (Glenn-Lewin and van der Maarel, 1992, p. 31).

The "vital attributes" concept (Noble and Slatyer, 1977, 1980; Noble, 1981) targeted those characteristics of a species that were vital in determining its role in vegetation replacement sequences. These included methods of recovery from disturbance, ability to establish in the face of competition, and time needed to reach critical life-history stages. Also, the basic trait of differential longevity is critical (Egler, 1954, 1977; Drury and Nisbet, 1973) and may be the major explanation of some successional sequences. Short-lived pioneer species may be replaced by long-lived, late-successional species that simply outlive them—providing periodic disturbance fails to regenerate the pioneers. Notice that these characteristics (e.g., seed size, nutrient requirements, vital attributes, longevity, etc.) evolved in response to site-specific ecosystem conditions that conferred persistence to the respective species (Chapter 4). However, broad generalizations without an ecosystem context can be misleading. For example, not all small-seeded species are short-lived pioneers, although in certain ecosystems these traits are key factors in explaining successional pathways in which such species are replaced by heavier-seeded, more understory-tolerant, long-lived species. However, hemlocks (*Tsuga*) and cedars (*Thuja*) are small-seeded, very tolerant, long-lived groups. Their persistence is related to these adaptations in certain kinds of ecosystems whose seedbed conditions (e.g., moist to wet sites; bare mineral soil) and disturbance regimes (e.g., fire, erosion and deposition) favor them. In contrast, walnuts are heavy-seeded but intolerant and fast-growing; their vital characteristics evolved and are expressed in specific kinds of ecosystems wherever they persist today in North America and Asia. In summary, the key life-history attributes and regeneration strategies are inherent parts of many overlapping and sequential *processes*—regeneration (Chapter 5), competition and mutualisms (Chapter 15), and biomass accumulation and growth (Chapters 18, 19)—that are ecosystem specific. These simultaneous and sequential processes, involving site and plant, explain succession ecosystem by ecosystem. Walker and Chapin (1987) describe this process model for primary succession in Alaska, and details are available in a number of sources (Walker and Chapin, 1986; Walker et al., 1986; Matthews, 1992).

AVAILABILITY AND ARRIVAL SEQUENCE OF SPECIES

The availability of species and their arrival at the disturbed site strongly affect the course of succession. Obviously, species may invade at different times and in varying numbers depending on their occurrence in adjacent ecosystems, site conditions of the disturbed area, species life-history traits, disturbance history, and position and distance of the disturbed site in relation to adjacent ecosystems providing the seed source. The factor of differential arrival sequence often explains why similar sites may have different successional pathways. In secondary successions following windstorm or fire, surviving understory plants, clonal plants, and those germinating from buried seed have immediate access to the site. For example, the differential arrival sequence of species was instrumental in determining three pathways of primary succession in Alaska where a 220-year glacial retreat progressively exposed parent material for colonization (Fastie, 1995). This differential was related to the distance from each site to the closest seed source of Sitka spruce at the time of deglaciation. In secondary succession following logging and fire in the Great Lakes region, disturbance history allowed some eastern white pine seed trees to survive logging and post-logging fires. The variable successional pathways on a given area today demonstrate the influence of these remnant white pine seed-sources (Palik and Pregitzer, 1994).

FACILITATION, TOLERANCE, AND INHIBITION

Three alternative models of succession as defined by the mechanisms of facilitation, tolerance, and inhibition were proposed by Connell and Slatyer (1977) and are shown in Figure 17.5. Ecologists have found these models restrictive (Huston and Smith, 1987) and

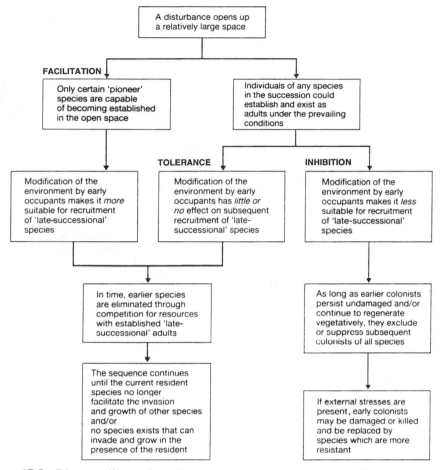

Figure 17.5. Diagram illustrating three autogenic mechanisms—facilitation, tolerance, and inhibition—that determine successional trends. (After Connell and Slatyer, 1977, as modified by Begon et al., 1990. Reprinted with permission from *American Naturalist,* © 1977 by the University of Chicago. Modification reprinted with permission of Blackwell Science.)

oversimplified (Christensen and Peet, 1981) because each was developed to represent a single, alternative pathway (Pickett et al., 1987). In nature there is an enormous variety of pathways, and specific mechanisms and pathways are not tied to one another. Also, several mechanisms may influence successional change in complex ways.

Facilitation is the model of relay floristics (Figure 17.3*a*) whereby early successional species modify the environment to favor later successional species. Actually, facilitation may operate through: (1) enhanced invasion ability, (2) amelioration of environmental stress, and (3) increasing the availability of resources (Pickett et al., 1987). Thus a species may establish and modify environmental conditions in a manner that favors individuals with other combinations of life history traits (Huston and Smith, 1987). In this sense, facilitation has been reported in many instances (Cooper, 1923, 1931, 1939; Jenny, 1941; Christensen and Peet, 1981; Matthews, 1992; Walker and Chapin, 1986; Wood and del Moral, 1987; Fastie, 1995). With special reference to facilitation, Chapin et al. (1994) investigated succession following deglaciaton at Glacier Bay, Alaska. The influence of facilitative and inhibitory effects at four stages in succession are illustrated in Figure 17.6. Life history traits determine the pattern of succession, and initial site conditions and facilita-

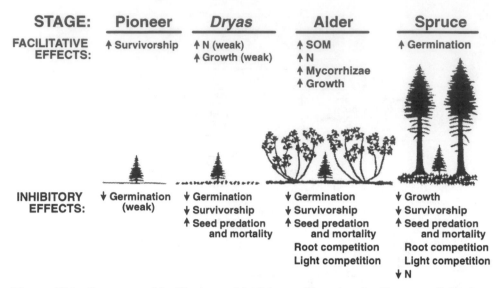

STAGE: **Pioneer**   ***Dryas***   **Alder**   **Spruce**

FACILITATIVE
EFFECTS:

↑ Survivorship | ↑ N (weak) | ↑ SOM | ↑ Germination
↑ Growth (weak)	↑ N
	↑ Mycorrhizae
	↑ Growth

INHIBITORY
EFFECTS:

↓ Germination (weak) | ↓ Germination | ↓ Germination | ↓ Growth
 | ↓ Survivorship | ↓ Survivorship | ↓ Survivorship
 | ↑ Seed predation and mortality | ↑ Seed predation and mortality | ↑ Seed predation and mortality
 | | Root competition | Root competition
 | | Light competition | Light competition
 | | | ↓ N

Figure 17.6. Summary of facilitative and inhibitory effects (weak effects noted) for four successional stages on establishment and growth of spruce seedlings at Glacier Bay, Alaska, as determined from field observations, field experiments, and greenhouse studies. SOM = Soil Organic Matter. (After Chapin et al., 1994. Reprinted with permission of the Ecological Society of America.)

tion and inhibition, where present, influence the rate of change and late-successional community composition and productivity. Facilitative effects were strongly related to the increase in nitrogen that is enhanced by all plants but especially alder.

The mechanism of "tolerance" relates to situations where the initial species modify the environment, but this change has little or no effect on subsequent recruitment and growth of later successional species. Tolerance often means the ability of a species to tolerate low resource levels. Sooner or later such species replace early successional species either by active competition or by simply living longer. Active and passive concepts of tolerance are discussed by Pickett et al. (1987). In "inhibition," the first-established species inhibit subsequent colonization of all species by pre-empting space and other resources or suppress the growth of those already present. Examples of inhibition are shrubs resisting invasion by trees (Niering and Egler, 1955; Webb et al., 1972; Niering and Goodwin, 1974; Damman, 1975; Chapin et al., 1994).

Huston and Smith (1987) observe that facilitation, tolerance, and inhibition describe processes that are relative, not absolute. They almost always occur simultaneously, with varying degrees of importance and may occur in most primary and secondary successions. Various combinations of them are reported for primary successions of glacier forelands (Matthews, 1992; Chapin et al., 1994; Fastie, 1995), in old-field succession in New York (Gill and Marks, 1991) and North Carolina piedmont, and in Alaskan floodplains (Walker et al., 1986; Walker and Chapin, 1986; Walker and Chapin, 1987). Walker and Chapin (1987) concluded that on Alaskan floodplains many different processes and factors (life history, chance, facilitative, competitive, herbivory) affect the interaction between alder and spruce during succession; no single successional process or model adequately describes successional change in these ecosystems.

DIAGRAMMATIC COMPARISON OF MODELS

Figure 17.7 illustrates several models, how they are related, and some simplified pathways for dominant species. Figure 17.7*a* shows the classical succession by facilitation of each

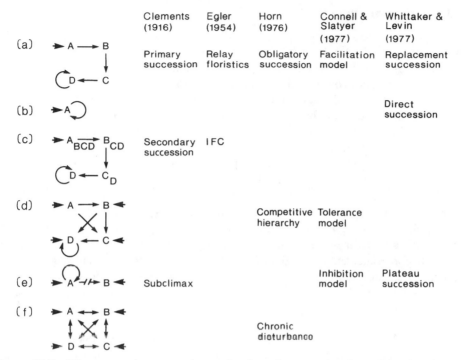

	Clements (1916)	Egler (1954)	Horn (1976)	Connell & Slatyer (1977)	Whittaker & Levin (1977)
(a)	Primary succession	Relay floristics	Obligatory succession	Facilitation model	Replacement succession
(b)					Direct succession
(c)	Secondary succession	IFC			
(d)				Competitive hierarchy	Tolerance model
(e)	Subclimax			Inhibition model	Plateau succession
(f)				Chronic disturbance	

Figure 17.7. Diagrammatic comparison of selected successional models showing replacement sequences (small arrows) of dominant species or community types (A–D). Large arrows represent alternative starting points after disturbance. Subscript letters in (c) indicate species present as subordinates or propagules (omitted from d–f). (After Noble, 1981 as modified by Miles, 1987.)

stage to "climax" (stage D) and its perpetuation. Figure 17.7*b* occurs in tundra and desert vegetation where a single species or community dominates. It might rarely operate in pure stands (pines, aspens) where fire recycles the dominant type. Figure 17.7*c* illustrates old-field succession where several species are present in the first stage (IFC, Figure 17.3*b*). Figure 17.7*d* shows species invasions occurring after succession is initiated (in contrast to Figure 17.7*c*), and one species may follow another depending on life history attributes and disturbance. The inhibition model is shown in Figure 17.7*e* where for some reason vegetation temporarily resists invasion of later-successional species. Figure 17.7*f* illustrates the potential for chronic disturbances to initiate all possible sequences among species with no stabilization. Considerations of stabilization (resistance or resilience to change) and the end point of succession, if any, have led to several viewpoints in reaction to the "climax" concept of Clements. These viewpoints are based, in part, on an understanding of change in forest ecosystems which is considered in the following section along with climax viewpoints.

CHANGE IN ECOSYSTEMS

Ecosystems change constantly in time and in space: from night to day, seasonally, and year to year as climate and soil change, and as the cycle of activity of each of the organisms changes. Thus a regional or local ecosystem never does and never can reach a complete balance or permanence.

End Point of Succession?

The process of ecosystem development or change is inconvertible, and forest succession has been recognized from the earliest days of natural history study. However, the question of what the last stage of succession is, if any, has been the subject of much discussion and debate, a controversy confused by semantic difficulties. Already in 1901, Cowles observed: "The condition of equilibrium is never reached, and when we say that there is an approach to the mesophytic forest, we speak only roughly and approximately. As a matter of fact we have a variable approaching a variable rather than a constant." However, Clements formalized the term, climax, as an organismal entity or "adult organism," defined it as the end point of a given succession, and used it as the basis for his classification of vegetation. Despite efforts to rehabilitate the term as "polyclimax," "polyclimatic climax," "site climax," "climax pattern," or position of relative stability, the term, climax, carries with it, and in any modified form, the implied ideas of vegetation as organism and an end point of succession. Basically, it epitomizes the vegetation context of succession rather than the landscape-ecosystem context. Using an ecosystem approach, the term, climax, is inappropriate, and its use should be discontinued. Instead, to characterize vegetation change, it is useful to recognize pioneer or early-successional, mid-successional, and late-successional species or groups for position along the temporal scale for site-specific ecosystems. These terms (early, mid, late) are not absolute but are relative to the specific ecosystem in the landscape.

Because climax terminology permeates the literature, it is useful to examine how the climax terms are used. In Clements monoclimax theory, macroclimate was the dominant community-forming factor, whereas other factors (physiography, soil, fire, biota) were of secondary importance. Arguing that other factors may be of equal importance, one can theorize a different climax community for different physiographic settings, soil types, and disturbance factors. This is the **polyclimax** theory (Tansley, 1939), a theory that holds that for any combination of environment and organisms, biotic succession will take place toward a climax, but that the specific nature of the climax will vary with the specific environmental factors and biotic conditions. The polyclimax theory has its roots in the contributions of Cowles (1899, 1901) and Moss (1913), but especially Nichols (1923), who argued for a different **physiographic climax** on each site, that differed more or less from the regional **climatic climax**.

For example, in Cowles' classic studies of plant succession on the sand dunes of lower Lake Michigan, it was thought at first that succession on all sand dune sites would eventually reach the same mixed mesophytic hardwood forest stage. Yet, in a later study of the same area, Olson (1958) concluded that the drier and more exposed sites would never support such a community but would be more or less permanently clothed with a black oak–blueberry type because of the dryness and low fertility of the soil itself (Figure 17.1). Furthermore, he felt that although vegetational changes would continue with time, they would become progressively slower and never reach complete stability:

> *Vegetation, soil and other properties of the ecosystem usually change rapidly at first and more slowly later on. If they approach some limit asymptotically or fluctuate around it, this limit should describe the climax community on mature soil . . . The limit itself may vary with time and place. Ideally, it describes a gradational "climax pattern" of communities or ecosystems in any region—generally not a uniform "climatic climax."*

Over the years, the concept of climax became broader, less rigid, and more in line with the polyclimax approach. Tansley (1939) was instrumental in broadening the concept by adopting a dynamic viewpoint in recognizing that "positions of relative equilibrium are reached in which the conditions and composition of the vegetation remain approximately constant for a longer or shorter time."

However, the polyclimax theory was criticized for its own terminology of climaxes: climatic, edaphic, topographic, topoedaphic, fire, zootic, salt, etc. (see descriptions in Oosting, 1956 and Daubenmire, 1968). Furthermore, many of these are the same as or are merely more specific distinctions of Clementsian climaxes: in the polyclimax approach the Clementsian "climax" was simply renamed "climatic climax;" Clements' subclimax may be either an edaphic or topographic climax; a disclimax becomes a fire or a zootic climax. Also, similar to Clements, only one climatic climax was recognized. To avoid confusion, a **polyclimatic climax** theory was developed (see Meeker and Merkel, 1984) in which more than one climatic climax was recognized for a macroclimatic region.

Like Olson (1958), Whittaker (1953, 1975) recognized a gradational **climax pattern** of vegetation corresponding to the pattern of environmental gradients—a continuum of climaxes. Whittaker (1953) wrote that climax composition was determined by:

all "factors" of the mature ecosystem—properties of each of the species involved, climate, soil and other aspects of site, biotic interrelations, floristic, and faunistic availability, chances of dispersal and interaction, etc. There is no absolute climax for any area, and climax composition has meaning only relative to position along environmental gradients and to other factors."

Recognizing the practical problems of the "climax pattern" continuum, Dyksterhuis (1949, 1958) advocated the use of units of site in characterizing climax for range classification. This approach is termed **site climax** by Meeker and Merkel (1984). Similar to the "climax pattern" in theory it, however, ties vegetation change to site-specific units that occur along a gradient.

Viewpoints and details of the climax concept have been presented by many authors (Phillips 1931, 1934-35; Tansley, 1935; Clements, 1936; Cain 1939; Whittaker, 1953, 1975; Daubenmire, 1968; Langford and Buell, 1969; White 1979; Matthews, 1992).

SUCCESSION AS A LANDSCAPE ECOSYSTEM PROCESS

Evident from the foregoing sections is that a general model of succession using a species or community context is unrewarding. A framework of stages based on stand or population development is useful, but the duration, distinctness, and predictability of stages is dependent upon specific site conditions as well as on the species involved. Species' life history traits are important but are inseparably linked to the sites and disturbance regimes to which they are adapted. Disturbance itself, a powerful successional force, is an ecosystem process. Therefore, the landscape ecosystem approach to forest succession provides the functional and terrain-specific framework for understanding the multiple mechanisms and processes that drive succession.

In adopting a landscape ecosystem approach to forest succession, the first-order focus is the regional and local ecosystem rather than their component parts, such as vegetation or soil. Succession is a multidimensional process involving many different factors and processes at different scales (Table 17.1) which can be assessed for specific regional and local ecosystems. Recall from Chapter 2 that the best starting place is from above (i.e., "top down") with an understanding of the geographic hierarchy of regional and local ecosystems, their physical factors, biotic interrelationships, and past histories. Each regional ecosystem, depending on scale, will exhibit many different successional pathways and mechanisms, but their character and pattern will be different among climatic-physiographic regions.

For example, the kinds of successions in the Southern Appalachian Mountains are different than those of the Yellowstone Plateau, slopes of the Sierra Nevada Mountains, or the southeastern Coastal Plain. In Michigan, the four ecosystem Regions (Figure 2.11) exhibit markedly different kinds and patterns of succession due to differences in climate, physiography, soil, and vegetation. Similarly, within a given Region in Michigan, ecosystem Districts (Figure 2.11) have their own characteristic patterns of successional trends for local ecosystems occurring as sand plain, acid swamp, calcareous swamp, rocky ridge, or moist and fertile lower slope. Even for the local ecosystems, not one but several successions may be expected (Simpson, 1990). This variation may be due to many factors, including past history of disturbance or management (logging, drainage) disturbance frequency and severity, seed source availability, and microsite variation, among others. Nevertheless, by specifically defining the geographic framework and working with more or less homogeneous site conditions, we reduce significantly the factors that obscure successional trends.

A general example of multiple successional pathways, based on the site characteristics and disturbance regimes of different ecosystems for the boreal forest of Alaska and Canada, is illustrated in Figure 17.8. Black spruce is the focus, and on many wet, mesic, and dry sites black spruce replaces itself with or without fire (Viereck, 1983). Invasion of

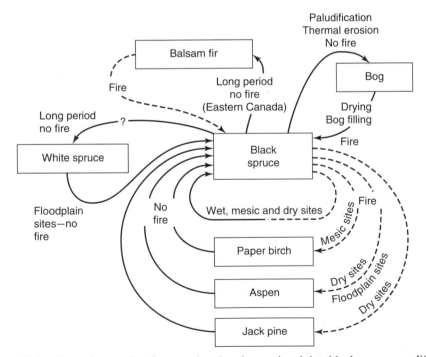

Figure 17.8.   General example of successional pathways involving black spruce on different ecosystems, defined by their site conditions, of the boreal forest in Alaska and Canada. (After Viereck, 1983. Reprinted with permission from SCOPE 18, The Effects Fire in Northern Circumpolar Ecosystems, edited by Wein, R. W. and D. A. MacLean, 1983 © SCOPE, John Wiley & Sons, Ltd., Chichester, UK.)

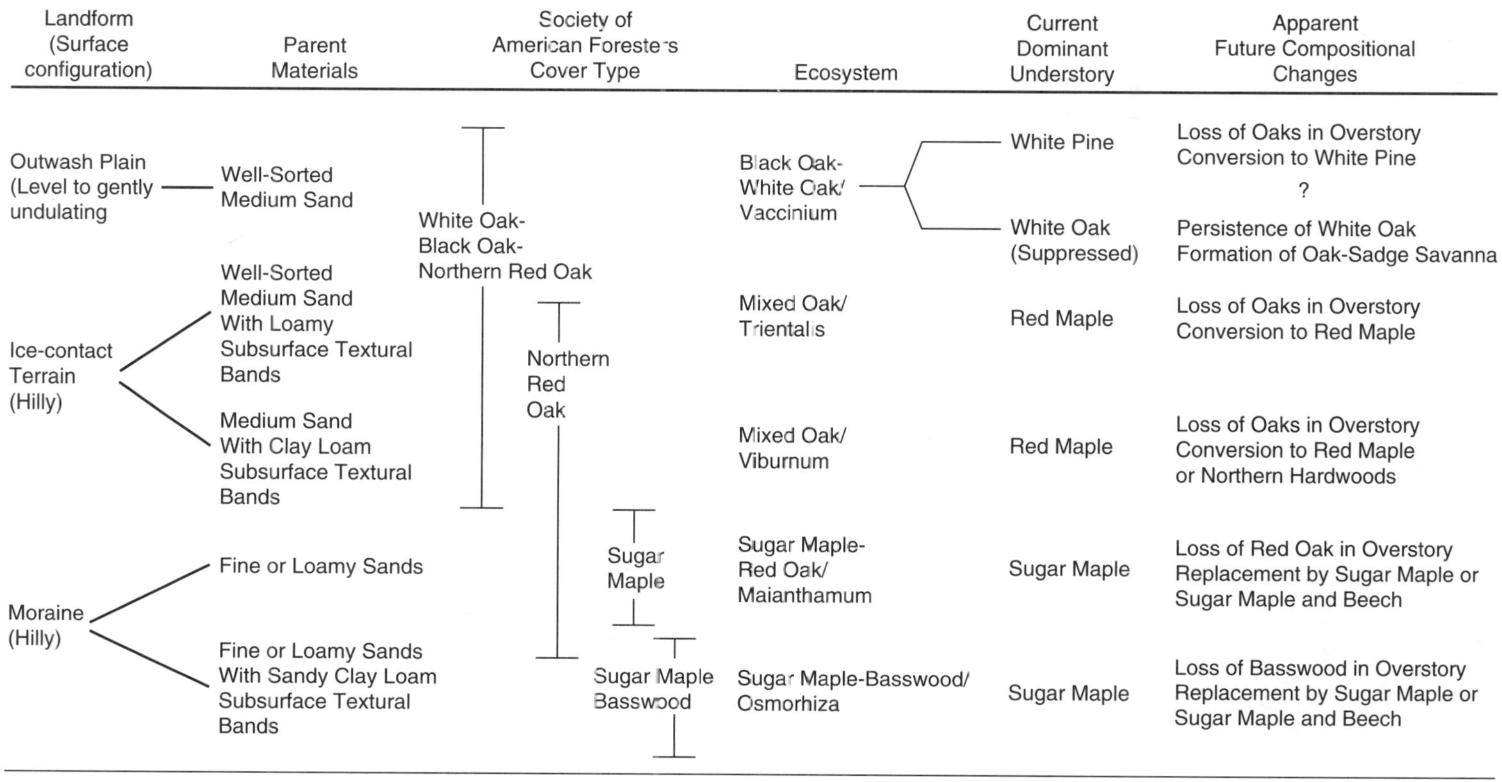

Figure 17.9. Estimated future compositional change associated with two levels of land-scape ecosystems (landforms and ecosystem types) in northwestern lower Michigan. (After Host et al., 1987. Reprinted with the permission of the Society of American Foresters.) Site conditions, overstory cover type (Eyre, 1980), and current dominant understory species are also shown.

burned black spruce forest by birch and aspen is more common in the southern regions of the boreal forest than in the north. In eastern Canada, long fire-free periods enable balsam fir to replace black spruce (Figure 17.8). The many other ecosystem-specific pathways are described by Viereck (1983).

A specific example of the ecosystem approach comes from research on ecosystem processes (forest succession, biomass accumulation, nutrient cycling) in northwestern lower Michigan (Figure 17.9; Zak et al., 1986; Host et al., 1987; Host et al., 1988). To estimate the likely pathways of succession, the site-specific ecosystems were first distinguished at two levels: *landforms* (outwash plain, ice-contact drift, and moraine) and within landforms the *ecosystem types* (named by overstory and ground-cover species). Then, based on sampling of the understory and ground-cover layers, and understanding site conditions and life history traits of the species, estimates of the successional trends for both landforms and ecosystems were made (Figure 17.9).

First, notice that the oak cover types fail to distinguish not only the landforms and ecosystem types, but the predicted successional pathways as well. This illustrates an important general point: communities identified by dominant species are not appropriate bases to ascertain successional pathways; an understanding of entire ecosystems is required. Second, observe that different pathways were predicted for ecosystem types within each landform. Finally, notice that these pathways, based on multiple factors (landform, soil, microclimate, and ground-cover, understory, and overstory vegetation) are the best estimate we have of successional trends for making management decisions. Since they are specific to ecosystems that recur throughout the landscape, the proposed trends, i.e., successional hypotheses, can be monitored and tested over time. Because landscape ecosystem maps are now available at several scales for most public and large private lands in North America, we have the opportunity to apply this powerful approach for improved management, conservation, and restoration of ecosystems.

BIOMASS AND DIVERSITY

Up to this point we have only examined compositional change in forest succession. Changes of biomass production have also been reported (Peet, 1992). Many studies show that biomass increases in a smooth, linear, or logistic fashion toward a late-successional maximum. Alternatively, biomass may increase to a maximum and then decrease slowly or rapidly depending on site and disturbance conditions. The process of biomass accumulation and alternate changes over time are discussed in Chapter 18.

An example of changes in five critical variables in a black spruce ecosystem following fire, and over a time span of about 300 years, is shown in Figure 17.10 (Zasada et al., 1977). The variables are: (A) overstory biomass and living forest floor, (B) dead and decaying forest floor biomass, (C) available pool of nitrogen and phosphorus, (D) soil temperature, and (E) soil water. Before fire (time stages 4–5 in Figure 17.10) the biomass of overstory and forest floor is high but the available nutrient pool is low. Decomposition is slow and soil temperature during the growing season is low; permafrost is as close as 30 cm below the surface. Following fire, there is a drastic change in all variables (Viereck, 1983). Overstory and forest-floor biomass are greatly reduced. Partial or complete burning of the forest floor releases large quantities of available nutrients (stage 1, Figure 17.10). Soil temperatures increase and the permafrost layer recedes; the depth of thaw may be two to three times greater following fire than it was before the burn.

In primary succession, plant diversity measured in numbers of species, i.e., richness,

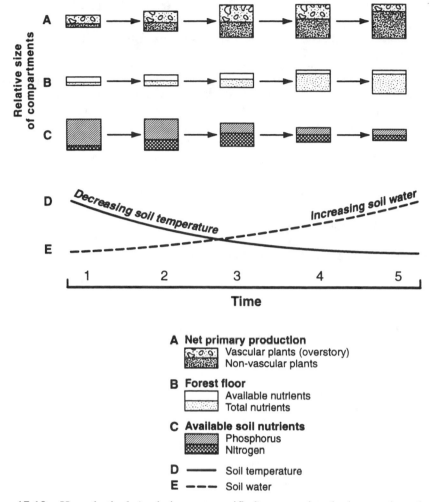

Figure 17.10. Hypothetical (and, in part, verified) successional changes in a burned black spruce-dominated ecosystem being revegetated again by black spruce. The time axis represents the stages of succession following fire, starting with black spruce–feathermoss forest to a more open black spruce–*Sphagnum* type, a time span of about 300 years. (After Zasada et al., 1977.)

may be low if site conditions are so severe that few species are able to colonize it. In contrast, this stage in secondary succession may exhibit higher richness because many species may sprout following fire or survive windstorm and flooding. Considering just secondary succession, richness is strongly related to ecosystem type. Where disturbance is recurrent, richness may be relatively high where ecosystems with mixtures of species of varying understory tolerance are maintained. Where mesic ecosystems lack major disturbance for 100 years or more species diversity may be very low as relatively few species are adapted to the shaded environment of the forest floor. Recurrent disturbance in such systems therefore tends to maintain diversity; richness may be highest at the stage in the disturbance regime where a combination of early-, mid-, and late-successional species are present (White, 1979; see Chapter 20 for discussion of diversity).

EXAMPLES OF FOREST SUCCESSION

Recently Deglaciated Terrain—A Geoecology Approach

The newly-formed land in front of a glacier, the glacier foreland, provides a unique opportunity to study primary succession; detailed studies have been underway since the late 1800s. John Matthews (1992) provides a comprehensive and worldwide review of primary succession on recently deglaciated terrain. He examines physical and ecological processes, successional stages, and models. From his geoecology (landscape ecology) viewpoint, ecosystem succession in front of the retreating glacier is part of the developing landscape in which all parts of the ecosystem develop together. In the glacier foreland, the most important visible feature of the landscape are the landforms and their associated parent materials. The developing vegetation is a function of the dominant factor *terrain age*—the other factors are subordinate: climate, relief (physiography), parent material, organisms, and the influx variables of physical site change and disturbance. Vegetation chronosequences have limitations, primarily the existence of differences between sites, both initial environment and their subsequent environmental histories. Nevertheless, judicious use of chronosequences can provide approximations of succession at particular sites. For example, the chronosequence at Glacier Bay, Alaska is a series of eight intergrading stages:

    I. early pioneer stage: mountain avens (*Dryas drummondii*) and prostrate willows, mosses, and herbs;

    II. *Dryas*-mat: coalescing mats of *Dryas*;

    III. late pioneer stage (after 15–20 years): young Sitka alder and black cottonwood less than 2 to 3 m tall;

    IV. open thicket stage (after 20–25 years): clumps of alder;

    V. closed thicket stage (after 25–30 years): dense alder with individual erect specimens of balsam poplar, shrubby willows, and Sitka spruce;

    VI. balsam poplar line stage (after 30–40 years) where poplars reach height of 4 to 5 m and form a canopy and visible line above the alder;

    VII. spruce forest stage (at 75–90 years) when spruce replaces alder, and possibly,

    VIII. spruce-western hemlock stage.

Succession past stage VII is speculative as even after 200 years the forest may be dominated by spruce. As indicated above, such "stages" must be viewed with care.

A general view of the horizontal and vertical structure of vegetation on the glacier foreland, based on permanent plots in front of the Grand Glacier d'Aletsch in the Swiss Alps, is illustrated in Figure 17.11. Typically, the sequence of herbs to shrubs to trees is parallel to the sequence of plant size, longevity, and competitive power. However, trees may invade much earlier in the sequence. Their establishment depends on seedbed conditions and seed availability. In the Canadian Rocky Mountains, establishment time ranges from 10 to 15 years to 80 years (Luckman, 1986).

Matthews' comprehensive review of primary succession in the glacier foreland context reveals that succession may proceed in strongly divergent, strongly convergent, as well as parallel chronosequences. Generally, divergent pathways are associated with early stages and relatively severe physical environments; convergence is favored by relatively strong biotic controls. Biomass and species richness tend to increase during succession;

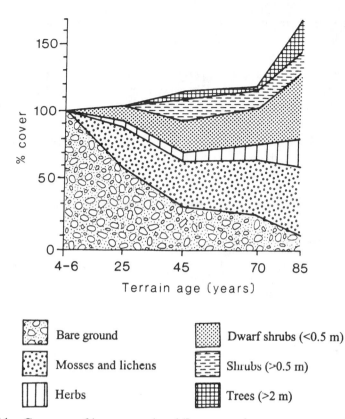

Figure 17.11. Coverage of bare ground and five vegetative strata on recently deglaciated terrain, which is on terrain of increasing age, Grand Glacier d'Aletsch, Swiss Alps. (After Lüdi, 1945. Reprinted with permission of the Geobotanisches Institut ETH Zürich, Stiftung Rübel, Zürich, Switzerland.)

patterns of diversity vary considerably from site to site and are reviewed by Matthews (1992).

In presenting his geoecological model, Matthews notes that most of the general models (as previously described) have only limited applicability on recently deglaciated terrain. The geoecological model emphasizes two distinctive features. First, physical environmental processes, as well as biological processes, form the model. Second, models of succession, as a landscape process, must pay attention to spatial variation, i.e., site position and environmental characteristics of the site through time. This approach emphasizes the vertical (functional) and horizontal (ecological) aspects of landscape ecology as described in Chapter 1 (Figure 1.5). Matthews strongly emphasizes the physical landscape, its properties (landforms and glacial sediments, climate, soil) and processes (erosion by water; frost weathering, heaving, and sorting; solifluction; wind deposition and erosion; and soil development).

Matthews' summary points concerning succession on recently deglaciated terrain provide real-world insights to the mechanisms and models described:

Allogenic and autogenic processes interact during succession and are often inseparable.

Differential seed arrival has been overemphasized; glacier forelands appear to be well supplied with propagules from the full range of species characteristic of the available flora.

Ecesis (establishment) is more important than migration (species arrival) in determining species composition. In the harsh environment of glacial forelands, pioneers probably require some special physical, chemical, or microbial conditions as well as lack of competition.

Multiple processes are characteristic in replacement of pioneers by later colonizers. Although biological processes are important (life history traits, competition, facilitation), the physical processes are also very important and often overlooked. The decline of pioneer species may be more related to physio-chemical changes associated with leaching (nutrient depletion, balance of available nutrients, toxicity associated with pH changes) and other physical changes than to the rise of species of the next successional stage.

The harsh physical environment of the glacier forelands markedly affects the nature and rate of succession by the action of allogenic factors (physical processes and disturbances) as compared with autogenic action of the plants, i.e., the importance of life history traits influencing succession. In Figure 17.12, a comparison is made of the effects of allogenic and autogenic factors in a range of environments. In favorable environments (high in resources; low in stresses), the steeply increasing curve (1 in Figure 17.12) illustrates the

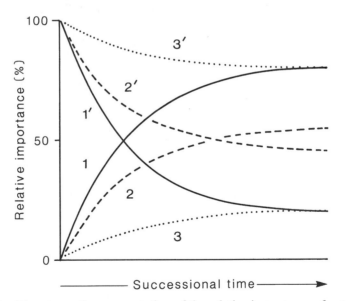

Figure 17.12. Diagrammatic representation of the relative importance of autogenic (1, 2, 3) and allogenic (1′, 2′, 3′) processes under increasing environmental severity (1 and 1′ with solid line = least severe; 2 and 2′ with dashed line = moderately severe; 3 and 3′ with dotted line = most severe) in the context of recently deglaciated terrain. The model assumes species/community context with autogenic processes relating only to vegetational processes; allogenic processes are a combination of physical processes and disturbance. (After Matthews, 1992. Reprinted from *The Ecology of Recently Deglaciated Terrain,* © 1992 by Cambridge University Press, Reprinted with the permission of Cambridge University Press.)

rapid increase in the importance of autogenic processes through time. The perceived increase may reflect the rapid accumulation of biomass due to increasingly favorable soil water and nutrients in a favorable climate or a relatively undisturbed site. In contrast, allogenesis, as a determinant of vegetational change, is reduced sharply (1′) due to the reduction of the intensity of physical processes and disturbances and the concomitant increase in plant biomass and vegetative control of the environment. As environmental harshness increases in severity, biological production is constrained, allogenic processes become more important in any successional stage, and the control of succession by allogenic processes becomes longer. The relative importance of allogenesis in Figure 17.12 (curves 2′ and 3′) is seen increasing with more severe environments. That is, in very severe sites, plants themselves modify the environment to a lesser extent than in favorable sites.

This relationship can be applied at the broad scale of regional ecosystems (increasing latitude for example: boreal forest, tundra, polar desert) or at the local scale within the different landforms of the glacier foreland. In addition, many facets of the nature and complexity of succession can be related to the severity of sites and the relative importance to autogenic or allogenic processes: (1) variety of growth forms, number of species, and plant biomass in the successional sequence, (2) number of successional stages recognized, (3) extent of divergence and/or convergence in successional pathways, and (4) rate of succession and time required to attain a late-successional condition. Finally, it is important to reiterate the caveat to the above discussion that autogenic and allogenic processes are not mutually exclusive. The above relationships apply to the extent that these processes can be distinguished. It appears likely that they are most distinct in the harshest environments.

Succession Following Fire in Ponderosa Pine Forests of Western Montana

Fire plays a predominant role in ecosystem change throughout forests of the western United States (Chapters 12 and 16). In western Montana, the major role of fire in determining successional pathways in grasslands and forests, from low to high elevations, is described by Fischer and Bradley (1987). As one example, the role of fire is illustrated for ponderosa pine, which is strongly fire-dependent for its establishment and widespread occurrence throughout the West (Chapter 4). In western Montana, ponderosa pine competes with grass-dominated ecosystems at low elevations and gradually gives way to Douglas-fir in the cooler, moister, and less fire-prone ecosystems of the montane zone. General and specific successional pathways for ponderosa pine communities (six habitat types) on warm and dry sites are illustrated in Figure 17.13. In this submontane zone, the natural role of fire is threefold: (1) to maintain grasslands, (2) to maintain open ponderosa pine stands, and (3) to encourage ponderosa pine regeneration. In the generalized concept (Figure 17.13a), frequent fires tend to maintain the grassland community by killing pine seedlings (stage 1). Ponderosa pine seedlings become established gradually over a long fire-free period as the result of a seedbed-preparing fire (stage 2). In absence of further burning, the seedlings develop into saplings. Fires during this period may kill young trees (stage 3) or thin them (stage 4). In time, light ground fires tend to produce an open stand of mature trees (stage 5) that may accumulate enough fuel to produce a stand-destroying fire (stage 6) which is followed by grassland.

In Figure 17.13b, multiple successional pathways are shown with the starting point of open, parklike, old-growth ponderosa pine forest. The various states of pine stands (e.g., A, B2, C3, E2, etc.) and pathways reflect the interaction of fire (timing and severity) or absence of fire and the life history traits of pine and grasses. For example, starting with an open, parklike, old-growth pine stand (A), frequent fires (pathway 33) maintain grass-

a

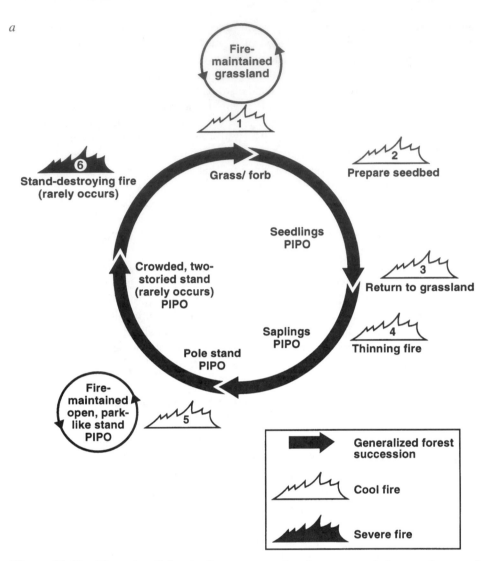

Figure 17.13*a*. The role of fire in forest succession in warm and dry ponderosa pine ecosystems. *a*, generalized diagram of succession illustrating situations where grassland is maintained or where ponderosa pine is favored by a less frequent fire regime. *b*, specific diagram of multiple successional pathways starting from an open, park-like old growth ponderosa pine stand. Numbers (1, 2, etc.) indicate successional pathways; letters (A, B) indicate states of pine forest or grassland. (After Fischer and Bradley, 1987.)

land (K). In contrast, moderate-severity fire (pathway 5) leads to stage D1, a closed canopy, multi-storied pine stand. If no fires occur, a more crowded stand occurs (E1), which is maintained by low-severity fire (pathway 7) or changed by severe fire (pathway 9) to grassland (F). The complexity of succession and the lack of any end point of the process is well illustrated in this simple example. Similar examples of fire ecology for the diverse range of habitat types of eastern Idaho and western Wyoming are also available (Bradley et al., 1992).

b

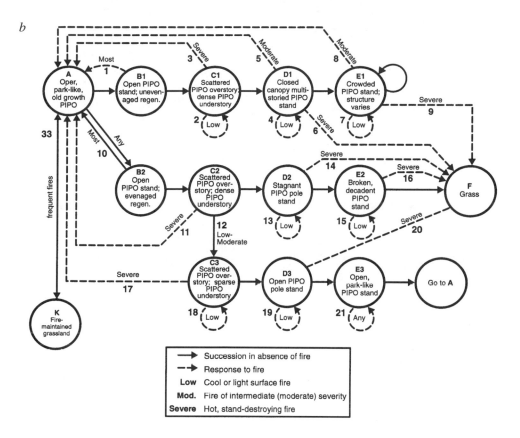

Figure 17.13*b*. Continued from previous page

A variety of successional simulation models have been developed (Huston et al., 1988; Botkin, 1993), and a mechanistic, ecological process model simulating fire succession of coniferous forests of the North Rocky Mountains is available (Keane et al., 1996). However, quantitative modeling of succession is beyond the scope of our treatment.

Gap Dynamics

Forest trees may establish and successfully reach the overstory canopy via three general situations, or modes (Veblen, 1992): following catastrophic disturbance, by establishing in overstory gaps, or, rarely, by continuous regeneration in a disturbance-free understory, and recruitment into an overstory that lacks discrete gaps. Understory intolerant species typically utilize the major disturbance mode, and mid-tolerant and tolerant species often achieve successful regeneration in relatively small treefall gaps. Tolerant species may use the third mode if the main canopy is not too dense. Usually however, they may establish in the disturbance-free understory (Lorimer et al., 1988; Busing, 1994) but probably require at least periodic episodes of overstory crown thinning. In this section, we examine gap-phase regeneration.

Once an opening or gap occurs in the forest overstory, advanced regeneration and new stems invading the gap form a patch of vegetation that responds to changed environ-

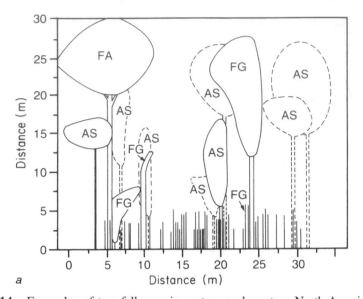

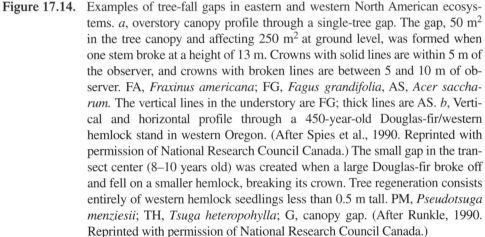

a

Figure 17.14. Examples of tree-fall gaps in eastern and western North American ecosystems. *a*, overstory canopy profile through a single-tree gap. The gap, 50 m² in the tree canopy and affecting 250 m² at ground level, was formed when one stem broke at a height of 13 m. Crowns with solid lines are within 5 m of the observer, and crowns with broken lines are between 5 and 10 m of observer. FA, *Fraxinus americana*; FG, *Fagus grandifolia*, AS, *Acer saccharum.* The vertical lines in the understory are FG; thick lines are AS. *b*, Vertical and horizontal profile through a 450-year-old Douglas-fir/western hemlock stand in western Oregon. (After Spies et al., 1990. Reprinted with permission of National Research Council Canada.) The small gap in the transect center (8–10 years old) was created when a large Douglas-fir broke off and fell on a smaller hemlock, breaking its crown. Tree regeneration consists entirely of western hemlock seedlings less than 0.5 m tall. PM, *Pseudotsuga menziesii*; TH, *Tsuga heteropohylla*; G, canopy gap. (After Runkle, 1990. Reprinted with permission of National Research Council Canada.)

mental conditions—leading to succession termed **gap or patch dynamics** (Pickett and White, 1985). Watt (1925, 1947) described four phases (Table 17.3) in the development of pure European beech forest in England, one of which was the "gap phase" that occurred in old-growth beech stands. Gaps are initiated by major or minor disturbances, and most emphasis has been placed on catastrophic disturbances caused by windstorm, fire, flooding, and landslides (Chapter 16). However, the small or tree-fall gaps significantly affect the structure and composition of mesic forests around the world, especially those dominated by understory tolerant trees.

Examples of small gaps that are characteristic of mesic forests of eastern and western North America are illustrated in Figure 17.14. In these forests, major disturbances are uncommon, and the interval between severe disturbances is long. Thus small gaps and slight but chronic disturbances to overstory crowns play a major role in tree regeneration. Lorimer (1977, 1980) found that mesic presettlement forests in the eastern North America and the Appalachian mountains were characterized by all-aged stands—stands with an inverse J-shaped size class distribution (Figure 15.12). In mesic old-growth forests of eastern North America, Runkle (1982) estimated average annual disturbance rates of only 0.8 to 0.9 percent per year, for an average canopy rotation time of about 110 to 125 years. For

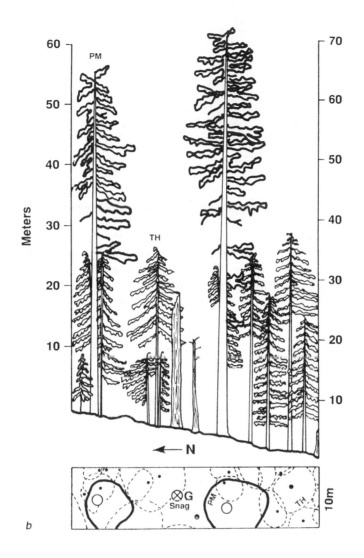

b

both tropical and temperate forests Veblen (1992) estimates an average rotation time of approximately 100 years and a range of 50 to 575 years.

In and around gaps caused by the death of one or several trees, late-successional species can perpetuate themselves more or less indefinitely in the absence of major disturbances. These species (beeches, sugar maple, western and eastern hemlocks, firs, basswoods) regenerate by persistent seedlings in the understory that fill the gaps to the virtual exclusion of more intolerant species that must first establish in the newly formed gap. A canopy profile through a single-tree gap in the sugar maple-beech forest is illustrated in Figure 17.14*a*. The gap, 50 m^2 in the tree canopy and affecting 250 m^2 at ground level, was formed when one stem broke at 13 m about 3 years before the profile was made. Such small gaps may be filled either by canopy trees expanding branches into it or by already-established saplings that will fill it from below. In temperate forests, the gap-size threshold is approximately 0.04 ha since the abundance of intolerant species increases markedly in gaps at or larger than this size (Runkle, 1985; Stewart et al., 1991; Busing, 1994). Many authors report that two or more gaps are required for most saplings of tolerant species to reach the overstory. Mean residence time in the understory for tolerant species is often long; for sugar maple it is often over 100 years and can be as long as 230 years (Barden, 1981; Canham, 1985).

Gap dynamics in mesic Douglas-fir–western hemlock forests of the Pacific Northwest appear to operate more slowly than in other forests (Spies et al., 1990). Gap formation is at the low end or below the range of 0.5 to 2.0 percent reported for temperate deciduous forests (Runkle, 1985) and tropical forests (Brokaw, 1985a,b). As illustrated in Figure 17.14*b*, vertical forest structure provides an important context for gap dynamics. In this figure, the small gap occurs in old-growth Douglas-fir/western hemlock forest whose overstory trees are over twice the height of those in the beech-sugar maple forest in Figure 17.14*a*. Unlike eastern deciduous forests, the death of one or a few narrow-crowned conifers in tall Douglas-fir/western hemlock forests may not transmit enough light for the regeneration of mid- or intolerant species (Spies et al., 1990; Canham et al., 1990). However, gaps may play this role for Douglas-fir and other early successional species in other ecosystems with shorter and less dense canopies, such as mixed-conifer forests (Spies and Franklin, 1989).

If a gap is large enough, mid-tolerant and intolerant species (opportunists or gap-phase species) may establish and reach the overstory. These opportunists have light, wind-disseminated seeds with the capacity of germinating on small patches of exposed mineral soil, such as that created by the uptorn roots of a windthrown tree. Juvenile growth under conditions of partial shade and partial root competition is rapid so that seedlings, or at least some of them, can outgrow and overtop the advanced growth of more tolerant species already established in the openings. Typical gap-phase-replacement species include yellow birch and white ash, and with increasingly larger gaps: tuliptree, northern red oak, black cherry, and sweet birch.

Small gaps form a shifting mosaic pattern in the transition and old-growth stages of mesic forests (Table 17.3). However, the forest is not just a matrix of closed canopy punctuated by gaps—the forest is not Swiss cheese (Lieberman et al., 1989)! As trees age their crowns thin; branches die and are ripped off by wind. As a result, light reaching the forest floor is a shifting patchwork of intensity. Gaps represent an extreme condition where light may penetrate to the forest floor. However, depending on latitude and height of trees, light penetrates outside the perimeter of the gap itself and under the branches of surrounding overstory trees (Canham et al., 1990). Actually, the forest understory and floor are spatially heterogeneous not only in light intensity but in the amount of vegetative cover (mosses, herbs, shrubs, seedling and understory trees), the distribution of soil water and nutrients, occurrence of large decomposing woody debris, and bare soil. Thus the succession in tree-fall gaps is dependent on many factors besides light. They include those associated with different regional and local ecosystems: latitude, physiography, soil properties, drainage, water and nutrient cycling, community composition, tree height, disturbance regime, mutualisms with animals and mycorrhizae, and those that may or may not be ecosystem dependent such as chance events affecting gap size and location, and periodicity of local disturbances. Of primary consideration in assessing gap dynamics is the cause of the gap—windstorm, fire, pest attack. The type of wind damage can affect succession. Wind uprooted trees may form larger gaps and provide bare soil for invading species, whereas stem breakage may favor existing advanced regeneration in a smaller, shadier gap.

Research by Runkle (1981, 1982, 1985, 1990) in a variety of different mesic forests of the eastern United States reveals that small gaps 50 to 200 m^2 on average strongly favor tolerant species but allow opportunists to persist in low densities. Only in low-diversity forests (especially beech–sugar maple stands) is there a possibility for gaps caused by the death of one species to favor seedlings of another species, i.e., reciprocal replacement (Fox, 1977; Woods, 1979, 1984; Runkle, 1981). However, long-term studies at one of the study sites rejects this concept (Poulson and Platt, 1996). In the Southern Appalachian Mountains, tree replacement in single and multiple-tree gaps was primarily by understory tolerant species (Barden, 1981). However, four mid-tolerant to intolerant species—northern red oak, white ash, black cherry, and tuliptree—were able to dominate three percent of the canopy area by infrequent success in gaps, especially multiple-tree gaps.

GAP SPECIALISTS: AMERICAN BEECH AND SUGAR MAPLE

American beech and sugar maple, two of the most widely distributed species in eastern North America, are long-lived, very tolerant, late-successional species of mesic forests where catastrophic disturbance is rare. With their ability to regenerate in shaded understories and utilize treefall gaps to gain the overstory canopy, they are the most likely of all deciduous species to perpetuate themselves for long periods of time. Where the ranges of the two species coincide, beech occurs over a wider variety of sites because it is less nutrient demanding and tolerates wetter sites (Crankshaw et al., 1965; Leak, 1978). For example, in presettlement forests of Indiana, beech occurred on 256 soil types, second only in distribution to white oak (Crankshaw et al., 1965). Furthermore, beech is the more shade tolerant of the two. Sugar maple is less competitive than beech in many ecosystems, not only because it requires more light, but because of its greater requirements for nutrients and its greater sensitivity to wet sites. Because the two species often are late-successional co-associates in mesic forests, many investigators have sought to explain their persistence and coexistence by their performance in relation to the light relations of forest understories.

A long-studied and instructive case is a 16-ha beech–sugar maple-dominated ecosystem that is part of Warren Woods, southwestern Michigan. Beech was the predominant overstory species in 1933 as it was in presettlement time in the surrounding area (Kenoyer, 1933). In 1933, beech had 10 times the number of trees (15 cm dbh and larger) and 82 percent of the basal area (Cain, 1935). It accounted for three-fourths of the treefalls from 1946 to 1976 (Brewer and Merritt, 1978). Sugar maple was more prevalent in the seedling and sapling layers. Long-term studies of the understory show both beech and sugar maple maintaining themselves for the area as a whole without marked replacement of one by the other; this relationship is termed "allogenic coexistence" by Poulson and Platt (1996). Allogenic is emphasized because wind disturbance causes treefall gaps, which appear to explain the observed pattern of species interaction. However, autogenic ecosystem factors such as old (300 yr), tall (30 to 40 m) trees with broad crowns, hollow trunks, brittle wood, and shallow rooting also are significant in gap formation, which has increased markedly from 1975 to the present in this aging stand. Thus the allogenic–autogenic dichotomy is inadequate to characterize succession in this and other forest ecosystems.

Beech and sugar maple understory individuals coexist in this ecosystem, in part, due to species differences in understory tolerance, tree architecture, and their differential response to light levels in the understory and treefall gaps (Runkle, 1985; Canham, 1988; Poulson and Platt, 1996). The slightly more tolerant beech is more competitive (beech replaces maple) in the deeply shaded understory and in small gaps where saplings can outsurvive those of sugar maple via its horizontal (plagiotropic) growth to obtain light. Multiple treefall gaps favor sugar maple (maple replaces beech), which has more rapid vertical (orthotropic) extension growth enabling it to colonize the gap. Thus if a mix of treefall gaps of different sizes occur in the appropriate frequencies both species may coexist in more or less equal numbers over the forest as a whole. This pattern of species interaction is illustrated in Figure 17.15, and the scenarios shown provide an excellent example of the species trade-offs based on light relations for this ecosystem. However, the Warren Woods example is incomplete without consideration of soil properties and nutrient and water relations that are specific to this local ecosystem.

25 YR BEFORE TREEFALL   25 YR AFTER TREEFALL

no gaps nearby

local gaps every
100–200 yr

< 1% full sun

B_C 6–8 to M_C 1
canopy tree ratio

10×10
metres

(1)

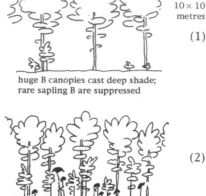

huge B canopies cast deep shade;
rare sapling B are suppressed

multiple M winners in a huge B treefall

nearby gaps every
50–100 yr

local gaps every
60–100 yr

1–3% full sun

B_C 3–4 to M_C 1

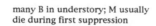

(2)

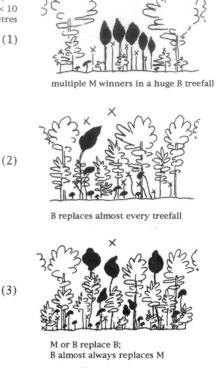

many B in understory; M usually
die during first suppression

B replaces almost every treefall

nearby gaps every
20–50 yr

local gaps every
40–60 yr

1–6% full sun

B_C 2–3 to M_C 1

(3)

more B than M in understory;
M endure 1–4 suppression cycles

M or B replace B;
B almost always replaces M

nearby gaps every
0–20 yr

local gaps every
0–40 yr

1–30% full sun

B_C1 to 1–2 M_C

(4)

more M than B in understory;
M rarely completely suppressed

gaps grow with time; M replaces most
B and some M; B replaces some M

Figure 17.15. A conceptual model of coexistence of beech (B) and sugar maple (M) at Warren Woods, southwestern Michigan. The sequence of four panels shows that infrequent and small treefall gaps (panel 1) favor beech (open silhouettes), whereas frequent and large gaps (panel 4) favor sugar maple (closed silhouettes) in the understory. The left panel of each pair, 25 years before treefall, shows how the understory percentages of full sunlight levels, due to branch gaps and nearby treefall gaps, affect forest structure and the relative success of beech and sugar maple. The right panel, 25 years after a treefall (x marks the former position of the crown top of a fallen tree), shows responses of species to gaps. Vertical and horizontal axes are drawn to a 10-m scale, and only individuals >1 cm dbh and >1–2m tall are shown. B_c = beech canopy tree; M_c = sugar maple canopy tree. (After Poulson and Platt, 1996. Reprinted with permission of the Ecological Society of America.)

The soils are sandy and strongly acid, and the high spring water table tends to favor shallow rooting in the acid topsoil. Because beech is less nutrient demanding than sugar maple the low nutrient status of this site should favor beech over sugar maple in understory performance and contribute significantly to beech dominance with minimum canopy disturbance. This relationship is borne out by the overwhelming dominance of beech overstory in 1933 (Cain, 1935). With the increasing windstorm disturbance the forest is now experiencing, the balance of understory dominance may shift to sugar maple, as shown in Figure 17.15, panel 4. However, in other ecosystems, where soil and nutrient cycling favor sugar maple, the proportion of the species in understory and overstory may reflect this advantage—provided sufficient light is available for sugar maple regeneration. For example, in the White Mountains of New Hampshire, sugar maple only dominates in two favorable ecosystems, whereas beech is more abundant in eight others (Leak, 1978). In the heavy-textured, calcareous morainal soils, where beech–sugar maple forest once dominated presettlement forests in central and southeastern Michigan (Quick, 1923; Veatch, 1953, 1959), sugar maple appears to be much more vigorous and the tree replacement pattern and process are different from those at Warren Woods. For example, in an old-growth forest on clayey, calcareous, nutrient-rich moraine in southeastern Michigan, sugar maple dominates the overstory (49 to 23 percent in basal area), and its saplings overwhelmingly dominate the understory (62 to 13 percent in number of stems). Furthermore, experimental evidence demonstrates that growth of seedlings of sugar maple under low light is enhanced two to four fold on rich versus poor sites in part because of higher nitrate N and water availability (Walters and Reich, 1997). Therefore, the landform and soil characteristics of site-specific ecosystems, as well as light relationships, are of potential significance in understanding the dynamics of beech and sugar maple coexistence.

Old-Field Succession in the Eastern United States

Agricultural use of land was at its most extensive development in the New England states about 1815 to 1830, and began to decline with the opening of the West and its ready access via the Erie Canal (1815) and the trans-Appalachian railroads. In the Atlantic-facing Piedmont of the southeastern states, agricultural use of lands reached its peak at the beginning of the Civil War. Throughout the entire eastern seaboard, most upland sites were cleared and were farmed until the 1815–1860 period, when the industrialization of the Northeast, coupled with the opening of the farm lands of the Midwest, initiated a long decline in agricultural acreage; a decline that is still in existence.

Secondary succession of forest on the abandoned upland fields and pastures involved great areas, and was instrumental in reforesting much of the eastern landscape. Since Thoreau's essay on the topic in 1860, many ecological studies have documented the major stages of forest succession on these sites, including eastern redcedar in Connecticut and New Jersey (Lutz, 1928; Bard, 1952) and pine-hardwood succession, primarily in the North Carolina piedmont (Billings, 1938; Oosting, 1942; Keever, 1950; Bormann, 1953; Peet and Christensen, 1980a,b, 1987; Christensen and Peet, 1981, 1984). Not surprisingly, successional processes are complex due to regional and local differences in: (1) climate, physiography, and soil throughout such a broad area, (2) the history of old-field use and abandonment, (3) the proximity and composition of surrounding forests that are the seed source, and (4) the life history traits of the species involved. Nevertheless, key features are the early dominance of conifers (often pines), the trend of their replacement by hardwoods, operation of successional processes (facilitation, tolerance, inhibition) simultaneously or singly at different stages, and different patterns of divergence or convergence of vegetation depending on stand age and site conditions (Christensen and Peet, 1981). Four stages of succession characterize old-field succession: establishment, thinning, transition, and steady state (see above discussion and Table 17.3). Several aspects of these are examined next.

In many cases, fields were abandoned as grass-bearing hayfields or pastures. In such cases, conifers form the initial old-field tree invaders (establishment stage). The old-field conifer in northern New England is red spruce, with white pine coming in on these sites in central New England and New York State. The old-field white pine succession in central New England produced much white pine of commercial importance. It has been detailed in many studies at the Harvard Forest (Spurr, 1956b; Foster, 1988a) and is pictorially depicted in the Harvard Forest models, which present in three-dimensional dioramas the history of land use in central Massachusetts (Figure 17.16*a-f*). Moving south, the dominant old-field conifer in southern New England and the mid-Atlantic states is eastern redcedar, with Virginia pine important in the upper South, and loblolly pine and shortleaf pine dominant throughout most of the rest of the Southeast. Only when fields were abandoned as fallow cultivated croplands did hardwoods—such as gray birch in the Northeast and sweet gum in the Southeast—predominate in the pioneer state of forest succession.

The greater ability of conifers to invade and establish themselves on old grasslands seems to be due to several factors: (1) the large numbers of relatively heavy wind-disseminated seed, which can work down through the sod to make contact with the soil; (2) the presence of enough stored food in the seed to develop a seedling sufficiently large to compete with the grass; (3) the drought resistance of conifer seedlings, which permits them to survive summer droughts occasioned by root competition with grass; and (4), the ability of pines and their ectomycorrhizae to obtain nutrients and water from the soil.

The conifer forest is not usually pure but is mixed with varying proportions of hardwood pioneers that become established mostly on small bare patches, in brush patches, or

Figure 17.16*a*.   The Harvard Forest models illustrate the sequence of events in forest history from the presettlement forest of the 17th century through agriculture and land abandonment of the 1800s to the old-field forests of the 1900s. *a*, A reconstruction of the mixed precolonial forest, with hemlock, tolerant hardwoods, and occasional white pine. (Model 1, courtesy of Harvard Forest, Harvard University.)

Figure 17.16*b.* The same view in central Massachusetts in 1740 shortly after settlement. (Model 2, courtesy of Harvard Forest, Harvard University.)

Figure 17.16*c.* The same view at the height of farming development, 1830. (Model 3, courtesy of Harvard Forest, Harvard University.)

other non-grassy areas in the old field. Given abundant seed supply, the proportion of initially establishing hardwoods is directly related to the density of the conifers and hence resource availability. In the North, gray and paper birch, pin and black cherry, and bigtooth and trembling aspen are the principal hardwood pioneers, although some white ash, northern red oak, and other mid-tolerants may come in with the first wave of tree invaders. In the South, sweetgum, red maple, and many other hardwoods may come in under these conditions although their early establishment is no guarantee of later success (Christensen and Peet, 1981). Heavy-seeded species, such as oaks and hickories, are absent from many young pine stands in the North Carolina piedmont, yet they dominate the mature forest—suggesting that early invasion is not a necessary prerequisite for eventual colonization and

Figure 17.16d. Farm abandonment and the seeding-in of old-field white pine, 1850. (Model 4, courtesy of Harvard Forest, Harvard University.)

Figure 17.16e. Harvesting the old-field pine in 1909, which seeded-in after farm abandonment in the same view. (Model 5, courtesy of Harvard Forest, Harvard University.)

dominance. The establishment of heavy-seeded species may depend heavily upon substantial animal populations which, in turn, may depend on development of the pine forest. Thus neither relay floristics nor the IFC model (Figure 17.3) are necessarily characteristic of old-field successions.

Intense competition and mortality of pines (Figure 15.14) characterizes the thinning stage. Dense pine stands inhibit establishment of hardwoods, but their break-up in the transition phase at 60 to 70 years in North Carolina piedmont facilitates invasion by a moderate number of hardwoods (oaks, hickories, red maple) (Peet, 1992). As most of the dominant overstory pines die, from 75 to 125 years after establishment, relatively long-lasting gaps are formed (steady-state phase) favoring establishment of new seedlings and

Figure 17.16ƒ. The young stand matures into the second-growth characteristic of central New England of 1930. (Model 7, courtesy of Harvard Forest, Harvard University.)

the release of already existing understory trees. Therefore, old-field succession is best described as the episodic establishment of species correlated with changes in availability of light, water, and nutrients—high in the establishment phase, low in the thinning phase, and higher in the transition and steady state phases. By the end of the second century, in the absence of any further major disturbance, replacement of conifers by more tolerant species, mostly hardwoods, is complete.

Fire and Oak Dominance—Oaks at Risk

Upland oaks dominate the overstory of dry and dry mesic sites in many parts of eastern North America today. However, on all but extremely dry sites the current generation may be their last. Succession to more mesophytic and understory tolerant species is progressing over large areas, and has been documented for many different regions and ecosystems (Christensen, 1977; Lorimer, 1984; McCune and Cottam, 1985; Host et al., 1987; Pallardy et al., 1988; Hammitt and Barnes, 1989; Abrams and Downs, 1990; Nowacki et al., 1990; Nowacki and Abrams, 1991; Abrams, 1992; Shotola et al., 1992). Although oak seedlings are sometimes found, there is virtually no recruitment to the sapling understory although oaks dominate in the overstory. Should this trend continue, major differences in ecosystem structure and function, including marked shifts in animal populations as well as plant communities, are foreseen.

Upland oaks evolved in relatively dry, fire-prone ecosystems and are characterized by a suite of fire and dry-site adaptations (Chapters 12, 22; Abrams, 1990). Depending on regional and local site conditions, the relative timing, frequency, intensity, and severity of fire is instrumental in the maintenance of oak regeneration and in patterning the distribution and occurrence of oak species. With the advent of forest fire control and fragmentation of landscapes, wildfires have been largely excluded throughout much of the East, and this is perceived as the major reason for lack of oak regeneration. Besides fire exclusion, herbivory by deer is a major contributing factor in many areas (Chapter 14).

The understory tolerant, mesophytic species replacing oaks in the understory on all but the driest sites are primarily red maple and sugar maple. However, in the mid-Atlantic region (from Pennsylvania to Virginia) and elsewhere, where oak advance regeneration fails and logging or wind disturbance to the oak overstory provide sufficient light, opportunistic species such as black cherry, sweet birch, blackgum, sassafras, and white ash can establish and develop in the absence of fire.

The successful regeneration of oak typically depends on their existence as saplings in the forest understory before disturbance, i.e., presence of advanced regeneration. Although there are many factors related to oak regeneration (seed production and dispersal, animal damage to seeds and seedlings; Lorimer, 1993), fire is critically important for successful regeneration for two interlinked reasons—it kills competitors and promotes resprouting of young oaks. In addition, by consuming organic matter fire provides a source of readily available nutrients for the developing oak sprouts. Being generally mid-tolerant species, young upland oaks are at a competitive disadvantage from early-successional species in open understories and from the species that tolerate shaded understory conditions. Mesophytic species are always invading the understories of oak forests by wind and animal dispersal.

Light and moderate surface fires may occur at all stages of stand development (Figure 17.4a), and they indiscriminately kill tree seedlings and saplings and shrubs. However, few species have the propensity of oaks (and their usual associates, the hickories) to resprout in the seedling or sapling stage following fire. Therefore, repeated fires purge the oak–hickory forest understory of mesophytic competitors. If young oak stems are killed by fire, they resprout vigorously from the well-established root system with its supply of stored food. However, with thicker bark than most mesophytic associates, many young saplings may survive surface fires. In the stand initiation and stem exclusion stages of stand development (Figure 17.4a), the density of the oak overstory may be sufficient to deter development of a competing understory of mesophytic species. However, from the understory reinitiation stage onward, when oaks are themselves becoming established and forming advanced regeneration, fire is of critical importance to kill invaders and reduce shrub competition. When crown breakage or gaps form in the old-growth stage, the advanced regeneration of oaks and hickories is well positioned to dominate the understory, often in a patchy pattern (Figure 15.15).

No one scenario of stand development exists; differences in soil water stress and fire regime determine the specific combinations of oak and other species across the landscape. Although oaks are typically transitional to mesophytic, understory-tolerant tree species on mesic and dry-mesic sites, on the driest sites of ridges and fire-prone barrens, oaks (blackjack, post, chestnut, black, northern pin) may be self-replacing or show a slow rate of replacement to other species (Abrams, 1992). In these ecosystems, fire, wind, soil movement, or other disturbance factors, together with other species adaptations to extreme sites, tend to maintain the intolerant oak species and associates of hickory or pine.

Suggested Readings

Abrams, M. D. 1992. Fire and the development of oak forests. *BioScience* 42:346–353.

Chapin, F. S., III, L. R. Walker, C. L. Fastie, and L. C. Sharman. 1994. Mechanisms of primary succession following deglaciation at Glacier Bay, Alaska. *Ecol. Monogr.* 64:149–175.

Christensen, N. L. 1988. Succession and natural disturbance: paradigms, problems, and preservation of natural ecosystems. In J. K. Agee and D. R. Johnson (eds.), *Ecosystem management for Parks and Wilderness*. Univ. Wash. Press, Seattle.

Glenn-Lewin, D.C., and E. van der Maarel. 1992. Patterns and processes of vegetation dynamics. In D. C. Glenn-Lewin, R. K. Peet, and T. T. Veblen (eds.), *Plant Succession: Theory and Prediction*. Chapman and Hall, London.

Grubb, P. J. 1985. Plant populations and vegetation in relation to habitat, disturbance and competition: problems of generalization. In J. White (ed.), *The Population Structure of Vegetation* (Handbook of Vegetation Science, Part III), W. Junk, Dordrecht.

Matthews, J. A. 1992. *The Ecology of Recently-deglaciated Terrain*. Cambridge Univ. Press, Cambridge. 386 pp. Chapter 6. Plant succession: processes and models.

Olson, J. S. 1958. Rates of succession and soil changes on southern Lake Michigan sand dunes. *Bot. Gaz.* 119:125–170.

Pickett, S. T. A., S. L. Collins, J. J. Armesto. 1987. Models, mechanisms and pathways of succession. *Bot. Rev.* 53:335–371.

White, P. S. 1979. Pattern, process and natural disturbance in vegetation. *Bot. Rev.* 45:229–299.

CHAPTER 18

CARBON BALANCE OF TREES AND ECOSYSTEMS

Although forest ecosystems cover only 21 percent of the Earth's land surface, they constitute a disproportionately large share of terrestrial plant mass (75 percent) and its annual growth (37 percent). Because plants are largely constructed of carbon (47 percent), the extent to which forests are altered by human activities (i.e., harvesting or conversion to other types of vegetation) has a substantial influence on the pattern in which carbon (C) is cycled and stored at local, regional, and global scales. Recent attention has focused on factors influencing the fixation and release of C by trees and forest ecosystems (i.e., their C balance), because the rising CO_2 concentration of the Earth's atmosphere and projected changes in global climate have the potential to alter the present-day geographic distribution of forests and the rate at which they sequester C from the atmosphere (Pastor and Post, 1986; Melillo et al., 1993). Such a change is of both ecological and economic importance, because the production of plant matter in forest ecosystems not only influences the global cycling of C but also represents an indispensable source of food, fuel, fodder, and fiber for human populations around the Earth. Globally, forests also are home to countless species of animals, and herbaceous and non-vascular plants.

The C balance of plants (i.e., growth) is controlled by the quantity of CO_2 fixed through photosynthesis and the rate at which fixed C is returned to the atmosphere by respiring plant tissues. Light, temperature, and the availability of soil water and nutrients constrain these physiological processes and thus influence the productivity of plants and the ecosystems in which they occur. One need only examine desert, grassland, and forest ecosystems to understand that the processes controlling the productivity of individual plants also have a substantial influence on the productivity of terrestrial ecosystems. This situation is particularly true in forests, where the growth and metabolism of plants far outweighs that of any other organism.

In this chapter, we focus on the physiological processes of plants that control the productivity of terrestrial ecosystems, especially forests. We discuss how climate and soil nutrient availability influence the C balance of plants, and hence their productivity. We then build upon this information to understand how environmental factors influence the C balance of ecosystems at local, regional, and global scales.

CARBON BALANCE OF TREES

Only a small fraction of the radiant energy reaching the Earth's surface (2 percent) is used by green plants to assimilate atmospheric CO_2 into organic compounds. These compounds are used to construct new plant tissue, maintain existing tissue, create storage reserves, or provide defense against insects and pathogens. Simply stated: trees grow (or gain C) when the amount of C fixed through photosynthesis exceeds the amount of C lost from respiring leaves, branches, stems, and roots. Understanding the factors influencing the balance between photosynthesis and respiration is of ecological importance, because plants with the greatest net C gain under a specific set of environmental conditions (i.e., light, water, and nutrient availability) are often the best competitors. It follows that fast-growing tree species require rapid rates of photosynthesis, but the converse is not always true. As you will see, the net C gain of plants is not influenced by photosynthetic rate alone (Ceulemans and Saugier, 1991). In the following section, we trace the flow of photosynthetically-fixed C to the construction and maintenance of plant tissue, focusing on how environmental factors influence the net C gain of plants.

Photosynthesis, Dark Respiration, and Leaf C Gain

The processes of photosynthesis and respiration occur in the leaves of all plants, and the balance of these more-or-less opposing physiological processes controls the net C gain of leaves. **Gross photosynthesis** is the total amount of C plants assimilate from the atmosphere. However, a portion of that fixed C is returned to the atmosphere from the leaves as CO_2 during **dark respiration**. This process is so termed because leaf respiration can only be determined in the absence of C fixation and is hence measured on un-illuminated leaves. **Net photosynthesis** is the balance between gross photosynthesis and leaf dark respiration, and it represents the amount of C available for growth and tissue maintenance. Light intensity, temperature, and the availability of water and soil nutrients (especially N) all influence photosynthesis, respiration, and the C gain of leaves.

Rates of net photosynthesis measured under saturating light, optimum air temperature, low vapor pressure deficit, and ambient atmospheric CO_2 represent the **maximum photosynthetic capacity** of plants (see Table 18.1). Care should be taken in interpreting the values in Table 18.1, because maximum photosynthetic capacity varies with canopy position (shade versus sun leaves), as trees develop from a juvenile to an adult phase (see ponderosa pine), soil nutrient availability, and measurement technique. Also be aware that maximum photosynthetic capacities are often poorly correlated with the C gain of forest trees (Ledig and Perry, 1969), because environmental factors (light, temperature, water, and soil nutrients) constrain photosynthesis under field conditions. Nevertheless, Table 18.1 illustrates several important physiological differences among forest trees that have important ecological implications.

The maximum photosynthetic capacity of forest trees ranges from 2 to 25 μmol m^{-2} sec^{-1} (Table 18.1), generally lower than that of most agronomic crops (Ceulemans and Saugier, 1991). Shade-intolerant deciduous species have some of the highest photosynthetic capacities of all forest trees (up to 25 μmol m^{-2} sec^{-1}; Table 18.1), whereas those of shade-tolerant deciduous species are much lower (3 to 6 μmol m^{-2} sec^{-1}; Table 18.1).

Recognize that shade tolerance is only one of several characteristics enabling a species to persist in the understory (i.e., understory tolerance; Chapter 15). Although some conifers have high photosynthetic capacities (see Scots and Monterey pine), most are modest when compared to those of deciduous trees.

A canopy architecture efficient at intercepting solar radiation can outweigh moderate photosynthetic rates, allowing some trees to attain large C gains despite the relatively low net photosynthetic rates of individual leaves. The rapid growth and modest net photosynthetic capacities (2–10 µmol m^{-2} sec^{-1}) of some conifers (e.g., Douglas-fir) illustrate this point (Table 18.1). Coniferous canopies disperse incoming solar radiation over a larger number of leaves than broad-leaved canopies, enabling them to capture more energy at high light intensities. The orientation of coniferous leaves and branches also produces less shading of leaves lower in the canopy, enabling them to photosynthesize more efficiently. Also note that the photosynthetic capacities in Table 18.1 are expressed on an area basis (µmole CO_2 per square meter of leaf surface per second). Coniferous species often maintain higher leaf areas than deciduous species, creating a larger surface area to capture solar radiation. For example, Douglas-fir and many true firs maintain leaf areas 2- to 6-

Table 18.1 The photosynthetic capacity of trees under conditions of saturating light, optimum temperature and water regimes, and ambient atmospheric CO_2 (350 µmol mol^{-1}). Rates are expressed in micromoles of CO_2 assimilated by 1 square meter of leaf during 1 second.

Shade-Understory Tolerance	Maximum Photosynthetic Capacity µmole CO_2 m^{-2} s^{-1}
Tolerant	
Deciduous Trees	
sugar maple	3–4
American beech	4–5
hophornbeam	6
red maple	3–5
American basswood	3
flowering dogwood	4–6
Coniferous Trees	
European silver fir	2–4
white fir	8
grand fir	4–5
balsam fir seedling	2
Sitka spruce	3–9
Norway spruce	2–7
subalpine fir	9
Midtolerant	
Deciduous Trees	
white oak	2
northern red oak	4–5
white ash	4–5
yellow birch	9
Engleman spruce	3–5
Coniferous Trees	
Douglas-fir	2–6

Table 18.1   *continued*

Shade-Understory Tolerance	Maximum Photosynthetic Capacity μmole CO_2 m^{-2} s^{-1}
Intolerant	
Deciduous Trees	
black oak	7
sassafras	6
black cherry	4
sweet gum	6–11
tuliptree	7–17
paper birch	10
black locust	13–17
red ash	20–25
big-tooth aspen	14
cottonwood	15–19
trembling aspen	20–22
black poplar hybrid	20–25
Coniferous Trees	
lodgepole pine	3–9
Ponderosa pine	
seedlings	25
mature trees	5–14
old growth	4
Scots pine	10–16
Monterey pine	17
loblolly pine	3–6

All values are expressed on a one-sided leaf area basis and are summarized from Bazzaz (1979), Ceulemans and Saugier (1991), Jurik et al. (1988), Larcher (1969), Korol and Marshall *unpublished data,* Wallace and Dunn (1980), and Walters et al. (1993).

times greater than that of some broad-leaved species. As such, low net photosynthetic rates in some conifers are offset by an efficient light-gathering canopy architecture, enabling them to achieve relatively large C gains.

Light and Leaf C Gain

Net photosynthesis varies substantially with light intensity, and the response of trembling aspen (intolerant) and northern red oak (midtolerant) leaves to increasing light intensity is illustrated in Figure 18.1. Sun leaves of trembling aspen attain a greater light-saturated rate of net photosynthesis (i.e., max. photosynthetic capacity; see right-hand portion of Figure 18.1) than do to the sun leaves of red oak. The shade leaves of trembling aspen also have greater light saturated rates of net photosynthesis than do the shade leaves of northern red oak, but rates in the shade leaves of both species are lower than those of sun leaves. Although shade leaves photosynthesize at a lower rate than sun leaves, they can contribute as much as 40 percent of the C assimilated by tree canopies (Schulze et al., 1977).

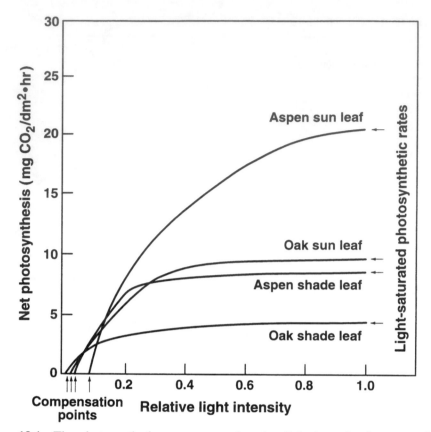

Figure 18.1. The photosynthetic response to changing light intensity for trees of contrasting tolerance. Trembling aspen is an intolerant species, whereas northern red oak is a mid-tolerant species that can persist in the forest understory. (After Loach, 1967.)

Plants attaining high rates of net photosynthesis, like trembling aspen and other intolerant species, typically contain large amounts of chlorophyll and the enzymes used for C-fixation (Field and Mooney, 1986), which represent the biochemical machinery plants use to convert light into chemical energy. High respiration rates are associated with plant tissues containing large amounts of metabolically-active enzymes (Amthor, 1984); such high rates of respiration are needed to replace and repair enzymes. Consequently leaves with high net photosynthetic rates also respire rapidly. Leaf dark respiration represents a relatively-constant proportion (7 to 10 percent) of the maximum photosynthetic capacity of many forest trees (Figure 18.2; Ceulemans and Saugier, 1991). However, rates of net photosynthesis rarely attain their maximum under field conditions, because light, soil nutrients, or water are often in short supply. Under these conditions, dark respiration can account for 10 to 20 percent of annual net photosynthesis in a wide range of coniferous and deciduous tree canopies (Ryan et al., 1994).

Because rates of net photosynthesis vary with light intensity while dark respiration remains constant, the ability of a species to persist in low light environments is related to the balance of these physiological processes. In Figure 18.1, the light intensities at which net photosynthesis is zero (i.e., light compensation point) for the sun and shade leaves of trembling aspen are substantially higher than those of northern red oak. This relationship results from

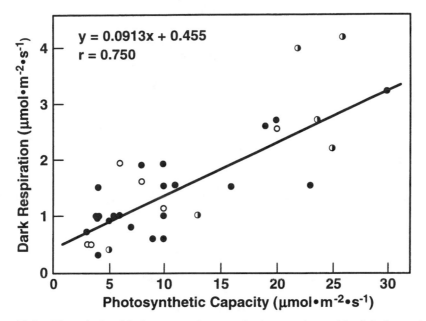

Figure 18.2. The relationship between photosynthetic capacity and leaf dark respiration for deciduous broad-leaved (closed circles), evergreen broad-leaved (half-open circles), and coniferous (open circles) tree species. (After Ceulemans and Saugier, 1991. Reprinted from *Physiology of Trees,* © 1991 by John Wiley & Sons, Inc. Reprinted with permission of John Wiley & Sons, Inc.)

high dark respiration rates that maintain rapid rates of C assimilation in aspen leaves. Dark respiration is lower in tolerant species, thus enabling them to maintain positive rates of net photosynthesis (i.e., positive leaf C gain) at much lower light intensities. This ability illustrates the point that high photosynthetic rates alone do not insure a high C gain in all light environments.

Temperature and Leaf C Gain

Air temperature directly controls leaf C gain by influencing rates of gross photosynthesis and dark respiration (Fig 18.3). The difference between these two physiological processes (i.e., net photosynthesis) reaches a maximum between 15 and 25 °C for most temperate trees (Kozlowski and Keller, 1966; Mooney, 1972). Rates above and below this range rapidly decline because low and high temperatures limit the process of C assimilation. In Figure 18.3 we see that gross photosynthesis increases with rising temperature to an optimum and then rapidly declines with a further increase in temperature. Dark respiration also increases with rising temperature, attains a maximum, and then declines markedly as temperature continues to rise. However, the response of dark respiration to temperature differs from that of gross photosynthesis in several important ways (Figure 18.3). First, the maximum rate of dark respiration occurs at a much higher temperature than does the maximum rate of gross photosynthesis. Also, the slope of the gross photosynthesis and respiration curves differ at high temperatures (i.e., 35 to 50 °C). Gross photosynthesis declines rapidly, while dark respiration continues to increase. Note also that leaf C gain approaches zero as temperature rises between 35 and 43 °C. At temperatures above 43 °C, rapid rates of dark respiration combined with slow rates of gross photosynthesis result in a

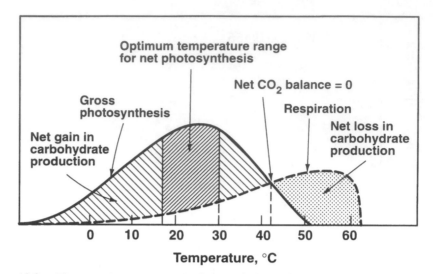

Figure 18.3. The temperature response of gross photosynthesis, dark respiration, and leaf C gain. Gross photosynthesis and dark respiration in leaves are temperature dependent processes, and leaf C gain is maximized at the temperature where the difference between these processes is greatest. (After Daniel et al., 1979. Reprinted from Daniel, T., J. A. Helms, and F. S. Baker, *Principles of Silviculture,* 2nd ed., © 1979 by McGraw-Hill, Inc. Reprinted by permission of The McGraw-Hill Companies.)

negative C balance and the net loss of C from the leaf. Thus, you can see that changes in rates of gross photosynthesis and dark respiration in response to temperature have a great impact on leaf C gain (Berry and Bjorkman, 1980; Berry and Downton, 1982).

Water and Leaf C Gain

Water is important for plant growth and its availability directly influences photosynthesis and leaf C gain. Plants are faced with a fundamental trade-off because C fixation and water loss occur simultaneously in leaves. Trees open their stomata allowing CO_2 to diffuse from the atmosphere to sites of C fixation in the leaf mesophyll. At the same time, water vapor diffuses out of stomata in response to differences in water potential between the leaf and atmosphere (see Chapter 11). Declines in leaf water content from 5 to 10 percent often have little influence on photosynthesis (Hanson and Hitz, 1982), but a drop in leaf water potential can sharply reduce C fixation. Low leaf water potentials cause stomata to close, restricting the loss of water to the atmosphere and the flow of CO_2 into the leaf. At leaf water potentials of -1.0 to -2.5 MPa (Figure 18.4), stomatal closure begins to limit the diffusion of CO_2, thus slowing rates of C fixation. As leaf water potentials fall below a critical level, stomata completely close and C fixation ceases.

Although the species in Figure 18.4 inhabit a wide range of sites (dry-mesic uplands for white pine and northern red oak; wet lowlands for speckled alder, red ash), leaf water potential limits C fixation over a surprisingly narrow range of values. Plants occurring in dry sites have several physiological and morphological adaptations enabling them to maintain high leaf water potentials at relatively low leaf water contents. Physiological changes in cell wall elasticity, membrane permeability, and the concentration of solutes within plant cells can maintain high leaf water potentials (i.e., less negative), allowing C fixation at relatively low leaf water contents (Bradford and Hsiao, 1982). Trees inhabiting

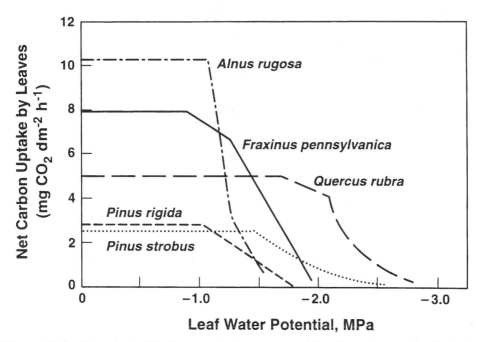

Figure 18.4. The relationship between leaf water potential and net photosynthesis for deciduous and coniferous trees occurring in different habitats. (After Hinckley et al., 1981. Reprinted from *Water Deficits and Plant Growth*, © 1981 by Academic Press, Inc. Reprinted with permission of Academic Press, Inc.)

dry environments also have a higher rate of gas exchange, lose turgor at a lower leaf water potential, and have greater root:shoot ratios, leaf thicknesses, guard cell lengths, and stomatal densities than trees from mesic sites (Abrams et al., 1994). Such adaptations provide dry-site species with increased water-use efficiency (μmole CO_2 fixed/mmole of H_2O transpired), allowing them to fix more C per unit of water lost.

Soil Nitrogen Availability and Leaf C Gain

Photosynthesis as well as other biochemical processes in the leaf are influenced by the supply of soil nutrients. Nitrogen is a constituent of chlorophyll, other light-harvesting pigments, and the enzymes involved with C fixation. Their presence makes leaves rich in N compared to most other plant tissues. Thus, the supply of N can have a large influence on a leaf's ability fix C from the atmosphere. For example, leaf N concentration is directly related to the photosynthetic rate of a wide array of herbaceous plants (Field and Mooney, 1986). Net photosynthesis in forest trees also increases with leaf N concentration, but coniferous and deciduous species respond much differently (Brix, 1971; Linder et al., 1981). In deciduous species, net photosynthesis is particularly responsive to leaf N concentration (Linder et al., 1981). For example, the highest rates of net photosynthesis for European white birch occur at the greatest leaf N concentration (5 percent), and rates sharply drop as leaf N concentrations decline (Figure 18.5). Coniferous species, on the other hand, lack the ability to increase photosynthetic enzyme contents to the same extent as deciduous trees, making photosynthesis much less responsive to changes in leaf N concentration. Photosynthetic rates in conifers like Douglas-fir initially increase up to 25 percent following N fertilization, but rates rapidly decline with time (Brix, 1971). However,

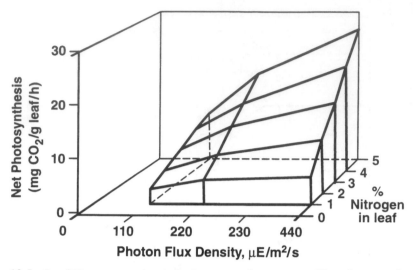

Figure 18.5. Leaf N concentration is an important factor controlling the rate of net photosynthesis in deciduous species. The highest rates of net photosynthesis for European white birch occur at saturating light intensities and high leaf N concentrations. (After Linder et al., 1981.)

conifers often increase C gain following fertilization by producing more foliage, thus creating a larger photosynthetic surface area (Brix and Ebell, 1969; Brix, 1971). In deciduous trees, N fertilization often increases both the photosynthetic rate of individual leaves and leaf production, resulting in a larger and more photosynthetically-active canopy.

Construction and Maintenance Respiration

Forest trees are unique compared to other types of terrestrial vegetation due to their size and longevity. As a consequence of their large mass, forest trees require substantial amounts of C to maintain the function of living cells located in the non-photosynthetic tissues of branches, boles, and reproductive structures. Carbon also is required for the construction of new leaves to capture light, roots to forage for water and nutrients, and reproductive structures to insure regeneration. The construction and maintenance of these tissues require C supplied by products of photosynthesis. Respiration during the biosynthesis of new tissue is termed **construction respiration**, whereas respiration used to maintain already existing living tissue is termed **maintenance respiration**. These physiological processes, which return fixed C to the atmosphere, have an important influence on the overall C balance of trees.

Construction Respiration: Respiration provides the energy used to convert the products of photosynthesis into the biochemical constituents of plant tissue. Thus photosynthetically-fixed C is lost to the atmosphere (as CO_2) during the synthesis of new plant tissue. This loss (i.e., construction respiration) is calculated knowing the biochemical constituents of a particular plant tissue and the biosynthetic pathways by which they are formed. This information can be used to determine the amount of CO_2 produced during the biosynthesis of a particular plant compound or organ. Glucose, a product of photosynthesis, is often considered the substrate for biosynthesis when calculating the construction respiration of plant tissue (Penning de Vries, 1975; Chung and Barnes, 1977).

Look at Table 18.2 and note that the needles and shoots of loblolly pine contain large proportions of carbohydrates, lignin, and phenolic compounds, whereas nitrogenous compounds, lipids, and organic acids are present in smaller amounts. The biosynthetic pathways giving rise to these compounds require different amounts of substrate (i.e., glucose) and produce different amounts of CO_2. For example, the synthesis of 1 g of carbohydrate in cell wall construction requires 1.2 g of glucose as substrate, of which 11 percent is lost as CO_2 (Table 18.2). Although lipids are present in relatively small amounts, their construction produces relatively large amounts of CO_2 compared to other biosynthetic pathways.

By knowing the constituents of a particular plant tissue and the amount of CO_2 lost during their synthesis, one can calculate construction respiration for entire plant organs. During the construction of 1 g of loblolly pine needles, approximately 19 percent of substrate is respired during biosynthesis. Similarly, construction respiration consumes approximately 16, 18, and 17 percent of the glucose used to synthesize loblolly pine shoots, cambium and roots, respectively (Chung and Barnes, 1977). Clearly, accounting for these losses of C is essential for understanding the C balance of whole plants.

Maintenance Respiration: Carbohydrates produced by photosynthesis are used to sustain the function of living plant tissues during the process of maintenance respiration. Protein synthesis and replacement, membrane repair, and the maintenance of ion gradients across cell membranes all require the expenditure of C initially fixed through photosynthesis (Ryan, 1991). The rate of maintenance respiration, like all other enzymatic processes, increases predictably with temperature. For most plants, maintenance respiration roughly doubles for every 10 °C increase in ambient temperature. That is, maintenance respiration increases exponentially with a linear increase in temperature.

Maintenance respiration also is strongly influenced by tissue N concentration (Waring et al., 1985), wherein tissues with high N concentrations also have high rates of maintenance respiration. In fact, leaf maintenance respiration for a wide range of plant species can be predicted from N concentration using the following equation (Ryan, 1991):

$$R_M = 0.0106\,N$$

where R_M is leaf maintenance respiration (mmoles of CO_2 mole tissue C^{-1} h^{-1}) and N is the total nitrogen concentration of the particular tissue (mmole N mole tissue C^{-1}). This

Table 18.2 The biochemical constituents of the needles and shoots of loblolly pine, and the proportion of carbon respired during their biosynthesis.

Compound	Needles	Shoots	Construction Respiration[1]
		percent	
Carbohydrates	35.7	38.0	11
Lignin	23.1	23.3	14
Phenolic compounds	21.0	20.0	29
Nitrogenous compounds	9.3	8.4	25
Lipids	5.6	5.3	49
Organic Acids	3.8	3.5	32

Source: After Chung and Barnes, 1977. Reprinted with permission of National Research Council Canada.

[1]Calculated as the percent of glucose lost as CO_2 during the synthesis of a compound in the left-hand column.

relationship arises because the majority of N present in plant tissue occurs in proteins, and a large proportion (60 percent) of maintenance respiration is allocated to their repair and replacement (Penning de Vries, 1975). Such a relationship also contributes to the high rates of leaf dark respiration that accompany rapid photosynthetic rates (see Fig 18.2). Herein lies an important link between soil N availability, leaf N concentrations and the physiological processes that control plant C gain. Greater soil N availability can increase the N concentration of plant tissues. This can lead to greater rates of photosynthesis in leaves and more rapid rates of maintenance respiration in non-photosynthetic tissue. Increases in maintenance respiration of non-photosynthetic tissues (i.e., plant roots) are offset by greater rates of net photosynthesis in leaves with higher N concentrations. In combination, these responses often lead to higher plant C gains when soil N availability increases.

As trees develop from juvenile to adult phases, the ratio of photosynthetic to non-photosynthetic tissues decreases, a factor that substantially influences maintenance respiration and the C balance of whole trees. Canopy leaf area reaches a maximum relatively early in tree development, thus imposing a limit on canopy photosynthesis and the supply of carbohydrate for tissue construction and maintenance. Although the supply of carbohydrate is constrained by canopy photosynthesis, the demand for carbohydrate rises as the proportion of non-photosynthetic tissue continues to increase. For example, stemwood accounts for a large proportion of total tree mass (ca. 50–70 percent in mature trees) and contains both living (sapwood) and dead tissue (heartwood). Although the proportion of living, non-photosynthetic tissue in stems decreases as trees mature, the absolute mass of living tissue continues to increase. Consequently, the amount of C allocated to the maintenance of living cells in stemwood dramatically increases as trees develop from a juvenile to an adult phase (Agren et al., 1980; Waring and Schlesinger, 1985; Ryan, 1988). Remember that C used to maintain the mass of live, non-photosynthetic tissue in mature trees cannot be used for the construction of new tissue. Consequently, the greater use of C for tissue maintenance in mature trees results in a decline in annual growth. Annual growth also can be slowed by reductions in the net photosynthetic rate of adult trees. As you will read later in this chapter, increases in maintenance respiration and declines in growth as trees mature have important implications for the C balance of forest ecosystems.

Allocation to Structure, Storage, and Defense

In the previous paragraphs, we have discussed physiological processes influencing the fixation and loss of C by forest trees. Accounting for these processes, one can construct a C budget illustrating how plants partition C into the growth and maintenance of leaves, stem, roots, reproductive structures, and the production of defense and storage compounds. Such an approach can shed light onto how environmental factors of light, water, and nutrients influence the allocation of C into organs that aid in gathering light or the acquisition of water and soil nutrients.

Summarized in Table 18.3 is the annual C budget for a young Scots pine tree (Agren et al., 1980). This budget quantifies the partitioning of photosynthetically-fixed C into the growth of various plant tissues and accounts for their respiration during construction and maintenance. Approximately 88 percent of fixed C in this young tree was used to produce new tissue, whereas only 10 percent was consumed during construction and maintenance respiration. The production of new roots to forage for water and nutrients in soil represented a large investment of C, accounting for 56 percent of annual net photosynthesis. In

Table 18.3 The carbon budget of a 14-year-old Scots pine tree.

	Assimilation	Allocation	Percent
	— g C year^{-1} —		of Total
Net Photosynthesis	1723		
Growth			
Current Needles		286	16.6
Branch Axes		132	7.7
Stem		145	8.4
Roots		960	55.6
Total Growth		1523	88.4
Construction and Maintenance Respiration			
Stem		49	2.8
Branch Axes		15	0.9
Roots		109	6.3
Total Respiration		173	10.0
Growth + Respiration		1696	98.5
Unaccounted Net Photosynthesis		27	1.5
TOTAL		1723	100.0

Source: After Agren et al., 1980.

combination, root growth and respiration comprised almost 62 percent of annual net photosynthesis, indicating a substantial proportion of fixed C was allocated to belowground growth and metabolism.

The amount of C plants allocate to growth, storage, and the production of defense compounds is under strong genetic control. Patterns of C allocation varies among species and within a species seasonally and with increasing age. Environmental factors also play an important role influencing the allocation of C into structure, storage, and defense. Figure 18.6 illustrates the C allocation priorities in lodgepole pine. Allocation of C to new foliage and buds represents the highest priority for carbohydrate, followed by the production of new roots. Tree growth often is limited by light, water, and nutrients, making it necessary for the plant to first meet the carbohydrate demand of tree crowns and roots. The storage of carbohydrate in leaves, stem, and coarse roots has a lower priority than allocation to new leaves and roots. Allocation of carbohydrate to stem growth follows that to storage and the production of chemicals to defend against herbivory, which is the lowest priority.

Patterns of C allocation vary substantially over the growing season, wherein certain tissues function as sinks or sources for carbohydrate. Nowhere is this more evident than during the phenological development of deciduous trees. Storage reserves in buds, branches, stems, and roots function as a source of carbohydrate to build new leaves early in the growing season. Once the expanding canopy attains a positive C balance, the priorities of C allocation shift. In sugar maple seedlings, for example, large amounts of C are allocated to the production of new leaves early in the growing season (Table 18.4). The proportion of C allocated to stems and coarse roots increased following canopy development (i.e., mid growing season), and represented the greatest sink for C late in the growing season. The amount of carbohydrate stored in stems, coarse roots, and fine roots also increases over the growing season, attaining the highest concentrations late in the growing season. These carbohydrate stores are then used to construct new leaves at the start of the next growing season.

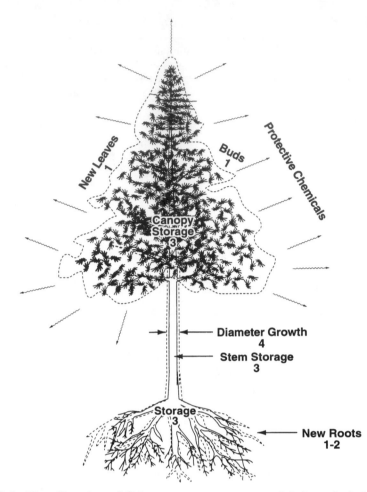

Figure 18.6. The allocation of C into structure, storage, and defense in lodgepole pine. Tree components with low numbers (e.g., 1) represent strong priorities for C allocations, whereas components with high numbers indicate a low priority for C allocation. (After Waring and Pitman, 1985. Reprinted with permission of the Ecological Society of America.)

Table 18.4   The seasonal C allocation to leaves, stem, and roots in sugar maple seedlings.

	Percent of Growth Allocated to:			
	Leaves	**Stem**	**Coarse Roots**[1]	**Fine Roots**[2]
Early Growing Season	85	5	0	10
Mid Growing Season	30	30	25	15
Late Growing Season	0	5	60	35

Source: After Burke et al., 1991. Reprinted with permission of National Research Council Canada.

[1]Coarse roots are 5 to 2 mm.

[2]Fine roots are < 2 mm.

Light and C Allocation

By knowing the C allocation pattern of trees, one can begin to understand how the availability of light, water, and nutrients influence allocation to structure, storage, and defense. Carbohydrate produced during photosynthesis is first allocated to new leaves and then new roots (Figure 18.6). If the carbohydrate demand of new leaves and roots is satisfied, then remaining carbohydrate can be allocated to storage, the growth of stemwood, and the production of chemicals to defend against herbivory. Such a pattern has been observed for both coniferous (Waring and Pitman, 1985) and deciduous species (Burke et al., 1991).

In very dense stands of lodgepole pine, low light levels limit photosynthesis and carbohydrate production, even when soil resources are abundant. In such a situation, carbohydrate allocated to meet the demand of new leaves and roots occurs at the expense of allocation to lower priorities of storage, stem growth, and the production of defense chemicals. As a consequence, stem diameter growth declines and the susceptibility to mountain pine beetle attack increases; trees with low diameter growth are those most susceptible to beetle attack (Waring and Pitman, 1985). Thinning these forests increases light availability, decreases the competition for light, and increases the photosynthesis of remaining trees. Thus carbohydrate limitation within the remaining trees lessens, leading to increases in diameter growth and a reduction in susceptibility to mountain pine beetle attack. Nitrogen fertilization produces a similar response, wherein greater soil N availability increases the production of new foliage, which in turn increases canopy photosynthesis and the supply of carbohydrate within the tree.

Soil Nitrogen Availability and C Allocation

The availability of soil nutrients is a potent modifier of C allocation in forest trees, substantially influencing the proportion of carbohydrate allocated to leaves, stems, and roots. Fine roots (< 2 mm in diameter) are the water- and nutrient-absorbing organs of forest trees, and their amount usually declines as water or nutrient availability increases. In contrast, the amount of foliage often increases when soil water and nutrients are in abundant supply, suggesting that a trade-off occurs between above- and belowground plant growth as soil resource availability increases. That is, the root to shoot ratio (root:shoot) of plants declines along a gradient of increasing soil nutrient availability. In sugar maple, for example, N fertilization increases the mass of leaves and stem by 12 percent, while root mass (coarse + fine) declines by an equivalent proportion (calculated from Burke et al., 1991). Scots pine and a black poplar hybrid respond in a similar manner to N fertilization, but these species exhibit an even more dramatic shift from roots to leaves (Linder and Axelsson, 1982; Pregitzer et al., 1995). Herbaceous species also apparently decrease the amount of roots and increase leaf mass along gradients of increasing soil N availability (Tilman, 1988). Such a pattern suggests that the plants "invest" smaller amount of carbohydrate into roots when soil resources are relatively abundant. As a result, carbohydrate is available to produce relatively more foliage, a response that allows the plant to capture more light energy on N-rich soils.

Nevertheless, there is considerable debate regarding the extent to which soil N availability alters the *total* allocation of carbohydrate to the production of belowground plant tissues. This debate stems from our limited understanding of fine-root production (i.e., birth) and mortality (i.e., death), factors that greatly influence the total amount of carbohydrate that plants allocate to belowground growth. Unlike the leaves of deciduous trees, which are initiated in spring and abscise during autumn, the birth and death of fine roots occurs simultaneously throughout the growing season (Hendrick and Pregitzer, 1992). Consequently, relatively little can be inferred about patterns of C allocation by comparing the standing crops of leaves and roots across gradients of soil nutrient availability.

Theoretically, the proportion (i.e., percent) of plant mass composed by fine roots could decline in N-rich soil by three alternative mechanisms: 1) fine-root production could remain constant while mortality increases (Figure 18.7*a*), 2) production could decline while mortality remained constant (Figure 18.7*b*), or 3) both production and mortality could increase with rates of mortality surpassing those of production (Figure 18.7*c*). The first mechanisms suggests that the life span of fine roots decreases, but root production (i.e., birth) remains constant; such a response would not alter the total amount of carbohydrate allocated to fine roots. The second mechanism argues for a decline in the total

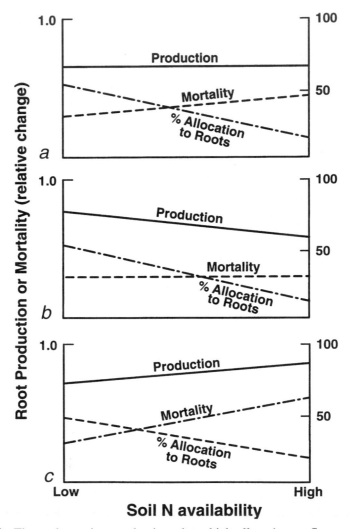

Figure 18.7.   Three alternative mechanisms by which allocation to fine roots could decline with increasing levels of soil N availability. Each alternative has very different implications for the total amount of carbohydrate allocated to fine-root production.

amount of carbohydrate allocated to fine roots, wherein fewer roots are produced with the same life span. In the third alternative, fine root production increases, but greater N availability decreases the life span of fine roots, which thereby increases the total amount of carbohydrate "invested" into these structures. In other words, fine roots "live fast and die young" in the third alternative. Clearly, determining which of these views is correct will rest on our understanding of the processes controlling fine-root production and mortality.

Attaining such an understanding has been difficult, largely because of the numerous problems associated with studying live plant roots in soil. Most methods of studying the production and mortality of fine roots modify the soil in ways that may invalidate the observations. However, new technology has recently enabled ecologists to actually view and record the growth and death of individual tree roots. Clear plastic tubes (mini-rhizotrons) can be placed in the soil of a forest stand, and a miniaturized video camera can be used to capture images of individual roots from their birth to death (Figure 18.8; for details see Hendrick and Pregitzer, 1992). The root images can be digitized, processed by computer,

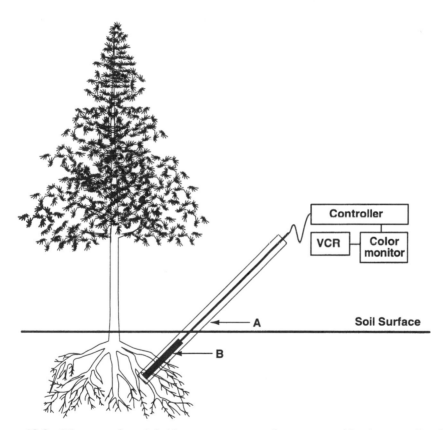

Figure 18.8. Diagram of a mini-rhizotron system used to capture video images of actively growing plant roots. The clear plastic mini-rhizotron tube (A) is placed into the soil using an auger, and the micro-video camera (B) is passed down the tube to record video images of the plant roots with the use of a video cassette recorder (VCR). The controller allows the operator to focus the camera and adjust light levels in the mini rhizotron tube. Recorded images are subsequently digitized with the use of computer software. (After Hendrick and Pregitzer, 1992. Reprinted with permission of the Ecological Society of America.)

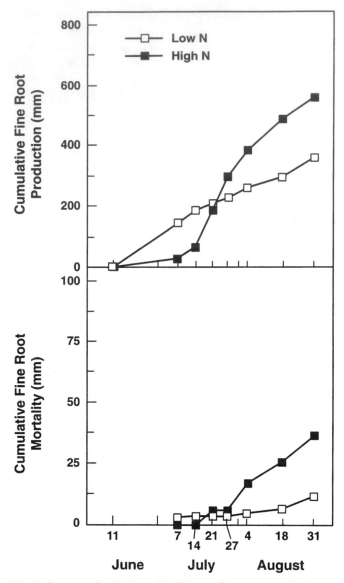

Figure 18.9. The influence of soil N availability on fine-root production and mortality of a black poplar hybrid. Note that both fine-root production and mortality increase with greater levels of soil N availability. Although production increases in the high N soil, the standing crop of fine roots is kept small by rapid rates of fine root mortality. Such a pattern suggests that the total amount of carbohydrate allocated to fine roots increases in N-rich soil. (After Pregitzer et al., 1995.)

and production can be calculated as the cumulative increase in root length over one growing season. Similarly, mortality can be determined as disappearance of roots that had previously appeared in the video image (i.e., a negative change in length). The results of such a study are shown in Figure 18.9, which illustrates the cumulative change in root production and mortality for a black poplar hybrid (cv Eugenei, a fast-growing intolerant tree) growing in soils of low and high N availability (Pregitzer et al., 1995).

The leaf mass of black poplar hybrid in the high N soil of this experiment increased by 35 percent, and fine root mass decreased by 25 percent, relative to plants in low N soil. The decline in fine roots and increase in leaves are consient with traditional views of changes in plant C allocation and with increases in soil N availability. In the high-N soil, however, fine-root production increased, whereas fine-root life span decreased (i.e., increase in mortality; Figure 18.9*a* and *b*), supporting the idea that more carbohydrate was allocated to fine root growth in N-rich soil (Pregitzer et al., 1995).

The underlying mechanisms for an increase in fine-root mortality can be found in the relationships among soil N availability, the N concentration of fine roots, maintenance respiration, and the longevity of plant tissue. Tissue N concentrations in the black poplar cultivar also increased in response to greater soil N availability, but the rate of maintenance respiration was not measured. Fine roots growing in N-rich soil should have relatively high tissue N concentrations and high rates of maintenance respiration, compared to fine roots growing in N-poor soil (Ryan, 1991). As a consequence, plants growing on N-rich soil should allocate proportionately more carbohydrate to fine roots, which have a greater maintenance "cost." Although we know little about the influence of tissue N concentration on the life span of fine roots, leaf life span is inversely related to photosynthetic rate, tissue N concentration, and metabolic activity (Larcher 1969; Chabot and Hicks, 1982; Reich et al., 1991, 1992). If the same relationship holds for fine roots, then one would expect fine-root life span to decline as soil N availability increases. Taken together, these results support the view that the allocation of C to fine root growth and maintenance increases in N-rich soil, but greater rates of fine root mortality (i.e., decreased life span) produce a proportionately smaller mass of fine roots.

We still have a great deal to learn about the C allocation patterns and fine root dynamics of forest trees. Certainly, the processes controlling fine root production and life span are of great importance for understanding the C allocation patterns of plants. Without a clear understanding of plant C allocation, we cannot fully understand how changes in light and soil resources influence the C balance of ecosystems. In the following section, we will build on the processes controlling the C balance of plants to understand how climate, soil resources, and disturbance influence the C balance of entire ecosystems.

CARBON BALANCE OF ECOSYSTEMS

Although the individual tree processes discussed in the previous section have an important influence on the C balance of forest ecosystems, we also must consider the activities of organisms other than trees to understand the flow of C into, within, and out of forest ecosystems. For example, most soil microorganisms use dead plant tissue (i.e., leaves, branches, stems, and roots) as an energy source. They incorporate a portion of the plant-derived C into microbial cells, producing organic by-products that aid in humus formation and return the remainder to the atmosphere as CO_2 during respiration. Leaf feeding (phytophagous) insects annually consume from 5 to 30 percent of the canopy leaves in most forests. Although such a loss of photosynthetic surface area often has little influence on tree growth (Mattson and Addy, 1975), it represents a transfer of C from one ecosystem component to another. That is, C once residing in plants has been transferred to insects, portions of which are lost during respiration and in excrement. To understand the C balance of forest ecosystems, we clearly must consider the growth and metabolism of organisms other than forest trees. In the sections that follow, we consider how climate, soil nutrient availability, and disturbance influence the processes controlling the cycling and storage of C in terrestrial ecosystems. We will later draw on this information to understand the cycling and storage of plant nutrients within forests, a topic covered in Chapter 19.

Biomass and Productivity of Forest Ecosystems

Several concepts are important in discussing the C balance of ecosystems, whether they are dominated by forest trees, prairie grasses, or desert shrubs. **Biomass** is the dry mass of living organisms and dead organic matter contained in a defined area, usually one square meter or a hectare (g m^{-2} or kg ha^{-1}, respectively). In some instances, biomass is reported in its C or energy equivalent (i.e., g C m^{-2} or cal m^{-2}), a necessary tool for understanding the flow of C (or energy) within ecosystems. In forest ecosystems, biomass is located in 5 major pools: 1) the above- and belowground tissues of over- and understory plants, 2) woody debris consisting of dead, fallen tree stems, 3) forest floor, 4) mineral soil, and 5) the tissues of heterotrophic organisms (decomposers and consumers). Although biomass in woody debris, forest floor, and mineral soil are composed of dead organic matter, it is convention to consider these compartments when discussing the distribution of biomass within forest or other terrestrial ecosystems.

Compare the amount and distribution of biomass among boreal, temperate and tropical forest ecosystems in Table 18.5. Notice that the total biomass of a tropical rain forest is almost five times that of a black-spruce-dominated boreal forest; the temperate coniferous and deciduous forest lies between these two extremes. Differences in the proportion of total biomass located in overstory trees, forest floor, and mineral soil also occur among the forests in Table 18.5. For example, forest floor and mineral soil compose the largest pool of biomass in the boreal forest, whereas the majority of biomass in the tropical forest resides in overstory trees. The temperate coniferous and deciduous forests in Table 18.5 also differ in biomass distribution, wherein overstory trees compose a larger proportion of total biomass in the coniferous forest.

Caution should be used in interpreting Table 18.5, because there is substantial variation in the amount and distribution of biomass within forests from boreal, temperate, and tropical regions. For example, overstory biomass in tropical forests ranges from 10 Mg

Table 18.5 The distribution of biomass in selected boreal, temperate, and tropical forest ecosystems.

	Boreal	Temperate		Wet Tropical
	Alaska	Washington	New Hampshire	Amazon
Location	USA	USA	USA	Brazil
Overstory				
Dominant	black	Douglas-	sugar maple	mixed
Species	spruce	fir	–beech	species[1]
Age (yrs)	95	60	55	mature
Biomass Pools		**Mg ha^{-1}**		
Overstory	50	410	165	990
Woody Debris	—	9	29	18
Forest Floor	76	15	48	7
Mineral Soil	152	119	173	250
Heterotroph	<1	<1	<4	<1
Total	278	553	419	1265

Source: Data have been summarized from Borman and Likens (1979), Harmon et al. (1990), Klinge et al. (1975), Schowalter (1989), Van Cleve et al. (1983), and Zak et al. (1994).

[1]The overstory of this tropical rain forest is composed of 50 species, but the total number of plant species exceeds 600 species per hectare. Species in the Leguminosae, Euphorbiaceae, and Sapotaceae families compose the majority of plant biomass.

ha^{-1} in dry forests to 540 Mg ha^{-1} in moist forests (Brown and Lugo, 1982), similar to the range of overstory biomass found in temperate deciduous (2 to 578 Mg ha^{-1}; Grier et al., 1989) and coniferous (5 to 810 Mg ha^{-1}; Grier et al., 1989) forests.

Primary production is the annual growth of all living plants within an ecosystem, and it can be subdivided into several basic components. **Gross primary production** (GPP) is the total amount of C fixed in an ecosystem during the process of photosynthesis (Fig 18.10). In many ways, GPP is the ecosystem equivalent of the gross photosynthesis of an individual plant. As with an individual plant, a portion of the total C fixed by an

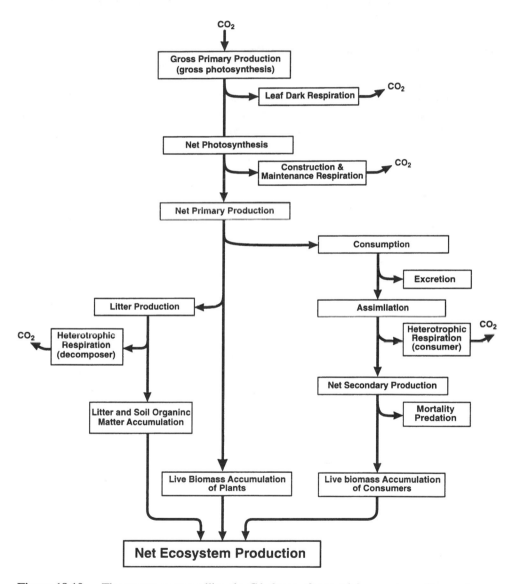

Figure 18.10. The processes controlling the C balance of terrestrial ecosystems. Gross primary production is the process responsible for C entering terrestrial ecosystems, whereas the construction and maintenance respiration of plants and animals is responsible for returning the C fixed through photosynthesis back to the atmosphere as CO_2. (After Aber and Melillo, 1991. Figure from TERRESTRIAL ECOSYSTEMS by John D. Aber and Jerry M. Melillo, copyright © 1991 by Saunders College Publishing, reprinted by permission of the publisher.)

ecosystem is lost to the atmosphere via leaf dark respiration and the construction and maintenance respiration of non-photosynthetic plant tissue (Fig 18.10). The difference between gross primary production and plant respiration is termed **net primary production** (NPP = GPP–R_A, where: R_A = total plant respiration consisting of leaf dark respiration, construction respiration—R_C, and maintenance respiration—R_M). Gross and net primary productivity can vary by an order of magnitude in forest and grassland ecosystems. However, the proportion of GPP lost via plant respiration (R_A) is surprisingly constant among the ecosystems in Table 18.6, varying from 50 to 60 percent. The majority of that loss arises from the maintenance respiration (R_M) of non-photosynthetic tissues (75 to 88 percent of R_A), whereas construction respiration (R_C) represents a much smaller proportion of R_A (12 to 24 percent). Much as the rate of net photosynthesis controls the C gain of an individual plant, the relationship between GPP and Ra directly influences NPP, and hence the rate at which C accumulates within or is lost from terrestrial ecosystems.

The growth and metabolism of decomposing organisms in soil and that of animals foraging within forest ecosystems is fueled by organic matter produced through NPP (Figure 18.10). The proportion of NPP remaining after accounting for losses to herbivory and litterfall is the **live biomass accumulation** of plants (Fig 18.10), and it can be either a positive or negative value. For example, a forest that has suffered a devastating windstorm, an insect outbreak, or has been harvested can experience a negative live biomass accumulation.

Net secondary production is the annual increase in the biomass of all organisms obtaining their energy from net primary production, including decomposers and consumers (i.e., heterotrophic organisms). Realize that not all organic matter assimilated by decomposers or consumers is used to build biomass; portions are lost during heterotrophic respiration (R_H) and in excrement (Figure 18.10). The live biomass of consumers accumulates only when the rate of net secondary production is greater than the rate at which consumer biomass is lost via mortality and predation.

Litter production and its decomposition directly control organic matter or biomass accumulation in the forest floor and mineral soil. Organic matter accumulates in these ecosystems pools when its rate of production (leaves and roots) exceeds the rate at which it is decomposed by soil microorganisms and returned to the atmosphere as CO_2 (Figure 18.10). Clearly, changes in either the rate of litter production or decomposition have the

Table 18.6   Estimates of total plant (R_A), construction (R_C), and maintenance (R_M) respiration in relationship to gross (GPP) and net primary productivity (NPP) in forest and grassland ecosystems.

Ecosystem	Plant Biomass (Mg C ha^{-1})	GPP	NPP	R_A	R_C	R_M	$R_A/$ GPP	$R_M/$ R_A
				Mg C ha^{-1} y^{-1}				
Forest								
deciduous	88	21.7	7.3	14.4	1.8	12.6	0.66	0.88
oak-pine	71	12.8	6.0	6.8	1.5	5.3	0.53	0.78
Grassland								
tallgrass	8	9.5	3.7	5.8	0.9	4.9	0.61	0.84
shortgrass	–	1.5	0.7	0.8	0.2	0.6	0.53	0.75

Source: After Ryan, 1991. Reprinted with permission of the Ecological Society of America.

Recall that NPP = GPP–R_A, and R_A = R_C + R_M; estimates of GPP do not include leaf dark respiration.

potential to alter the accumulation of organic matter in soil, a topic we will consider later in this section.

The annual rate of biomass accumulation in live plants, live animals, and soil organic matter is termed **net ecosystem production** (NEP), and it represents the summed change in all ecosystem biomass pools (Figure 18.10), and is defined in the following expression:

$$NEP = GPP - (R_A + R_H)$$

Net ecosystem production, or the C gain of ecosystems, can be either positive or negative depending on the change in the biomass of live plants, live animals and soil organic matter. Biomass accumulates within terrestrial ecosystems only when the amount of C fixed during GPP exceeds the rate at which C is lost via the respiration of plants (R_A) and heterotrophic organisms (R_H). That is, NEP > O when GPP > ($R_A + R_H$).

In forests, the biomass and respiration of organisms that consume plant material and their predators is relatively small compared to that of the overstory trees, forest floor, and mineral soil. Consequently, NEP in forests is largely controlled by the difference between NPP and the rate at which CO_2 is returned to the atmosphere during organic matter decomposition (R_H). Any disturbance that reduces or eliminates the live biomass accumulation of plants and increases rates of soil organic matter decomposition could lead to a net loss of C from an ecosystem. Under such a condition, NEP is negative and total ecosystem biomass declines.

Understanding the C balance of forest ecosystems has clear management implications—any management practice that tilts the balance toward negative NEP has the potential to alter the amount of C sequestered by forest ecosystems. Nevertheless, it is important to realize that the C balance of forest ecosystems is dynamic and can rapidly become positive following destruction of plants by harvest, fire, disease, or insect outbreak.

Measurement of Biomass and Productivity

Ecologists have long been interested in understanding how and why the productivity of ecosystems differs from one geographic location to another. Clearly, biomass and productivity differ markedly among forest, agricultural, grassland or desert ecosystems. Nonetheless, measuring the biomass and productivity of an ecosystem is not an easy task, particularly in forests where tremendous amounts of biomass reside in the above- and belowground portion of overstory trees.

Three approaches provide ecologists with insight into the biomass and productivity of forest ecosystems. One frequently-used approach is based on relationships that exist between the dimensions of forest trees and their weight. These species-specific, or allometric, relationships are typically expressed in the form of a mathematical equation in which the DBH and height of a particular tree is used to predict the weight of leaves, branches, bolewood, and sometimes roots. Development of **allometric biomass equations** is a very labor intensive task in which individual trees spanning a range of DBH and height are harvested, divided into components, and weighed. Allometric equations have been developed for most North American forest trees and a significant number of shrub species (Whittaker, 1966; Tritton and Hornbeck, 1982; Smith and Brand, 1983). Because of the great difficulty of collecting tree roots from the soil, most allometric biomass equations are used to predict the aboveground biomass and productivity of forest ecosystems.

With the use of allometric biomass equations, aboveground net primary productivity can be estimated with the following expression:

$$ANPP = \Delta B + L$$

In this equation, the annual change in biomass (ΔB) of all aboveground plant tissue plus the amount of leaves, seeds, flowers (L) produced in that year is equal to the aboveground net primary productivity of the ecosystem. Estimates of root biomass and root litter are often unavailable and hence are often absent from estimates of net primary productivity.

Aboveground net primary productivity is estimated by making measurements of tree or shrub heights and diameters in successive years, computing the biomass for each year using allometric equations, and subtracting the biomass estimate in year 1 from that estimated in year 2. Estimates of net aboveground primary productivity obtained using this technique are readily available for a wide array of forest ecosystems distributed throughout North America and Europe (see Grier et al., 1989; Reichel, 1981)

A second approach estimates net primary productivity by directly measuring the physiological processes controlling the C balance of ecosystems. This approach entails measuring photosynthesis and respiration for representative ecosystem components (i.e., leaves, branches, stems, soil) and then extrapolating their CO_2 flux to the entire ecosystem. Ecologists have used this approach to obtain estimates of gross primary productivity, ecosystem respiration, and net primary productivity for a late-successional oak-pine ecosystem at the Brookhaven National Laboratory on Long Island, New York (Botkin et al., 1970; Woodwell and Botkin, 1970).

In this study, leaves, branches, stems, and the soil surface were enclosed within small chambers that were connected to gas analyzers which measured CO_2 concentration over

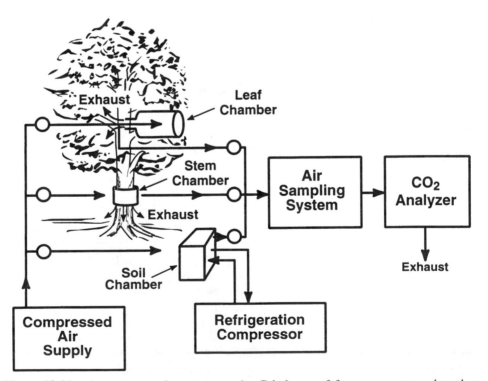

Figure 18.11.   A system used to measure the C balance of forest ecosystems based on quantifying the physiological processes responsible for C fixation and loss. (After Woodwell and Botkin, 1970. Reprinted from *Analysis of Temperate Forest Ecosystems,* © by Springer-Verlag Berlin, Heidelberg 1970. Reprinted by permission of Springer-Verlag New York, Inc.)

time (Fig 18.11). Because the entire forest canopy cannot be practically enclosed with a chamber, estimates of photosynthesis and dark respiration from a known quantity of leaves must be scaled to the entire forest canopy. Care must be taken to obtain estimates of photosynthesis and dark respiration for sun and shade leaves, thus accounting for differences in photosynthesis from the top to the bottom of the canopy. Similarly, estimates of branch, stem, and soil respiration must be extrapolated to a flux from the entire ecosystem. Given that rates of CO_2 exchange from a small portion of the entire ecosystem are actually measured, great care must be taken to insure that data are collected on representative samples of the ecosystem. Otherwise, estimates could be greatly in error when they are extrapolated to the entire ecosystem.

Using this approach, gross primary productivity at the Brookhaven forest was estimated to be 14.0 megagrams of C per hectare (Mg C ha^{-1}) during the growing season, whereas leaf dark respiration (4.0 Mg C ha^{-1}), stem respiration (5.5 Mg C ha^{-1}), and soil respiration (4.5 Mg C ha^{-1}) totaled 14.0 Mg C ha^{-1}. These results indicate that net ecosystem productivity was 0 Mg C ha^{-1}, suggesting that this relatively late-successional forest is neither gaining nor losing C (i.e., GPP = R_A + R_H).

One drawback of the small chamber approach is that, despite great effort, estimates of NPP are often inaccurate because root respiration cannot be separated from the respiration of soil microorganisms (i.e., soil respiration = root maintenance, root construction, and microbial respiration). This currently remains a challenge for ecologists, because it places great uncertainty on the actual belowground C budget of forest ecosystems.

Recently, micrometeorological techniques have provided a third approach to study the exchange of CO_2 between terrestrial ecosystems and the atmosphere. The **eddy covariance** method quantifies the net exchange of CO_2 by measuring vertical gradients of CO_2 from the forest floor to above the canopy. In combination with wind speed, wind direction and temperature, the vertical CO_2 profile can be used to estimate the flux of C into and out of forested ecosystems on hourly, daily, and yearly time intervals (Fig 18.12). This technique has recently been used to study the C balance of a second-growth northern hardwood ecosystem at the Harvard Forest in Massachusetts. Net ecosystem production estimated using the eddy correlation was 3.7 Mg C ha^{-1} y^{-1}. Ecosystem respiration, calculated from the CO_2 flux during nighttime hours, was 7.4 Mg C ha^{-1} y^{-1}, suggesting

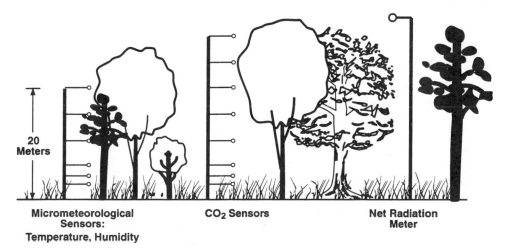

20 Meters

Micrometeorological Sensors: Temperature, Humidity

CO$_2$ Sensors

Net Radiation Meter

Figure 18.12. Instrumentation of a forest ecosystem to measure the net exchange of C with the atmosphere. The eddy covariance approach uses the vertical gradient of CO_2 from above the canopy to the forest floor to estimate the net uptake and release of CO_2 by terrestrial ecosystems.

that gross primary productivity was 11.1 Mg C ha^{-1} y^{-1} (Wofsy et al., 1993). In contrast to the late-successional forest at Brookhaven, New York, this second-growth northern hardwood ecosystem is accumulating C based on the fact that NEP is positive i.e., GPP > R$_A$ + R$_H$.

It is important to realize that the eddy correlation technique also suffers from the fact that soil respiration cannot be broken into separate contributions of plant roots and soil microorganisms—making it difficult to estimate the proportion of net ecosystem production and net primary production occurring below the soil surface. Clearly, the greatest uncertainty associated with estimating the C balance of ecosystems occurs in measuring rates of root production and the amounts of CO$_2$ separately respired by plant roots and soil microorganisms.

Climate and Productivity

Although there is still much to learn about the C budget of forests and other terrestrial ecosystems, striking patterns and predictable differences in aboveground net primary productivity (ANPP) occur over the Earth's surface. Geographic differences in ANPP are not random, but appear to be strongly linked to patterns of climate. It is no surprise that the productivity of desert or shrub steppe ecosystems is much less than that of prairie or forests.

Differences in temperature and precipitation over the Earth's surface influence the distribution of vegetation and the processes controlling ANPP. Figure 18.13 summarizes the relationship between temperature, precipitation, and the ANPP of a wide array of terrestrial ecosystems, including tundra, deserts, grasslands, and forests from tropical, temperate, and boreal regions. Aboveground net primary productivity is relatively low in cold (Figure 18.13a) and dry (Figure 18.13b) climates, and it rapidly rises as both temperatures precipitation increase. This relationship is striking, given the wide diversity of plants and ecosystems used to compile this relationship. Why are global-scale patterns of ANPP so closely related to climatic factors like temperature and precipitation?

The connection between climate and ANPP arises due to a fundamental trade-off during photosynthesis—plants lose water when they open their stomates to fix C from the atmosphere. At a global scale, the amount of C fixed by plants and terrestrial ecosystems is limited by the quantity of water available for transpiration. Actual evapotranspiration is an ecologically-important climatic factor that incorporates precipitation and temperature to estimate the amount of water lost to the atmosphere from evaporation (soil surface) and plant transpiration. It can be calculated with knowledge of mean monthly temperature, precipitation and the available water content of soil. Actual evapotranspiration (AET) often is compared with potential evapotranspiration (PET), which estimates the amount of evapotranspiration that would occur if the supply of water was unlimited. In regions where temperatures are high and precipitation is infrequent (i.e., deserts), PET greatly exceeds AET because low amounts of precipitation limit evapotranspiration even when ample energy is available to evaporate water from the surfaces of soil and leaves.

Figure 18.14 illustrates the relationship between the ANPP of a wide array of terrestrial ecosystems from different climatic regimes and AET. Desert ecosystems occur in regions where little water is available for evapotranspiration and hence can be found at the low end of the AET axis in Figure 18.14. Forests, in contrast, occur where annual amounts of precipitation exceed AET (upper end of the axis); a situation where the availability of water does not often limit evapotranspiration. Notice that a positive relation exists between ANPP and AET—an increase in the amount of water returned to the atmosphere via AET results in greater ANPP. Again, this relationship exists because plants open their sto-

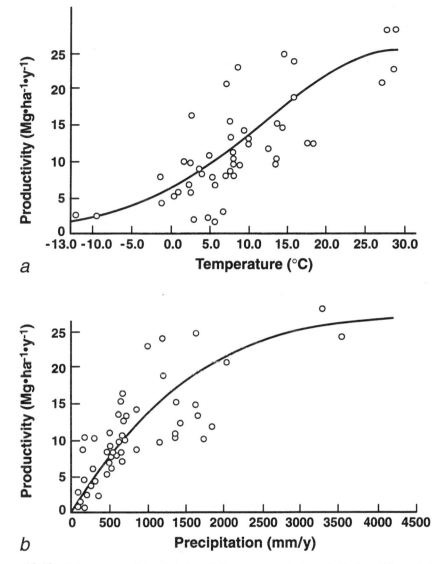

Figure 18.13. Mean annual temperature (*a*), mean annual precipitation (*b*), and the net aboveground primary productivity (ANPP) of terrestrial ecosystems. Data used to construct these relationships were collected from a wide array of terrestrial ecosystems including deserts, grasslands, forests, and tundra. Increasing temperature and precipitation generally result in greater amounts of ANPP. (After Lieth, 1973.)

mates to fix C from the atmosphere and simultaneously lose water during transpiration. At a global scale, the amount of water available for transpiration sets the upper bounds on plant C gain, and hence is responsible for differences in ANPP among ecosystems from different climatic regions.

Mountainous regions exhibit dramatic temperature and precipitation gradients that create differences in ecosystem productivity equivalent to those depicted in Figure 18.14, albeit at a much smaller spatial scale. The Coastal and Cascade Mountain ranges in the Pacific Northwest help create marked changes in climate, vegetation, and productivity

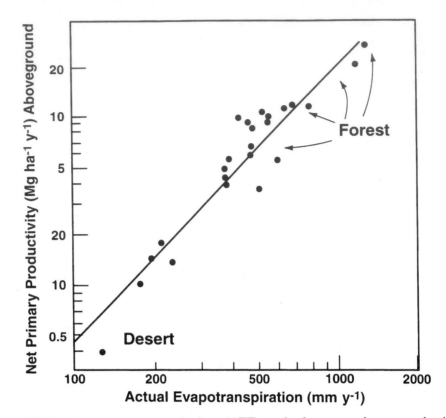

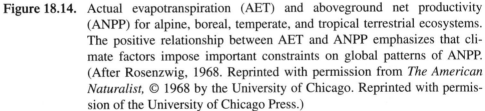

Figure 18.14. Actual evapotranspiration (AET) and aboveground net productivity (ANPP) for alpine, boreal, temperate, and tropical terrestrial ecosystems. The positive relationship between AET and ANPP emphasizes that climate factors impose important constraints on global patterns of ANPP. (After Rosenzwig, 1968. Reprinted with permission from *The American Naturalist,* © 1968 by the University of Chicago. Reprinted with permission of the University of Chicago Press.)

across relatively small distances (Chapter 10). At low elevations in the Coastal range (200 m) near the Pacific Ocean, relatively warm and wet (246 cm y^{-1}) conditions give rise to tall-statured forests dominated by Sitka spruce and western hemlock (Franklin and Dyrness, 1973). Inland lies the Cascade range where relatively warmer and drier conditions at lower elevation give rise to forests ecosystems dominated by Douglas-fir and grand fir. Moving eastward and up in elevation, orographic cooling produces progressively colder and wetter climatic conditions where Douglas-fir, mountain hemlock, and Pacific silver fir dominate the forest overstory. Rainshadow conditions occur on the east slope of the Cascade range, creating progressively warmer and drier climates as one moves further eastward and down in elevation. Ponderosa pine ecosystems occur on the east slope (870 m) of the Cascade range in regions receiving 40 cm y^{-1} of precipitation; western juniper (25 cm y^{-1}) and sagebrush (20 cm y^{-1}) occur in the drier climates at lower elevations on the east slope.

Along this relatively-broad climatic gradient in the Pacific Northwest, ANPP ranges from 0.3 Mg ha^{-1} y^{-1} in dry, sagebrush ecosystems to 15 Mg ha^{-1} y^{-1} in wet, coastal western hemlock ecosystem (Gholz, 1982), a range of values equivalent to that in Figure 18.14. In this mountainous region, patterns of ANPP are strongly related to differences in temperature, precipitation, and soil water storage. Gholz (1982) combined these physical site factors to estimate the water availability in Pacific Northwest ecosystems using the following expression:

$$WB = P - E + SWC$$

In this equation, P is monthly precipitation (cm), E is monthly potential evaporation (cm), SWC is the average monthly soil water content (cm), and WB is the monthly water balance. Aboveground annual net primary productivity is relatively low in juniper and sagebrush ecosystems where annual losses of water to evaporation greatly exceed precipitation and soil water content (i.e., a negative water balance; Figure 18.15). As either precipitation increases or evaporation decreases, the annual water balance in this region becomes less negative, which relates to a linear increase in ANPP (Figure 18.15). Again, this reinforces the idea that water availability sets a fundamental limit on ecosystem productivity by controlling the amount of C fixed through photosynthesis.

Although evapotranspiration is strongly related to global gradients of productivity, it is not the sole determinant of ecosystem productivity. In Figure 18.14, study the variation in ANPP where AET equals 500 mm; note that it ranges from 3 to 10 Mg ha^{-1} y^{-1}. Such variation is not uncommon in temperate forests where ANPP differs by a factor of 10 (Grier et al., 1989), even though AET varies to a lesser extent (500 to 700 mm). Also realize that ecosystems with very low productivity can be found in warm, tropical regions receiving relatively large amounts of precipitation, something that would not be predicted from the relationships depicted in Figures 18.13 and 18.14. What factors cause productivity to differ so widely among ecosystems within a particular climatic region? In the following sections we explore the extent to which soil properties, species composition, and stage of ecosystem development modify NPP and NEP within a particular climatic region.

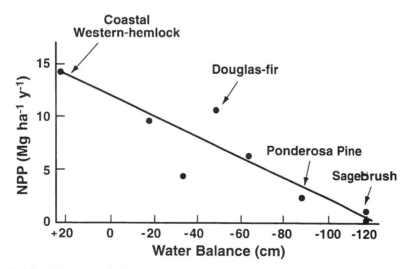

Figure 18.15. The water balance and aboveground net primary productivity (ANPP) of ecosystems in the Pacific Northwest. The relationship between water balance and ANPP emphasizes the idea that increases in water availability along climatic gradients in mountainous regions relate to an increase in net primary productivity. (After Gholz, 1982. Reprinted with permission of the Ecological Society of America.)

Soil Properties, Forest Biomass, and ANPP

Within a particular climatic region, soil parent material can exert a substantial influence on the productivity of forest ecosystems through its influence on soil water holding capacity. Recall that sandy-textured soils hold relatively lower amounts of plant available water compared to soils of loam or clay loam texture (Chapter 11). In the Lake States, glacial activity (10,000 to 14,000 yr BP) created local landscapes where parent materials range from coarse outwash sands to relatively fine-textured glacial till; such changes occur over relatively short distances, e.g., 100 m to 1 km. Because these contrasting parent materials occur within a particular climatic regime, they present the opportunity to examine how landform-mediated differences in soil parent material (i.e., plant available water) modify the productivity of forest ecosystems.

One example occurs in northern Lower Michigan where productive northern hardwood forests growing on loamy till occur in the same landscape as relatively-unproductive oak-dominated (black and white) ecosystems of the sandy outwash plains. Overstory biomass (208 Mg ha^{-1}) and productivity (3.2 Mg ha^{-1} y^{-1}) of a northern hardwood ecosystem is twice that of a xeric oak ecosystem (105 Mg ha^{-1} and 1.5 Mg ha^{-1} y^{-1}; Host et al., 1988). Thus soil texture can greatly modify the influence of regional climate on the productivity of forest ecosystems.

Differences in soil parent material also give rise to differences in soil N availability, which is thought to control the productivity of many boreal, temperate, and a selected

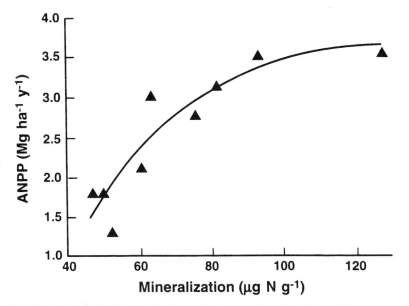

Figure 18.16. In the Lake States Region, aboveground productivity of forest ecosystems is well correlated with the rate at which soil microorganisms release N from soil organic matter. Illustrated in this figure is the ANPP (overstory only) and net mineralization rate of nine upland forest ecosystems in northern Lower Michigan. The positive relationship between these variables suggests that soil nitrogen availability imposes an important constraint on ecosystem productivity within the climate of the Great Lakes Region. (After Zak et al., 1989. Reprinted with permission of National Research Council Canada.)

number of tropical forests. Unlike agricultural ecosystems that receive N additions from chemical fertilizer, soil N availability in forests and other less-intensively managed ecosystems is controlled by the activity of soil microorganisms. Most soil microorganisms obtain energy and essential nutrients during the decomposition of plant litter. During this process, only a portion of the N contained in litter is used by soil microbes to maintain or build biomass. The remainder is released into soil solution as ammonium (NH_4^+) where it can be assimilated by plant roots (see Chapter 19 for details). In the Upper Lake States, variation in ANPP among forest ecosystems is related to differences in soil N availability, wherein relatively-high rates of ANPP (overstory only) correspond with high rates of N mineralization (Figure 18.16). It is important to note that fine-textured soils, which hold relatively-large amounts of plant available water, also release the greatest amounts of N for plant growth. The relationship between ANPP and soil N availability is not unique to Lake States forests. ANPP also is related to soil N availability in grassland (Schimel et al., 1985), shrub steppe (Burke, 1989), and desert (Fisher et al., 1987) ecosystems.

Soil texture can modify the influence of climate on ANPP by controlling the amount of water and N that is available for plant growth. In other words, climate sets the regional potential for productivity through patterns of temperature and precipitation, but that potential is constrained by landform-mediated differences in soil texture which control the availability of water and N within the local landscape.

Biomass Accumulation During Ecosystem Development

Following an event that destroys forest vegetation, living plant biomass increases from near zero to that of a mature forest ecosystem (Figure 18.17). Concomitant changes also occur in the activities of decomposing organisms and in the biomass stored in forest floor and mineral soil. The rate at which biomass accumulates in forest ecosystems is controlled by the amount of C fixed through photosynthesis (i.e., GPP) and the amount of C returned to the atmosphere via maintenance and construction respiration (i.e., $R_A + R_H$). These processes are illustrated for an idealized forest following the destruction of living plant biomass (Figure 18.17).

In the newly established forest, living plant biomass (Mg ha^{-1}) is nearly zero and it rapidly increases during the first 20 to 25 years of ecosystem development. After this period of time, the increase in living plant biomass begins to slow and it eventually attains a maximum value near year 50 (Figure 18.17b). The biomass of leaves initially increases and attains a maximum value after approximately 20 years (Figure 18.17b), the point at which canopy closure occurs. Recall that net canopy photosynthesis, and hence the amount of photosynthate available for plant growth and maintenance, is a product of leaf photosynthetic rate (μmol m^{-2} sec^{-1}) and total canopy leaf area. Leaf biomass (Mg ha^{-1}) and area (m^2 ha^{-1}) are well correlated in most forest trees, indicating that the amount of leaf area (or biomass) at canopy closure sets the upper limit on canopy photosynthesis and GPP. It follows that this relationship also sets a limit on the amount of C that can be allocated to growth and maintenance. Notice that total biomass (non-photosynthetic tissue) continues to increase well after leaf biomass has attained its maximum (Fig 18.17b).

The pattern of total biomass accumulation in this forest ecosystem indicates the balance between GPP and $R_A + R_H$ changes during ecosystem development. In Figure 18.17a, GPP is relatively consistent after approximately 20 years, whereas R_A represents a variable C cost that continues to increase for a longer period of time. This pattern suggests that an increasingly greater amount of fixed C is allocated to tissue maintenance, diminishing the amount allocated to the construction of new tissue. This is consistent with

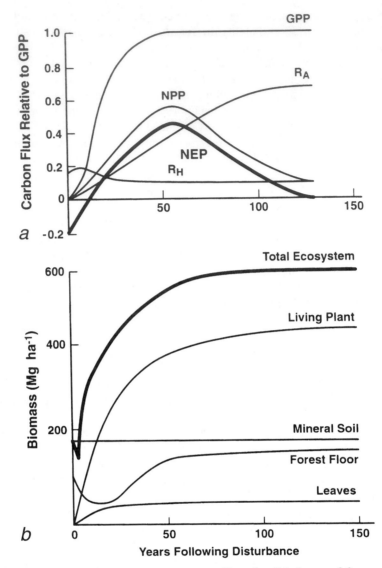

Figure 18.17. Changes in the processes controlling the C balance of forest ecosystems (*a*) and the rate of biomass accumulation following a disturbance that destroys living plant biomass (*b*). Such a disturbance includes intensive harvesting of overstory trees, fire or windstorm. Change in net secondary production is not illustrated and accounts for a relatively minor flux of C in forest ecosystems.

the increase in non-photosynthetic biomass (total) long after the biomass of photosynthetic tissue has reached its maximum at canopy closure. The initial increase and subsequent decline in NPP (i.e., GPP $-$ R_A) reflects this fundamental trade-off: C allocated to maintenance cannot be allocated to growth. Clearly, the C balance of plants can exert direct control on the C balance of entire ecosystems.

Changes in forest floor biomass following a disturbance reflect a balance between litter production (i.e., abscised leaves, fine roots, twigs, reproductive structures) and the amount of litter consumed by decomposing organisms (Fig 18.17*b*). Early in ecosystem development (0 to 15 years), plant litter production is low, but activities of decomposing organisms are often enhanced by the relatively warmer, moister conditions at the soil surface. Note that R_H increases during the first several years following disturbance. Low litter production in combination with enhanced decomposition rates result in an initial decline in forest floor biomass. Biomass begins to accumulate in the forest floor when litter production exceeds decomposition, a situation that occurs between 20 and 50 years following disturbance. The relatively constant biomass of the forest floor late in ecosystem development suggests that litter inputs to forest floor are balanced by rates of decomposition.

The amount of organic matter stored in mineral soil reflects the long-term (1,000 to 1,000,000 years) balance between plant litter production and the rate at which decomposing organisms return it to the atmosphere as CO_2. Biomass or organic matter stored in mineral soil often changes little following a disturbance the destroys a forest overstory (Figure 18.17b; Bormann and Likens, 1979). In some situations, the amount of organic matter lost from the forest floor and mineral soil are quickly regained by inputs from rapidly developing vegetation (Gholz and Fisher, 1982). However, soil erosion following disturbance can result in a substantial reduction in soil organic matter, particularly in areas of high rainfall and steep topography.

Net ecosystem production measures the C balance of terrestrial ecosystems, and it can dramatically change during ecosystem development. Immediately following a disturbance, high rates of heterotrophic respiration (R_H) and low rates of NPP ($GPP - R_A$) result in a net flux of C from the ecosystem depicted in Figure 18.17a. This negative C balance is marked by a reduction in total ecosystem biomass, which produced a negative NEP. Recall that any disturbance that diminishes NPP and accelerates rates of organic matter decomposition (R_H) has the potential to reduce NEP below 0. Total ecosystem biomass increases only when GPP exceeds the respiratory loss of C from plants and heterotrophic organisms, a situation occurring from 5 to 75 years following disturbance. Note that NEP declines to 0 late in ecosystem development (130 yrs), as a result of the relationship $GPP = R_A + R_H$. Because the fixation and loss of C on an ecosystem basis are in balance, total ecosystem biomass reaches a relatively constant value (Figure 18.17b).

The generalized patterns illustrated in Figure 18.17 occur in all terrestrial ecosystems, albeit at different magnitudes and over different periods of time. After an initial decline, for example, total ecosystem biomass rapidly accumulates in young slash pine plantations in Florida (Figure 18.18; Gholz and Fisher, 1982). Maximum leaf biomass and canopy closure occur after 5 years, which sets an upper limit on canopy leaf area and the amount of photosynthate available for growth and maintenance. Notice that total biomass continues to accumulate long after canopy closure, indicating that greater amounts of GPP are allocated to maintain the increasing amount of non-photosynthetic tissue (Figure 18.18b). This trade-off in C allocation from growth to maintenance is further illustrated by the peak in tree and understory biomass at 25 years ($GPP \approx R_A$; Figure 18.18a). After 25 years of growth, overstory trees are commercially harvested and seedlings are planted to re-establish a new plantation. These activities again reduce living plant biomass to near zero and re-start the process of biomass accumulation in the newly developing plantation.

Natural disturbance regimes of fire and windstorms also alter the C balance of forest ecosystems. Prior to European settlement, fire was an important component of ecosystem development in the hardwood forests of southern Wisconsin (Figure 18.19; Loucks, 1970). Repeated, random fires altered the C balance of these ecosystems by destroying living plant and reducing GPP. Biomass oxidized by fire to CO_2 also represents a loss of C and causes a decline in total ecosystem biomass. In combination, these responses cause NEP to fluctuate in a cyclic matter that reflects the magnitude and frequency of fire. Notice that the period between some fires in Figure 18.19 is sufficient for R_A and R_H to equal

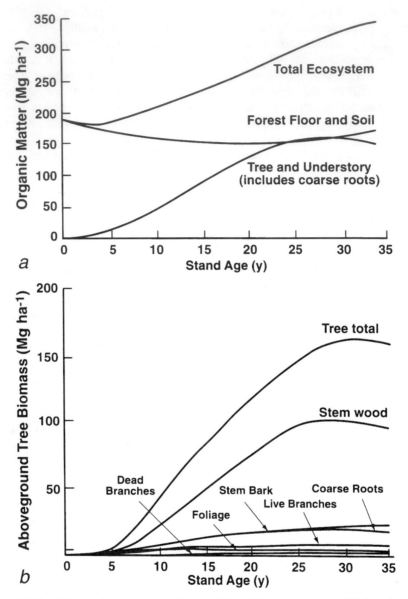

Figure 18.18.   Changes in ecosystem biomass (*a*) and the components of living plant biomass (*b*) in a developing slash pine ecosystem in Florida. (After Gholz and Fisher, 1982. Reprinted with permission of the Ecological Society of America.)

GPP, producing a NEP that is near 0 late in ecosystem development. In some ecosystems, fire may occur during a period of total biomass accumulation (GPP > R_A + R_H), thus preventing total biomass from attaining a "steady state" that might occur late in ecosystem development.

Natural disturbance that destroys living plant biomass can occur over relatively long periods of time (i.e., 100s to 1000s yrs.) in some ecosystems, thus allowing them to reach a "steady state" in total biomass. In the Pacific Northwest, old-growth ecosystems dominated by Douglas-fir and western hemlock can attain ages of 450 years or more. Although

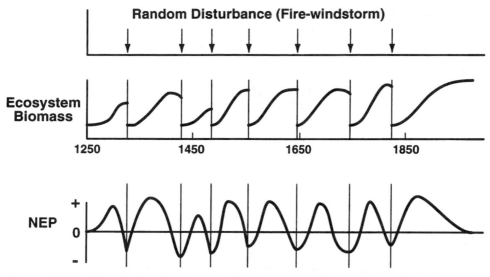

Figure 18.19. The relationship between fire frequency, total ecosystem biomass and net ecosystem productivity for forests in southern Wisconsin (modified from Loucks, 1970). Fire was an important disturbance prior to European settlement that functioned to alter ecosystem C balance in a cyclic manner. The present suppression of fire in Lake States forests has greatly altered this relationship. Note the lack of fire following 1850 and the relatively long period of time over which NEP is greater than 0.

NEP is low (or 0) in these old-growth forests, they have accumulated large amounts of C relative to second-growth forests (Table 18.7). Total ecosystem biomass of old-growth forest is over twice that of the second-growth forest, suggesting that GPP exceeds $R_A + R_H$ for a substantial period of time between 60 and 450 years following disturbance. The accumulation of biomass in stem wood and coarse woody debris accounts for 77 percent of the increase in total biomass between the second- and old-growth forest.

Table 18.7 Biomass pools in second- and old-growth ecosystems dominated by Douglas-fir and western hemlock in the Pacific Northwest.

Component	60-yr-old forest	450-yr-old forest
	Mg ha^{-1}	
Foliage	12	13
Branch	15	56
Stem	308	687
Coarse Roots	62	151
Fine Roots	12	12
Forest Floor	15	55
Coarse Woody Debris	8	206
Mineral Soil	119	119
TOTAL	551	1299

Source: After Harmon, M.E., W.K. Ferrell, and J.F. Franklin, 1990. Effects on carbon storage of conversion of old-growth forests to young forests. Reprinted with permission from *Science,* p. 700, © 1990 by the American Association for the Advancement of Science.

Although NPP and NEP decline through time, substantial amounts of biomass can accumulate in old-growth forests. Clearly, the conversion of old-growth to second-growth forests has the potential to greatly alter the cycling and storage of C in the Pacific Northwest.

SOIL N AVAILABILITY AND BELOWGROUND NET PRIMARY PRODUCTIVITY

Allocation of C to belowground plant growth accounts for a substantial proportion of NPP in many forest ecosystems. Studies conducted in coniferous and deciduous forests suggest 50 percent of NPP can be allocated to the growth and maintenance of roots. Earlier in this chapter (see C Balance of Plants), we observed that soil N availability substantially alters the proportion of carbohydrate allocated to the leaves, stem, and roots of individual plants. In this section, we extend this idea to an ecosystem level and summarize patterns of belowground NPP in forest ecosystems. We focus on the extent to which soil N availability modifies the proportion of NPP allocated to fine roots and associated mycorrhizae, the nutrient and water absorbing organs of forest trees (Chapter 15).

Quantifying fine-root production in forest ecosystems remains one of the most important, but difficult challenges to ecologists. Unlike leaves that senesce and abscise during a defined time of the year, root production and mortality are continuous processes occurring throughout the entire growing season. Most often, belowground NPP is estimated by measuring the monthly change in root biomass over one year. Other techniques include the use of mini-rhizotrons and several methods that employ a nutrient budget. These methods of quantifying belowground production have led ecologists to very different conclusions regarding the influence of soil N availability on belowground NPP. Regardless of this uncertainty, it is clear that the production of plant tissue below the soil surface in forest ecosystems is a substantial component of NPP.

Two alternative views regarding the influence of soil N availability on fine-root dynamics have emerged (Figure 18.20). The one hypothesis argues that fine-root life span (i.e., mortality) does not change along gradients of soil N availability. It also predicts that fine-root production declines in N-rich soil, enabling the plant to allocate proportionately more carbohydrate to foliage and stems (Hypothesis 1 in Figure 18.20). Such a view contends that fewer roots are needed to forage for soil resources when they are in abundant supply. Evidence for this view comes from a comparison of two 40-year-old Douglas-fir ecosystems occurring on soils that differ in N availability (Table 18.8). Sequential measurements of fine-root biomass were used to estimate fine-root production and allometric equations were used to estimate the change in coarse root biomass. The proportion of NPP allocated to fine-root production declined from 36.4 percent in the N-poor soil to 7.9 percent in the N-rich soil, but rates of fine-root mortality were unchanged (Keyes and Grier, 1981). At the same time, allocation to leaves rose from 33.1 percent in the N-poor soil to 55.5 percent in the N-rich soil (Table 18.8). In combination, these data suggest that the proportion of NPP allocated belowground declined in N-rich soil, facilitating a decline in fine-root biomass.

Another view contends that C allocation to fine roots remains constant along a gradient of N availability, but the life span of fine roots declines when soil N availability is relatively high (Figure 18.20; Hypothesis 2). Therefore, the proportion of NPP allocated to fine-root growth remains constant, but an increase in mortality facilitates a reduction in fine-root biomass when soil N availability is high. This view of fine-root dynamics has developed from the use of an N budget to estimate the amount forest trees allocate to the annual production of fine roots. It assumes that the allocation of N to fine roots equals the difference between the amount of N assimilated by forest trees and the amount of N allo-

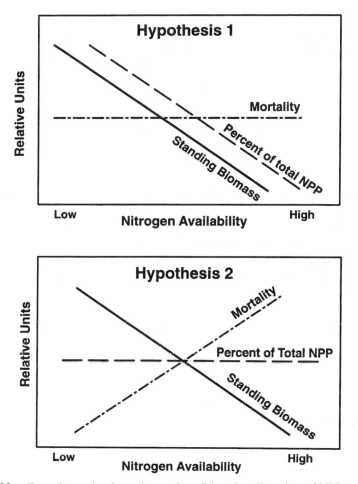

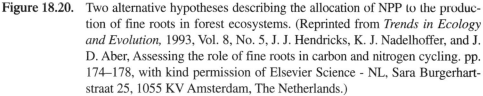

Figure 18.20. Two alternative hypotheses describing the allocation of NPP to the production of fine roots in forest ecosystems. (Reprinted from *Trends in Ecology and Evolution,* 1993, Vol. 8, No. 5, J. J. Hendricks, K. J. Nadelhoffer, and J. D. Aber, Assessing the role of fine roots in carbon and nitrogen cycling. pp. 174–178, with kind permission of Elsevier Science - NL, Sara Burgerhartstraat 25, 1055 KV Amsterdam, The Netherlands.)

Table 18.8 The Allocation of Net Primary Productivity to Above- and Belowground Plant Tissues in 40-year-old Douglas-fir Stands Growing on Soils of Low and High Fertility.

Component	Low N Soil (Mg ha^{-1} y^{-1})	Percent of Total	High N Soil (Mg ha^{-1} y^{-1})	Percent of Total
Stem	5.1	33.1	9.9	55.5
Branch	0.2	1.3	0.6	3.4
Leaves	2.0	13.0	3.2	18.0
Coarse Roots	2.5	16.2	2.7	15.2
Fine Roots	5.6	36.4	1.4	7.9
Total	15.4	100	17.8	100

Source: After Keyes and Grier, 1981. Reprinted with permission of National Research Council Canada.

cated to aboveground growth. This approach has been applied to a series of hardwood and coniferous forests in Wisconsin that occur along a gradient of soil N availability (Nadel-hoffer et al., 1985; Figure 18.21) and has yielded results in direct contrast to the aforementioned study of NPP in Douglas-fir. Using the N budget approach, the biomass N content (g N m^{-2}) of fine roots declined along an increasing gradient of soil N availability, but the total amount of N allocated to fine-root production increased; NPP allocated to fine roots varied little and averaged 27 percent. The inverse relationship between fine-root biomass N and the total amount of N allocated to fine-root production indicates that increased mortality at high soil N availability maintained a relatively low fine-root biomass. Recall that plant tissues with high N concentrations often have high maintenance respiration costs and a short life span. These attributes may partially explain the increase in fineroot mortality, because roots in high-N-availability soil often have a high N concentration (percent N).

A third approach, based on a short-term C budget, has also produced evidence for a positive relationship between above- and belowground NPP (Raich and Nadelhoffer, 1989). It is based on the principle that all above- and belowground litter production is respired by decomposing organisms on an annual basis. Under that assumption, the allocation of C to root growth and maintenance is proportional to the flux of C from soil respiration minus the C in aboveground leaf litter production (for details see Raich and Nadelhoffer, 1989). Soil respiration and leaf litter production have been quantified in many forest ecosystems, and Figure 18.22 summarized the relationship between C allocated to leaf litter and fine roots (fine-root production, construction respiration, and maintenance respiration) for 30 forest ecosystems occurring in tropical, temperate and boreal regions. Notice that leaf litter and the amount of C allocated to fine roots appear to be positively related on a global basis. Such a relationship provides further evidence that a relatively constant proportion of NPP is allocated belowground, which provides additional support for Hypothesis 2 in Figure 18.20.

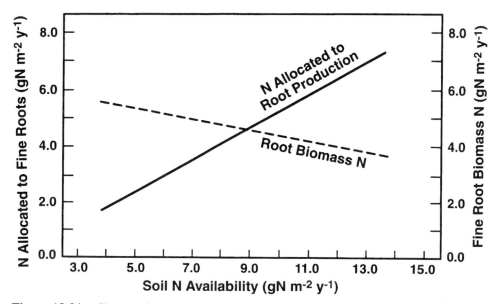

Figure 18.21.   Changes in the N content of fine-roots (dashed line) and the total amount of N allocated to fine-root production (solid line) along a gradient of increasing soil N availability. (After Nadelhoffer et al., 1985. Reprinted with permission of the Ecological Society of America.)

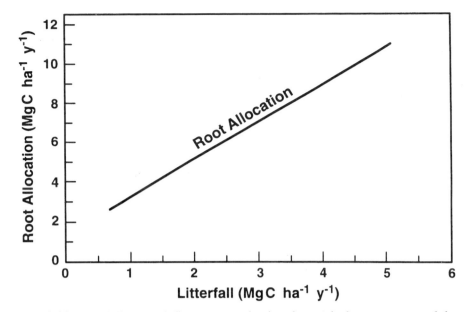
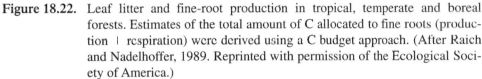

Figure 18.22. Leaf litter and fine-root production in tropical, temperate and boreal forests. Estimates of the total amount of C allocated to fine roots (production + respiration) were derived using a C budget approach. (After Raich and Nadelhoffer, 1989. Reprinted with permission of the Ecological Society of America.)

All of the approaches discussed use indirect methods of quantifying fine-root dynamics. That is, none directly observe the production and death of individual roots, which is the only approach that has the potential to resolve our uncertainty of the belowground C budget of forest ecosystems. We still have a great deal to learn regarding the ecosystem-level allocation of C to fine-root production in forest ecosystems before we can resolve the disparity in observations of fine-root production and mortality. This fact clearly limits the ability of ecologists to accurately estimate the NPP of forest ecosystems. Nevertheless, it is clear that the growth and maintenance of plant tissues below the soil surface account for a significant proportion of C fixed via photosynthesis in many forest ecosystems.

Suggested Readings

Bormann, F.H., and G.E. Likens. 1994. Pattern and Process in a Forested Ecosystem. Springer-Verlag, New York. 266 pp.

Givnish, T.J. 1988. Adaptation to sun and shade: a whole-plant perspective. *Australian J. of Plant Physiol.* 15:63–92.

Mooney, H.A. 1972. The carbon balance of plants. *Ann. Rev. of Ecol. Syst.* 3:315–346.

Pearcy, P.W., J. Ehleringer, H.A. Mooney, and P.W. Rundel (eds.). 1991. *Plant Physiological Ecology: Field Methods and Instrumentation.* Chapman and Hall, New York. 457 pp.

Smith, W.R. and T.M. Hinkley (eds.). 1995. *Resource Physiology of Conifers: Aquisition, Allocation, and Utilization.* Academic Press, New York. 396 pp.

Raghavendra, A.S. (ed.) 1991. *Physiology of Trees.* John Wiley, New York. 509 pp.

CHAPTER *19*

NUTRIENT CYCLING

Although carbon (C), hydrogen (H), and oxygen (O) form the basic building blocks of all biological tissue, plants require a suite of 14 other elements, termed nutrients, in order to maintain existing tissue and build new biomass (see Table 11.3). These nutrients mostly enter terrestrial ecosystems from the atmosphere (wet and dry precipitation) or through the weathering of soil minerals. Plants assimilate nutrients from soil, incorporate them with photosynthetically-fixed C to form living biomass, and eventually return them to soil in dead leaves, roots, branches, and stems (i.e., plant litter). After dead plant material enters the forest floor (O horizons) it is subject to decomposition, a microbially-mediated process that releases organically-bound nutrients into forms that can again be assimilated by plant roots. The uptake of nutrients by plant roots, their incorporation into living tissue, and the release of nutrients during organic matter decomposition cause nutrients to flow or *cycle* within terrestrial ecosystems. Nutrient cycles are biogeochemical processes, so named because they are controlled by the physiological activities of plants and soil microorganisms, and the geochemical processes in soil that control nutrient supply (Chapter 11).

Recall that net primary productivity differs among boreal, temperate, and tropical forests (Chapter 18). Because nutrients are assimilated along with C to form living biomass, differences in net primary productivity among ecosystems suggest that rates of nutrient cycling also must differ. In this chapter, we discuss processes controlling the input of nutrients to forest ecosystems, the redistribution of nutrients within forest ecosystems by the metabolic activities of plants and soil microorganisms, and the loss of nutrients from forest ecosystems to streams, groundwater, and the atmosphere. These processes are illustrated in Figure 19.1, which outlines the organization of this chapter. Nutrient absorption by plant roots is central to the cycling of nutrients in forest ecosystems, and we place emphasis on the physiological and morphological mechanisms by which plants forage for

524

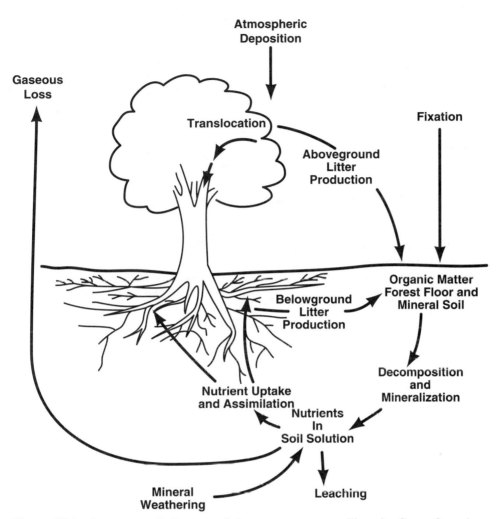

Figure 19.1. A conceptual diagram of the processes controlling the flow of nutrients into, within, and out of forest ecosystems. Nutrients enter forest ecosystems through atmospheric deposition, N_2 fixation, and mineral weathering. The flow of nutrients within forest ecosystems is controlled by nutrient uptake, the translocation of nutrients from senescent tissue, the return of nutrients in litter, the decomposition of litter on the forest floor and mineral soil, and the mineralization of nutrients from organic matter. Leaching and gaseous losses (i.e., denitrification) are processes by which nutrients are lost from forest ecosystems.

nutrients in soil. We also discuss the extent to which the C balance of terrestrial ecosystems influences nutrient retention and loss within forests, particularly during forest development. Our treatment of nutrient cycling focuses on nitrogen (N), because the supply of this nutrient in soil most often limits the productivity of boreal, temperate, and several types of tropical forests (i.e., dry and montane tropical forests).

NUTRIENT ADDITIONS TO FOREST ECOSYSTEMS

Nutrient additions to forest ecosystems are relatively small compared to the annual requirements of actively-growing overstory trees, typically contributing 5 to 30 percent of the annual need for most plant nutrients. Nevertheless, input processes serve as important conduits by which nutrients flow into forests and other terrestrial ecosystems, and over relatively-long periods of time (100 to 200 years) they contribute significantly to the nutrient capital of an ecosystem. Nutrients enter terrestrial ecosystems through geological, hydrological, and biological processes. These are high-

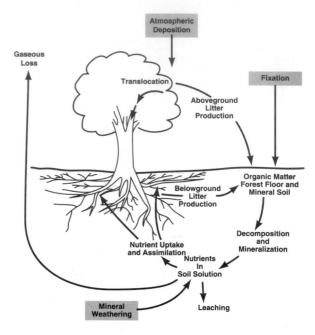

lighted in the reduced version of Figure 19.1, which lies next to this paragraph. In the paragraphs that follow, we discuss the magnitude of nutrient inputs to forest ecosystems from mineral weathering, atmospheric deposition, and biological fixation, the primary mechanisms by which nutrients flow into forest ecosystems.

Mineral Weathering

Plant nutrients contained in the chemical structure of rocks and soil minerals are largely unavailable to plants. Through physical abrasion and chemical dissolution, nutrients are slowly released from geologic materials into forms that plants take up from soil solution. The rate at which geologic materials weather to yield plant nutrients is strongly influenced by climate; temperature, and precipitation regimes directly control rates of chemical reactions in soil. Rates of mineral weathering are quite slow, and because not all geologic materials are chemically similar, they weather at different rates and yield different plant nutrients. With the exception of N, all other plant nutrients are found in the chemical structure of rocks, and therefore can be supplied to the plant from mineral weathering. In Chapter 11, we discussed mineral weathering as it pertains to soil horizon development, soil acidity, and base saturation. Here, we focus on aspects of mineral weathering that control the input of plant nutrients to forest ecosystems.

The most-well known attempt at quantifying mineral weathering in ecosystems began in 1963, when a team of scientists quantified the nutrient cycles of northern hardwood forests in the White Mountains of New Hampshire (Likens et al., 1977). In this region, a number of comparable watersheds (i.e., similar in size, geology, topography, soil parent material, and vegetation) are underlain by a layer of impermeable bedrock, preventing rainfall from reaching groundwater. Because these watersheds are sealed, streamflow represents the main process by which nutrients can exit. Therefore, the difference between the atmospheric deposition of nutrients and streamwater loss of nutrients should approximate the annual release of nutrients from mineral weathering. Using this approach, a large number of watershed-level studies of mineral weathering have been completed, allowing a comparison of weathering rates for different climates and parent materials. In areas dominated by silicate rocks, plant nutrients weather from soil parent materials in the following order:

$$Ca > Na > Mg > K > Fe$$

This sequence partially reflects mineral solubility and partially the degree to which these elements participate in the formation of secondary minerals.

The products of mineral weathering are particularly important to plant growth and the cycling of nutrients within terrestrial ecosystems. Table 19.1 summarizes nutrient inputs from mineral weathering for forest ecosystems occurring in different climates and on different parent materials. Notice that the weathering of dolomitic parent materials ($CaCO_3$ and $MgCO_3$) beneath limber and bristlecone pines yield relatively large amounts of Ca compared to weathering of adamellite (SiO_2:$NaAlSi_3O_8$:$CaAl_2Si_2O_8$). The weathering of Ca- and Mg-bearing minerals represents an important input of these nutrients in many terrestrial ecosystems, satisfying 35 percent of Ca and Mg required for the annual growth of some temperate forests (Table 19.1; Likens et al., 1977). Similarly, the weathering Ca-, Fe-, and Al-phosphate minerals provides an important source of P for plant growth. However, the situation is very different for other plant nutrients like N. With the exception of mica schist (Dahlgren, 1994), only a few geologic materials contain N. Consequently, mineral weathering is not an important process by which N enters terrestrial ecosystems.

As you will read later in this section, inputs from weathering are small relative to the nutrient demand of forest ecosystems that are accumulating biomass. Nonetheless, the weathering of nutrients from soil minerals represents an important source of nutrients over long periods of time (100 to 200 yrs).

Atmospheric Deposition

Over the past two decades, ecologists have become increasingly interested in understanding the geographic pattern and amount of nutrients entering terrestrial ecosystems from the atmosphere. In many portions of the Earth, human activity has altered rainfall chemistry and hence the atmospheric deposition of nutrients. In the northeastern U.S. and cen-

Table 19.1 Rates of mineral weathering for forest ecosystems occurring in different climates and on different soil parent materials.

ECOSYSTEM	Parent Material	Location	N	P	K	Ca	Mg
			kg ha^{-1} y^{-1}				
Wet Tropical	Alluvium	Venezuela	–	–	–	6	1
Douglas-fir	Volcanic Tuff	Oregon	–	–	2	47	12
Limber and	Dolomite	California	–	–	4	86	52
bristlecone	Adamellite	California	–	–	8	17	2
pines							
Trembling aspen	Glacial Till	Wisconsin	–	1	4	7	–
Northern hardwood	Glacial Till	New Hampshire	0	13	7	21	4

Source: Data have been summarized from Boyle and Ek (1973), Fredricksen (1972), Hase and Foelster (1983), Likens et al. (1977), Marchand (1971), and Wood et al. (1984).

tral Europe, for example, terrestrial and aquatic ecosystems receive enhanced inputs of NO_3^-, SO_4^{2-}, and H^+ (i.e., acidic deposition or acid rain) from the burning of fossil fuels. Both NO_3^- and SO_4^{2-} contain essential plant nutrients, and elevated inputs of these ions have the potential to alter the rate and pattern by which N and S are cycled and stored within terrestrial ecosystems.

Atmospheric deposition occurs through three separate processes: 1) **wet deposition**— the addition of nutrients or other materials contained in rain or snow, 2) **dry deposition**— the direct deposition of atmospheric particles and gases to vegetation, soil or water surfaces, and 3) **cloud deposition**—the input of small, non-precipitating water droplets (in clouds and fog) to terrestrial surfaces (Fowler, 1980; Lovett and Kinsman, 1990; Lovett 1994). Wet and dry deposition occur in all terrestrial ecosystems, but their importance as a source of nutrients varies greatly from place to place. In contrast, cloud deposition is generally restricted to coastal and mountainous regions that are frequently immersed in clouds or fog (Lovett, 1994).

As discussed in Chapter 7, wet deposition occurs when solid particles (0.2 to 2 mm diameter) and gases in the atmosphere dissolve in water droplets that fall to Earth as rain or snow (Lovett, 1994). This process is controlled by several factors that vary markedly from one region to another and result in a broad range of nutrient inputs from wet deposition. In North America, continental-scale patterns of wet deposition are relatively well understood (Figure 19.2). For example, anthropogenic emissions of N and S in the Midwest, Southeast, and Northeast result in widespread patterns of increased NO_3^- and SO_4^{2-} deposition over the eastern U.S. In contrast, NH_4^+ and Ca^{2+} inputs from wet deposition are relatively low in the eastern U.S., but are elevated in the eastern Great Plains region and the Midwest. The plowing of agricultural fields and traffic on unpaved roads causes Ca^{2+} associated with dust particles to be carried eastward from the Great Plains region by the prevailing winds. Ammonium (NH_4^+) released into the air from heavily-fertilized agricultural fields and animal manure is transported eastward in a similar manner.

Wet deposition entering forest ecosystems can pass through the canopy as throughfall or portions can flow down tree stems in stemflow. The chemical composition of wet deposition can be greatly altered by the "wash off" of dry material accumulated on plant surfaces, the leaching of nutrients from plant tissues, and the direct assimilation of nutrients into leaves. Potassium, Ca^{2+}, and Mg^{2+}, are easily leached from leaves (Tukey, 1970), and their concentrations in throughfall can be greater than that of the original wet deposition. This may be particularly important in tropical forests where species-specific differences in canopy leaching are related to the abundance of epiphytes (Schlesinger and Marks, 1977). In England, annual throughfall contains two- to eight-times more K^+, Ca^{2+}, and Mg^{2+} than wet deposition collected in adjacent, treeless openings (Madgwick and Ovington, 1959). In some cases, however, forest canopies directly assimilate water-soluble nutrients (i.e., N) from wet precipitation, thus lowering their concentration in throughfall (Olsen et al., 1981). Stemflow is significant in that it returns a relatively concentrated supply of nutrients directly to the base of the tree (Gersper and Holowaychuck, 1971).

Determining the extent to which the "wash off" of dry material and leaching influence the nutrient concentration of throughfall and stemflow can be difficult. However, it appears that 85 percent of the SO_4^{2-} in throughfall originates from the "wash off" of dry material deposited on leaves (Lindberg and Garten, 1988), making the dry deposition of nutrients an important process in forest ecosystems.

Dry deposition is a complex process involving atmospheric chemistry, wind velocity, and canopy characteristics like leaf shape, orientation, spatial arrangement, and area (Lovett, 1994). Solid particles (< 5 μm diameter) enter terrestrial ecosystems through gravitational sedimentation, the processes by which the force of gravity pulls particles

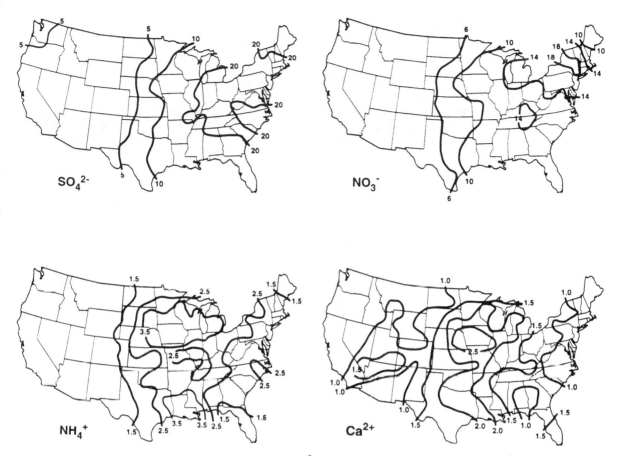

Figure 19.2. Geographic patterns of sulfate (SO_4^{2-}), nitrate (NO_3^-), ammonium (NH_4^+), and calcium (Ca^{2+}) in wet deposition across the continental U.S. Values are in kilograms per hectare per year ($kg\ ha^{-1}\ y^{-1}$). (Modified from Lovett, 1994. Reprinted with permission by the Ecological Society of America.)

suspended in the atmosphere toward the Earth's surface. Some particles are small enough to be kept aloft by wind turbulence and enter terrestrial ecosystems only when they impact the surface of vegetation. Even smaller dry particles and gases directly enter the plant through stomata and lenticles.

The movement of atmospheric particles toward leaf surfaces is determined by the change in particle concentration from the atmosphere to the leaf surface and the resistance to flow along that path (Garland, 1977). Atmospheric scientists estimate the input of particles to terrestrial ecosystems by determining their deposition velocity:

$$\text{Deposition velocity} = \frac{\text{Rate of Particle Dryfall (mg} \cdot \text{cm}^{-2}\ \text{of leaf} \cdot \text{sec}^{-1})}{\text{Concentration in Air (mg} \cdot \text{cm}^{-3})}$$

With knowledge of deposition velocity ($cm\ sec^{-1}$), atmospheric concentration ($mg\ cm^{-3}$), and leaf area ($m^2\ m^{-2}$), atmospheric scientists estimate the amount nutrient-containing dry particles entering a particular forest ecosystems. Total dry deposition ($mg\ m^{-2}$) is simply the mathematical product of atmospheric concentration, deposition velocity, and canopy leaf area.

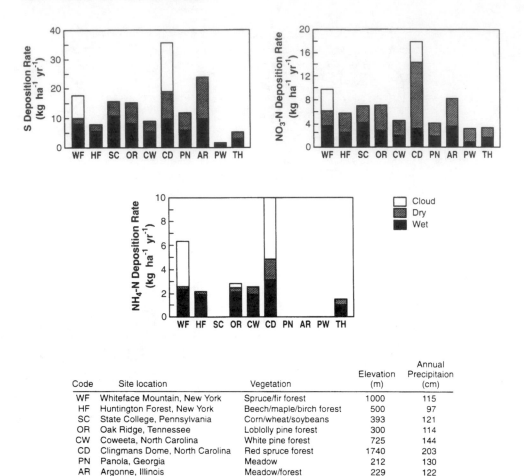

Figure 19.3. Atmospheric deposition of S, NO_3^--N, and NH_4^+-N in agricultural, prairie, and forest ecosystems in North America. The relative contribution of wet, dry, and cloud deposition is depicted. Note that cloud deposition is an important process in coastal areas and at high elevations. (Modified from Lovett, 1994. Reprinted with permission of the Ecological Society of Amrica.)

Dry deposition of nutrients is somewhat more variable than wet deposition for several reasons (Figure 19.3). The ratio of wet to dry deposition increases with distance from the source, because readily dry-deposited materials are depleted from the air as it travels downwind (Ollinger et al., 1993). Second, differences in canopy structure and leaf morphology among ecosystems cause nutrients in dry material to be captured with different efficiencies. Nonetheless, dry deposition can bring significant amounts of nutrients into forest ecosystems relative to those contained in wet deposition.

Figure 19.3 summarizes wet, dry, and cloud deposition for a series of forest, prairie, and agricultural ecosystems distributed across North America. Notice that the ratio of dry to wet deposition is highly variable among these ecosystems, reflecting differences in the distance to nutrient sources and canopy characteristics. The prairie ecosystem (AR) located downwind from Chicago, Illinois, receives relatively greater inputs of S and N in dry deposition than would be expected for an ecosystem in the Midwest (Figure 19.3) because it is relatively close to a major source of anthropogenic NO_3^- and SO_4^{2-} production.

Cloud deposition results when small, non-precipitating water droplets containing dissolved nutrients come in contact with terrestrial surfaces such as tree canopies (Chapter 7). In contrast to wet deposition, tree canopies exert a substantial influence on the amount of nutrients entering forest ecosystems from cloud deposition. This process is a particularly important nutrient input to ecosystems that lie in coastal and mountainous regions (WF and CD in Figure 19.3), whereas in continental areas (e.g., Midwest) cloud deposition contributes relatively minor amounts of nutrients to terrestrial ecosystems. For example, relatively large quantities of S, NO_3^-, and NH_4^+ enter spruce/fir ecosystems in the northeastern US and red spruce ecosystems at high elevation in the Appalachian Mountains through cloud deposition (Figure 19.3). Over 50 percent of the atmospheric deposition of NH_4^+ in these ecosystems occurs through cloud deposition.

Biological Fixation of Nitrogen

Biological inputs of N to forest ecosystems occur through symbiotic associations of soil bacteria and tree roots, free-living bacteria in litter and mineral soil, and epiphytic lichens living on the external surfaces of trees and decaying wood (Melillo, 1981; Waughman et al., 1981; Boring et al., 1988). Several genera of soil bacteria, actinomycetes, and cyanobacteria (blue-green algae) have the ability to reduce or "fix" atmospheric N_2 into NH_4^+. Depending on the fixation process, NH_4^+ can be directly used by the microbial cell or it can be transported from the microorganism to the host plant as N-containing organic compounds. The process of N_2 fixation is catalyzed by the nitrogenase enzyme system, which requires substantial quantities of energy (i.e., ATP). The following equation summarizes this process:

$$N_2 + 16\,ATP + 8\,e^- + 10\,H^+ \xrightarrow[\text{Nitrogenase}]{} 2\,NH_4^+ + H_2 + 16\,ADP + P_i$$

The high energy cost (i.e., ATP) associated with N_2 fixation has important ecological implications and gives rise to widely different rates of fixation depending on the fixation process and the microorganisms involved. Soil organic matter provides small amounts of energy for fixation by free-living soil bacteria, whereas carbohydrates from photosynthesis can fuel rapid rates of fixation by symbiotic bacteria inhabiting plant roots. As a result, quantities of N entering terrestrial ecosystems from N_2 fixation are highly variable, and range from as little as 1 kg N ha^{-1} y^{-1} for free-living soil bacteria to 200 kg N ha^{-1} y^{-1} from symbiotic fixation in the roots of legumes, alder and other higher plants (Cole and Rapp, 1981).

Rates of free-living N_2 fixation by free-living organisms are low and equivalent to rates of atmospheric deposition (Figures 19.2 and 19.3) in most terrestrial ecosystems. Free-living heterotropic bacteria of the genera *Azotobacter* and *Clostridium* fix atmospheric N_2 into forms that plants can assimilate after their cells die and cellular constituents are decomposed by other soil microorganisms. Because soil organic matter supplies the energy for free-living fixation, this process is greatest in soils with relatively high organic matter contents (Granhall, 1981). In the forests of the Pacific Northwest and northeastern United States, free-living N_2 fixation has been observed in decaying logs (Roskoski, 1980; Silvester et al., 1982), a relatively rich energy source for heterotropic bacteria. In most cases, the input of N to forest ecosystems from free-living N_2 fixation is relatively small, approximately 1 to 5 kg N $ha^{-1} y^{-1}$ (Boring et al. 1988).

Free-living cyanobacteria (blue-green algae) are common soil microorganisms that can, in some instances, be a source of N for some terrestrial ecosystems. Fixation by these photosynthetic microorganisms is associated with high-light environments, and in most forests fixation by cyanobacteria is limited by low light levels at the soil surface. However, algal crusts that form on the surface of some desert soils can have exceptionally-high rates of N_2 fixation (Rychert et al., 1978). Nitrogen-fixing cyanobacteria also enter into mutualistic associations with some fungi to form N_2-fixing lichens that bring N into forest ecosystems. *Nostoc,* an N_2-fixing cyanobacteria, is one of two species of algae present in the lichen *Lobaria oregana* that is found inhabiting the crowns of old-growth Douglas-fir trees. *Lobaria* is estimated to fix approximately 8 to 10 kg N ha^{-1} y^{-1}, an amount comparable to the input of N via atmospheric deposition (see Figure 19.3).

Nitrogen-fixing actinomycetes and unicellular bacteria can infect the root hairs of certain higher plants and induce the formation of root nodules, which are the location of symbiotic N_2 fixation. These small (< 5 mm), round, hollow structures are formed on individual roots and enclose large numbers of bacterial cells, the agents of N_2 fixation. Nitrogen fixed by the microbial symbiont is transferred to the plant, and carbohydrate supplied by the host in turn fuels the microbial fixation of N_2 in the root nodule. Because the cost of N_2 fixation is subsidized by a supply of carbohydrate from the plant, symbiotic fixation can bring substantial quantities of N into terrestrial ecosystems, far surpassing fixation by free-living bacteria and cyanobacteria.

Legumes are perhaps the most important plants with N_2-fixing bacteria inhabiting their roots. They occur as overstory species in both temperate and tropical forests. However, not all legumes have the ability to fix N_2 in association with *Rhizobium.* Black locust, honey locust, acacias, and mesquite are trees of the legume family that occur in temperate forests; there are hundreds of similar species in the tropics. Also many herbs and shrubs of this family inhabit the forest floor throughout much of the Earth, many becoming dominant following forest harvesting or fire. Soil bacteria belonging to the *Rhizobium* genus exclusively infect the roots of legumes and enter into a symbiotic N-fixing relationship.

In the southeastern United States, black locust readily establishes following a major disturbance, like fire or clearcut harvest. Fixation in the root nodules of this leguminous tree represents a substantial input of N to the forests of the southern Appalachians, particularly during the early to intermediate stages of ecosystem development (Waide et al., 1988). Four years following clearcut harvest of black locust, rates of N_2 fixation equaled 48 kg N ha^{-1} y^{-1}, a substantial input of N relative to atmospheric inputs (see Figures 19.2 and 19.3). Rates subsequently increased to 75 kg N ha^{-1} y^{-1} at 17 years following harvest, and after 38 years of ecosystem development rates declined to 33 kg N ha^{-1} y^{-1} (Waide et al., 1988).

Actinomycetes (filimentous bacteria) of the genus *Frankia* form a symbiotic relationship with the roots of *Alnus, Ceanothus, Casuarina, Eleagnus, Comptonia,* and *Myrica.* This symbiotic relationship is particularly important in the Pacific Northwest, where N_2 fixation by red alder and *Frankia* can substantially increase the amount of N entering forest ecosystems. One example comes from a comparison of 38-year-old red alder and Douglas-fir ecosystems that initially established adjacent to one another on the same soil type. Differences in N pools between these ecosystems resulted directly from N_2 fixation by the alder–*Frankia* symbiosis (Table 19.2). An additional 85 kg N ha^{-1} y^{-1} have accumulated in the red alder–dominated ecosystem, an input of N far exceeding atmospheric deposition (see Figure 19.3). Much of the N_2 fixed by the red alder–*Frankia* resides in forest floor and mineral soil, where it is eventually released during organic matter decomposition. The shrub *Ceonothus velutinous* also occurs in the Pacific and Inland Northwest and enters into an N_2-fixing symbiosis with *Frankia* that can contribute as much as 100 kg N ha^{-1} y^{-1} (Youngberg and Wollum, 1976).

Greater N availability resulting from N_2 fixation on N-poor soils can dramatically increase rates of biomass accumulation and nutrient cycling during ecosystem development. In the Hawaiian Islands, *Myrica faya* is an exotic N_2-fixing tree that has greatly

Table 19.2 The accumulation of N in forest ecosystems dominated by nitrogen-fixing (red alder) and non-nitrogen fixing (Douglas-fir) tree species. The overstory of each ecosystem is 38-years-old and differences in the amount of N contained in each ecosystem pool reflects the fixation of nitrogen by red alder.

ECOSYSTEM POOL	NITROGEN Douglas-fir	Red Alder	Annual Accumulation
	——— kg N ha^{-1} ———		kg N ha^{-1} y^{-1}
Overstory	320	590	7.1
Understory	10	100	2.4
Forest Floor	180	880	18.4
Mineral Soil	3270	5450	57.4
Total Ecosystem	3780	7020	85.3

Source: After Cole and Rapp, 1981. Reprinted from *Dynamic Properties of Forest Ecosystems,* © 1981 by Cambridge University Press. Reprinted with the permission of Cambridge University Press.

altered the N cycle of native forests (Vitousek et al., 1987). This invasive, early-successional species establishes on volcanic ash flows that were formerly colonized by *Metrosideros polymorpha*, a native, non-N$_2$ fixing tree. In the absence of *Myrica faya*, N accumulates in ash-flow soils at a very slow rate, constraining rates of ecosystem development and productivity (ca. 5 kg N ha^{-1} y^{-1} from free-living fixation and atmospheric deposition). Nitrogen fixation by *Myrica faya* brings an additional 18 kg N ha^{-1} y^{-1} into the ecosystem, increasing the annual rate of N accumulation by a factor of three. The N$_2$ fixed by *Myrica* eventually enters the soil via leaf and root litter, and during decomposition relatively greater amounts of N are released into soil solution where they are assimilated by plant roots.

NUTRIENT CYCLING WITHIN FOREST ECOSYSTEMS

Nutrients entering forest ecosystems from mineral weathering, atmospheric deposition, and biological fixation can enter soil solution where they are absorbed by plant roots. Within the plant, absorbed nutrients participate in a wide array of physiological processes, and in some cases, nutrients are mobilized (i.e., translocated) prior to the shedding of some plant tissues. The production of plant litter above and below the soil surface eventually returns nutrients to forest floor and mineral soil, where they are decomposed by soil microorganisms. During

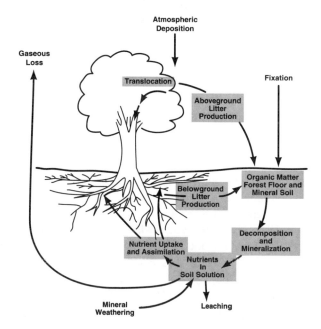

the process of litter decomposition, soil microorganisms incorporate organically-bound nutrients (and C) into their biomass and release excess nutrients into soil solution where they can again be absorbed by plant roots. Thus the rate at which nutrients flow within forest ecosystems is controlled by the physiological activities of plants and soil microorganisms, and their requirement for growth-limiting nutrients.

In the sections that follow, we trace the flow of nutrients within forest ecosystems. The cycling of nutrients within ecosystems occurs through: 1) root nutrient absorption via the processes of uptake and assimilation, 2) nutrient allocation to biomass construction and maintenance, 3) nutrient translocation from senescent tissue, 4) the return of nutrients in above- and belowground litter, and 5) the microbially-mediated release of inorganic nutrients into soils solution (i.e., mineralization) during organic matter decomposition. These processes are highlighted in the reduction of Figure 19.1, which is located on the previous page. Central to this discussion is the ability of plants to forage for nutrients in soil solution, a process mediated by physiological and morphological adaptations of plant root systems. Also of importance is the growth dynamics of soil microorganisms during litter decomposition, because they control the release of organically-bound nutrients from plant litter. As you will learn latter in this chapter, soil microorganisms are the "gate keepers" of N availability within forest ecosystems. We emphasize the belowground processes of plants and soil microorganisms that control the flow of nutrients within terrestrial ecosystems.

Nutrient Transport to Roots

Nutrients must move to the root surface before they can be taken up and incorporated into biologically-active compounds. The uptake of nutrients by plant roots is initially constrained by rates of organic matter decomposition, mineral solubility, and cation/anion exchange reactions. Although these processes control nutrient availability in soil, the fact that nutrients are present in soil solution or on exchange sites does not ensure uptake by plant roots. In the paragraphs that follow, we discuss the processes controlling the flow of nutrients to the root surface, uptake from soil solution, and their incorporation into biologically-active compounds within the plant.

Nutrients move toward root surfaces in response to two processes: **mass flow** and **diffusion**. Mass flow is a passive process in which ions move with the flow of water for transpiration. Diffusion occurs when ions move from a region of high concentration to one of low concentration, e.g., from soil solution to the root surface. In most soils, the concentration of nutrients in solution is low enough that mass flow does not meet the demand of actively-growing plants. Rapid uptake by plant roots also causes some nutrients to be virtually absent near their surface, thus forming zones of depletion around individual roots. Consequently, the supply of nutrients to roots is constrained by the rate at which ions diffuse toward root surfaces from areas of higher concentration in soil (Nye, 1977). Nutrients like PO_4^{3-}, K^+, and NH_4^+ diffuse slowly in water and rapid uptake depletes their concentration near root surfaces, so that concentrations increase as the distance from the root surface increases. Although NO_3^- is very mobile in soil solution and diffuses rapidly toward roots, a high demand by most plants maintains low NO_3^- concentrations throughout soil solution. Because of its high rate of diffusion, mass flow may meet the NO_3^- demand of some forest trees (Marchner et al., 1991).

The size of the zone of depletion surrounding an individual root is directly proportional to its radius and the rate at which a particular ion diffuses in water. The radius of the zone of depletion ($R_{depletion}$ in cm) for any ion is:

$$R_{depletion} = a + 2\sqrt{Dt}$$

where; a is the root radius (cm), D is the diffusion coefficient for an ion ($cm^2 sec^{-1}$) and t is time in seconds (Nye and Tinker, 1977). One also can compare root size (cm^3) to the zone of depletion (cm^3) it creates using the following expression:

$$\left[a + 2\sqrt{Dt}\right]^2/a^2$$

Given this relationship, smaller roots exploit a greater volume of soil than larger roots (Fitter, 1987). Recall that large roots have higher construction and maintenance costs than small roots, indicating that plants "invest" relatively greater amounts of photosynthate into the production of large-diameter roots (Chapter 18). By forming small-diameter roots, plants increase the volume of soil exploited for water and nutrients while reducing the amount of C allocated to root production. It is likely that this is why roots generally less than 1 mm in diameter are the nutrient and water absorbing organs of forest trees.

Once nutrients arrive at the root surface they must be taken up from soil solution and incorporated into biologically-active compounds. In some soils, the supply of some nutrients is excessive, and they are actively excluded from uptake. It is not uncommon to observe accumulations of $CaCO_3$ adjacent to the roots of desert shrubs growing on highly calcareous soils (Klappa, 1980). In most instances, however, nutrient supply is low (especially for N and P), and plants actively forage for nutrients in soil. Nutrient foraging occurs through the physiological mechanisms of nutrient absorption and through morphological changes in root architecture that increase the absorptive area of plant root systems.

Nutrient Uptake and Assimilation by Roots

Nutrients are incorporated into biologically-active compounds in a two-step process consisting of uptake and assimilation. **Uptake** is the physiological process by which nutrients in soil solution are actively transported across cell membranes. **Assimilation** occurs when inorganic nutrients transported into plant cells are biochemically incorporated into organic compounds such as amino acids, nucleic acids, lipids or other biologically-active compounds.

High rates of enzymatic uptake are one physiological means by which plants can maximize acquisition of growth-limiting nutrients. For example, the uptake of nutrients from soil solution is mediated by enzymes associated with the membranes of very-fine roots (<0.5 mm diameter). The rate of activity of enzymes involved with nutrient uptake increases with increasing nutrient concentrations in soil solution until the capacity of the enzyme system is saturated. Uptake capacity for nutrients varies widely among tree species, and this process can be highly responsive to soil temperature (Figure 19.4). In a comparison of trees of the taiga, rapidly-growing species like balsam poplar and trembling aspen have relatively-high uptake capacities for NH_4^+, NO_3^-, and PO_4^{3-}, compared to the slower-growing paper birch and green alder (Chapin et al., 1986). Balsam poplar and trembling aspen also occur on relatively-warm soils in the taiga and exhibit rapid increases in nutrient uptake with rising soil temperature (Figure 19.4).

Although the productivity of many temperate forests is constrained by soil N availability, we understand relatively little regarding the ecological importance of NH_4^+ versus NO_3^- uptake by overstory trees. Ammonium (NH_4^+) often is the dominant form of N in forest soils, and some trees exhibit a physiological preference for NH_4^+ over NO_3^-. This relationship can be observed in Figure 19.4 for taiga trees, and it also has been documented for several overstory trees in temperate forests (Figure 19.5). In the fine roots of Douglas-fir and sugar maple, rates of NH_4^+ uptake far exceed rates of NO_3^- uptake, suggesting that these widely-distributed trees have a physiological preference for NH_4^+.

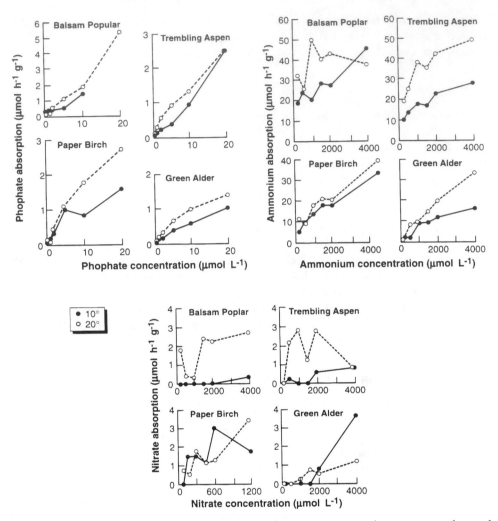

Figure 19.4. Nutrient uptake by four taiga trees in response to nutrient concentration and soil temperature. Uptake rates of NH_4^+, NO_3^-, and PO_4^{3-} were measured on excised roots of seedling growing under laboratory conditions. (Modified from Chapin et al., 1986. Reprinted from *Oecologia,* © Springer-Verlag Berlin, Heidelberg 1986. Reprinted with permission of Springer-Verlag, New York, Inc.)

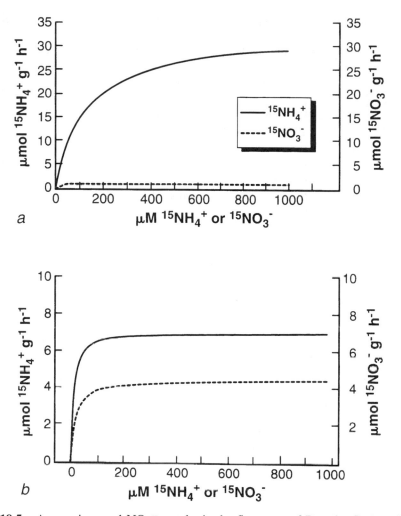

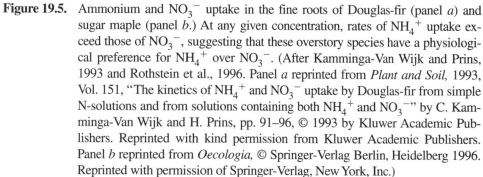

Figure 19.5. Ammonium and NO_3^- uptake in the fine roots of Douglas-fir (panel *a*) and sugar maple (panel *b*.) At any given concentration, rates of NH_4^+ uptake exceed those of NO_3^-, suggesting that these overstory species have a physiological preference for NH_4^+ over NO_3^-. (After Kamminga-Van Wijk and Prins, 1993 and Rothstein et al., 1996. Panel *a* reprinted from *Plant and Soil,* 1993, Vol. 151, "The kinetics of NH_4^+ and NO_3^- uptake by Douglas-fir from simple N-solutions and from solutions containing both NH_4^+ and NO_3^-" by C. Kamminga-Van Wijk and H. Prins, pp. 91–96, © 1993 by Kluwer Academic Publishers. Reprinted with kind permission from Kluwer Academic Publishers. Panel *b* reprinted from *Oecologia,* © Springer-Verlag Berlin, Heidelberg 1996. Reprinted with permission of Springer-Verlag, New York, Inc.)

Western hemlock and jack pine commonly occur on acidic soils with extremely low NO_3^- availability. These overstory species also have a limited capacity for NO_3^- uptake and appear to satisfy their demand for N through rapid NH_4^+ uptake (Lavoie et al., 1992; Knoepp et al., 1993). In jack pine, NO_3^- uptake enzymes are present at very low levels in fine roots (Lavoie et al., 1992), and when seedlings are supplied with NH_4^+ or NO_3^- as a sole source of N, those supplied with NH_4^+ attain twice the biomass of seedling grown on NO_3^-. (Lavoie et al., 1992). This observation suggests that jack pine has a low physiological capacity for NO_3^- uptake, even when concentrations in soil solution are relatively high. The inability of this species to use NO_3^- suggests that it would be a poor competitor in NO_3^- rich soils, relative to plants that have rapid rates of NO_3^- uptake and assimilation.

There are several reasons why NH_4^+ uptake is more rapid than NO_3^- uptake in many forest trees. First, NH_4^+ is often the dominant form of N in soil solution and small amounts can strongly inhibit enzymes involved with NO_3^- uptake and assimilation. Secondly, plants must reduce NO_3^- to NH_4^+ before it can be assimilated into biologically-active compounds, a process that requires substantial amounts of energy. When this reaction occurs in leaves, it is subsidized by energy from the light-harvesting reactions of photosynthesis. However, many woody plants have the ability to reduce NO_3^- in roots, which requires the transport of reducing compounds from the leaf. Most often, plants with rapid rates of NO_3^- uptake and assimilation occur in high-light habitats where excess energy from photosynthesis can be allocated to NO_3^- reduction.

A second physiological mechanism for maximizing the uptake of limiting nutrients is via the production of root enzymes that release nutrients from soil organic matter. This is particularly true of PO_4^{3-}, which diffuses slowly in soil solution. Phosphatases are enzymes produced by plants and microorganisms that act upon soil organic matter to yield PO_4^{3-} from organically-bound P. The release of these enzymes into soil appears to be an important mechanism for increasing PO_4^{3-} supply to roots in PO_4^{3-}-poor soils. Phosphatase activity associated with the roots of arctic tundra plants can supply up to 65 percent of their annual PO_4^{3-} demand (Kroehler and Linkins, 1988).

Root Architecture, Mycorrhizae, and Nutrient Acquisition

ROOT ARCHITECTURE

In addition to physiological mechanisms, plants also forage for nutrients by altering root architecture and growth. Ecologists have become increasingly interested in the structure and function of plant root systems, because plants appear to alter root branching patterns, or architecture, in response to soil nutrient availability (Fitter, 1987; Fitter and Stickland, 1991). Several analyses suggest that fine-root systems consisting of only a main axis and primarily laterals (herringbone architecture; Figure 19.6) are the most efficient at nutrient acquisition (Fitter and Stickland, 1991; Berntson and Woodward, 1992). The herringbone arrangement maximizes the volume of soil exploited by roots, while minimizing the overlap between the zone of depletion created by adjacent roots; however, the construction and maintenance cost of this type of root system is high. It requires the production of larger diameter roots (e.g., higher construction and maintenance respiration), each of which transports a greater proportion of the total flow of water and nutrients into the aboveground portion of the plant (Figure 19.6). In contrast, the mean root diameter of a dichotomously-branched root system is much smaller than that of a herringbone architecture, resulting in relatively-lower construction and maintenance costs. This spatial arrangement of roots allows the depletion zones of adjacent roots to overlap, and thus it is less efficient at foraging for nutrients than the herringbone architecture.

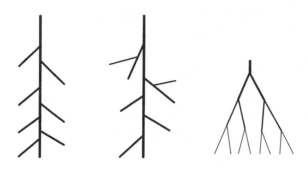

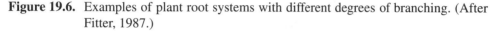

Herringbone   **Intermediate**   **Dichotomous**

Figure 19.6. Examples of plant root systems with different degrees of branching. (After Fitter, 1987.)

Differences in the cost of constructing and maintaining these root systems, and the efficiency with which they explore soil for nutrients, suggest that root-system architecture should vary in response to soil nutrient availability. Plants growing in nutrient-poor soils should allocate relatively more C to a highly efficient root system for nutrient uptake (i.e., herringbone branching pattern), whereas those growing in nutrient-rich soil should invest relatively less C into a dichotomously-branched root architecture. These predictions hold for different herbaceous plants (dicots) from nutrient-rich and nutrient-poor soil, and also hold for the same species of herbaceous plants grown in soils of different fertility (Fitter and Stickland, 1991; Berntson and Woodward, 1992; Taub and Goldberg, 1996). The fine-root system of forest trees appears to respond in a similar manner, but unfortunately few studies have focused on these organisms. The fine roots of pin cherry rapidly proliferate within experimental patches (125 cm^3) of N-rich soil by increasing rates of production and the degree of branching, i.e., their fine-root system became more dichotomous (Pregitzer et al., 1992).

MYCORRHIZAE

Forest trees also increase their ability to forage for nutrients in soil by entering into a symbiotic relationship with soil fungi. This form of mutualism, termed a **mycorrhiza** or literally a *root-fungus*, increases the volume of soil exploited by an individual plant. In Chapter 15, we discussed the importance of mycorrhizal fungi for the successful establishment of some tree seedlings. Mycorrhizae facilitate an important nutrient cycling process—plant nutrient uptake. In this section, we further explore the extent to which mycorrhizal fungi increase the ability of plant roots to forage for nutrients within soil.

Root hairs are not common in nature, because tree roots are almost universally colonized by mycorrhizal fungi. In pines, spruces, firs (and all other genera of the Pinaceae), birches, beeches, oak, basswoods, and willows, ectomycorrhizal fungi (*ecto* meaning outside) form a sheath or mantle surrounding fine roots giving them a characteristic swollen appearance. Fungal hyphae penetrate the space between the outer cortical cells, but do not enter individual root cells. Over 2,400 species of fungi are known to form ectomycorrhizae on North America trees (Marx and Beattie, 1977). A less conspicuous group, vesicular-arbuscular mycorrhizae (VAM) form no sheath, but individual hyphae grow within and between epidermal and cortical cells of roots. Many plant species form VAM, including cultivated crops, grasses, and most tropical tree species (Janos, 1987). Temperate species forming VAM include redwood and many hardwoods: maples, ashes, tuliptree, sweet gum, sycamore, black walnut, and black cherry (Marx and Beattie, 1977; Harley and Smith, 1983).

Ectomycorrhizae and VAM are effective accumulators of nutrients (and water) because they increase the absorbing surface area of tree roots, thus allowing the plant to forage for nutrients in a greater volume of soil. The hyphae of mycorrhizae can extend great distances into the forest floor and mineral soil. For example, 1 mm^3 of soil can contain 4 m of hyphae that transport water and nutrients back to the host plant. In a 450-year-old Douglas-fir ecosystem in Oregon, 5000 kg ha^{-1} of ectomycorrhizae occur in the surface soil (0–10 cm), composing over 11 percent of the total root biomass (Trappe and Fogel, 1977). Moreover, a single Douglas-fir-*Cenococcum* ectomycorrhizae can form 200 to 2,000 individual hyphae, some of which extend over 2 m into soil and form more than 120 lateral branches or fusions with other hyphae.

As compared to non-mycorrhizal roots, those infected with mycorrhizal fungi tend to be more metabolically active, exhibiting rapid rates of respiration and ion uptake. Mycorrhizae are widely known to increase the PO_4^{3-} uptake of forest trees growing in P-poor soils; however, mycorrhizae also facilitate the uptake of other plant nutrients like N (Bowen and Smith, 1981). The mycorrhizae formed by the fungus *Paxillus involutus* are able to take up substantial amounts of NH_4^+ from soil solution, assimilate the NH_4^+ into amino acids, and transfer the newly-formed amino acids to European beech (Finlay et al., 1989). *Paxillus involutus* also is able to assimilate NO_3^-, but it does so at a substantially lower rate. In addition to the direct uptake of ions from soil solution, mycorrhizae have the ability to produce extracellular enzymes (e.g., cellulases and phosphatases) which aid organic matter decomposition (Antibus et al., 1981; Dodd et al., 1987) and also produce organic acids which release plant nutrients from soil minerals (Bolan et al., 1984).

Because mycorrhizal fungi are heterotrophic microorganisms, they require a supply of carbohydrate for growth, for uptake of nutrients from soil, and for translocation of nutrients to the host plant. The cost of forming mycorrhizae is not inconsequential. Carbohydrate formed by the plant that could otherwise be used for growth and maintenance must be transferred to the mycorrhizal fungi in return for a greater nutrient supply. In a Pacific silver fir ecosystem, mycorrhizae compose only 1 percent of total ecosystem biomass but their growth and maintenance consume over 15 percent of net primary production (Vogt et al., 1983). Because of the high C cost of forming mycorrhizae, plants growing in nutrient-poor soil tend to form more mycorrhizae than those growing in nutrient rich-soils. Thus mycorrhizae facilitate nutrient acquisition by plants, which in turn can influence the return of nutrients to soil in plant litter.

Plant Litter and the Return of Nutrients to Forest Floor and Soil

Plant litter production above and below the soil surface directly controls the amount of nutrients returned to the forest floor and mineral soil, and therefore constitutes important processes controlling the cycling of nutrients within forest ecosystems. Nutrients absorbed by roots and mycorrhizae are used for a wide variety of physiological functions, including the growth of new tissue, the maintenance of existing tissue, storage, and the production of defense compounds—fates similar to that of fixed C (Chapter 18). Prior to abscission, both coniferous and deciduous trees withdraw nutrients from leaf tissue and transport them into the nearby branches or bole for storage. However, it is not clear whether nutrients are withdrawn from fine roots prior to their death. Consequently, both the production of litter and the concentration of nutrients contained within it control the rate at which nutrients absorbed by plants are returned to the soil from which they came. In the paragraphs that follow, we explore geographic patterns of above- and belowground plant litter production and the extent to which plants withdraw nutrients prior to leaf abscission, a process controlling the nutrient concentration of plant litter.

LEAF AND ROOT LITTER PRODUCTION

On a global basis, leaves account for approximately 60 to 75 percent of total aboveground litterfall in forest ecosystems; the remaining proportion is composed of woody material (ca. 30 percent) and reproductive structures (1 to 20 percent). In Chapter 18, we found that global patterns of aboveground net primary production were strongly related to climate, particularly temperature and actual evapotranspiration. Leaf litter production in forest ecosystems also is strongly related to global patterns of climate. In Figure 19.7, leaf litter production is low at high latitudes where short growing seasons limit plant growth; it increases toward the equator where plant growth can occur throughout the entire year. Also notice that substantial variation occurs at any particular latitude (Figure 19.7). This regional-scale variability in leaf litter production undoubtedly results from the modification of climate by physiography (i.e., slope and aspect), differences in soil water and nutrient availability, or disturbance.

Recall that leaf biomass (or area) places the upper limit on the amount of photosynthetically-fixed C available for net primary productivity (Chapter 18). As a result, leaf biomass and aboveground net primary productivity (ANPP) are positively related across a

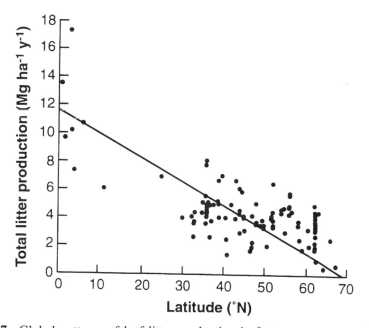

Figure 19.7. Global patterns of leaf litter production in forest ecosystems. (After Bray and Gorham, 1964 and O'Neill and DeAngelis, 1981. Modified from *Advances in Ecological Research*, ©, 1964 by Academic Press, Ltd. London. Reprinted with permission of Academic Press, Ltd. London and from *Dynamic Properties of Forest Ecosystems*, © Cambridge University Press 1981. Reprinted with permission of Cambridge University Press.)

wide array of terrestrial ecosystems, including deserts, grasslands, and forests (Webb et al., 1983). Because leaves are temporary plant tissues, even in coniferous forests, one would expect that patterns of leaf litterfall should be related to ANPP. The solid line in Figure 19.8 indicates a 1:1 relationship between ANPP and leaf litter production, wherein all ANPP is allocated to leaf production. At low values of ANPP (i.e., <7.5 Mg ha^{-1} y^{-1}), note that leaf litter production represents a large proportion of ANPP. At higher rates of ANPP, leaf litter constitutes a smaller proportion, reflecting a greater allocation of C to other plant tissues.

Although we have a clear understanding of global patterns of leaf litterfall, our knowledge of the production of plant litter below the soil surface is incomplete. It is clear, however, that the death of fine roots represents a significant proportion of net primary productivity (NPP) in many forest ecosystems, often exceeding leaf litter production. Notice that fine-root litter represents 40 to 330 percent of leaf litter in the temperate coniferous and deciduous forests summarized in Table 19.3. In the Pacific Northwest, the production and death of roots can account for 59 to 67 percent of NPP in Pacific silver fir ecosystems,

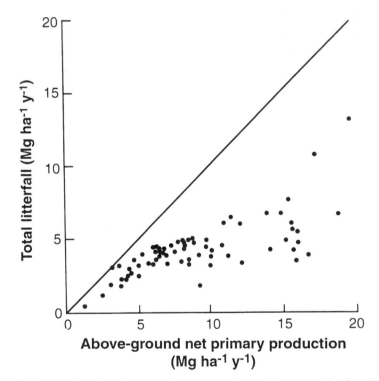

Figure 19.8. The relationship between aboveground net primary production (ANPP) and aboveground litter production in temperate and boreal forest ecosystems. The solid line indicates a 1:1 relationship between ANPP and the production of aboveground litter. At low levels of ANPP, virtually all production is allocated to leaves, and the departure from the 1:1 line at high levels of ANPP indicates a greater allocation of production to non-photosynthetic tissues. (After O'Neill and DeAngelis, 1981. Reprinted from *Dynamic Properties of Forest Ecosystems,* © Cambridge University Press 1981. Reprinted with permission of Cambridge University Press.)

Table 19.3   Leaf- and root-litter production in forest ecosystems from different geographic locations in temperate North America. The ratios of fine-root litter to leaf litter indicate that the production of fine-root litter in forest ecosystems ranges from 40 to 330 percent of leaf litter.

| | LITTER | | |
| | Leaf | Fine-Root | Root:Leaf |
ECOSYSTEM	—— $Mg\ ha^{-1}\ y^{-1}$ ——		
Temperate Coniferous			
Pacific silver fir	1.5–2.2	4.4–7.4	2.9–3.3
Douglas-fir	2.8–3.0	2.4–2.7	0.9
Scots pine	2.1	2.0	1.0
white pine	2.9	2.6	0.9
mixed pine	3.1	2.6	0.8
white spruce	2.7	1.6	0.6
red pine	2.5–5.3	1.3–2.0	0.4–0.5
Temperate Deciduous			
black oak	4.1	5.9	1.4
northern red oak	4.2	5.2	1.2
white oak	3.6	4.1	1.1
sugar maple	2.9	4.0	1.4
paper birch	2.8	3.2	1.1
mixed hardwoods	4.4	2.7	0.6
American beech	3.2	1.3	0.4

Source: After Vogt et al., 1986 and Nadelhoffer et al., 1985. Modified from *Advances in Ecological Research*, © 1986 by Academic Press, Inc. Reprinted with permission of Academic Press, Inc. and from *Ecology* by the permission of the Ecological Society of America.

whereas leaf litter production ranges from 6 to 8 percent of NPP (Vogt et al., 1983). Fine roots often have nutrient contents (N and P) greater than or equivalent to leaf litter, suggesting that fine-root litter represents a substantial transfer of nutrients from plant tissue to forest floor and mineral soil.

Recall that the return of nutrients to the forest floor and mineral soil is controlled by the production of plant litter and its nutrient concentration. From the previous paragraphs, it should be clear that forest ecosystems dramatically differ in the production of litter, both above and below the soil surface. In the following paragraphs, we discuss the withdrawal, or **retranslocation**, of nutrients prior to the shedding of plant litter. This process directly influences the concentration (i.e., percent) of nutrients in litter, which together with litter production, control the total return of nutrients in plant litterfall.

NUTRIENT RETRANSLOCATION

The biochemical C-fixing machinery of leaves requires substantial quantities of N and P, as do the nutrient uptake and assimilation mechanisms of fine roots. As a result, green leaves and live fine roots contain relatively high concentrations of nutrients compared to other plant tissues, like stems and branches. Because low quantities of N and P in soil often limit plant growth, plants have evolved mechanisms to conserve growth-limiting nutrients within their tissues. As noted above, in both coniferous and deciduous trees substantial quantities of N and P (and other nutrients) are retranslocated from leaves prior to abscission and are mobilized into storage. Much of the N and P removed from leaves resides within adjacent branches, bole, and large structural roots, where they can be remobilized to meet the demand of newly forming leaves when growth resumes. However, the extent to which forest trees retranslocate nutrients from senescent leaves is highly variable (Killingbeck, 1996); it changes from year to year within an individual species and changes among species from different habitats.

The retranslocation of many nutrients reflects their availability in soil and their demand within the plant for subsequent growth. In Brazilian rainforest trees, 17 to 73 percent of N and 41 to 83 percent of P are translocated from leaves prior to abscission. Although retranslocation efficiency (i.e., percent removed) varies among rainforest species, they all reduce P to similar concentrations in abscised leaves (i.e., 0.04 to 0.06 percent P; Scott et al., 1992). The retranslocation of N is not as complete and relatively large amounts of N (>1.0 percent) remain in abscised leaves (Scott et al., 1992). Differences in the amount of P and N remaining in abscised leaves of rainforest trees reflects the fact that P availability in soil, not N, limits the productivity of wet tropical forests (Vitousek, 1984).

The extent to which forest trees retranslocate nutrients from senescing leaves directly influences the **nutrient-use efficiency** of litter, which is defined by the following expression:

$$\text{Nutrient use-efficiency} = \frac{\text{Dry Weight of Leaf Litterfall (kg ha}^{-1}\text{ y}^{-1})}{\text{Nutrient Content of Leaf Litterfall (kg ha}^{-1}\text{ y}^{-1})}$$

Plants that retranslocate large quantities of nutrients prior to leaf abscission have high nutrient-use efficiencies and return relatively few nutrients to the forest floor in litterfall.

On a global basis, there is a remarkable relationship between the N- and P-use efficiency and nutrient availability in tropical, temperate, and boreal forests. In Figure 19.9, the amount of nutrients returned in litterfall (x axis) is used as a surrogate for nutrient availability—large amounts of nutrients are contained in litterfall where their availability in soil is high. Notice that N use-efficiency is inversely related to the amount of N contained in litterfall, with the exception of tropical forests (Figure 19.9*a*). The inverse relationship between N-use efficiency and litterfall N in temperate coniferous and deciduous forests indicates that N-use efficiency is high in ecosystems where relatively small quantities of N are annually cycled (i.e., those with low litterfall N contents).

In tropical ecosystems, litterfall N varies widely (10 to 180 kg N ha^{-1} y^{-1}), but N-use efficiency is relatively constant (ca. 80; Figure 19.9*a*). The fact that N-use efficiency changes little while the amount of N annually cycled to the forest floor varies by a factor of three indicates that N availability does not influence the N-use efficiency of tropical forest trees. In contrast to this pattern, P-use efficiency and the amount of P returned in litterfall are inversely related in tropical forests (Fig 19.9*b*). This pattern suggests P translocation from senescent leaves is high (i.e., high P-use efficiency) where P availability is relatively low within the ecosystem. Also notice that there is a weak relationship between P-use efficiency and litterfall P in other forest ecosystems. These patterns are consistent with the observation that N generally limits the productivity of boreal and temperate forests, whereas a limited supply of P constrains productivity in wet tropical forests.

Taken together, these observations suggest that limited supplies of N or P result in greater translocation prior to leaf abscission, thus increasing N- or P-nutrient-use efficiency of leaf litter. This is likely an evolved mechanism that permits plants growing in low N or P soils to conserve these growth-limiting nutrients within their tissue. Forest trees that conserve nutrients through a high nutrient-use efficiency should have a competi-

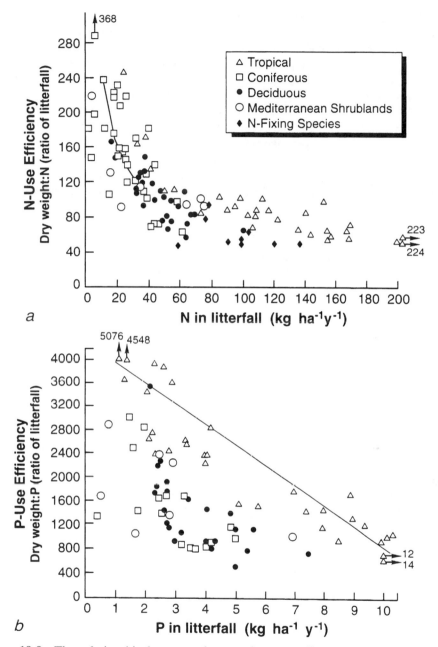

Figure 19.9. The relationship between plant nutrient-use efficiency and the amount of nutrients returned in leaf litterfall for tropical, coniferous, temperate deciduous, and Mediterranean shrublands. Forests dominated by N-fixing plants are denoted with a diamond. Plant nutrient-use efficiency is estimated as the ratio of litterfall mass to its N (panel *a*) or P (panel *b*) content. The amount of N and P returned in litterfall (kg N or P ha^{-1} y^{-1}) is a relative index of soil N or P availability. (After Vitousek, 1982. Reprinted with permission from *American Naturalist,* © 1982 by the University of Chicago. Reprinted with permission of the Chicago University Press.)

tive advantage on nutrient-poor sites, and it is not uncommon to observe species replacement along gradients of nutrient availability (Pastor et al., 1984; Zak and Pregitzer, 1990). Individual species also are able to alter nutrient-use efficiency in response to nutrient supply, which may allow them to persist in soils of markedly different fertility. Northern red oak, for example, occurs on soils with a wide range of N availability in the northern Lake States; N-use efficiency in this species drops as soil N availability increases; similar responses have been observed in sugar maple, basswood, and American beech (Zak et al., 1986). Although fine-root mortality represents a significant input of litter to forest floor and mineral soil, we do not yet understand the extent to which soil nutrient availability influences the nutrient-use efficiency of fine roots.

The combined influence of plant litter production and nutrient-use efficiency on the amount of N returned to forest floor and mineral soil is summarized in Table 19.4. Notice that N inputs from fine-root litter are equivalent to or greater than the amount of N contained in leaf litter, regardless of whether forests occur in tropical, temperate, and boreal regions (i.e., root:leaf N ratio >1). The combined amounts of N entering forest floor and mineral soil are generally greatest in tropical forests. Recall rates of N input from atmospheric deposition (see Figure 19.3) and free-living fixation (*ca.* 1 to 5 kg N ha^{-1} y^{-1}) and realize that leaves and fine roots annually contribute 10 times more N to forest floor and mineral soil. Thus above- and belowground litter production and its nutrient-use effi-

Table 19.4   The nitrogen content of leaf and fine root litter in forest ecosystems from different geographic locations.

ECOSYSTEM	Leaf	Fine-Root	Root:Leaf N
	LITTER — kg N ha^{-1} y^{-1} —		
Tropical			
broadleaf evergreen	119	255	1.9
Warm temperate			
broadleaf deciduous	36	44	1.2
Cold temperate			
needleleaf evergreen			
white pine	21	40	2.0
mixed pine	16	36	2.2
white spruce	28	22	0.8
red pine	12	19	1.6
broadleaf deciduous			
black oak	31	79	2.5
northern red oak	30	62	2.0
white oak	26	47	1.8
sugar maple	23	47	2.0
paper birch	25	43	1.7
Boreal			
needleleaf evergreen	24	26	1.1

Compare the magnitude of nitrogen entering the soil from leaf and fine-root litter, and notice that the N content of fine-root litter is equivalent to or greater than that of leaf litter.

Source: After Vogt et al., 1986 and Nadelhoffer et al., 1985. Modified from *Advances in Ecological Research,* © 1986 by Academic Press, Inc. Reprinted with permission of Academic Press, Inc. and from *Ecology* by the permission of the Ecological Society of America.

ciency represent an important process influencing the internal cycling of nutrients within terrestrial ecosystems.

Nutrients in the Forest Floor

Large amounts of organic matter and nutrients can accumulate on the soil surface in boreal and cool temperate forests, emphasizing the importance of the forest floor in controlling the flow of nutrients within these ecosystems (Table 19.5). Nevertheless, forest floor organic matter and nutrient content do not vary in a predictable manner among tropical, subtropical and warm-temperate forests. In fact, the quantity of organic matter and nutrients in the forest floor of dry and montane tropical forests differ greatly from wet tropical forests and are more similar to that in temperate forests (Anderson and Swift, 1983). Nevertheless, several important generalizations can be drawn from the information in Table 19.5. First, the accumulation of organic matter and nutrients in the forest floor of cold-temperate and boreal evergreen forests is much greater than that in warm-temperate, subtropical or tropical forests. In temperate and tropical regions, deciduous forest floors contain half the organic matter of evergreen forest floors, regardless of whether they are dominated by needleleaf or broadleaf evergreen species. Although the greatest forest floor N and P contents occur in boreal evergreen forests, there is substantial variation among the N and P contents of tropical, subtropical, and temperate forest floors.

The quantity of organic matter and nutrients contained in the forest floor of tropical, temperate, and boreal forests reflect two opposing processes: the production and the de-

Table 19.5   Forest floor mass and nutrient content in tropical, temperate, and boreal forest ecosystems. These values have been summarized for a wide range of forest ecosystems occurring in tropical, temperate, and boreal regions of the Earth.

ECOSYSTEM	FOREST FLOOR		
	Organic Matter (Mg ha^{-1})	N kg ha^{-1}	P
Tropical			
broadleaf evergreen	22.6	325	8
broadleaf semideciduous	2.2	35	–
broadleaf deciduous	8.8	–	14
Subtropical			
broadleaf evergreen	22.1	121	5
broadleaf deciduous	8.1	–	–
Warm temperate			
broadleaf evergreen	19.1	60	4
needleleaf evergreen	20.0	362	25
broadleaf deciduous	11.5	163	12
Cold temperate			
needleleaf evergreen	44.6	200	10
broadleaf deciduous	32.2	624	50
Boreal			
needleleaf evergreen	44.6	875	81

Source: Vogt et al., 1986. Modified from *Advances in Ecological Research,* © 1986 by Academic Press, Inc. Reprinted with permission of Academic Press, Inc.

composition of plant litter. Consequently, it is not surprising that patterns of leaf and root litter production alone do not well reflect differences in forest floor mass among different forest ecosystems (compare Tables 19.4 and 19.5). Temperature, precipitation, and the chemical constituents of plant litter have a substantial influence on the rate at which leaf and root litter is decomposed by soil microorganisms; differences in these factors contribute to the variation in forest floor accumulation among ecosystems in Table 19.5. Earthworm activity is also of importance, because these organisms incorporate fresh leaf litter into surface mineral soil horizons which accelerates the decomposition process (Edwards and Bohlen, 1996). In the absence of earthworms, relatively thick, distinct organic horizons develop over thin A horizons. In contrast, discontinuous and thin organic horizons form over thick, organic-matter-rich A horizons in the presence of earthworms.

A simple model considering plant litter production and decomposition has been used to gain insight into forest floor dynamics (Olson, 1963). This mass-balance approach assumes that the annual rate of litter decomposition equals the annual rate of plant litter production, such that the amount of organic matter in the forest floor is unchanged. Under this assumption, a constant proportion (k) of forest floor organic matter decomposes on an annual basis:

$$\text{Litter Production} = k(\text{Forest Floor Mass})$$

In this equation, k represents the decomposition rate constant for the entire forest floor. With knowledge of litter production and forest floor mass, values of k are derived using the following equation:

$$k\,(\text{y}^{-1}) = \frac{\text{Litter Production (Mg ha}^{-1}\,\text{y}^{-1})}{\text{Forest Floor Mass (Mg ha}^{-1})}$$

When decomposition is rapid, relatively small amounts of organic matter accumulate in the forest floor, and values for k are typically greater than 1 y^{-1}. This situation occurs in tropical forests in which microbial activity has the potential to respire more C than that contained in above- and belowground litter production (Cuevas and Medina, 1988). In contrast, decomposition is slow in cold, boreal forests where substantial amounts of organic matter and nutrients accumulate in the forest floor; values of k are approximately 0.01 y^{-1}.

If the forest floor is neither gaining nor losing organic matter (i.e., it is in equilibrium), then k can be used to estimate the mean residence of time of organic matter or nutrients ($1/k$), which is the average time organic matter and nutrients stay in the forest floor. Values range widely among the forest ecosystems listed in Table 19.6, primarily reflecting differences in temperature, precipitation, and plant-litter chemistry. The mean residence time for organic matter, N, and P, in the forest floor of tropical forests is less than 1 year (Table 19.6), whereas these constituents can reside within the forest floor of boreal forests for 10s to 100s of years.

Relatively large amounts of N and P (Table 19.6) in the forest floor of cool-temperate and boreal forests, combined with long residence times, emphasizes the importance of the forest floor as a site for nutrient storage in these forest ecosystems. Also, compare the mean residence times of coniferous (i.e., needleleaf) and deciduous forests, and notice that organic matter and nutrients reside in the forest floor of coniferous forests for relatively longer periods of time. This pattern reflects the fact that conifers contain greater proportions of organic compounds that are not easily metabolized by soil microorganisms, thus they have slower rates of litter decomposition.

Table 19.6 The mean residence time of organic matter, nitrogen and phosphorus in the forest floor of tropical, temperate and boreal forest ecosystems. Mean residence time of forest floor organic matter is calculated by dividing the forest floor mass (kg ha^{-1}) by the annual leaf litterfall mass (kg ha^{-1} y^{-1}); residence times for nutrients are calculated in the same manner.

| ECOSYSTEM | MEAN RESIDENCE TIME[1] | | |
| | Organic Matter | N | P |
	——————————— yrs ———————————		
Tropical			
broadleaf evergreen	2	2	1
broadleaf semideciduous	0.4	0.2	–
broadleaf deciduous	0.9	–	2
Subtropical			
broadleaf evergreen	6	3	2
broadleaf deciduous	2	–	–
Warm Temperate			
broadleaf evergreen	3	1	2
needleleaf evergreen	5	14	11
broadleaf deciduous	3	5	4
Cold Temperate			
needleleaf evergreen	18	33	22
broadleaf deciduous	10	19	11
Boreal			
needleleaf evergreen	60	138	225

[1]Mean residence times are $1/k$.

Source: After Vogt et al., 1986. Modified from *Advances in Ecological Research,* © 1986 by Academic Press, Inc. Reprinted with permission of Academic Press, Inc.

Mean residence times in Table 19.6 are calculated using rates of aboveground litter production alone, which comprise only 28 to 56 percent of total litter input to forest floor (see Table 19.4). Using leaf litter production as the primary input of plant litter, the mean residence time of organic matter in the forest floor of cool temperate forests ranges from 8 to 67 years. The range of values is much lower when fine-root litter is included in the calculation of mean residence time (e.g., 5 to 15 years; Vogt et al., 1986), further emphasizing the importance of belowground litter production in studying the storage and cycling of nutrients in forest ecosystems.

Differences in the mean residence time of organic matter and nutrients in the forest floor reflect the influence of temperature, water availability, and plant litter chemistry on the metabolic activity of soil microorganisms. Rates of microbially-mediated processes exponentially increase with rising temperature. At a given temperature, rates also attain maximum values when soil is near field capacity (i.e., at a matric potential of -0.01 MPa). Given the warm temperatures and ample soil water in tropical forests, it is not surprising that the forest floor mean residence times are generally less than 1 year. Meentemeyer (1978a) used the relationship between climatic factors and microbial activity to predict litter decomposition rates in arctic, subarctic, temperate (cool and warm), and tropical ecosystems. Actual evapotranspiration (Chapter 7) was used in a simulation model predicting rates of organic matter decomposition. Although this model generally agreed with observations of plant litter decomposition in the field, its predictive ability increased when information on plant litter chemistry (i.e., lignin and N concentration) was included (Meentemeyer, 1978b; Melillo et al., 1982). In the following sections, we discuss the chemical properties of plant litter that control the rate at which plant litter decomposes in the forest floor and mineral soil.

Organic Matter Decomposition and Nutrient Mineralization

To fully understand the process of decomposition and the release of nutrients from plant litter and organic matter, one needs to consider the factors that control microbial growth and maintenance in soil. Once abscised leaves, dead roots, and other plant tissues enter the forest floor and mineral soil, they become a substrate for decomposing microorganisms. During the decomposition process, soil bacteria, actinomycetes, and fungi assimilate the organic compounds contained in plant litter into their cells for biosynthesis (i.e., growth and maintenance), albeit at different rates depending on the types of compounds contained in plant litter. Whether soil microorganisms release nutrients into, or assimilate nutrients from, soil solution depends on chemical constituents of plant litter and their suitability for microbial growth and maintenance. Previously, we described the influence of temperature and soil water potential on microbial activity and organic matter decomposition. In the paragraphs that follow, we discuss how plant litter functions as a substrate for microbial growth and maintenance in soil, and in turn, how these microbial processes influence the supply of growth-limiting nutrients to plants. As you will read later in this section, the amount of energy that soil microorganisms derive from plant litter control the processes of decomposition and soil N availability.

CHEMICAL CONSTITUENTS OF PLANT LITTER

The leaves, stems, and roots of plants are constructed of a remarkably small set of organic compounds, regardless of differences in phylogeny and growth form. Plant litter is largely composed of cellulose (15 to 60 percent), hemicellulose (10 to 30 percent), lignin (5 to 30 percent), protein (2 to 15 percent), fats (1 percent) and soluble compounds like sugars, amino acids, nucleic acids and organic acids (10 percent). These organic compounds carry out essential metabolic functions, provide physical support, and defend plants against herbivores and pathogens. Upon entering soil, however, they become substrates that yield different amounts of energy for microbial growth and maintenance.

The process of decomposition is mediated by the microbial production of enzymes that harvest the chemical energy contained in compounds formed during the biosynthesis of plant tissue. In most terrestrial ecosystems, microbial growth in soil is limited by usable forms of energy, and amounts contained in annual litter production are only sufficient to maintain (i.e., no growth) microbial populations in soil. The microbial production of enzymes that make the energy contained in plant compounds available is expensive in terms of C and N. Consequently, energy produced during the microbial metabolism of a particular plant compound must surpass the cost of the enzymes used to harvest its energy in order for microorganisms to use it as a substrate.

The use of a plant-derived organic compound as a substrate for microbial growth is determined by: 1) the amount of energy released by breaking different types of chemical bonds, 2) the size and three-dimensional complexity of molecules, and 3) its nutrient content (N and P). Look at Figure 19.10 and notice that the complexity of chemical bonds, three-dimensional structure, and nutrient content varies dramatically among the primary constituents of plant litter. Simple sugars (i.e., carbohydrates) like glucose are among some of

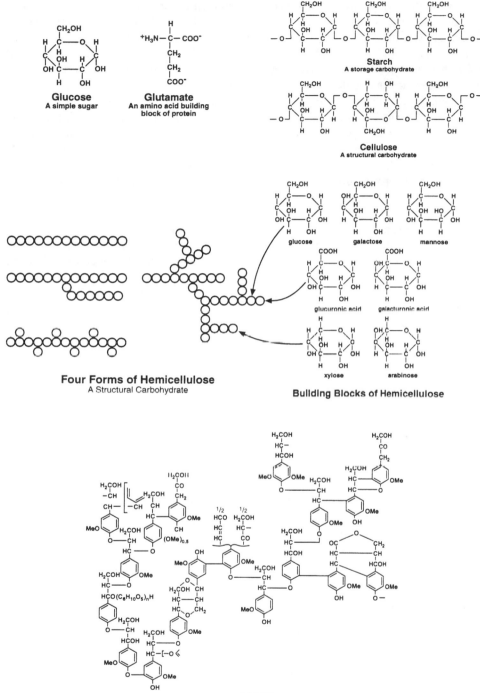

Figure 19.10. The chemical constituents of plant litter that influence the rate of organic matter decomposition in forest floor and mineral soil. Notice that nitrogen is a component of amino acids and proteins, but it is not found in the chemical structure of the other plant compounds illustrated in this figure.

the first products of photosynthesis, and although their concentration in fresh litter is low (<5 percent), these water-soluble compounds yield substantial amounts of energy for microbial growth. In soil, glucose and other simple sugars can be directly taken up and used to fuel energy-producing biochemical reactions (i.e., glycolysis and tricarboxylic acid cycle) within microbial cells. Given these characteristics, it should not be surprising that simple sugars are rapidly decomposed in plant litter.

Amino acids are the building block of proteins, molecules that carry out all biochemical reactions. A single protein contains thousands of amino acid subunits, making them too large to be directly taken up by microbial cells. Proteases released by microbial cells enzymatically-cleave proteins into their amino acid subunits. These relatively small, high-energy-yielding molecules can then be taken up by microbial cells and used for protein synthesis or energy production. Notice that proteins and amino acids are the only plant compounds in Figure 19.10 that contain N, an essential nutrient also required for microbial growth and maintenance. This fact has important implications for the release of N from plant litter during the process of mineralization, which we discuss later in this section.

Starch is a storage carbohydrate (Figure 19.10) and its synthesis occurs when the supply of photosynthate exceeds the carbohydrate requirements for the construction and maintenance of plant tissues. The concentration of this energy-storing glucose polymer is typically low in many types of plant litter, because starch is used for maintenance respiration prior to the senescence of leaves and fine roots. Starch is broken down by many soil bacteria and fungi, which synthesize extracellular enzymes (i.e., amalyses) that cleave starch into subunits consisting of several glucose molecules; these subunits can then be taken up and assimilated by microbial cells where they are used to produce energy. Although the decomposition of starch is somewhat slower than that of simple sugars, its metabolism yields relatively large amounts of energy for microbial growth.

Cellulose is the primary component of plant cell wall (Figure 19.10), and is the most abundant compound in plant litter. It is often associated with hemicellulose and lignin, which also are present in the plant cell wall. Cellulose is formed by linking glucose molecules into an unbranched, semicrystaline polymer. Notice that starch and cellulose differ in the type of bond joining adjacent glucose subunits (Figure 19.10). Unlike starch, the glucose subunits of cellulose cannot be mobilized by plants for use in energy production. Cellulose is a structural polymer, whereas starch is an energy storage polymer.

The decomposition of cellulose occurs as a two step process. First, the bacterial production of cellulases repeatedly cleave off two glucose subunits from the end of the molecule. Second, the resulting disaccharide, cellobiose, is degraded into glucose by the enzyme cellobioase. As with starch degradation, the glucose subunits can be directly assimilated by the decomposing organism. Unlike proteins and amino acids that contain N, both starch and cellulose are exclusively composed of C, H, and O.

Hemicellulose is a heterogeneous group of polymers composed of several types of plant sugars (Figure 19.10). The linking of these sugar subunits produces branched as well as unbranched molecules, which in pure state are often broken down quite rapidly. In nature, however, the association of hemicellulose with other substances such as lignin in the plant cell wall make their breakdown more difficult. Pectin (polyglacturonic acid) is a type of hemicellulose molecule found in the middle lamella of plant cell walls, and it is broken down by several enzymes collectively know as pectinases. These enzymes appear to be primarily produced by soil fungi and actinomycetes, which use the sugar subunits as a source of energy. The initial entry of mycorrhizal fungi into plant roots is thought to be facilitated by the production of pectinases. There is little difference between cellulose and hemicellulose in energy yield and hence their rate of decay by microorganisms.

Lignin is by far the most chemically-complex constituent of plant litter, containing a wide variety of chemical bonds and three-dimensional conformations (Figure 19.10). It is an abundant component of leaf litter, and high concentrations occur in most woody tissues. Lignin molecules surround cellulose and hemicellulose microfibrils in the plant cell wall, providing rigidity and protecting them against pathogenic fungi and bacteria that break down plant cell walls. Because of its chemical complexity and the fact that most soil microorganisms cannot enzymatically harvest the energy contained within its chemical bonds, lignin decomposition occurs at a very slow rate. White- (*Pleurotus, Phanerochaete*) and brown-rot fungi (*Poria, Gloephyllum*) are primarily responsible for the degradation of lignin in plant litter. In order to degrade lignin, these organisms require the presence of an alternative energy source, which functions as the primary growth substrate. As a result, only small portions of the C contained in lignin is actually incorporated into fungal cells; the majority enters into reactions that eventually form humus.

DYNAMICS OF DECOMPOSITION

The chemical constituents of plant litter and their use for microbial biosynthesis directly control the rate at which plant litter decays on and in the soil. The microbial breakdown of the chemical constituents in plant litter can be described using the following first-order rate equation:

$$A_t = A_o e^{-kt}$$

In this equation, the amount of substrate remaining (A_t in µg g^{-1}) at any point during the decomposition process is proportional to its initial concentration (A_o in µg g^{-1}) and its first-order rate constant for decomposition (k in days^{-1}, months^{-1}, or years^{-1}). First-order decomposition rate constants are experimentally-derived values obtained by measuring the decline in a substrate during its decomposition. As such, they directly reflect how rapidly a particular compound can be use for microbial biosynthesis. Rapidly decomposed substrates have a high k, whereas substrates that provide little energy for microbial growth and maintenance have low values. Decomposition rate (k) constants also are sensitive to temperature and soil water potential, doubling for a 10 °C increase in temperature and attaining maximum values near field capacity.

In Table 19.7, we summarize decomposition rate constants for the primary constituents of plant litter. Notice that glucose, a small, high-energy yielding molecule has a very-rapid decomposition rate constant, as do other simple carbohydrates and proteins.

Table 19.7 Rate constants for the decomposition of organic compounds contained in plant litter.

COMPOUND	Rate Constant for Decay k (day^{-1})
Glucose	0.500 to 1.000
Protein and Simple Sugars	0.200
Cellulose	0.036 to 0.080
Hemicellulose	0.030 to 0.080
Lignin	0.003 to 0.010

Values were determined by decomposing purified compounds under laboratory conditions.
Source: After Alexander, 1977, Paul and Clark, 1996, and van Veen et al., 1984.

Contrast these values with the low rate constant of lignin. Complex chemical bonds, subunits that provide little energy for microbial growth, and a folded three-dimensional structure that protects the inner portion of the lignin molecule against enzymatic attack all contribute its slow rate of microbial degradation. Also notice that cellulose and hemicellulose have similar decomposition rate constants, reflecting the fact that these molecules provide equivalent amounts of energy for microbial growth (Table 19.7).

The chemical attributes of plant molecules that influence their use for microbial biosynthesis are clearly reflected in the decomposition rate constants listed in Table 19.7. However, plant litter is not composed of a single organic compound, but instead contains varying proportions of simple sugars, protein, cellulose, hemicellulose, and lignin. How do different proportions of these compounds in plant litter influence the rate at which it is metabolized by soil microorganisms?

Consider the decomposition of an abscised leaf composed of 10 percent protein and simple carbohydrates, 25 percent lignin, and 65 percent cellulose and hemicellulose (Figure 19.11a). The overall decline in leaf mass seen in Figure 19.11a results from microbial respiration for growth and maintenance, which returns CO_2 to the atmosphere. Notice that the decomposition of this leaf is initially rapid (0 to 20 days), but that rate slows during the later stages of decay. During the initial stages of decomposition, the leaching of water- soluble compounds and the metabolism of proteins and simple carbohydrates (high decomposition rate constants) contribute to a rapid loss of C from the leaf. In this example, proteins and simple carbohydrates are almost totally degraded after only 15 days of decomposition. Metabolism of cellulose and hemicellulose influence the intermediate stages of decomposition (10 to 50 days), while the latter stages (50 to 100 days) of decomposition are dominated by lignin breakdown. The concentration and decomposition rate constants of these compounds additively yield the overall pattern of leaf decomposition in Figure 19.11a.

The situation is much different for an abscised leaf containing lower amounts of lignin (5 percent) and higher proportions of protein and simple sugars (25 percent), and cellulose and hemicellulose (70 percent; Figure 19.11b). Compared to the leaf in Figure 9.11a, the leaf containing 5 percent lignin (Figure 19.10b) exhibits a more rapid overall decline in C than the leaf containing 25 percent lignin (Figure 19.11a). Moreover, a higher proportion of protein and simple carbohydrates (Figure 19.11b) results in a rapid initial decline in leaf C, much greater than that of the leaf material lower in protein and simple carbohydrates (Figure 19.11a). Clearly, the relative proportion of easily-metabolized, high-energy yielding plant compounds like simple sugars and of chemically-complex molecules that function as poorer substrates (i.e., lignin) for microbial growth have a profound influence on the overall decomposition of plant litter entering forest floor and mineral soil.

Decomposition is not solely controlled by differences in the chemical constituents of plant litter. It also is controlled by the quantity of N and other nutrients available for microbial biosynthesis. In the Alaskan taiga, N and lignin concentration in leaf litter has an important influence on its decomposition rate constant and the rate at which it decomposes in the forest floor. High concentrations of lignin, which reduce the overall energy yield of litter for microbial biosynthesis, result in substantial declines in decomposition rate constants (Figure 19.12a). In contrast to this relationship, decomposition rate constants are rapid in leaf litter with high concentrations of N (Figure 19.12b). A similar relationship can be observed in the northeastern U.S., where high lignin:N ratios in the leaf litter of deciduous trees are reflected in slow rates of decomposition (Melillo et al., 1982).

The relationship between decomposition rate and litter N concentration results from the fact that soil microorganisms, like plants, require N to form new cells and maintain existing biomass. Fungal biosynthesis generally requires 1 atom of N for every 5 to 15

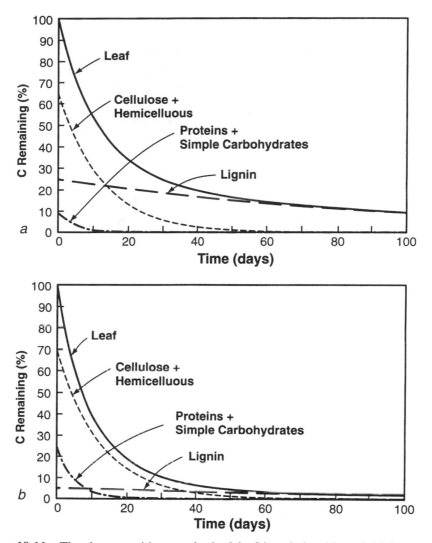

Figure 19.11. The decomposition an abscised leaf in relationship to initial concentrations of protein, simple carbohydrates, hemicellulose, cellulose and lignin. The amount of leaf C remaining is controlled by the concentration and decomposition rate constant of protein and simple sugars, cellulose and hemicellulose, and lignin. The amount of leaf C remaining in panels *a* and *b* were calculated using the following series of first-order decay equations (percent leaf $C = A_1 e^{-k1t} + A_2 e^{-k2t} + A_3 e^{-k3t}$; where A_1 = percent protein and simple carbohydrates, A_2 = percent cellulose and hemicellulose, and A_3 = percent lignin. The decomposition rate (k) constant are 0.2 days^{-1} for protein and simple carbohydrates, 0.08 days^{-1} for cellulose and hemicellulose, and 0.01 days^{-1} for lignin. Note that differences in initial concentrations have a profound influence on leaf decomposition. High lignin concentrations and low concentrations of other constituents result in relatively slow decomposition rates.

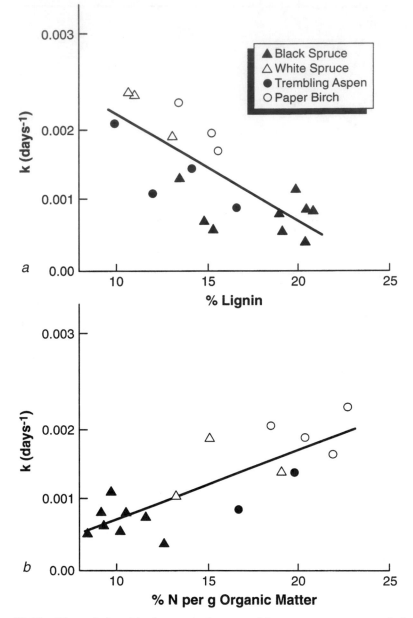

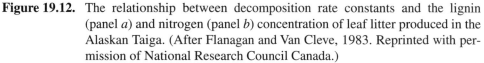

Figure 19.12. The relationship between decomposition rate constants and the lignin (panel *a*) and nitrogen (panel *b*) concentration of leaf litter produced in the Alaskan Taiga. (After Flanagan and Van Cleve, 1983. Reprinted with permission of National Research Council Canada.)

atoms of C assimilated during the decomposition of a particular substrate. Bacteria, on the other hand, require 1 atom of N for every 3 to 5 atoms of C. As a result, soil fungi have a C:N ratio ranging from 15:1 to 5:1, whereas those of soil bacteria are much lower (i.e., 5:1 to 3:1; Paul and Clark, 1996). Look again at Figure 19.10 and realize that proteins and amino acids are the primary N-containing constituents of plant litter. Because protein and amino acid concentrations are low (2 to 15 percent) in most plant tissues, newly abscised leaves and dead fine roots are C-rich and N-poor substrates for microbial metabolism.

Once litter enters forest floor and mineral soil, its C-based constituents can only be used for microbial biosynthesis if there is a sufficient source of N and other nutrients. Leaf litter always has a C:N ratio wider (e.g., 50:1) than that of microbial cells, and its decomposition requires the presence of an additional source of N. Otherwise, soil microorganisms are unable to use the compounds contained in litter for biosynthesis. The positive relationship between litter N concentration and decomposition rate constants illustrate this point (Figure 19.12*b*); low concentrations of N in plant litter limit the rate at which C-based compounds are used for microbial growth and hence slow its rate of decomposition. The inverse relationship between lignin and decomposition rate indicates that increasing proportions of poor-energy yielding substrates in plant litter also slow the process of decomposition (Figure 19.12*a*).

In summary, decomposition is controlled by the energy yield of C-based compounds (e.g., proportion of cellulose vs. lignin) in plant litter and the amount of N and other nutrients available for microbial biosynthesis (i.e., growth and maintenance). Herein lies an important ecological link between N-use efficiency, litter N concentration, and microbial activity during the process of decomposition. Plants with a high N-use efficiency withdraw a relatively large proportion of N prior to leaf abscission, thus lowering the N concentration of leaf litter. The production of leaf litter with a low N concentration slows decomposition and the subsequent release of inorganic N and other nutrients. Thus nutrient-use efficiency and plant litter chemistry can exert an important feedback on the rate at which nutrients cycle between plants and soil. In the following paragraphs, we further explore the microbially-mediated release of N during organic matter decomposition—processes that have a substantial influence on the productivity of many forest ecosystems.

NITROGEN IMMOBILIZATION AND MINERALIZATION

Although forest floor and mineral soil represent the largest pools of N in forest ecosystems, the majority (90 percent) of N contained within them resides in organically-bound forms that are unavailable to plants. Soil microorganisms control soil N availability within terrestrial ecosystems, because their growth and maintenance control the amount of inorganic N (i.e., NH_4^+) that is released from litter and soil organic matter. The amount of N available to plants is controlled by the simultaneous assimilation and release of NH_4^+ during microbial biosynthesis. The balance of microbial NH_4^+ assimilation and release supplies approximately 90 percent of the N that is annually taken up and assimilated by plants. In many ecosystems, the microbial release of inorganic N during organic matter decomposition is well correlated with the net primary productivity of forests (see Figure 18.16) and other terrestrial ecosystems.

In the previous section, we saw that litter decomposition was controlled by the types of organic compounds and the N required to use them for microbial biosynthesis. In the following paragraphs, we further explore the extent to which these factors control microbial requirements for N and the release of NH_4^+ from soil organic matter, the rate-limiting step in soil controlling the supply of N for plant growth.

When plant litter enters the soil, it is colonized by microorganisms, and high-energy-yielding constituents (i.e., organic acids, simple sugars and carbohydrates) are preferentially used for microbial biosynthesis. Because this material is a C-rich and N-poor substrate for microbial growth and maintenance, NH_4^+ must be assimilated from soil solution to form new N-containing compounds within microbial cells. **Nitrogen immobilization** is the microbial uptake and assimilation of NH_4^+ into organic compounds that are used for biosynthesis. This process is illustrated by the following reaction in which an organic acid combines with NH_4^+ to produce an amino acid:

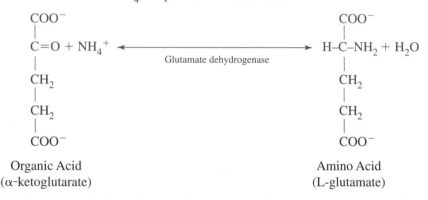

Amino acid synthesis, facilitated by the glutamate dehydrogenase enzyme, is the first step in the production of proteins needed for the growth and maintenance of microorganisms. Nitrate (NO_3^-) also can be used by soil microorganisms, but it first must be reduced to NH_4^+ before it can be used to form an amino acid. Nitrogen immobilization is the result of microbial growth and maintenance on a C-rich and N-poor substrate, and it characterizes the initial stages of litter decomposition. The hyphal networks of fungi, which extend from forest floor into mineral soil, are particularly effective at transporting NH_4^+ into freshly decaying litter.

Nitrogen mineralization is the release of NH_4^+ during the microbial breakdown of proteins, amino acids and other N-containing organic compounds. Thus N mineralization is the reverse of the N immobilization reaction illustrated above, and it is carried out by all microorganisms involved with organic matter decomposition. Whether an amino acid produced during protein degradation is used as an energy source or as a building block for microbial protein synthesis depends on a series of feedback controls. For example, carbohydrates contained in litter are preferentially used to generate energy in microbial cells, thus preserving any amino acids present for microbial protein synthesis. Nonetheless, this relationship can change depending on the availability and energy yield of substrates remaining in plant litter.

During the initial stages of decomposition, carbohydrate availability is relatively high and amino acids contained in litter are directly assimilated for microbial protein synthesis. However, carbohydrate supply (i.e., energy) begins to limit microbial biosynthesis during the latter stages of litter decomposition (Figure 19.11). When quantities are insufficient to meet the maintenance energy requirements of microbial cells, mortality occurs and the constituents of the dead cells begin to serve as substrates for surviving microorganisms. Under energy-limited conditions, the C-skeletons of amino acids are used by surviving microbial cells to generate energy. During this process, NH_4^+ is released into soil solution where it can be assimilated by plant roots, participate in cation exchange reactions (Chapter 11), or enter into other microbially-mediated processes.

Nitrogen mineralization and immobilization occur simultaneously during the process of litter decomposition, albeit at different rates depending on the organic compounds present and the amount of N available to use those compounds for microbial biosynthesis. At what stage during litter decomposition is there a net release of NH_4^+ into soil solution where it can be taken up and assimilated by plant roots? Nitrogen is released from decomposing litter only when the gross rate of N mineralization exceeds the gross rate at which NH_4^+ is incorporated into microbial biomass (i.e., gross N immobilization).

Figure 19.13 illustrates the relationship between the C and N contents of leaf litter and changes in microbial metabolism that influence the gross immobilization and mineralization of N. During the initial phases of decomposition, plant litter is relatively rich in energy-yielding compounds, poor in N, and has a C:N ratio (50:1; Figure 19.13*a*) wider than that of microbial cells (*ca.* <8:1). In order to use litter constituents for growth or maintenance, soil microorganisms need to assimilate NH_4^+ or NO_3^- from soil solution. The initial demand for N for microbial biosynthesis causes high rates of gross N immobilization and an initial accumulation of N in decomposing litter (compare Figures 19.13*a* and 19.13*c*). Comparing the decline in leaf C over time with rates of microbial respiration illustrates that litter mass loss results from the return of leaf C to the atmosphere as CO_2 during microbial respiration (Figures 19.13*a* and 19.13*b*). The initial increase in litter N from gross N immobilization, in combination with the loss of C during microbial respiration, causes leaf C:N to decline over time.

Later in the decomposition process, decomposing microorganisms are limited by substrates that provide little energy (i.e., energy- or C-limited phase), and microbial respiration and biosynthesis subsequently decline (Figure 19.13*b*). Proteins and other N-containing compounds released from cell death during this energy-limited phase of decomposition are then metabolized by surviving microbes. These organisms release NH_4^+ and use the resulting organic compounds for energy production. As a result, gross rates of N immobilization fall below gross rates of N mineralization, and there is a net release of NH_4^+ from decomposing litter (net N mineralization). These dynamics illustrate an important point—the microbially-mediated processes of gross N mineralization and gross N immobilization control the transfer of N from soil organic matter to soil solution where it can be taken up by plant roots. Although the biomass of soil microorganisms contains only 1.5 percent of the C and 3.0 percent of the N within forest ecosystems (Wardle, 1992), their metabolic activities truly make them the "gate keepers" of soil N availability! We direct the reader to Staaf and Berg (1982) for a discussion of P, S, K, Ca, Mg, and Mn mineralization during plant litter decomposition.

NITROGEN AVAILABILITY IN FOREST ECOSYSTEMS

It is important to realize that net N immobilization only characterizes the initial phases of plant litter decomposition (Figure 19.13*c*). Soil organic matter or humus is the end product of litter decomposition; it has a low C:N (10:1) and contains few compounds that can be used for microbial biosynthesis. It is characterized by the net release of NH_4^+, which supplies the majority of N that is taken up and assimilated by plants on an annual basis. Generally, net N mineralization rates increase with the organic matter content of mineral soil (Marion and Black, 1988). We direct the reader to Binkley and Hart (1989) and Hart et al. (1994) for a review of current techniques for measuring soil N transformations.

In North American temperate forests, rates of net N mineralization range from 30 to 120 kg N ha^{-1} y^{-1} (Pastor et al., 1984; Zak and Pregitzer, 1990; Binkley, 1995). This process supplies approximately 90 percent of the N available for plant uptake on an annual basis in forest ecosystems. For example, compare rates of net N mineralization with inputs from atmospheric deposition (Figures 19.2 and 19.3) and free-living N_2 fixation (*see* Biological Fixation of Nitrogen). Notice that rates of net N mineralization in most

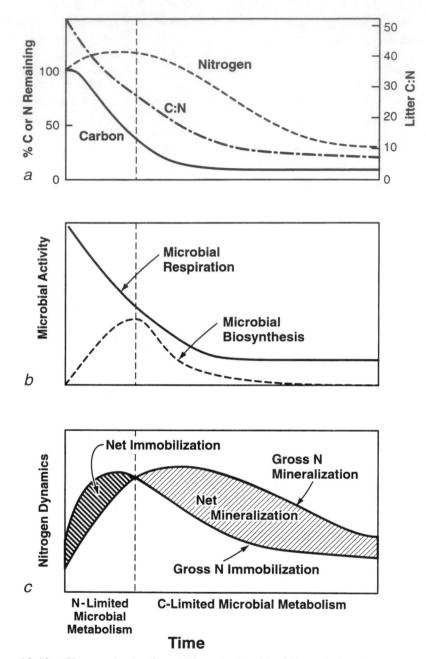

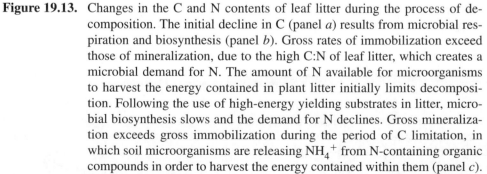

Figure 19.13. Changes in the C and N contents of leaf litter during the process of decomposition. The initial decline in C (panel *a*) results from microbial respiration and biosynthesis (panel *b*). Gross rates of immobilization exceed
those of mineralization, due to the high C:N of leaf litter, which creates a
microbial demand for N. The amount of N available for microorganisms
to harvest the energy contained in plant litter initially limits decomposition. Following the use of high-energy yielding substrates in litter, microbial biosynthesis slows and the demand for N declines. Gross mineralization exceeds gross immobilization during the period of C limitation, in
which soil microorganisms are releasing NH_4^+ from N-containing organic
compounds in order to harvest the energy contained within them (panel *c*).

forest ecosystems are approximately 10 times greater than the combined annual inputs from atmospheric deposition and free-living N_2 fixation.

NITRIFICATION

The NH_4^+ produced during the process of net N mineralization can be retained by cation exchange reactions, taken up by plant roots, or it can be oxidized to NO_3^- during the process of **nitrification.** Nitrification is an ecologically-important N-cycling process because, in the absence of plant uptake or microbial immobilization, NO_3^- can rapidly move in the downward flow of water through the soil profile. The high mobility of this ion, in combination with the low anion exchange capacity of most temperate soils, make NO_3^- the form of N most often lost from terrestrial ecosystems. Once NO_3^-, or any other nutrient, has passed below the majority of plant roots, it is lost from the ecosystem and eventually enters groundwater, streams, or lakes. Consequently, nitrification is of great importance for understanding patterns of N loss from forest ecosystems.

Nitrification is the two-step oxidation of NH_4^+ to NO_3^- by chemoautotrophic bacteria. These organisms use NH_4^+ or NO_2^- as an energy source to fix CO_2 from the soil atmosphere. Nitrifying bacteria are sensitive to pH, water potential, and aeration. In general, nitrification diminishes below pH 6.0, is most rapid at matric potentials of -0.01 to -1.0 MPa, and requires the presence of O_2 (Paul and Clark, 1996). However, nitrification has been observed in some acidic forest soils. There is some evidence to suggest that several heterotrophic bacteria (*Arthrobacter* spp.) and fungi (*Aspergillus* spp.) can produce small amounts of NO_3^- from NH_4^+, but their ecological significance in forest soils is largely unknown (Schimel et al., 1984).

The first step in the process of nitrification is usually carried out by *Nitrosomonas europaea, Nitrosospira briensis, Nitrosococcus mobilis,* and *Nitrosovibrio tenuis* in the following manner:

$$NH_4^+ + 1.5\,O_2 \longrightarrow NO_2^- + 2\,H^+ + H_2O$$

The NO_2^- produced in the above reaction is used by *Nitrobacter winogradskyi* and *Nitrobacter agilis* to obtain energy in the second reaction in the nitrification process:

$$NO_2^- + H_2O \longrightarrow H_2O{\cdot}NO_2^- \longrightarrow NO_3^- + 2H$$
$$2H + 0.5\,O_2 \longrightarrow H_2O$$

Notice that NH_4^+ and NO_2^- oxidation require O_2, making nitrification unlikely in very poorly drained soil. The production of H^+ during NH_4^+ oxidation (first reaction) can lower the pH of soil with high NH_4^+ concentrations and rapid nitrification rates.

The potential for NO_3^- loss following nitrification has drawn the attention of ecologists and soil microbiologists, who have sought to understand the factors controlling the activity of nitrifying bacteria. Competition appears to be an important controlling factor, because plants, decomposing microorganisms, and nitrifying bacteria all require NH_4^+ as either a biosynthetic building block or as an energy source. Most plant roots and decomposing microorganisms have NH_4^+ uptake rates more rapid than those of NH_4^+-oxidizing bacteria, making plant uptake and microbial immobilization important controls constraining nitrification in many ecosystems (Robertson and Vitousek, 1981; Robertson, 1982; Zak et al., 1990). Nitrification rates can greatly increase in the absence of plant uptake, further suggesting that plant-microbe competition can limit nitrification. There is some evidence that plant compounds synthesized to deter herbivory (tannins and terpenoides) can inhibit nitrifying bacteria (White, 1986; 1988).

Nitrification rates dramatically differ among intact forest ecosystems, ranging from 0 to 100 percent of net N mineralization (Pastor et al., 1984; Zak and Pregitzer, 1990; Binkley, 1995). The highest rates of nitrification often occur in forest soils with rapid rates of net N mineralization. Because of its high rate of diffusion and low retention by anion exchange reactions in soil, one would expect that NO_3^- loss following disturbance should vary according to initial nitrification rate and the extent to which it increases following the removal of plant uptake. We build on our understanding of nutrient uptake by plants, organic matter decomposition, and nitrification to understand patterns of N loss prior to and following disturbance later in this chapter.

NUTRIENT LOSS FROM FOREST ECOSYSTEMS

In the previous sections, we saw that relatively large quantities of nutrients annually circulate between overstory trees, forest floor and mineral soil. In intact forest ecosystems, a small proportion of the nutrients that annually cycle among these ecosystem pools are lost through physical export via the hydrologic cycle or biological export to the atmosphere. Highlighted in the reduced version of Figure 19.1 (to the right) are leaching and gaseous loss, major pathways of nutrient loss.

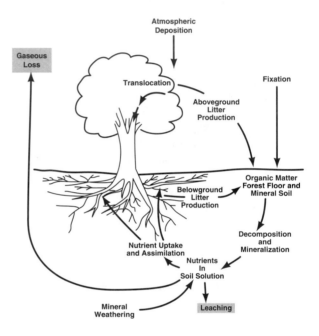

Nutrients in soil solution that exceed plant and microbial demands can be transported below the rooting zone by the downward flux of water through the soil profile. Nutrients moving in this way eventually enter ground or surface waters where they then become a nutrient input for aquatic ecosystems. Thus the nutrient cycles of terrestrial and aquatic ecosystems are intimately linked through the flow of water in the landscape. Nutrient loss from forest ecosystems also influences water quality and the nutrient dynamics of aquatic ecosystems. In the paragraphs that follow, we discuss the physical and biological loss of nutrients from intact forest ecosystems, thus completing the journey of nutrients into, through, and out of forest ecosystems. We later build upon this foundation to understand the extent to which natural disturbance and human activities influence nutrient loss from forest ecosystems.

Nutrient Leaching from Forest Ecosystems

Leaching is the physical process by which nutrients exit terrestrial ecosystems in the downward flow of water through the soil profile. This process can occur when precipitation exceeds the amount of water lost through transpiration and evaporation from the soil surface. Water moving downward through the soil profile has the potential to entrain and transport nutrients dissolved in soil solution. Nutrients are subject to leaching *only* when water is moving downward through the soil profile. Thus the hydrologic cycle provides the transport mechanism for the leaching of nutrients from terrestrial ecosystems. Once below the rooting zone, nutrients are lost from the forest ecosystem.

The high demand for growth-limiting nutrients (N or P) by overstory trees typically maintains low concentrations in soil solution (*also see* Nutrient Uptake and Assimilation) and thus lessens the potential for nutrient leaching in forest ecosystems. The leaching loss of N is often much less than N inputs via atmospheric deposition and free-living N_2 fixation (Vitousek and Melillo, 1979). On the other hand, non-limiting nutrients (i.e., Ca or K) can accumulate in soil solution because the rate of supply from weathering or the atmosphere exceeds their biological demand. Under these conditions, leaching can export relatively large quantities of non-limiting nutrients from terrestrial ecosystems.

It should be remembered, however, that excess water must be present to serve as the transport mechanism for nutrient leaching. In temperate deciduous forests, the greatest potential for nutrient leaching occurs in early spring prior to canopy development and in late autumn after leaf fall. During these periods, a low biological demand for nutrients and water, and the resulting ample amounts of water moving through the soil profile, provide the potential for leaching.

The potential for nutrient leaching, particularly the loss of limiting nutrients like N and P, has drawn the attention of forest ecologists over the past 30 years. In 1963, one of the first attempts to quantify nutrient leaching from intact forests began at the Hubbard Brook Experimental Forest in New Hampshire (Likens et al., 1977). Using small, bedrock-scaled watersheds drained by a single stream, a team of ecologists began measuring nutrient leaching from northern hardwood forests (*also see* Mineral Weathering). In this unique situation, leached nutrients could only exit the bedrock-sealed watershed in streamflow. Many studies have been conducted subsequent to this pioneering research; the reader is directed to Henderson et al. (1978), Vitousek and Melillo (1979), and Swank and Crossley (1988).

Table 19.8 summarizes leaching losses of limiting (N, P, S), non-limiting (K, Ca) and non-essential (Si) plant nutrients from the second-growth northern hardwood forests at the Hubbard Brook Experimental Forest; also provided are nutrient inputs from atmospheric deposition and biological N_2 fixation. Notice that leaching losses of N, P, and S are much

Table 19.8 Nutrient leaching from an intact northern hardwood-dominated watershed in the Hubbard Brook Experimental Forest, New Hampshire.

NUTRIENT	Leaching Loss	Input	Export (−) or Retention (+)
	kg ha^{-1} y^{-1}		
N	4.0	20.7	16.7
P	0.019	0.036	0.017
S	17.6	18.8	1.2
K	2.4	0.9	−1.5
Ca	13.9	2.2	−11.7
Si	23.8	0	−23.8

Nutrient inputs via atmospheric deposition and biological fixation are also summarized. Nutrients required for plant growth in large quantities are accumulating within these second-growth forests. Leaching losses are greater than additions for nutrients in excess of plant demand (i.e., Ca) or those not required for plant growth (i.e., Si).
Source: After Likens et al., 1977. Reprinted from *Biogeochemistry of a Forested Ecosystem,* © 1977 by Springer-Verlag New York, Inc. Reprinted with permission from Springer-Verlag.

lower than their input, suggesting that these nutrients are being taken up and assimilated by forest vegetation. Contrast this pattern to the losses of K, Ca, and Si, which exceed inputs from atmospheric deposition. These nutrients weather from the K-, Ca- and Si-rich bedrock of the Hubbard Brook Experimental Forest at a rate surpassing plant demand, giving rise to their net export form the forest.

Similarly, Henederson et al. (1978) compared leaching losses of N, K, and Ca from watersheds dominated by old-growth Douglas-fir in the Pacific Northwest with those dominated by oaks and hickories in the southeastern United States. Despite ample amounts of precipitation in these widely different forests (158 to 233 cm y^{-1}), leaching losses of N were low (<0.5 kg N ha^{-1} y^{-1}) and were much less than the annual input of N from atmospheric deposition (1 to 6 kg N ha^{-1} y^{-1}). These ecosystems also leached higher amounts of K and Ca than they received in atmospheric deposition, again reflecting mineral weathering rates in excess of plant demand. In summary, these examples illustrate that leaching is controlled by the biological demand for nutrients and the availability of excess water to transport nutrients out of the rooting zone.

In the next section, we will consider how these factors change following natural and human-induced disturbances which initiate the process of ecosystem development (i.e., secondary succession). The harvesting of the overstory, which also constitutes a loss of nutrients, will be treated in our discussion of forest harvesting and nutrient loss later in this chapter. In the following section, we discuss the biological loss of N from forest ecosystems through the process of denitrification.

Denitrification

Nitrogen is the only plant nutrient that enters terrestrial ecosystems through a biological process. It is also one of the few plant nutrients that leaves terrestrial ecosystems via a biological process. **Denitrification** is the microbially-mediated reduction of NO_3^- to nitrous oxide (N_2O) or N_2, which returns N to the atmosphere (Tiedje, 1988). This process has received considerable attention for several reasons. Denitrification results in the loss of a limiting nutrient, potentially influencing the productivity of terrestrial ecosystems. Moreover, N_2O produced during denitrification is a "greenhouse gas" and its concentration in the atmosphere continues to rise (Elkins and Rosen, 1989). Consequently, atmospheric scientists, ecosystem ecologists, and microbial ecologists have become increasingly interested in understanding the environmental and biological controls on this process (Davidson and Swank, 1987).

The denitrification pathway is used by soil bacteria (facultative anaerobes) as an alternate form of respiration when O_2 is absent in the soil atmosphere. Species within 13 genera of bacteria have the ability to use NO_3^- for respiration. Those common in soil (*Pseudomonas*, *Bacillus*, and *Alcaligenes*) carry out the following overall reaction:

$$5CH_2O + 4H^+ + 4NO_3^- \longrightarrow 2N_2 + 5CO_2 + 7H_2O$$

The reduction of NO_3^- is accomplished in several steps ($NO_3^- \rightarrow NO_2^- \rightarrow NO \rightarrow N_2O \rightarrow N_2$) that are mediated by a series of enzymes collectively know as reductases (Knowles, 1981). Some organisms lack the enzymatic capacity to carry out the final reductive step ($N_2O \rightarrow N_2$) and produce N_2O as the end product of denitrification. In some situations, N_2O can account for one third of N lost during denitrification (Robertson and Tiedje, 1984). Whether the end product of denitrification is N_2O or N_2, the majority of denitrifying bacteria require organic compounds (CH_2O) in order to generate energy. Field techniques for measuring denitrification in terrestrial ecosystems are summarized by Tiedje et al. (1989).

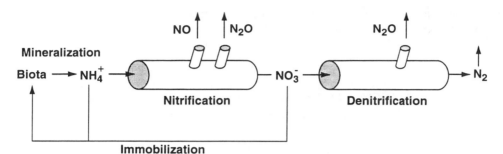

Figure 19.14. Transformation of N by soil microorganisms resulting in the gaseous loss of N from forest ecosystems. Gaseous loss of N can occur during the process of nitrification as well as during denitrification. (Reprinted from "Microbial basis of NO and N_2O production and consumption in soil" by M. K. Firestone and E. A. Davidson, In M. O. Andreae and D. S. Schimel (eds.), *Exchange of Trace Gasses between Terrestrial Ecosystems and the Atmosphere,* © 1989 by John Wiley & Sons Limited. Reproduced with permission.)

Summarized in Figure 19.14 are the combined sources of gaseous N loss during denitrification and nitrification. There is a substantial body of evidence suggesting that gaseous N loss can also occur during nitrification. Nitric oxide (NO) and N_2O are gaseous byproducts that are released by nitrifying bacteria as they oxidize NH_4^+ and NO_2^- (Firestone and Davidson, 1989). Apparently, NO and N_2O are chemical intermediates that can dissociate from the NH_4^+- and NO_3^-- oxidizing enzyme systems—processes that may be an important source of gaseous N loss in some forest soils (Robertson and Tiedje, 1984).

The physiological requirements for denitrification place considerable restrictions on where this process occurs in the landscape. Most often, denitrification is associated with poorly-drained soils in which the water table is near or at the soil surface for a considerable portion of the year. Soils in these landscape positions are often rich in organic-matter and experience periodic low water levels during which nitrifying bacteria can produce NO_3^-. Once the soil becomes inundated, O_2 diffusion slows, and the soil becomes anaerobic. Denitrification can ensue, provided that organic substrates and NO_3^- are present.

Groffman and Tiedje (1989) studied denitrification in northern temperate forests occurring on soils of sand, loam, and clay loam texture. Soils of each textural class were also located in well drained, moderately well drained and poorly drained landscape positions. Denitrification increased from well drained to poorly drained landscape positions, and were generally higher on finer-textured soils. For example, losses were substantial (40 kg N ha^{-1} y^{-1}) in lowland forests occurring on poorly drained clay loam soil, whereas denitrification was minimal in upland forests located on well drained sandy soils (< 1 kg N ha^{-1} y^{-1}). The lack of available NO_3^-, the result of rapid plant uptake, was the primary factor limiting denitrification during midsummer.

Nitrate uptake by plants also appears to be an important constraint on denitrification in wet tropical forests. Denitrification is rapid in lowland tropical rainforests that were recently cleared of vegetation, but only for the first 6 months following harvesting (Robertson and Tiedje, 1988). After 6 months, denitrification was substantially lower than in intact rainforests, apparently the result of N uptake by rapidly regrowing forest and accelerated decomposition, which lessened organic substrate availability to denitrifying bacteria. In combination, these results suggest that physical factors like soil drainage and aeration directly control the activity of denitrifying bacteria. However, denitrification is also controlled by the activities of plants and heterotropic microorganisms that influence NO_3^- and C availability in soil, factors that can dramatically change following the removal of overstory trees.

THE CYCLING AND STORAGE OF NUTRIENTS IN FOREST ECOSYSTEMS

In the preceding sections, we have discussed the physical, chemical, and biological processes controlling the flow of nutrients into, within, and out of forest ecosystems. Recall that nutrients enter forest ecosystems from mineral weathering, atmospheric deposition, and biological N_2 fixation. They cycle within forest ecosystems due to root uptake, above- and belowground litter production, decomposition, and nutrient mineralization. And nutrients are lost from forest ecosystems through leaching and denitrification. Here, we synthesize the specific processes of nutrient input, internal flow, and loss into overall patterns of nutrient cycling and storage in forest ecosystems. We discuss the N and Ca cycles of forest ecosystems to compare and contrast the overall cycling of limiting and nonlimiting, essential nutrients within temperate forest ecosystems.

Nutrient Storage in Boreal, Temperate, and Tropical Forests

The uptake of nutrients by roots, the return of nutrients in above- and belowground litter, and the release of nutrients during decomposition result in the continual movement of nutrients between plants and soil, albeit at different rates depending on the particular ecosystem. It is a widely held misconception that overstory trees contain the majority of nutrients within wet tropical forests, and that the soils beneath them are relatively poor in organic matter and nutrients. The origin of this misconception is obscure, but our current understanding of tropical forests clearly refutes this notion. For example, the organic matter and nutrient content of many tropical forest soils is equivalent to that of temperate forest soils (Klinge and Rodrigues, 1968a, b; Sanchez 1976). In some instances, as much as 60 percent of N and P in wet tropical forests resides in the organic matter-rich surface and subsurface horizons of Spodosols (Klinge and Rodriques, 1968a, b). What differs between temperate and tropical forests is the rate at which nutrients cycle from soil to plants and back to soil again. Recall the rapid mean residence time of organic matter and nutrients in forest floor of some tropical forests (Table 19.6). Declines in soil fertility following the harvesting of tropical forests are more likely related to a rapid decomposition and loss of nutrients than to the removal of nutrients in the harvested trees (Cole and Johnson, 1980).

In Table 19.9, we summarize the distribution of N within boreal, temperate and tropical forests. Notice that overstory trees in the wet tropical forest contain 10-times more N than those of the boreal forest and 3 to 5 times more N than the overstory of temperate forests. Regardless of this large difference, comparable quantities of N reside within the soil beneath tropical, temperate, and boreal forests. For example, mineral soil contains 62 percent of the N in the wet tropical forest and 69 to 84 percent of the N within the temperate and boreal forests in Table 19.9. When forest floor and mineral soil are combined, these proportions increase to 70 percent of total ecosystem N in the wet tropical forest, and 90 percent of total ecosystem N in the boreal and temperate forests.

From this analysis, it should be clear that forest management practices that erode or remove (i.e., windrowing or scalping) nutrient-rich O and A soil horizons have the potential to greatly reduce the nutrient capital of forest ecosystems, regardless of whether they

Table 19.9 Distribution of N in tropical temperate, and boreal forest ecosystems.

	Boreal		Temperate		Wet Tropical
Location	Alaska USA	Washington USA	New Hampshire USA	Tennessee USA	Amazon Venezuela
Dominant Species	black spruce	Douglas-fir	sugar maple –beech	oak–hickory	mixed species
Age (yr)	55	42	55	30-80	mature
Nitrogen			kg N ha^{-1}		
Overstory	134	316	491	497	1670
Understory	51	21	9	–	–
Forest Floor	657	233	1100	334	406
Mineral Soil	2200	2476	3600	4500	3507
Total Ecosystem	3042	3046	5200	5331	5583

The amount of N in the overstory (leaves, branches, stems, large roots) increases from boreal to tropical forests, but the largest amount of N resides within the forest floor and mineral soil of these markedly different forest ecosystems.
Source: After Cole and Rapp 1981, Jordan et al., 1982, and Likens et al., 1977.

occur in boreal, temperate, or tropical regions. Recall the rates of nutrient input from atmospheric deposition and mineral weathering earlier in this chapter (Figure 19.3 and Table 19.1) and realize that substantial reductions in forest floor and mineral soil nutrient pools can only be replenished over relatively long periods of time. Also realize that the loss of organic matter-rich surface horizons reduces rates of nutrient mineralization.

The Nitrogen and Calcium Cycle of a Temperate Forest Ecosystem

The complete N cycle of a second-growth northern hardwood forest is depicted in Figure 19.15a. In this diagram, boxes represent the storage of N in ecosystems pools (kg N ha^{-1}) and the arrows depict processes (kg N ha^{-1} y^{-1}) by which nutrients flow from one ecosystem pool to another. The N cycle of this forest ecosystem is dominated by storage within soil (3600 kg N ha^{-1}), forest floor (1100 kg N ha^{-1}), and overstory trees (532 kg N ha^{-1}). Look at Figure 19.15a and compare nutrients stored in overstory trees, forest floor and mineral soil with rates of aboveground litter production (54.2 kg N ha^{-1} y^{-1}), belowground (6.2 kg N ha^{-1} y^{-1}) litter production, and net N mineralization (69.6 kg N ha^{-1} y^{-1}). Notice that relatively small amounts of N are annually transferred from one ecosystem pool to another, compared to amounts stored in trees, forest floor, and mineral soil. Also notice that the total input of N is greater than the total loss of N, indicating an overall retention of N in this rapidly-growing forest ecosystem.

The Ca cycle of the same northern hardwood ecosystem is depicted in Figure 19.15b; it is similar to and different from the N cycle in several important aspects. The Ca cycle is dominated by storage in mineral soil, forest floor, and overstory biomass with relatively small amounts of Ca annually flowing between these pools, similar to the pattern of N cycling and storage. However, the relatively large amounts of Ca stored in soil occur in the chemical structure of minerals, whereas N is stored in soil organic matter not soil miner-

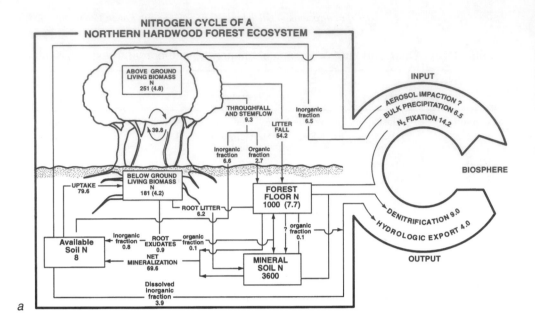

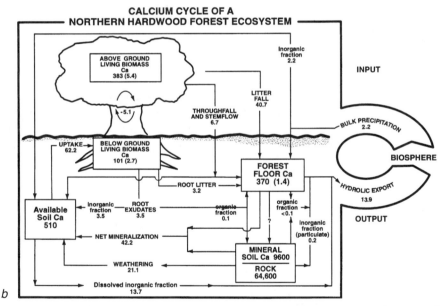

Figure 19.15. The nitrogen (panel *a*) and calcium (panel *b*) cycle of a second-growth northern hardwood forest at the Hubbard Brook Experimental Forest. Boxes in each diagram represent ecosystem pools in which N or Ca is stored, and the arrows represent processes by which nutrients are transferred from one ecosystem pool to another. Nutrient pools are measured in kg ha^{-1} and the processes of transfer between pools is presented in kg ha^{-1} y^{-1}. (Panel a reprinted from Bormann et al., 1977. Reprinted with permission from *Science* 196, p. 982. © 1977 by American Association for the Advancement of Science. Panel b reprinted from Likens et al., 1977, *Biogeochemistry of a Forested Ecosystem,* © Springer-Verlag Berlin, Heidelberg 1977. Reprinted with permission of Springer-Verlag, New York, Inc.)

als. The large mineral pool of Ca provides weathering inputs (21.1 kg Ca ha^{-1} y^{-1}) that are relatively large. Also note that Ca translocation (arrow from canopy to stem) during leaf senescence is not an important process; in fact, the Ca increases in leaves prior to abscission! Contrast this with the large amount of N translocated from leaves prior to senescence. Also contrast patterns of nutrient input and loss for the Ca cycle with that of the N cycle where losses are much less than inputs. Calcium losses are six times greater than Ca inputs from atmospheric deposition, suggesting that Ca is not being retained by this ecosystem (compare Figure 19.15a and 19.15b).

Why is N retained by this forest ecosystem, and why is there a net export of Ca? Recall that biomass accumulates early in ecosystem development, because net ecosystem productivity (NEP) is positive (NEP > 0, when GPP > R_A + R_H; Chapter 18). Growth-limiting nutrients like N are rapidly taken up to build new biomass, whereas non-limiting nutrients like Ca can accumulate in soil solution—their supply exceeding the biological demand to build new biomass. A positive NEP and the resulting accumulation of biomass within this forest ecosystem (Gosz et al., 1978), combined with the fact that N limits the rate of biomass accumulation and Ca does not, cause the net retention of N and a net export of Ca from this developing forest ecosystem. In the following section, we build upon these principles to understand patterns of nutrient retention and loss during ecosystem development.

ECOSYSTEM C BALANCE AND THE RETENTION AND LOSS OF NUTRIENTS

In 1975, Peter Vitousek and William Reiners proposed and tested a hypothesis integrating patterns of nutrient loss and retention with the C balance of forest ecosystems (Vitousek and Reiners, 1975). It was based on the idea that patterns of net ecosystem production (NEP) control nutrient loss and retention in forest ecosystems. Notice in Figure 19.16a that net ecosystem production is low or negative during the initial stages of succession, increases during the intermediate stages of succession, and declines in the latter stages of succession (i.e., ecosystem development). In Chapter 18, we saw that patterns of NEP during ecosystem development result from changes in gross primary productivity (GPP), the respiration of plants (R_A) and heterotrophic organisms (R_H). When net ecosystem production is positive (GPP > R_A + R_H), biomass and nutrients should accumulate within ecosystems.

Vitousek and Reiners (1975) reasoned that the loss of growth-limiting nutrients should be greatest early or late in succession when NEP is low and there is little or no demand for nutrients. Examine Figure 19.16a and b and observe the proposed inverse relationship between NEP and the loss of limiting nutrients. When NEP is less than zero, total ecosystem biomass declines (not shown in Figure 19.16), and the loss of limiting nutrients should exceed their input. When NEP is zero, total ecosystem biomass is at an equilibrium, and the loss of limiting nutrients should equal their input. And when NEP is greater than zero, total ecosystem biomass increases, and the loss of limiting nutrients should be less than their input. That is, limiting nutrients should be retained in ecosystems when NEP is positive and biomass is accumulating. They also predicted these same patterns should hold for non-limiting nutrients, but to a much smaller extent (Figure 19.16a and 19.16b). Because non-essential nutrients are not required to build plant biomass, changes in NEP during ecosystem development should have little influence on their retention or loss (Figure 19.16a and 19.6b).

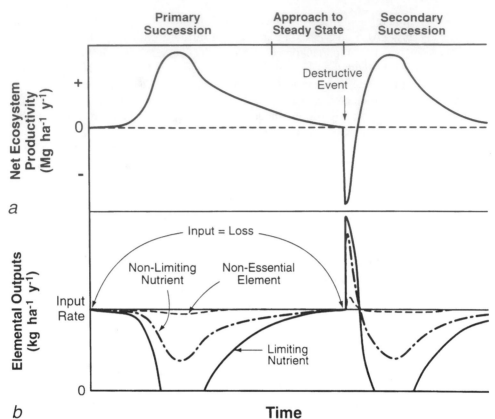

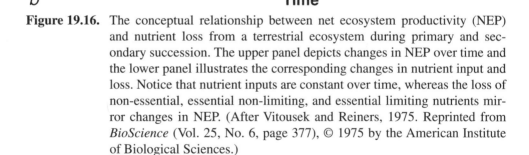

Figure 19.16. The conceptual relationship between net ecosystem productivity (NEP) and nutrient loss from a terrestrial ecosystem during primary and secondary succession. The upper panel depicts changes in NEP over time and the lower panel illustrates the corresponding changes in nutrient input and loss. Notice that nutrient inputs are constant over time, whereas the loss of non-essential, essential non-limiting, and essential limiting nutrients mirror changes in NEP. (After Vitousek and Reiners, 1975. Reprinted from *BioScience* (Vol. 25, No. 6, page 377), © 1975 by the American Institute of Biological Sciences.)

To test this hypothesis, Vitousek and Reiners (1975) located a series of bedrock-sealed watersheds that were entirely dominated by mid- (high NEP) or late-successional (low or 0 NEP) forests. If their hypothesis was correct, differences in NEP between mid- and late-successional forests should be reflected in the loss of limiting, non-limiting, and non-essential nutrients. These watersheds were located in close proximity to one another and received an equivalent input of nutrients in precipitation. Because each was sealed by underlying bedrock, nutrients that leached below the rooting zone could only exit in streamwater. Over a two year period, they compared the concentration of NO_3^- (limiting nutrient), K^+, Mg^{2+}, Ca^{2+} (non-limiting) and Na^+ (non essential) in the streams draining the watersheds dominated by the mid- and late-successional forests.

By comparing nutrient concentrations in streamwater, it was clear that NO_3^- loss from late-successional forests was much greater than NO_3^- loss from mid-successional forests (Table 19.10 and Figure 19.17), supporting the contention that loss is low in

Table 19.10 The mean concentration of limiting, non-limiting, and non-essential plant nutrients in the streamwater draining watersheds dominated by mid- and late-successional forests in New Hampshire.

| Nutrient | Streamwater Concentration µeq L^{-1} | | |
	Late Successional	Mid-Successional	Late:Mid Ratio
NO_3^-	53	8	6.6
K^+	13	7	1.8
Mg^{2+}	40	24	1.7
Ca^{2+}	56	36	1.6
Na^+	29	28	1.0

Source: After Vitousek and Reiners, 1975. Reprinted from *BioScience* (Vol. 25, No. 6, page 378), © 1975 by the American Institute of Biological Sciences.

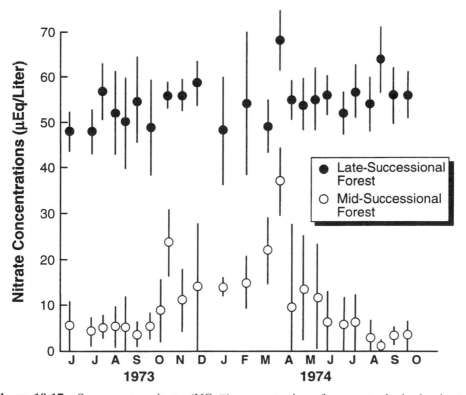

Figure 19.17. Streamwater nitrate (NO_3^-) concentrations from watersheds dominated by mid- and late-successional forests in the White Mountains of New Hampshire (After Vitousek and Reiners, 1975. Reprinted from *BioScience* (Vol. 25, No. 6, page 379), © 1975 by the American Institute of Biological Sciences.)

ecosystems where NEP is positive. The export of non-limiting nutrients (K^+, Mg^{2+}, Ca^{2+}) was slightly greater from late-successional forests, whereas the loss of Na^+, a nonessential nutrient, did not differ between mid- and late-successional forests. These results support the idea that the C balance of terrestrial ecosystems exerts a substantial influence on the cycling and storage of nutrients. Clearly, the C balance of forest ecosystems exerts a substantial influence on nutrient retention and loss, especially on growth-limiting nutrients like N.

FOREST HARVESTING AND NUTRIENT LOSS

Following the removal of overstory trees, there are several fundamental changes that drastically alter the C balance of forest ecosystems and patterns of nutrient loss. Gross primary productivity immediately declines, because there is little leaf area to convert solar radiation into plant biomass. Greater amounts of solar radiation reaching the soil warm its surface and increase the rates of organic matter decomposition. Large declines in transpiration increase soil water contents, which also increases organic matter decomposition and the return of CO_2 to the atmosphere. These changes in ecosystem C balance are illustrated in Figure 19.16 as the negative values of NEP (i.e., GPP $< R_A + R_H$) following a disturbance that initiates secondary succession. Low rates of nutrient uptake by overstory trees, in combination with a greater nutrient supply from accelerated rates of organic matter decomposition, may lead to the loss of nutrients from forest ecosystems.

One of the first and probably most well-known attempts to quantify the influence of forest harvesting on nutrient loss occurred at the Hubbard Brook Experimental Forest in the White Mountains of New Hampshire (Likens et al., 1977; Bormann and Likens, 1979). To quantify nutrient loss, scientists located a series of similar watersheds (i.e., size, topography, geology, forest vegetation, and harvesting history) that were drained by a single stream. Nutrients leached below the rooting zone encounter shallow bedrock and flow laterally toward the stream draining each watershed. For a period of several years (1957 to 1965), researchers measured the volume and nutrient concentration of streamwater draining two watersheds dominated by 55-year-old northern hardwood forests. Notice in Figure 19.18 that precipitation, streamflow, evapotranspiration, and nutrient concentrations are nearly identical during this period. In 1965, however, the overstory of one watershed was clearcut and herbicide was applied for three subsequent years to preclude revegetation (shown by shaded portion of Figure 19.18). This treatment resulted in a decrease in evapotranspiration, an increase in streamflow and the greater export of NO_3^-, K^+, and Ca^{2+} (compare control versus clearcut in Figure 19.18). Related studies demonstrated that clearcut harvest also increased denitrification (Melillo et al., 1983). As plants recolonized the clearcut site, NEP began to increase, biomass accumulated on the site, and streamwater nutrient concentrations returned to pre-harvest concentrations (Figure 19.19). This example illustrates the importance of NEP in regulating the loss of nutrients from clearcut forest ecosystems.

It is important to point out, however, that not all forest management is conducted in such a manner nor are these northern hardwood forests typical of all temperate forests. Silvicultural systems ranging from single-tree selection to whole-tree harvest, soils of broadly different fertility, and differences the balance between precipitation and evapotranspiration (i.e., excess water) all contribute to variation in nutrient loss from managed forest ecosystems. Moreover, not all temperate forests experience large losses of NO_3^- following disturbance. In the southeastern United States, for example, NO_3^- loss from intensively-harvested loblolly pine plantations was reduced by the addition of logging slash to the forest floor, which increased soil C availability, net N immobilization, and N reten-

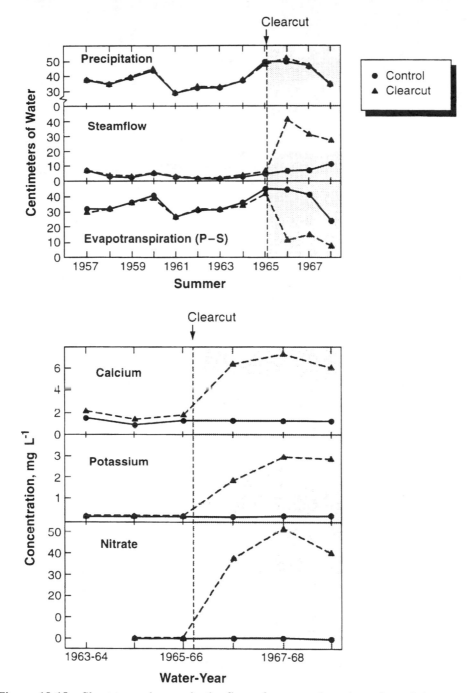

Figure 19.18. Short-term changes in the flow of water and nutrients through intact and clearcut northern hardwood forest ecosystems at the Hubbard Brook Experimental Forest in New Hampshire. The shaded portion of the figure indicates the period of time during which regeneration was prevented with the use of herbicide. Notice that harvesting increases streamflow and the export of nutrients from the harvested watershed. (After Borman and Likens, 1979. Reprinted from *Pattern and Process in a Forested Ecosystem,* © 1979 by Springer-Verlag New York Inc. Reprinted with permission of Springer-Verlag, New York, Inc.)

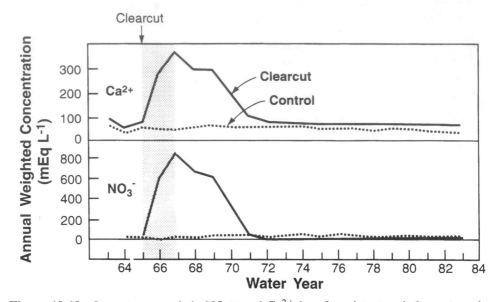

Figure 19.19. Long-term trends in NO_3^- and Ca^{2+} loss from intact and clearcut north-ern hardwood forest ecosystems. The shaded portion of the figure indi-cates the period of time during which regeneration was prevented with the use of herbicide. The decline in nutrient loss from the clearcut forest coin-cides with the accumulation of biomass (i.e., GPP > R_A + R_H, therefore NEP >0) and a renewed demand for nutrients. (After Likens et al., 1978. Reprinted with permission from *Science* 199, p. 493, © 1978 by American Association for the Advancement of Science.)

tion (Vitousek and Matson, 1984). Temperate forests prone to NO_3^- loss are generally those with high rates of net N mineralization and nitrification prior to disturbance (Vi-tousek et al., 1982).

In addition to the export of nutrients to surface waters and the atmosphere, nutrients can also be lost from forest ecosystems in harvested biomass. One approach for quantify-ing the impact for forest harvesting on nutrient cycles uses a "balance sheet" to keep track of all nutrient inputs and losses (Silkworth and Grigal, 1982). Using this technique, one can estimate the influence of repeated harvesting on nutrient storage in forest ecosystems. If the loss of nutrients during harvest is not met by nutrient additions during the next rota-tion, then forest harvesting can cause an overall decline in nutrient storage.

Table 19.11 summarizes the nutrient balance for trembling aspen ecosystems in which the overstory was removed by whole-tree harvest (removal of stems, branches and leaves). Observe that whole-tree harvesting accelerated Ca leaching, the only nutrient for which annual inputs matched normal leaching losses. For other nutrients, annual inputs were greater than normal leaching losses, and harvesting did not result in accelerated leaching losses. Although a substantial amount of N was lost through biomass removal, it can be replaced by annual inputs after 48 years. This period of time is generally less than the rotation age for trembling aspen in the Lake States region. Phosphorus, K, and Mg be-haved in a similar manner. However, harvest export and accelerated Ca leaching suggest that whole-tree harvest could deplete amounts stored within the ecosystem. It is unlikely that such losses would reduce forest productivity on these Ca-rich soils, because Ca is not a limiting nutrient (Silkworth and Grigal, 1982).

Table 19.11 The nutrient balance for whole-tree harvested trembling aspen ecosystems in northern Minnesota.

	N	P	Nutrient K kg ha^{-1}	Ca	Mg
Annual Input					
Precipitation	6.9	2.6	9.5	5.0	1.7
Weathering	0	0.4	8.7	20.8	10.4
N$_2$ Fixation	3.0	–	–	–	–
Ecosystem					
Storage[†] (kg ha^{-1})	4834.0	147.8	642.5	9081.0	1866.3
Output					
Normal Annual Leaching	0.4	0.6	3.6	28.8	11.3
Accelerated Leaching*	0	0	0	62.3	0
Removal in Biomass	452.0	43.1	354.6	1034.0	94.5
Years to Replenish					
Harvest Losses[§]	48	18	24	–[#]	118

Source: After Silkworth and Grigal, 1982. Reprinted with the permission of the Soil Science Society of America, Segoe, Wisconsin, USA.

[†]Includes aboveground biomass, forest floor and storage in organic mineral soil and on cation exchange sites.

*Increases in leaching immediately following whole-tree harvest

[§]Calculated as (removal in biomass + accelerated leaching)/(inputs—normal leaching)

[#]The normal leaching loss of Ca is approximately balanced by inputs, therefore the loss of Ca in harvested biomass and accelerated leaching cannot be replenished.

The time it would take to replenish harvest-associated nutrient losses can be greatly reduced for some nutrients by altering the amount and type of biomass removed from forests. The leaves of trembling aspen are relatively rich in N and P, and leaving them on the site dramatically reduces harvest losses. Similarly, trembling aspen has Ca-rich bark, and leaving this biomass component on site can reduce the net export of Ca from harvesting. This example illustrates the need to assess nutrient inputs and harvest-associated exports to assess the long-term influence of harvesting on the storage of nutrients within forest ecosystems.

Suggested Readings

Anderson, J. M., and M. J. Swift. 1983. Decomposition in tropical forests. In S .L. Sutton, T. C. Whitmore, and A. C. Chadwick (eds.). *Tropical Rain Forest: Ecology and Management.* Blackwell, Oxford. 498 pp.

Bormann, F. H., and G. E. Likens. 1979. *Pattern and Process in a Forested Ecosystem.* Springer–Verlag, New York. 253 pp.

Hedin, L. O., J. J. Armesto, and A. H. Johnson. 1995. Patterns of nutrient loss from unpolluted, old-growth temperate forests: evaluation of biogeochemical theory. *Ecology* 76:493–509.

Likens, G. E., F. G. Bormann, R. S. Pierce, J. S. Eaton, and N. M. Johnson. 1977. *Biogeochemistry of a Forested Watershed.* Springer–Verlag, New York. 146 pp.

Paul, E. A., and F. E. Clark. 1996. *Soil Microbiology and Biochemistry* 2nd edition. Academic Press, New York. 340 pp.

Roy, J. and E. Garnier. 1994. *A Whole Plant Perspective on Carbon–Nitrogen Interactions,* SPB Academic Pub. bv. The Hague, The Netherlands, 313 pp.

Schlesinger, W. H. 1991. *Biogeochemistry: an analysis of global change.* Academic Press, New York. 443 pp.

Staaf, H., and B. Berg. 1982. Accumulation and release of plant nutrients in decomposing Scots pine needle litter: Long-term decomposition in a scots pine forest II. *Can. J. Bot.* 60:1561–1568.

Vitousek, P. M., and R. L. Sanford, Jr. 1986. Nutrient cycling in moist tropical forests. *Ann. Rev. Ecol. Syst.* 17:137–167.

Vogt, K. A., C. C. Grier, and D. J. Vogt. 1986. Production, turnover, and nutrient dynamics of above- and belowground detritus in world forests. *Adv. Ecol. Res.* 15:303–377.

CHAPTER *20*

DIVERSITY

The diversity of organisms, measurement of diversity, and hypotheses about the causes of diversity have long interested ecologists. Exploring natural forests, making lists of species, hunting rare species, and posing questions such as why a species is present, rare, or abundant at some sites but not at others, is fundamental to the forest ecologist. Diversity is a simple concept; in ecology it means variety. However, diversity grows rapidly in complexity with attempts to describe and measure it and compare it from one area to another. At its simplest level, the study of diversity is a descriptive pursuit involving counting or listing species.

The study of diversity may also involve a range of foci from genes to ecosystems. It becomes closely intertwined with theoretical and practical questions related to ecosystem productivity and conservation. Numerous questions arise: Is species diversity a product of a site and its physical attributes? Is it a product of biotic interactions of competition and mutualism within the community residing on that site? Are ecosystems with many species more or less stable than those with only a few? How do we best identify and preserve the variety of organisms and ecosystems that exist locally, regionally, and globally? How do we manage forests in a manner that maintains their diversity? We have many good insights to these questions, but clearly we know too little about the genetic, species, and ecosystem aspects of diversity and their causes.

This chapter is organized into several sections. First, we consider concepts and values of species and ecosystem diversity. Second, we examine the measurement of diversity and provide examples at species and ecosystem levels. Then we discuss controls and patterns of forest diversity, beginning first at a continental scale and progressively focusing on more local landscapes. Finally, we discuss important issues related to management and conservation. While our treatment is primarily at organism and ecosystem levels, it is important to recognize throughout that, ultimately, diversity is closely intertwined with many interacting factors, including paleogeology and ecology, ecosystem geography, plant physiology, and human activities. Each chapter of this book can be thought of as presenting one or more aspects of the physical and ecological processes whose ultimate result includes the diversity of ecosystems and organisms that inhabit them. The scope of our overview of diversity is necessarily limited, and readers are encouraged to delve deeply into books that treat organismal diversity (Wilson, 1992; Ricklefs and Schluter, 1993; Huston, 1994), measurement of species diversity (Peet, 1974; Pielou, 1975; Magurran, 1988: Krebs, 1989), and conservation biology (Soulé, 1986; Primack, 1993; Meffe and Carroll, 1997).

CONCEPTS OF BIOLOGICAL AND ECOSYSTEM DIVERSITY

Diversity can be examined from many different viewpoints. For example, the diversity of physical features of bedrock, landform, geomorphic processes, parent material, and climate is of ultimate importance to the distribution and growth of plants. The diversity of the biota—plants and animals, their gene pools and their arrangement in aggregates of populations and communities— is of great interest and concern. Because biota are fundamentally different from the physical environmental factors and the ecosystems that support them, we define **biological diversity** (**biodiversity**) as the kinds and numbers of organisms and their patterns of distribution. This definition also includes properties of biota, including the genetic diversity of their gene pools and other types of diversity related to groupings of biota, such as the diversity of families and genera, populations and communities, and "structural" properties of communities (vertical layering, plant density, or patchiness, etc.).

Although it is popular today to include landscapes and ecosystems as part of "biodiversity," biota are not ecosystems and biodiversity should be distinguished from ecosystem diversity. Because biodiversity depends on ecodiversity (Rowe, 1992b, 1997), the diversity of ecosystems deserves separate and special consideration. **Ecosystem diversity** is defined as the kind and number of ecosystems in an area and includes the patterns of association of ecosystems with one another and the recurrence of these patterns in a given landscape. Ecosystem diversity can be estimated at multiple spatial scales by using the same diversity indices that are used to measure species diversity. Because species are integral parts of ecosystems, understanding ecosystem diversity is fundamental to maintaining and conserving biodiversity.

SPECIES DIVERSITY AND ITS VALUE

The statement by Solbrig (1991) that scientists have a very rudimentary knowledge of biodiversity accurately characterizes our understanding of the Earth's biota. Of the 10 to 100 million species that are estimated to exist, only about 1.4 million species of plants, animals, and microorganisms have been described (Wilson, 1992). The breakdown of the organisms known to exist (Figure 20.1) shows that insects alone comprise 53 percent of the total. In the plant kingdom, angiosperms have been the dominant group for the past 150 million years and constitute about 18 percent of the total with dicots the dominant group by far (Figure 20.2). The immense combined diversity of insects and flowering plants is no accident because these groups are united by intricate symbioses (Wilson, 1992). The insects live on and in plants in their every nook and cranny, consume every anatomical part, and, as we recall from Chapter 14, a large fraction of plants depend on insects for pollination and dispersal.

Despite this seemingly large number of organisms, virtually all plant and animal

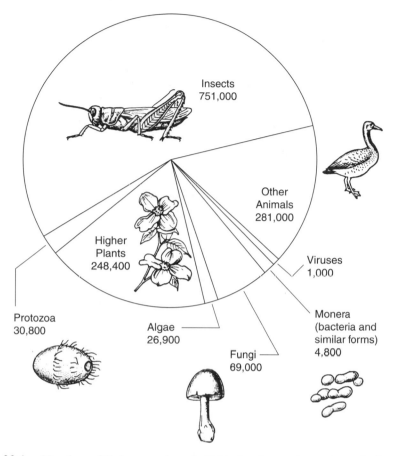

Figure 20.1. Number of living species of all kinds of organisms currently known. The total number of all species is 1,413,000. Organisms are shown by major group. Insects and higher plants dominate the diversity of living organisms known to date, but vast arrays of species remain to be discovered in bacteria, fungi, and other poorly studied groups. Most flowering plants live on the land; algae prevail in the sea. (After Wilson, 1992. Reprinted by permission of the publisher from THE DIVERSITY OF LIFE by E. O. Wilson, Cambridge, Mass.: Harvard University Press, Copyright © 1992 by the President and Fellows of Harvard College.)

species that have ever lived on the Earth are extinct (Raup, 1986). Up to four billion species of plants and animals are estimated to have lived at some time in the geologic past (Simpson, 1952), yet there are far fewer living today. Thus extinction of species has been almost as common as origination. Major extinction events are regularly spaced in geologic time (Raup, 1986). There are several major mass extinctions, like that at the end of the Cretaceous, but relatively sudden and rapid turnovers occur at lesser scales as well. For the past 600 million years, the course of biodiversity, despite extinctions, has been upward (Figure 20.3). Global biodiversity of plants (Figure 20.4) reached its peak in the Cenozoic by: (1) creation of the aerobic environment, (2) fragmentation of land masses, and (3) increased packing of species into regional and local ecosystems of the developing landscape (Wilson, 1992). Note in Figure 20.4 that the number of plant species in local floras has more than tripled in the past 100 million years. Although this brief overview illustrates the immense increase in biodiversity, Wilson (1992, p. 343) observes that: "The

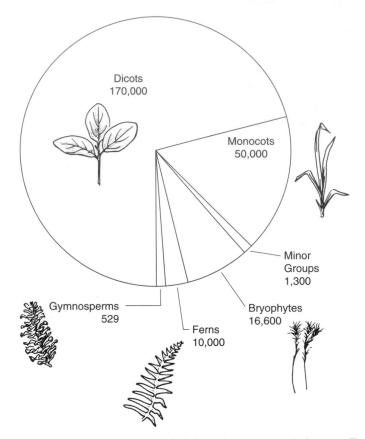

Figure 20.2.   Number of living species of higher plants currently known. Total number of higher plant species is 248,000. Species are shown by major group. Plant diversity of the world consists primarily of angiosperms (flowering plants), which in turn make up grasses and other monocots and also a huge variety of dicots, from magnolias to asters and roses. (After Wilson, 1992. Reprinted by permission of the publisher from THE DIVERSITY OF LIFE by E. O. Wilson, Cambridge, Mass.: Harvard University Press, Copyright © 1992 by the President and Fellows of Harvard College.)

sixth great extinction spasm of geological time is upon us, grace of mankind. Earth has at last acquired a force that can break the crucible of biodiversity." Therefore, it is useful next to consider the value of biodiversity; later in this chapter we consider the increasing role of biodiversity in shaping the science of ecosystem conservation, management, and restoration.

Value of Biodiversity

The reasons for conserving, promoting, and managing a rich variety of organisms are many (Burton et al., 1992). These include:

The intrinsic value of the existence of life forms and ecosystems (Leopold, 1949; Regan, 1981; Norton, 1982; Naess, 1986; Rowe, 1990; Mosquin et al., 1994).

Aesthetic or anthropocentric values; nature preserves, wilderness, and biological diversity can all play a role in promoting human well-being (Kaplan and Kaplan, 1989; Easley et al., 1990; Kaplan, 1992; Thompson and Barton, 1994).

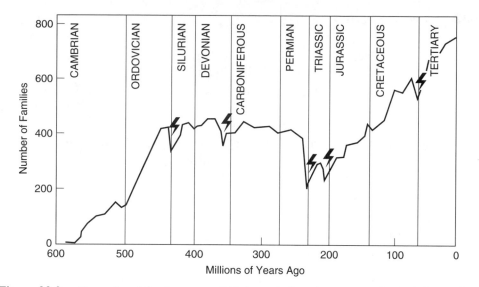

Figure 20.3. Example of the increase of biological diversity over geological time using data from families of marine organisms. A slow increase is seen, with occasional setback through mass global extinctions. There have been five such extinctions so far, indicated here by lightning flashes. A sixth major decline is now underway as the result of human activity. (After Wilson, 1992. Reprinted by permission of the publisher from THE DIVERSITY OF LIFE by E. O. Wilson, Cambridge, Mass.: Harvard University Press, Copyright © 1992 by the President and Fellows of Harvard College.)

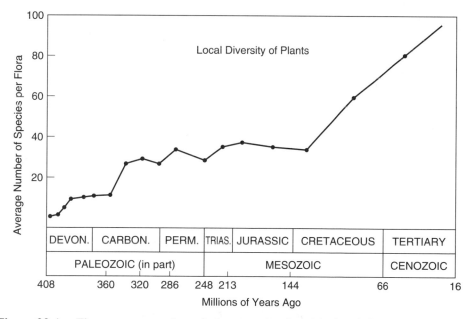

Figure 20.4. The average number of plant species found in local floras over geologic time. The number of plants found in local floras has risen steadily since the invasion of the land by plants 400 million years ago. The increase reflects a growing complexity in terrestrial ecosystems around the world. (After Wilson, 1992. Reprinted by permission of the publisher from THE DIVERSITY OF LIFE by E. O. Wilson, Cambridge, Mass.: Harvard University Press, Copyright © 1992 by the President and Fellows of Harvard College.)

581

Important products other than fiber may be derived from forest organisms, e.g., wild species for food and medicine (McNeely, 1989).

Certain species indicate the integrity of ecosystem processes, whereas others act as warnings of critical stress thresholds of pesticides or atmospheric pollution that may endanger ecosystem function (Burton et al., 1992).

Many species are important ecological indicators of site productivity (Chapter 13).

Diverse organisms, and especially their gene pools, provide a sound resource base as the best insurance for coping with a rapidly changing and uncertain future environment (Burton et al., 1992).

Many species play mutualistic roles in the regeneration and function of forest ecosystems.

Many species play key roles in the resiliency of ecosystems following normal disturbance events.

The importance of plant species for a variety of uses besides timber or fiber has been widely emphasized. Opportunities for identifying natural products for future pharmaceutical, agricultural, or industrial value increase with species diversity. Pharmaceutical compounds derived from wild plants are often of global importance. The Madagascar periwinkle (*Catharanthus roseus*), for example, is the source of at least 60 alkaloids that are used in the treatment of childhood leukemia and Hodgkin's disease, and were reported worth about US$160 million in sales each year (Shiva, 1990). Another medicinal example is the western yew tree, the source of taxol, a compound found to have strong activity against a number of cancers (Wani et al., 1971). Wild species are not only used for food, medicine, and crafts in subsistence societies, but some are brought into cultivation such as was done with kiwi fruit (*Actinidia deliciosa*). As Burton et al. (1992) point out, the valuable use of this and other species could not have been anticipated decades ago. A number of equally useful species are doubtless being destroyed today by short-sighted land management. The evolution of these species and their useful attributes is, in part, a function of their membership in diverse and complex ecosystems.

In addition to many specific values for biota already discussed, Costanza et al. (1997) have estimated the market value of the world's ecosystem services (ecosystem processes and functions such as food production, waste treatment, pollination, and recreation) to human welfare to average about US$33 trillion ($10^{12}$). Global gross national product total is around US$18 trillion. Forests contribute 14 percent of the $33 trillion, and wetlands (in part forested swamps and floodplains) contribute another 15 percent. This bottom line approach is striking and provides a powerful incentive for conservation of the natural capital stock that produces these services. Many ecologists and others would additionally, or alternatively, cite very real and "priceless" ethical and cultural reasons for sustaining the Earth's ecosystems.

MEASURING DIVERSITY

Levels of Diversity

Diversity encompasses the two different concepts of variety: richness and evenness. Richness refers to the number of units (alleles, species, families, communities, ecosystems) per unit area, and evenness refers to their abundance, dominance, or spatial distribution. These two concepts can be applied and measurements made at many levels of organization: Ecosphere and continental-level ecosystems (macroecosystems, see Chapter 2), regional or mesoecosystems, local or microecosystems, species, and the genetic diversity of allele frequencies within populations of a species. The focus of biodiversity measurement is typically the species, because species are easily observed and the most common taxonomic unit used in ecological studies of forest ecosystems.

Species diversity can refer to all organisms in a community, but usually discussions are restricted on practical grounds to certain groups of species related to the interest of the ecologist. A single forest ecosystem type may contain many trees and shrubs and a great number of herbs, bryophytes, birds, mammals, insects, and microorganisms. However, most studies ignore whole groups of organisms, especially those that are difficult to locate, let alone to count or measure. Thus when studying diversity or comparing it between ecosystems or regions, it is important to identify explicitly the set of organisms being considered and the site-specific area of their occurrence. This point is especially important because forest ecosystems are remarkably different and complex; the diversity pattern of trees may not reflect that of ground-cover vegetation, birds, or microorganisms.

Most studies of biodiversity occur at the local level in areas of <1 to ca 25 ha (microecosystem scale, Chapter 2). Local site factors of microclimate, landform, and soil, as well as interspecific interactions determine the observed patterns of richness and evenness. Diversity at the broad meso- and macroecosystem scales is sometimes referred to as landscape, or regional diversity in the literature. Here the focus may be on relatively broad vegetative cover types such as those in Yellowstone National Park (Romme, 1982; see Chapter 21) or the White Mountains of New Hampshire (Reiners and Lang, 1979). The pattern of diversity observed in such cases results from the superimposition of multiple patterns of: (1) species distribution along gradients of limiting factors, (2) recurrent spatial patterns of local ecosystems, and (3) communities in different stages of succession following disturbance. In continental-level studies (Currie and Paquin, 1987; Latham and Ricklefs, 1993a,b), diversity of taxonomic units above the species level (i.e., genera, family, order) are often examined. At this scale, diversity is related to macroclimate, physiography, continental drift, and the historical development of plant floras.

Whittaker (1960, 1972, 1975, 1977) was one of the first to recognize that species diversity needs to be measured in several ways and at several scales. First, he recognized **inventory diversity** that could be estimated at four increasingly larger spatial scales:

Point diversity: at a microsite scale with samples taken from within a homogeneous site, i.e., microsite within a local landscape ecosystem type (e.g., 100 to 500 m^2).

Alpha diversity: at a homogeneous site, i.e., for a given landscape ecosystem type or community in a homogeneous site (microecosystem level, e.g., < 1 to several hundred hectares; Table 2.4).

Gamma diversity: at a larger landscape unit, i.e., for a regional ecosystem at the mesoecosystem level, i.e., a group of alpha-level ecosystems within a region (e.g., 625 to 2,500 ha; Table 2.4).

Epsilon diversity: total diversity for an area encompassing a group of areas of gamma diversity, i.e., macroecosystem level (e. g., > 2,500 ha).

Whittaker also recognized that species composition changes across the landscape, both along environmental gradients, and from one generally homogeneous site to the next. This change from one area to the next, regardless of scale, he referred to as **differentiation diversity** or **beta diversity**.

Besides community composition, diversity can also be used to describe the "structural" diversity of communities. In forested systems, this analysis focuses on the spatial pattern of species occurrence (random, systematic, contagious or patchy), stand density and diameter-class distribution, and the vertical layering of a forest from the tallest tree to the shortest plant on the forest floor. For example, Franklin et al. (1981) used a diversity index to confirm differences in spatial heterogeneity (between-tree spacing intervals) that had been observed between old-growth Douglas-fir forests (450-year-old trees) and young-growth stands (125-year-old trees).

Structural diversity of communities is of considerable importance in wildlife ecology where the number of animals capable of making use of a forested system may depend heavily on the presence of appropriate areas for nesting, feeding, resting, and hiding. Structural diversity for old-growth, mixed-age or size, and mixed-species stands is typically high, whereas that for plantations or single-age, single-species stands is low. For example, the Kirtland's warbler, a rare and endangered neotropical migrant nesting in ecosystems dominated by jack pine in northern Lower Michigan, prefers the patchy distribution of jack pine individuals following fire to the systematic arrangement of jack pines in plantations (see Chapter 21).

Measurement

INVENTORY DIVERSITY: ALPHA DIVERSITY

There are various ways to measure the "diversity" of species. Whittaker (1977) described inventory diversity as having two components, richness and equability (evenness), and, as we have described, cited four levels of scale, each reflecting successively larger land areas varying from a microsite to the macroecosystem level.

The simplest way to measure species diversity is to count the number of species present in a designated area. This number, or species **richness**, is the oldest, most fundamental, and perhaps the least ambiguous of the diversity measurements (Peet, 1974). It depends on the area sampled, so richness is often expressed as number of species per unit area. Richness varies with the size of the area sampled. In an area of homogeneous composition, such as a single ecosystem type with a relatively little microtopographic variation, relatively small gaps or disturbance features, and uniform climate, richness will increase with increasing size of sampled area up to a point where all, or almost all, species have been captured in the sample. At this point, increasing the sample size results in no increase in species number. In determining the richness for a given ecosystem by measuring sample plots within it, it is important to use a plot size large enough, or enough plots, to encompass most of the species. In any sample of a large forested area, this total number is very difficult to determine for most major groups of organisms. Further, as the sampled area expands beyond a single local ecosystem, environmental heterogeneity increases, and different ecosystems with different species are included in the sample. Therefore, to obtain a reliable estimate of alpha diversity of species for a given ecosystem, it is important to recognize that ecosystem's boundaries.

Diversity also depends on the distribution of species within an area, termed **evenness** or equability, in other words, the degree to which all species share dominance in an area. For example, forest stands A and B in Figure 20.5 both have eight species; therefore, they have the same richness. However, stand A is clearly dominated by one or two species, with the other species being relatively uncommon. Stand B exhibits a more equal abundance of all species and thus exhibits greater evenness. Which stand is more "diverse?" Stand B—because with the same number of species, it has the more equal abundance of the species. Note that each of the two diversity indices shown in Figure 20.5 has a higher value for stand B than for stand A.

The two diversity indices illustrated in Figure 20.5 *combine* richness and evenness in one index value. This dual-concept diversity was introduced into the ecological literature by Simpson (1949), and although generally accepted, it can be a source of confusion (Peet,

1974). Most workers have considered that diversity should include both an evenness and a richness component. Nevertheless, as Peet (1974) observed, retention of diversity as a broad term encompassing all of the subordinate concepts, with a different terminology for duel-concept measures, seems more desirable. Thus the term **heterogeneity** is often used to denote diversity indices that combine richness and evenness in a single measure.

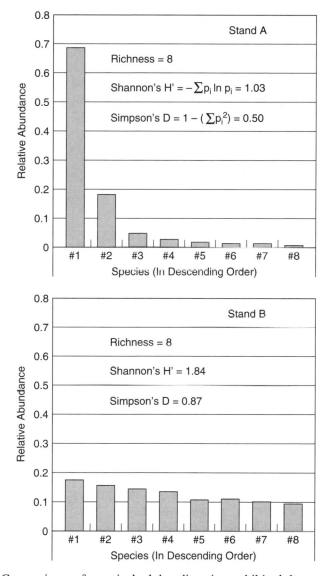

Figure 20.5. Comparison of species' alpha diversity exhibited by two hypothetical stands having the same number of tree species. Species richness for each stand is eight. Two diversity indices are given, where p_i is the proportional abundance of each species, i, and the summation is performed over all eight species. Because stand B exhibits greater evenness, with less dominance by any one species, it has greater overall diversity than stand A, as measured by the indices. (Reprinted with permission by the *Forestry Chronicle* from: Burton, P. J., A. C. Balisky, J. P. Coward, S. G. Cumming, and D. D. Kneeshaw. "The values of managing for biodiversity." *Forestry Chronicle,* 1992 Vol 68, p. 227.)

There exists a wide variety of species diversity indices that attempt to summarize richness and evenness in a single number; each has its theoretical advantages under specific circumstances. Two of the most commonly used indices for measuring species diversity are Simpson's index and the Shannon-Wiener (or Shannon's) index. The Shannon-Wiener index is probably the most commonly used diversity index. It is computed as:

$$H' = -\sum_{i=1}^{S} p_i \ln(p_i)$$

where S is the number of species in the sample, and p_i is the proportion of individuals that belong to species i, which is estimated by dividing the number of individuals in species i by the total number of individuals in the sample. H' (for heterogeneity) generally ranges from a low of around 1.5 to a high of 4.5, rarely exceeding this value (Margalef, 1972). The Shannon-Wiener index is most sensitive to the number of species in a sample, so it is usually considered to be biased toward measuring species richness (Figure 20.5). The evenness component of H' can be computed as (Magurran, 1988):

$$E = \frac{H'}{H'\ \max} = \frac{H'}{\ln S}.$$

In this computation, the Shannon-Wiener index is scaled by its maximum possible value (H' max), which would occur if all species had equal abundance and happens to be the natural log of S.

Simpson (1949) developed an index of diversity that is computed as:

$$D = \sum_{i=1}^{S} \left(\frac{n_i(n_i - 1)}{N(N-1)} \right),$$

where n_i is the number of individual in species i and N is the total number of species in the sample. Diversity is inversely related to D, so Simpson's index is usually expressed as its complement, $1-D$ or $1/D$. Simpson's index is strongly influenced by the most common or dominant species in a community and relatively insensitive to rare species. For this reason, Simpson's index, or its complement (1-D) is sometimes used as a measure of evenness (Figure 20.5).

Indices of diversity vary in computation and interpretation, and it is inappropriate to compare diversity measured by one index with that measured by another. The values will also vary depending on what organisms or what layers of the forest (overstory; ground cover, all layers) are sampled, the specific area sampled (i.e., number of different ecosystems included), and sometimes on sampling methods. However, given reasonable standardization, when multiple indices are computed for the same data, they tend to be correlated. That is, the sampled areas will rank similarly in diversity as measured by various indices. Given that agreement is not complete, however, studies using diversity indices should use an index sensitive to the factors of interest in the study.

Species abundance models also are a way of expressing commonly observed abundance distributions of species and thereby assessing diversity. These models include the log normal distribution, the geometric series, the logarithmic series, and MacArthur's brokenstick model (Magurran, 1988). One of the most commonly used techniques is to plot some measure of abundance or dominance of species against their rank order of abundance, or dominance, or coverage. These so-called species-abundance plots can be used to compare not only the relative richness of ecosystem communities, but also to gain an insight into evenness. Figure 20.6 shows species-abundance curves for ground-cover plants sampled in two very different ecosystem types on glaciated terrain in northern Lower Michigan: (1) an excessively drained, infertile, sandy outwash plain supporting a sparse,

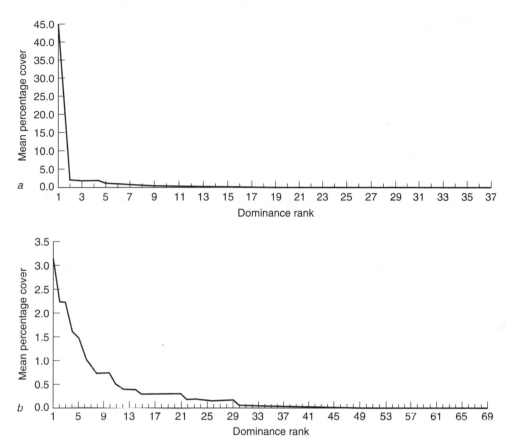

Figure 20.6. Comparison of ground-cover species diversity of two ecosystems using plots of species-abundance. *a,* ecosystem type 1 occurs on an outwash plain landform with excessively drained, infertile, sandy soil; *b,* type 116 occurs on a moraine with moderately well fertile, loamy soil. Abundance is measured by percentage cover. Ground-cover species for each ecosystem are ranked in order by their percent coverage along the x-axis (1=species with greatest abundance). Richness is indicated by the number of species shown at the far right on the x-axis. Evenness is indicated by the shape of the curve, with ecosystem 116 being more "even" in distribution than ecosystem type 1. (After Pearsall, 1995.)

short overstory of small bigtooth and trembling aspens (ecosystem type 1), and (2) a moderately well drained, loamy, fertile moraine supporting a moderately dense canopy of tall bigtooth and trembling aspens, white ash, paper birch, and basswood (ecosystem type 116). In this example, percent coverage is used as the measure of species abundance. The species-abundance plot for type 1 shows lower ground-cover species richness (37 versus 69 species, Figure 20.6) and a sharp drop in dominance from the most dominant species (bracken fern, mean percentage cover: 44) to the second most dominant species (mean percentage cover of about 2). Such a plot is often observed in ecosystems that are extreme in some way, such as in microclimate, fire frequency and severity, light level, or fertility, and in an early successional stage. In contrast, the plot for type 116 illustrates nearly twice the species richness as type 1. It exhibits a much more gradual drop in dominance, from about 3.3 percent to <1 percent, from the most to least dominant species. This plot is more typical of ecosystems that are not at the extremes of microclimate, fertility, light, or disturbance, and in mid-successional stage.

DIFFERENTIATION OR BETA DIVERSITY

Besides species diversity within a given area, one may measure how different or similar a range of samples is in terms of the variety (and sometimes the abundance) of species found in them (Magurran, 1988). For example, communities of a given area that have very few species in common, collectively have a high beta diversity, and vice versa. This change or difference in species variety may be determined for ecosystems or communities that occur along an environmental gradient or within an area supporting diverse sites. The often-used term beta diversity gives little clue to the spatial or ecological context of the diversity that is measured. Thus beta diversity may be usefully understood as differentiation diversity (Whittaker, 1977; Magurran, 1988)—a measure of species diversity along a gradient containing different ecosystems or throughout an area where multiple ecosystems occur.

Beta diversity is typically measured by determining the similarity in species composition between pairs of communities or ecosystems; six methods are assessed and discussed by Wilson and Shmida (1984) and Magurran (1988). Two basic approaches for measuring beta diversity are similarity coefficients and a variety of indices developed for measuring change along gradients. Percentage Similarity and the Jaccard and Sorenson similarity coefficients are widely used where the presence or absence of species is known. These methods relate the number of shared species to the total number of species in two samples along a gradient or in a given geographic area. For example, in simplest form, if two samples are taken and each has the same 10 species, the similarity is 100 percent. However, if only five species are shared then the similarity is only 50 percent (2 x 5/ 10 + 10 = 10/20 = .5). This example of Percentage Similarity illustrates a qualitative (presence/absence) estimate of beta diversity. There is a quantitative form of the Sorensen index that can be used when abundance data are available. Another quantitative measure, the Morista–Horn coefficient, is preferred among all similarity measures because it is independent of sample size (Wolda, 1981).

A new quantitative approach by Loehle and Wein (1994) using a multifactor "information dimension" (d_i) facilitates the measurement of community and ecosystem diversity not only by the species that comprise them, but by their relative abundance in the different communities and the arrangement of communities in the landscape. This new method provides a powerful tool for both displaying and assessing community or landscape ecosystem diversity at multiple scales.

Diversity of Landscape Ecosystems

Studies of diversity have focused primarily on the diversity of species, especially of the alpha and beta types defined by Whittaker (1977). Ecosystem diversity, however, is concerned with higher levels of organization: local and regional landscape ecosystems. Diversity exists and can be treated at a variety of levels of ecological organization, but few researchers have applied the concepts of diversity to the hierarchically nested series of ecosystems or to vegetation types throughout a large landscape. Exceptions are Romme's (1982) study of forest and landscape-level processes in Yellowstone National Park (Chapter 21), Loehle and Wein's (1994) study of forest vegetation types in western Tennessee, and Pearsall's (1995) study of ecosystem diversity in northern Lower Michigan. Because we can distinguish and map landscape ecosystems at multiple scales, we have the spatial framework for determining entire-ecosystem diversity at multiple spatial scales. In the following section, we consider examples of species diversity among landscape ecosystems and an example of ecosystem diversity.

Examples of Diversity

GROUND-COVER SPECIES DIVERSITY IN NORTHERN LOWER MICHIGAN

In the following sections, we examine measures of species diversity in a large landscape in northern Lower Michigan. The purpose is not only to compare various measures of alpha and beta diversity but to examine diversity at several spatial ecosystem scales (major and minor landforms) and soil types, by groups of ecosystems, and by individual ecosystem types. In addition, ecosystem diversity is also considered.

Ecosystem Groups

Diversity was examined for landscape ecosystems of multiple scales that were distinguished and mapped at the 4,000-hectare landscape of the University of Michigan Biological Station in northern Lower Michigan (Lapin and Barnes, 1995; Pearsall, 1995; Zogg and Barnes, 1995). The terrain was shaped by two late-Wisconsinan glacial advances and retreats between 14,000 and 11,000 years ago. The area has been highly disturbed by logging and repeated post-logging fires. Early successional species, primarily bigtooth and trembling aspens, now dominate the overstory of upland ecosystems, whereas many kinds of upland and lowland conifers and hemlock-northern hardwood forests prevailed in the presettlement forest over 200 years ago. Landscape ecosystems were distinguished (using multiple factors of physiography, microclimate, soil, and ground-cover vegetation) and mapped at four hierarchical scales: major landforms, minor landforms, ecosystem groups, and local ecosystem types.

Initially, the three major landforms of outwash plain, ice-contact terrain, and interlobate moraine were subdivided into 21 ecosystem groups (Lapin, 1990). Six of the 21 ecosystems groups, representing extremes along gradients of moisture and nutrient availability, were used in analysis of alpha and beta diversity of ground-cover vegetation. Figure 20.7 illustrates the general location of the ecosystem groups on the Biological Station terrain, and their key site features are given in Table 20.1.

A comparison of alpha diversity indices among the groups is presented in Table 20.2. Ground-cover species richness ranged from 27 species to 16 per 100 m^2 plot, and Shannon–Wiener heterogeneity ranged from 2.21 to 0.61. Ecosystem group 20 (moist and nutrient-rich) was the richest and most heterogeneous, and ecosystem group 4 (dry and nutrient-poor) was the least diverse. The general pattern of ecosystem groups in alpha-diversity measurement, from more diverse to less diverse, was moist and nutrient-rich (group 20), moderately moist and moderately nutrient-rich (group 17), dry and nutrient-rich (group 11), moist and nutrient-poor and climatically extreme (group 2), dry and nutrient-poor and climatically extreme (group 1), and dry and nutrient-poor (group 4) (Tables 20.1 and 20.2). Multiple-comparison methods often showed significant pairwise differences (Table 20.2).

As might be expected, species richness and heterogeneity tended to increase as moisture and nutrient availability increased, as long as other factors (temperature, light, and herbivory) were not limiting. The fire-regenerated aspens that dominate the overstory of all plots of these ecosystem groups provide a favorable light environment for many vascular plants. Survey records indicate that the presettlement forests—dominated by eastern white pine, American beech, and eastern hemlock (Kilburn, 1960)—were probably more dense than today's relatively open canopies of aspens. Just the reverse of the results of ground-cover diversity might have been the case in presettlement time.

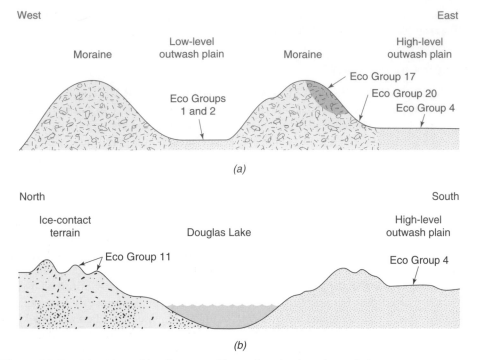

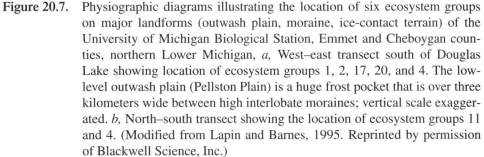

Figure 20.7. Physiographic diagrams illustrating the location of six ecosystem groups on major landforms (outwash plain, moraine, ice-contact terrain) of the University of Michigan Biological Station, Emmet and Cheboygan counties, northern Lower Michigan, *a,* West–east transect south of Douglas Lake showing location of ecosystem groups 1, 2, 17, 20, and 4. The low-level outwash plain (Pellston Plain) is a huge frost pocket that is over three kilometers wide between high interlobate moraines; vertical scale exaggerated. *b,* North–south transect showing the location of ecosystem groups 11 and 4. (Modified from Lapin and Barnes, 1995. Reprinted by permission of Blackwell Science, Inc.)

The ground-cover diversity results reveal a much higher level of diversity than might be expected from examining only the overstory composition. Also, many of the herb and shrub species are shared among different ecosystem groups. Yet, distinguishing the landscape ecosystems based on geologic landform, soil, and climatic factors, and comparing species diversity of these groups, reveals many significant differences in species richness and heterogeneity of the groundflora. That is, there are significant differences in ground-cover composition and pattern within a landscape having an overstory that is dominated only by two closely related tree species—bigtooth and trembling aspen.

This relationship calls attention to the important point that alpha and beta species diversity depend greatly on how sample plots are defined and bounded. If plot boundaries are set by community type (i.e., by cover type such as the "forest of bigtooth aspens"), it may be assumed that all subdivisions will share the diversity of the whole cover type and that any one will be representative. If one sets the plot boundaries by ecosystem type (multifactor landform-climate-soil-vegetation unit), however, then a pattern of diversity emerges within the cover type from which the idea of representativeness can be better gauged.

Table 20.1 Comparative summary of site conditions of six landscape ecosystem groups of the University of Michigan Biological Station, Emmet and Cheboygan counties, northern Lower Michigan.

Ecosystem Group[1]	Site Condition	Physiography	Soil and Drainage
20	moist, nutrient rich	interlobate moraine; flat and moderate slopes	loamy sand over clayey calcareous till; well drained
17	moderately moist, moderately nutrient rich	interlobate moraine; flat	sandy soil with many heavy-textured bands; noncalcareous; well drained
11	dry, nutrient rich	ice-contact terrain	calcareous, gravelly medium sand, somewhat excessively drained
2	dry to seasonally moist; nutrient poor; wide daily and seasonal temperature extremes	low-lying outwash plain between two moraines; flat	deep, noncalcareous medium sand; moderately well drained
1	dry, nutrient poor; wide daily and seasonal temperature extremes	low lying outwash plain between two moraines; flat	deep, noncalcareous medium sand; excessively drained
4	dry, nutrient poor	high-level outwash plain; flat	deep, noncalcareous medium sand; excessively drained

Source: Modified from Lapin and Barnes, 1995. Reprinted with permission of Blackwell Science, Inc.)

[1]Presettlement cover type: groups 20, 17, 11: hemlock northern hardwoods; groups 1, 2, 4: eastern white pine, red pine, northern red oak.

[1]Present cover type: groups 20, 17, 4: bigtooth aspen; groups 1 and 2: trembling and bigtooth aspen; group 11: northern hardwood species and bigtooth aspen.

Analysis of beta diversity further demonstrates the compositional differences in ground-cover vegetation among ecosystems groups. Two measures of beta diversity are presented in Table 20.3—Jaccard's Coefficient seen outside parentheses and Percentage Similarity (within parentheses). Jaccard's Coefficient is a qualitative measure, based on species presence/absence, whereas Percentage Similarity (*PS*) is the similarity of the percentage coverage of the ground-cover species (Krebs, 1989). Jaccard's Coefficient shows that Ecosystem group 20 (moist, nutrient-rich) and 11 (dry, nutrient-rich) share the most species (C_j=56 percent). A number of herb species that are not present in other groups are characteristic of these two groups. Three nutrient-rich groups (20, 17, 11), all with presettlement vegetation of hemlock–northern hardwood forest, are the most similar ecosystem groups based on presence–absence data. Yet, the most similar two groups are only slightly over 50 percent alike. The least similar groups (less than one-third similar) are the various combinations of nutrient-rich (20, 17, 11) and the nutrient poor, climatically-extreme ecosystem group (1).

Percentage Similarity (*PS*) indicates the extreme differences in species coverage among the ecosystem groups (Table 20.3, values in parentheses). Percentage similarity

Table 20.2 Comparison of alpha diversity indices for ground-cover plant species among six ecosystem groups of the University of Michigan Biological Station, Emmet and Cheboygan Counties, northern Lower Michigan.

Index[2]	Ecosystem Group[1] 20	17	11	2	1	4
S_{GC}	27.2	18.4	17.2	23.6	19.4	15.8
	a	bc	bc	ac	bc	b
S_H	14.2	7.8	9.0	10.2	9.4	7.2
	a	b	b	b	b	b
S_W	13.0	10.6	8.2	13.4	10.0	8.6
	a	abc	c	ab	ac	c
H'	2.21	1.66	1.15	0.99	0.70	0.61
			a	ab	b	b
$1\text{-}D$	0.827	0.704	0.537	0.420	0.268	0.241
	a	ab	bc	cd	cd	d
E	0.68	0.57	0.41	0.31	0.23	0.22
	a	a	b	bc	c	c

Source: Modified from Lapin and Barnes, 1995. Reprinted with permission of Blackwell Science, Inc.)

[1]Number of plots=5 for all groups. There was no significant difference between groups that share a letter, alpha=0.1

[2]S_{GC}=ground-cover species richness, S_H=herbaceous species richness, S_W=woody species richness, H'=Shannon–Wiener heterogeneity, $1\text{-}D$=Simpson's heterogeneity, E=Shannon–Wiener evenness.

Table 20.3 Comparison of beta diversity measures for ecosystem groups of the University of Michigan Biological Station, Emmet and Cheboygan counties, northern Lower Michigan.

Group	20	17	11	2	1	4
20	———					
17	53 (57)	———				
11	56 (28)	47 (36)	———			
2	42 (23)	36 (14)	31 (2)	———		
1	33 (20)	30 (12)	28 (2)	39 (56)	———	
4	36 (23)	46 (16)	34 (3)	39 (55)	37 (92)	———

Source: Modified from Lapin and Barnes, 1995. Reprinted with permission of Blackwell Science, Inc.)

Jaccard's Coefficient for all combination-pairs of six ecosystem groups (20, 17, 11, 2, 4, 1) is shown outside parentheses, and the Percentage Similarity *(PS)* is shown within parentheses.

Jaccard's Coefficient (species presence/absence): $C_j = j/(a + b - j)$, where j equals the number of species present in both samples, a equals the number of species present in sample 1, and b equals the number of species present in sample 2 (Magurran, 1988). Percentage similarity, PS=minimum $(p1i, p2i)$, based on the coverage of a species in a sample where $p1i$ equals the proportion of species i in sample 1 and $p2i$ equals the proportion of species i in sample 2 (Krebs, 1989).

ranged from 2 to 92 percent, with most of the pairs of ecosystem groups from 12 to 36 percent similar. Ecosystem group 11 (dry, nutrient rich) is very dissimilar to the three nutrient-poor groups (2, 4, and 1; PS=2–3 percent). The pair with greatest PS (92) are groups 1 and 4, both of which are dry and nutrient poor. However, they are not spatially adjacent but occur in different landforms (Figure 20.7). Although ecosystem groups may share many species, their relative coverage and the role that they play in different ecosystems may be very different. For example, ecosystem groups 20 and 11 share 56 percent of the species (Table 20.3), but their PS in coverage is only 28 percent. Conversely, groups 1 and 2 share only 39 percent of the same species but are 56 percent similar in species coverage.

Ecosystem Types

The classification and mapping of 125 local landscape ecosystems types (units within ecosystem groups) for the 4,000-hectare tract described previously facilitated the study of alpha and beta diversity of ground-cover species within and among diverse major and minor landforms. For example, several measures of alpha diversity in Table 20.4 illustrate differences in species richness, heterogeneity, and evenness for major landforms (outwash plains and moraines), groups of ecosystem types within these landforms, and ecosystem types with calcareous versus non-calcareous soils. Notice also in Table 20.4 that the diversity values are presented in two ways: for all ecosystems and plots within a given unit and the average per unit area (within parentheses). For example, total species richness for outwash plains with 41 ecosystem types is 182, whereas the average per 150 m^2 plot is 20.2. At the major landform level, moraine landforms exhibit greater richness and heterogeneity of ground cover species per unit area than outwash plains (26 versus 20, respectively) despite the greater number of ecosystem types in the outwash plains. Ecologically, this finding is related to the greater soil water and nutrient availability on moraines than on the outwash plains.

Within moraines, species richness, heterogeneity, and evenness are markedly higher in the bigtooth aspen-dominated interlobate moraine as compared with the hemlock-northern hardwood-dominated Colonial Point moraine (Table 20.4). These landforms that are very similar in their high soil water and nutrient availability. However, low ground-cover richness occurs on the Colonial Point moraine primarily due to the dense overstory and understory of beech, sugar maple, hemlock, and other species that shade the forest floor at Colonial Point. In contrast, bigtooth aspens, with relatively open crowns, dominate ecosystems of the Interlobate moraine. The relationship between species richness and canopy cover for these two moraine landforms is illustrated in Figure 20.8. Richness is greatest at intermediate levels of canopy coverage, whereas under a dense canopy the low light reaching the forest floor tends to be limiting to many species.

Within outwash plains, two contrasts are shown in Table 20.4—between (1) the broad, climatically severe Pellston outwash plain (giant frost pocket, Figure 20.7) and the warmer, high-level outwash plains, and (2) the non-calcareous versus calcareous ecosystem types of the outwash plains. The contrast of the climatically severe with warmer high-level plains shows their similar heterogeneity (H') of species per unit area. Note that the diversity measures for the total high-level outwash landform are higher, due primarily to the nearly double number of ecosystem types in the high-level outwash landform. The contrast of calcareous versus non-calcareous ecosystems in outwash plains (with similar overstory coverage of aspens) shows that the more nutrient rich calcareous outwash exhibits greater diversity, as one might expect. Nutrient requiring and non-requiring plants grow in calcareous ecosystems, whereas some plants requiring relatively high nutrient levels are excluded in the dry, non-calcareous outwash soils.

Table 20.4 Alpha diversity measure of ground-cover diversity for upland ecosystem groups on outwash plain and moraine landforms of the University of Michigan Biological Station, Emmet and Cheboygan counties, northern Lower Michigan.

Ecosystem Group	Number of Ecosystem Types	Number of Plots[1]	Index[2]			
			S	H'	1-D	E
Major landforms						
Outwash plain	41	175	182 (20.2)	1.77 (1.22)	0.54 (0.48)	0.34 (0.41)
Moraine	9	48	141 (25.8)	3.06 (1.84)	0.93 (0.71)	0.62 (0.58)
Minor landforms						
Interlobate moraine	3	26	123 (31.5)	3.05 (21.3)	0.93 (0.80)	0.63 (0.62)
Colonial Point moraine	6	22	76 (19.1)	2.29 (1.49)	0.81 (0.60)	0.53 (0.52)
Low-level outwash plain	11	34	132 (23.6)	1.53 (1.16)	0.46 (0.45)	0.31 (0.36)
High-level outwash plain	21	121	142 (18.9)	1.75 (1.17)	0.56 (0.47)	0.35 (0.40)
Soil type in outwash plains						
Calcareous outwash	18	70	142 (22.4)	1.89 (1.33)	0.60 (0.52)	0.38 (0.43)
Non-calcareous outwash	23	105	149 (18.3)	1.63 (1.05)	0.50 (0.43)	0.32 (0.36)

Source: Courtesy of Douglas Pearsall.

[1] plot size = 150m²

[2] S=ground-cover species richness, H'=Shannon-Wiener heterogeneity, 1-D=compliment of Simpson's index, E=Shannon-Wiener eveness. Diversity values *outside parentheses* are based on all ecosystem types and plots for a given unit. Diversity values *inside parentheses* are expressed on a per unit area basis, in this study 150 m². For example, there is a total of 182 different ground-cover species in the 175 plots and 41 ecosystem types in outwash plain landform, whereas the average number of species per unit area (150 m²) is 20.2. See text for discussion.

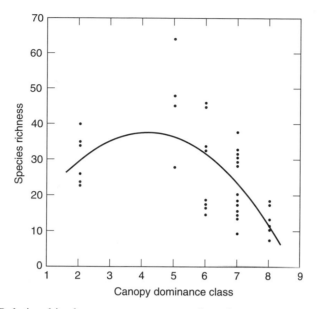

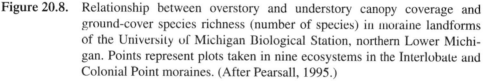

Figure 20.8. Relationship between overstory and understory canopy coverage and ground-cover species richness (number of species) in moraine landforms of the University of Michigan Biological Station, northern Lower Michigan. Points represent plots taken in nine ecosystems in the Interlobate and Colonial Point moraines. (After Pearsall, 1995.)

Beta diversity of ground-cover species, based on paired comparisons of ecosystems in upland landforms, is essentially the same for ground-cover vegetation of ecosystem types of outwash plains and moraines. Beta diversity is relatively low in the Interlobate and Colonial Point moraines where there are relatively few ecosystem types and they have similar species richness and coverage.

Overall, both alpha and beta diversity analyses and relationships reveal the usefulness of hierarchical classification and mapping of landscape ecosystems in examining biodiversity at multiple spatial levels. Although much of the outwash plain landscape appears relatively flat and strongly dominated by aspens, the multifactor-multiscale landscape ecosystem approach reveals site differences across the landscape that are ecologically meaningful and that are reflected in differences in ground-cover vegetation. These differences are also reflected in ecosystem processes such as nutrient cycling (Zak et al., 1986) and succession (Host et al., 1987).

Ecosystem Diversity

In the previous examples, we have considered species diversity. In landscapes where whole ecosystems of a given scale have been distinguished and mapped, the map itself (see Figures 2.12 and 13.11) provides a visual estimate of ecosystem type richness and evenness. For example, in the Huron Mountains in Marquette County of Upper Michigan, the Pine River flows through a large expanse of Lake Superior beach (Figure 20.9). In this landscape, six riverine ecosystem types are tightly clustered along the river, and their number, size, and spatial distribution are in marked contrast to the single, flat, fire-prone beach ecosystem dominated by jack pine. The riverine ecosystems exhibit diverse soil conditions that range from extremely acid and infertile peat to very fertile muck and a remarkable richness of tree, shrub, herb, and bryophyte species (Simpson et al., 1990). Sim-

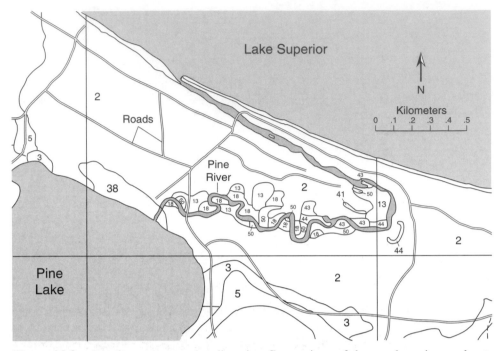

Figure 20.9.   Landscape ecosystem diversity. Comparison of the number, size, and pattern of landscape ecosystem types of an area along the south shore of Lake Superior, Huron Mountain Club, Marquette Co., upper Michigan. In contrast to the extensive beach terrace (ecosystem type 2) six small ecosystem types (13, 18, 41, 43, 44, 50) border the Pine River where it flows through the beach ecosystem. (After Simpson et al., 1990.)

ilar "hotspots" of high ecosystem diversity were easily identified on the ecosystem map of the 4,000-hectare Biological Station tract described above (Pearsall, 1995). High biodiversity is associated with these hotspots of ecosystem diversity in the glacial terrain of Michigan (Simpson et al., 1990; Pearsall, 1995). Where ecosystem maps are available, land managers have the opportunity to efficiently conserve biodiversity by identifying areas of high ecosystem diversity.

The diversity of landscape ecosystems and landforms can be assessed quantitatively using conventional alpha and beta diversity indices. Also, ecosystem diversity can be quantitatively determined by using the information dimension (d_i) of Loehle and Wein (1994). Using this approach, Pearsall (1995) used the number of ecosystems, multiple characteristics of their physiography, soil, and vegetation, and their pattern in the landscape to determine the ecosystem diversity of landforms and ecosystems of the Biological Station tract. Diversity of ecosystems was highest in outwash plains, and the centers of highest diversity that were apparent from ecosystem maps were confirmed by this multivariate diversity measure.

CAUSES OF SPECIES DIVERSITY AT MULTIPLE SCALES

In the following sections, we examine biological diversity from the broad scale of macroecosystems and continental events to the local scale of microecosystems. The objective is to consider some of the main factors that work together in creating the pattern of diversity that we observe.

Diversity at Continental and Subcontinental Scales

PALEOGEOGRAPHY AND CONTINENTAL RELATIONSHIPS

At the macroecosystem scale, the age of continents, their relative isolation or contiguity with other land masses, and their movements across the Earth's surface have provided opportunity for speciation and extinction. Continental drift has had a major influence on the range of species present on any continent. Collision of continents permitted species from one continent to colonize another. About 100 million years ago, the modern continents were in place, and major floras and faunas developed in a state of increasing isolation. Remarkable floras and faunas developed in Australia and New Zealand in contrast to Asia. South American, African, and North American floras differentiated in unique ways, but in all cases were strongly influenced by their historical origins and development.

A fascinating example is the history of the beeches of the northern and southern hemispheres that share the family Fagaceae but are in different genera. Beeches of the genus *Fagus* ("true" or northern beeches) are deciduous and differ in many other respects from the southern or "false" beeches of the genus *Nothofagus*. Today, southern beech forests (genus *Nothofagus*) are found in South America, Australia, New Zealand, New Caledonia, and New Guinea (Figure 20.10). The plant fossil record of these countries and of Antarctica shows that they might have descended from similar forest that was once thought to have been extensive in the cool and temperate latitudes of the gigantic land mass, Gondwanaland, that existed 150 to 100 million years ago (Poole, 1987; Barlow and

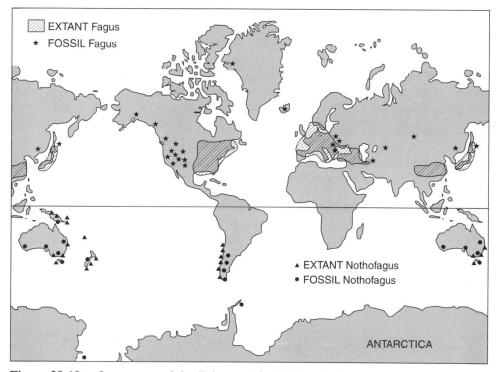

Figure 20.10. Occurrence of the living and fossil taxa of the northern beeches (*Fagus*) and the southern beeches (*Nothofagus*). (After Poole, 1987.)

Hyland, 1988). What the origin of those forests was and how they came to be separated from the northern hemisphere *Fagus*, so-called true beech—presumably true because they were first named by botanists in that part of the world—remain mysteries. Figure 20.10 illustrates the continental and hemispheric separation of living and fossil beeches of the Earth. Consideration of the movements of continental plates, isolation, and evolution of temperate and tropical forests of Australia's flora by Barlow and Hyland (1988) provide insights to the complex paleogeographic and paleoclimatic derivation of the remarkable diversity of this region.

For nearly two centuries the causes for the striking differences in tree species diversity among regions of the Earth has interested biogeographers and ecologists. Wet tropical forests have 10 times as many trees species as moist forest of the northern and southern temperate zones at several spatial scales. Various explanations have been offered. Yet no one mechanism has gained general acceptance, and several may contribute to the complex patterns. An outstanding review of regional tree diversity and comparison of temperate-zone diversity is presented by Latham and Ricklefs (1993b). They observe that ecologists generally have invoked local-scale deterministic processes of competition and other interactions to explain global-scale diversity patterns. It is a large step from local-scale processes to such regional-scale predictions. Nevertheless, differences in diversity between regions with similar climate suggest that local and regional processes contribute separately to regional species richness. Their analyses indicate that contemporary patterns of tree species diversity owe much to historical and evolutionary events, as noted for the beeches.

Latham and Ricklefs (1993b) provide a comprehensive analysis of present-day taxonomic diversity among major moist temperate forest regions of the Northern Hemisphere: (1) northern, central, and eastern Europe, (2) east-central Asia, (3) Pacific slope of North America, and (4) eastern north America. A total of 1,166 species make up the characteristic tree flora of these regions, approximately in the ratio 2:12:1:4 (Table 20.5); eastern Asia has by far the most species and genera.

A comparison of genera that occur in either two or three of the regions (i.e., those that are neither endemic nor cosmopolitan), is presented in Figure 20.11. Of these 63 genera, 59 (94 percent) occur in eastern Asia. Regions sharing the most genera (20) are

Table 20.5   Summary of Moist Temperate Forest Trees in the Northern Hemisphere by Taxonomic Level and Geographic Region.

Taxonomic level	Number of tree taxa characteristic of moist temperate forests in:				
	Northern, central, & Eastern Europe	East-central Asia	Pacific slope of North America	Eastern North America	Northern Hemisphere (total)
Subclasses	5	9	6	9	10
Orders	16	37	14	26	39
Families	21	67	19	46	74
Genera	43	177	37	90	213
Species	124	729	68	253	1,166

Source: After Latham and Ricklefs, 1993b. Reprinted with permission from *Species Diversity in Ecological Communities,* Ricklefs and Schluter, eds., © 1993 by the University of Chicago. Reprinted with permission of the University of Chicago Press.)

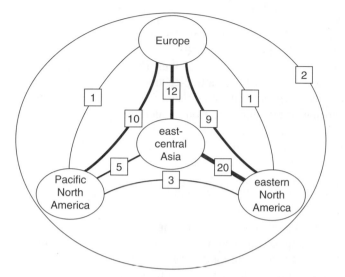

Figure 20.11. Numbers of tree genera native to either two or three of the four north temperate forest regions. Straight lines and the middle curved lines (bowing out) indicate genera that occur in two regions. The inner curved lines (bowing in) indicate genera that occur in three regions including east-central Asia. The outer ellipse indicates genera that occur in the three regions excluding east-central Asia. (After Latham and Ricklefs, 1993b. Reprinted with permission from *Species Diversity in Ecological Communities,* Ricklefs and Schluter, eds., © 1993 by the University of Chicago. Reprinted with permission of the University of Chicago Press.)

eastern Asia and eastern North America, reflecting the well-known range disjunction of many plants inhabiting both regions (Boufford and Spongberg, 1983). Despite their proximity, the temperate forests of eastern and western North America share few genera and none uniquely. Eastern Asia emerges overwhelmingly as the core area of temperate tree genus distributions of the Northern Hemisphere.

Why the high species richness in eastern Asia? The marked differences in diversity of temperate tree floras between east-central Asia and other regions of the Northern Hemisphere is probably ancient (Latham and Ricklefs, 1993b). It appears to have arisen from differences between regions in colonization history and perhaps in subsequent rates of proliferation of endemic taxa. Colonization appears to have occurred more frequently in Asia, giving rise to new native taxa and their geographical spread among the temperate regions. The species proliferation in Asia is reflected in the occurrence of markedly more species at the local level than found in temperate forests elsewhere. Reasons for this include the enormous site differentiation provided by ubiquitous mountainous terrain and the lack of significant influence of continental glaciation. Current geographical distributions and the fossil record suggest that most cosmopolitan taxa of temperate trees originated in eastern Asia and dispersed to Europe and North America. Furthermore, differences between temperate and tropical flora reflect a physiological barrier to colonization of temperate zones that can be crossed only by the evolution of freezing tolerance mechanisms. Overall, explanations for latitudinal gradients in richness can be traced primarily to historical and evolutionary factors affecting land masses and floras rather than to present-day ecological interactions.

GLACIATION

Continental glaciation has had marked effects on species diversity, and many of these effects are described in Chapters 3 and 22. Extinctions have occurred in areas where species were unable to migrate to their former locations. Elimination of species was severe in central Europe where the Alps acted as a barrier to southerly migration of species. The massive continental ice sheets altered the climate of areas that were located at great distances from the advancing or retreating ice fronts. Such differences for eastern North America are discussed in Chapters 3 and 22 and seen in Figure 3.5.

Glaciation has also affected sea levels around the world. Low sea levels during the Pleistocene Era caused formation of a land bridge between Asia and North America allowing immigration of humans into the Americas. The presence of humans has been associated with elimination of many of the large North American mammals such as the sabertooth tiger and the American horse. As we have seen in Chapters 16 and 17 on disturbance and succession, human alteration of disturbance regimes of fire and flooding has been widespread. The extent to which humans have influenced diversity in the Americas is unknown, but their introduction during the Pleistocene certainly had a remarkable effect on the structure and diversity of North American forests.

LATITUDE AND ELEVATION

Plant richness decreases with both increasing elevation and increasing latitude toward the Arctic (Billings, 1973, 1995). Reasons for this decrease are very long-term evolutionary and geographic processes and local ecological conditions such as decreases in mean annual and growing season temperatures, drought stress, and ultraviolet-B irradiation at high elevations (Billings, 1995).

The latitudinal gradient in diversity of many organisms from the tropics to the poles has generated great scientific interest. Wilson (1992) illustrates this gradient by numbers of breeding bird species by a slice through the northern hemisphere using land areas of roughly the same size: Greenland, 56; Labrador, 81; Newfoundland, 118, New York State, 195, Guatemala 469, and Columbia, 1,525. He notes that 30 percent of the Earth's bird species occur in the Amazon Basin and another 16 percent in Indonesia. Tree species show an even more striking and ecologically-significant gradient of decreasing richness with increasing latitude. The remarkable tree and plant richness of the wet tropics is well known and is discussed in Chapter 22. However, the latitudinal relationship is complex due to marked differences in physiography and climate of regional ecosystems within any continent.

For example, the geographical pattern of tree species richness for much of North America, illustrated in Figure 20.12, shows that factors other than latitude affect richness. Tree richness is greatest in the southeastern United States and decreases steadily northward. It also decreases in the arid parts of the continental interior showing that latitude (i.e., primarily temperature-related factors) is not the only determinant of richness. A strong relationship was found between tree richness and annual actual evapotranspiration and mean annual temperature when large geographic quadrats were used in computation (Figure 20.12, Currie and Paquin, 1987). However, at the fine scale, no significant relationship was found between actual evapotranspiration and local tree richness (0.5 to 10-ha plots) of broad-leaved deciduous trees (Latham and Ricklefs, 1993a,b). The tree richness pattern in North America reflects the evolutionary history of broad-leafed trees and the relative newness of continental arctic climates compared to the unglaciated areas of southeastern United States (Latham and Ricklefs, 1993a,b) as well as local ecological interactions.

Plant diversity is strongly related to elevational gradients that are steeper up the sides of mountains than latitudinal gradients. Generally, plant diversity decreases as mountain

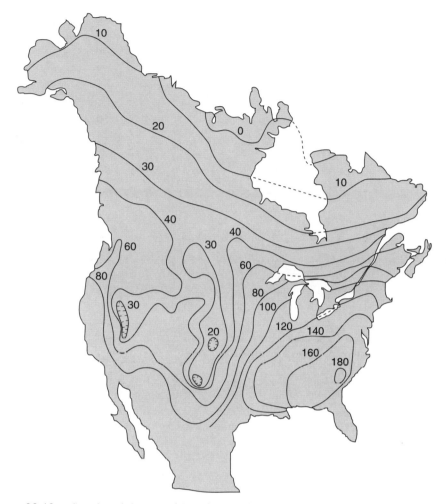

Figure 20.12.   Species richness of North American trees north of the Mexican border. Contours connect points with the same approximate number of species per quadrat. Data is based on 620 native tree species in quadrats each 2 1/2° × 2 1/2° south of 50° N, and 2 1/2° × 5° (longitude) north of 50° N. (Reproduced with permission from *Nature* (Currie, D. J., and V. Paquin. Large-scale biogeographical patterns of species richness.), ©1987 Macmillan Magazines Limited.)

elevations increase (Billings, 1987). This relationship is especially true when elevation is combined with increasing latitude as illustrated in Figure 20.13 for relatively high mountain ranges in Europe and Asia.

An excellent example contrasting patterns of plant species diversity in forests of two mountain ranges and two markedly different regional ecosystems of the western United States is presented in Figure 20.14. In the relatively humid, maritime climate of the Siskiyou Mountains of southwestern Oregon, the slopes are forested at all elevations that were sampled. Richness of woody plant species decreases with increasing elevation; herb diversity is highest at middle elevations and lower in the dense stands with a sclerophyll

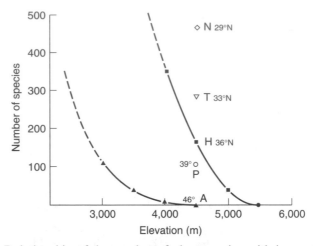

Figure 20.13.   Relationship of the number of plant species with increasing elevation in the European Alps (left) compared with high mountain ranges in central Asia (right) having many peaks above 6,000 meters. The letters show the increasing number of species with decreasing latitude: A = Alps, P = Pamirs, H = Hindu Kush, T = Tibet, N = Nepal Himalaya. (From Breckle, 1974. Reprinted with permission by the Council for Nordic Publications in Botany, Copenhagen, Denmark.)

tree stratum at low elevations (Whittaker, 1960). The combination of these two trends produces the pattern of maximum richness in mesic forests at middle elevations (ca 1,200 m), with a secondary maximum at lowest elevations. Notice that at all elevations, richness increases from xeric (ridges and SSW aspects) to mesic and wet ravines. In the drier, continental climate of the Colorado Front Range of the Rocky Mountains, the forests are more open and without sclerophyllous trees. Forests give way to shrubland and grassland at low elevations and to alpine meadows at high elevations. Plant richness is low in forests of middle elevations and middle topographic positions of exposed, open slopes. Richness is higher in ravine forests, the transition to alpine meadow, and the more open stands of driest topographic situations and low elevations. Again, richness increases from xeric aspects to mesic and wet ravines.

   These examples emphasize the importance of considering biodiversity through studies of regional ecosystems (southwestern Oregon mountains versus Colorado Rocky Mountains) and the local ecosystems (mesic ravines, open mid-elevation slopes, xeric ridges, subalpine woodland etc.) that occur at positions along gradients of temperature and moisture in the respective mountain systems.

Diversity at Local Scales

There are few sure or "global" generalities or principles that we can offer about biodiversity. For example, fertile soil may or may not promote diversity, and biomass productivity may be positively or negatively correlated with diversity. However, two concepts seem reasonably clear. Heterogeneity promotes diversity at all scales, and extreme conditions tend to constrain diversity. For example, a heterogeneous pattern of ecosystems or their components (climate, physiography, soil, hydrology) and heterogeneous disturbance regimes and stand structures tend to promote diversity. In contrast, the most extreme phys-

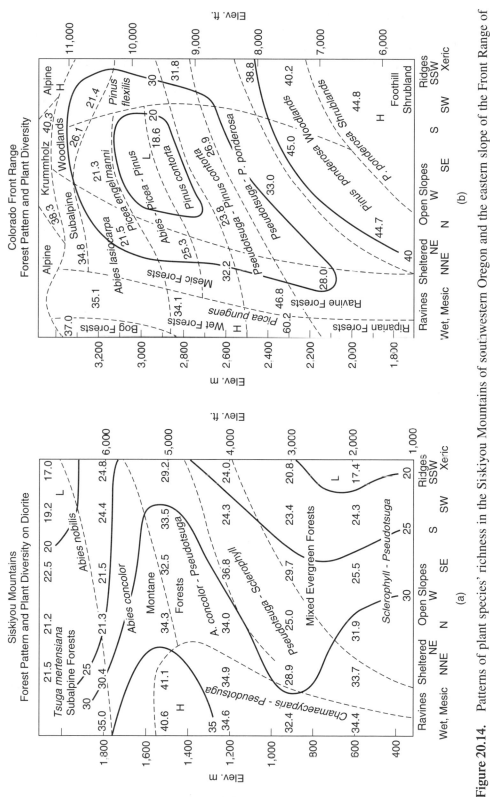

Figure 20.14. Patterns of plant species' richness in the Siskiyou Mountains of southwestern Oregon and the eastern slope of the Front Range of the central Rocky Mountains in Colorado. Numbers are for vascular plants species in 0.1-hectare quadrats averaged for several stands representing a given combination of elevation and physiographic position. (After Whittaker, 1977. Reprinted from *Environmental Biology* with permission of the Plenum Publishing Corp.)

ical conditions of temperature, water, soil reaction, and combinations thereof tend to limit biodiversity. Above all, however, biodiversity is most closely related to the spatial pattern, diversity, and key characteristics (physiography, climate, soil, and disturbance regimes) of regional and local ecosystems. It is beyond the scope of this chapter to examine the enormous array of factors and multiple-factor determinants of species diversity so that only selected ecological relationships are discussed in the following sections.

PHYSIOGRAPHY AND SOIL

At meso- and microecosystem scales, physiography and soil, as described in Chapters 10 and 11, strongly influence the distribution and diversity of plants. Physiography, in particular, because of its elements of form and parent material markedly affects plant diversity by its effects on microclimate and soil development. Diversity in mountainous regions in the temperate zone and in unglaciated areas is often very high because of the dissection (i.e., heterogeneity) of terrain and the diverse sites and niches available for plant occupancy. Figures 2.12, 10.2, 10.4, 10.12, 11.8, 13.11, 15.1–3, and 6; and 20.14 all illustrate the striking effect of physiography as it mediates diversity in microclimate and soil water and nutrients.

Microtopography increases species diversity by providing highly localized environmental heterogeneity (Chapter 10, Figure 10.15; Foster, 1988a). Regeneration of some understory tolerant species depends on microtopography, either of mounds of treefalls or from coarse woody debris. For example, yellow birch and eastern hemlock seedlings establish preferentially on moss-covered logs and mounds in hemlock–northern hardwood forests (Chapter 5) as do seedlings of western hemlock and sitka spruce in old-growth forests of the Pacific Northwest (Harmon, 1987; Harmon and Franklin, 1989).

Soil conditions of water, nutrient, and O_2 availability account for many observed differences in diversity. Soils that are extreme in some attribute—low water or nutrient availability, high in specific mineral content (e.g., soil derived from serpentine parent material), or very shallow—often support forests low in tree richness (e.g., pine barrens, black spruce swamps). However, such soils may support many herbs or rare and unusual herbs that have adaptations giving them a competitive advantage.

Soils that are high in nutrient availability may support high tree species richness. For example, for a broad range of ecosystems in the Piedmont Province of North Carolina, vascular plant richness was strongly related to soil nutrient availability (Figure 20.15). High richness was also associated with abundant moisture, and among sites with low soil fertility, the moister sites had more species.

Soils that are deep and also high in water and nutrient availability may support either high- and low-diversity forests depending on the history and amount of disturbance and successional stage. Dominance by one or a few species often occurs more rapidly under favorable conditions, where all organisms can potentially grow rapidly (Huston (1993, 1994). Under such conditions, the best competitors are able to eliminate most other species by monopolizing an essential resource such as light. Under poorer soil water and nutrient conditions, the superior competitors are not able to dominate as rapidly, and many species are likely to coexist, leading to individuals that are smaller in size, fewer in number, or both.

In mesic beech–sugar maple and hemlock–northern hardwood forests of the western Great Lakes region, canopy coverage in late-successional stands may be very dense: Relatively few species survive in the understory and ground-cover layers other than young individuals of the canopy species (Figure 15.16). This relationship for a mesic hemlock–northern hardwood forest is illustrated in Figure 20.8. Ecosystems in this fertile and moist Colonial Point moraine exhibit relatively low ground-cover diversity (Table 20.4). How-

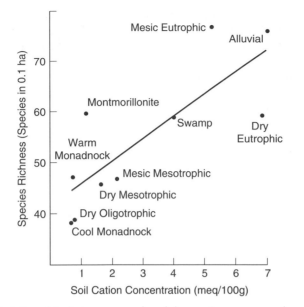

Figure 20.15. Relationship between species richness (average number of plant species in a 0.1-hectare sample) and average exchangeable cation (Ca, Mg, K) for a diverse range of ecosystem types, representing wet, mesic, and dry sites, of the North Carolina Piedmont. (After Peet and Christensen, 1980b. Reprinted with permission of the ETH Zürich, Finanzdienste, CH8092 Zürich, Switzerland.)

ever, recall from the discussion above that a bigtooth aspen community, with relatively-open crowns, on a similarly fertile and mesic soil (Table 20.4, interlobate moraine) has approximately 40 percent greater ground-cover richness due primarily to favorable light conditions. In this case, catastrophic fire replaced the northern hardwood forest with aspens. Ground-cover diversity observed in summer in ecosystems with a closed canopy may be somewhat misleading because of a relatively diverse flora of early spring ephemerals that complete much of their aboveground growth before the canopy trees leaf-out in the spring.

COMMUNITY COMPOSITION AND STRUCTURE

The diversity of birds, mammals, and other forest animals may be affected by: (1) the species composition of forest stands, and (2) community heterogeneity or structure/age differences within and among communities, vertical layering of vegetation, and horizontal arrangement of trees and other vegetation within and among stands (e.g., random versus patchy stem distribution). A classic case is the relationship of bird species richness and structural complexity of vegetation or foliage height diversity. Bird species diversity has been shown to be positively correlated with foliage height diversity in northwestern North America (MacArthur and MacArthur, 1961) but negatively correlated with it in northern Patagonia, Argentina (Ralph, 1985; Huston, 1994, p. 41). The northern spotted owl (*Strix occidentalis*), a rare inhabitant of old-growth forests of the Pacific Northwest, requires large hollow trees for nesting and open subcanopy areas for foraging. The spotted owl controversy, a watershed event of American environmental policy (Yaffee, 1994), revolves around conserving structurally-heterogeneous old-growth forests of the Pacific

Northwest to retain habitat of the owl and many other organisms of these unique forests (Franklin et al., 1981; Franklin et al., 1997; see Chapter 21). In contrast, the red-cockaded woodpecker (*Picoides borealis*), a rare bird of southeastern United States, is associated with old-growth pine communities with low structural diversity. This woodpecker nests in colonies in cavities that are carved over generations in large, living, fungally infected, resin-rich, longleaf, slash, or loblolly pines (Lennartz, 1988; McFarlane, 1992). The birds preferentially reside in ecosystems having large trees with little or no mid-story, few tall shrubs, and generally sparse ground-cover. Protection from predators, hot fires, and availability of insect prey appear associated with the bird's habitat—one with simplified vertical structure.

DISTURBANCE AND SUCCESSION

It is difficult to generalize about the effect of disturbance on species diversity. Ecosystems are so diverse in composition, structure, and function that they defy generalized textbook examples. Disturbance, being ecosystem specific, typically maintains a diversity level of trees, ground-cover, and animals consistent with the geographic setting of regional ecosystems and the characteristic physiography, soil, and vegetation of local ecosystems (see Chapters 16 and 21). Diversity patterns of hurricane-prone ecosystems are different than fire- or flood-prone ecosystems. Overall, many ecologists have found that low species diversity occurs at both very high and very low frequencies and the highest diversity at intermediate frequencies of disturbance (Huston, 1994). This concept applied to tree species of a tropical rain forest in Uganda is illustrated in Figure 20.16.

Depending on ecosystem type, disturbance can be perceived to maintain either high or low species diversity. For example, frequent fires in presettlement time maintained some dry, outwash-plain ecosystems in the Lake States with jack pine as the only dominant tree species, whereas tree species richness increases markedly in the adjacent mesic moraines in which fires are virtually absent. Simultaneously, periodic fire in the outwash-plain ecosystems maintains high herbaceous richness because the herbs die out as the jack pine canopy closes as the trees develop in crown density with age.

A universal example of fire disturbance reducing temporal species richness is observed in ecosystems where mesophytic, fire-sensitive species colonize the site during the period of fire absence. In these situations, richness increases, but then decreases at the occurrence of the next fire. For example, in fire-prone ecosystem types in northern Florida and southern Georgia, diversity of canopy trees is low in the sandhill, flatwoods, and cypress ecosystems, each being nearly monotypic (Monk, 1967). Diversity of saplings and seedlings is higher than that of canopy trees and includes hardwood species that colonize the understory—only to be eliminated or reduced in size by the next fire. In contrast, the southern mixed hardwoods, mixed hardwood swamps, and bayheads that seldom burn, the same richness of species is found in both the canopy and the understory.

In some ecosystems, especially those characterized by periodic fires, frequent, less severe disturbances are necessary for the maintenance of diversity. For example, lack of periodic fires in prairie or savanna ecosystems typically leads to the encroachment of woody plants (Curtis, 1959) and a reduction in herbaceous diversity. In an attempt to restore presettlement oak savanna structure and composition to a northern pin oak community in central Minnesota, annual prescribed burning was begun in 1964 (White, 1983; 1986). After 13 years of burning, the basal area of oaks was significantly reduced in burned versus unburned plots. Shrub cover was eliminated in burned plots, and the richness of herbs significantly increased in burned plots by about 60 percent.

Overall, fire is an extremely important factor affecting biodiversity because pyrodiversity promotes biodiversity (Martin and Sapsis, 1991). Pyrodiversity, the variety in in-

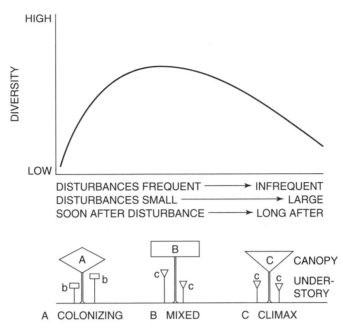

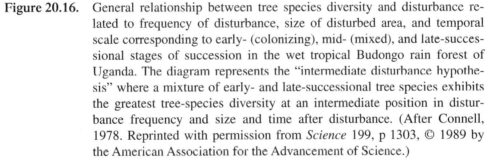

Figure 20.16. General relationship between tree species diversity and disturbance related to frequency of disturbance, size of disturbed area, and temporal scale corresponding to early- (colonizing), mid- (mixed), and late-successional stages of succession in the wet tropical Budongo rain forest of Uganda. The diagram represents the "intermediate disturbance hypothesis" where a mixture of early- and late-successional tree species exhibits the greatest tree-species diversity at an intermediate position in disturbance frequency and size and time after disturbance. (After Connell, 1978. Reprinted with permission from *Science* 199, p 1303, © 1989 by the American Association for the Advancement of Science.)

tervals between fires, seasonality, size, and fire characteristics, has produced biodiversity at all ecosystem levels and worldwide from prehistoric times onward.

Disturbance markedly affects species diversity in successional sequences because these are ecosystem specific. Overall, total plant species diversity is closely associated with the four general stages of secondary succession following disturbance (Chapter 17; Figure 17.4): (1) stand initiation or establishment, (2) stem exclusion or thinning, (3), understory reinitiation or transition, and (2) old-growth. We may characterize diversity of vascular plants in layers for the respective stages as follows: (1) high diversity in an open site soon after disturbance as species resprout or colonize by seedlings, (2) low diversity as dense overstory may "shade-out" all but the most understory tolerant species, (3) low but increasing diversity as gaps appear in the overstory (Figure 17. 14), and (4) low to high diversity depending on canopy density and number of gaps, as minor and major disturbances affect the declining old-growth forest more than younger stands of earlier stages. Examples of this general pattern of diversity with successional stage are given for Rocky Mountain forests (Peet, 1978) and for old-field mixed conifer–hardwood forests of the North Carolina piedmont (Christensen and Peet, 1981; Peet and Christensen, 1987). Obviously, disturbances occurring during these arbitrary stages would affect diversity in different ways depending on ecosystem type and disturbance characteristics—kind (wind, fire insect, etc.), origin (natural or human), and severity.

HUMAN EFFECTS AND CONCERNS FOR DIVERSITY

Species extinction has escalated with increasing human populations and domestication of natural ecosystems by intensive agriculture, forest management, and urbanization. Maintaining the local, regional, and global diversity of species has become an increasing focus of ecologists, managers, and the public. An entire science of conservation biology has arisen in response to a desire to maintain species richness. Of special concern is the management of species whose ranges or habitats are highly limited in area, conservation of species whose presence or absence greatly influences the ecosystems in which they reside, and reduction of invasive exotic species that have the potential to outcompete native species or alter ecosystems in which native species occur (see Chapters 14 and 19). In the following sections, we examine several of these topics.

Endemics and Rare and Endangered Species

Much of the concern about diversity arises around species that are considered rare, in danger of extinction, or endemic. **Endemic** species are those whose ranges are extremely limited in size, being restricted to an ecological region, a single mountain range, serpentine soils or other spatially-limited sites, or to politically-bounded units such as states or provinces. Endemics occur in areas that may or may not be considered diverse. However, endemics often contribute substantially to regional diversity.

Endemism is not uniform across North America. In the United States, California, Texas, and Florida have the highest numbers of endemic species (Gentry, 1986). Not all endemics, of course, occur in forest ecosystems, but many do. The majority of Florida's endemic plants are associated with swamps, flatwoods, and hardwood forests in northern and central Florida. A large number of species is associated with a spatially-limited area known as "scrub" (Myers, 1990). Many plants with very limited distributions apparently arose on isolated islands during periods when most of the current peninsula was below sea level. These species persist today in isolated scrub "islands." An array of animals also evolved on these islands, and they include the Florida scrub jay (*Aphelocoma coerulescens* var. *coerulescens*), a showy and sociable bird, and the remarkable sand skink (*Neoseps reynoldsi*) which "swims" just below the sand surface. While these species are well adapted to survive on very specific sites and on an archipelago of scrub islands, they are also regionally rare, and many are considered to be endangered, due to encroachment of agriculture and urbanization into the ecosystems that support them.

Keystone Species

What is a keystone? A keystone is the central, topmost stone of an arch, popularly thought of as holding the others in place, or more generally: that one of a number of assorted parts that supports or holds together the others (Neufeldt and Guralnik, 1988). This concept can't be taken too literally, however—as did the writer who observed that if all the pines of a forest were wiped out, "the ecosystem would totally disintegrate." But ecosystems are more than species; ecosystems change, slowly or drastically, but they don't "fall apart" for loss of a particular species. Instead, our perception is that the pine community, which we once knew intimately, upon catastrophic change is gone (i.e., had "disintegrated"), and the new community is a major force in building a new ecosystem. Dominant species are not necessarily considered "keystone" species.

In ecology, the definition of **keystone species** has been expanded from the original animal ecology usage of Paine (1969) to identify a keystone species as one whose impact on its community or ecosystem is large, and disproportionately large relative to its abundance (Power et al., 1996). Ecologists recognize that the most abundant species play major roles in controlling the rates and directions of many ecosystem processes. These dominants are often crucial for maintaining their communities because they typically provide the major energy flow and the physical structure that supports and shelters other organisms (Gentry and Dodson, 1987; Ashton, 1992). An excellent example is old-growth Douglas-fir trees that dominate their community. They are significant in ecosystem function (Franklin et al., 1981) and support multitudes of organisms on their boles and branches (Chapter 21, Figure 21.1). However, many experiments have shown that some less abundant species have major community and ecosystem effects. Thus keystone species differ from dominant species in that their effects are much larger than would be predicted from their abundance (Power et al., 1996). Identifying keystone species is difficult because we have learned to think of dominant species as playing key roles in forest ecosystems. Nevertheless, identifying keystone species will lead to a better understanding of how loss of species will affect ecosystem function.

The Rosetta stone for identifying keystones is provided by Power et al. (1996). Keystone species are often, but not always of high trophic status. They can exert effects, not only through consumption (e.g., pocket gophers, elephants, wolves, predatory ants and trout, kangaroo rats, etc.) but also by such interactions and processes as competition, mutualism, dispersal, pollination, disease, and by modifying abiotic factors (Bond, 1993; Mills et al., 1993). Possible examples of ecosystem modifiers or "ecosystem engineers" (Lawton and Jones, 1995) include beavers that swamp-out meadows and forests and badgers whose mounds maintain diversity in prairie floras (Platt, 1975). Few woody plants have been identified as keystone species by Power et al. (1996); fig trees (*Ficus* spp.) that provide food for animals was the only woodland keystone identified.

The importance of the keystone concept is in helping identify the most suitable areas for biodiversity preserves and in understanding the complex linkages among ecosystem biota and site-biota interactions. Power et al. (1996) gave three useful insights from the keystone concept: (1) land managers should carefully consider the consequences of the loss of species for which no obvious role in ecosystems has been discovered, (2) introduced alien species may, like keystone species, have potential strong effects disproportionate to their biomass, and (3) there is a lack of well-developed protocol of identifying keystone species; the field is littered with far too many untested anecdotal "keystone species."

Effects of Forest Management on Diversity

Silviculture and forest management include many operations, such as regenerating, tending, and protecting forests (Smith et al., 1997). Traditional management procedures, however, are typically very different from the processes occurring in natural stands. The creation of even-aged, even-sized stands using clearcutting and artificial reforestation (Oliver and Larson, 1990; Smith et al., 1997) has become the most common managerial system. Such stands are highly simplified and lack many components of community structure, such as snags and logs, as well as stand-level structural complexity, such as multiple canopy layers, irregular tree spacing, and gaps (Franklin, 1995). Forest ecosystems may be simplified at several levels: genetic, individual stand, landscape, and temporal (succession). Managed forests have been simplified in response to economic criteria—efficient management (e.g., whole-tree harvesting) and high productivity of commercially-important tree species—in the belief that much of the structural complexity found in natural stands is not essential to sustained tree productivity of the site. Research on the importance of stand structure and biological legacies as they affect ecosystem function, biodiversity, and long-term site productivity are discussed by Franklin (1995, 1997) and Franklin et al., (1997) for Pacific Northwest forests (see Chapter 21).

Clearcutting, because of the enormous changes to ecosystems and their biota, has been the subject of many studies, and a good review is provided by Boyle (1992). Diversity of animals and plants may increase or decrease depending on the specific sites, scale and pattern of clearcuts, and the methods employed, including post-cutting burning or site preparation. The effect of clearcutting is typically not a simple reduction in species diversity, but rather a differential effect. Species dependent on old-growth or near-natural forests are often lost from the system, presumably until suitable environmental conditions are restored, as illustrated by the following example.

A detailed study of understory species diversity in natural and managed Douglas-fir forest of the Pacific Northwest, revealed that changes in understory diversity are fairly short-lived following clear-cut logging and slash burning (Halpern and Spies, 1995). Populations of most vascular plants recover to original levels prior to canopy closure. However, diversity may remain depressed for decades on severely burned sites, and some species may experience local extinction. Silvicultural prescriptions that maintain or foster spatial and temporal diversity of stand structure and site conditions will be most effective in maintaining diversity. Practices associated with intensive, short-rotation plantation forest, that preclude or delay the development of old-growth attributes, may result in long-term loss of diversity.

Fragmentation of forests by roads, agricultural fields, and clearcut patches also markedly influences biodiversity. The consequences of ecosystem fragmentation are not only the creation of isolated "islands" of various sizes but also include large changes in the physical environment that markedly influence species and gene pool diversity. The significance of fragmentation and edge effects have been considered by many ecologists and geneticists (Harris, 1984; Saunders et al., 1991, Ledig, 1992), and we examine it in Chapter 21 at the landscape-level.

Genetic diversity of forests (i.e., species diversity and gene diversity within species) has been greatly influenced by human activities that affect evolutionary processes of extinction, selection, drift, gene flow and mutation (Ledig, 1992). In some cases, diversity may be increased by exploitation and harvest practices (e.g., exposing recessive genes and increasing the occurrence of novelties and mutants), but these practices can also reduce diversity. Most forest harvest practices and associated road building cause fragmentation, result in drastic swings in population sizes, and simplify age structure—all of which affect the breeding systems of forest trees. Ledig (1992) describes how habitat alteration, environmental deterioration, and domestication of forest species may all lead to reduced biodiversity and, therefore, the urgent need to inventory genetic diversity of forest species and monitor their changes.

CONSERVING ECOSYSTEM AND BIOLOGICAL DIVERSITY

As landscape ecosystems have become increasingly dominated by humans, the list of species considered to be in danger of extinction has grown (Vitousek et al., 1997). Currently, more than 950 species or subspecies are considered to be in peril of extinction in the United States by the year 2,000 (Wilson, 1992). However, the concern is deeper than rare and endangered species; it is for the immense number of unknown species and for ecosystem processes that are basic to long-term sustainability of ecosystems. Endangered organisms *per se* cannot be preserved. However, ecosystems of which organisms are notable parts can be preserved using an ecosystem approach (Rowe, 1990, 1997).

Landscape Ecosystem Approach

At the outset, a key problem in the conservation of biodiversity is locating and inventorying all the species, including the endemic and endangered ones. Rowe (1997) observes that the biodiversity of any area depends on its "taxodiversity:" the number of different kinds of taxonomists that have packed or been packed into it! Although we don't know how many kinds of organisms share the Ecosphere, we know that they are inseparable from their landscape or waterscape ecosystems. Therefore, the conservation of forest biodiversity requires the conservation of landscape ecosystems that support organisms.

Recall from the previous discussion that species tend to be the focus of conservation and it typically proceeds on a species-by-species basis. However, ecologists increasingly realize that an ecosystem approach is required because organisms *per se* cannot be preserved. Franklin (1993) concludes that efforts to preserve biological diversity must focus increasingly at the ecosystem level because of the immense number of species, the majority of which are currently unknown. A species-by-species approach will fail because it will quickly exhaust: (1) financial resources, (2) the time available, (3) societal patience, and (4) scientific knowledge. To even come close to attaining the goal of preserving biodiversity, broad-scale approaches—at the levels of regional and local ecosystems—are the only way to conserve the overwhelming mass—millions of species—of existing biodiversity (Franklin, 1993). Fortunately, we have the ability to distinguish and map ecosystems at multiple spatial scales as described in Chapters 2, 13, and 21. Also, as discussed earlier in the chapter, qualitative and quantitative procedures are available to determine ecosystem diversity and facilitate conservation of ecosystems and their biota.

Public interest and concern over biodiversity and the conservation of forest ecosystems is reflected in many ways: in the passing of the Endangered Species Act, development and widespread interest in the discipline of Conservation Biology (Soulé, 1986; Primack, 1993; Meffe and Carroll, 1997), interest in conserving existing old-growth forests by ecotourism and bioprospecting for medicinal uses, and efforts by public and private organizations to develop programs in conservation and sustainability of ecosystems. In addition, the following initiatives illustrate the widespread interdisciplinary interest in the management of ecosystems to sustain biodiversity: (1) detailed documentation of the most endangered ecosystems in the United States (Noss et al., 1995), (2) publication of the Canadian guide to living sustainability using an ecosystem approach (Gray et al., 1995), (3) the insightful analysis of the spotted owl controversy (Yaffee, 1994), and the emphasis on forest ecosystem management for the 21st century (Kohm and Franklin, 1997).

Conservation of biological and ecosystem diversity may be furthered by establishing parks and preserves, but it primarily rests on innovative management of semi-natural ecosystems and cut-over lands that form the vast majority of the world's forested lands. Therefore, a variety of procedures have been developed for the conservation of biodiversity, and they are available in the works of authors of diverse viewpoints (Harris, 1984, Part 4; Norse, 1990, Chapter 8; Hunter, 1990; Hammond, 1991; Meffe and Carroll, 1997; and Rowe, 1997). Specific guidelines for conserving forest biodiversity are provided by Burton et al. (1992) and by Probst and Crow (1991). Key recommendations emphasize: (1) ecosystem management at a landscape scale using a regional perspective, (2) thinking beyond the boundaries of specific ownerships and avoiding managing using a stand-by stand approach, (3) stressing multispecies and ecosystem management, (4) conducting ecological surveys and inventories to know the land and what is on it, and (5) monitoring problem ecosystems and problem species. Furthermore, Janzen's (1997) perceptive consideration of wildland biodiversity management in the tropics emphasizes innovative approaches and, above all, the relevance of the pivotal professions of taxonomy and natural history.

Conservation of biological and ecosystem diversity may take place locally, but as we have seen, a landscape-level ecosystem approach is required. Landscape ecology is considered in the following chapter, and examples of ecosystem management for structurally and ecologically heterogeneous systems are presented.

Suggested Readings

Burton, P. J., A. C. Balisky, L. P. Coward, S. G. Cumming, and D. D. Kneeshaw. 1992. The value of managing for biodiversity. *For. Chronicle* 68:225–237.

Franklin, J. F. 1993. Preserving biodiversity: species, ecosystems, or landscapes? *Ecological Applications* 3:202-205.

Huston, M A. 1994. *Biological Diversity.* Cambridge Univ. Press. 681 pp.

Janzen, D. H. 1997. Wildland biodiversity management in the tropics. In M. J. Reaka-Kudla, D. E. Wilson, and E. O. Wilson (eds.), *Biodiveristy II: Understanding and Protecting Our Biological Resources.* Joseph Henry Press, Washington, D.C.

Latham, R. E., and R. E. Ricklefs. 1993. Continental comparisons of temperate-zone tree species diversity. In R. Ricklefs and D. Schluter (eds.), *Species Diversity in Ecological Communities: Historical and Geographical Perspectives.* Univ. Chicago Press, Chicago.

Ledig, F. T. 1992. Human impacts on genetic diversity in forest ecosystems. *Oikos* 63:87–108.

Magurran, A. E. 1988. *Ecological Diversity and Its Measurement.* Princeton Univ. Press, Princeton, NJ. 179 pp.

Rowe, J. S. 1997. The necessity of protecting ecoscapes. *Global Biodiversity* 7:9–12.

Wilson, E. O. 1992. *The Diversity of Life.* Belknap Press, Harvard Univ. Press, Cambridge, MA. 424 pp.

CHAPTER 21

LANDSCAPE ECOLOGY

Landscape ecology is a complex interaction of disciplines. It developed in central and eastern Europe through a combination of geography and ecology. In Europe, it is strongly human-centered and used as the scientific basis for landscape planning, management, conservation, and restoration (Naveh and Lieberman, 1993). Naveh and Lieberman (1993) define landscape ecology as the interrelationship between humans and their open and built-up landscapes. It includes not only the classical ecology–biology disciplines but emphasizes human-centered fields—the sociopsychological, economic, geographic, and cultural sciences that are intimately connected with modern land uses. All these considerations are beyond the scope of our treatment but are examined in detail by Zonneveld and Forman (1990), Naveh and Lieberman (1993), and Forman (1995). The history of the landscape ecology concept is reviewed by Schreiber (1990) and Naveh and Lieberman (1993), and an excellent review of the discipline by Turner (1989) is available.

Landscape ecology, in the broad sense, is the study of anything situated in a landscape, often identified as "elements" or "patches," at scales above those of the individual organism or a single ecosystem type. It considers the spatial pattern and heterogeneity of these landscape features, their development and dynamics, exchanges between them, and their management (Turner and Gardner, 1991). Landscape ecology also examines the influences of spatial heterogeneity on ecological processes. Turner (1989) observes that the explicit effect of spatial patterns on ecological processes is what distinguishes landscape ecology from other disciplines of ecology. Quantitative approaches and modern technologies of remote sensing and geographic information systems are considered by Hobbs and Mooney (1990), Turner and Gardner (1991), and Sample (1994).

Given the great diversity of possible topics, we wish to emphasize the landscape ecosystem perspective by considering spatial relationships of ecosystems, as well as disturbance at the landscape level, landscape fragmentation, and the conservation and management of ecosystems.

613

CONCEPTS OF LANDSCAPE ECOLOGY

A landscape is a spatially heterogeneous area that may be conceived in several ways. The term landscape in world literature, first referenced in the Book of Psalms, gives the visual-aesthetic connotation of landscape that is typically adopted in literature and art and used by people in landscape planning and design (Naveh and Lieberman, 1993). Early in the 19th century, von Humboldt defined landscape as the total character of an area of the Earth (Troll, 1950). The oldest meaning of landscape contains the visual-aesthetic element—an expanse of natural scenery seen by the eye in one view (Neufeldt and Guralnik, 1988). A second meaning is the pattern and interrelationships of surface features—landforms, soil bodies, and vegetation (Zonnefeld, 1990). This view has been broadened to include features created by humans such as agricultural fields, tree plantations, hedgerows, roads, and buildings.

The third point of view is that of landscape as ecosystems—three-dimensionally structured segments of the Earth with interacting physical and biotic components. The German geographer Carl Troll (1939, 1968) introduced the name landscape ecology and defined it in this ecological framework ". . . to denote the analysis of the physio-biological complex of interrelations which govern the different areal units of a region" (Troll, 1963a). As described in Chapter 1 and illustrated in Figure 1.5, Troll emphasized both a horizontal (geographic–ecological) and vertical (functional) approach to landscape ecology.

Troll (1939, 1968) noted that research using aerial photographs was to a great extent landscape ecology. Actually seeing the modern landscape from an airplane at 50 m, 500 m, or 5,000+ m reveals features at multiple scales in a variety of natural and human-derived patterns that form networks and corridors. In the presettlement landscape, or in the rare contiguous forests of today, the forest patches would appear as a contiguous matrix with "soft" edges between them (Crow, 1991). In contrast, the patches of the human-disturbed landscape of today show fragmented forests with sharp boundaries and geometric shapes of woodlots together with fields, roads, and houses.

Patches (e.g., forest woodlot, clearcut, field) are often the focus in landscape ecology because they can be easily identified and monitored by aerial photographs and satellites, and they can be mapped by geographic information systems. The term, patch, is convenient conceptually since it requires no necessary assumptions or preconceived notions of ecosystem, community type, or successional stage. A patch is simply an area that differs from its surroundings in appearance. However, all vegetated patches *are* interconnected to the air–landform–soil substrate that supports them. Thus for many practical management and conservation issues the understanding of the specific regional and local landscape ecosystems in which patches are embedded is increasingly important because of the air and soil heterogeneity that accompanies each patch.

Geographers, like Troll (1971) and Bailey (1996), regard landscapes as integral units very much like ecosystems. Thus understood, landscape ecology is the study of land systems (including rivers and wetlands) and of their relationships to one another, for each is set in the "environment" of its surrounding neighbors (Rowe, 1988). Therefore, landscape ecology is the science of landscapes, understood concretely as spatial and volumetric ecosystems in their regional contexts. The sections that follow emphasize the spatial structure of ecosystems, exchanges between them, and their inherent disturbance regimes. As discussed in Chapter 2, ecosystems occur at many different spatial scales. The following sections illustrate some of these patterns and exchanges, from very broad to local scales, and involving both terrestrial and aquatic landscape interactions.

OLD-GROWTH FORESTS OF THE PACIFIC NORTHWEST

One of the best examples of the coupling of spatial pattern and functional interactions among terrestrial and aquatic ecosystems comes from the Pacific Northwest in which old-growth forests of the Cascade Mountains of Oregon and Washington are interspersed with riparian and stream ecosystems. A unique feature of the Pacific Northwest is the occurrence of long-lived conifers, often growing on steep mountain terrain (Chapter 10; Waring and Franklin, 1979). The dense forests are dominated almost totally by the conifers whose biomass accumulation far exceed that in almost every other temperate forest region (Chapter 18). Douglas-fir trees typically dominate the old-growth forests; individual trees 80 m tall and 450 years old are not uncommon. Four key features of old-growth stand structure—large live trees, large snags, and large logs on land and in streams—markedly influence ecosystem composition and function and distinguish old-growth ecosystems from those of plantations or young natural stands (Franklin et al., 1981; Franklin and Spies, 1991).

Each large tree is itself an ecological landscape with variations in topography, climate, vegetation, and an associated fauna (Figure 21.1). For example, six zones supporting epiphytic lichens and bryophytes were recognized, and striking differences were found between the lower moist and dry sides of the trunk (Figure 21.1; Pike et al., 1975). Generally, the epiphytic biomass ranges between 10 and 20 percent of the foliage of the host tree. Nitrogen plays a key role in coniferous forests, and lichens, particularly *Lobaria oregana,* fix atmospheric nitrogen (N_2) at an average rate of about 8 to 10 kg ha^{-1} yr^{-1} (Chapter 19). In contrast, epiphytes play a negligible role in young second-growth stands (Grier et al., 1974). The large canopy of old-growth Douglas-fir trees also acts as a site for decomposition since much of the dead needle, twig, and bark litter becomes lodged in the crown and creates a perched soil habitat (Figure 21.1*b*). Also, fungi and bacteria infect living needles and decompose needles and other detritus lodged in the crown. Associated with the epiphytes and lodged litter are over 900 taxa of invertebrates, many of which feed on the foliage. Thus the epiphytic biomass contributes significantly to the functioning of these ecosystems by fixation of N_2, interception of water and nutrients, and provision of crown habitat for decomposer flora and invertebrates.

Coarse Woody Debris and Ecosystem Interactions in Riparian Zones

Another feature, unique to old-growth forests with huge quantities of biomass, is the presence of great amounts of coarse woody debris on the forest floor and in small streams (Harmon et al., 1986). In a 450-year-old Douglas-fir stand in the H. J. Andrews Experimental Forest of western Oregon's Cascade Range, 13 percent of the forest floor was covered with logs amounting to 172 Mg (megagrams) of biomass per hectare (Figure 21.2). In contrast, a 300-year-old yellow birch–red spruce forest in New York had 43 Mg ha^{-1}, and values of 17 to 29 Mg ha^{-1} were reported for old-growth hardwood forests in Kentucky and Tennessee (McMinn and Hardt, 1996). In addition to woody debris on the forest floor, the Douglas-fir forest had 109 Mg ha^{-1} of standing dead trees over 4 m tall. Decomposition of trunks on the forest floor is slow; 120 to 140 years are required to reduce Douglas-fir logs to 50 percent of their original weight. Therefore, logs constitute a long-term reservoir for nutrients that are slowly available to the soil and are more readily available to plants with roots in the litter layer. The logs also provide an important habitat for small mammals that spread mycorrhizal fungi. These characteristics not only influence the function of upland ecosystems but affect riparian and aquatic systems as well.

Figure 21.1 Old-growth Douglas-fir, H. J. Andrews Experiment Forest, Blue River Ranger District, Willamette National Forest, western Oregon. (*a*) Lower trunk showing sharp distinction between moist and dry sides. Tree is on a north-facing slope and leans toward the north, which is to the right in the picture. (*b*) View of two axes in the canopy showing mosses and accumulation of trapped needles. (*c*) Moist side of the lower trunk as viewed from the ground. (*d*) Dry side of lower trunk of the same tree shown in (*c*). (After Pike et al., 1975. Photos courtesy of William C. Denison.)

Figure 21.2 Large masses of decaying logs are a dominant feature of Pacific Northwest old-growth forests. This 450-year-old stand of Douglas-fir, western hemlock, and Pacific silver fir illustrates maximal accumulations. H. J. Andrews Experiment Forest, western Oregon. (U.S. Forest Service Photo. Courtesy of Jerry F. Franklin.)

An example of the spatial and functional linkage between the old-growth coniferous forest, riparian, and aquatic ecosystems is illustrated in Figure 21.3. Trees of ecosystems adjacent to streams are important in regulating their food base and habitat structure (Swanson and Franklin, 1992). The trees shade the stream and modify its temperature as well as providing a source of coarse woody debris and fine foliage litter. Notice in Figure 21.3 that both streamside ecosystems, often characterized by hardwoods, shrubs, and conifer-dominated ecosystems of adjacent terraces and slopes, significantly affect the stream.

Figure 21.4 illustrates the amount of coarse woody debris that may come to rest in small- and intermediate-sized streams. For example, 56 Mg of organic debris were measured in a 3-m-wide belt along a small stream in the Andrews Experimental Forest (Froehlich, 1973). Such debris, in the form of tree tops, limbs, root wads, and whole trees, is a principal factor determining the biological and physical character of these streams (Figures 21.5 and 21.6; Swanson et al., 1976; Swanson and Lienkaemper, 1978). Forest streams and their biota developed through a long history of high concentrations of debris. Wood moves through the stream channels by flotation during high-flow events and as dissolved and fine particulate materials following breakdown by wood-processing organ-

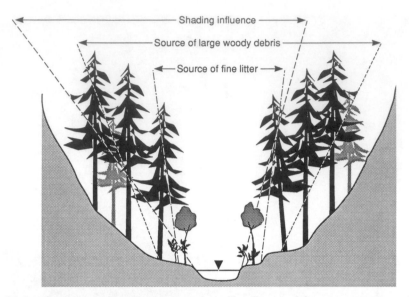

Figure 21.3   Physiographic cross section of a valley floor and forest ecosystems adjacent to a stream, showing several of the major ecological effects of forest vegetation on stream systems. The asymmetry of the zone of shading influence is intended to represent the dependence of this function on the orientation of the valley floor relative to the solar path. The *functional riparian zone*, shown here adjacent to the stream, includes those ecosystems whose vegetation contributes to the food base, shading, and habitat structure of the stream. (After Swanson and Franklin, 1992. Reprinted with permission of the Ecological Society of America.)

isms. The debris may reside in the channel over 100 years, acting to retain organic and inorganic sediment. The logs and associated gravel bars form a stream profile of steps—long, low-gradient sections separated by relatively short falls or cascades. This pattern can account for up to 100 percent of the total stream drop (Swanson et al., 1976), and much of the resulting stream bed may have a gradient less than that of the valley bottom. Energy is dissipated in short, steep falls and results in less energy available for erosion of stream bed and banks, more sediment storage in the channel, slower routing of organic materials, and greater habitat diversity than in straight, even-gradient channels (Swanson and Lienkaemper, 1978). Except in periodic flushing episodes, the debris in the channel tends to route the sediment through the stream in a slow trickle, reducing the rate of downstream movement. The woody debris, through its very presence and the formation of the stepped profile, creates relatively stable habitats for a variety of stream biota. Consumer organisms (i.e., those feeding on woody debris) tend to be concentrated in the wood and wood-created habitat, which in small undisturbed streams may exceed 50 percent of total stream area. A food base for these organisms is constantly available (fine organic material and large debris), even during periods when leaves or needles are not available.

In streams adjacent to logged-over areas, coarse and fine debris may be largely eliminated (Figure 21.6) and the stream significantly altered physically and biologically. For example, on Mack Creek (Figure 21.6), extremely high winter storm flows moved logs, slash, and roots downstream, leaving only about four percent of the coarse debris, compared to the uncut area upstream, seven years after logging (Froehlich et al., 1972). The consequences of removing coarse debris are several (Swanson et al., 1976): (1) a more

Figure 21.4 Coarse woody debris in Lookout Creek, H. J. Andrews Experimental Forest, Oregon. Note the person in upper left of picture. (After Swanson et al., 1976; U.S. Forest Service photo. Courtesy of F. J. Swanson.)

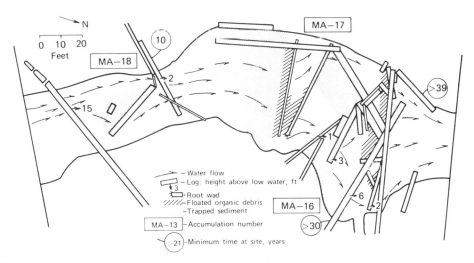

Figure 21.5 Map of coarse woody debris and other material in a 60-m forest section of Mack Creek upstream from the section in Figure 21.6. H. J. Andrews Experimental Forest, Oregon. (After Swanson et al., 1976.)

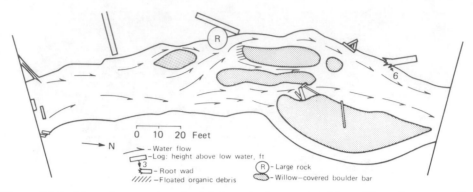

Figure 21.6 Map of coarse woody debris and other material in a 60-m clearcut section of Mack Creek downstream from the section in Figure 21.5. H. J. Andrews Experimental Forest, Oregon. (After Swanson et al., 1976.)

uniformly steep stream profile is formed, (2) diversity of stream habit is reduced as debris-created depositional pools are eliminated, (3) increased water velocity contributes to the accelerated transport of fine organic matter through the channel system, thereby decreasing the opportunity of stream organisms to process the material, and ultimately (4) reduction of long-term biological productivity and increased rate of sediment transfer from headwater streams to downstream areas.

As we have seen, the small streams in coniferous forests are a part of the spatial complex of forest and aquatic ecosystems. A consideration of streams in coniferous forests has been presented by Triska et al. (1982). The riparian zone, shown in Figure 21.3, serves as the interface between the old-growth forest farther upslope and stream ecosystems. Its important functions have been reviewed by Swanson et al. (1982b) and Gregory et al. (1991). Erosional processes, described in the next section, also illustrate the influence of spatially adjacent terrestrial and stream ecosystems upon one another.

Erosion and Geomorphology in Riparian Systems

Steep forest terrain combined with heavy winter precipitation, weakly developed soils, and massive trees make linkages between physical and ecological processes in terrestrial and aquatic ecosystems more apparent than in gently sloping terrain of geologically-older regions. Thus studies of erosion and forest geomorphology have demonstrated the significance of physiography in ecosystem interactions (Dyrness, 1967; Fredriksen, 1970; Swanson et al., 1982a; Gregory et al., 1991).

Past studies of erosion have identified the hillslope and stream channel processes that transfer organic matter, soil, and vegetation in old-growth forests (Figure 21.7). For example, subtle, slow mass movement processes of creep, slip, and earth flow operate simultaneously and sequentially with rapid soil mass movements on hillslopes and debris torrents in stream channels (Swanson and Swanston, 1977). Tree-ring analyses suggest that mass soil movement has spanned centuries. These mass movements are responsible for microenvironments affecting the composition, growth, and functioning of forests and streams. An analysis of erosion processes for a watershed of the H. J. Andrews Experimental Forest in Oregon revealed that on a long-term basis, the most episodic and infrequent processes (debris avalanches on hillslopes occurring approximately once every 370 years, and debris torrents in streams, occurring about once every 580 years) transferred more mineral matter than the persistent, pervasive, and continuous processes or creep and

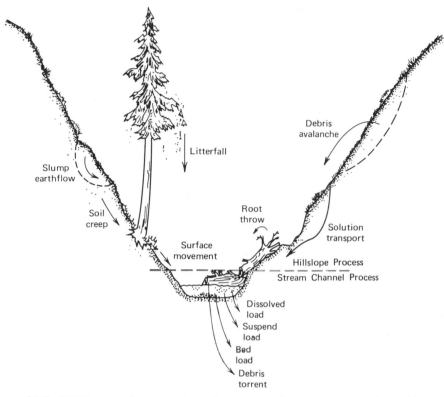

Figure 21.7 Hillslope and stream channel processes that transfer organic and inorganic material among coniferous forest ecosystems and between terrestrial and aquatic systems. (Modified from Swanson et al., 1982a. Courtesy of the Coniferous Forest Biome and Oregon State University. Drawing by M. F. Orlando.)

surface erosion. Through such studies, the impact of erosion on other ecosystem processes are evaluated, and management recommendations are given to minimize damage to riparian ecosystems. For example, current riparian management policies focus on many issues including retention of live trees and snags, width of riparian management zone, road crossings, and floodplain protection (Gregory, 1997).

Forest-Stream Linkages at Multiple Spatial Scales

In the above sections we have examined just part of a complex system—the small, first-order mountain stream and its riparian zone. The stream connects forest and sea (Maser and Sedell, 1994). Thus a landscape-level ecosystem consideration is needed that integrates habitat structure, food resources for aquatic and terrestrial wildlife, disturbance regimes, and nutrient cycling (Gregory et al., 1991). This synthesis can be accomplished by examining linkages between forest and aquatic ecosystems at multiple hierarchical spatial scales, as illustrated in Figure 21.8 (Swanson and Franklin, 1992). These range from the broad stream network (scale 6) of a watershed to a small stream channel (scale 3) and to the boulder or particle of the smallest scale. The local ecosystems and processes we

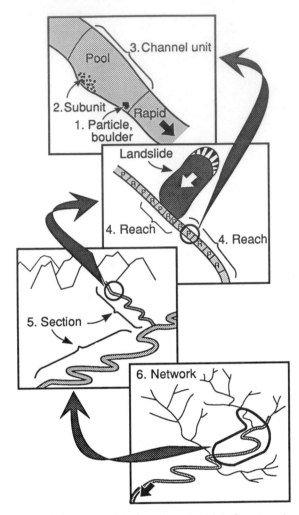

Figure 21.8 Diagram of six hierarchical scales at which forest and aquatic ecosystems interact. River sections (scale 5) reflect the geomorphic effect on riparian zones as the stream flows through mountains or broad valleys. "Reaches" (scale 4) are segments of stream, and "P" and "R" refer to pool or riffle units of the channel (scale 3). (After Swanson and Franklin, 1992 as modified from Swanson et al., 1982.)

have considered in Figures 21.3 to 21.7—characteristic of first-order streams in mountain terrain—are encompassed in scales of the upper two panels. The landslide of Figure 21.7 is shown in Figure 21.8 as affecting several pool and rapid channel units of a segment or "reach" (scale 4) of the stream. At a broader scale the "section" illustrates geomorphic influences as a high-gradient, mountain stream section joins a low-gradient river section in a broad valley. The terrestrial and aquatic ecosystems and their processes in a broad valley section are markedly different than those of the small mountain stream (Gregory et al., 1991). Thus the landscape-scale hierarchy is the basis for understanding riparian ecosystems, interactions of terrestrial and aquatic systems, and management efforts on stream and riparian systems.

LANDSCAPE-LEVEL DISTURBANCE

Disturbances of catastrophic proportions have been significant in shaping the physical features of landscape ecosystems and instrumental in the evolution of biota. Only recently have we been able to study the effects of these rare events to gain insights on a broad scale. Two examples are considered next.

Catastrophic Windstorm in New England

The forests of central New England have been repeatedly affected by wind and ice storms, fires, and pathogens. Hurricanes have had the greatest effect, and their ecological effects and damage to the vegetation at several spatial scales have been studied (Foster, 1988a,b; Boose et al., 1994; Foster and Boose, 1992; 1995). Records show that hurricanes strike southern and central New England every 20 to 40 years, and the most destructive storms (1625, 1788, 1815, 1938) provide compelling evidence that this, of all disturbances, most strongly affects the pattern of vegetation in this region. Figure 21.9 illustrates the hurricane tracks and the frequency of their occurrence over the New England landscape. The 1938 hurricane, with winds in excess of 200 km hr^{-1} destroyed forest along a 100-km^2 path. At the broad level, the post-hurricane landscape was a mosaic of differentially-damaged stands controlled by physiography, wind direction, and the nature of the pre-hurricane vegetation. The strongest winds came from the southeast. Overall, severe damage occurred on south- to east facing slopes and the northwestern shores of lakes. Little damage occurred to forests in the lee of broad hills (north-west exposure), on the south-east and east shores of lakes, or in valleys. The spatial position of damage classes related to physiography and forest type is illustrated in Figure 21.10. Notice that severe damage is immediately adjacent to boundaries of the pond and large field where exposure is greatest. Also at the broad scale, severity of damage is also related to historical factors such as the time since the last major storm and the previous pattern of human clearing and agriculture which, in turn, affect the structure (spatial and canopy) and composition of the pre-hurricane vegetation.

For example, landscapes of central New England were ripe for catastrophe because of the dominance on old fields of 30- to 100-year-old white pines that were highly susceptible to wind damage. In contrast, a similar hurricane in 1815 caused much less damage because it occurred at the height of the agricultural period when most of the land was cleared. A major storm today would leave yet another imprint on the landscape due to the legacy of the 1938 storm which destroyed the older conifers and tall pioneer species, leaving more storm-resistant hardwoods and young understory conifers (Figure 21.10).

Landscape-level study of the 1938 hurricane (Foster and Boose, 1992) did not bear out the perceived expectation that major storms cause very large openings across a region, whereas small storms create small gaps. Although the latter is undoubtedly true, this hurricane produced a continuum of effects ranging from damage to individual trees to uprooting of entire stands. The sight of entire windthrown stands is spectacular, as illustrated in Figure 21.11. However, a heterogeneous mosaic of small areas, most less than 2 ha, was created.

Overall, the heterogeneity of abrupt changes in damage intensity across short distances is due, in part, to landscape-level physiography and land-use history and, in part, to the interrelated factors of stand structure and species composition. At the stand level, damage was strongly related to topography and increased with increasing stand age and height; it decreased with stand density as stand opening increased air turbulence. Conifer

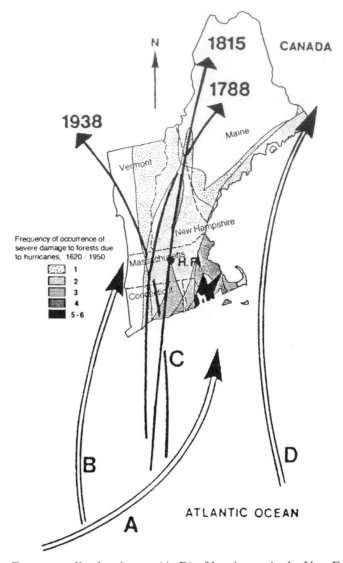

Figure 21.9 Four generalized pathways (A–D) of hurricanes in the New England region superimposed over a map of the frequency of occurrence of the major hurricanes from 1620 to 1950. In addition, the historical tracks of the hurricanes of 1788, 1815, and 1938. Hurricane damage ratings range from: 1 = light to 5–6 = severe. HF = location of the Harvard Forest, Petersham, MA. (Modified from Foster and Boose, 1992 and Whitney, 1994. Reprinted from the *Journal of Ecology* with permission of the British Ecological Society.)

forests, dominated by white pine, were significantly more susceptible than hardwood forests. Also susceptible were fast-growing species that tended to occupy dominant canopy positions, such as red pine, aspens, and white birch. The slower growing oaks, hickories, red maple, and hemlock occupied subordinate canopy positions and thus were more protected from wind.

At the local level, the 1938 hurricane was notable for the catastrophic uprooting of trees, primarily white pine. Uprooting was attributed to the 15 to 35 cm of rain that pre-

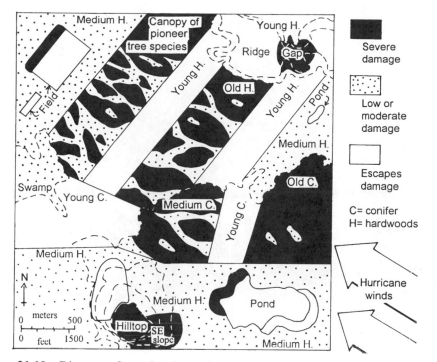

Figure 21.10 Diagram of post-hurricane damage classes from a characteristic landscape in southern New England. Young, medium-age, and old stands of conifers (C) or hardwoods (H) represent a hypothetical pre-hurricane vegetation mosaic. Conifer = white pine; hardwoods = oaks, red maple, birch, beech, hickories (and hemlock); pioneers = pines, aspens, white birch. Predicted damage distribution is primarily based on studies of Foster (1988a,b) and Foster and Boose (1992). (After Forman, 1995. Reprinted from Land Mosaics: The Ecology of Landscapes and Regions, © Cambridge University Press 1995. Reprinted with permission of Cambridge University Press.)

ceded the storm, causing soil saturation and reduced rooting strength. The pit and mound topography subsequently produced affects succession at the landscape level because of the microtopographic and soil variation created. The exposed soil favors establishment of small-seeded, pioneer species such as birches and aspens, as well as white pine and the tolerant hemlock that require moist, competition-free environments.

Following the 1938 hurricane, succession followed an even-aged pattern of stand development with the composition of the new stand on a given area dependent upon: (1) soil type, (2) composition of the former stand, (3) advanced regeneration present at the time of blowdown, and (4) whether the down timber was logged or not (Spurr, 1956; Hibbs, 1983; Foster 1988a). Where present, advanced regeneration, especially beech and hemlock, was released. New regenerations of pioneer species and understory tolerants occurred where disturbance by blowdown or logging created favorable sites. Most hardwoods established within 10 years after the hurricane and grew rapidly into the canopy. The striking lack of white pine following the 1938 hurricane is attributed to the absence of fire that had accompanied the 1635 hurricane and facilitated the establishment of white pine–hemlock–hardwood forests that persisted until 1938 (Foster, 1988b).

Figure 21.11   The Harvard Tract of the Pisgah National Forest in 1942, four years after the hurricane. (Photo by Stephen H. Spurr. Reprinted courtesy of the Harvard Forest, Harvard University.)

In summary, for the areas studied in the New England landscape, catastrophic windstorm occurs quite predictably and specifically with profound consequences on the vegetation at several spatial scales. Post-hurricane forests in this region are mediated strongly by physiography and associated ecosystem characters, by historical factors of human land use, and by related structural and compositional features of the vegetation.

Vegetational Diversity Mediated by Fire in Yellowstone National Park

In contrast to the long and heavily settled central New England region, Yellowstone National Park is an example of a large wilderness area that can provide the best and probably the only place for studying landscape changes that occurred for millennia in presettlement times (Romme and Knight, 1982). In contrast to the catastrophic disturbance of windstorm in New England, fire affects vegetation and associated animals in the Yellowstone Plateau over vast areas (Chapter 16). William Romme's studies of landscape-level vegetation diversity and fire patterns in Yellowstone watersheds provide an excellent example of fire effects at the landscape scale.

Romme (1982; Romme and Knight, 1982) studied the interrelationship of fire history and upland plant communities in the 73-km^2 Little Firehole River subalpine watershed prior to the catastrophic fires in 1988. Fire history during a 350-year period was determined using fire-scar methods (Chapter 12). The major fires of 1739, 1755, and 1795 burned over half of the area, avoiding young forests and topographically protected sites. Large, stand-regenerating fires (over 1,000 ha) were absent until 1988; from 1795 to 1978 there were only three fires > 4 ha, and these were all small (<100 ha). The absence of catastrophic fires was due primarily to lack of suitable fuel conditions over most of the watershed, not to fire suppression. Major fires occur naturally at very long intervals,

300–400 years, because of very slow forest regrowth and fuel accumulation after fire (Romme, 1982). Such a long fire cycle for large fires occurs on the high subalpine plateaus because geologic substrate, soils, and vegetation are very similar over much of the area; forests over large contiguous areas grow and develop a fuel accumulation approximately at the same rates; the plateau topography has low relief and few natural barriers to the spread of fire.

Forest vegetation is dominated by lodgepole pine, and three main stages of succession following fire (early, 0–40 yr; middle, 40–250 yr; late, >250 yr) were recognized on upland sites (Romme and Knight, 1982). In the stand reestablishment and old-growth stages, the even-aged pine canopy deteriorates and is replaced by trees developing in the understory. On moist sites, lodgepole pine shares dominance with subalpine fir and whitebark pine. This relatively homogeneous site-vegetation pattern is in marked contrast to the diverse mosaic of physiography, human-use history, and disturbance regime of central New England, and the landscape-level vegetation mosaic reflects these differences.

Using fire history and the rates and patterns of forest succession, vegetation mosaics for the last 240 years were reconstructed. Figure 21.12 illustrates the percentage of the area covered by early, middle, and late stages of forest succession from 1738 to 1978. The high initial proportion of old-growth forest was markedly reduced by the fires of 1739, 1755, and 1795 and replaced with early-successional forests. Middle-successional stages became most abundant around 1800 and gradually declined as late-successional forest increased. How elegantly simple is this pattern compared to the complexity in New England described above.

In addition to vegetation coverage, community diversity from 1778 to 1978 was characterized using indices for measuring species diversity (Chapter 20). Figure 21.13 illustrates changes in community diversity over the landscape using (1) a weighted average of richness (number of community types), evenness (relative amount of landscape occupied by each community type), and patchiness (Romme, 1982); and (2) Shannon-Wiener

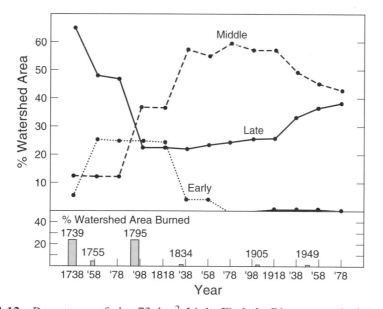

Figure 21.12 Percentage of the 73 km^2 Little Firehole River watershed covered by early, middle, and late stages of forest succession from 1738 to 1878 in Yellowstone National Park, northwestern Wyoming. (After Romme and Knight, 1982. Reprinted with permission from BioScience (vol. 32, no. 8, p. 666), © 1982 American Institute of Biological Sciences.)

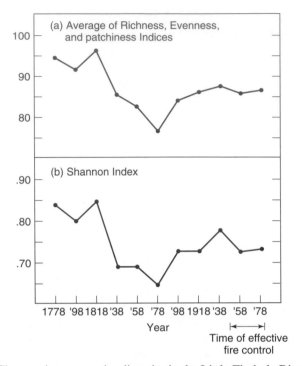

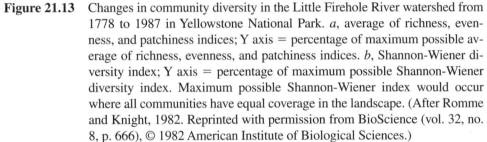

Figure 21.13 Changes in community diversity in the Little Firehole River watershed from 1778 to 1987 in Yellowstone National Park. *a*, average of richness, evenness, and patchiness indices; Y axis = percentage of maximum possible average of richness, evenness, and patchiness indices. *b*, Shannon-Wiener diversity index; Y axis = percentage of maximum possible Shannon-Wiener diversity index. Maximum possible Shannon-Wiener index would occur where all communities have equal coverage in the landscape. (After Romme and Knight, 1982. Reprinted with permission from BioScience (vol. 32, no. 8, p. 666), © 1982 American Institute of Biological Sciences.)

index. Both measures indicate a similar pattern: high diversity in the late 1700s and early 1800s after the extensive fires of 1739, 1755, and 1795, and low diversity in the late 1800s during a 70-year period with no major fires. It increased during the twentieth century due to differential rates of succession related to two small fires and mountain beetle attack (Romme, 1982). The effects of these landscape-level differences over time in relation to bird and elk populations are described by Romme and Knight (1982).

LANDSCAPE PATTERN AND PROCESS USING THE LANDSCAPE ECOSYSTEM APPROACH

Forest patches are often the focus of landscape-level studies. Another approach focuses on the identification and mapping of landscape ecosystems at multiple spatial scales using multiple factors of climate, geology, physiography, soil, and vegetation (Chapters 2, 13). This approach emphasizes spatial landscape hierarchy (e.g., Figure 2.11) and a strong physical site basis for understanding site-species relationships; community composition and diversity; ecosystem processes; and for providing the cartographic framework for the management, conservation, and restoration of ecosystems. Illustrations of this approach in the following examples emphasize: (1) the tight linkage between geology/physiography and community composition and diversity, (2) the great heterogeneity of landscapes at fine scales, (3) nutrient cycling differences among local ecosystems, and (4) the behavior of a rare and endangered species to spatial patterns of ecosystems.

Geomorphic Processes and Ecosystem Patterns

In Michigan, a landscape ecosystem approach has been and is being used to distinguish and map regional and local ecosystems following the integrated, multiscale and multifactor approach developed in southwestern Germany (Chapter 13). Multiple factors of climate, physiography, soil, vegetation, and hydrology were used simultaneously in the field to distinguish and map ecosystems (Barnes et al., 1982; Pregitzer and Barnes, 1984; Spies and Barnes, 1985a). In every regional and local setting, the nature and pattern of ecosystem types and their communities are strongly related to geological processes and landforms and reflect their heterogeneity. For example, in the Huron Mountains of Michigan's Upper Peninsula along the southern shore of Lake Superior, the interaction of retreating glacial ice about 10,000 years ago, bedrock, and water action created distinctive landforms that now support a repeating pattern of ecosystems (Simpson et al., 1990). Figure 21.14 illustrates the sequence of geological processes. As the ice slowly retreats, it traps meltwater between it and the crystalline mountain bedrock (Figure 21.14*a*). Here, the outwash stream deposits sand and gravel layers over glacial drift, which rest on sandstone bedrock. As catastrophic meltwater torrents develop with warming and further ice retreat, they strip off the glacial drift, rip out sections of the relatively soft sandstone bedrock, and leave extensive scoured flats of sandstone with little covering material (Figure 21.14*b* and *c*). Later, outwash sands cover these scoured areas. Vegetation develops on these diverse substrates, and ecosystem change (i.e., in soil, microclimate, and vegetation) occurs over time. Today we distinguish and map an array of eight ecosystems that differ in microclimate, soil, structure and composition of vegetation, and ecosystem processes that reflect their geologic–physiographic origin (Figure 21.14*d*).

Because of their spatial position in the landscape and soil conditions, fire and windstorm have also affected the forest communities such that the scoured sandstone flats in exposed topographic positions (Ecosystem 26) are dominated by old-growth hemlock-white pine forest. In contrast, Ecosystem 22 is underlain by till, and being more moist and fertile and less fire-prone, it supports northern-hardwood forests. This and other distinctive ecosystem patterns are repeated in the area and reflect the geomorphic processes of their origin.

Similarly, landforms strongly influence groundflora composition and ecosystem function. For example, Host and Pregitzer (1992), again using a multifactor approach, identified two ecosystem levels, *landform* and *ecosystem type,* in upland forests of northwestern Lower Michigan. Landforms are markedly different in overstory dominants and ground-cover species. For example, Figure 21.15*a* illustrates the major landforms of the area that include moraines, outwash plains, ice-contact terrain, and alluvial plains. The striking association of overstory dominants and ecological species groups of groundflora (Chapter 13) to landform is illustrated in Figure 21.15*b*. Black oak and the *Vaccinium* group dominate the matrix of dry, outwash plains. Northern red oak and the *Viburnum* group are relatively uncommon on outwash plains but dominate the dry-mesic, ice-contact hills and Port Huron Moraine. In contrast, sugar maple and the *Osmorhiza* group dominate the mesic and more fertile Interlobate Moraine. This example illustrates again a key point made in Chapter 10 that the largest part of the repetitive patterns detectable in vegetation can be traced directly to the repetitive pattern of topography (Rowe, 1984b).

Landforms not only affect community composition but play a significant role in controlling ecosystem processes such as succession (Figure 17.9; Host and Pregitzer, 1992),

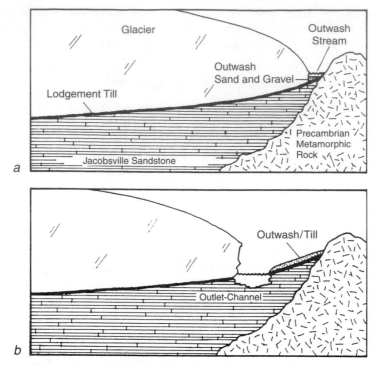

Figure 21.14   Diagram of geomorphic events that immediately followed the retreat of the glacier from the north slope of the Huron Mountain. *a*, Initial retreat of the ice front, 10,000 years ago; *b*, Initial outlet channel erosion by glacial meltwater;

biomass production (Host et al., 1988) and nutrient cycling (Zak et al., 1986). For example, three local ecosystems types that occur on different landforms exhibit markedly different patterns of nitrogen cycling. The black oak/*Vaccinium* ecosystem on outwash plains was contrasted with two sugar maple-dominated ecosystems that occurred on morainal landforms but differed in diversity and abundance of groundflora species. In Figure 21.16, we see that nitrogen mineralization differs by a factor of two between the oak and sugar maple-dominated ecosystems, and that the two maple ecosystems are significantly different. In addition, the species-rich sugar maple ecosystems exhibited a fourfold increase in potential nitrification compared with the species-poor sugar maple ecosystem, suggesting that NO_3^- loss following disturbance is an ecosystem-specific process in this landscape. A key point is that because these ecosystems and landforms repeat throughout the regional landscape, the pattern of nutrient cycling for a given ecosystem can also be expressed throughout the landscape. Thus understanding patterns of landforms and ecosystems across local and regional landscapes should enable managers to predict the response of ecosystems to management treatments, disturbances, and climatic change.

Landscape Ecosystems and Endangered Spaces

Focusing on spaces as well as species, the landscape ecosystem approach provides the basis for conserving rare and endangered species through understanding of the landscapes that sustain them. A case in point is the Kirtland's warbler (*Dendroica kirtlandii*), the rarest member of the wood warbler family in North America. It is a part of the sand outwash plain, jack pine–oak forest landscapes in northern Lower Michigan (District 8.2 in

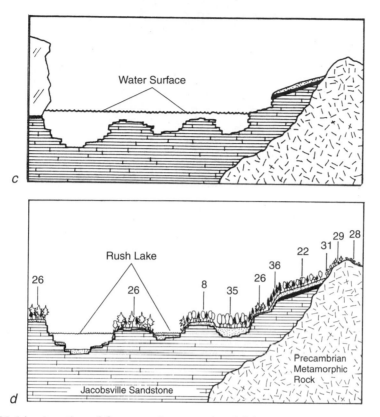

Figure 21.14 *(continued from previous page)* *c*, Melt-water channel location and configuration when the glacial front was about 400 m north of the Huron Mountain. *d*, Occurrence of ecosystem types in relation to physiography. Ecosystem types are: 26, Flood-scoured sandstone; 8, Flats and slopes; 35, Footslopes and intermittent stream valleys; 36, Steep terrace risers; 22, Mountain slopes; 31, Thin stony soil; 29, Broad ridge tops; 28, Narrow ridge tops, rocky knolls. The deep and shallow parts of Rush Lake reflect the intensity of channel erosion by glacial meltwater. (After Simpson et al., 1990.)

Figure 2.11). Overwintering in the Bahamas, the warblers fly to their summer breeding grounds in Michigan that are characterized by flat to gently rolling, dry, glacial outwash plains that support fire-prone stands of jack pine and to a lesser extent, northern pin oak. The population level of Kirtland's warblers is determined by an annual spring census of singing male warblers. Warblers were maintained for centuries by wildfire, which periodically creates young pine–oak stands. However, Kirtland's warbler populations declined from an estimated 1,000 birds in 1961 (Mayfield, 1962) to about 400 in 1972 (Mayfield, 1972), primarily as a result of lack of habitat (Probst and Weinrich, 1993). Forest fragmentation and lack of fire to regenerate jack pine-oak stands are the primary causes for lack of habitat. Furthermore, brown-headed cowbirds parasitize the warbler nests, and from 1972 to 1996, 101,600 cowbirds were destroyed to minimize loss of warbler fledglings (Deloria and DeCapita, 1997).

The warbler nests on the ground in young pine stands characterized by dense patches of pine interspersed by numerous small openings that are created by the irregular pattern

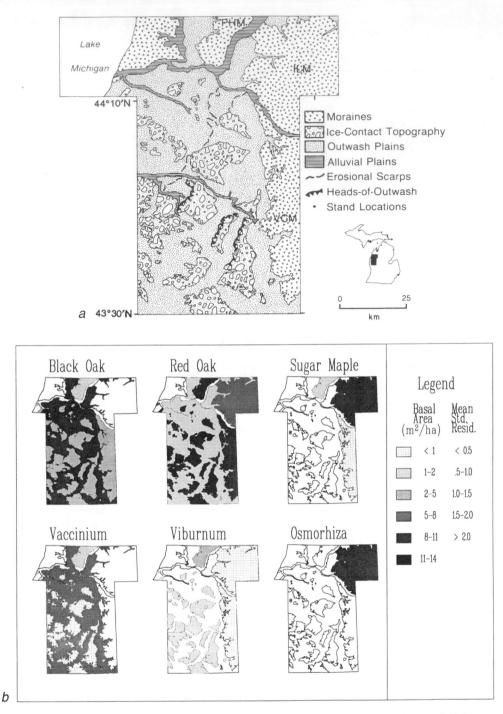

Figure 21.15 Relationships of landform and species in northwestern Lower Michigan. *a*, Major glacial landforms of study area; moraines include: PHM, Port Huron Moraine, ILM, Interlobate Moraine, VCM, Valparaiso–Charlotte Moraine; *b*, Maps of spatial distribution and abundance of selected overstory species and ecological species groups across the landforms. Overstory species (top row) are shaded according to mean basal area; ecological species groups are defined by one characteristic species and are shaded according to mean standard residuals that indicate the relative degree of association with the landform. (After Host and Pregitzer, 1992. Reprinted with permission of Canadian Research Council Canada.)

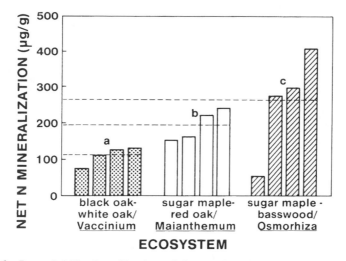

Figure 21.16 Potential N mineralization of three upland forest ecosystems. Ecosystems are named by dominant, late-successional trees and characteristic ecological species group. Four stands in each ecosystem were sampled. Values represent stand means for the amount of ammonium + nitrate N produced during an 8-week aerobic laboratory incubation. Ecosystem means are represented by the broken lines. Ecosystem means with different letters are significantly different ($\alpha = 0.05$). (After Zak et al., 1986. Reprinted with permission of Canadian Research Council Canada.)

of wildfire. Warblers delay colonization of an area until the trees reach a height of 1.5 to 3 meters and leave an area when the tree crowns begin to close over the openings (height 5 to 7 meters). Although considerable research has been done on the bird itself, little detailed information was available on the landforms and ecosystem types occupied by the warbler. However, on May 5, 1980 one of the most intensive fires ever recorded burned about 9,700 ha across the sandy, outwash-plain basin surrounding Mack Lake (Simard et al., 1983). This burn provided the opportunity for studying the pattern of warbler occupancy on a landscape ecosystem basis (Zou et al., 1992). Figure 21.17 illustrates several landforms as giant stair steps from high-elevation outwash and ice-contact terrain (385 m) to the low-elevation outwash and glacial meltwater channels (365 m) that once drained the basin. Air drainage follows this high-to-low pattern; the high-elevation landform is warmer in the growing season than the low-elevation outwash and the channels. Soils follow the same gradient; finer-textured sand and sand with heavy-textured bands occur in the high-level outwash and ice-contact ecosystems, but they are nearly absent in the low-level outwash and channels.

The growth of pines (and oak in the warmer ecosystems) follows the landform-based climate and soil pattern described—relatively fast in the high-elevation sites and slow in the low-elevation sites. Trees reach acceptable heights (ca. 2 meters) for warbler colonization in the high-elevation landform before those in the low-elevation landform. Not surprisingly, the pattern of warbler colonization and occupancy has followed this pattern of tree height. Overall, the warbler population expanded dramatically from 14 singing males in 1986 to 305 males 1994, thereafter declining to 276 in 1995 (Figure 21.18*a*). Warblers first colonized the area of taller trees and were for several years more abundant in ecosystems of the high-level area than the lower landforms. However, by 1988 populations in the high- and low-elevation landforms were equal, and after 1990 warblers were much more abun-

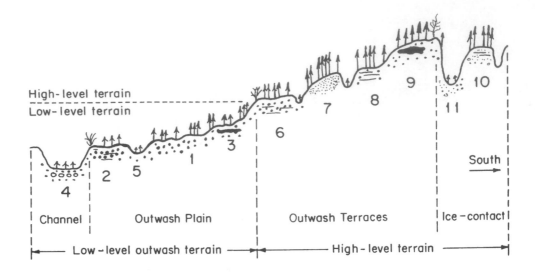

PHYSIOGRAPHIC DIAGRAM –MACK LAKE BURN
N-S Transect

Figure 21.17   Physiographic diagram of part of the Mack Lake Basin illustrating the spatial arrangement of landforms and ecosystems. The boundary between high-elevation and low-elevation landforms was arbitrarily distinguished at 372 m. Poorer soils (primarily medium sand; gravel and cobbles), colder microclimate, and shorter, slow-growing trees characterize the low-elevation pitted outwash plain. Heavier textured soils, warmer microclimate, and taller trees (except in ice-contact depressions, ecosystem 11) characterize the high-elevation landform. Warblers colonize high-level ecosystems first and over time predominate in the low-level ecosystems. Numbers indicate diverse ecosystem types characterized by differences in soil, microclimate, topography, and vegetation. For example, types 1, 4, and 5 occupy the driest, nutrient-poor soils; type 7 occurs on fine sand; types 4, 5, and 11 occupy sites with relatively cold microclimates; types 3, 6, 8, 9, and 10 occupy sites with fine-textured-soil bands of various thicknesses.

dant in the low-elevation landforms (Figure 21.18*b*). Once the warblers saturated the Mack Lake basin, the "overflow" birds colonized new areas, often young jack pine plantations.

As shown in Figure 21.19, the Kirtland's warbler is an inseparable part of landscapes where physiography and related landforms, soil, fire regime, and vegetation all are closely interconnected. Conserving this rare and endangered species requires identifying landscapes like the Mack Lake basin where the duration of warbler occupancy will be 12 to 15 years or longer. Landscape ecology research in over 50 other areas of current and former warbler occupancy shows that we can identify those specific landform-level ecosystems that have favorable conditions of microclimate, soil, and vegetation for long-duration warbler occupancy.

FOREST FRAGMENTATION

Human activities throughout the Ecosphere have produced great changes that affect the composition and diversity of biota. These include a reduction in total forest area, conver-

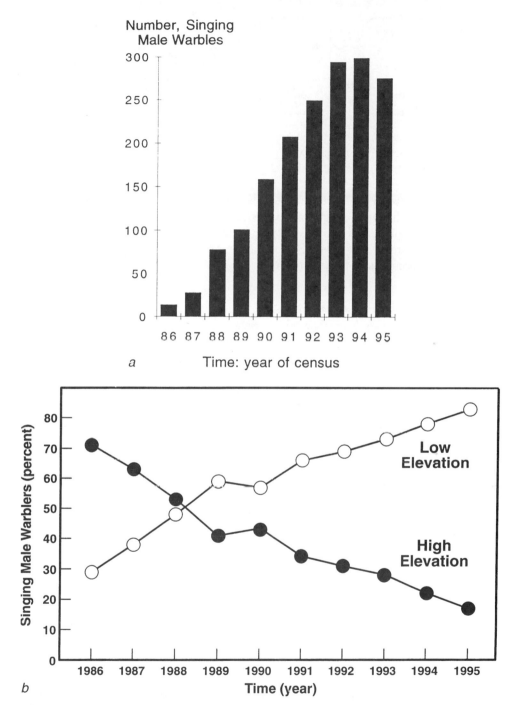

Figure 21.18   Abundance of the Kirtland's Warbler from 1986 to 1995 in the Mack Lake burn based on annual census of singing male warblers. *a*, Number of warblers occupying the basin from initial colonization in 1986 through 1995; *b*, Percentage of warblers occupying high-elevation (>372 m) low-elevation (<372 m) ecosystems from 1986 to 1995.

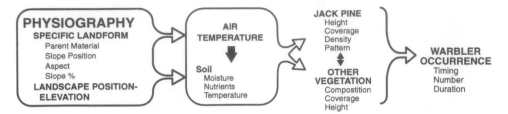

Figure 21.19   Diagram illustrating the interconnected relationships of physiography, microclimate, soil, and vegetation upon the initial time of colonization, the number of warblers, and the duration of their occupancy of landforms and ecosystems of the Mack Lake basin. Specific landforms refer to pitted outwash plain, ice-contact terrain, and channels; landscape position refers to the relative high- or low-elevation position of these landforms. For example, low-level outwash plains and depressions are frost pockets whose extreme late spring frost affect the growth of pine and may preclude the occurrence of northern pin oak.

sion of natural forest ecosystems to biotically simplified and often even-aged monocultures of plantations, and fragmentation of remaining forests into progressively smaller patches isolated by adjacent plantations, roads, or agricultural and urban development (Harris, 1984; Saunders et al., 1991). Forest fragmentation leads to (1) reduction of ecosystem diversity and concomitant loss of habitat heterogeneity for animals and plants, (2) surrounding of forest ecosystem remnants by inhospitable conditions for biota, (3) creation of abrupt forest edges, and (4) increased likelihood of decreased biotic diversity and extinction of animals and plants.

Forested land in the 48 contiguous states is estimated to have occupied 400 million ha in the 1500s (Harrington, 1991), and it was reduced by about 53 percent to approximately 188 million ha by the 1920s. Only three to five percent of pristine old-growth forest remains today, principally in the Pacific Northwest (Miller, 1992). Fragmentation of Pacific Northwest forests and its effects are described by Harris (1984). The remaining forests in the East are relatively young, dating from the 1920s and 1930s, and forest fragmentation is widespread (Rudis, 1995; Vogelmann, 1995). Small, isolated woodlots interspersed in farmlands and suburbia comprise about 40 percent of the deciduous forest (Terbough, 1989). Figure 21.20 illustrates a township in southern Wisconsin where land clearing and farming led to a fragmented forest in 1882 (30 percent of original) and forest fragments in 1950. By 1954, Curtis (1956) noted that only 3.6 percent of the original forest remained and most of that (77 percent) was heavily grazed so that no regeneration was taking place. However, landscape ecosystems are dynamic, and as much as 20 percent of this area can be naturally reforested (Dunn et al., 1991). This example is one of forest fragments or islands where the forest becomes functionally isolated from other and larger areas of forest. Fragmentation is often less severe than that pictured in Figure 21.20, and in many situations organisms and ecological processes that are characteristic of forests continue to dominate.

Ecological Effects

Natural fragmentation of forest communities by disturbance occurs in all forest ecosystems (Chapter 16). Disturbances characteristic of regional and local ecosystems (fire, windstorm, flood, avalanche) provided openings of various sizes in the presettlement forest and created habitat heterogeneity for diverse animals and plants. However, distur-

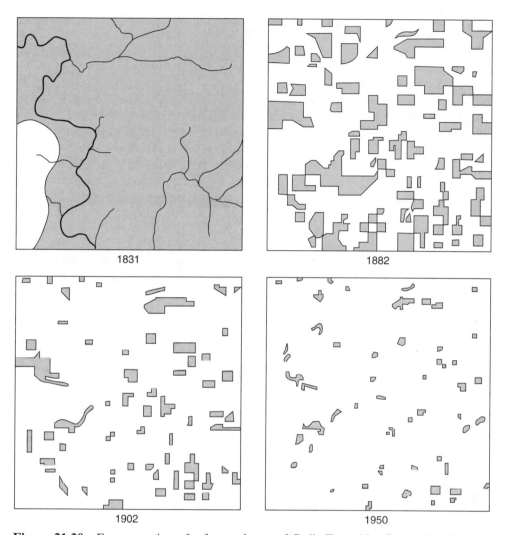

1831

1882

1902

1950

Figure 21.20 Fragmentation of a forested area of Cadiz Township, Green Co., Wisconsin, during the period of European settlement. The township is six miles on a side. The shaded areas represent land remaining in, or reverting to, forest in 1882, 1902, and 1950. (After Curtis, 1956. Reprinted from *Man's Role in Changing the Face of the Earth,* W. L. Thomas, ed., © 1956 by the University of Chicago Press. Reprinted with permission of the University of Chicago Press.)

bances in presettlement time were typically not as systematic and extensive, and forest edges were generally not so abrupt as they are today.

Fragmentation by human activities results in several major features that affect the diversity and abundance of animal species. First, the forest is subdivided into relatively small, more or less isolated patches of differing sizes and shapes. This reduction may yield an area too small to provide adequate habitat heterogeneity for territory size, food supply (Whitcomb et al., 1981), or other required features such as streams or wetlands. As we have seen (Figures 2.12 and 13.11), a fine-scale mosaic of ecosystem types and their diverse plant communities provide faunal "habitat." This fine-scale pattern is destroyed or simplified by fragmentation. Second, an extensive forest-edge habitat is created that rings the perimeter of forests adjacent to disturbed areas. Forest edges favor species adapted to edge habitats and select against those requiring interior habitats (Whitcomb et al., 1981; Yahner, 1988, 1995). Third, the forest edge is characterized by relatively sharp boundaries with microclimatic, vegetational, and biotic conditions (e.g., presence of predators) markedly different than in the interior of the forest.

As a result of these effects, changes in species richness may occur. **Forest-edge** bird species, those that feed and nest near forest edges, are relatively independent of forest size. They are either unaffected or even positively affected by fragmentation. In contrast, **forest interior** species—those that depend on large forest tracts and nest away from edges—may decline in numbers with fragmentation. For example, the probability of encountering a breeding pair of an area-dependent species, such as the scarlet tanager, is much greater (>70 percent) in a forest of 100 ha than in a tract less (< 50 percent) than 10 ha (Robbins et al., 1989). The decline in abundance of some neotropical migratory bird species that overwinter in the forests of Central and South America may be due to forest fragmentation in summer breeding grounds or loss of wintering habitat (Yahner, 1995).

The values of edge-effect to favor edge-dwelling game species have been emphasized by wildlife managers, dating back to the writings of Aldo Leopold (1933). However, it is clear that the characteristics of forest edge have a marked negative effect on many other fauna and flora (Wilcove et al., 1986; Yahner, 1995). Many game species (white-tailed deer, red fox), raptors (great horned owl, red-tailed hawk), and songbirds (gray catbird, brown-headed cowbird) are favored by edge habitat (Yahner, 1995). However, predation and parasitism in edge habitat are suggested as a main source of the decline of neotropical migrant bird species (Robinson, et al., 1995). For example, nest predation and parasitism by the brown-headed cowbird (*Molothrus ater*) increased with forest fragmentation in nine midwestern landscapes located in Missouri, Indiana, Illinois, and Wisconsin. Cowbirds lay their eggs in the nests of other "host" species, which then raise cowbirds at the expense of their own. Nest parasitism significantly increased with decreasing amount of forest cover. In fact, the reproductive success for some host species is so low that they may depend for persistence on immigration from source populations in landscapes with more extensive forest cover. The results suggest that a good conservation strategy for migrant songbirds in the Midwest is to restore and maintain large tracts of unfragmented forest in each regional landscape (Gustafson and Crow, 1994).

The human-created, abrupt forest edges along fields or clearcuts are not representative of forest disturbance edges to which many forest species have adapted. However, they may attract animals to nest or reproduce near the forest edge, precisely where predation rates are highest. This edge, or so-called "ecological trap" (Gates and Gysel, 1978), has microclimatic conditions markedly different from the forest interior. In studies of microclimatic variables from recent clearcut edges into the old-growth Douglas-fir forests of the Pacific Northwest, edge effects, on variables such as air temperature, humidity, wind speed, and radiation, were found to extend from 30 to over 240 m into the forest (Chen et al., 1995). Less crown cover, greater tree mortality, and a different tree species composition was found in edge versus interior areas (Chen et al., 1992). To maintain forest interior conditions and biodiversity, new management guidelines are suggested to reduce the amount and depth of forest edge.

One of the most severe examples of fragmentation occurs in Florida where the consequences of large highways are devastating. Harris and Silva-Lopez (1992) observe that, whereas Florida's native species evolved in a setting surrounded by sea, none of them evolved in the presence of high-density, high-speed traffic. In Florida, hard-surface highways were built at a rate of over 6 km/day for the last 50 years to accommodate human population growth rate and the tourist industry. Collisions between motor vehicles and animals represent the primary known source of mortality for most of Florida's endangered large wildlife species, including the panther, black bear, key deer, manatee, and American crocodile. Harris and Silva-Lopez (1992) conclude that roads are perhaps America's number-one fragmenting force.

Forest fragmentation is one result of a suite of human activities that have intensified concerns of ecologists and the public at large over diminishing forest lands, declining native plant and animal diversity, concomitant increase in exotic and invasive species, and sustaining natural ecological processes. These and other factors have led to an examination of forest and resource management practices and basic changes in policies and practices to sustain ecosystems.

NEW PARADIGM IN FORESTRY: SUSTAINABLE ECOSYSTEMS

Changing societal values have ushered in a new paradigm shift for the 21st century as the focus in forestry and land management becomes "ecosystems first" rather than "people first" (Rowe, 1994). In the past, the forest was conceived as trees and other organisms, with a production emphasis on single factors such as timber and wildlife. Now, society values other roles of forest ecosystems, such as ecological and biological diversity, forest health, and aesthetics, derived by seeking the integrity or sustainability of ecosystems and their processes (Christensen et al., 1996; Franklin, 1997; Kohm and Franklin, 1997; Callicott and Mumford, 1997).

For centuries, people have generally shared the view of subduing nature, of exercising their right to use "their" forests. Such community-shared political and economic beliefs set the standards by which good or bad forestry has been judged. A typical view is that forests were meant to provide people with timber, wildlife, water, and recreation. On the other hand, ecology suggests that landscape ecosystems are the life-giving segments of Earth upon which organisms depend (Rowe, 1994). Therefore, ecosystems are more important than the species they support, and increasingly, ecosystem-based management, or **ecosystem management** is being implemented. Definitions of ecosystem management emphasize that the goal is to *sustain ecosystem composition, structure, and function, i.e., protect native ecosystem integrity, over the long term* (Rowe, 1992b; Grumbine, 1994; Christensen et al., 1996). This objective is accomplished not only by our best understanding of multiscale ecological interactions and processes but also by the integration of scientific knowledge of ecological interactions with a complex sociopolitical and value framework (Grumbine, 1997). From ecological and practical perspectives, ecosystem management necessarily focuses on spatial units of the landscape at multiple scales and a detailed understanding of the complexity of their composition, structure and function. Several examples of ecosystem management are presented next.

Examples of Ecosystem Management

In a comprehensive survey of ecosystem-management projects being carried out in the United States, Yaffee et al. (1996) identified a set of 619 sites over remarkably diverse landscapes representing all regions and states. Sites selected were those that included elements of ecosystem-based land management in a significant way—where management was extended across property or political boundaries to incorporate ecological boundaries or where managers sought to shift management priorities away from emphasis on a single resource or species to consider ecosystem processes and the landscape as a whole. Of the 619 candidate sites, 105 were selected for detailed analysis of the participants' experiences.

Results of the assessment revealed that efforts to define and implement ecosystem management are under way in all regions of the United States, and that the early experience with these efforts is very positive. Participation and enthusiasm are widespread, and public and private sectors are involved at all levels. Managers are dealing with real problems at the ground level whose resolution requires more understanding and collaboration and greater consideration of larger spatial and temporal scales than was typical in the past. Collaborative groups have succeeded in dealing with complex ecological and social systems by managing adaptively, that is, by undertaking activities experimentally while investing in information that will determine whether the strategies are effective over time. Ecosystem management is therefore not only a set of goals, but a long-term process of understanding and decision making that requires the involvement of multiple sources of expertise and numerous stakeholders. Based on the assessment of the experiences at 105 sites, the overall message from the practitioners about the process of ecosystem management is to "know your land and know your neighbors." An outstanding example of adaptive ecosystem management comes from the Douglas-fir forest region of the Pacific Northwest.

21st-Century Ecosystem Management in Pacific Northwest Forests

In introducing the book on creating 21st-century forestry, Franklin and Kohm (1997) state: "if the 20th-century forestry was about simplifying systems, producing wood, and managing at the stand level, 21st-century forestry will be defined by understanding and managing complexity, providing a wide range of ecological goods and services, and managing across broad scales." In applying ecosystem management to sustain ecosystems and maintain their complexity, one of the key practical applications is a new set of silvicultural practices associated with forest harvest and regeneration.

Silviculture is traditionally concerned with the control of forest composition, establishment, and growth (Spurr, 1945). Traditional silvicultural regeneration and harvest methods—clearcut, seed-tree, and shelterwood in even-age management systems and selection used in uneven-aged systems (Smith et al., 1997)—were created for a singular objective: regeneration and subsequent growth of a commercially-important tree species. Today, however, multiple ecosystem management objectives typically include maintenance of specific levels of ecosystem processes and provision of habitats for diverse biota. Tree regeneration and its subsequent growth are often still concerns, although these objectives—especially for rapid growth of regeneration—often are subordinated to other goals. In Douglas-fir forests of the western slopes of the Cascade Mountains in Washington and Oregon, harvest objectives may include such diverse goals as providing for specified levels of snags and standing live trees of various species and sizes, maintaining tree root strength, managing for special forest products (fungi, moss, native understory plants), and fulfilling specific aesthetic criteria (Franklin et al., 1997; Molina et al., 1997).

Research on Douglas-fir–western hemlock forests of the western Cascades has emphasized the structural complexity of old-growth forest stands that range in age from 250 to over 600 years (Franklin et al., 1981; Harmon et al., 1986; Franklin et al., 1987; Maser et al., 1988; Spies and Franklin, 1991; Franklin, 1993; Spies, 1997). Many of these stand-structure features were described in the first section of this chapter, and they include varied sizes and conditions of live trees (Figure 21.1), snags, woody debris on the forest floor (Figure 21.2) and in streams (Figure 21.4), multiple canopy layers, and presence of canopy gaps (Figure 17.14*b*). These structural features markedly influence biotic composition of the forest, ecosystem function (biomass production, nutrient cycling, and succession; Franklin et al., 1981), and regeneration following natural disturbances. Therefore, they form the centerpiece for creating new managed forests.

Studies of regeneration following natural disturbances have emphasized the importance of biological legacies—surviving organisms and organically-derived structures, such as snags, logs, and soil organic horizons—to the rapid reestablishment of forests that have highly diverse composition, function, and stand structure (Chapter 16). A consideration of biological legacies illustrate some critical differences between natural disturbances and traditional even-aged cutting methods. For example, the effects of clearcutting are not ecologically comparable to the effects of most natural disturbances (Franklin, 1997). The high levels of biological legacies that typically follow natural disturbances lead to rapid redevelopment of compositionally, structurally, and functionally complex forest ecosystems. Traditional approaches to clearcutting purposely eliminate most of the structural and much of the compositional legacy in the interest of efficient harvest and wood production. A fragmented landscape in western Oregon, created by dispersed patch clearcutting, is illustrated in Figure 21.21.

As a result of new knowledge about forest structure and ecosystem function, structurally-complex managed stands are being developed in the United States, and independently in many countries on several continents, to manage forests for multiple and complex objectives, including wood production and biodiversity (Franklin et al., 1997). Three proposed approaches to create structurally complex stands are: (1) use of long rotations, (2) retention of structural features at the time of harvest, and (3) silvicultural treatment of established stands. These approaches can be combined effectively, although each has specific circumstances where it is particularly appropriate to address different environmental issues and economic implications.

Using long rotations, forests are managed on time frames often 50 to 300 percent longer than those currently in use. For example, sites supporting Douglas-fir forests traditionally managed on an 80-year rotation would be extended to 120 to 240 years. The advantages and limitations of long rotations are discussed by Franklin et al. (1997) and Curtis

Figure 21.21 Fragmented landscape due to dispersed patch clearcutting in Douglas fir forests on Warm Springs Indian Reservation (formerly Mt. Hood National Forest) in western Oregon. Individual clearcut patches range in size from 6 to 16 ha. (Photo courtesy of Jerry F. Franklin.)

(1997). Silvicultural treatment of established stands seeks to "restore" structurally-simplified stands created by traditional even-aged methods (e.g., clearcutting) to more structurally-complex and compositionally-rich forests (DeBell et al., 1997; Tappeiner et al., 1997). Methods employed in such young stands include early density control in order to vary spacing and create patches and thinning in older stands to encourage regeneration that may produce structures similar to old-growth stands. In addition, fertilization, pruning, and managing dead wood (e.g., creation of snags, down wood, and cavities for wildlife) are methods that may increase financial returns while creating more compositional and structural diversity.

VARIABLE-RETENTION HARVEST SYSTEM

The variable-retention harvest system attempts to create in managed stands, as far as possible, elements of the rich structural, compositional, and functional diversity that are characteristic of old-growth forests. It is based upon the concept of biological legacies—retaining structural elements of the harvested stand for at least the next rotation. Variable-retention harvest prescriptions have three major purposes: (1) "lifeboating" species and processes immediately after logging and before forest cover is reestablished, (2) "enriching" reestablished forest stands with structural features that would otherwise be absent, and (3) "enhancing connectivity" in the managed landscape (Franklin et al., 1997). Strategies of structural retention focus on both living and dead organic structures, including living trees of various species, sizes, and conditions and their derivatives such as snags and logs on the forest floor. Such structures provide habitats for many species (Carey and Johnson, 1995) and ecosystem functions (Franklin et al., 1981; Spies, 1997). The term "variable retention" is used because the structures retained may vary widely in amount (percent of area covered), kind, and spatial pattern.

Lifeboating: Refugia and Inocula

Lifeboating provides refugia for biota that might otherwise be lost from the harvested area. The primary objective of structural retention is achieved in three ways: (1) providing structural elements that fulfill habitat requirements for various organisms, (2) ameliorating microclimatic conditions in relation to those that would be encountered under clearcutting, and (3) providing energetic substances to maintain heterotrophic organisms. Figure 21.22 illustrates dispersed retention of living trees following partial cutting in a 135-year-old Douglas-fir forest. Woody debris was left on the forest floor. Microclimatic conditions are often critical for survival of some organisms, and such conditions are most likely retained in small forest patches by aggregating the retained structures, a topic discussed in the following section. Structural retention will almost always result in harvested areas that have less stressful microclimatic regimes than those that are found on clearcuts. In addition, retention of live trees and shrubs provides critical habitat to maintain populations of soil organisms, such as mycorrhizae (Amaranthus et al., 1994).

Structural Enrichment and Enhanced Connectivity

Besides serving as refugia immediately after logging, structural retention is a technique for enriching the complexity of managed forest stands for an entire rotation. Suitable conditions for species can thereby be reestablished much earlier in the rotation than would otherwise be possible. Retention can provide suitable habitat for species that are generally rare or absent in young stands of simple and homogeneous structure. For example, many forest stands 80 to 200 years old in the Douglas-fir region provide suitable nesting and foraging habitat for the northern spotted owl and other species associated with late-successional forests. By retaining some old-growth Douglas-fir trees, managed stands may provide suitable nesting and foraging habitat for spotted owls within 50 or 60 years of

Figure 21.22   Variable-retention harvest system in old-growth Douglas-fir forest, western Oregon. Silvicultural prescription is for dispersed structural retention of large live trees (135-yr old), snags, and logs (20 to 25 trees per hectare) to meet minimal goals for coarse woody debris. H. J. Andrews Experimental Forest, western Oregon. (Photo courtesy of Jerry F. Franklin.)

harvest. Without retention, it may take 120 years or more to create the necessary structural elements (Franklin et al., 1997).

Facilitating the movement of organisms within a managed landscape is a third value of structural retention. Connectivity is enhanced in the harvest unit by retaining structures that facilitate the dispersal of organisms, in addition to creating corridors between intact stands. The objective here is to make the traditionally "non-habitat," managed stands more hospitable for dispersion by retaining, for example, well-spaced logs, trees, and shrub patches for protective cover or habitat.

Designing a Variable-Retention Harvest System

In developing a variable-retention harvest system, three major issues are: (1) what structures to retain on the harvest site, (2) how much of each of these structures to retain, and (3) the spatial pattern for the retention, i.e., *dispersed* or *aggregated*, or some combination of these. Structures to retain may include live trees, especially large-diameter trees, snags in varying states of decay, logs and other debris on the forest floor, forest understory, and undisturbed layers of forest floor. How much to retain depends on the management objective for the harvest unit, but little data are available on how specific ecological objectives respond to various levels of structural retention.

Dispersed retention, as illustrated in Figure 21.22, creates an even distribution of structures in a harvest unit. Well-dispersed structures provide wildlife habitats and future sources of energy across a large area. Aggregated retention, illustrated in Figures 21.23 and 21.24, focuses on small patches of forest ("green trees") left in a harvest unit. Aggregated structures provide more microclimatic modification and retain more diversity by canopy structure and species composition. In one of the first trials (Figure 21.23), foresters of the Plum Creek Timber Company retained 15 percent of the green trees as aggregates in clearcut unit. Figure 21.24 illustrates aggregated retention with teardrop-shaped islands of green trees left between radiating clearcut strips. These compact aggregates are better at modifying microclimate and may be less susceptible to windthrow, whereas linear aggregates can provide better visual screening, wildlife corridors, and stream protection.

Figure 21.23   One of the first examples of aggregated retention in which 15 percent of green trees were left in the clearcut area which lies in the white polygon above. Aggregated retention is often advantageous from standpoints of logging costs, safety concerns, and management efficiency. Plum Creek Timber Company lands in southwestern Washington State. (Photo courtesy of Jerry F. Franklin.)

Figure 21.24 Variable-retention harvest system in Douglas-fir forest illustrating the aggregated method of retaining live trees following harvest. Teardrop-shaped patches of trees are left, providing corridor-like connections of live-tree groups to the intact stand. Note individual trees left between the islands of green trees. City Of Seattle, Cedar River Watershed, western Washington State. (Photo courtesy of Jerry F. Franklin.)

Figure 21.25 illustrates an experiment where both dispersed and aggregated methods are applied in combination together with the shelterwood system. Whether to have dispersed retention, aggregated retention, or some combination of both involves many issues of forest protection (wind, wildfire, insects, and disease) and forest harvest and management (worker safety, costs of transportation and logging, impact on forest product receipts). Common objectives for many public and some private timberlands include provision of moderate to high levels of timber production, regeneration of understory-intolerant tree species, and maintenance of minimal structural levels to fulfill basic lifeboat and stand-enrichment functions.

Silvicultural prescriptions to achieve this mix of objectives typically involve relatively low retention and leave 10 to 20 percent of the area in live trees, using both aggregated and dispersed approaches. Experiments involving 15, 40, 75, and 100 percent retention are underway to study the effects on small mammal, bird, and fungal populations, forest understory communities, growth of tree regeneration, and growth and mortality of retained trees. Overall, variable-regeneration harvest prescriptions are emerging as a major strategy for integrating ecological and economic objectives throughout the temperate forest regions of the world.

Simulating Harvests and Forest Fragmentation

In the previous section, we learned that managers can create biological linkages from one generation to another by variable-retention forest harvest and thereby mimic natural disturbances to a certain degree. The harvest patches themselves—their size and dispersion

Figure 21.25   Experimental application of the variable-retention harvest system on MacMillan-Bloedel Company lands near Campbell River, British Columbia, Canada. This coastal-montane forest includes Pacific silver fir, western hemlock, Alaska-cedar, and Douglas-fir. Note the aggregated dispersal of tree groups and several densities of dispersed retention, including shelterwood (left-center). This experiment was designed to address regeneration, wildlife habitat, and aesthetic concerns in managing forests, and considers both ecological and economic issues. (Photo courtesy of Jerry F. Franklin.)

in the landscape—are also of concern in ecosystem management. As we observed in Figures 21.21 and 21.25, clearcutting creates patches of various sizes and causes forest fragmentation and increased edge effect to one degree or another. Therefore, in practicing ecosystem management, the landscape ecologist is concerned with the size, shape, and dispersal pattern of harvest units because of their ecological, aesthetic, and economic consequences. Generally, it is thought that forest harvesting has the following effects on fragmentation: (1) large cutting units cause less fragmentation than small units, (2) aggregated and progressive cutting patterns create less fragmentation than dispersed cutting patterns, and (3) low harvest rates produce less fragmentation than high cutting rates (Crow and Gustafson, 1997).

To test these assumptions and evaluate the effects of different harvest strategies on the amount of forest interior and edge, Crow and Gustafson (1997) examined the interaction between the size and distribution of harvest units and the rate of harvest. They applied a simulation model to a 23,000-ha area in southern Indiana that is largely forested with some patches of nonforest cover. They sought to determine the relative impact of three sources of variation: (1) *size* of harvest unit, (2) *intensity* of harvest, and (3) *spatial dispersion* of openings on the total area of forest interior and total linear forest edge over the eight decades. Thus for the simulation, two levels of patch size (18- and 0.18-ha harvest units), intensity of cut (1 and 7 percent per decade), and spatial dispersion (randomly dispersed and aggregated) were used. The results, illustrated in Figure 21.26*a*, show that for-

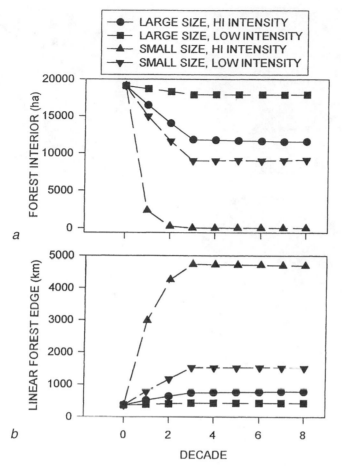

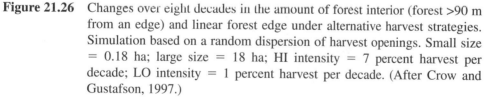

Figure 21.26 Changes over eight decades in the amount of forest interior (forest >90 m from an edge) and linear forest edge under alternative harvest strategies. Simulation based on a random dispersion of harvest openings. Small size = 0.18 ha; large size = 18 ha; HI intensity = 7 percent harvest per decade; LO intensity = 1 percent harvest per decade. (After Crow and Gustafson, 1997.)

est interior area was greatest when harvest opening size was large, harvest intensity was low, and openings were clustered; it was lowest when openings were small, harvest intensity was high, and openings were dispersed. Notice that harvesting only 1 percent of the forest each decade using small openings (low intensity, small size in Figure 21.26*a*) leaves less forest interior than harvesting 7 percent of the forest using large openings (high intensity, large size in Figure 21.26*a*). Small cuttings simply perforate the forest, eliminating interior. They would also require a greater road network, further fragmenting the area. The amount of linear forest edge was lowest when harvest opening size was large, harvest intensity was low, and openings were clustered; it was highest when openings were small, harvest intensity was high, and openings were dispersed (Figure 21.26*b*). Again, even a low rate of harvest using small openings produces more edge than a high rate of harvest using large openings.

A visual illustration of one of the simulations is shown in Figure 21.27. The initial landscape pattern (Figure 21.27*a*) is compared with those projected after 80 years using

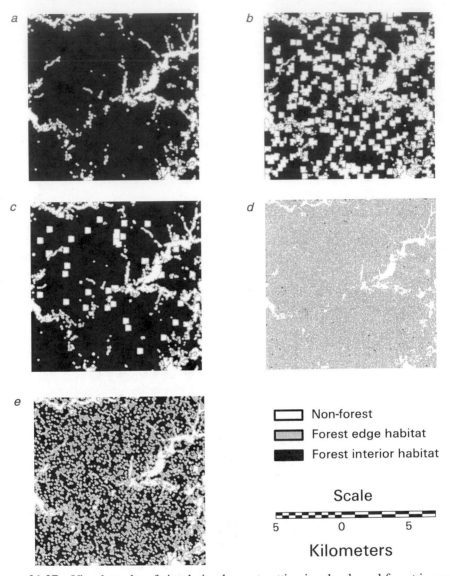

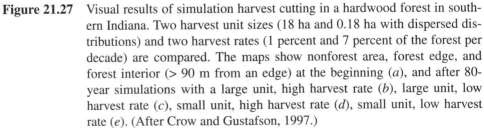

Figure 21.27 Visual results of simulation harvest cutting in a hardwood forest in southern Indiana. Two harvest unit sizes (18 ha and 0.18 ha with dispersed distributions) and two harvest rates (1 percent and 7 percent of the forest per decade) are compared. The maps show nonforest area, forest edge, and forest interior (> 90 m from an edge) at the beginning (*a*), and after 80-year simulations with a large unit, high harvest rate (*b*), large unit, low harvest rate (*c*), small unit, high harvest rate (*d*), small unit, low harvest rate (*e*). (After Crow and Gustafson, 1997.)

two harvest unit sizes (18 ha and 0.18 ha with dispersed distributions) and two harvest rates (1 percent and 7 percent of the forest per decade). The marked reduction in forest interior by dispersing small harvest units across the landscape is obvious in this visual format (Figures 21.27*d* and 21.27*e*). Thus a combination of larger harvest units, the aggregation of those units, and lower harvest rates may be necessary to reduce forest fragmentation and its ecological consequences.

Simulation modeling can increase a manager's ability to evaluate the effects of alternative management strategies across the landscape. In addition, to develop the models and apply them successfully on the ground, one must know intimately the spatial patterns of landscape ecosystems of the area, their above- and belowground dynamics, and the forest land-use history. The tools for evaluating ecosystem management alternatives and for monitoring the landscape dynamics over time, such as geographic information systems (GIS) and remote sensing, are widely available and applied by ecosystem managers. The role of remote sensing is considered next.

REMOTE SENSING

Remote sensing is a modern technology that links spatial patterns and ecological processes at multiple spatial and temporal scales. The sensing of ecosystem components, such as landforms (Pickup, 1990) and vegetation (Hobbs, 1990), has the potential to extend measurements over spatial scales from microscopic at short wavelengths to the global scale over a broad range of wavelengths (Ustin et al., 1993). Remote sensing, especially the use of aerial photographs, has been long an integral part of forest and wildlife management (Johnson, 1969). The main advantages of remote sensing are that: (1) observation and measurement are above the landscape, thereby preventing observer interference in the system, (2) local, regional, and global-scale measurement may be done repeatedly, and (3) a wide variety of spectral ranges and sensors are available to provide remotely-sensed data (Lulla and Mausel, 1983).

Landscape ecology and ecosystem-based management are intimately concerned with landscape heterogeneity, especially forest patchiness due to disturbances by human activities, fire, wind, and water. Remote sensing techniques are eminently suited to providing data in several key areas related to forest landscape patterns (Johnson, 1969): (1) inventory and mapping of ecosystem components (landforms, waterforms, and vegetation), (2) quantification of environmental characteristics (soil water, land-use cover), (3) inferences regarding flow of energy and nutrients within and among ecosystems, and (4) evaluating ecosystem change (e.g., fragmentation of forests over time). Above all, determining the meaning and validity of remotely sensed data for these applications requires familiarity with the ecosystems being measured, along with the basic understanding of relationships expressed in the data.

For any of these application areas, identifying and characterizing the landscape features is essentially the starting point. In any given landscape, every object has a vertical and horizontal spatial structure and also a characteristic multispectral response. Such a "signature" is due to the way objects reflect, emit, or transmit electromagnetic energy. Thus identification and boundary delineation in the landscape is accomplished, in part, by obtaining (with airborne or satellite sensors) and analyzing spectral signatures. The landscape seen in Figure 21.28 illustrates the use of several different spectral bands to distinguish among a variety of features, including broadleaved and coniferous forests, lakes and wetlands, roads, and "hot-spots." Decreased leaf absorption and increased leaf reflectance in the wavelength region between 0.7 and 1.0 μm are more striking with broadleaved than coniferous species. This fact is the basis for the use of infrared-sensitive films for forest aerial photography when separation of broadleaved and coniferous species is desired. Figure 21.28 demonstrates that reflection from these groups is essentially the same in the visible spectrum (Figure 21.28*a*; spectral range 0.4 to 0.7 μm), but that the broadleaved species are significantly more reflective than conifers (dark tones) in the near-infrared spectral band (Figure 21.28*b*; spectral range 0.7–0.98 μm).

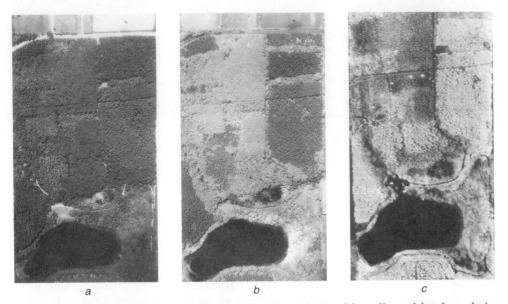

a   b   c

Figure 21.28   Differences in the reflectance of conifers and broadleaved hardwoods in the visible and near-infrared spectral bands and emittance in the thermal infrared band, Saginaw Forest, Ann Arbor, Michigan. *a*, Visible spectral band (0.4–0.7μm); hardwood plantations not distinctly different in tone from conifers in midsummer. *b*, Near-infrared spectral band (0.7–0.98μm); hardwoods markedly lighter in tone than conifers in midsummer. *c*, Thermal infrared band (4.5–5.5μm); small charcoal fires detected; conifers warmer (lighter tone) than leafless hardwoods in midwinter. (Courtesy of University of Michigan.)

The use of optical-electronic detectors instead of photographic film makes it possible to monitor or "sense" landscape phenomena over a broader spectrum of wavelengths. A multispectral sensor may collect data at several discrete wavelength ranges, e.g., 0.45–0.49 μm, 0.5–0.59 μm, 0.6–0.69 μm, 0.7–0.79 μm, and 4.5–5.5 μm. The image interpreter then may choose to look at each of these "channels" as separate images or combined in a composite image. Thermal remote sensors with detectors capable of recording infrared energy with wavelengths too long to be recorded photographically (4.5–5.5 μm) can detect small fires from altitudes as high as 7,000 m. For example, the eight bright spots in Figure 21.28*c* were caused by small charcoal fires with a surface area of approximately 80 cm^2 each. These hot spots are detectable in the 4.5–5.5 μm band because this is the wavelength of maximum emittance for an object whose temperature is approximately 300 °C. Remote sensors with electronic detectors operating at still longer thermal wavelengths (8.0–14.0 μm) can detect the effects of several kinds of insect and disease attacks, in some instances before visual symptoms appear. The affected trees are consistently 1 to 3 °C warmer than unaffected trees.

Use of black and white panchromatic, color, and color-infrared aerial photography or airborne multispectral or thermal sensors, such as those described, have been basic working tools of the ecologist and manager for many years. In the late 1970s, space-borne optical remote sensors began to provide ecologists with multispectral imagery whereby large regions or all of the globe may be studied. The American Landsat and AVHRR, and French SPOT systems are three of the most significant satellite sensor programs that are still actively acquiring data.

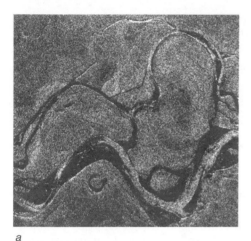

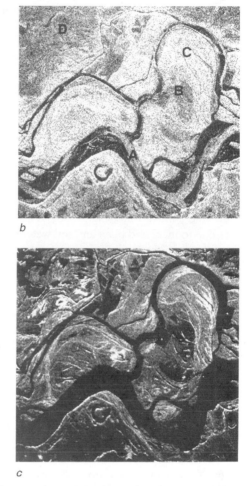

Figure 21.29 Radar images illustrating differences in topography and moisture and phenological states of several forest communities, Bonanza Creek research site, near Fairbanks, Alaska. Plant communities adjacent to the Tannana River (A) include black spruce (B), white spruce–balsam poplar (C), and shrub bog (D). *a*, March image at sub-freezing temperatures when water in plants is frozen; *b*, March image several days later during a thaw with liquid water in vegetation and soil; *c*, May image with more advanced phenological stage. (Images courtesy of The University of Michigan Radiation Laboratory and the NASA Jet Propulsion Laboratory.)

Remote sensors are not limited to sensing reflected or emitted solar thermal radiation, as in the previous examples. Synthetic aperture radar (SAR) does not record reflected or emitted solar radiation as do passive optical sensors. It is an active sensor that transmits its own energy and then records the signal reflected back to the sensor from the target. SAR uses the microwave portion of the electromagnetic spectrum with wavelengths ranging from 1 to 60 cm. Optical instruments operating in the visible and infrared ranges primarily record energy reflected off the top of a closed forest canopy, responding to internal leaf structure and pigments. At its longer wavelengths, SAR effectively may penetrate the forest canopy to provide information on tree bole biomass, bole moisture, soil water, understory vegetation, and topographic features. Radar is used for diurnal monitoring and for gathering information in frequently cloud-covered boreal or tropical regions because it is independent of solar illumination, and because precipitation and clouds are invisible to SAR's longer wavelengths.

SAR applications in forest ecology include land-cover mapping (Rignot et al., 1994), estimation of biomass (Dobson et al., 1995), and estimation of carbon storage and aboveground net primary production (Bergen et al., 1997). For example, Figure 21.29 illustrates the use of radar to distinguish differences in plant and soil water and phenological states of forest communities at a site near Fairbanks, Alaska. Plant communities adjacent to the Tannana River (A) include black spruce (B), white spruce–balsam poplar (C), and shrub bog (D). In March (Figure 21.29*a*), water in plants and soil is frozen, and only major differences between main river channels and vegetation as a whole are observed. Several days later during a thaw (Figure 21.29*b*), the much brighter tone of forested areas results from higher reflectance of SAR energy due to presence of liquid water in vegetation and saturated soil. In May (Figure 21.29*c*), good discrimination is evident among communities due to increased plant and soil water differences and more advanced phenology.

Remote sensing methods, therefore, provide efficient ways to gather data on the spatial patterns of ecosystems and communities, biomass production, insect and disease outbreaks over large areas, and to monitor changes in these parameters over time—something heretofore impossible to accomplish at such scales. A variety of books and papers describe the scope and detail of the use of remote sensing in landscape ecology and ecosystem management (Hobbs, 1990; Hobbs and Mooney, 1990; Luvall and Holbo, 1990; Pickup, 1990; Quatrochi and Pelletier, 1990; Dunn et al., 1991; Ustin et al., 1993; Sample, 1994).

Suggested Readings

Christensen, N. L., A. M. Bartuska, J. H. Brown, S. Carpenter, C. D'Antonio, R. Francis, J. F. Franklin, J. A. MacMahon, R. R. Noss, D. J. Parsons, C. H. Peterson, M. G. Turner, and R. G. Woodmansee. 1996. *Ecol. App.* 6:665–691.

Crow, T. R., and E. J. Gustafson. 1997. Ecosystem management: managing natural resources in time and space. In K. A. Kohm and J. F. Franklin (eds.), *Creating a Forestry for the 21st Century*. Island Press, Washington, D.C.

Forman, R. T. T. 1995. *Land Mosaics, The Ecology of Landscapes and Regions*. Cambridge Univ. Press. 632 pp.

Foster, D. R., and E. R. Boose. 1992. Patterns of forest damage resulting from catastrophic wind in central New England, USA. *J. Ecology* 80:79–98.

Franklin, J. F. 1997. Ecosystem management: an overview. In M Boyce (ed.), Proc. Symp. *Ecosystem Management: Applications for Sustainable Forest and Wildlife Resources*. Yale Univ. Press. New Haven, CT.

Franklin, J. F., D. R. Berg, D. A. Thornburgh, and J. C., Tappeiner. 1997. Alternative silvicultural approaches to timber harvesting: variable retention harvest systems. In K. A. Kohm and J. F. Franklin (eds.), *Creating a Forestry for the 21st Century*. Island Press, Washington, D.C.

Grumbine, R. E. 1997. Reflections on "What is ecosystems management?" *Conserv. Biol.*11:41–47.

Kohm, K. A., and J. F. Franklin. 1997. *Creating a Forestry for the 21st Century*. Island Press. Washington D.C. 475 pp.

Rowe, J. S. 1988. Landscape ecology: the ecology of terrain ecosystems. In M. R. Moss (ed.), *Landscape Ecology and Management,* Proc. First Symp. Can. Soc. Landscape Ecology and Management. Univ. Guelph Polysci. Publ., Montreal.

Rowe, J. S. 1994. A new paradigm for forestry. *For. Chron.* 70:565–568.

Swanson, F. J., and J. F. Franklin. 1992. New forestry principles from ecosystem analysis of Pacific Northwest forests. *Ecol. App.* 2:267–274.

PART 6

Although ecological principles and concepts are basic to the forest ecologist's education, hands-on experience with factors of the physical environment is required to fully understand their application. Furthermore, ecologists must know the plants and associated animals to fully understand ecological processes and predict their outcomes. Usually, the forest ecologist's field training is local and often necessarily provincial. Therefore, it is critically important that one explore not only the rich ecological literature but travel widely—studying forest ecosystems at regional and local scales in North America and around the world.

The diversity of the Earth's landscape ecosystems is truly amazing. Thus a comparative study of forests inhabiting regions of continental or subcontinental scale reveals striking similarities in plant occurrence—and equally remarkable differences! Exposure to a broad range of ecosystems means that preconceived ideas are challenged and to the keen observer ecosystem complexity is revealed at multiple scales. New combinations of climate, geology and physiography, and forest history continually challenge our thinking about the structure and function of forest ecosystems, and they raise new and exciting questions. Thus travel and comparative study of the Earth's ecosystems is immensely rewarding.

In Chapter 22, we can only gain a brief insight of forest diversity, its origins and complexity. Nevertheless, at this point in our journey in forest ecology, it is useful to step back and examine the overall history and relationships of forest organisms to the landscape ecosystems that support and sustain them.

CHAPTER *22*

FORESTS OF THE WORLD

In this final part, we provide a brief description of the present-day forests of the world with particular reference to forests of temperate North America. Description, however, is not enough. An understanding also must be provided by summarizing briefly the evolutionary development of today's forests, and, in that description, bringing in as much as possible of the climatic, edaphic, and dynamic relationships that go far toward explaining why a particular type of forest is growing in a particular geographic location. Therefore, we present in the first part of this chapter a brief overview of the woody plant evolution and development. Next, we examine the general kinds of world forests arranged in either geographic-climatic, physiognomic-structural, or floristic classifications. This discussion is followed by a description of the major ecological features of tropical, boreal, and temperate forests. Emphasis is placed on the groups of plants that have similar morphology and physiological and ecological requirements and distributions.

EVOLUTION OF MODERN TREE SPECIES

Aside from a few tree ferns and cycads that reach tree proportions, modern trees may be grouped as gymnosperms and angiosperms. The appearance of their prototypes and the gradual development of modern forms are recorded in fossils with which the study of paleobotany is concerned (Beck, 1976; Behrensmeyer et al., 1992; Stewart and Rothwell, 1993; Taylor and Taylor, 1993). A few salient points may be summarized briefly.

The oldest lineage of modern trees is that of the conifers, which can be traced back to late Paleozoic (Figure 22.1). Some early conifers had leaves and branches quite like those of modern species of *Araucaria*, a Southern Hemisphere genus that persists today with species such as Norfolk Island pine, hoop pine of Queensland, Parana pine of Brazil, and

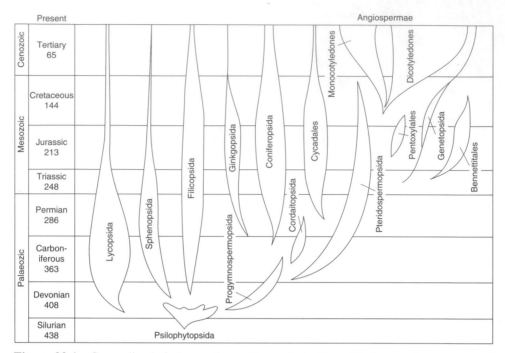

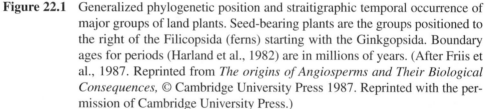

Figure 22.1 Generalized phylogenetic position and straitigraphic temporal occurrence of major groups of land plants. Seed-bearing plants are the groups positioned to the right of the Filicopsida (ferns) starting with the Ginkgopsida. Boundary ages for periods (Harland et al., 1982) are in millions of years. (After Friis et al., 1987. Reprinted from *The origins of Angiosperms and Their Biological Consequences,* © Cambridge University Press 1987. Reprinted with the permission of Cambridge University Press.)

the monkey-puzzle tree of Chile. These primitive conifers, however, constituted but a small part of the forests that formed the coal beds of Carboniferous (Mississippian and Pennsylvanian) times. Seed ferns (Pteridospermospida), horse-tails (Sphenopsida), club-mosses (Lycopsida), and two lines of primitive conifers (Coniferopsida and Cordaitopsida, now extinct) formed the dominant tree flora. These lowland, mild-climate plants, along with many marine organisms and insects, were largely eliminated at the Permian-Triassic boundary due to large-scale glaciation, volcanic activity, and continental uplift. This left the land open to colonization by the developing modern arborescent groups.

In Mesozoic times, the conifers evolved into many forms and achieved great abundance (Figure 22.1). By Jurassic times, forms similar to modern *Libocedrus, Thuja, Sequoia,* and *Agathis* (the kauri of New Zealand and nearby lands) had appeared. Because identifications frequently must be made from impressions made by fragments of leaves and other vegetative parts, many are highly tentative. The general predominance of conifers in the fossil record of Mesozoic times is clear, however, as is the fact that most present-day genera of conifers are much less widely distributed than they were in late Jurassic and early Cretaceous times of 150 to 65 million years ago (Li, 1953; Stuart and Rothwell, 1993). The genus *Sequoia,* for example, is known from fossil records across Europe, central Asia, and North America as well as from Greenland, Spitsbergen, and the Canadian Arctic but is restricted today to one species with local distributions in California. The dawn redwood, also, was abundant at high northern latitudes during Cretaceous and early Tertiary times but now exists only locally over a very restricted range in the interior of China (Chaney, 1949; Boufford and Spongberg, 1983; Tallis, 1991). *Cupressus* is still another once widespread coniferous genus, that now occurs only in widely scattered relict stands. The now widespread pines were less abundant during the Mesozoic period. However, a Cretaceous pine described from the Cretaceous of Minnesota is similar to red pine, which occupies the same region today (Chaney, 1954)

One of the earliest and most primitive forms that has survived in a closely related form through to the present day is the ginkgo, the sole surviving member of a once numerous group of gymnosperms. Ginkgo is widely introduced in North America from individuals growing in China, and it was once common in Triassic floras, over 200 million years ago (Figure 22.1; Friis et al., 1987; Stewart and Rothwell, 1993).

The angiosperms developed later than the conifers; estimates placed their origin as early Cretaceous, about 130 million years ago (Scott et al., 1960). However, new estimates place their origin as early as 220 million years ago in the late Triassic age (Cornet, 1993; Crane, 1993). By the late Cretaceous, angiosperms had increased explosively (Figure 22.1; Friis et al. 1987; Graham, 1993; Wing and Sues, 1993). Pollen and macrofossil evidence indicates that the primitive angiosperms of the early Cretaceous or late Jurassic showed little taxonomic diversity and were probably not closely similar to modern flowering plants. The first angiosperms apparently developed slowly on disturbed, semiarid areas—including loose talus slopes and stream banks—of the ancient landmass of Gondwanaland (Axelrod, 1970; Doyle and Hickey, 1976; Stebbins, 1976). Thereafter, angiosperms evolved together with insects, birds, and mammals during periods of great ecosystem change, which probably elicited increasing diversity in plant forms (Chapter 14). Continents changed in position, configuration, size, and altitude through ocean-floor spreading, continental drift, and the fragmentation and joining of plants of the earth's crust. Mountains elevated as the plates collided (Himalayas and American Cordillera), and some desert lands became moist tropics; climates changed drastically. As summarized by Axelrod (1970):

> *The great diversity of taxa in numerous families in the tropics and subtropics, as well as the evolution of unique floras of arid to semiarid regions, and those of the temperate climates as well seems directly related to the breakup of Gondwanaland following the medial Cretaceous and subsequent evolution in isolation.*

During the Tertiary, flowering plants became widely distributed and well differentiated throughout the world. In Eocene times, the warmest Tertiary epoch, tropical plants were intermingled with warm temperate types as far north as London, England, and the coastal plain of eastern United States. In the Eocene coal beds of London, Nipa palm was one of the commonest species—it is now a saltwater swamp species in southeastern Asia. In the northeast, North America was connected to Europe by a landbridge, and the early Eocene was a time of rapid species migration from Europe to North America (Graham, 1993). The vegetation around the north Pacific basin, from Oregon to Alaska and southwest to Japan, was tropical to subtropical in character. Then a major deterioration in climate took place in the Oligocene, and broadleaved deciduous forests replaced subtropical forests. Except for minor fluctuations, the cooling trend continued in the Miocene and Pliocene and culminated in the Pleistocene ice age.

In Miocene times, a forest of temperate mesophytic species (with many genera similar to those of the modern Mixed Mesophytic Forest) replaced the warm temperate and subtropical types in much of the Northern Hemisphere. A continuous band of this broadleaved deciduous forest existed around the north Pacific from Oregon to Japan. The resemblance of forests in these areas is remarkable even on the species level. Furthermore, as late as 15 million years ago (middle Miocene) floristic continuity existed between eastern North America and eastern Asia. The mixed mesophytic forest included several species that ranged across the broad land bridge (Beringia) in the Bering Sea area. This land bridge, open throughout much of Cenozoic times, aided in securing a broad interchange between Asiatic and American floras and faunas (Hopkins, 1967a,b). Human settlement of Beringia and the peopling of the New World from eastern Asia began about 12,000 years ago, coinciding with a major interstadial (warmer climate between two cold maxima or stadia) that may have caused the reappearance of trees in river valleys and other areas (Hoffecker et al., 1993). From 12,000 to 11,000 years ago the human population expanded rapidly into other parts of the New World.

By late Miocene, coniferous forests began to occupy large upland areas in Siberia and northern North America as the mesophytic hardwoods retreated southward (Wolfe and Leopold, 1967; Graham, 1993). A coniferous forest of spruce, fir, and hemlock for the first time extended from the uplands of Oregon northward through British Columbia into Alaska. The mesophytic forest of the northwest gradually became extinct with the coming of a cooler climate and the rise of the Cascades, which brought dry and even arid conditions to vast areas of the interior. A southern extension of the eastern American forest in the cloud forest of Mexico and Central America (Röhrig, 1991)—represented by such trees as sweetgum, white pine, blue beech, American beech, the sugar maple group, black walnut, and black gum—apparently originated about this time (Martin and Harrell, 1957).

In Europe, the mesophytic flora was largely eliminated (Ellenberg, 1988; Jahn, 1991). Examples of genera that became extinct in Europe with progressive cooling are the following (Van der Hammen et al., 1971):

Epoch	*Extinct Genera in Europe*
Miocene	*Castanopsis, Clethra, Libocedrus, Metasequoia*
Pliocene	*Aesculus, Diospyros, Elaeagnus, Liquidambar, Nyssa, Palmae, Pseudolarix, Rhus* (and probably *Sequoia* and *Taxodium*)
Pleistocene	*Carya, Castanea, Celtis, Juglans, Liriodendron, Magnolia, Ostrya, Tsuga*

In addition, other extinctions included: *Sassafras, Torreya, Chamaecyparis,* and *Pseudotsuga.* That many of these genera survived in North America and eastern Asia is ascribed to the north–south orientation of the mountain chains of these areas, permitting the temperate species to migrate southward to warmer areas and then expand northward in the interglacial intervals. In Europe, the Pleistocene continental glaciations, combined with east–west chains of mountains (Alps, Pyrenees, Carpathian, and Caucasian Mountains), which blocked the southward migration of species and then the northward migration of plants during interglacial intervals, are thought to account for the extinction of these genera in Europe and western Asia. The extensive glaciations of the Pleistocene, which denuded much of the northern portions of the Northern Hemisphere, were discussed in Chapter 3.

PRESENT-DAY FORESTS OF THE WORLD

That phase of science in which geographical problems of plant distribution are emphasized is termed plant geography or **phytogeography**. Although much of the concern and knowledge is shared with plant ecology, phytogeography differs in that it has arisen from studies of the distribution of particular taxa of plants, rather than from studies of the environment. In other words, plant geography is primarily floristically oriented, whereas plant ecology is, or should be, concerned with Earth space and everything in it, including the sites of landscape ecosystems that support plants.

Plant geography is concerned with historical and present-day distributions of plant taxa, the location of their origins, studies of dispersal and migration, and in general with the evolution and present distribution of our flora. Only those phases pertinent to a brief survey of the present distribution of the world's forests can be summarized here. Excellent textbooks representing different perspectives are available for those wishing to delve more deeply (Polunin, 1960; Walter, 1973; Good, 1974; Cox et al., 1976; Daubenmire, 1978; Pielou, 1979; Tallis, 1991). Furthermore, several volumes in the series *Ecosystems of the World* have been published that treat such diverse assemblages as: temperate deciduous forests (Röhrig and Ulrich, 1991), temperate broad-leaved evergreen forests (Ovington, 1983), tropical rain forests (Golley, 1983; Lieth and Werger, 1989), tropical savannas (Bourliére, 1983) and forested wetlands (Lugo et al., 1990).

The description of the forests of the world on either a physiognomic-structural or a floristic basis involves the adoption of some type of classification within a geographical framework. The choices are many, for there are approximately as many classifications of vegetation types as there are writers on the subject. The fact that no single approach has achieved widespread recognition and adoption testifies to the different ways vegetation is perceived and, in part, to the arbitrariness of most classifications. Nevertheless, there are broad floristic provinces that correspond to broad climatic and physiographic conditions that support different groups of plants (Walter, 1973). Within a broad regional ecosystem, vegetation is conventionally described either on a physiognomic-structural or a floristic basis. Several approaches are described briefly.

Geographic-Climatic Classification

To even the most casual observer, the vegetation of the North Temperate Zone is characterized by common genera that set it apart from the South Temperate Zone. Similarly, the tropical flora of the New World is quite different floristically from that of the Old World. Plant geographers, therefore, recognize four plant domains: (1) **boreal** and **north polar** (North Temperate Zone), (2) **paleotropical** (Old World Tropics), (3) **neotropical** (American Tropics), and (4) **southern oceanic** and **subantarctic** (South Temperate Zone).

These units may be subdivided into geographical regions, which represent major subdivisions of continents on the basis of broad vegetation differences. Here there is less consistency, but North America, for example, can be divided into: (1) Arctic or tundra zone, (2) boreal conifer forest, (3) western temperate conifer forest, (4) eastern deciduous forest, (5) southern mixed hardwood–conifer forest, (6) central prairie belt (shortgrass and tallgrass prairies), and (7) subtropical and tropical forests (Figures 2.6 and 2.8).

The regions can be further subdivided various ways into plant formations or vegetation types (Daubenmire, 1978; Barbour and Christensen, 1993). The further the subdivision is carried, the more arbitrary it becomes and the less is the agreement between authors. In Canada, useful classifications based on an integration of climate and physiography are available in the works of Rowe (1972) and Wicken (1986).

An instructive example of a geographic-climatic classification of selected plant formations is illustrated in Figure 22.2 (Stephenson, 1990). In this diagram, the relationship of these broad vegetation types to annual actual evapotranspiration and annual water deficit are shown. Tundra and forest formations occur where annual water deficit is lower than about 400 mm. Within this band of low water deficit, formations are differentiated by increasing actual evapotranspiration in the sequence: tundra, coniferous forest, deciduous forest, and southern mixed coniferous–deciduous forest. The line joining an actual evapotranspiration of 1,000 mm to an annual water deficit of 1,000 mm represents an east-to-west transect along latitude 40 °N (Washington, D.C.; Indianapolis, IN; Denver, CO; Salt

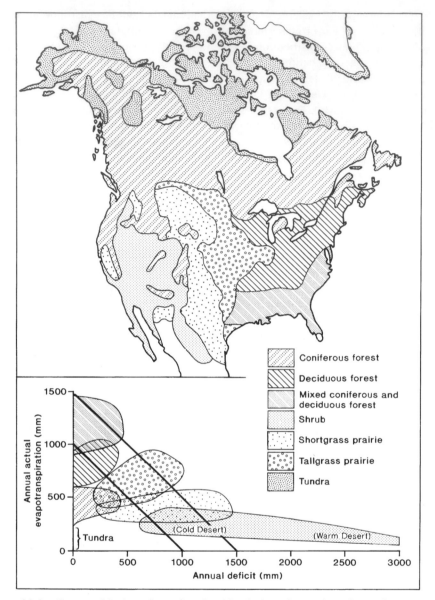

Figure 22.2 Geographic location of major North American plant formations (top) and their relationship to annual actual evapotranspiration and the annual water deficit (bottom). (After Huggett, 1995 as modified from Stephenson, 1990. Reprinted from the *American Naturalist,* © 1990 by The University of Chicago. Reprinted with permission of the University of Chicago Press.)

Lake City, UT). The vegetational sequence following the humidity–aridity gradient runs from deciduous forest, tallgrass prairie, shortgrass prairie, to cold-desert shrub. The line joining 1,500 mm co-ordinates represents a lower latitude transect at about 30 to 35 °N latitude (Charleston, SC; Dallas, TX; Phoenix, AZ) where southern coniferous–hardwood forest replaces the deciduous forest of the more northerly transect. Although we may describe the relationship shown in Figure 22.2 as a geographic-climatic classification, the vegetation types exhibit major differences in physiognomy and structure, the focus of the next section.

Physiognomic-Structural Classification

The outward appearance of vegetation, that is, its physiognomy (closed forest, open woodland, grassland), and structural features (evergreen forest, deciduous forest, thorn forest) have been widely used to describe and map vegetation on a world scale. The classification by Mueller-Dombois and Ellenberg (1974) is a good example, and a modified portion of this classification is shown in Table 22.1. As seen in Table 22.1, the physiognomic-structural classification is closely related to environmental factors, especially temperature and precipitation. For example, "Closed Forests" are not exclusively characteristic of any one geographic region and are further subdivided on the basis of geographic differences of temperature and precipitation. Webb (1959, 1968) has provided a useful physiognomic-structural classification of the forests of eastern Australia. The change in structural types along three environmental gradients is clearly depicted in Figure 22.3.

Physiognomic-structural features are not only suitable for classification at the broad landscape level but have proved useful at lower levels as well, as in the tropical rain forests of eastern Australia (Webb et al., 1970). Here they were demonstrated to be as efficient as floristic classification in assessing the environmental conditions of an area. Structural classification is particularly useful in areas such as tropical rain forests where the

Table 22.1 A Portion of a Physiognomic-Structural Classification of Forests and Related Woody-Plant Vegetation.

I. ***Closed Forests***
    A. Mainly Evergreen Forests
        1. Tropical rain forests
        2. Tropical and subtropical evergreen seasonal forests
        3. Tropical and subtropical semideciduous forests
        4. Subtropical seasonal rain forests
        5. Mangrove forests
        6. Temperate and subpolar evergreen rain forests
        7. Temperate evergreen seasonal broadleaved forests
        8. Winter-rain evergreen broadleaved sclerophyllous forests
        9. Temperate and subpolar evergreen coniferous forests
    B. Mainly Deciduous Forests
        1. Drought-deciduous forests (tropical, subtropical)
        2. Cold-deciduous forests with evergreens
        3. Cold-deciduous forests without evergreens
    C. Extremely Xeromorphic Forests
        1. Sclerophyllous-dominated forests
        2. Thorn forests
        3. Mainly succulent forests
II. ***Woodlands***
    A. Mainly Evergreen Woodlands
        1. Evergreen broadleaved woodlands
        2. Evergreen needle-leaved woodlands
    B. Mainly Deciduous Woodlands
        1. Drought-deciduous woodlands
        2. Cold-deciduous woodlands with evergreens
        3. Cold deciduous woodlands
    C. Extremely Xeromorphic Woodlands
III. ***Scrub***

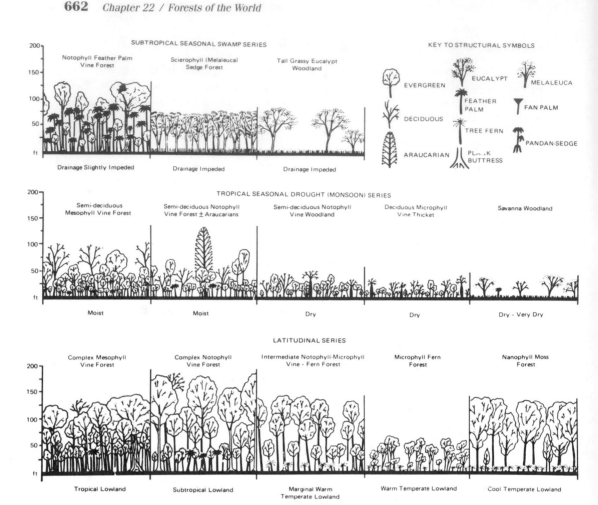

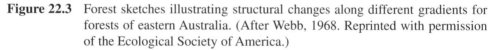

Figure 22.3 Forest sketches illustrating structural changes along different gradients for forests of eastern Australia. (After Webb, 1968. Reprinted with permission of the Ecological Society of America.)

flora is not well known and structural data may be rapidly collected, even by inexperienced personnel.

Floristic Classification

Plant sociologists have long attempted to develop systems for classifying plant communities along lines similar to those that have been developed for plant species. Starting with the analogy between the plant species and the plant association, elaborate schemes of community classification have been proposed for various geographic regions but not on a worldwide basis (Chapter 15). In central Europe, where the greatest effort at systematization has occurred, the schools of Braun-Blanquet, Schmid, DuRietz, Gaussen, and Aichinger (Küchler, 1967) are in conflict on various points. In the conterminous United States, Küchler (1967) has developed a classification of potential natural vegetation.

Multifactor Classification

A major result of vegetational classification is usually a vegetation map. To prepare a reliable and useful map multiple factors of climate, physiography, and soil must eventually be integrated with vegetative physiognomy and floral composition. Such approaches are rare, but we would cite one interesting example. Prentice et al. (1992) have developed a model predicting patterns of global vegetation (biomes). The resulting map was based on climatic and soil variables that affect plant physiology and therefore plant form and function. This model was developed from studies of physiological considerations of plant function by Box (1981) and Woodward (1987).

A Classification of Forests

Since it is necessary to adopt some system of classification in order to discuss logically the forests of the world, the following loose and approximate scheme will be followed.

As a basic major division, forests characteristic of freeze-free regions (i.e., the Tropics) will be considered separately from those annually subject to frequent and severe frosts (i.e., the Temperate Zone and temperate belts within the Tropics). Within tropical ecosystems, forest formations are recognized on the basis of soil–water relationships arising from the amount and seasonal pattern of precipitation and upon drainage conditions of the soil.

Within the Temperate Zone, forest formations are recognized on the basis of dominant tree genera, and within these formations, forest types are recognized as characterized by dominant tree species. These in turn may be subdivided into communities on the basis of total floristic composition.

TROPICAL FORESTS

In the freeze-free tropical zone, many species of trees live and thrive, forming a great complex of vegetational formations and communities. The species are numbered in the hundreds and even in the thousands, and classification of forests on the basis of floristic composition is not only extremely difficult but results in the recognition of hundreds of taxa. Forests on similar sites, however, tend to have a similar physiognomy and structure regardless of their species composition. An ecological classification of vegetation based on site has proved much simpler and more useful than one based upon species. Such an approach will suffice for the present brief description of tropical forests.

A few tropical plants occur in sites protected from frost north of the Tropic of Cancer and south of the Tropic of Capricorn. These are almost entirely confined to windward ocean sites where moderating maritime influences create a freeze-free—but not necessarily a warm—climate. Palms are grown in Devon and Cornwall in southwestern England. Both the southern half of the Florida peninsula and the southern coast of California support frost-susceptible plants in large numbers and can be considered at least subtropical, if not actually tropical. Similar conditions occur in North Africa, northern New Zealand, and other areas of mild oceanic climate.

Classification of Tropical Forests

Within the freeze-free zone of the world, the physiognomy of vegetation varies primarily according to the amount and seasonal pattern of rainfall and secondarily in response to human activity, primarily land clearing, burning, and plant introduction. Many classifications of tropical forests have been proposed (Beard, 1955; Haden-Guest et al., 1956; Cain and Castro, 1959; Webb, 1968). All recognize a gradation ranging from swamp forests to tropical desert, a moisture cline which may be divided arbitrarily into few or many vegetational types. The following is perhaps the simplest possible classification:

Swamp forest

    Salt water swamp

    Fresh water swamp

Rain forest, climate wet the year around

    Lowland rain forest, grading with increasing altitude into

    Montane rain forest and cloud forest

Monsoon forest, alternating wet and dry seasons

Dry forest, no pronounced wet periods

    Closed dry forest, grading with decreasing precipitation into

    Savanna woodland

Swamp Forest

Poorly drained and undrained sites in the Tropics support the most hydrophytic formations. These range from saltwater swamps and low beaches through brackish-water swamps to fresh-water swamps. Swamp forests may be classified further according to the depth of the water, aeration of the water, the type of substrate, and the seasonal variation in the water level.

SALT-WATER TROPICAL SHORELINES

The vegetation closely associated with sea water is quite distinct from that associated with fresh water. The nature of the vegetation differs with geographical location, nearness to the shore, and soil—whether mud, sand, or coral.

    Mangroves are evergreen shrubs and trees belonging to several unrelated families that form communities very similar in physiognomic appearance on tidal mud flats throughout freeze-free climates of the world. The term "mangal" has been increasingly used to designate the distinctive community of which mangroves and other species occur. Thus mangal is a community that contains mangrove plants. The two main centers of mangrove diversity are in the Eastern Hemisphere (East Africa, India, Southeast Asia, Australia, and the Western Pacific) with 40 species and the Western Hemisphere (West Africa, Atlantic South America, the Caribbean, Florida, Central America, and Pacific North and South America) with 8 species (Barth, 1982; Tomlinson, 1986). Mangrove swamps grow to latitude 32° north, as far as Bermuda, the Gulf of Aqaba in the Red Sea, and southern Japan; and in the Southern Hemisphere as far south as Natal (Brazil) and northern New Zealand. *Rhizophora* is the commonest and most widespread genus of mangroves. The species occurring in mangrove swamps of the Indian and western Pacific Oceans differ from those of America, the West Indies, and Africa, but the two formations are very similar in appearance and ecology. Mangroves are of considerable local importance for fuel wood, tannin, and even for timber. The strongly developed zonation in mangrove swamps is indicative of physiological responses to different tidal levels, salt concentrations, and in some situations propagule sorting.

    In brackish swamps inland, other species do occur. Nipa palm, important for thatch, is widespread in occurrence in southeastern Asia and the western Pacific.

    On sandy beaches, still other species are found. Species of *Barringtonia* (Indian and Pacific Oceans), *Pandanus* (Asia and Africa), and *Coccoloba* (New World) are particularly widespread trees and shrubs. *Casuarina* is a characteristic littoral tree that has been planted even more widely throughout the tropics for its shelterbelt and fuel value. Most important of the shoreline trees, however, is the coconut, a species so widely spread by humans that its natural range is still in dispute.

FRESH-WATER SWAMPS

Both the site and the vegetation of fresh-water swamps grade into upland tropical rain forest as soil drainage improves. The swamp vegetation, because of limiting soil aeration, however, is more open and irregular in structure and consists of fewer species than nearby upland types. Various palms reach their optimum development and abundance in fresh-water swamps and are perhaps the most characteristic plants of these sites. Cabbage palm (*Sabal palmetto*) is such a tree in the fresh-water swamps of Florida. Others are Euterpe and Maurita of South America, Phoenix in Africa, and sago palm (*Metroxylon*) in the western Pacific. Various *Cecropias* are important in developing forest successions in tropical America, where the largest late-successional fresh-water swamp species is the huge emergent, *Ceiba pentandra*.

Rain Forest

Under conditions of adequate precipitation, lush rain forests develop in the Tropics. The extent, however, is limited by climatic patterns, much of the Tropics being too dry. The rain forest is concentrated in three broad regions: (1) northern South and adjacent Central America, particularly in the Amazon Basin; (2) western Equatorial Africa from Sierra Leone to the Congo (Zaire) Basin; and (3) the Indo-Malayan region, including the west coast of India, much of the Indo-Chinese Peninsula, Indonesia, Papua New Guinea, and the northeast coast of Australia. The vegetation differs greatly not only between the three main belts of tropical rain forest but in different portions of each belt. The number of species represented is not known. Individual regions such as Borneo and the Congo Basin have been estimated to contain about 10,000 species of vascular plants each. In the Malay Peninsula alone, about 2,500 species of trees are known, while the great rain forest or **hylaea** of the Amazon contains at least that many species of large trees. The richest single plant community on Earth is reported to be a 100 m^2 plot in wet lowland rain forest in Costa Rica where 233 vascular plant species (including 73 tree species, mostly seedlings) were recorded (Whitmore, 1991). And in Peru, a 1,000 m^2 plot contained 580 trees over 10 cm in diameter, of which every second tree was a different species.

Tropical rain forests differ from temperate forests not only in species richness, but in many other characteristics. These include: (1) much greater proportion of epiphytes and lianas, (2) greater proportion of buds lacking bud scales, (3) greater proportion of leaves with entire margins and "drip tips," (4) multiple flushes of shoot growth, (5) many more species with buttressed trunks, (6) a far greater part of precipitation reaching the ground is from throughfall, and (7) more nutrients occur in the above- and belowground biomass. One of the most common and widely held assumptions is that, in contrast to other forests, almost all of the inorganic nutrient capital is held in the living plants, mainly above ground (Whitmore, 1984, 1991). However, studies in a variety of areas indicate that this assumption is not always true. Nutrients and their cycling are dependent on a number of factors, including the geographic and altitudinal location of the forest, the nature of the substrate, and the age of the soil.

Richards (1952, 1973, 1996) stimulated great interest through his comprehensive study and interpretation of the tropical rain forest. Many other treatments have followed since his classic book in 1952 (2nd edition, 1996). These include an excellent overview of tropical rain forests (Whitmore, 1991), descriptions of tropical rain forests of the Far East

(Whitmore, 1984) and rain forests of eastern Asia (Lovett and Wasser, 1993), as well as the diversity of tropical rain forests (Terborgh, 1992).

Fortunately, the general appearance and structure of the tropical rain forest is much the same everywhere. A description of its structure in tropical Africa by Aubréville (Haden-Guest et al., 1956) is typical and worth quoting:

> *The rain forest is very dense, with a tightly closed canopy. Three stories may be distinguished. The upper story is discontinuous, composed of a relatively few gigantic and usually isolated trees with mighty crowns rising 40 to 45 meters above the ground. At a height of about 25 to 30 meters a continuous middle story of crowns pressed one against another gives the forest, as seen from an airplane, a characteristically undulating and unbroken appearance and hides the trunks from view. The lowest story is made up of small trees and bushes whose crowns fill almost all of the remaining space.*

The great number of tree species and the complexity of tropical rain forests have long intrigued ecologists. Their very complexity and relative inaccessibility have limited detailed analytical studies of their nature. Nevertheless, the large number of species and the low density of adults of each species in the rain forest as compared to that of the temperate forest may be explained at least in part by several interacting factors: (1) favorable temperature, precipitation, and soil conditions permitting expression of many and varied mechanisms of plant growth and reproduction, (i.e., relatively free release and expression of genetic variation); (2) mutualisms of exceeding complexity and frequency among plants and animals; (3) proportionally greater interspecific competition as individuals approach maturity (contrasted to greater intraspecific competition in temperate forests) leading to the evolution of mutually avoiding, hence ecologically complementary, species (Ashton, 1969); (4) site conditions favoring a rich fauna, particularly insect, bird, and bat pollinators which, being many and specialized, effect pollination among widely spaced individuals and promote speciation of plants (Chapter 14); and (5) the action of predators destroying seeds and seedlings, which tends to increase the distance between adult breeding trees of many species (Janzen, 1970). Summarizing his work in the species-rich southeast Asian mixed dipterocarp forests, Ashton (1969) stated that the complexity of the rain forest may be explained in terms of:

> *(i) The seasonal and geological stability of the climate which had led to selection for mutual avoidance, and through increased specialization, to increasingly narrow ecological amplitudes, leading to complex integrated ecosystems of high productive efficiency. As the complexity increases, the numbers of biotic niches into which evolution can take place increase but become increasingly narrow.*
> *(ii) Their great age.*

The tropical rain forest in any one geographical area represents a multidimensional continuum with gradual changes in composition and structure occurring with distance from the ocean, distance from rivers, increasing altitude, and changing geographical position. Riparian forest communities are usually different from the forest communities away from streams, and both gradual and relatively sharp ecotones may be observed.

With increasing altitude, the rain forest becomes shorter in stature, simpler in floristic composition, and increasingly characterized by luxuriant epiphytes, particularly mosses and lichens. At elevations above about 2,000 m in the equatorial Tropics, this trend culminates in the **montane rain forest,** also termed the **cloud forest** (Roth, 1990) or **mossey forest**. If dwarfed, it may be called **elfin woodland**. As the names imply, the mountain climate around such forests is apt to be mostly cloudy with the air saturated with moisture and fog-drip providing daily precipitation. The climate is cool, constantly damp, and persistently misty. Whereas tree heights in the lowland rain forest frequently exceed 30 m (maximum 84 m for *Koompassia excelsa* in Borneo), heights in lower montane belts commonly are only 20 to 25 m, and in the high mossy forest may be only 6 m. The number of tree stories and the number of tree species become fewer with increasing elevation, but the number of individual trees increases.

Monsoon Forest

Since evapotranspiration in the Tropics (outside of the cloud zone in the mountains) is extremely high, evergreen broadleaf forests are confined to the rain belts, where at least 100 mm of rain normally fall each month of the year and at least 2,000 mm or more of rain fall during the year at least. In regions with 1,000 to 2,000 mm annual precipitation that are characterized by a dry period of a month or more, some of the dominant trees tend to lose their leaves, especially toward the end of this period. The term **monsoon forest** is given to the deciduous and semideciduous forests of southeastern Asia developed under a climate characterized by a very dry period of two to six or more months, broken by the very wet monsoon which comes in June or thereabouts depending upon the locality. Not too accurately, the term has been extended to refer to smaller areas of deciduous and semideciduous forests elsewhere.

Changes in the seasonal foliation of forests tend to occur gradually with climatic gradients in the Tropics. A marked deciduous period occurs in forests growing in areas with an equally marked dry season. A general lowering of annual precipitation without a strongly marked dry season, however, results in the forest being simpler in vertical structure, smaller in size, and more xerophytic in character, but not necessarily deciduous.

The tropical deciduous forest is extremely important in southeast Asia, both for its extent, the commercial importance of its tree species, and its species diversity. Dipterocarps constitute the most important genus. In the wetter zone, the dipterocarps are tall and shed their leaves annually only after the new leaves are expanded. In drier zones, the dominants may have a leafless period while the understory shrubs and trees may remain evergreen. Teak is commercially the most important of the monsoon forest trees. *Xylia xylocarpa*, a deciduous leguminous tree, however, is more abundant. *Terminalia* and *Shorea* are other common genera.

Dry Forest

Although major emphasis is usually placed on tropical rain forest, most ecosystems of subtropical and tropical latitudes are seasonally affected by drought. With less than about 1,000 mm of rain, forests in the Tropics tend to be very xerophytic. Dry forests occur where there are several months of severe, even absolute, drought, which is the only unifying characteristic of such forests (Bullock et al., 1995). Depending upon the amount and distribution of annual precipitation, such forests vary from low and rather simply structured closed forest to open woodland, thorn woodland, and open wooded savannas. These dry forests occupy most of tropical Africa south of the Sahara except for the equatorial rain forest (Menaut et al., 1995), much of tropical Australia, and a good deal of South America both north and south of the Amazon Basin.

As with other vegetation types, composition and structural changes are graded to climatic changes, there being no sharp cleavage between dry closed forest, dry open forest, and savanna forest. Fire is the dominant factor in these dry tropical types, having been widespread and frequent for many thousands of years. Vast wooded savannas have been degraded by repeated burning into sparsely wooded savannas (Bartlett, 1956; Bourliére, 1993). On the driest sites, thorny shrubs and small trees with thin parasol-like crowns replace the larger arborescents. *Acacia* is a particularly common dry tropical genus of wide distribution.

Dry forests are unremarkable in species richness (Gentry, 1995). The average species richness (plants > 3 cm in diameter on a 0.1 ha plot) is about 65 for dry forests compared with 152 for lowland wet and moist forest. Surprisingly, there is no significant change in species richness with increasing precipitation. The exceptions are the most arid regions that tend to have open, single species-dominated forest (Gentry, 1995). Apparently, once the critical rainfall threshold needed to maintain a closed-canopy forest is reached, increases in precipitation have little effect on species richness until rainfall is high enough to support moist forest. Dry forests are dominated by trees that shed their foliage for major portions of each year. Leaves, stems, and roots have many adaptations to deciduousness, and because water is the primary limiting factor, leaf abscission is the most significant drought response of dry forest trees (Holbrook et al., 1995). A great variety of phenological responses are evident due to the interactions between water availability and structural and physiological characteristics, such as rooting depth, stem water storage, hydraulic architecture, and sensitivity to soil water stress.

FREEZE-HARDY FORESTS

As with tropical forests of freeze-free climates, freeze-hardy forests subject to annual freezing temperatures are marked by discontinuity. The distribution of temperate genera and species is broken, however, not only by oceanic barriers, but by the barrier of the equatorial belt as well. Although high elevations with temperate or near-temperate climates occur in the tropics of all continents, these areas are not continuous. In many cases, however, closely related "vicarious" species occupy similar sites in widely separated areas. In still other places, vicarious species are missing through historical factors governing their distribution. Their place in the ecological scheme is taken by unrelated trees that can occupy the site in the absence of better competitors.

The other major factor governing the distribution of freeze-hardy forests is the existence of a circumpolar boreal forest belt populated around the world with various closely related species of spruces, firs, larches, birches, and aspens. This belt is of relatively new origin, much of it growing on sites that have been glaciated up to within the last 10,000 years. As a result, it is inhabited by northward migrants of various populations that survived the Pleistocene in cool, moist refugia. Many tree genera are represented in several of the temperate forest centers in the Northern Hemisphere, and others are similarly represented in temperate forest centers of the Southern Hemisphere. Because the species of each genus tend to occupy similar ecological habitats and niches wherever they occur, the homologous nature of the temperate forests of the world can perhaps best be approached by considering the world distribution and ecological habits of the most important tree taxa. These may, for convenience, be considered in two groups: boreal trees that comprise the circumboreal forest ecosystems of the high latitudes and that also occur at high altitudes in the temperate zone; and temperate forest trees that form discontinuous distributions between the boreal forest and the Tropics.

Boreal Forest Taxa

The principal boreal forest groups include the spruces, firs, larches, birches, and aspens (section Populus of genus *Populus*). Of these, the spruces are the most widespread in distribution and most characteristic of the boreal forest. The spruce–fir, larch, and birch–

aspen forests everywhere have a similar physiognomy, similar ecological relationships, and respond to similar silvicultural treatment.

SPRUCES

The circumpolar boreal forest is characterized by spruces, except for eastern Siberia, where larches are more numerous. The principal species in North America are white and black spruce, which have similar ranges from Alaska to Newfoundland. Norway spruce in western Europe and the closely related Siberian spruce in Russia and Siberia occupy similar sites in similar latitudinal belts from Norway to eastern Siberia (Schmidt-Vogt, 1977).

South of the boreal belt, about 25 species of spruce occur in isolated mountain ranges as far south as the 25th Parallel in Mexico (*Picea chihuahuana*) and the 23rd Parallel in Taiwan (*P. morrisonicola*) (Wright, 1955). All are similar in appearance and in ecological habit, and they show close taxonomic relationships with widely-spaced species in the different continents. They are characterized by a shallow root system, often largely confined to the organic humus layers of the typical mor-type soil, which develops under all spruces. Species of the genus are particularly well adapted to grow in cold soils and even on permanently frozen ground and on glaciers covered with detritus, for it is this shallow top layer that warms up and dries out sufficiently under arctic climatic conditions to make tree growth possible. On lighter-textured and warmer soils, however, spruces, particularly Norway spruce and white spruce, develop moderately deep root systems. This habit suggests considerable plasticity in their rooting depth, depending on the local soil and organic-matter environment. Partly because of their shallow root systems, and partly because of their low nutrient requirements, spruces are tolerant of acid, undrained soil conditions and are able to survive and even grow in northern bogs and (as in the case of Sitka spruce) in cold, wet, rain forest, or cloud forest conditions. Being characterized by sharply conical crowns, the spruces are well suited to bear winter snows and to shed them when slight warming or wind occurs. It is not by accident, therefore, that spruce is the characteristic genus of the boreal forest, and that its worldwide distribution may be taken as approximately delineating both the circumpolar forest and its outliers in the high mountains of the Northern Hemisphere.

FIRS

Being similar to the spruces in general form and appearance, and occupying much the same ecological niche, it is not surprising that the genus *Abies* has much the same distribution as the genus *Picea*. The fact that spruces and firs are frequently lumped together to characterize the boreal forest, however, should not obscure the fact that the firs are less tolerant of poorly-drained conditions, less tolerant of fire, and more tolerant of warmer and drier climates. Therefore, although firs are common upland species in the boreal forest of North America, they do not occur to any extent in the boreal forests of Eurasia. Also, they occur farther south and in somewhat drier mountain ranges than the spruces.

The principal boreal forest firs are the balsam fir of eastern North America and the subalpine fir of western North America. In Eurasia, fir is primarily a high mountain genus of the Temperate Zone. Important species are European silver fir in the mountains of central Europe and as far south as the Pyrenees, where it is the major tree species, *A. nordmanniana* in the Caucasus mountains and the Black Sea region of Turkey, *A. nephrolepsis* in northeastern China, and *A. pindrow* in the Himalayas and mountains of northern Indo-China. The genus occurs in several high mountain ranges far south, with *A. guatemalensis* and *A. religiosa* growing in the mountains of Mexico and Guatemala, *A. pinapso* and *A. numidica* in the mountains of North Africa, and *A. cilicia* in the mountains of Asia Minor.

LARCHES

In contrast to Abies, which is more a boreal group of the New World than the Old, *Larix* is the dominant boreal genus in Siberia. Larches also occur in boreal and northern forests of eastern and western North America. Being deciduous, the larches are better adapted to survive in cold, dry climates than the spruces and firs, which are better suited to cold, wet climates. Being highly understory intolerant as a group, however, the larches require an open site and a dominant crown position throughout life and cannot compete successfully with spruces, firs, and other understory tolerant trees under moister climatic conditions.

Around the world, larches are characteristic of regions with very short growing seasons and their deciduous foliage is an important adaptation in such extreme areas. Young conifer needles may not be capable of maturing properly, and the cuticle cannot attain the required final thickness to survive large water losses in the strong sunlight of early spring when the ground is frozen (Walter, 1973). However, deciduousness in larches, coupled with the ability to produce large masses of thin, rapidly photosynthesizing leaves under full sun, gives species of this genus a competitive advantage over other conifers in open habitats. Western larch and subalpine larch occur in cold, dry climates of the northern Rocky Mountains. Tamarack is a pioneer on mountain sites in northeastern North America and in wetlands of boreal Canada. However, at the edge of its range in southern Michigan, for example, it is relegated to frost pocket–wetland bogs and deciduous swamps as relict populations of the boreal forests that were once widespread there 12,000 to 14,000 years ago.

In eastern Siberia, larch forms extensive stands in cold and dry regions, from longitude 90° to 150° and almost as far south as the southern border of the former U.S.S.R. Elsewhere, with the exception of the boreal swamp-inhabiting tamarack, larch is primarily a mountain tree of cold, dry climates. Such are western larch of northern Idaho and western Montana, European larch of the mountains of central Europe, *L. griffithii* in the Himalayas, *L. gmelini* in Korea and northeastern China, and *L. leptolepis* in Japan. None of these species, however, extends as far to the south as species of spruces and firs in the same localities, nor do they occur as far down the slopes when they do occur in the same latitudes.

PINES

Jack pine and the closely related lodgepole pine are the boreal pines of fire-prone ecosystems of eastern and western North America, respectively. Lodgepole pine, with its several varieties (Critchfield, 1957, 1985; Chapter 4) spans a more geographically and altitudinally diverse range of ecosystems than does jack pine. Their fire adaptations were described in Chapter 12. In Eurasia, the related short-needled Scots pine occurs in boreal ecosystems in Scandinavia, Finland, and Siberia. It is highly genetically variable (Chapter 4), and its geographic distribution in Eurasia is greater than that of any pine in the world.

BIRCHES AND ASPENS

Associated with spruces and firs, but growing into drier climates than the conifers, are the birches and aspens. Both groups feature light-seeded pioneer trees that can colonize large burns, and that can regenerate from undamaged rootstocks after fire has killed their aboveground parts—the birches by root-collar sprouts and the aspens by root suckers. Together with the jack pine, aspens and birches are the fire species of the boreal forest *par excellence*, and their abundance is a direct measure of the severity and frequency of past forest fires.

Birches form circumpolar populations composed of separate migrations from various Pleistocene refugia that have merged and interbred to form a complex of closely related species. The *Betula papyrifera* complex in North America and a similar complex in Europe (including, *B. pubescens* and *B. verrucosa*) form a group of white-barked birches that are similar in morphology and ecological habitat and are closely related genetically. The birches and aspens extend farther east in Siberia than the spruces and larches and are the principal forest trees of the Kamchatka Peninsula. Birches and European aspen occur abundantly throughout the dry larch forests of eastern Siberia and commonly, though less abundantly, in the moist spruce–fir forests of Europe and western Siberia. In North America, birches and aspens are most abundant in the drier sections of the Canadian north and in the most heavily burned regions of the western Great Lakes region.

In the Arctic tundra, dwarf birches form a characteristic part of the vegetation, whereas south of the boreal forest, dark-barked birches (Section Costata) occur throughout the Allegheny and Appalachian Mountains of the eastern United States, the Pyrenees, the Himalayas, and mountains of northern Korea and northeastern China. These temperate species vary from midtolerant to intolerant, and in their site preferences they vary from riparian to high-elevation mountain slopes. They reach their optimum development in the northeastern United States and Appalachian Mountains, where birches grow on disturbed sites from low elevation (sweet birch) to high elevations where yellow birch is one of the highest-elevation deciduous associates of red spruce and Fraser fir at 2,000 m in the Southern Appalachians.

The aspens similarly form a circumboreal complex of intergrading populations that have migrated north from various Pleistocene centers of refuge. *Populus tremula* in Europe and Asia, *P. davidiana* and *P. rotundifolia* in China (Barnes and Han, 1993), and *P. tremuloides* in North America are names given to the Old and New World portions of this complex. Trembling aspen has the greatest range of any native North American tree species, spanning the continent from Atlantic shores in the East to western Alaska and southward in the montane zone throughout the Rocky Mountains and into the mountains of northern Mexico. Individual clones can be very long-lived, include thousands of stems, and extend over 40 ha in area (Figure 5.14; Barnes, 1975; Kemperman and Barnes, 1976).

Temperate Forest Taxa

A belt of freeze-hardy forests occupies the temperate zones of the Northern Hemisphere between the boreal forest and the freeze-free forest of the Tropics and all of the Southern Hemisphere south of the Tropics (there being no austral equivalent of the boreal forest). It consists of many disjunct tree populations separated by oceanic and desert barriers of long geologic standing. In both the Northern and Southern Hemispheres, however, closely related species of the same genera (i.e., vicariads) frequently occupy similar ecological niches in the different regions. Thus a discussion of the temperate forest on the basis of the dominant forest taxa serves to indicate the large common denominator of these forests throughout the world. Pines, oaks, beeches, maples, and eucalypts are the principal forest-forming trees, but other conifers and angiosperms must be included in the discussion.

PINES

The genus *Pinus* is one of the most diverse and important of all forest-tree genera. There are approximately 100 species and most of them are locally or nationally important in the timber or fiber economy. They occur from the boreal forest to south of the Equator in the equatorial mountains of Sumatra and Java (*Pinus merkusii*). An excellent monograph of the genus is available (Mirov, 1967) as well as succinct descriptions and maps of their occurrence (Critchfield and Little, 1966).

Despite the taxonomic variety and the wide distribution of pines in the Northern Hemisphere, all have a great deal in common in their ecological place in the world's forest ecosystems. Virtually all occur on coarse dry soils, especially sands, gravels, and rock outcrops; and most owe their dominance to the frequent burning of such sites and their ability to regenerate abundantly in the ashes of the blackened site. The closed-cone pines are the most dependent on fire, but even the five-needled white pines, the most mesophytic of this group, are commonly found in pure, even-aged stands most often after destruction of the prior forest by fire. Their many adaptations to fire are discussed in Chapter 12.

As a group, the pines have a deep root system, with the growing root tips requiring large soil interstices for penetration. Because of the deep root system, they do not grow well on frozen or poorly drained soils, and because of the poor penetrating ability of the roots, they grow best on deep, coarse soils. They have a high ability to withstand hot, dry conditions, whether in the Tropics or in northern continental climates as in the interior of Canada and Siberia. They are widely used for lumber and edible nuts (pinyons and stone pines) throughout much of the Northern Hemisphere.

In the boreal forest, the closely related species of lodgepole pine, jack pine, and Scots pine form a circumpolar belt of forest across North America, Europe, and Asia. These species do not extend far into the northern zones of permanently frozen soils, but do extend quite far south along high mountain ranges, lodgepole occurring as far south as southern California and Mexico and Scots pine growing in Spain, Italy, Greece, and Turkey.

Soft pines

Pines of the north temperate zone, for the most part, are the five-needled white pines, of which closely related vicarious species occur in relatively cool, moist temperate regions around the world. The subsection Strobi include: sugar pine and western white pine in the western United States; eastern white pine in eastern North America and Chiapas; Mexican white pine in Mexico; Japanese white pine; Formosan white pine; Macedonian white pine; and Himalayan white pine in the Himalayas. The other groups of soft pines include the stone pines and the foxtail pines, both also widely distributed and often occurring on rocky mountain slopes and ridges.

Hard pines

The hard pines, usually with two or three needles in a fascicle, are more xerophytic than the soft pines and generally grow in warmer, drier temperate regions (except for the boreal pines just mentioned which do, however, endure a hot, dry continental summer in much of their ranges). These pines have a similar physiognomic appearance and occupy similar sites.

In western North America, ponderosa pine is the principal hard pine, with lodgepole, and Jeffrey pines being also of major importance. The closed-cone relict pines of the Pacific Coast, particularly Monterey and knobcone pines, however, thrive extremely well when transferred to Maritime climates elsewhere.

In eastern North America, red pine and the four principal southern pines: shortleaf, loblolly, longleaf, and slash are the major hard pines, but, as in the west, many other species occur locally.

Mexico and Central America are main centers of evolution and speciation of the genus *Pinus* (Mirov, 1967). Nowhere in western North America is there greater variation, and Mexico has the greatest number of pine species of any country in the world (Perry,

1991). The pines occupy rocky, gravelly, and sandy soils on a variety of geologic substrates and occur from moderate to high elevations. Caribbean pine, however, reaches the seacoast as far south as Honduras and Nicaragua. Caribbean and patula pines have proven particularly well suited for extensive planting in other continents.

The dry mountain ranges of southern Europe, the Mediterranean region in general, and Asia Minor shelter many species of pines, although their ranges and quality have been severely limited by centuries of grazing and burning. Perhaps the most important is black pine, which grows in moderate temperate climates from the Pyrenees east to Greece, Turkey, and Cyprus. Several races have been distinguished, the most important in reforestation being Corsican pine and Austrian pine. Other species of the region of widespread distribution are: French maritime pine, an important sandy coastal species growing from Portugal to Italy and reaching its optimum development along the Bay of Biscay in northern Spain and southwestern France; Aleppo pine, which occurs on both sides of the Mediterranean Sea from Spain to Jordan, Israel, and Turkey, enduring hot and dry climates; Calabrian pine in the mountains of the Near East from Turkey to Iraq; and Italian stone pine, a nut or pinyon pine bearing edible seeds and ranging from Spain to Turkey. A purely tropical European pine is the Canary Island pine, found only on those islands.

In Asia, aside from the great northern extent of Scots pine, temperate pines occur in the mountains of India and Pakistan, China, Korea, and Japan. In the Himalayan region, chir pine grows in the hot and dry lower elevations, with Himalayan white pine (*P. griffithii*) being the principal species of the higher zones. In China, *P. massoniana* and *P. tabulaeformis* are the principal pines, while in Japan, Japanese black pine and Japanese red pine are both important species. Two of the important Asiatic pines are tropical, growing widely in the southwest Pacific in the same regions as podocarps and other south-temperate genera. Merkus pine occurs in the mountains of Burma, Thailand, the Indo-Chinese peninsula, the Philippines, and one locality in Sumatra, where it extends south of the equator. *Pinus khasia* has a very similar range and frequently grows in conjunction with Merkus pine but at higher elevations. The geography of each pine species is discussed by Mirov (1967).

OAKS

In the Northern Hemisphere, the members of the genus *Quercus* are the angiosperm equivalents of the pines. They are a widely-distributed group of deep-rooted, xerophytic trees that occupy dry sites from the southern edge of the boreal forest well into the Tropics. As with the pines, oak bark is thick and fire-resistant. When the bole is killed, the root system remains alive almost indefinitely, sending up generation after generation of coppice stump sprouts that perpetuate both the individual tree and the dominance of oaks in the forest. Throughout the North Temperate Zone and much of the high country of the north Tropics, oaks are the characteristic late-successional dominants in the drier forest zones. They also occur as early- to mid-successional species on mesic sites and some occur as dominants in floodplains and swamps. Oaks do not extend as far north as the pines, and thrive on drier sites in southern latitudes. For example 80 percent of the oaks occur between 30° north latitude and the equator, and only 2 percent are found between 30° and 40° north latitude (Axelrod, 1983). Otherwise, the ecological requirements of upland oaks are much the same as those of the pines. Fire-dependent oak types in much of the world will invade as understory plants and assume dominant position when the pines die or are overthrown by wind. In other words, the oaks are more tolerant of understory conditions than the pines as a group, and follow the pines in natural succession.

The oaks show a great deal of genetic variation. More than 500 species have been recognized, but the merging of morphological characteristics between adjacent populations makes species identification difficult and nomenclature uncertain in many instances. With this wide diversity in morphology goes a similar wide tolerance of ecological conditions. Although primarily xerophytes, there are species of oaks that have become adapted to mesic and even hydric conditions. In the eastern United States, for instance, cherrybark oak is the most important of many bottomland oaks in the Mississippi Valley floodplain; the northern and southern red oaks grow over a wide range of sites and regions but reach their best development under mesic conditions; white and black oaks similarly have a wide ecological amplitude but are most characteristic of drier sites, especially black oak. Chestnut oak is dominant on very dry hillsides and mountain slopes; while a whole series of scrub or dwarf oaks, (such as *Q. illicifolia*, *Q. laevis*, and *Q. marilandica*) grow on the very driest, fire-prone, and most infertile sand plains.

In the North Temperate Zone, the oaks are deciduous, but in the South Temperate Zone, oak species are frequently evergreen or have but a brief leafless season. Most oaks are **sclerophylls**, having a hard leaf with thick-walled structures suitable for preventing wilting of the leaves, even under protracted drought conditions. Mesophytic species of oaks, however, may have quite succulent leaves, particularly under shade-grown conditions. On the very driest sites, on the other hand, oaks tend not only to be highly sclerophyllous, but also have narrow or deeply-cut leaves with small leaf surfaces.

The Temperate Zone has dry forest ranging from dry-closed through dry-open to dry scrub just as in the Tropic Zone. Oaks are frequently the characteristic species of these zones. Open oak woodlands are widespread in the South Temperate Zone, from California east throughout the Southwest to Texas, throughout the Mediterranean region, Asia Minor, and southern Asia to New Guinea, and in the south of eastern Asia and Japan in the north. Scrub woodland stands, frequently with oak as a major species, are similarly distributed on the drier sites. The **chaparral** ("little oak") of California, the **scrub oak** of the southeastern United States, and the **maquis** of the Mediterranean are examples. *Arbutus* and *Pistacia* are common associates of oak in the Mediterranean scrub.

Some of the major oak species around the world should be mentioned. In eastern North America, oaks are the most abundant and widely distributed tree genus, with the white oak, northern and southern red oaks, and black oak being perhaps the most important. Along the Gulf Coast, in the Southwest, and in Mexico, many of the oaks are evergreen or have only a brief leafless period. These live oaks include several species in both the Southeast (*Q. virginiana*) and the Southwest (*Q. agrifolia*, *Q. chrysolepis*, *Q. emoryi*, and *Q. wislizenni*). As far south as Costa Rica, *Q. copeyensis* forms magnificent stands in the mountains of Central America, while other species occur as far south as Columbia and Ecuador.

White oak is the most important and widely distributed hardwood in Europe with two closely related species, *Q. robur* (= *Q. pedunculata*) and *Q. petraea* (= *Q. sessiliflora*) extending from Portugal and Ireland on the west to central Russia and the Caspian Sea on the east and from southern Scandinavia south to the Mediterranean (Schoenicher, 1933). The Mediterranean forests are dominated by other oak species including cork oak in the west from southern France to Morocco; holm oak in the maquis of the entire Mediterranean; and *Q. macrolepis* or Vallonia oak, important for the tannin from its acorn cups in the eastern Mediterranean. In Asia Minor, *Q. aegylops* is important; and *Q. aegylops*, *Q. castaneaefolia*, and *Q. infectoria* occur widely in Iraq, Iran, and surrounding regions. Farther east, the Himalayan oaks are numerous and range widely both in elevation and extent. To the north, oaks are important in China, Korea, and Japan; while to the south they grow in Assam, Indo-China, Indonesia, and the mountains of New Guinea. The Mongolian oak is widely distributed throughout southeastern Siberia, Mongolia, northeastern China, Japan and Korea.

BEECHES AND MAPLES

Several genera of angiosperms typically occupy the more mesic sites in the Temperate Zone of the Northern Hemisphere. Of these, the beeches and maples are particularly widespread and important. The species of these genera are commonly highly shade tolerant, as a result of which they frequently succeed oak in the absence of fire by seeding into the understory and growing slowly many decades until they are released by the death or decadence of the oaks. Due to the lack of stand-regenerating fires in North America, maples, beeches, and other mesophytic species are replacing the oaks on all but the driest sites (Chapter 16). In the moister and cooler sites, beech and maple occur in all stages of succession and characterize the hardwood forest.

The principal beech species are: American beech, which grows in eastern North America from Cape Breton Island and Ontario to Florida and east Texas on the Gulf of Mexico, with outlying relict colonies in Mexico; European beech, which occurs from England and southern Sweden and southward to northern Spain, the Italian mountains, and Greece; the closely allied Oriental beech, from Greece through the high mountains of Asia Minor and Caucasia to the Caspian forest of northern Iran; and five species in eastern Asia. These are all very similar in appearance and in ecological requirements, characteristic of the moist, cool, temperate forest, and forming late-successional communities whether in pure stands or mixed with other tolerants in areas long unburned.

Beeches as a group typically thrive in oceanic or suboceanic climates (western Europe, northeastern North America, Japan) where rainfall and soil water are high. Their numbers decline in the relatively dry and cold interiors of both continents. Whereas, a mixed deciduous forest of maples, ashes, lindens, elms, and many other species are typical of the northeastern parts of North America and Asia, beech is notably lacking from this group in northeastern China (Barnes et al., 1992). Completely absent, also are its eastern North America associates, eastern hemlock and hop-hornbeam. The primary reasons for the absence of these species are apparently the relatively low annual precipitation and relative dryness of the spring months, together with a severely cold winter. The "Mongolian-Siberian Cold High" air mass invades southward during winter and spring in eastern China (Chang, 1983). The spring drought and cold winter caused by this air mass are the most important limiting conditions for species such as beech that require relatively high precipitation. However, several beech and hemlock species do occur in the rainy, montane zone of mountains in humid, east-central China.

The maples have similar ecological preferences to the beeches, but being more variable and consisting of many more species (70 as compared to 8 or 9), have a wide distribution and a wider ecological amplitude, including drier climates. Some maples are large trees, but others are primarily shrubs. In eastern North America, the sugar maple group (sugar maple, black maple, Florida maple) has a very similar distribution to beech, from Canada to Mexico, and occurs in mixed beech–sugar maple stands on cool, moist, fertile or infertile, soils that are rarely disturbed by fire. The red maple has even a wider range and an even wider ecological amplitude, growing on virtually all sites, and frequently more prevalent on wet-mesic areas and in deciduous and conifer-hardwood swamps. Silver maple is a species of river floodplain forests throughout much of eastern North America. The book by van Gelderen et al. (1994) provides an excellent survey of the world's maples and their cultivated varieties.

In western North America, bigleaf and vine maples are fire-susceptible tolerants that form a minor part of the conifer forest of the Pacific Northwest, especially in the moist ravines and draws, while in the Rocky Mountains, the bigtooth maple occurs on hillsides and canyons.

In Europe, the maples are less important than beech, but do occur widely as minor components of forest ecosystems. Sycamore maple, Norway maple, and field maple occur throughout much of central Europe, with Norway maple extending as far north as southern Norway and Finland, and field maple as far south as North Africa and Turkey.

Other maples are associated with oak forests throughout Asia, forming minor portions of the stands, especially on the cooler, moister sites of the Middle East, the Himalayas, China, southern Siberia, Korea, and Japan. They extend into the tropics and south of the equator to the mountains of Indonesia.

MISCELLANEOUS HARDWOODS

Other hardwood or angiosperm tree taxa are also widely distributed in the North Temperate Zone. As with the more abundant genera, these are frequently represented with vicarious species in North America, Europe, southwestern Asia, the Himalayan region, and eastern Asia. The mixed hardwood complexes of the eastern United States, western Europe, the Himalayas, eastern Siberia, northeastern China, and Japan have much in common. Genera that occur with disjunct species pretty much around the Northern Hemisphere include the basswoods and lindens (*Tilia*), elms (*Ulmus*), ashes (*Fraxinus*), walnuts (*Juglans*), blue beeches and hop-hornbeams (*Carpinus* and *Ostrya*), chestnuts (*Castanea*), sycamores or planes (*Platanus*), willows (*Salix*), and alders (*Alnus*). Still others are common to both the eastern American and east Asian forests, including *Liriodendron* (Parks et al., 1983), *Liquidambar*, *Magnolia*, and *Nyssa*; only a few are confined to a single continent.

The extent of the southern temperate land zone is very limited and what little there is is sparsely forested. Relatively few hardwoods, therefore, are abundant or widespread south of the Tropics. *Nothofagus*, the Antarctic or southern beech, is the most important, with deciduous and evergreen species of southern beeches characteristic of the cool wet forests of southern Chile and Argentina, and evergreen species characteristic of similar zones in New Zealand (Poole, 1987; Wardle, 1991) and southeastern Australia. (Chapter 20). In New Guinea, a region forming a bridge between the north-temperate and south-temperate Fagaceae, species of *Nothofagus* and *Quercus* occur together.

The other southern hemisphere hardwood genus of great importance is *Eucalyptus*, with several hundred species in Australia and the islands to its north (Brooker and Kleinig, 1990a, b, c). It occurs both in tropical and temperate climates, and in both xeric and mesic zones. In fact, there is in Australia a eucalyptus community in every type of forest niche from cool, wet, rain forest through closed mesic forest to open xeric forest and sparsely wooded semidesert. The large, fast-growing eucalypts, however, typically grow in the higher rainfall belts under temperature climatic conditions. These eucalypt species include: blackbutt and flooded gum in the east; alpine ash, mountain ash, and messmate stringybark in the southeast; and jarrah and karri in the far west. These trees reach heights of 60 to 98 meters and are among the fastest-growing trees in the world.

MISCELLANEOUS CONIFERS

The principal north-temperate conifers mentioned thus far—the spruces, firs, larches, and pines—all belong to the Pinaceae. Others of the pine family are of local importance. These include Douglas-fir of western North America, hemlocks in western and eastern North America and China, and the true cedars: Lebanon cedar in the near east and Deodar cedar in the Himalayas.

The members of the family Cupressaceae are also of great importance in the Northern Hemisphere, with some genera occurring locally in the Southern Hemisphere. The junipers, in particular, occur around the Northern Hemisphere, occupying dry and infertile sites and forming open woodlands at the drier edges of the forests. Other members of the family of local importance include sha mu, or Chinese fir, the most important timber tree of China; sugi in Japan; *Cupressus lusitanica* in the mountains from Mexico to Honduras; cypress pine in the dry interior of Queensland; southern bald cypress and pondcypress in the swamps of the American Southeast, and redwood in coastal northern California.

In the Southern Hemisphere, however, Araucariaceae and Podocarpaceae are the two important gymnosperm families. The former includes *Araucaria* and *Agathis*, and the latter, *Phyllocladus*, *Podocarpus*, and *Dacridium*.

The araucarias are important timber trees in Australia, islands east and north of Australia, Chile, and Argentina. *Agathis*, including the kauri of New Zealand, is a tropical genus of the southwestern Pacific with a number of commercially important species, frequently reaching gigantic size (Figure 22.4). It extends from Borneo and the Philippines in the North to Queensland and northern New Zealand in the South.

The various podocarps are widely distributed in the Southern Hemisphere, occurring in Africa, southern Asia, the southwest Pacific, and the Americas from the West Indies to southern Chile. They constitute the native conifers of east and south Africa, and also occur in Indo-China, Papua New Guinea, the Philippines, Australia, and New Zealand. In the

Figure 22.4   Giant kauri (*Agathis australis*) in Waipoua Forest, North Island, New Zealand. Diameters may exceed 3 m with little stem taper. (New Zealand Forest Service photo by J. H. Johns, A.R.P.S.)

latter country, rimu (Dacrydium cupressinum), totara (*Podocarpus totara*), and kahikatea (*Dacrycarpus dacrydioides*) are the principal native conifers. Among the American species is *Podocarpus coriaceous*, found in the uplands of the West Indies, Venezuela, and Columbia.

Suggested Readings

Axelrod, D. I. 1983. Biogeography of oaks in the Arcto-Tertiary Province. *Ann. Missouri Bot. Gard.* 70:423–439.

Beard, J. S. 1955. The classification of tropical American vegetation-types. *Ecology* 36:89–100.

Mirov, N. T. 1967. *The Genus Pinus.* Ronald Press, New York. 602 pp.

Richards, P. W. 1996. *The Tropical Rain Forest: An Ecological Study.* 2nd ed. Cambridge Univ. Press, Cambridge. 575 pp.

Röhrig, E., and B. Ulrich. 1991. *Temperate Deciduous Forests.* Vol. 7 of Ecosystems of the World. Elsevier, Amsterdam. 635 pp.

Terborgh, J. 1992. *Diversity and the Tropical Rain Forest.* Sci. Amer. Library, HPHLP, New York. 242 pp.

Walter, H. 1973. *Vegetation of the Earth: In Relation to Climate and the Eco-Physiological Conditions.* Springer–Verlag, New York. 237 pp.

Webb, L. J. 1968. Environmental relationships of the structural types of Australian rain forest vegetation. *Ecology* 49:296–311.

Whitmore, T. C. 1991. *An Introduction to Tropical Rain Forests.* Clarendon Press, Oxford. 226 pp.

LITERATURE CITED

Abbot, C. G. 1929. *The Sun.* Appleton-Century-Crofts, New York. 447 pp.

Aber, J. D., and J. M. Melillo. 1991. *Terrestrial Ecosystems.* Saunders, Philadelphia. 429 pp.

Abrams, M. D. 1990. Adaptations and responses to drought in *Quercus* species of North America. *Tree Physiol.* 7:227–238.

Abrams, M. D. 1992. Fire and the development of oak forests. *BioScience* 42:346–353.

Abrams, M. D., and J. A. Downs. 1990. Successional replacement of old-growth white oak by mixed-mesophytic hardwoods in southwest Pennsylvania. *Can. J. For. Res.* 20:1864–1870.

Abrams, M. D., M. E. Kubiske, and S. A. Mostoller. 1994. Relating wet and dry year ecophysiology to leaf structure in contrasting temperate tree species. *Ecology* 75:123–133.

Abrams, M. D., and M. L. Scott. 1989. Disturbance-mediated accelerated succession in two Michigan forest types. *For. Sci.* 35:42–49.

Abrams, M. D., D. G. Sprugel, and D. I. Dickmann. 1985. Multiple successional pathways on recently disturbed jack pine sites in Michigan. *For. Ecol. and Manage.* 10:31–48.

Adams, W. T. 1992. Gene dispersal within forest tree populations. *In* W. T. Adams, S. H. Strauss, D. L. Copes, and A. R. Griffin (eds.), *Population Genetics of Forest Trees.* Kluwer, Boston, MA.

Adams, W. T., S. H. Strauss, D. L. Copes, and A. R. Griffin (eds.) 1992. *Population Genetics of Forest Trees.* Kluwer, Boston, MA. 420 pp.

Agee, J. K. 1993. *Fire Ecology of Pacific Northwest Forests.* Island Press, Washington D.C. 493 pp.

Agren, G. I., B. Axelsson, J. G. K. Flower-Ellis, S. Linder, H. Persson, H. Staaf, and H. Troeng. 1980. Annual carbon budget for a young Scots pine. *In* T. Persson (ed.), *Structure and Function of Northern Coniferous Forests: An Ecosystem Study.* Swedish Natural Sci. Res. Council, Stockholm.

Ahlgren, C. E. 1960. Some effects of fire on reproduction and growth of vegetation in northeastern Minnesota. *Ecology* 41:431–445.

Ahlgren, C. 1974. Effects of fires on temperate forests: north central United States. *In* T. T. Kozlowski and C. E. Ahlgren (eds.), *Fire and Ecosystems.* Academic Press, New York.

Ahlgren, I. F. 1974. The effect of fire on soil organisms. *In* T. T. Kozlowski and C. E. Ahlgren (eds.), *Fire and Ecosystems.* Academic Press, New York.

Ahuja, M. R., and W. J. Libby (eds). 1993. *Clonal Forestry. I, Genetics and Biotechnology,* 277 pp. *II, Conservation and Application,* 240 pp. Springer-Verlag, New York.

Alban, D. H. 1969. The influence of western hemlock and western redcedar on soil properties. *Soil Sci. Soc. Amer. Proc.* 33:453–457.

Alban, D. H. 1977. Influence on soil properties of prescribed burning under mature red pine. USDA For. Serv. Res. Paper NC-139. North Central For. Exp. Sta., St. Paul, MN. 8 pp.

Albert, D. A. 1995. Regional landscape ecosystems of Michigan, Minnesota, and Wisconsin: a working map and classification. USDA For. Serv. Gen. Tech. Report NC-178. North Central For. Exp. Sta., St. Paul, MN. 250 pp. + map.

Albert, D. A., S. R. Denton, and B. V. Barnes. 1986. Regional landscape ecosystems of Michigan. School of Natural Resources, University of Michigan, Ann Arbor. 32 pp.

Albertson, F. W., and J. E. Weaver. 1945. Injury and death or recovery of trees in prairie climate. *Ecol. Monogr.* 15:395–433.

Alexander, M. 1977. *Introduction to Soil Microbiology,* 2nd ed. John Wiley, New York. 467 pp.

Alexander, N. L., H. L. Flint, and P. A. Hammer. 1984. Variation in cold-hardiness of *Fraxinus americana* stem tissue according to geographic origin. *Ecology* 65:1087–1092.

Alexander, R. R. 1964. Minimizing windfall around clear cuttings in spruce-fir forests. *For. Sci.* 10:130–142.

Alexander, R. R. 1966. Sites indexes for lodgepole pine, with corrections for stand density; instructions for field use. USDA For. Serv. Res. Paper RM-24. Rocky Mountain For. and Rge. Exp. Sta., Fort Collins, CO. 7 pp.

Alexander, R. R. 1985. Major habitat types, community types, and plant communities in the Rocky Mountains. USDA For. Serv. Gen. Tech. Report RM-123. Fort Collins, CO. 105 pp.

Alexander, R. R. 1986. Classification of the forest vegetation of Wyoming. USDA. For. Serv. Res. Note RM-466. Rocky Mountain For. and Rge. Exp. Sta., Fort Collins, CO. 10 pp.

Alexander, R. R., G. R. Hoffman, and J. M. Wirsing. 1986. Forest vegetation of the Medicine Bow National Forest in southeastern Wyoming: a habitat type classification. USDA For. Serv. Res. Paper RM-271. Rocky Mountain For. and Rge. Res. Sta. Fort Collins, CO. 39 pp.

Alexander, R. R., D. Tackle, and W. Dahms. 1967. Sites indexes for lodgepole pine, with corrections for stand density; methodology. USDA For. Serv. Res. Paper RM-29. Rocky Mountain For. and Rge. Exp. Sta., Fort Collins, CO. 18 pp.

Allen, G. S., and J. N. Owens. 1972. The life history of Douglas-fir. Environment Canada, For. Serv., Ottawa. 139 pp.

Amaranthus, M. P., and D. A. Perry. 1987. The effect of soil transfers on ectomycorrhizal formation and the survival and growth of conifer seedlings on old, nonreforested clear-cuts. *Can. J. For. Res.* 17:944–950.

Amaranthus, M. P., J. M. Trappe, L. Bednar, and D. Arthur. 1994. Hypogeous fungal production in mature Douglas-fir forest fragments and surrounding plantations and its relation to coarse woody debris and animal mycophagy. *Can. J. For. Res.* 24:2157–2165.

Amthor, J. S. 1984. The role of maintenance respiration in plant growth. *Plant, Cell and Environment* 7:561–569.

Anderson, E. 1948. Hybridization of the habitat. *Evolution* 2:1–9.

Anderson, E. 1949. *Introgressive Hybridization.* John Wiley, New York. 109 pp.

Anderson, E. 1953. Introgressive Hybridization. *Biol. Rev.* 28:280–307.

Anderson, H. W., G. B. Coleman, and P. J. Zinke. 1959. Summer slides and winter scours—dry-wet erosion in southern California mountains. USDA For. Serv. Tech. Paper 36. Pacific Southwest For. and Rge. Exp. Sta., Berkeley, CA. 12 pp.

Anderson, H. W., M. D. Hoover, and K. G. Reinhart. 1976. Forests and water: effects of forest management on floods, sedimentation, and water supply. USDA For. Serv., Gen. Tech. Report PSW-18. 115 pp.

Anderson, J. M., and M. J. Swift. 1983. Decomposition in tropical forests. *In* S. L. Sutton, T. C. Whitmore, and A. C. Chadwick (eds.), *Tropical Rain Forest: Ecology and Management.* Blackwell, Oxford.

Andersson, E. 1963. Seed stands and seed orchards in the breeding of conifers, World Consult. For. Gen. and For. Tree Imp. Proc. II FAO-FORGEN 63– 8/1:1–18.

Andresen, J. W. 1959. A study of pseudo-nanism in *Pinus rigida* Mill. *Ecol. Monogr.* 39:309–332.

Antibus, R. K., J. G. Croxdale, O. K. Miller, and A. E. Linkins. 1981. Ectomycorrhizal fungi of *Salix rotundifolia.* III. Resynthesized mycorrhizal complexes and their surface phosphatase activities. *Can. J. Bot.* 59:2458–2465.

Antonovics, J. 1971. The effects of a heterogeneous environment on the genetics of natural populations. *Amer. Sci.* 59:593–599.

Arbeitsgemeinschaft "Oberschwäbische Fichtenreviere," 1964. *Standort, Wald und Waldwirtschaft in Oberschwaben.* Verein forstl. Standortsk. Forstpflz., Stuttgart. 323 pp.

Arbeitsgruppe Biozönosenkunde. 1995. Ansätze für eine Regionale Biotop-und Biozönosenkunde von Baden-Württemberg. Mitt. Forst. Versuchs u. Forschungsanst. Baden-Württemberg. Freiburg, Germany. 166 pp.

Archambault, L., B. V. Barnes, and J. A. Witter. 1989. Ecological species groups of oak ecosystems of southeastern Michigan, USA. *For. Sci.* 35:1058–1074.

Archambault, L., B. V. Barnes, and J. A. Witter. 1990. Landscape ecosystems of disturbed oak forests of southeastern Michigan, USA. *Can J. For. Res.* 20:1570–1582.

Armand, A. D. 1992. Sharp and gradual mountain timberlines as a result of species interaction. *In* A. J. Hansen and F. di Castri (eds.), *Landscape Boundaries: Consequences for Biotic Diversity and Ecological Flows.* Springer, New York.

Arno, S. F. 1976. The historical role of fire on the Bitterroot National Forest. USDA For. Serv. Res. Paper INT-187. Intermountain For. and Rge. Exp. Sta., Ogden, UT. 29 pp.

Arno, S. F. 1980. Forest fire history of the northern Rockies. *J. For.* 78:460–465.

Arno, S. F., and W. C. Fischer. 1989. Using vegetation classifications to guide fire management. *In* D. E. Ferguson, P. Morgan, and F. D. Johnson (comps.), *Proceedings—Land Classifications Based on Vegetation: Applications for Resource Management.* USDA For. Serv. Gen. Tech. Report INT-257, Intermountain Res. Sta, Ogden, UT.

Arno, S. F., J. H. Scott, and M. G. Hartwell. 1995. Age-class structure of old growth ponderosa pine/Douglas-fir stands and its relationship to fire history. USDA, For. Serv. Res. Paper INT-RP-481, Intermountain Res. Sta., Ogden, UT. 25 pp.

Arno, S. F., D. G. Simmerman, and R. E. Keane. 1985. Forest succession on four habitat types in western Montana. USDA For. Serv. Gen. Tech. Report INT-177, Intermountain For. and Rge. Exp. Sta., Ogden, UT. 74 pp.

Arnold, M. L. 1992. Natural hybridization as an evolutionary process. *Ann. Rev. Ecol. Syst.* 2:237–261.

Arnold, M. L. 1994. Natural hybridization and Louisiana irises. *BioScience* 44:141–147.

Ashton, P. S. 1969. Speciation among tropical forest trees: some deductions in light of recent evidence. *Biol. J. Linn. Soc.* 1:155–196.

Ashton, P. S. 1992. Species richness in plant communities. *In* P. L. Fiedler and K. S. Jain (eds.), *Conservation Biology.* Chapman and Hall, London.

Assmann, E. 1970. *The Principles of Forest Yield Studies.* Pergamon Press, Oxford. 506 pp.

Aston, J. L., and A. D. Bradshaw. 1966. Evolution in closely adjacent plant populations. II. *Agrostis stolonifera* in maritime habitats. *Heredity* 21:649–664.

Augenbaugh, J. E. 1935. Replacement of chestnut in Pennsylvania. Pa. Dept. For. Waters, *Water Resour. Bull.* 54:1–38.

Aulitzky, H. 1967. Significance of small climatic differences for the proper afforestation of highlands in Austria. *In* W. E. Sopper and H. W. Lull (eds.), *Forest Hydrology.* Pergamon Press, New York.

Austin, M. P. 1985. Continuum concept, ordination methods, and niche theory. *Ann. Rev. Ecol. Syst.* 16:39–61.

Austin, M. P. 1990. Community theory and competition in vegetation. *In* J. B. Grace and D. Tilman (eds.), *Perspectives on Plant Competition.* Academic Press, New York.

Austin, M. P., and T. M. Smith. 1989. A new model for the continuum concept. *Vegetatio* 83:35–47.

Avers, P. E., D. T. Cleland, and W. H. McNab. 1994. National hierarchical framework of ecological units. *In* L. H. Foley (comp.), Workshop Proc. *Silviculture: From the Cradle of Forestry to Ecosystem Management.* USDA For. Serv. Gen. Tech. Report SE-88. Southeastern For. Exp. Sat., Asheville, NC.

Avery, T. E., and H. E. Burkhart. 1994. *Forest Measurements.* McGraw-Hill, New York. 408 pp.

Axelrod, D. I. 1970. Mesozoic paleogeography and early angiosperm history. *Bot. Rev.* 36:277–319.

Axelrod, D. I. 1983. Biogeography of oaks in the Arcto-Tertiary Province. *Ann. Missouri Bot. Gard.* 70:629–657.

Ayala, F. J. 1982. *Population and Evolutionary Genetics: A Primer.* Benjamin/Cummings, Menlo Park, CA. 268 pp.

Bahari, Z. A., S. G. Pallarady, and W. C. Parker. 1985. Photosynthesis, water relations, and drought adaptations in six woody species of oak-hickory forests in central Missouri. *For. Sci.* 31:557–569.

Bailey, R. G. 1983. Delineation of ecosystem regions. *Env. Manage.* 7:365–373.

Bailey, R. G. 1988. Ecogeographic analysis: a guide to the ecological division of land for resource management. USDA For. Serv. Misc. Publ. No. 1465. Washington, D.C. 18 pp.

Bailey, R. G. 1994. Ecoregions of the United States (map). USDA For. Serv., Washington, D.C.

Bailey, R. G. 1995. Description of the ecoregions of the United States. USDA For. Serv. Misc. Publ. 1391. Washington, D.C. 108 pp. + map

Bailey, R. G. 1996. *Ecosystem Geography.* Springer-Verlag, New York. 204 pp + 2 maps.

Bailey, R. G., and H. C. Hogg. 1986. A world ecoregions map for resource reporting. *Env. Conservation* 13:195–202.

Bailey, R. G., S. C. Zoltai, and E. B. Wiken. 1985. Ecological regionalization in Canada and the United States. *Geoforum* 16:265–275.

Baker, F. S. 1918. Aspen as a temporary forest type. *J. For.* 16:294–300.

Baker, F. S. 1929. Effect of excessively high temperatures on coniferous reproduction. *J. For.* 27:949–975.

Baker, F. S. 1944. Mountain climates of the western United States. *Ecol. Monogr.* 14:223–254.

Baker, F. S. 1945. Effects of shade upon coniferous seedlings grown in nutrient solutions. *J. For.* 43:428–435.

Baker, F. S. 1950. *Principles of Silviculture*. McGraw-Hill, New York. 414 pp.

Baker, H. G., K. S. Bawa, G. W. Frankie, and P. A. Opler. 1983. Reproductive biology of plants in tropical forests. *In* F. B. Golley (ed.), *Tropical Rainforest Ecosystems*, Elsevier, New York.

Baldwin, K. A., J. A. Johnson, R. A. Sims, and G. M. Wickware. 1990. Common landform toposequences of northwestern Ontario. Forestry Canada, Ontario Min. Nat. Res. Publ. 5311. Sault Ste. Marie, Ontario. 26 pp.

Bannister, M. H. 1965. Variation in the breeding system of *Pinus radiata*. *In* H. G. Baker and G. L. Stebbins (eds.), *The Genetics of Colonizing Species*. Academic Press, New York.

Bao, Y., and E. T. Nilsen. 1988. The ecophysiological significance of thermotropic leaf movements in *Rhododendron maximum*. *Ecology* 69:1578–1587.

Barber, H. N. 1955. Adaptive gene substitutions in Tasmanian eucalyptus: I. genes controlling the development of glaucousness. *Evolution* 9:1–14.

Barber, H. N., and W. D. Jackson. 1957. Natural selection in action in Eucalyptus. *Nature* 179:1267–1269.

Barbour, M. G. 1988. Californian upland forests and woodlands. *In* M. G. Barbour and W. D. Billings (eds.), *North American Terrestrial Vegetation*. Cambridge Univ. Press, Cambridge.

Barbour, M. G., and W. D. Billings. 1988. *North American Terrestrial Vegetation*. Cambridge Univ. Press, Cambridge. 434 pp.

Barbour, M. G., and N. L. Christensen. 1993. Vegetation. *In* R. R. Morin (conv. ed.), *Flora of North America, North of Mexico. Vol. 1*. Oxford Univ. Press, New York.

Bard, G. E. 1952. Secondary succession on the Piedmont of New Jersey. *Ecol. Monogr.* 22:195–215.

Barden, L. S. 1981. Forest development in canopy gaps of a diverse hardwood forest of the Southern Appalachian Mountains. *Oikos* 37:205–209.

Barlow, B. A., and B. P. M. Hyland. 1988. The origins of the flora of Australia's wet tropics. *In* R. Kitching (ed.), *The Ecology of Australia's Wet Tropics*. Ecol. Soc. Australia, Surrey Beatty & Sons, Chipping Norton, Australia.

Barnes, B. V. 1967. The clonal growth habit of American aspens. *Ecology* 47:439–447.

Barnes, B. V. 1969. Natural variation and delineation of clones of *Populus tremuloides* and *P. grandidentata* in northern Lower Michigan. *Silvae Genetica* 18:130–142.

Barnes, B. V. 1975. Phenotypic variation of trembling aspen in western North America. *For. Sci.* 21:319–328.

Barnes, B. V. 1976. Succession in deciduous swamp communities of southeastern Michigan. formerly dominated by American elm. *Can. J. Bot.* 54:19–24.

Barnes, B. V. 1984. Forest ecosystem classification and mapping in Baden-Württemberg, Germany. *In* J. G. Bockheim (ed.), *Symp. Proc. Forest Land Classification: Experience, Problems, Perspectives*. NCR-102 North Central For. Soils Com., Soc. Am. For., USDA For. Serv., and USDA Soil Cons. Serv. Madison, WI.

Barnes, B. V. 1991. Deciduous forests of North America. *In* E. Röhrig and B. Ulrich (eds.), *Ecosystems of the World, Temperate Deciduous Forests. Vol. 7*. Elsevier, New York.

Barnes, B. V. 1993. The landscape ecosystem approach and conservation of endangered spaces. *Endangered Species Update* 10:13–19.

Barnes, B. V. 1996. Silviculture, landscape ecosystems, and the iron law of the site. *Forstarchiv* 67:226–235.

Barnes, B. V., and B. P. Dancik. 1985. Characteristics and origin of a new birch species, *Betula murrayana*, from southeastern Michigan. *Can. J. Bot.* 63:223–226.

Barnes, B. V., and F. Han. 1993. Phenotypic variation of Chinese aspens and their relationships to similar taxa in Europe and North America. *Can. J. Bot.* 71:799–815.

Barnes, B. V., K. S. Pregitzer, T. A. Spies, and V. Spooner. 1982. Ecological forest site classification. *J. For.* 80:493–498.

Barnes, B. V., Z. Xü, and S. Zhao. 1992. Forest ecosystems in an old-growth pine—mixed hardwood forest of the Changbai Shan Preserve in northeastern China. *Can. J. For. Res.* 22:144–160.

Barnes, W. J., and E. Dibble. 1988. The effects of beaver in riverbank forest succession. *Can. J. Bot.* 66:40–44.

Barnosky, C. W. 1987. Response of vegetation to climatic changes of different duration in the Late Neogene. *Trends Ecol. Evol.* 2:247–250.

Barry, R. C. 1967. Seasonal location of the Arctic Front over North America. *Geographical Bull.* 9:79–95.

Barth, H. 1982. The biogeography of mangroves. *In* E. N. Sen and K. S. Rajpurohit (eds.), *Tasks for Vegetation Science, Vol. 2.* Junk, The Hague.

Barthelemy, D., C. Edelin, and F. Hallé. 1991. Canopy architecture. *In* A. S. Raghavendra (ed.), *Physiology of Trees.* John Wiley, New York.

Bartlett, H. H. 1956. Fire, primitive agriculture, and grazing in the tropics. *In* W. L. Thomas (ed.), *Man's Role in Changing the Face of the Earth.* Univ. Chicago Press, Chicago.

Basey, J. M., S. H. Jenkins, and P. E. Busher. 1988. Optimal central-place foraging by beavers: tree-size selection in relation to defensive chemicals of quaking aspen. *Oecologia* 76:278–282.

Basey, J. M., S. H. Jenkins, and G. C. Miller. 1990. Food selection by beavers in relation to inducible defenses of *Populus tremuloides. Oikos* 59:57–62.

Bassett, J. R. 1964. Diameter growth of loblolly pine trees as affected by soil moisture availability. USDA For. Serv. Res. Note SO–9, Southern For. Exp. Sta., New Orleans, LA. 7 pp.

Bates, C. G., and J. Roeser, Jr. 1928. Light intensities required for growth of coniferous seedlings. *Amer. J. Bot.* 15:185–244.

Baumgartner, A. 1958. Nebel und Nebelniederschlag als Standortsfacktoren am grossen Falkenstein (Bayr. Wald). *Forstw. Centralbl.* 77:257–272.

Bazzaz, F. A. 1979. The physiological ecology of plant succession. *Ann. Rev. Ecol. Syst.* 10:351–372.

Beale, E. F. 1958. Wagon road from Fort Defiance to the Colorado River. 35 Cong. 1 Sess., Sen. Exec. Doc. 124.

Beard, J. S. 1953. The savanna vegetation of northern tropical America. *Ecol. Monogr.* 23:149–215.

Beard, J. S. 1955. The classification of tropical American vegetation-types. *Ecology* 36:89–100.

Beaufait, W. R. 1956. Influence of soil and topography on willow oak sites. U.S. For. Serv., Southern For. Exp. Sta. Occ. Paper 148. 12 pp.

Beck, C. B. (ed.). 1976. *Origin and Early Evolution of Angiosperms.* Columbia Univ. Press, New York. 341 pp.

Beck, D. E. 1971. Polymorphic site index curves for white pine in the southern Appalachians. USDA For. Serv. Res. Note SE-80. Southeastern For. Exp. Sta., Asheville, NC. 8 pp.

Beck, D. E. 1977. Twelve-year acorn yield in southern Appalachian oaks. USDA For. Serv. Res. Note SE-244, Southeastern For. Exp. Sta., Asheville, NC. 8 pp.

Becking, R. W. 1957. The Zürich-Montpellier school of phytosociology. *Bot. Rev.* 23:411–488.

Begon, M., J. L. Harper, and C. R. Townsend. 1990. *Ecology, Individuals, Populations, and Communities.* Blackwell Sci. Publ., Boston. 945 pp.

Behrensmeyer, A. K., J. D. Damuth, W. A. DiMichele, R. Potts, H.-D. Sues, and S. L. Wing (eds.). 1992. *Terrestrial Ecosystems Through Time.* Univ. Chicago Press, Chicago. 568 pp.

Bell, A. D. 1991. *Plant Form, An Illustrated Guide to Flowering Plant Morphology.* Oxford Univ. Press, New York. 341 pp.

Belsky, A. J., and C. D. Canham. 1994. Forest gaps and isolated savanna trees. *BioScience* 44:77–84.

Bendell, J. F. 1974. Effects of fire on birds and mammals. *In* T. T. Kozlowski and C. E. Ahlgren (eds.), *Fire and Ecosystems.* Academic Press, New York.

Benson, L. B. 1962. *Plant Taxonomy: Methods and Principles.* Ronald Press, New York. 494 pp.

Bergen, K. M., M. C. Dobson, L. E. Pierce, and F. T. Ulaby. 1997. Characterizing carbon in a northern forest using SIR-C/X-SAR imagery. *Remote Sensing Env. 62(3).*

Bernabo, J. C., and T. Webb, III. 1977. Changing patterns in the Holocene pollen record of northeastern North America: a mapped summary. *Quat. Res.* 8:64–96.

Berntson, G. M., and F. I. Woodward. 1992. The root system architecture and development of *Senecio vulgaris* in elevated CO_2 and drought. *Functional Ecology* 6:324–333.

Berry, J., and O. Bjorkman. 1980. Photosynthetic response and adaptation to temperature in higher plants. *Ann. Rev. Plant Physiology* 31:491–543.

Berry, J., and W. J. S. Downton. 1982. Environmental regulation of photosynthesis. *In* R. Govindjee (ed.), *Photosynthesis, Development, Carbon Metabolism, and Plant Productivity*. *Vol. 2*. Academic Press, New York. 580 pp.

Bidwell, R. G. S. 1974. *Plant Physiology*. Macmillan. New York. 643 pp.

Bigelow, R. S. 1965. Hybrid zones and reproductive isolation. *Evolution* 19:449–458.

Billings, W. D. 1938. The structure and development of old field shortleaf pine stands and certain associated properties of the soil. *Ecol. Monogr.* 8:437– 499.

Billings, W. D. 1973. Arctic and alpine vegetations: Similarities, differences, and susceptibility to disturbance. *BioScience* 23:697–704.

Billings, W. D. 1987. Constraints to plant growth, reproduction, and establishment in Arctic environments. *Arctic and Alpine Res.* 19:357–365.

Billings, W. D. 1995. The effects of global and regional environmental changes on mountain ecosystems. *In* D. G. Despain (ed.), *Plants and their Environment: Proceedings of the First Biennial Scientific Conference on the Greater Yellowstone Ecosystem*. Tech. Report NPS/NRYELL/ NRTR-93XX. US Dept. Interior, Nat. Park Serv., Nat. Res. Publ. Office, Denver. CO.

Binkley, D. 1995. The influence of tree species on forest soils: processes and patterns. *In* Mead, D. J., and I. S. Cornforth (eds.), *Proc. Trees and Soils Workshop*. Agronomy Society of New Zealand Special Publication No. 10. Lincoln Univ. Press, Canterbury.

Binkley, D., and S. Hart. 1989. The components of nitrogen availability assessments in forest soils. *Adv. Soil Sci.* 10:57–115.

Birkeland, P. W. 1974. *Pedology, Weathering and Geomorphological Research*. Oxford Univ. Press, New York. 285 pp.

Biswell, H. H. 1961. The big trees and fires. *Nat. Parks Mag.* 35:11–14.

Biswell, H. H. 1974. Effects of fire on chaparral. *In* T. T. Kozlowski and C. E. Ahlgren (eds.), *Fire and Ecosystems*. Academic Press, New York.

Blais, J. R. 1952. The relationship of the spruce budworm (*Choristoneura fumiferana* Clem.) to the flowering condition of balsam fir (*Abies balsamea* (L.) Mill.). *Can. J. Zool.* 30:1–29.

Bliss, L. C., and J. E. Cantlon. 1957. Succession on river alluvium in northern Alaska. *Am. Midl. Nat.* 58:452–469.

Bloomberg, W. J. 1950. Fire and spruce. *For. Chron.* 26:157–161.

Boettcher, S. E., and P. J. Kalisz. 1990. Single-tree influence on soil properties in the mountains of eastern Kentucky. *Ecology* 71:1365–1372.

Boggess, W. R. 1964. Trelease Woods, Champaign County, Illinois: woody vegetation and stand composition. *Trans. Ill. Acad. Sci.* 57:261–271.

Boggess, W. R., and L. W. Bailey. 1964. Brownfield Woods, Illinois: woody vegetation and changes since 1925. *Am. Midl. Nat.* 71:392–401.

Boggess, W. R., and J. W. Geis. 1966. The Funk Forest Natural Area, McLean County, Illinois: woody vegetation and ecological trends. *Trans. Ill. Acad. Sci.* 59:123–133.

Bolan, N. S., A. D. Robson, N. J. Barrow, and L. A. G. Aylmore. 1984. Specific activity of phosphorus in mycorrhizal and non-mycorrhizal plants in relation to the availability of phosphorus to plants. *Soil Biol. Biochemistry* 6:299–304.

Bonan, G. B., and H. H. Shugart. 1989. Environmental factors and ecological processes in boreal forests. *Ann. Rev. Ecol. Syst.* 20:1–28.

Bond, W. J. 1993. Keystone species. *In* E. D. Schulze and H. A. Mooney (eds.), *Ecosystem Function and Biodiversity*. Springer-Verlag, New York.

Bond, W. J., and B. W. van Wilgen. 1996. *Fire and Plants*. Chapman & Hall, London. 263 pp.

Bonga, J. M., and P. von Aderkas. 1993. Rejuvenation of tissues from mature conifers and its implications for propagation in vitro. *In* M. R. Ahuja and W. J. Libby (eds.), *Clonal Forestry I, Genetics and Biotechnology*. Springer-Verlag, Berlin.

Books, D. J., and C. H. Tubbs. 1970. Relation of light to epicormic sprouting in sugar maple. USDA For. Serv. Res. Note NC-93. North Central For. Exp. Sta., St. Paul, MN. 2 pp.

Boose, E. R., D. R. Foster, and M. Fluet. 1994. Hurricane impacts to tropical and temperate forest landscapes. *Ecol. Monogr.* 64:369–400.

Booysen, P. de V., N. M. Tainton. 1984. *Ecological Effects of Fire in South African Ecosystems* (Ecological Studies 48). Springer-Verlag, New York. 426 pp.

Borchert, J. R. 1950. The climate of the central North American grassland. *Ann. Assoc. Amer. Geog.* 40:1–39.

Borchert, R. 1991. Growth periodicity and dormancy. *In* A. S. Raghavendra (ed.), *Physiology of Trees.* John Wiley, New York.

Boring, L. R., W. T. Swank, J. B. Waide, and G. S. Henderson. 1988. Sources, fates, and impacts of nitrogen inputs to terrestrial ecosystems: review and synthesis. *Biogeochemistry* 6:119–159.

Bormann, F. H. 1953. Factors determining the role of loblolly pine and sweetgum in early old-field succession in the Piedmont of North Carolina. *Ecol. Monogr.* 23:339–358.

Bormann, F. H. 1966. The structure, function, and ecological significance of root grafts in *Pinus strobus* L. *Ecol. Monogr.* 36:1–26.

Bormann, F. H., and G. E. Likens. 1979. *Pattern and Process in a Forested Ecosystem.* Springer-Verlag, New York. 253 pp.

Bormann, F. H., G. E. Likens, and J. M. Melillo. 1977. Nitrogen budget for an aggrading northern hardwood forest ecosystem. *Science* 196:981–982.

Botkin, D. B. 1993. *Forest Dynamics: An Ecological Model.* Oxford Univ. Press, New York. 309 pp.

Botkin, D. B., G. M. Woodwell, and N. Tempel. 1970. Forest productivity estimated from carbon dioxide uptake. *Ecology* 51:1057–1060.

Botting, D. 1973. *Humboldt and the Cosmos.* Harper and Row, London. 295 pp.

Boucher, D. H. 1985. *The Biology of Mutualism.* Croom Helm, London. 388 pp.

Boucher, D. H., S. James, and K. H. Keeler. 1982. *Ann. Rev. Ecol. Syst.* 13:315–347.

Boufford, D. E., and S. A. Spongberg. 1983. Eastern Asian–eastern North American phytogeographical relationships—a history from the time of Linnaeus to the twentieth century. *Ann. Missouri Bot. Gard.* 70:423–439.

Bourliére, F. (ed.) 1983. *Tropical Savannas.* Elsevier, New York. 730 pp.

Bowen, G. D., and S. E. Smith. 1981. The effects of mycorrhizas on nitrogen uptake by plants. *In* F. E. Clark and T. Rosswall (eds.), *Terrestrial Nitrogen Cycles.* Ecol. Bull. 33. Swedish Natural Sci. Res. Council, Stockholm.

Box, E. O. 1981. *Macroclimate and Plant Forms: an Introduction to Predictive Modelling in Phytogeography.* Junk, The Hague, 258 pp.

Boyce, S. G., and M. Kaeiser. 1961. Why yellow-poplar seeds have low viability. USDA For. Serv. Central States For. Exp. Sta., Tech. Paper 186. 16 pp.

Boyle, T. J. B. 1992. Biodiversity of Canadian forests: current status and future challenges. *For. Chronicle* 68:444–453.

Boyle, J. R., and A. R. Ek. 1973. Whole tree harvesting: nutrient budget evaluation. *J. For.* 71:760–762.

Boyle, T., C. Liengsiri, and C. Piewluang. 1990. Genetic structure of black spruce on two contrasting sites. *Heredity* 65:393–399.

Bradford, K. J., and T. C. Hsiao. 1982. Physiological responses to moderate water stress. *In* O. L. Lange, P. S. Nobel, C. B. Osmond, and H. Ziegler. (eds.), *Physiological Plant Ecology II: Water Relations and Carbon Assimilation.* Encyclopedia of Plant Physiology, Volume 12B. Springer-Verlag, New York.

Bradley, A. F., W. C. Fischer, and N. V. Noste. 1992. Fire ecology of the forest habitat types of eastern Idaho and western Wyoming. USDA For. Serv. Gen. Tech. Report INT-290. Intermountain Res. Sta., Ogden, UT. 92 pp.

Bradshaw, A. D. 1971. Plant evolution in extreme environments. *In* R. Creed (ed.), *Ecological Genetics and Evolution.* Blackwell Press, Oxford.

Bradshaw, K. E. 1965. Soil use and management in the national forest of California. *In* C. T. Youngberg. (ed.), *Forest–Soil Relationships in North America.* Oregon State Univ. Press, Corvallis.

Brady, N. C. 1990. *The Nature and Properties of Soils*, 10th ed., Macmillan, New York. 621 pp.

Brady, N. C., and R. R. Weil. 1996 *The Nature and Properties of Soils,* 11th ed. Prentice Hall, Upper Saddle River, NJ. 740 pp.

Braun, E. L. 1950. *Deciduous Forests of Eastern North America.* McGraw-Hill, New York. 596 pp.

Braun-Blanquet. J. 1921. Prinzipien einer Systematik der Pflanzengellschaften auf floristischer Grundlage. *Jahrb. St. Gllen Naturw. Ges.* 57:305–351.

Braun-Blanquet. J. 1964. *Pflanzensoziologie, Grundzüge der Vegetationskunde.* 3rd ed. Springer-Verlag, New York. 865 pp.

Bray, J. R. 1956. Gap phase replacement in a maple basswood forest. *Ecology* 37:598–600.

Bray, J. R., and E. Gorham. 1964. Litter production in forests of the world. *Adv. Ecol. Res.* 2:101–157.

Brayshaw, T. C. 1965. Native poplars of southern Alberta and their hybrids. Canada, Dept. of Forestry Publ, No. 1109. 40 pp.

Brayton, R., and G. A. Mooney. 1966. Population variability of *Cercocarpus* in the White Mountains of California as related to habitat. *Evolution* 20:383–391.

Breckle, S.-W. 1974. Notes on alpine and nival flora of the Hindu Kush, East Afganistan. *Bot. Notises* 127:278–284.

Brewer, R., and P. G. Merritt. 1978. Wind throw and tree replacement in a climax beech-maple forest. *Oikos* 30:149–152.

Brinson, M. M. 1990. Riverine forests. *In* D. Goodall, A. Lugo, M. Brinson, and S. Brown (eds.), *Ecosystems of the World, Forested Wetlands, Vol. 15.* Elsevier, New York.

Brix, H. 1971. Effects of nitrogen fertilization on photosynthesis and respiration in Douglas-fir. *For. Sci.* 17:407–414.

Brix, H., and L. F. Ebell. 1969. Effects of nitrogen fertilization on growth, leaf area, and photosynthesis rate of Douglas-fir. *For. Sci.* 15:189–196.

Broadfoot, W. M., and H. L. Williston. 1973. Flooding effects on southern forests. *J. For.* 71:584–587.

Brokaw, N. V. L. 1985a. Treefalls, regrowth and community structure in tropical forests. *In* S. T. A. Pickett and P. S. White (eds.), *The Ecology of Natural Disturbance and Patch Dynamics.* Academic Press, New York.

Brokaw, N. V. L. 1985b. Gap-phase regeneration in a tropical forest. *Ecology* 66:682–687.

Bronstein, J. L. 1994. Our current understanding of mutualism. *Quart. Rev. Biol.* 69:31–51.

Brooker, M. I. H., and D. A. Kleinig. 1990a. *Field Guide to Eucalypts, Southeastern Australia. Vol. 1.* Inkata Press, Melbourne. 299 pp.

Brooker, M. I. H., and D. A. Kleinig. 1990b. *Field Guide to Eucalypts, Northern Australia. Vol. 2.* Inkata Press, Sidney. 383 pp.

Brooker, M. I. H., and D. A. Kleinig. 1990c. *Field Guide to Eucalypts, Southwestern and southern Australia. Vol. 3.* Inkata Press, Melbourne. 428 pp.

Brooks, K. N., P. F. Ffolliott, H. M. Gregersen, and J. L. Thames. 1991. *Hydrology and the Management of Watersheds.* Iowa State Univ. Press, Ames. 392 pp.

Brouillet, L., and R. D. Whetstone. 1993. Climate physiography. *In* R. R. Morin (conv. ed.), *Flora of North America, North of Mexico. Vol. 1.* Oxford Univ. Press, New York.

Brown, C. L., R. G. McAlpine, and P. P. Kormanik. 1967. Apical dominance and form in woody plants: a reappraisal. *Amer. J. Bot.* 54:153–162.

Brown, J. M. B. 1955. Ecological investigations: shade and growth of oak seedlings. Rep. For. Res. Comm., London, 1953–54.

Brown, S., and A. E. Lugo. 1982. The storage and production of organic matter in tropical forests and their role in the global carbon cycle. *Biotropica* 14:161–187.

Brunig, E. F. 1976. Tree forms in relation to environmental conditions: an ecological viewpoint. *In* M. G. R. Cannell and F. T. Last (eds.), *Tree Physiology and Yield Improvement.* Academic Press, New York.

Bryant, J. M. 1981. Phytochemical deterrence of snowshoe hare browsing by adventitious shoots of four Alaskan trees. *Science* 213:889–890.

Bryant, J. P., F. S. Chapin III, P. Reichardt, and T. Clausen. 1985. Adaptation to resource availability as a determinant of chemical defense strategies in woody plants. *In* G. A. Cooper-Driver, T. Swain, and E. E. Conn (eds.), *Chemically Mediated Interactions Between Plants and Other Organisms.* Recent Adv. Phytochemistry 19. Plenum Press, New York.

Bryson, R. A. 1966. Air masses, streamlines, and the boreal forest. *Geographical Bull.* 8: 228–269.

Bryson, R. A., and F. K. Hare. 1974. The climates of North America. *In* R. A. Bryson and F. K. Hare (eds.), *Climates of North America* (World Survey of Climatology, Vol. 11). Elsevier, New York.

Bryson, R. A., W. N. Irving, and J. A. Larsen. 1965. Radiocarbon and soil evidence of former forest in the southern Canadian tundra. *Science* 147:45–48.

Buckley, J. L. 1959. Effects of fire on Alaskan wildlife. *Proc. Soc. Amer. For.* 1958:123–126.

Budyko, M. I. 1974. *Climate and Life.* Academic Press, New York. 508 pp.

Budyko, M. I. 1986. *The Evolution of the Biosphere.* D. Reidel, Dordrecht. 423 pp.

Buffo, J., L. J. Fritschen, and J. L. Murphy. 1972. Direct solar radiation on various slopes from 0 to 60 degrees north latitude. USDA For. Serv. Res. Paper PNW-142. Pacific Northwest For. and Rge. Exp. Sta., Portland, OR. 74 pp.

Bullock, S. H., Mooney, H. A., and E. Medina (eds.). 1995. *Seasonally Dry Tropical Forests.* Cambridge Univ. Press, Cambridge. 450 pp.

Buol, S. W., F. D. Hole, and R. J. McCracken. 1980. *Soil Genesis and Classification.* Iowa State Univ. Press, Ames. 404 pp.

Burger, D. 1972. Forest site classification in Canada. *Mitt. Vereins forstl. Standortsk. Forstpflz.* 21:20–36.

Burger, D. 1993. Revised site regions of Ontario: concepts, methodology and utility. Ontario For. Res. Inst., Sault Ste. Marie, For. Res. Report No. 129. 24 pp.

Burke, I. C. 1989. Control of nitrogen mineralization in a sagebrush steppe landscape. *Ecology* 70:1115–1126.

Burke, M. J., L. V. Gusta, H. A. Quamme, C. J. Weiser, and P. H. Li. 1976. Freezing and injury in plants. *Ann. Rev. Plant Physiol.* 27:507 528.

Burke, M. K., D. Raynal, and M. J. Mitchell. 1991. Soil nitrogen availability influences seasonal carbon allocation patterns in sugar maple (*Acer saccharum*). *Can. J. For. Res.* 22:447–456.

Burley, J. 1966. Provenance variation in growth of seedling apices of Sitka spruce. *For. Sci.* 12:170–175.

Burns, G. P. 1923. Studies in tolerance of New England forest trees IV. Minimum light requirements referred to a definite standard. Univ. Vermont Agr. Exp. Sta. Bull. 235.

Burns, G. R. 1942. Photosynthesis and absorption in blue radiation. *Amer. J. Bot.* 29:381–387.

Burton, P. J., A. C. Balisky, L. P. Coward, S. G. Cumming, and D. D. Kneeshaw. 1992. The value of managing for biodiversity. *For. Chronicle* 68:225–237.

Büsgen, M., and E. Münch. 1929. *The Structure and Life of Forest Trees*, 3rd ed. Trans. by Thomas Thomson. Chapman & Hall, London. 436 pp.

Busing, R. T. 1994. Canopy cover and tree regeneration in old-growth cove forests of the Appalachian Mountains. *Vegetatio* 115:19–27.

Buttrick, P. L. 1914. Notes on germination and reproduction of longleaf pine in southern Mississippi. *For. Quart.* 12:532–537.

Byers, H. R. 1953. Coast redwoods and fog drip. *Ecology* 34:192–193.

Byram, G. M., and G. M. Jemison. 1943. Solar radiation and forest fuel moisture. *J. Agr. Res.* 676:149–176.

Cain, S. A. 1935. Studies on virgin hardwood forests: III. Warren's Woods, a beech-maple climax forest in Berrien County, Michigan. *Ecology* 16:500– 513.

Cain, S. A. 1939. The climax and its complexities. *Am. Midl. Nat.* 21:146–181.

Cain, S. A., and G. M. de Oliveira Castro. 1959. *Manual of Vegetation Analysis.* Harper & Row, New York. 325 pp.

Cajander, A. K. 1926. The theory of forest types. *Acta For. Fenn.* 29. 108 pp.

Caldwell, M. M. 1968. Solar ultraviolet radiation as an ecological factor for alpine plants. *Ecol. Monogr.* 38:243–268.

Callaham, R. Z. 1962. Geographic variability in growth of forest trees, *In* T. T. Kozlowski (ed.), *Tree Growth.* Ronald Press, New York.

Callicott, J. B., and K. Mumford. 1997. Ecological sustainability as a conservation concept. *Conserv. Biol.* 11:32–40.

Campbell, R. K. 1979. Genecology of Douglas-fir in a watershed in the Oregon Cascades. *Ecology* 60:1036–1050.

Campbell, R. K. 1986. Mapped genetic variation of Douglas-fir to guide seed transfer in southwest Oregon. *Silvae Genetica* 35:85–96.

Campbell, R. K. 1991. Soils, seed-zone maps, and physiography: guidelines for seed transfer in Douglas-fir in southwestern Oregon. *For Sci.* 37:973–986.

Campbell, R. K., W. A. Pawuk, and A. S. Harris. 1989. Microgeographic genetic variation of Sitka spruce in southeastern Alaska. *Can. J. For. Res.* 19:1004–1013.

Campbell, R. K., and F. C. Sorensen. 1978. Effect of test environment on expression of clines and on delimitation of seed zones in Douglas-fir. *Theor. Appl. Genet.* 51:233–246.

Campbell, R. K., and A. I. Sugano. 1979. Genecology of bud-burst phenology in Douglas-fir: response to flushing temperature and chilling. *Bot. Gaz.* 140:223–231.

Campbell, R. S. 1955. Vegetational changes and management in the cutover longleaf pine-slash pine area of the Gulf Coast. *Ecology* 36:29–34.

Canada Inst. Forestry. 1992. *The Forestry Chronicle* 68(1):21–120.

Canham, C. D. 1985. Suppression and release during canopy recruitment in *Acer saccharum. Bull Torrey Bot. Club.* 112:134–145.

Canham, C. D. 1988. Growth and canopy architecture of shade-tolerant trees: response to canopy gaps. *Ecology* 69:786–795.

Canham, C. D., J. S. Denslow, W. J. Platt, J. R. Runkle, T. A. Spies, and P. S. White. 1990. Light regimes beneath closed canopies and tree-fall gaps in temperate and tropical forests. *Can. J. For. Res.* 20:620–631.

Canham, C. D., and O. L. Loucks. 1984. Catastrophic windthrow in the presettlement forests of Wisconsin. *Ecology* 65: 803–809.

Cannell, M. G. R., M. B. Murray, and L. J. Sheppard. 1985. Frost avoidance by selection for late budburst in *Picea sitchensis. J. App. Ecol.* 22: 931–941.

Cantor, L. F., and T. G. Whitham. 1989. Importance of belowground herbivory: pocket gophers may limit aspen to rock outcrop refugia. *Ecology* 70:962– 970.

Caplenor, C. D., R. E. Bell, J. Brook, D. Caldwell, C. Hughes, A. Regan, A. Scott, S. Ware, and M. Wells. 1968. Forests of west central Mississippi as affected by loess. *Miss. Geol. Econ. Topogr. Survey Bull.* 111:205–267.

Carey, A. B., and M. L. Johnson. 1995. Small mammals in managed, naturally young, and old-growth forests. *Ecol. Applications* 5:336–352.

Carmean, W. H. 1967. Soil refinements for predicting black oak site quality in southeastern Ohio. *Proc. Soil Sci. Soc. Amer.* 31:805–810.

Carmean, W. H. 1975. Forest site quality evaluation in the United States. *Adv. Agronomy* 27:209–269.

Carmean, W. H. 1977. Site classification for northern forest species. *In* Proc. Symp *Intensive Culture of Northern Forest Types*, USDA For. Serv. Gen. Tech. Report NE-29. Northeastern For. Exp. Sta., Upper Darby, PA.

Carmean, W. H. 1970a. Site quality for eastern hardwoods. *In: The Silviculture of Oaks and Associated Species.* USDA For. Serv. Res. Paper NE-144. Northeastern For. Exp. Sta., Upper Darby, PA.

Carmean, W. H. 1970b. Tree height-growth patterns in relation to soil and site. *In* Youngberg, C. T., and C. B. Davey (eds.), *Tree Growth and Forest Soils*. Oregon State Univ. Press, Corvallis.

Carmean, W. H. 1975. Forest site quality evaluation in the United States. *Adv. Agronomy* 27:209–269.

Carmean, W. H. 1977. Site classification for northern forest species. *In* Symp. Proc. *Intensive Culture of Northern Forest Types*. USDA For. Serv. Gen. Tech. Report NE-29. Northeastern For. Exp. Sta., Upper Darby, PA.

Carmean, W. H. 1996. Site-quality evaluation, site-quality maintenance, and site-specific management for forest land in northwest Ontario. Ontario Ministry Nat. Res., Northwest Sci. and Technology Unit, NWST Tech. Report TR-105, Thunder Bay, ON. 121 pp.

Carter, G. S. 1934. Reports of the Cambridge expedition to British Guiana, 1933. Illumination in the rain forest at the ground level. *J. Linn. Soc., London. Zool.* 38:579–589.

Castello, J. D., D. J. Leopold, and P. J. Samllidge. 1995. Pathogens, patterns and processes in forest ecosystems. *BioScience* 45:16–24.

Ceulemans, R. J., and B. Saugier. 1991. Photosynthesis. *In* A. S. Raghavendra (ed.), *Physiology of Trees*. John Wiley, New York.

Chabot, B. F., and D. J. Hicks. 1982. The ecology of leaf life spans. *Ann. Rev. Ecol. Syst.* 11:233–257.

Chabot, B. F., and H. A. Mooney (eds.) 1985. *Physiological Ecology of North American Plant Communities*. Chapman and Hall, New York. 351 pp.

Chambless, L. F., and E. S. Nixon. 1975. Woody vegetation—soil relations in a bottomland forest of west Texas. *Texas J. Sci.* 26:407–416.

Chaney, R. W. 1949. Redwoods—occidental and oriental. *Science* 110:551–552.

Chaney, R. W. 1954. A new pine (*Pinus clementsii*) from the Cretaceous of Minnesota and its palaeoecological significance. *Ecology* 35:145–151.

Chang, D. H. S. 1983. The Tibetal plateau in relation to the vegetation of China. *Ann. Missouri Bot. Gard.* 70:564–570.

Chapin, F. S., III, A. J. Bloom, and C. B. Field. 1987. Plant response to multiple environmental factors. *BioScience* 37:49–57.

Chapin, F. S., III, K. Van Cleve, and P. R. Tryon. 1986. Relationship of ion absorption to growth rate in taiga trees. *Oecologia* 69:238–242.

Chapin, F. S., III., L. R. Walker, C. L. Fastie, and L. C. Sharman. 1994. Mechanisms of primary succession following deglaciation at Glacier Bay, Alaska. *Ecol. Monogr.* 64:149–175.

Chapman, H. H. 1932. Is the longleaf pine a climax? *Ecology* 13:328–335.

Chazdon, R. L., and R. W. Pearcy. 1986a. Photosynthetic responses to light variation in rain forest species. I. Induction under constant and fluctuating light conditions. *Oecologia* 69:517–523.

Chazdon, R. L., and R. W. Pearcy. 1986b. Photosynthetic responses to light variation in rain forest species. II. Carbon gain and light utilization during sunflecks. *Oecologia* 69:524–531.

Chen, J., J. F. Franklin, and T. A. Spies. 1992. Vegetation responses to edge environments in old-growth Douglas-fir forests. *Ecol. Applications* 2:387–396.

Chen, J., J. F. Franklin, and T. A. Spies. 1995. Growing-season microclimatic gradients from clearcut edges into old-growth Douglas-fir forests. *Ecol. Applications* 5:74–86.

Christensen, K. M., and T. G. Whitham. 1991. Indirect herbivore mediation of avian seed dispersal in pinyon pine. *Ecology* 72:536–542.

Christensen, K. M., and T. G. Whitham. 1993. Impact of insect herbivores on competition between birds and mammals for pinyon pine seeds. *Ecology* 74:2270–2278.

Christensen, N. L. 1977. Changes in structure, pattern and diversity associated with climax forest maturation in Piedmont, North Carolina. *Am. Midl. Nat.* 97:176–188.

Christensen, N. L. 1988a. Vegetation of the southeastern Coastal Plain. *In* M. G. Barbour and W. D. Billings (eds.), *North American Terrestrial Vegetation*. Cambridge Univ. Press, New York.

Christensen, N. L. 1988b. Succession and natural disturbance: paradigms, problems, and preservation of natural ecosystems. *In* J. K. Agee and D. R. Johnson (eds.), *Ecosystem Management for Parks and Wilderness*. Univ. Washington Press, Seattle.

Christensen, N. L., A. M. Bartuska, J. H. Brown, S. Carpenter, C. D'Antonio, R. Francis, J. F. Franklin, J. A. MacMahon, R. R. Noss, D. J. Parsons, C. H. Peterson, M. G. Turner, and R. G. Woodmansee. 1996. The report of the Ecological Society of America committee on the scientific basis for ecosystem management. *Ecol. Applications* 6:665–691.

Christensen, N. L., and R. K. Peet. 1981. Secondary forest succession on the North Carolina piedmont. *In* D. C. West, H. H. Shugart, and D. B. Botkin (eds.), *Forest Succession: Concepts and Applications*. Springer-Verlag, New York.

Christensen, N. L., and R. K. Peet. 1984. Convergence during secondary forest succession. *J. Ecol.* 72:25–36.

Christy, H. R. 1952. Vertical temperature gradients in a beech forest in central Ohio. *Ohio J. Sci.* 52:199–209.

Chung, H. H., and R. L. Barnes. 1977. Photosynthate allocation in *Pinus taeda*. I. Substrate requirements for synthesis of shoot biomass. *Can. J. For. Res.* 7:106–111.

Cieslar, A. 1895. Über die Erblichkeit des Zuwachsvermögens bei den Waldbäumen. *Centralbl. gesam. Forstw.* 21:7–29.

Cieslar, A. 1899. Neues aus dem Gebiete der fostlichen Zuchtwahl. *Centbl. gesam. Forstw.* 25:99–117.

Clark, J. S. 1989. Ecological disturbance as a renewal process: theory and application to fire history. *Oikos* 56:17–30.

Clark, J. S. 1990. Fire and climate change during the last 750 years in northwestern Minnesota. *Ecol. Monogr.* 60:135–169.

Clausen, J., D. D. Keck. and W. M. Hiesey. 1940. Experimental studies on the nature of species. I. Effect of varied environments on western North American plants. Carnegie Inst. Washington , Publ. 520. 452 pp.

Clausen, J., D. D. Keck, and W. M. Hiesey. 1958a. Experimental studies on the nature of species. III. Environmental responses of climatic races of *Achillea*. Carnegie Inst. Washington, Publ. 581. 129 pp.

Clausen, J., D. D. Keck, and W. M. Hiesey. 1958b. Experimental studies on the nature of species. IV. Genetic structure of ecological races. Carnegie Inst. Washington, Publ. 615. 312 pp.

Clements, F. E. 1905. *Research Methods in Ecology*. Univ. Publ. Co., Lincoln, NB. 334 pp. (reprinted 1977, Arno Press, New York).

Clements, F. E. 1910. The life history of lodgepole burn forests. *U.S. For. Serv. Bull.* 79. 56 pp.

Clements, F. E. 1916. Plant succession: an analysis of the development of vegetation. Carneg. Inst. Wash. Publ. 242. 512 pp.

Clements, F. E. 1936. Nature and structure of the climax. *J. Ecology* 24:252–284.

Clements, F. E. 1949. *Dynamics of Vegetation: Selections from the Writings of Frederic E. Clements, Ph. D.* The H. W. Wilson Co., New York. 296 pp.

Cline, A. C., and S. H. Spurr. 1942. The virgin upland forest of Central New England. *Harvard For. Bull.* 21. 51 pp.

Coe, J. J., and F. M. Isaac. 1965. Pollination of the baobab (*Adansonia digitata* L.) by the lesser bush baby (*Galego crassicaudatum* E. Geoffroy). *E. African Wildlife J.* 3:123–124.

Coile, T. S. 1937. Distribution of forest tree roots in North Carolina Piedmont soils. *J. For.* 35:247–257.

Coile, T. S. 1938. Forest classification: classification of forest types with special reference to ground vegetation. *J. For.* 36:1062–1066.

Coile, T. S. 1952. Soil and the growth of forests. *Adv. Agronomy* 4:330–398.

Cole, D. W., and D. W. Johnson. 1980. Mineral cycling in tropical forests. *In* C. T. Youngberg (ed.), *Forest Soils and Land Use*. Proc. 5th North American Forest Soils Conf. Colorado State Univ., Fort Collins.

Cole, D. W., and M. Rapp. 1981. Elemental cycling in forest ecosystems. *In* D. E. Reichle (ed.), *Dynamic Properties of Forest Ecosystems*, Int. Biol. Programme 23. Cambridge Univ. Press, Cambridge.

Cole, L. C. 1958. The ecosphere. *Sci. Amer.* 198:83–92.

Conkle, M. T. 1973. Growth data for 29 years from the California elevational transect study of ponderosa pine. *For. Sci.* 19:31–39.

Conkle, M. T., and W. B. Critchfield. 1988. Genetic variation and hybridization of ponderosa pine. *In* D. M. Baumgartner and J. E. Lotan (eds.), Symp. Proc. *Ponderosa Pine, The Species and Its Management*. Washington State Univ., Dept. Nat. Res. Sciences. Pullman.

Connell, J. H. 1978. Diversity in tropical rain forests and coral reefs. *Science* 199:1302–1310.

Connell, J. H., and R. O. Slatyer. 1977. Mechanisms of succession in natural communities and their roles in community stability and organization. *Am. Nat.* 111:1119–1144.

Cook, R. E. 1979. Asexual reproduction: a further consideration. *Am. Nat.* 113:769–772.

Cooper, C. F. 1960. Changes in vegetation, structure, and growth of southwestern pine forests since white settlement. *Ecol. Monogr.* 30:129–164.

Cooper, C. F. 1961. The ecology of fire. *Sci. American* 204(4):150–160.

Cooper, S. V., K. E. Neiman, and D. W. Roberts. 1991. Forest habitat types of northern Idaho: a second approximation. USDA For. Serv. Gen. Tech. Report INT-236. Int. Res. Sta., Ogden, UT. 143 pp.

Cooper, W. S. 1913. The climax forest of Isle Royale, Lake Superior, and it development. *Bot. Gazette.* 55:1–44. 189–235.

Cooper, W. S. 1923. The recent ecological history of Glacier Bay, Alaska. *Ecology* 4:93–128, 223–246, 355–365.

Cooper, W. S. 1931. A third expedition to Glacier Bay, Alaska. *Ecology* 12:61–95.

Cooper, W. S. 1939. A fourth expedition to Glacier Bay, Alaska. *Ecology* 20:130–155.

Cornet, B. 1993. Dicot-like leaf and flowers from the late Triassic Tropical Newark Supergroup Rift Zone, U.S.A. *Modern Geology* 19:81–99.

Cottam, G., and R. P. McIntosh. 1966. Vegetational continuum. *Science* 152:546–547.

Cottam, W. P. 1947. Is Utah Sahara bound? *Bull. Univ. Utah* 37:1–40.

Coutinho, L. M. 1990. Fire in the ecology of the Brazilian Cerrado. *In* J. G. Goldammer (ed.), *Fire in the Tropical Biota* (Ecological Studies 84). Springer-Verlag, New York.

Coutts, M. P., and J. Grace (eds.). 1995. *Wind and Trees*. Cambridge Univ. Press, Cambridge. 485 pp.

Coutts, M. P., and B. C. Nicoll. 1991. Development of the surface roots of trees. *In* C. Edelin (ed.), *L'Arbre, Biologie et Development*. Naturalia Monspeliensia, Montpellier.

Covell, R. R., and D. C. McClurkin. 1967. Site index of loblolly pine on Ruston soils in the southern Coastal Plain. *J. For.* 65:263–264.

Cowles, H. C. 1899. The ecological relations of the vegetation on the sand dunes of Lake Michigan. *Bot. Gaz.* 27:95–117, 167–202, 281–308, 361–391.

Cowles, H. C. 1901. The physiographic ecology of Chicago and vicinity; a study of the origin, development, and classification of plant societies. *Bot. Gaz.* 31:73–108, 145–182.

Cowles, H. C. 1909. Ecology of plants. *Bot. Gaz.* 48:149–152.

Cowles, H. C. 1911. The causes of vegetative cycles. *Bot. Gaz.* 51:161–183.

Cox, C. B., I. N. Healey, and P. D. Moore. 1976. *Biogeography: An Ecological and Evolutionary Approach*. 2nd ed. Blackwell, Oxford. 194 pp.

Cragg, J. B. 1953. Book review of natural communities by L. R. Dice. *Bull. Inst. Biol.* I:3.

Crane, P. R. 1993. Time for angiosperms. *Nature* 366:631–632.

Crankshaw, W. B., S. A. Qadir, and A. A. Lindsey. 1965. Edaphic controls of tree species in presettlement Indiana. *Ecology* 46:688–698.

Crawford, D. J. 1974. A morphological and chemical study of *Populus acuminata* Rydberg. *Brittonia* 26:74–89.

Crawford, H. S., R. G. Hooper, and R. F. Harlow. 1976. Woody plants selected by beavers in the Appalachian Ridge and Valley Province. USDA For. Serv. Res. Paper NE-346. Northeastern For. Exp. Sta., Upper Darby, PA. 6 pp.

Critchfield, W. B. 1957. Geographic variation in *Pinus contorta*. Maria Moors Cabot Found. Publ. No. 3. Harvard Univ., Cambridge, MA. 118 pp.

Critchfield, W. B. 1985. The late Quaternary history of lodgepole and jack pines. *Can. J. For. Res.* 15:749–772.

Critchfield, W. B., and E. L. Little. 1966. Geographic distribution of the pines of the world. USDA For. Serv. Misc. Publ. 991. Washington, D.C. 97 pp.

Crouch, G. L. 1969. Deer and reforestation in the Pacific Northwest. *In* H. C. Black (ed.), *Proc. Wildlife and Reforestation in the Pacific Northwest,* School For., Oregon State Univ., Portland.

Crouch, G. L. 1976. Wild animal damage to forests in the United States and Canada. *In Proc. XVI IUFRO World Congress, Div. II.* Oslo, Norway. pp. 468–478.

Crow, T. R. 1991. Landscape ecology: the big picture approach to resource management. *In* D. J. Decker, M. E. Krasny, G. R. Goff, C. R. Smith, and D. W. Gross (eds.), *Challenges in the Conservation of Biological Resources, A Practitioner's Guide*. Westview Press, Boulder, CO.

Crow, T. R., and E. J. Gustafson. 1997. Ecosystem management: managing natural resources in time and space. *In* K. A. Kohm and J. F. Franklin (eds.), *Creating a Forestry for the 21st Century*. Island Press, Washington, DC.

Cuddihy, L. W., and C. P. Stone. 1990. *Alteration of Native Hawaiian Vegetation: Effects of Humans, Their Activities and Introductions*. Univ. Hawaii Press, Honolulu. 138 pp.

Cuevas, E., and E. Medina. 1988. Nutrient dynamics within Amazonian forest ecosystems. I. Nutrient flux in fine litter fall and efficiency of nutrient utilization. *Oecologia* 68:466–472.

Currie, D. J., and V. Paquin. 1987. Large-scale biogeographical patterns of species richness of trees. *Nature* 329:326–327.

Curtis, J. D., and N. R. Lersten. 1974. Morphology, seasonal variation and function of resin glands on buds and leaves of *Populus deltoides* (Salicaceae). *Amer. J. Bot.* 61:835–845.

Curtis, J. T. 1956. The modification of mid-latitude grasslands and forests by man. *In* W. L. Thomas, Jr. (ed.), *Man's Role in Changing the Face of the Earth.* Univ. Chicago Press, Chicago.

Curtis, J. T. 1959. *The Vegetation of Wisconsin.* Univ. Wisconsin Press, Madison. 657 pp.

Curtis, J. T., and R. P. McIntosh. 1951. An upland forest continuum in the prairie-forest border region of Wisconsin. *Ecology* 32:476–496.

Curtis, R. O. 1972. Yield tables past and present. *J. For.* 70:28–32.

Curtis, R. O. 1997. The role of extended rotations. *In* K. A. Kohm and J. F. Franklin (eds.), *Creating a Forestry for the 21st Century.* Island Press, Washington, D.C.

Curtis, R. O., D. J. DeMars, and F. R. Herman. 1974a. Which dependent variable in site index—height-age regressions. *For. Sci.* 20:74–87.

Curtis, R. O., F. R. Herman, and D. J. DeMars. 1974b. Height growth and site index for Douglas-fir in high-elevation forests of the Oregon-Washington Cascades. *For. Sci.* 20:307–315.

Cwynar, L. C., and G. M. MacDonald. 1987. Geographical variation of lodgdepole pine in relation to population history. *Am. Nat.* 129:463–469.

Cypert, E. 1973. Plant succession on burned areas in Okefenokee Swamp following the fires of 1954 and 1955. *In Proc. Annual Tall Timbers Fire Ecology Conf.* 12:199–217. Tall Timbers Res. Sta., Tallahassee, FL.

Dahlgren, R. A. 1994. Soil acidification and nitrogen saturation from weathering of ammonium-bearing rock. *Nature* 368:838–840.

Damman, A. W. H. 1964. Some forest types of central Newfoundland and their relation to environmental factors. *For. Sci. Monogr.* 8. 62 pp.

Damman, A. W. H. 1975. Permanent changes in the chronosequence of a boreal forest habitat induced by natural disturbances. *In* W. Schmidt (ed.), *Int. Proc. Symp. Sukzessionsforschung. Int. Verein. Vegetationsk.* J. Cramer, Vaduz.

Daniel, T. W., J. A. Helms, and F. S. Baker. 1979. *Principles of Silviculture.* McGraw-Hill, New York. 500 pp.

Dansereau, P. (ed.). 1968. The continuum concept of vegetation: responses. *Bot. Rev.* 34:253–332.

Darley-Hill, S., and W. C. Johnson. 1981. Dispersal of acorns by blue jays (*Cyanocita cristata*). *Oecologia* 50:231–232.

Dasmann, R. F. 1984. *Environmental Conservation.* 5th ed. John Wiley, New York. 486 pp.

Daubenmire, R. F. 1936. The "Big Woods" of Minnesota: its structure, and relation to climate, fire, and soils, *Ecol. Monogr.* 6:223–268.

Daubenmire, R. F. 1952. Forest vegetation of northern Idaho and adjacent Washington, and its bearing on concepts of vegetation classification. *Ecol. Monogr.* 22:301–330.

Daubenmire, R. F. 1954. Alpine timberlines in the Americas and their interpretation. *Butler Univ. Bot. Stud.* 11:119–136.

Daubenmire, R. F. 1956. Climate as a determinant of vegetation distribution in eastern Washington and northern Idaho. *Ecol. Monogr.* 26:131–154.

Daubenmire, R. F. 1960. A seven year study of cone production as related to xylem layers in *Pinus ponderosa. Am. Midl. Nat.* 64:187–193.

Daubenmire, R. F. 1966. Vegetation: identification of typal communities. *Science* 151:291–298.

Daubenmire, R. F. 1968. *Plant Communities.* Harper & Row. New York. 300 pp.

Daubenmire, R. F. 1970. Steppe vegetation of Washington. Tech. Bull 62. Wash. State Agr. Exp. Sta., Pullman, WA. 131 pp.

Daubenmire, R. F. 1976. The use of vegetation in assessing the productivity of forest lands. *Bot. Rev.* 42:115–143.

Daubenmire, R. F. 1978. *Plant Geography.* Academic Press, New York. 338 pp.

Daubenmire, R. F. 1989. The roots of a concept. *In* D. E. Ferguson, P. Morgan, F. D. Johnson (comp.), *Proceedings—Land Classifications Based on Vegetation: Applications for Resource Management.* USDA For. Serv. Gen. Tech. Report INT-257, Int. Res. Sta., Ogden, UT.

Daubenmire, R. F., and J. B. Daubenmire. 1968. Forest vegetation of eastern Washington and northern Idaho. Washington Agric. Exp. Sta., Tech. Bull. 60. 104 pp.

Davidson, E. A., and W. T. Swank. 1987. Factors limiting denitrification in soils from mature and disturbed southeastern hardwood forests. *For. Sci.* 33:135–144.

Davies, W. J., and T. T. Kozlowski. 1974. Stomatal responses of five woody angiosperms to light intensity and humidity. *Can. J. Bot.* 52:1525–1534.

Davis, M. B. 1981. Quaternary history and the stability of forest communities. *In* E. C. West, H. H. Shugart, and D. B. Botkin (eds.), *Forest Succession: Concepts and Application.* Springer-Verlag, New York.

Davis, M. B. 1983a. Holocene vegetational history of the eastern United States. *In* H. E. Wright and S. Porter (eds.), *The Late Quaternary Environments of the United States. Vol. 2: The Holocene.* Univ. Minnesota Press, Minneapolis.

Davis, M. B. 1983b. Quaternary history of deciduous forests of eastern North America and Europe. *Ann. Missouri Bot. Gard.* 70:550–563.

Davis, M. B. 1986. Climatic instability, time lags, and community disequilibrium. *In* J. Diamond and T. J. Case (eds.), *Community Ecology.* Harper and Row, New York.

Davis, M. B. 1987. Invasions of forest communities during the Holocene: beech and hemlock in the Great Lakes region. *In* A. J. Gray, M. L. Crawley, and P. J. Edwards (eds.), *Colonization, Succession, and Stability.* Blackwell Sci. Publ., London.

Davis, M. B., K. D. Woods, S. L. Webb, and R. P. Futyma. 1986. Dispersal versus climate: expansion of *Fagus* and *Tsuga* into the Upper Great Lakes region. *Vegetatio* 67:93–103.

Davis, P. H., and V. H. Heywood. 1963. *Principles of Angiosperm Taxonomy.* D. Van Nostrand, New York. 558 pp.

Davis, R. B., and G. L. Jacobson, Jr. 1985. Late glacial and early Holocene landscapes in northern New England and adjacent areas of Canada. *Quat. Res.* 23:341–368.

Dawson, A. G. 1992. *Ice Age Earth, Late Quaternary Geology and Climate.* Routledge, New York. 293 pp.

Day, F. P., Jr., and C. D. Monk. 1974. Vegetation patterns on a southern Appalachian watershed. *Ecology* 55:1064–1074.

Day, G. M. 1953. The Indian as an ecological factor in the northeastern forest. *Ecology* 34:329–346.

De Steven, D. 1989. Genet and ramet demography of *Oenocarpus mapora* ssp. *mapora,* a clonal palm of Panamanian tropical moist forest. *J. Ecology* 7:579–596.

DeBano, L. F. 1969. The relationship between heat treatment and water repellency in soils. *In* L. F. DeBano and J. Letey (eds.), *Water-Repellent Soils.* Univ. Calif., Riverside.

DeBano, L. F., J. F. Osborn, J. S. Krammes, and J. Letey, Jr. 1967. Soil wettability and wetting agents ... our current knowledge of the problem. USDA For. Serv. Res. Paper PSW-43. Pacific Southwest For. and Rge. Exp. Sta., Berkeley, CA 13 pp.

DeBell, D. S., R. O. Curtis, C. A. Harrington, and J. C. Tappeiner. 1997. Shaping stand development through silvicultural practices. *In* K. A. Kohm and J. F. Franklin (eds.), *Creating a Forestry for the 21st Century.* Island Press, Washington, D.C.

DeByle, N. V. 1976. Soil fertility as affected by broadcast burning following clearcutting in Northern Rocky Mountain larch/fir forests. *In Proc. Annual Tall Timbers Fire Ecology Conf.* 14:447–464. Tall Timbers Res. Sta., Tallahassee, FL.

Decker, J. P. 1959. A system for analysis of forest succession. *For. Sci.* 5:154–157.

Del Vecchio, T. A., C. A. Gehring, N. S. Coob, and T. G. Whitham. 1993. Negative effects of scale insect herbivory on the ectomoycorrhizae of juvenile pinyon pine. *Ecology* 74:2297–2302.

Delcourt, H. R. 1978. Roland McMillan Harper, recorder of early Twentieth-Century landscapes in the South. *Pioneer America* 10:37–50.

Delcourt, H. R. 1980. Late quaternary vegetation history of the eastern Highland Rim and adjacent Cumberland Plateau of Tennessee. *Ecol. Monogr.* 49:225–280.

Delcourt, H. R., and P. A. Delcourt. 1974. Primeval magnolia—holly—beech climax in Louisiana. *Ecology* 55:638–644.

Delcourt, H. R., and P. A. Delcourt. 1988. Quaternary landscape ecology: relevant scales in space and time. *Landscape Ecology* 2:23–44.

Delcourt, H. R., and P. A. Delcourt. 1991. *Quaternary Ecology, A Paleoecological Perspective.* Chapman and Hall, New York. 242 pp.

Delcourt, H. R., P. A. Delcourt, and T. Webb, III. 1983. Dynamic plant ecology: the spectrum of vegetational change in space and time. *Quat. Sci. Rev.* 1:153–175.

Delcourt, P. A., and H. R. Delcourt. 1977. The Tunica Hills, Louisiana—Mississippi: late-glacial locality for spruce and deciduous forest species. *Quat. Res.* 7:218–237.

Delcourt, P. A., and H. R. Delcourt. 1979. Late Pleistocene and Holocene distributional history of the deciduous forest in the southeastern United States. *Veröff. Geobot. Inst. ETH, Stift. Rübel* 68:79–107.

Delcourt, P. A., and H. R. Delcourt. 1981. Vegetation maps for eastern North America: 40,000 yr B. P. to the present. *In* R. Romans (ed.), *Proc. 1980 Geobotany Conference,* Plenum, New York.

Delcourt, P. A., and H. R. Delcourt. 1987. *Long-Term Forest Dynamics of the Temperate Zone.* Springer-Verlag, New York. 439 pp.

Delcourt, P. A., and H. R. Delcourt. 1993. Paleoclimates, paleovegetation, and paleofloras during the late Quaternary. *In* N. Moran (conv. ed.), *Flora of North America North of Mexico,* Vol. 1, Introduction. Oxford Univ. Press, New York.

Delcourt, P. A., H. R. Delcourt, and T. Webb, III. 1984. Atlas of mapped distributions of dominance and modern pollen percentages for important tree taxa of eastern North America. *Am. Assoc. Strat. Palynol. Contrib. Ser.* 14:1–131.

Delevoryas, T. 1980. Polyploidy in Gymnosperms. *In* W. H. Lewis (ed.), *Polyploidy.* Plenum Press, New York.

Deloria, C. M., and M. E. DeCapita. 1997. Control of brown-headed cowbirds on Kirtland's warbler nesting areas of northern Michigan. USDI, U.S. Fish and Wildlife Serv., Report, Jan. 10, 1997. East Lansing, MI. 11 pp.

Demarchi, D. A., R. D. Marsh, A. P. Harcombe, and E. C. Lea. 1990. The environment (of British Columbia). *In* R. W. Campbell, N. K. Dawe, I. McTaggart-Cowan, J. M. Cooper, G. W. Kaiser, and M. C. E. McNall (eds.), *The Birds of British Columbia, Vol. I: Introduction and Loons Through Waterfowl.* Royal British Columbia Museum, Victoria, B.C. and Canadian Wildlife Service, Environment Canada, Delta B.C.

Denevan, W. M. 1992. The pristine myth: the landscape of the Americas in 1492. *Annals Assoc. Am. Geog.* 82:369–385.

Denton, S. R. 1985. *Ecological Climatic Regions and Tree Distributions in Michigan.* Ph.D. dissertation, University of Michigan, Ann Arbor. 321 pp.

Denton, S. R., and B. V. Barnes. 1987. Tree species distributions related to climatic patterns in Michigan. *Can. J. For. Res.* 17:613–629.

Denton, S. R., and B. V. Barnes. 1988. An ecological climatic classification of Michigan: a quantitative approach. *For. Sci.* 34:119–138.

Dice, L. R. 1943. *The Biotic Provinces of North America.* Univ. Michigan Press, Ann Arbor. 78 pp.

Dickerson, B. P. 1976. Soil compaction after tree-length skidding in northern Mississippi. *Proc. Soil Sci. Soc. Amer.* 40:965–966.

Dieterich, H. 1970. Die Bedeutung der Vegetationskunde für die forstliche Standortskunde. *Der Biologieunterricht.* 6:48–60.

Dieterich, H., S. Müller, and G. Schlenker. 1970. *Urwald von morgen.* Verlag Eugen Ulmer, Stuttgart, Germany. 174 pp.

Dieterich, J. H., and T. W. Swetnam. 1984. Dendrochronology of a fire-scarred ponderosa pine. *For. Sci.* 30:238–247.

Dimock, E. J., II. 1970. Ten-year height growth of Douglas-fir damaged by hare and deer. *J. For.* 68:285–288.

Dobbs, R. C. 1976. White spruce seed dispersal in central British Columbia. *For. Chron.* 52:225–228.

Dobson, M. C., F. T. Ulaby, L. E. Pierce, T. L. Sharik, K. M. Bergen, J. Kellndorfer, J. R. Kendra, E. Li, Y. C. Lin, A. Nashashibi, K. Sarabandi, and P. Siqueira. 1995. Estimation of forest biophysical characteristics in northern Michigan with SIR-C/X-SAR. *IEEE Trans. Geosci. Remote Sensing* 33:877–895.

Dobzhansky, T. 1951. *Genetics and the Origin of Species.* 3rd ed. Columbia Univ. Press, New York. 364 pp.

Dobzhansky, T. 1968. Adaptedness and fitness. *In* R. C. Lewontin (ed.), *Population Biology and Evolution.* Syracuse Univ. Press, Syracuse, New York.

Dodd, J. C., C. C. Burton, R. G. Burns, and P. Jefferies. 1987. Phosphatase activity associated with the roots and rhizosphere of plants infected with vesicular-arbuscular mycorrhizal fungi. *New Phytol.* 107:163–172.

Dodd, J. L., 1994. Desertification and degradation in Sub-Saharan Africa. *BioScience* 44:28–34.

Doolittle, W. T. 1957. Site index of scarlet and black oak in relation to southern Appalachian soil and topography. *For. Sci.* 3:114–124.

Douglass, A. E. 1919. Climatic cycles and tree growth; a study of the annual rings of trees in relation to climate and solar activity. Carnegie Inst. Wash. Publ. 289. Vol. 1, 127 pp; Vol. 2 (1928), 166 pp; Vol. 3 (1936), 171 pp.

Downs, A. A. 1938. Glaze damage in the birch-beech-maple-hemlock type of Pennsylvania and New York. *J. For.* 36: 63–70.

Downs, A. A., and W. E. McQuilken. 1944. Seed production of southern Appalachian oaks. *J. For.* 53:439–441.

Downs, R. J. 1962. Photocontrol of growth and dormancy in woody plants. *In* T. T. Kozlowski (ed.), *Tree Growth.* Ronald Press, New York.

Doyle, J. A., and L. J. Hickey. 1976. Pollen and leaves from the Mid-Cretaceous Potomac Group and their bearing on early angiosperm evolution. *In* C. B. Beck (ed.), *Origin and Early Evolution of Angiosperms.* Columbia Univ. Press, New York.

Drew, A. P., and E. J. Alyanak. 1981. Summer drought effects on the growth of mature balsam fir trees. *Mich. Botanist* 20:105–110.

Drew, T. J., and J. W. Flewelling. 1977. Some recent Japanese theories of yield-density relationships and their application to Monterey pine plantations. *For. Sci.* 23:517–534.

Drew, T. J., and J. W. Flewelling. 1979. Stand density management: an alternative approach and its application to Douglas-fir plantations. *For. Sci.* 25:518–532.

Drury, W. H. 1969. Discussion on concepts of ecosystem and landscapes (an excerpt). *In* K. N. Greenidge (ed.), *Essays in Plant Geography and Ecology.* Nova Scotia Museum, Halifax, Nova Scotia, Canada.

Drury, W. H., and I. C. T. Nisbet. 1973. Succession. *J. Arnold Arboretum* 54:331–368.

Duffield, J. W., and E. B. Snyder. 1958. Benefits from hybridizing American forest trees. *J. For.* 56:809–815.

Duffy, D. C., and A. J. Meier. 1992. Do Appalachian herbaceous understories ever recover from clearcutting? *Conserv. Biol.* 6:196–201.

DuMerle, P. 1988. Phenological resistance of oaks to the green oak leafroller, *Tortrix viridana* (Lepidoptera: Tortricidae). *In* W. J. Mattson, J. Levieux, and C. Bernard-Dagan (eds.), *Mechanisms of Woody Plant Defenses Against Insects: Search for Pattern.* Springer-Verlag, New York.

Dunn, C. P., D. M. Sharpe, G. R. Guntenspergen, F. Stearns, and Z. Yang. 1991. Methods for analyzing temporal changes in landscape pattern. *In* M. G. Turner and R. H. Gardner (eds.), *Quantitative Methods in Landscape Ecology.* Springer-Verlag, New York.

Duvigneaud, P. 1946. La variabilité des associations végétales. *Bull. Soc. R. Bot. Belg.* 78:107–134.

Dwyer, L. M., and G. Merriam. 1981. Influence of topographic heterogeneity on deciduous litter decomposition. *Oikos* 37:228–237.

Dyksterhuis, E. J. 1949. Condition and management of range land based on quantitative ecology. *J. Range Manage.* 2:104–115.

Dyksterhuis, E. J. 1958. Ecological principles in range evaluation. *Bot. Rev.* 24:253–272.

Dyrness, C. T. 1967. Mass soil movements in the H. J. Andrews Experimental Forest. USDA For. Serv. Res. Paper PNW-42. Pacific Northwest Forest Exp. Sta., Portland, OR. 12 pp.

Dyrness, C. T. 1976. Effect of wildfire on soil wettability in the high Cascades of Oregon. USDA For. Serv. Res. Paper PNW-202. Pacific Northwest For. and Rge. Exp. Sta., Portland, OR. 18 pp.

Easley, A. T., J. F. Passineau, and B. L. Driver (comps.). 1990. The use of wilderness for personal growth, therapy, and education. USDA. For. Serv., Gen. Tech. Rep. RM-193. Rocky Mountain Res. Sta., Ft. Collins, CO. 197 pp.

Eckenwalder, J. E. 1977. North American cottonwoods (*Populus*, Salicaceae) of sections *Abaso* and *Aigeiros*. *J. Arnold. Arbor.* 58:193–208.

Eckenwalder, J. E. 1996. Systematics and evolution of *Populus*. *In* R. F. Stettler, H. D. Bradshaw, Jr., P. E. Heilman, and T. M. Hinckley (eds.), *Biology of Populus and Its Implications for Management and Conservation*. NRC Res. Press. Ottawa. 539 pp.

Ecoregions Working Group. 1989. Ecoclimatic regions of Canada, first approximation. Ecoregions Working Group of the Canada Committee on Ecological Land Classification. Ecol. Land Classif. Series, No. 23, Can. Wildlife Serv., Env. Canada, Ottawa, Ontario. 119 pp. and map.

Edmunds, G. F., Jr., and D. N. Alstad. 1978. Coevolution in insect herbivores and conifers. *Science* 199:941–945.

Edwards, C. A. 1974. Macroarthropods. *In* C. H. Dickinson and G. J. F. Pugh (eds.), *Biology of Plant Litter Decomposition*. Academic Press, New York.

Edwards, C. A., and P. J. Bohlen. 1996. *Biology and Ecology of Earthworms*. Chapman and Hall, London. 426 pp.

Egler, F. E. 1954. Vegetation science concepts I. Initial floristic composition, a factor in old-field vegetation development. *Vegetatio* 4:412–417.

Egler, F. E. 1968. The contumacious continuum. *In* P. Dansereau (ed.), The continuum concept of vegetation: responses. *Bot. Rev.* 34:253–332.

Egler, F. E. 1977. *The Nature of Vegetation, Its Management and Mismanagement*. Conn. Cons. Assoc., Bridgewater, CO. 527 pp.

Ehleringer, J. R., and K. S. Werk. 1986. Modifications of solar-radiation absorption patterns and implications for carbon gain at the leaf level. *In* T. J. Givnish (ed.), *On the Economy of Plant Form and Function*. Cambridge Univ. Press, Cambridge.

Eis, S. 1976. Association of western white pine cone crops with weather variables, *Can. J. For. Res.* 6:6–12.

Eis, S., E. H. Garman, and L. F. Ebell. 1965. Relation between cone production and diameter increment of Douglas-fir (*Pseudotsuga menziesii* (Mirb.) Franco), grand fir (*Abies grandis* (Dougl.) Lindl.), and western white pine (*Pinus monticola* Dougl.). *Can. J. Bot.* 43:1553–1559.

Ekberg, I., G. Eriksson, and I. Gormling. 1979. Photoperiodic reactions in conifer species. *Holarctic Ecology* 2:255–263.

Elkins, J. W., and R. Rosen. 1989. Summary Report 1988: Geophysical monitoring for climate change. NOAA, ERL, Boulder, CO. 142 pp.

Ellenberg, H. 1963. *Vegetation Mitteleuropas mit den Alpen*. Eugen Ulmer, Stuttgart. 943 pp.

Ellenberg, H. 1968. Wege der Geobotanik zum Verständnis der Pflanzendecke. *Naturwissenschaften* 55:462–470.

Ellenberg, H. 1974. Zeigerwerte der Gefässpflanzen Mitteleuropas. *Scripta Geobotanica* 9:1–97.

Ellenberg, H. 1988. *Vegetation Ecology of Central Europe,* 4th ed. Cambridge Univ. Press, Cambridge. 731 pp.

Elliott-Fisk, D. L. 1988. The Boreal Forest. *In* M. G. Barbour and W. D. Billings(eds.), *North American Terrestrial Vegetation*. Cambridge Univ. Press, New York.

Ellis, J. A., W. R. Edwards, and K. P. Thomas. 1969. Responses of bobwhites to management in Illinois. *J. Wildl. Manage.* 33:749–762.

Ellis, M., C. D. von Dohlen, J. E. Anderson, and W. H. Romme. 1994. Some important factors affecting density of lodgepole pine seedlings following the 1988 Yellowstone fires. *In* E. G. Despain (ed.), *Conf. Proc. Plants and Their Environments*. Tech. Report NPS/NRYELL/NRTR-93/XX. USDI, Nat. Park Service.

Ellstrand, N. C. 1992. Gene flow among seed plant populations. *In* W. T. Adams, S. H. Strauss, D. L. Copes, and A. R. Griffin (eds.), *Population Genetics of Forest Trees*. Kluwer, Boston, MA.

Engler, A. 1905. Einfluss der Provenienz des Samens auf die Eigenschaften der forstlichen Holzgewächse. *Mitt. schweiz. Centralanst. forstl. Versuchsw.* 8:81–236.

Engler, A. 1908. Tatsachen, Hypothsen und Irrtümer auf dem Gebiete der Samenprovenienz-Frage. *Forstwiss. Centralbl.* 30:295–314.

Erdmann, G. G., and R. R. Oberg. 1974. Sapsucker feeding damages crown-released yellow birch trees. *J. For.* 72:760–763.

Esau. K. 1977. *Anatomy of Seed Plants.* John Wiley, New York. 550 pp.

Eschner, A. R., and J. H. Patric. 1982. Debris avalanches in eastern upland forests. *J. For.* 80:343–347.

Eshelman, K. R., R. E. Wagner, and F. M. Secrist. 1989. Vegetation classification—problems, principles, and proposals. *In* D. E. Ferguson, P. Morgan, and F. D. Johnson (comps.), *Symp. Proc. Land Classifications Based on Vegetation: Applications for Resource Management.* USDA For. Serv. Gen. Tech. Report INT-257. Intermountain Res. Sta., Ogden, UT. 315 pp.

Estrada, A., and T. H. Fleming (eds.). 1986. *Frugivores and Seed Dispersal.* Kluwer, Dordrecht. 392 pp.

Ewel, K. C. 1990. Swamps. *In* R. L. Myers and J. J. Ewel (eds.), *Ecosystems of Florida.* Univ. Central Florida Press, Orlando.

Eyde, R. H. 1985. The case for monkey-mediated evolution in big-bracted dogwoods. *Arnoldia* 45:3–9.

Eyre, F. H. 1980. *Forest Cover Types of the United States and Canada.* Soc. Am. For. Bethesda, MD. 148 pp.

Faegri, K., and L. Van der Pijl. 1979. *The Principles of Pollination Ecology.* Pergamon Press, New York. 244 pp.

Fahn, A., and E. Werker. 1972. Anatomical mechanisms of seed dispersal. *In* T. T. Kozlowski (ed.), *Seed Biology, Vol. 1.* Academic Press, New York.

Fairbridge, R. W. (ed.) 1972. *The Encyclopedia of Geochemistry and Environmental Sciences.* Van Nostrand Reinhold, New York. 1321 pp.

FAO, 1980. *Poplars and Willows in Wood Production and Land Use.* Food and Agr. Organization of the United Nations. Rome. 328 pp.

Fastie, C. 1995. Causes and ecosystem consequences of multiple pathways of primary succession at Glacier Bay, Alaska. *Ecology* 76:1899–1916.

Federer, C. A., and C. B. Tanner. 1966. Spectral distribution of light in the forest. *Ecology* 47:555–560.

Fenneman, N. M. 1931. *Physiography of Western United States.* McGraw-Hill, New York. 534 pp.

Fenneman, N. M. 1938. *Physiography of Eastern United States.* McGraw-Hill, New York. 691 pp.

Ferguson, D. E., P. Morgan, and F. D. Johnson (comps.) 1989. *Proceedings—Land Classifications Based on Vegetation: Applications for Resource Management.* USDA For. Serv. Gen. Tech. Report INT-257. Intermountain Res. Sta., Ogden, UT. 315 pp.

Field, C., and H. A. Mooney. 1986. The photosynthesis-nitrogen relationship in wild plants. *In* T. J. Givnish (ed.), *On the Economy of Plant Form and Function.* Cambridge Univ. Press, Cambridge.

Findlay, B. F. 1976. Recent developments in eco-climatic classifications. *In* J. Thie and G. Ironside (eds.), *Ecological (Biophysical) Land Classification in Canada.* Lands Directorate, Environment Canada, Ottawa.

Finegan, B. 1984. Forest succession. *Nature* 312:109–114.

Finlay, R. D., H. Ek, G. Odham, and B. Söderström. 1989. Uptake, translocation and assimilation of nitrogen from ^{15}N-labeled ammonium and nitrate sources by intact ectomycorrhizal systems of *Fagus sylvatica* and *Paxillus involutus. New Phytol.* 113:47–55.

Finney, M. A., and R. E. Martin. 1989. Fire history in a *Sequoia sempervirens* forest at Salt Point State Park, California. *Can. J. For. Res.* 19:1451–1457.

Finnigan, J. J., and Y. Brunet. 1995. Turbulent airflow in forests on flat and hilly terrain. *In* M. P. Coutts and J. Grace (eds.), *Wind and Trees.* Cambridge Univ. Press, Cambridge.

Firestone, M. K., and E. A. Davidson. 1989. Microbial basis of NO and N_2O production and consumption in soil. *In* M. O. Andreae and D. S. Schimel (eds.), *Exchange of Trace Gases between Terrestrial Ecosystems and the Atmosphere.* John Wiley, New York.

Fischer, W. C., and A. F. Bradley. 1987. Fire ecology of western Montana forest habitat types. USDA For. Serv. Gen. Tech. Report INT-223. Int. Res. Sta., Ogden, UT. 95 pp.

Fisher, F. M., L. W. Parker, J. P. Anderson, and W. G. Whitford. 1987. Nitrogen mineralizaiton in a desert soil: interacting effects of soil moisture and nitrogen fertilizer. *Soil Sci. Soc. Amer. J.* 1033–1041.

Fisher, H. M., and E. L. Stone. 1990a. Air-conducting porosity in slash pine roots from saturated soils. *For. Sci.* 36:18–33.

Fisher, H. M., and E. L. Stone. 1990b. Active potassium uptake by slash pine roots from O_2-depleted solutions. *For. Sci.* 36:582–598.

Fisher, H. M., and E. L. Stone. 1991. Iron oxidation at the surfaces of slash pine roots from saturated soils. *Soil Sci. Soc. Amer. J.* 55:1123–1129.

Fisher, W. C. 1989. The fire effects information system: a comprehensive vegetation knowledge base. *In* D. E. Ferguson, P. Morgan, and F. D. Johnson (comps.), *Proceedings—Land Classifications Based on Vegetation: Applications for Resource Management.* USDA For. Serv. Gen. Tech. Report INT-257. Intermountain Res. Sta., Ogden, UT.

Fitter, A. H. 1987. An architectural approach to the comparative ecology of plant root systems. *New Phytol.* 106(Supp.): 61–77.

Fitter, A. H., and R. K. M. Hay. 1987. *Environmental Physiology of Plants.* 2nd ed. Academic Press, New York. 423 pp.

Fitter, A. H., and T. R. Stickland. 1991. Architectural analysis of plant root systems 2. Influence of nutrient supply on architecture in contrasting plant species. *New Phytol.* 118:383–389.

Flanagan, P. W., and K. Van Cleve. 1983. Nutrient cycling in relation to decomposition and organic matter quality in taiga ecosystems. *Can. J. For. Res.* 13:795–817.

Fleming, T. H., and A. Estrada (eds.). 1993. *Frugivory and Seed Dispersal: Ecological and Evolutionary Aspects.* Kluwer, Dordrecht. 392 pp.

Flint, H. L. 1972. Cold hardiness of twigs of *Quercus rubra* L. as a function of geographic origin. *Ecology* 53:1163–1170.

Flint, H. R. 1930. Fire as a factor in the management of north Idaho national forests. *Northwest Sci.* 4:12–15.

Floate, K. D., M. J. C. Kearsley, and T. G. Whitham. 1993. Elevated herbivory in plant hybrid zones: *Chrysomela confluens*, *Populus* and phenological sinks. *Ecology* 74:2056–2065.

Foggin, G. T. III, and L. F. DeBano. 1971. Some geographic implications of water-repellent soils. *Prof. Geographer* 23:347–350.

Fonseca, C. R. 1994. Herbivory and the long-lived leaves of an Amazonian ant-tree. *J. Ecology.* 82:833–842.

Ford, E. D., and P. G. Jarvis (eds.). 1981. *Plants and Their Atmospheric Environment.* Blackwell. Oxford, England. 419 pp.

Ford-Robertson, F. C. (ed.). 1983. *Terminology of Forest Science Technology Practice and Products,* 2nd Printing. Soc. Am. Foresters, Washington, D.C. 370 pp.

Forman, R. T. T. 1995. *Land Mosaics, The Ecology of Landscapes and Regions.* Cambridge Univ. Press, Cambridge. 632 pp.

Forman, R. T. T., and M. Godron. 1986. *Landscape Ecology.* New York. John Wiley, New York. 619 pp.

Foster, D. R., 1988a. Disturbance history, community organization and vegetation dynamics of the old-growth Pisgah forest, south-western New Hampshire, U.S.A. *J. Ecology* 76:105–134.

Foster, D. R., 1988b. Species and stand response to catastrophic wind in central New England, U.S.A. *J. Ecology* 76:135–151.

Foster, D. R., and E. R. Boose. 1992. Patterns of forest damage resulting from catastrophic wind in central New England, USA. *J. Ecology* 80:70–98.

Foster, D. R., and E. R. Boose. 1995. Hurricane disturbance regimes in temperate and tropical forest ecosystems. *In* M. P. Coutts and J. Grace (eds.), *Wind and Trees.* Cambridge Univ. Press, Cambridge.

Fowells, H. A. 1948. The temperature profile in a forest. *J. For.* 46:897–899.

Fowler, D. 1980. Removal of sulphur and nitrogen compounds from the atmosphere in rain and by dry deposition. *In* D. Drablos and A. Tollan (eds.), *Ecological Effects of Acid Precipitation.* SNSF (Acid Precipitation-Effects on Forests and Fish) Project, Oslo, Norway.

Fowler, D. P. 1965a. Effects of inbreeding in red pine, *Pinus resinosa* Ait. II. Pollination studies. *Silvae Genetica* 14:12–23.

Fowler, D. P. 1965b. Effects of inbreeding in red pine, *Pinus resinosa* Ait. IV. Comparison with other Northeastern *Pinus* species. *Silvae Genetica* 14:76–81.

Fowler, D. P., and R. E. Mullin. 1977. Upland–lowland ecotypes not well developed in black spruce in northern Ontario. *Can. J. For. Res.* 7:35–40.

Fox, J. T. 1977. Alternation and coexistence of tree species. *Am. Nat.* 111:69–89.

Foy, C. D., O. Chaney, M. C. White. 1978. The physiology of metal toxicity of plants. *Ann. Rev. Plant Physiology* 29:511–566.

Fralish, J. S., F. B. Crooks, J. L. Chambers, and F. M. Harty. 1991. Comparison of presettlement, second-growth and old-growth forest on six site types in the Illinois Shawnee Hills. *Am. Midl. Nat.* 125:294–309.

Fralish, J. S., R. P. McIntosh, and O. L. Loucks (eds.). 1993. *John T. Curtis, Fifty Years of Wisconsin Plant Ecology.* Wisc. Acad. Sci., Arts & Letters, Madison. 339 pp.

Franklin, J. F. 1968. Cone production by upper-slope conifers. USDA For. Serv. Res. Paper PNW-60. Pacific Northwest For. and Rge. Exp. Sta., Portland, OR. 21 pp.

Franklin, J. F. 1988. Pacific Northwest Forests. *In* M. G. Barbour and W. D. Billings (eds.), *North American Terrestrial Vegetation.* Cambridge Univ. Press, Cambridge.

Franklin, J. F. 1990. Biological legacies: a critical management concept from Mount St. Helens. *In Trans. Fifty-fifth North American Wildlife and Natural Resources Conference.* Wildlife Manage. Inst., Washington, D.C.

Franklin, J. F. 1993. Preserving biodiversity: species, ecosystems, or landscapes? *Ecol. Applications* 3:202–205.

Franklin, J. F. 1995. Sustainability of managed temperate forest ecosystems. *In* M. Munasinghe and W. Shearer (eds.), *Defining and Measuring Sustainability: The Biogeophysical Foundations.* World Bank, Washington, D.C.

Franklin, J. F. 1997. Ecosystem management: an overview. *In* M. Boyce (ed.), *Proc. Symp. Ecosystem Management: Applications for Sustainable Forest and Wildlife Resources.* Yale Univ. Press, New Haven, CT.

Franklin, J. F., D. R. Berg, D. A. Thornburgh, and J. C. Tappeiner. 1997. Alternative silvicultural approaches to timber harvesting: variable retention harvest systems. *In* K. A. Kohm and J. F. Franklin (eds.), *Creating a Forestry for the 21st Century.* Island Press, Washington, D.C.

Franklin, J. F., K. Cromack, Jr., W. Denison, A. McKee, C. Maser, J. Sedell, F. Swanson, and G. Juday. 1981. Ecological characteristics of old-growth Douglas-fir forests. USDA For. Serv. Gen. Tech. Report PNW-118. Pacific Northwest For. and Rge. Exp. Sta., Corvallis OR. 48 pp.

Franklin, J. F., and C. T. Dyrness. 1973. Natural vegetation of Oregon and Washington. USDA For. Serv. Gen. Tech. Rep. PNW-8, Pacific Northwest For. Rge. Exp. Sta., Portland OR. 417 pp.

Franklin, J. F., and C. T. Dyrness. 1980. *Natural Vegetation of Oregon and Washington.* Oregon Univ. Press, Corvallis. 452 pp.

Franklin, J. F., P. M. Frenzen, and F. J. Swanson. 1988. Re-creation of ecosystems at Mount St. Helens: contrasts in artificial and natural approaches. *In* J. Cairns, Jr. (ed.), *Rehabilitating Damaged Ecosystems, Vol. 2.* CRC Press, Boca Raton, FL.

Franklin, J. F., and K. Kohm. 1997. Introduction. *In* K. A. Kohm and J. F. Franklin (eds.), *Creating a Forestry for the 21st Century.* Island Press, Washington, D.C.

Franklin, J. F., J. A. MacMahon, F. J. Swanson, and J. R. Sedell. 1985. Ecosystem responses to the eruption of Mount St. Helens. *National Geographic Research* 1985:196–215.

Franklin, J. F., H. H. Shugart, and M. E. Harmon. 1987. Tree death as an ecological process. *Bio-Science* 37:550–556.

Franklin, J. F., and T. A. Spies. 1991. Composition, function, and structure of old-growth Douglas-fir forests. *In* L. F. Ruggerio, K. B. Aubry, A. B. Carey, and M. H. Huff (tech. coords.), *Wildlife and vegetation of unmanaged douglas-fir forests.* USDA For. Serv. Gen. Tech. Report PNW-GTR-285. Pacific Northwest Res. Sta., Portland, OR.

Franklin, R. T. 1970. Insect influences on the forest canopy. *In* D. E. Reichle (ed.), *Analysis of Temperate Forest Ecosystems.* Springer-Verlag, New York.

Fredriksen, R. L. 1970. Erosion and sedimentation following road construction and timber harvest on unstable soils in three small western Oregon watersheds. USDA For. Serv. Res. Paper PNW–104. Pacific Northwest For. and Rge. Exp. Sta., Portland OR. 15 pp.

Fredriksen, R. L. 1972. Nutrient budget of a Douglas-fir forest on an experimental watershed in western Oregon. *In* J. F. Franklin, L. J. Dempster, and R. H. Waring (eds.), *Symp. Proc. Re-*

search on Coniferous Forest Ecosystems. USDA For. Serv., Pacific Northwest For. and Rge. Exp. Sta., Portland, OR.

Frelich, L. E., and C. G. Lorimer. 1985. Current and predicted long-term effects of deer browsing in hemlock forests in Michigan, USA. *Biol. Conserv.* 34:99–120.

Friis, E. M., W. G. Chaloner, and P. R. Crane. (eds.) 1987. *The Origins of Angiosperms and Their Biological Consequences.* Cambridge Univ. Press, Cambridge. 358 pp.

Frissell, S. S., Jr. 1973. The importance of fire as a natural ecological factor in Itasca State Park, Minnesota. *Quat. Res.* 3:397–407.

Fritts, H. C., D. G. Smith, and M. A. Stokes. 1965. The biological model for paleoclimatic interpretation of Mesa Verde tree-ring series. *Amer. Antiquity* 31:101–121.

Froehlich, H. A. 1973. Natural and man-caused slash in headwater streams. Loggers Handb. No. 33. Pac. Logging Congr., Portland, OR. 8 pp.

Froehlich, H. A., D. McGreer, and J. R. Sedell. 1972. Natural debris within the stream environment. Coniferous For. Biome. U. S. Int. Biol. Program. Int. Rep. 96, Univ. Washington, Seattle. 10 pp.

Frye, R. J., and J. A. Quinn. 1979. Forest development in relation to topography and soils on a floodplain of the Raritan River, New Jersey. *Bull. Torrey Bot. Club* 106:334–345.

Fuller, W. H., S. Shannon, and P. S. Burgess. 1955. Effect of burning on certain forest soils of northern Arizona. *For. Sci.* 1:44–50.

Furnier, G. R., P. Knowles, M. A. Clyde, and B. P. Dancik. 1987. Effects of avian seed dispersal on the genetic structure of whitebark pine populations. *Evolution* 4:607–612.

Furniss, M. 1972. Observations of resistance and susceptibility to Douglas-fir beetles. *In* Program Abstracts, Second North American For. Biol. Workshop, Oregon State Univ., Corvallis. p. 24.

Gaffney, W. S. 1941. The effects of winter elk browsing, south fork of the Flathead River, Montana. *J. Wildl. Manage.* 5:427–453.

Gail, M. R., and D. F. Grigal. 1987. Vertical root distributions of northern tree species in relation to successional status. *Can. J. For. Res.* 17:829–834.

Galston, A. W. 1994. *Life Processes of Plants.* W. H. Freeman, New York. 245 pp.

Gara, R. I., W. R. Littke, J. K. Agee, D. R. Geiszler, J. D. Stuart, and C. H. Driver. 1985. Influence of fires, fungi, and mountain pine beetles on development of a lodgepole pine forest in south-central Oregon. *In* D. M. Baumgartner, R. G. Krebill, J. T. Arnott, and G. F. Weetman (eds.), *Symp. Proc. Lodgepole Pine: The Species and Its Management.* Washington State Univ., Dept. Nat. Res. Sciences, Pullman, WA.

Garland, J. A. 1977. The dry deposition of sulphur dioxide to land and water surfaces. *Proc. Royal Soc. London* 12:245–268.

Garner, W. W. 1923. Further studies in photoperiodism in relation to hydrogen-ion concentration of the cell-sap and the carbohydrate content of the plant. *J. Agr. Res.* 23:871–920.

Garner, W. W., and H. A. Allard. 1920. Effect of the relative length of day and night and other factors of the environment on growth and reproduction in plants. *J. Agr. Res.* 18:553–606.

Gashwiler, J. S. 1967. Conifer seed survival in a western Oregon clearcut. *Ecology* 48:431–438.

Gates, D. M. 1968. Energy exchange between organism and environment. *In* W. P. Lowry (ed.), *Biometeorology.* Oregon State Univ. Press, Corvallis.

Gates, D. M. 1980. *Biophysical Ecology.* Springer-Verlag, New York. 611 pp.

Gates, J., and L. Gysel. 1978. Avian nest dispersion and fledging success in field-forest ecotones. *Ecology* 59:871–873.

Gauche, H. G. Jr. 1982. *Multivariate Analysis in Community Ecology.* Cambridge Univ. Press, Cambridge. 298 pp.

Gehring, C. A., and T. G. Whitham. 1991. Herbivore-driven mycorrhizal mutualism in insect-susceptible pinyon pine. *Nature* 353:556–557.

Gelderen, D. M. van, P. C. de Jong, and H. J. Oterdoom. 1994. *Maples of the World.* Timber Press, Portland, OR. 458 pp.

Gentry, A. H. 1986. Endemism in tropical versus temperate plant communities. *In* M. Soulé (ed.), *Conservation Biology: The Science of Scarcity and Diversity.* Sinauer Associates, Sunderland, MA.

Gentry, A. H. 1995. Diversity and floristic composition of neotropical dry forests. *In* S. H. Bullock, H. A. Mooney, and E. Medina (eds.), *Seasonally Dry Tropical Forests*. Cambridge Univ. Press, Cambridge.

Gentry, A. H., and C. H. Dodson. 1987. Diversity and biogeography of neotropical vascular epiphytes. *Annals Missouri Bot. Gardens* 74:205–233.

George, M. F., and M. J. Burke. 1976. The occurrence of deep supercooling in cold hardy plants. *Curr. Adv. Plant Sci.* 22:349–360.

George, M. F., and M. J. Burke. 1977. Supercooling in overwintering azalea flower buds. Additional freezing parameters. *Plant Physiol.* 59: 319–325.

George, M. F., M. J. Burke, H. M. Pellet, and A. G. Johnson. 1974. Low temperature exotherms and woody plant distribution. *Hort. Sci.* 6:519–522.

Gersper, P. L., and N. Holowaychuk. 1971. Some effects of stem flow from forest canopy trees on chemical properties of soils. *Ecology* 52:691–702.

Ghent, A. W. 1958. Studies of regeneration in forest stands devastated by the spruce budworm, II. Age, height, growth and related studies of balsam fir seedlings. *For. Sci.* 4:135–146.

Gholz, H. L. 1982. Environmental limits on aboveground net primary production, leaf area, and biomass in vegetation zones of the Pacific Northwest. *Ecology* 63:469–481.

Gholz, H. L., and R. L. Fisher. 1982. Organic matter production and distribution in slash pine (*Pinus elliottii*) plantations. *Ecology* 63:1827–1839.

Gill, A. M. 1977. Plant traits adapted to fires in Mediterranean land ecosystems. *In* H. Mooney and C. E. Conrad (eds.), *Proc. Symp. The Environmental Consequences of Fire and Fuel Management in Mediterranean Ecosystems*. USDA, For. Serv. Gen. Tech. Report WO-3, Washington, D.C.

Gill, A. M., R. H. Groves, and I. R. Noble. 1981. *Fire and the Australian Biota*. Australian Acad. Sci., Canberra. 582 pp.

Gill, A. M., J. R. L. Hoare, and N. P. Cheney. 1990. Fires and their effects in the wet-dry tropics of Australia. *In* J. G. Goldammer (ed.), *Fire in the Tropical Biota* (Ecological Studies 84). Springer-Verlag, New York.

Gill, C. J. 1970. The flooding tolerance of woody species—a review. *For. Abstracts* 31:671–688.

Gill, D. S., and P. L. Marks. 1991. Tree and shrub seedling colonization of old fields in central New York. *Ecol. Monogr.* 61:183–205.

Givnish, T. J. 1981. Serotiny, geography, and fire in the pine barrens of New Jersey. *Evolution* 35:101–123.

Givnish, T. J. 1986. Biomechanical constraints on crown geometry in forest herbs. *In* T. J. Givnish (ed.), *On the Economy of Plant Form and Function*. Cambridge Univ. Press, Cambridge.

Gleason, H. A. 1910. The vegetation of the inland sand deposits of Illinois. *Bull. Illinois Stat. Lab. Nat. History* 9:20–173.

Gleason, H. A. 1917. The structure and development of the plant association. *Bull. Torrey Bot. Club.* 44:463–481.

Gleason, H. A. 1926. The individualistic concept of the plant association. *Bull. Torrey Bot. Cl.* 53:7–26.

Gleason, H. A. 1939. The individualistic concept of the plant association. *Am. Midl. Nat.* 21:92–110.

Glenn-Lewin, D. C., R. K. Peet, and T. T. Veblen (eds.) 1992. *Plant Succession: Theory and Prediction*. Chapman and Hall, London. 352 pp.

Glenn-Lewin, D. C., and E. van der Maarel. 1992. Patterns and processes of vegetation dynamics. *In* D. C. Glenn-Lewin, R. K. Peet, and T. T. Veblen (eds.), *Plant Succession: Theory and Prediction*. Chapman and Hall, London.

Godwin, H. 1956. *The History of the British Flora*. Cambridge Univ. Press, Cambridge. 384 pp.

Goldammer, J. G. 1990a. *Fire in the Tropical Biota* (Ecological Studies 84). Springer-Verlag, New York. 497 pp.

Goldammer, J. G. 1990b. Tropical wild-land fires and global changes: prehistoric evidence, present fire regimes, and future trends. *In* J. S. Levine (ed.), *Global Biomass Burning*. Mass. Inst. Tech. Press, Cambridge.

Goldammer, J. G., and S. R. Penafiel. 1990. Fire in the pine-grassland biomes of tropical and sub-tropical Asia. *In* J. G. Goldammer (ed.), *Fire in the Tropical Biota*. (Ecological Studies 84). Springer-Verlag, New York.

Goldblatt, P. 1980. Polyploidy in angiosperms: monocotyledons. *In* W. H. Lewis (ed.), *Polyploidy*. Plenum Press, New York.

Golley, F. B. (ed.)1983. *Ecosystems of the World, Tropical Rain Forest Ecosystems. Vol. 14A*. Elsevier, New York. 382 pp.

Golley, F. B. 1993. *A History of the Ecosystem Concept in Ecology*. Yale Univ. Press, New Haven, CT. 254 pp.

Good, N. F. 1968. A study of natural replacement of chestnut in six stands in the Highlands of New Jersey. *Bull. Torrey Bot. Club* 95:240–253.

Good, R. O. 1974. *The Geography of Flowering Plants*. 4th ed. Longman, London. 557 pp.

Goodall, D., A. Lugo, M. Brinson, and S. Brown. 1990. *Ecosystems of the World, Forested Wetlands. Vol. 15*. Elsevier, New York. 527 pp.

Gosz., J. R., R. T . Holmes, G. E. Likens, F. H. Bormann. 1978. The flow of energy in a forest ecosystem. *Sci. Amer.* 328:92–102.

Gottlieb L. D., and S. K. Jain (eds.). 1988. *Plant Evolutionary Biology*. Chapman and Hall, New York. 414 pp.

Graber, R. E. 1970. Natural seed fall in white pine (*Pinus strobus* L.) stands of varying density. USDA For. Serv. Res. Note NE-119. Northeastern For. Exp. Sta., Upper Darby, Pa. 6 pp.

Grace, J. 1989. Tree lines. *Phil. Trans. R. Soc. Lond.* B 324: 233–245.

Gradmann, R. 1898. *Das Pflanzenleben der Schwäbischen Alb*. Verlag Schwäbischen Albvereins, Tübingen. Vol 1:1–376; Vol 2:1–376.

Graham, A. 1993. History of the vegetation: Cretaceous (Maastrichtian)—Tertiary. *In* N. Moran (conv. ed.), *Flora of North America, North of Mexico, Vol. 1*, Oxford Univ. Press, Oxford.

Graham, B. F., Jr., and F. H. Bormann. 1966. Natural root grafts. *Bot Rev.* 32:255–292.

Graham, S. A. 1954. Scoring tolerance of forest trees. *Michigan For.* Note 4. Univ. Michigan, Ann Arbor. 2 pp.

Graham, S. A., R. B. Harrison, Jr., and C. E. Westell, Jr. 1963. *Aspen: Phoenix trees of the Great Lakes Region*. Univ. Michigan Press, Ann Arbor. 272 pp.

Granhall, U. 1981. Biological nitrogen fixation in relation to environmental factors and functioning of natural ecosystems. *In* F. E. Clark and T. Rosswell (eds.), *Terrestrial Nitrogen Cycles*. Ecol. Bull. 33. Swedish Natural Sci. Res. Council, Stockholm.

Grant, M. C. 1993. The trembling giant. *Discover* 14:83–89.

Grant, M. C., J. B. Mitton, and Y. B. Linhart. 1992. Even larger organisms. *Nature* 360:216.

Grant, V. 1963. *The Origin of Adaptations*, Columbia Univ. Press, New York. 606 pp.

Grant, V. 1971. *Plant Speciation*. Columbia Univ. Press, New York. 435 pp.

Grant, V. 1977. *Organismic Evolution*. W. H. Freeman, San Francisco. 418 pp.

Grasovsky, A. 1929. Some aspects of light in the forest. *Yale Univ. School For. Bull.* 23. 53 pp.

Gratkowski, H. J. 1956. Windthrow around staggered settings in old-growth Douglas-fir. *For. Sci.* 2:60–74.

Gray, P. A., L. Demal, D. Hogg, D. Green, D. Euler, and D. DeYoe. 1995. *An Ecosystem Approach to Living Sustainability*. Queen's Printer, Ottawa, Ontario, Canada.

Green, D. C., and D. F. Grigal. 1979. Jack pine biomass accretion on shallow and deep soils in Minnesota. *Soil Sci. Soc. Amer. J.* 43:1233–1237.

Green, D. C., and D. F. Grigal. 1980. Nutrient accumulations in jack pine stands on deep and shallow soils over bedrock. *For. Sci.* 26:325–333.

Green, D. G. 1981. Time series and postglacial forest ecology. *Quat. Res.* 15:265–277.

Greenwood, M. S., and K. W. Hutchison. 1993. Maturation as a developmental process. *In* M. R. Ahuja and W. J. Libby (eds.), *Clonal Forestry I, Genetics and Biotechnology*. Springer-Verlag, Berlin.

Gregory, S. V. 1997. Riparian management in the 21st century. *In* K. A. Kohm and J. F. Franklin (eds.), *Creating a Forestry for the 21st Century*. Island Press, Washington, D.C.

Gregory, S. V., F. J. Swanson, W. A. McKee, and K. W. Cummins. 1991. An ecosystem perspective of riparian zones. *BioScience* 41:540–551.

Greller, A. M. 1989. Correlation of warmth and temperateness with the distributional limits of zonal forests in eastern North America. *Bull. Torrey Bot. Club* 116:145–163.

Grier, C. C. 1975. Wildfire effects on nutrient distribution and leaching in a coniferous ecosystem. *Can. J. For. Res.* 5:599–607.

Grier, C. C., D. W. Cole, C. T. Dyrness, and R. L. Fredriksen. 1974. Nutrient cycling in 37– and 450–year-old Douglas-fir ecosystems. *In* R. H. Waring and R. L. Edmonds (eds.), *Integrated Research in the Coniferous Forest Biome*. Conif. For. Biome Bull. No. 5. Coniferous Forest Biome, US/IBP, Univ. Washington, Seattle.

Grier, C. C., K. M. Lee, N. M. Nadkarni, G. O. Klock, and P. J. Edgerton. 1989. Productivity of forests of the United States and its relation to soil and site factors and management: a review. USDA For. Serv. Gen. Tech. Report PNW-GTR-222. Pacific Northwest Res. Sta., Corvallis, OR. 51 pp.

Griffin G. J. 1992. American chestnut survival in understory mesic sites following the chestnut blight pandemic. *Can. J. Bot.* 70:1950–1956.

Grigal, D. F. 1984. Shortcomings of soil surveys for forest management. *In* J. G. Bockheim (ed.), *Symp. Proc. Forest Land Classification: Experience, Problems, Perspectives*. NCR-102 North Central For. Soils Com., Soc. Am. For., USDA For. Serv., and USDA Soil Cons. Serv. Madison, WI.

Grime, J. P. 1965a. Comparative experiments as a key to the ecology of flowering plants. *Ecology* 46:513–515.

Grime, J. P. 1965b. Shade tolerance on flowering plants. *Nature* 208:161–163.

Grime, J. P. 1966. Shade avoidance and shade tolerance in flowering plants. *In* R. Bainbridge, G. C. Evans, and O. Rackham (eds.), *Light as an Ecological Factor*. Blackwell, Oxford, England.

Grime, J. P. 1977. Evidence for the existence of three primary strategies in plants and its relevance to ecological and evolutionary theory. *Am. Nat.* 111:1169–1194.

Grime, J. P. 1979. *Plant Strategies and Vegetation Processes*. John Wiley, New York. 222 pp.

Grime, J. P. 1988. The C-S-R model of primary plant strategies—origins, implications and tests. *In* L. D. Gottlieb and S. K. Jain (eds.), *Plant Evolutionary Biology*. Chapman and Hall, New York.

Grime, J. P., and D. W. Jeffrey. 1965. Seedling establishment in vertical gradients of sunlight. *J. Ecol.* 53:621–642.

Grimm, E. C. 1988. Data analysis and display. *In* B. Huntley and T. Webb, III (eds.), *Vegetation History*. Kluwer, Dordrecht.

Grimm, R. C. 1984. Fire and other factors controlling the Big Woods vegetation of Minnesota in the mid-nineteenth century. *Ecol. Monogr.* 54:291–311.

Groffman, P. M., and J. M. Tiedje. 1989. Denitrification in north temperate forests soils: spatial and temporal patterns at the landscape and seasonal scales. *Soil Biol. Biochemistry* 21:613–620.

Gross, H. L. 1972. Crown deterioration and reduced growth associated with excessive seed production by birch. *Can. J. Bot.* 50:2431–2437.

Gross, H. L., and A. A. Harnden. 1968. Dieback and abnormal growth of yellow birch induced by heavy fruiting. Canada Dept. For. and Rural Develop., For. Res Lab. Info. Report 0–X-79, Sault Ste. Marie, Ontario. 12 pp.

Grubb, P. J. 1977. The maintenance of species-richness in plant communities: the importance of the regeneration niche. *Biol. Rev.* 52:107–145.

Grubb, P. J. 1985. Plant populations and vegetation in relation to habitat, disturbance and competition: problems of generalization. *In* J. White (ed.), *The Population Structure of Vegetation* (Handbook of Vegetation Science, Part III). W. Junk, Dordrecht.

Grumbine, R. E. 1994. What is ecosystem management? *Conserv. Biol.* 8:27–38.

Grumbine, R. E. 1997. Reflections on "What is ecosystems management?" *Conserv. Biol.* 11:41–47.

Gullion, G. W. 1972. Improving your forested lands for ruffed grouse. Ruffed Grouse Soc. No. Am., Rochester, N. Y. 34 pp. (Minn. Agr. Exp. Sta., Misc. J. Ser., Publ. 1439).

Gustafson, E. J., and T. R. Crow. 1994. Modeling the effects of forest harvesting on landscape structure and the spatial distribution of cowbird brood parasitism. *Landscape Ecology* 9:237–248.

Gustafson, F. G. 1943. Influence of light upon tree growth. *J. For.* 41:212–213.

Haack, R. A. 1994. Insect-pollinated trees and shrubs in North America. *Mich. Ent. Soc. Newsletter* 39:8–9.

Haack, R. A., and J. W. Byler. 1993. Insects & pathogens, regulators of forest ecosystems. *J. For.* 91:32–37.

Haase, E. F. 1970. Environmental fluctuations on south-facing slopes in the Santa Catalina Mountains of Arizona. *Ecology* 51:959–974.

Habeck, J. R. 1958. White cedar ecotypes in Wisconsin. *Ecology.* 39:457–463.

Habeck, J. R., and R. W. Mutch. 1973. Fire-dependent forests in the northern Rocky Mountains. *Quat. Res.* 3:408–424.

Hack, J. T., and J. C. Goodlett. 1960. Geomorphology and forest ecology of a mountain region in the central Appalachians. U. S. Geol. Surv. Prof. Paper 347. 66 pp + map.

Haden-Guest, S., J. K. Wright, and E. M. Teclaff (eds.), 1956. *A World Geography of Forest Resources.* Amer. Geog. Soc. Spec. Publ. 33. Ronald Press, New York. 736 pp.

Hagen, J. B. 1992. *An Entangled Bank.* Rutgers Univ. Press, New Brunswick, NJ. 245 pp.

Hagglund, B. 1981. Evaluation of forest site productivity. *For. Abstr.* 42:515–527.

Hall, D. M., A. I. Matus, J. A. Lamberton, and H. N. Barber. 1965. Infra-specific variation in wax on leaf surfaces. *Aust. J. Biol. Sci.* 18:323–332.

Hall, F. C. 1989. Plant community classification: from concept to application. *In* D. E. Ferguson, E. E., P. Morgan, and F. D. Johnson (comps.), *Proceedings—Land Classifications Based on Vegetation: Applications for Resource Management.* USDA For. Serv. Gen. Tech. Report INT-257. Intermountain Res. Sta., Ogden, UT. 315 pp.

Hallé, F., R. A. A. Oldeman, and P. B. Tomlinson. 1978. *Tropical Trees and Forests.* Springer-Verlag, Berlin. 441 pp.

Haller, J. R. 1965. The role of 2–needle fascicles in the adaptation and evolution of ponderosa pine. *Brittonia* 17:354–382.

Halliday, W. E. D. 1937. A forest classification for Canada. *For. Serv. Bull.* 89; Can. Dept. Mines and Resources. 50 pp.

Halpern C. B., and T. A. Spies. 1995. Plant species diversity in natural and managed forest of the Pacific Northwest. *Ecol. Applications* 5:913–934.

Hambrey, M. J. 1994. *Glacial Environments.* Univ. British Columbia Press, Vancouver. 296 pp.

Hammerson, G. A. 1994. Beaver (*Caston canadensis*): ecosystem alterations, management, and monitoring. *Nat. Areas J.* 14:44–57.

Hammitt, W. E., and B. V. Barnes. 1989. Composition and structure of an old-growth oak-hickory forest in southern Michigan over 20 years. *In* G. Rink and C. A. Budelsky (eds.), *Proc. Seventh Central Hardwoods Conference.* USDA For. Serv. Gen. Tech. Report NC-132. North Central For. Exp. Sta., St. Paul, MN.

Hammond, H. 1991. *Seeing the Forest Among the Trees: The Case for Wholistic Forest Use.* Polestar, Vancouver, B.C. Canada. 309 pp.

Hampf, F. E. (ed.). 1965. Site index curves for some forest species in the eastern United States. USDA For. Serv., Eastern Region. Upper Darby, PA. 43 pp.

Hamrick, J. L. 1989. Isozymes and the analysis of genetic structure in plant populations. *In* D. E. Soltis and P. S. Soltis (eds.), *Isozymes in Plant Biology.* Dioscorides Press, Portland, OR.

Hamrick, J. L., and M. J. W. Godt. 1989. Allozyme diversity in plant species. *In* A. H. D. Brown, M. T. Clegg, A. L. Kahler, and B. S. Weir (eds.), *Plant Populations Genetics, Breeding, and Genetic Resources.* Sinauer, Sunderland, MA.

Hamrick, J. L., M. J. W. Godt, and S. L. Sherman-Broyles. 1992. Factors influencing levels of genetic diversity in woody plant species. *In* W. T. Adams, S. H. Strauss, D. L. Copes, and A. R. Griffin (eds.), *Population Genetics of Forest Trees.* Kluwer, Boston, MA.

Handley, C. O., Jr. 1969. Fire and mammals. *In Proc. Annual Tall Timbers Fire Ecology Conf.* 9:151–159. Tall Timbers Res. Sta., Tallahassee, FL.

Hanes, T. 1988. California chaparral. *In* M. G. Barbour and J. Major (eds.), *Terrestrial Vegetation of California.* California Native Plant Society, Special Publ. No. 9. Berkeley, CA.

Hanks, J. P., E. Fitzhugh, and S. R. Hanks. 1983. A habitat type classification system for ponderosa pine forests of northern Arizona. USDA For. Serv. Gen. Tech. Report RM-97. Rocky Mountain For. and Rge. Exp. Sta., Fort Collins, CO. 22 pp.

Hanover, J. W. 1975. Physiology of tree resistance to insects. *Ann. Rev. Entomol.* 20:75–95.

Hansen, H. P. 1947. Postglacial forest succession, climate, and chronology in the Pacific Northwest. *Trans. Amer. Philosoph. Soc.* 37(1). 130 pp.

Hansen, H. P. 1955. Postglacial forests in south central and central British Columbia. *Amer. J. Sci.* 253:640–658.

Hanson, A. D., and W. D. Hitz. 1982. Metabolic responses of mesophytes to plant water deficits. *Ann. Rev. Plant Physiology* 33:163–203.

Hanson, H. C. 1917. Leaf-structure as related to environment. *Amer. J. Bot.* 4:533–560.

Hardin, J. W. 1979. Patterns of variation in foliar trichomes of eastern North American *Quercus*. *Amer. J. Bot.* 66:576–585.

Hare, F. K., and J. C. Ritchie. 1972. The boreal bioclimates. *Geographical Review* 62:333–365.

Harland, W., A. V. Cox, P. G. Llewellyn, C. A. G. Pickton, A. G. Smith, and R. Walters. 1982. *A Geologic Time Scale*. Cambridge Univ. Press, Cambridge. 131 pp.

Harley, J. L. 1939. The early growth of beech seedlings under natural and experimental conditions. *J. Ecol.* 27:384–400.

Harley, J. L., and S. E. Smith. 1983. *Mycorrhizal Symbiosis*. Academic Press, New York. 483 pp.

Harlow, W. M., E. S. Harrar, J. W. Hardin, and F. M. White. 1996. *Textbook of Dendrology*. 8th ed. McGraw-Hill, New York. 501 pp.

Harmon, M. E. 1984. Survival of trees after low-intensity surface fires in Great Smoky Mountains National Park. *Ecology* 65:796–802.

Harmon, M. E. 1987. The influence of litter and humus accumulations and canopy openness on *Picea sitchensis* (Bong.) Carr. and *Tsuga heterophylla* (Raf.) Sarg. seedlings growing on logs. *Can. J. For. Res.* 17:1475–1479.

Harmon, M. E., W. K. Ferrell, and J. F. Franklin. 1990. Effects on carbon storage of conversion of old-growth forests to young forests. *Science* 247:699–702.

Harmon, M. E., and J. F. Franklin. 1989. Tree seedlings on logs in *Picea-Tsuga* forests of Oregon and Washington. *Ecology* 70:48–59.

Harmon, M. E., J. F. Franklin, F. J. Swanson, P. Sollins, S. V. Gregory, J. D. Lattin, H. H. Anderson, S. P. Cline, N. G. Aumen, J. R. Sedell, G. W. Lienkaemper, K. Cromack, Jr., and K. W. Cummins. 1986. Ecology of coarse woody debris in temperate ecosystems. *Adv. Ecol. Res.* 15:133–302.

Harper, J. L. 1961. Approaches to the study of plant competition. *In* F. L. Milthorpe (ed.), *Mechanisms in Biological Competition*. Symp. Soc. Exp. Biology 15:1–39.

Harper, J. L. 1977. *Population Biology of Plants*. Academic Press, New York. 892 pp.

Harper, R. M. 1906. A phytogeographical sketch of the Altamaha Grit region of the coastal plain of Georgia. *Ann. N. Y. Acad. Sci.* 17:1–415.

Harper, R. M. 1911. The relation of climax vegetation to islands and peninsulas. *Bull. Torrey Bot. Club*. 38:515–525.

Harper, R. M. 1914. Geography and vegetation of northern Florida. *Fla. Geol. Surv. Sixth Ann. Rep.* pp. 163–437.

Harper, R. M. 1962. Historical notes on the relation of fires to forests. *In Proc. Annual Tall Timbers Fire Ecology Conf.* 1:11–29. Tall Timbers Res. Sta. Talahassee, FL.

Harrington, T. C. 1986. Growth decline of wind-exposed red spruce and balsam fir in the White Mountains. *Can. J. For. Res.* 16:232–238.

Harrington, W. 1991. Wildlife: severe decline and partial recovery. *In* K. D. Frederick and R. A. Sedjo (eds.), *America's Renewable Resources: Historical Trends and Current Challenges*. Resources for the Future, Washington, D.C.

Harris, L. D. 1984. *The Fragmented Forest*. Univ. Chicago Press. Chicago, 211 pp.

Harris, L. D., and G. Silva-Lopez. 1992. Forest fragmentation and the conservation of biological diversity. *In* P. L. Fiedler and S. K. Jain (eds.), *Conservation Biology*, Chapman and Hall, New York.

Harris, T. M. 1958. Forest fire in the mesozoic. *J. Ecology*. 46:447–453.

Harshberger, J. W. 1903. An ecologic study of the flora of mountainous North Carolina. *Bot. Gaz.* 36:241–258,368–383.

Harshberger, J. W. 1899. Thermotropic movement of leaves of *Rhododendron maximum. Proc. Nat. Acad. Sci.,* U.S.A. 1899:214–222.

Hart, J. W. 1988. *Light and Plant Growth.* Unwin Hyman, Boston, MA. 204 pp.

Hart, S .C., J. M. Stark, E. A. Davidson, and M. K. Firestone. 1994. Nitrogen mineralization, immobilization, and nitrification. *In* R. W. Weaver, S. Angle, P. Bottomley, D. Bezdicek, S. Smith, A. Tabatabai, and A. Wollum (eds.), *Methods of Soil Analysis: Part 2. Microbiological and Biochemical Properties.* Soil Sci. Soc. Amer. Book Series, No. 5., Soil Sci. Soc. Amer., Segoe, WS.

Hartesveldt, R. J., and H. T. Harvey. 1967. The fire ecology of sequoia regeneration. *In Proc. Annual Tall Timbers Fire Ecology Conf.* 7:65–77. Tall Timbers Res. Sta., Tallahassee, FL.

Hase, H. and H. Foelster. 1983. Impact of plantation forestry with teak (*Tectonia grandis*) on the nutrient status of young alluvial soils in west Venezuela. *For. Ecol. Manage.* 6:33–57.

Hatch, A. B. 1937. The physical basis of mycotrophy in the genus *Pinus. Black Rock For. Bull.* 6:1–168.

Hatchell, G. E., and C. W. Ralston. 1971. Natural recovery of surface soils disturbed in logging. *Tree Planters' Notes* 22:5–9.

Hatchell, G. E., C. W. Ralston, and R. R. Foil. 1970. Soil disturbances in logging. *Tree Planters' Notes* 68:772–778.

Hawk, G. M., and D. B. Zobel. 1974. Forest succession on alluvial landforms of the McKenzie River valley, Oregon. *Northwest Science* 48:245–265.

Hebda, R. J. 1985. Museum collections and paleobiology. *In* E. H. Miller (ed.), *Museum Collections: Their Roles and Future in Biological Research.* Brit. Columbia Prov. Mus. Occ. Pap. 25.

Heikkenen, H. J., S. E. Scheckler, P. J. J. Egan, Jr., and C. B. Williams, Jr. 1986. Incomplete abscission of needle clusters and resin release from artificially water-stressed loblolly pine (*Pinus taeda*): a component for plant-animal interactions. *Amer. J. Bot.* 73:1384–1392.

Heinselman, M. L. 1973. Fire in the virgin forests of the Boundary Waters Canoe Area, Minnesota. *Quat. Res.* 3:329–382.

Heinselman, M. L. 1981a. Fire and succession in the conifer forests of northern North America. *In* D. C. West, H. H. Shugart, and D. B. Botkin. (eds.), *Forest Succession: Concepts and Application.* Springer-Verlag, New York.

Heinselman, M. L. 1981b. Fire intensity and frequency as factors in the distribution and structure of northern ecosystems. *In* H. A. Mooney et al. (eds.), *Fire Regimes and Ecosystems Properties.* USDA For. Serv. Gen. Tech. Report WO-26. Washington, D.C.

Heizer, R. F. 1955. Primitive man as an ecologic factor. *Kroeber Anthropological Soc. Papers* 13:1–31.

Helgath, S. F. 1975. Trial deterioration in the Selway-Bitterroot Wilderness. USDA For. Serv. Res. Note INT-193. Intermountain For. and Rge. Exp. Sta., Ogden, UT. 15 pp.

Hellmers, H. 1962. Temperature effect upon optimum tree growth. *In* T. T. Kozlowski (ed.), *Tree Growth.* Ronald Press, New York.

Hellmers, H. 1966a. Temperature action and interaction of temperature regimes in the growth of red fir seedlings. *For. Sci.* 12:90–96.

Hellmers, H. 1966b. Growth response of redwood seedlings to thermoperiodism. *For. Sci.* 12:276–283.

Hellmers, H., M. K. Genthe, and F. Ronco. 1970. Temperature affects growth and development of Engelmann spruce. *For. Sci.* 16:447–452.

Hellmers, H., and W. P. Sundahl. 1959. Response of *Sequoia sempervirens* (D. Don) Endl. and *Pseudotsuga menziesii* (Mirb. Franco) seedlings to temperature. *Nature* 184:1247–1248.

Henderson, G. S., W. T. Swank, J. B. Waide, and C. C. Grier. 1978. Nutrient budgets of Appalachian and cascade region watersheds: a comparison. *For. Sci.* 24:385–397.

Hendrick, R. L., and K. S. Pregitzer. 1992. The demography of fine roots in a northern hardwood forest. *Ecology* 73:1094–1104.

Hendricks, J. J., K. J. Nadelhoffer, and J. D. Aber. 1993. Assessing the role of fine roots in carbon and nitrogen cycling. *Trends in Ecology and Evolution* 8:174–178.

Henniker-Gotley, G. R. 1936. A forest fire caused by falling stones. *Indian Forester* 62:422–423.

Henry, J. A., J. M. Portier, and J. Coyne. 1994. *The Climate and Weather of Florida*. Pineapple Press, Sarasota, FL. 279 pp.

Herbertson, A. J. 1965. (reprint of 1913 article), The higher units—A geographical essay. *Geography* 50:332–342.

Hermann, R. K., and D. P. Lavender. 1968. Early growth of Douglas-fir from various altitudes and aspects in southern Oregon. *Silvae Genetica* 17:143–151.

Herms, D. A., and W. J. Mattson. 1992. The dilemma of plants: to grow or defend. *Quart. Rev. Biol.* 67:283–335.

Heslop-Harrison, J. 1964. Forty years of genecology. *Adv. Ecol. Res.* 2:159–247.

Heslop-Harrison, J. 1967. *New Concepts in Flowering-Plant Taxonomy*. Harvard Univ. Press, Cambridge, MA. 134 pp.

Hett, J. M. 1971. A dynamic analysis of age in sugar maple seedlings. *Ecology* 52:1071–1074.

Heusser, C. J. 1960. *Late-Pleistocene environments of North Pacific North America*. Amer. Geog. Soc. Spec. Publ. 35. 308 pp.

Heusser, C. J. 1965. A Pleistocene phytogeographical sketch of the Pacific Northwest and Alaska. *In* H. E. Wright, Jr., and David G. Frey (eds.) *The Quaternary of the United States*. Princeton Univ. Press, Princeton, NJ.

Heusser, C. J. 1983. Vegetational history of the northwestern United States including Alaska. *In* S. C. Porter (ed.), *The Late Pleistocene*, Vol. 1 of H. E. Wright, Jr., (ed.), *Late-Quaternary Environments of the United States,* Univ. Minnesota Press, Minneapolis.

Hewlett, J. D., and A. R. Hibbert. 1963. Moisture and energy conditions within a sloping soil mass during drainage. *J. Geophys. Res.* 68:1081–1087.

Hibbs, D. E. 1981. Leader growth and the architecture of three North American hemlocks. *Can. J. Bot.* 59:476–480.

Hibbs, D. E. 1983. Forty years of forest succession in central New England. *Ecology* 64:1394–1401.

Hills, G. A. 1952. The classification and evaluation of site for forestry. Ontario Dept. Lands and For., Res. Rept. 24. 41 pp.

Hills, G. A. 1953. The use of site in forest management. *For. Chron.* 29:128–136.

Hills, G. A. 1960. Regional site research. *For. Chron.* 36:401–423.

Hills, G. A. 1977. An integrated iterative holistic approach to ecosystem classification. *In* J. Thie and G. Ironside (eds.), *Ecological (Biophysical) Land Classification in Canada.* Ecological land classification series, No. 1, Lands Directorate, Env. Canada, Ottawa.

Hills, G. A., and G. Pierpoint. 1960. Forest site evaluation in Ontario. Ontario Dept. Lands and For., Res. Rept. 42. 64 pp.

Himelick, E. B., and D. Neely. 1962. Root-grafting of city-planted American elms. *Plant Dis. Rep.* 46:86–87.

Hinckley, T. M., R. O. Teskey, F. Duhme, and H. Richter. 1981. Temperate hardwood forests. *In* T. T. Kozlowski (ed.), *Water Deficits and Plant Growth. Vol. 6.* Academic Press, New York.

Hinds, T. E., F. G. Hawksworth, and R. W. Davidson. 1965. Beetle-killed Engelmann spruce: its deterioration in Colorado. *J. For.* 63:536–542.

Hix, D. M. 1988. Multifactor classification and analysis of upland hardwood forest ecosystems of the Kickapoo River watershed, southwestern Wisconsin. *Can J. For. Res.* 18:1405–1415.

Hobbs, R. J. 1990. Remote sensing of spatial and temporal dynamics of vegetation. *In* R. J. Hobbs and H. A. Mooney (eds.), *Remote Sensing of Biosphere Functioning*. Springer-Verlag, New York.

Hobbs, R. J., and H. A. Mooney (eds.). 1990. *Remote Sensing of Biosphere Functioning*. Springer-Verlag, New York. 312 pp.

Hocker, H. W., Jr. 1956. Certain aspects of climate as related to the distribution of loblolly pine. *Ecology* 37:824–834.

Hoffecker, J. F., W. R. Powers, and T. Goebel. 1993. The colonization of Beringia and the peopling of the New World. *Science* 259:46–53.

Holbrook, N. M., J. L. Whitbeck, and H. A. Mooney. 1995. Drought responses of neotropical dry forest trees. *In* S. H. Bullock, H. A. Mooney, and E. Medina (eds.), *Seasonally Dry Tropical Forests*. Cambridge Univ. Press, Cambridge.

Holderidge, L. R. 1947. Determination of world plant formations from simple climatic data. *Science* 105:367–368.

Holderidge, L. R. 1967. *Life Zone Ecology.* Tropical Sci. Center, San Jose, Costa Rica. 206 pp.

Holmsgaard, E. 1955. Tree-ring analyses of Danish Forest trees. *Det. Forstl. Forsøgsvaesen i Danmark* 22:1–246.

Holmsgaard, E. 1962. Influence of weather on growth and reproduction of beech. *Commun. Inst. Forst. Fenni.* 55:1–5.

Holmsgaard, E., and C. Bang. 1989. Loss of volume increment due to cone production in Norway spruce. *Forstl. Forsøgsvawsen i Danmark* 42:217–231.

Holthuijzen, A. M. A., T. L. Sharik, and J. D. Fraser. 1987. Dispersal of eastern red cedar (*Juniperus virginiana*) into pastures: an overview. *Can. J. Bot.* 65:1092–1095.

Hook, D. D. 1984. Waterlogging tolerance of lowland tree species of the South. *Southern J. Applied For.* 8:136–149.

Hoover, W. H. 1937. The dependence of carbon dioxide assimilation in a higher plant on wave length of radiation. Smithsn. Misc. Coll., Vol. 95. 13 pp.

Hopkins, D. M. (ed.). 1967a. *The Bering Land Bridge.* Stanford Univ. Press, Stanford, CA. 495 pp.

Hopkins, D. M. 1967b. The Cenozoic history of Beringia—a synthesis. *In* D. M. Hopkins (ed.), *The Bering Land Bridge.* Stanford Univ. Press, Stanford, CA.

Hopkins, D. M., J. V. Matthews, Jr., C. E. Schweger, and S. B. Young (eds.). 1982. *Paleoecology of Beringia.* Academic Press, New York. 489 pp.

Hori, T. (ed.). 1953. *Studies on Fogs in Relation to Fog-Preventing Forest.* Foreign Books Dept. Tanne Trading , Sapporo, Hokkaido. 399 pp.

Hornstein, F. von. 1958. *Wald und Mensch, Wald Geschichte des Alpen Vorlandes.* Otto Maier Verlag, Ravensburg, Germany. 283 pp.

Horton, K. W. 1956. The ecology of lodgepole pine in Alberta and its role in forest succession. For. Br. Can. Tech. Note 45. 29 pp.

Horton, K. W. 1959. Characteristics of subalpine spruce in Alberta. Can. Dept. Northern Affairs and Nat. Res. For. Res. Div. Tech. Note No. 76, 20 pp.

Hosie, R. C. 1969. *Native Trees of Canada,* 7th ed. Can. For. Serv., Ottawa, Canada. 380 pp.

Host, G. E., and K. S. Pregitzer. 1992. Geomorphic influences on ground flora and overstory composition in upland forests of northwestern Lower Michigan. *Can. J. For. Res.* 22:1547–1555.

Host, G. E., K. S. Pregitzer, C. W. Ramm, J. B. Hart, and D. T. Cleland. 1987. Landform-mediated differences in successional pathways among upland forest ecosystems in northwestern Lower Michigan. *For. Sci.* 33:445–457.

Host, G. E., K. S. Pregitzer, C. W. Ramm, D. P. Lusch, and D. T. Cleland. 1988. Variation in overstory biomass among glacial landforms and ecological land units in northwestern Lower Michigan. *Can. J. For. Res.* 18:659–668.

Hough, A. F. 1945. Frost pocket and other microclimates in forests of the northern Allegheny plateau. *Ecology* 26:235–250.

Hough, A. F., and R. D. Forbes. 1943. The ecology and silvics of forests in the high plateaus of Pennsylvania. *Ecol. Monogr.* 13:299–320.

Howe, H. F. 1977. Bird activity and seed dispersal of a tropical wet forest tree. *Ecology* 58:539–550.

Howe, H. F. 1993. Specialized and generalized dispersal systems: where does "the paradigm" stand? *Vegetatio* 107/108:3–13.

Huber, O. 1978. Light compensation point of vascular plants of a tropical cloud forest and an ecological interpretation. *Photosynthetica* 12:382–390.

Huberman, M. A. 1943. Sunscald of eastern white pine, *Pinus strobus* L. *Ecology* 24:456–471.

Huenneke, L. F., and R. R. Sharitz. 1986. Microsite abundance and distribution of woody seedlings in a South Carolina cypress-tupelo swamp. *Am. Midl. Nat.* 115:328–335.

Huggett, R. J. 1995. *Geoecology, An Evolutionary Approach.* Routledge, New York. 320 pp.

Hunter, M. L. 1990. *Wildlife, Forests and Forestry: Principles of Managing Forests for Biological Diversity.* Prentice-Hall, Englewood Cliffs, NJ. 370 pp.

Hunter, A. F., and M. J. Lechowicz. 1992. Predicting the timing of budburst in temperate trees. *J. Applied Ecology* 29:597–604.

Huntley, B., and I. C. Prentice. 1993. Holocene vegetation and climates of Europe. *In* Wright, H. E., Jr., J. E. Kutzbach, T. Webb III, W. F. Ruddiman, F. A Street-Perrott, and P. J. Bartlein, (eds.), *Global Climates Since the Last Glacial Maximum.* Univ. Minnesota Press, Minneapolis.

Huntly, N. 1991. Herbivores and the dynamics of communities and ecosystems. *Ann. Rev. Ecol. Syst.* 22:477–503.

Hupp, C. R., and W. R. Osterkamp. 1985. Bottomland vegetation distribution along Passage Creek, Virginia, in relation to fluvial landforms. *Ecology* 66:670–681.

Hursch, C. R. 1948. Local climate in the Copper Basin of Tennessee as modified by the removal of vegetation. U.S. Dept. Agr. Circ. 774. 38 pp.

Huston, M. A. 1993. Biological diversity, soils, and economics. *Science* 262:1676–1680.

Huston, M. A. 1994. *Biological Diversity.* Cambridge Univ. Press, Cambridge. 681 pp.

Huston, M., D. DeAngelis, and W. Post. 1988. New computer models unify ecological theory. *BioScience* 38:682–691.

Huston, M., and T. Smith. 1987. Plant succession: life history and competition. *Am. Nat.* 130:168–198.

Hutchins, H. E., S. A. Hutchins, and B.-W. Liu. 1996. The role of birds and mammals in Korean pine (*Pinus koraiensis*) regeneration dynamics. *Oecologia* 107:120–130.

Hutchins, H. E., and R. M. Lanner. 1982. The central role of Clark's nutcracker in the dispersal and establishment of whitebark pine. *Oecologia* 55:192–201.

Huxley, C. R., and D. F. Cutler. 1991. *Ant–Plant Interactions.* Oxford Univ. Press. Oxford. 601 pp.

Huxley, J. S. 1938. Clines, an auxiliary taxonomic principle. *Nature* 142:219–220.

Huxley, J. S. 1939. Clines, an auxiliary method in taxonomy. *Bijdragen tot de Dierkunde* 27:491–520.

Illick, J. S. 1914. *Pennsylvania Trees.* Pa. Dept. For. Bull. 11, 231 pp.

Illick, J. S. 1921. Replacement of the chestnut. *J. For.* 19:105 114.

Imbrie, J., and K. P. Imbrie. 1979. *Ice Ages: Solving the Mystery.* Enslow Publ., Hillside, NJ. 224 pp.

Irgens-Moller, H. J. 1968. Geographical variation in growth patterns of Douglas-fir. *Silvae Genetica* 17:106–110.

Irwin, L. L., J. G. Cook, R. A. Riggs, and J. M. Skovlin. 1994. Effects of long-term grazing by big game and livestock in the Blue Mountains forest ecosystems. USDA For. Serv. Gen. Tech. Report PNW-GTR-325. Pacific Northwest Res. Sta., Portland, OR. 49 pp.

Isaac, L. A. 1938. Factors affecting establishment of Douglas-fir seedlings. U.S. Dept. Agr. Circ. 486. 45 pp.

Isaac, L. A. 1946. Fog drip and rain interception in coastal forests. U. S. For. Serv., Pacific Northwest For. Rge. Exp. Sta. Paper 34. 16 pp.

Isaac, L. A. 1956. Place of partial cutting in old-growth sands of the Douglas-fir region. U.S. For. Serv., Pacific Northwest For. Rge. Exp. Sta. Res. Paper 16. Corvallis, OR. 48 pp.

Jackson, L. W. R. 1952. Radial growth of forest trees in the Georgia Piedmont. *Ecology* 33:336–341.

Jackson, L. W. R. 1967. Effect of shade on leaf structure of deciduous tree species. *Ecology* 48:498–499.

Jackson, L. W. R., and R. S. Harper. 1955. Relation of light intensity to basal area of short-leaf pine (*Pinus echinata*) stands in Georgia. *Ecology* 36: 158–159.

Jackson, R. C. 1976. Evolution and systematic significance of polyploidy. *Ann. Rev. Ecol. Syst.* 7:209–234.

Jacobs, D. F., D. W. Cole, and J. R. McBride. 1985. Fire history and perpetuation of natural coast redwood ecosystems. *J. For.* 83:494–497.

Jacobson, R. B., A. J. Miller, and J. A. Smith. 1989. The role of catastrophic geomorphic events in central Appalachian landscape evolution. *Geomorphology* 2:257–284.

Jacoby, G. C., P. L. Williams, and B. M. Buckley. 1992. Tree ring correlation between prehistoric landslides and abrupt tectonic events in Seattle, Washington. *Science* 258:1621–1623.

Jahn, G. 1991. Temperate deciduous forests of Europe. *In* E. Röhrig and B. Ulrich (eds.) *Ecosystems of the World, Temperate Deciduous Forests, Vol. 7.* Elsevier, Amsterdam.

James, S. 1984. Lignotubers and burls: their structure, function, and ecological significance in Mediterranean ecosystems. *Bot Rev.* 50:225–245.

Janos, D. P. 1987. VA mycorrhizas in humid tropical systems. *In* G. R. Safer (ed.), *Ecophysiology of VA Mycorrhizal Plants.* CRC Press, Boca Raton, FL. 224 pp.

Janzen, D. H. 1969. Seed-eaters versus seed size, number, toxicity and dispersal. *Evolution* 23:1–27.

Janzen, D. H. 1970. Herbivores and the number of tree species in tropical forests. *Am. Nat.* 104:501–528.

Janzen, D. H. 1971. Seed predation by animals. *Ann. Rev. Ecol. Syst.* 2:465–492.

Janzen, D. H. 1976. Why do bamboos wait so long to flower? *In* J. Burley and B. T. Styles (eds.), *Tropical Trees, Variation, Breeding and Conservation.* Linn. Soc. Symp. Series No. 2, Academic Press, New York.

Janzen, D. H. 1983. Dispersal of seeds by vertebrate guts. *In* D. J. Futuyma and M. Slatkin (eds), *Coevolution.* Sinauer Assoc., Sunderland, MA.

Janzen, D. H. 1985. The natural history of mutualisms. *In* D. H. Boucher (ed.), *The Biology of Mutualism*, Croom Helm, London.

Janzen, D. H. 1997. Wildland biodiversity management in the tropics. *In* M. L. Reada-Kudla, D. E. Wilson, and E. O. Wilson (eds.), *Biodiversity II: Understanding and Protecting Our Biological Resources.* Joseph Henry Press, Washington, D.C.

Janzen, D. H., and P. S. Martin. 1982. Neotropical anacronisms: the fruits the gomphotheres ate. *Science* 215:19–27.

Jarvis, P. G., and J. W. Laverenz. 1983. Productivity of temperate, deciduous, and evergreen forests. *In* O. L. Lange, P. S. Nobel, C. B. Osmond, and H. Ziegler (eds.), *Encyclopedia of Plant Physiology, New Series, Vol. 12D.* Springer-Verlag, New York.

Jenny, H. 1941. *Factors of Soil Formation.* McGraw-Hill, New York. 281 pp.

Jenny, H. 1961. Derivation of state factor equations of soils and ecosystems. *Soil Sci. Soc. Amer. Proc.* 25:385–388.

Jenny, H. 1980. *The Soil Resource: Origin and Behavior.* Springer-Verlag, New York. 377 pp.

Jenny, H., R. J. Arkley, and A. M. Schultz. 1969. The pygmy forest-podsol ecosystem and its dune associates of the Mendocino coast. *Madroño* 20:60–74.

Johnson, E. A. 1992. *Fire and Vegetation Dynamics: Studies from the North American Boreal Forest.* Cambridge Univ. Press, Cambridge. 129 pp.

Johnson, P. L. 1969. *Remote Sensing in Ecology.* Univ. Georgia Univ. Press, Athens. 244 pp.

Johnson, P. W. 1972. Factors affecting buttressing in *Triplochiton scleroxylon* K. Schum. *Ghana J. Agr. Sci.* 5:13–21.

Johnson, W. C., and C. S. Adkisson. 1985. Airlifting the oaks. *Nat. Hist.* 10:41–46.

Johnson, W. C., and T. Webb III. 1989. The role of blue jays (*Cyanocitta cristata* L.) in the postglacial dispersal of fagaceous trees in eastern North America. *J. Biogeogr.* 16:561–571.

Johnson, W. C., R. L. Burgess, and W. R. Keammerer. 1976. Forest overstory vegetation and environment on the Missouri River floodplain in North Dakota. *Ecol. Monogr.* 46:59–84.

Jones, J. R. 1969. Review and comparison of site evaluation methods. USDA For. Serv. Res. Paper RM-51. Rocky Mountain For. and Rge. Exp. Sta., Fort Collins, CO. 27 pp.

Jones, R. K., G. Pierpoint, G. M. Wickware, J. K. Jeglum, R. W. Arnup, and J. M. Bowles. 1983. Field guide to forest ecosystem classification for the Clay Belt. Site Region 3E. Ont. Min. Nat. Res. Toronto, ON. 123 pp.

Jones, S. M. 1991. Landscape ecosystem classification for South Carolina. *In* D. L. Mengel and D. T. Tew (eds.), *Symp. Proc., Ecological Land Classification: Applications to Identify the Productive Potential of Southern Forests.* USDA For. Serv. Gen. Tech. Report SE-68. Southeastern For. Exp. Sta., Asheville, NC.

Jones, S. M., and F. T. Lloyd. 1993. Landscape ecosystem classification: the first step toward ecosystem management in the southeastern United States. *In* G. H. Aplet, N. Johnson, J. T. Olson, and V. A. Sample (eds.), *Defining Sustainable Forestry.* Island Press, Washington D.C. 231 pp.

Jordan, C., W. Caskey, G. Escalante, R. Herrera, F. Montagnini, R. Todd, and C. Uhl. 1982. The nitrogen cycle in a "Terra Firme" rainforest on oxisol in Amazon territory of Venezuela. *Plant and Soil* 67:325–332.

Jorgensen, J. R., and C. S. Hodges, Jr. 1970. Microbial characteristics of a forest soil after twenty years of prescribed burning. *Mycologia* 62:721–726.

Jurik, T. W., J. A. Webber, and D. M. Gates. 1988. Effects of temperature and light on photosynthesis of dominant species of a northern hardwood forest. *Bot. Gaz.* 149:203–208.

Kalela E. K. 1957. Über Veränderungen in den Wurzelverhältnissen der Kiefernbestände im Laufe der Vegetationsperiode. *Acta For. Fenn.* 65:1–41.

Kamminga-Van Wijk, C., and H. Prins. 1993, The kinetics of NH_4^+ and NO_3^- uptake by Douglas-fir from single N-solutions and from solutions containing both NH_4^+ and NO_3^-. *Plant and Soil* 151:91–96.

Kaplan, R., and S. Kaplan. 1989. *The Experience of Nature.* Cambridge Univ. Press, Cambridge. 340 pp.

Kaplan, S. 1992. Environmental preference in a knowledge-seeking, knowledge-using organism. *In* J. H. Barkow, L. Cosmides, and J. Tooby (eds.), *The Adapted Mind: Evolutionary Psychology and the Generation of Culture.* Oxford Univ. Press, New York.

Kappen, L. 1981. Ecological significance of resistance to high temperature. *In* O. L. Lange, P. S. Nobel, C. B. Osmond, H. Ziegler (eds.), *Physiological Plant Ecology I, Responses to the Physical Environment*, New Series, Vol. 12A. Springer-Verlag, New York.

Kaszkurewicz, A., and P. J. Fogg. 1967. Growing season of cottonwood and sycamore as related to geographic origin and environmental factors. *Ecology* 48: 785–793.

Kauffman, J. B., and C. Uhl. 1990. Interactions of anthropogenic activities, fire, and rain forests in the Amazon Basin. *In* J. G. Goldammer (ed.), *Fire in the Tropical Biota.* (Ecological Studies 84). Springer-Verlag, New York.

Kaufman, Y. J., A Setzer, C. Justice, C. J. Tucker, M. G. Pereira, and I. Fung. 1990. Remote sensing of biomass burning in the tropics. *In* J. G. Goldammer (ed.). *Fire in the Tropical Biota* (Ecological Studies 84). Springer-Verlag, New York.

Kay, C. E. 1994. Aboriginal overkill, the role of Native Americans in structuring western ecosystems. *Human Nature* 5:359–398.

Kay, C. E. 1997. Is aspen doomed? *J. For.* 95:4–11.

Kay, C. E., and F. H. Wagner. 1993. Historic condition of woody vegetation on Yellowstone's northern range: a critical test of the "natural regulation" paradigm. *In* D. G. Despain (ed.), *Conf. Proc. Plants and Their Environments,* Yellowstone National Park, WY.

Kayll, A. J. 1974. Use of fire in land management. *In* T. T. Kozlowski and C. E. Ahlgren (eds.), *Fire and Ecosystems.* Academic Press, New York.

Keane, R. E., P. Morgan, and S. W. Running. 1996. FIRE-BGC—a mechanistic ecological process model for simulating fire succession on coniferous forest landscapes of the northern Rocky Mountains. USDA For. Serv. Res. Paper INT-RP-484. Int. Res. Sta., Odgen, UT. 122 pp.

Kearsley, M. J. C., and T. G. Whitham. 1989. Developmental changes in resistance to herbivory: implications for individuals and populations. *Ecology* 70:422–434.

Keeley, J. E., and S. C. Keeley. 1988. Chaparral. *In* M G. Barbour and W. D. Billings (eds.), *North American Terrestrial Vegetation.* Cambridge Univ. Press, Cambridge.

Keeling, C. D., and T. P. Whorf. 1996. Atmospheric CO_2 records from sites in the SIO air sampling network. *In Trends. A Compendium of Data on Global Change.* Carbon Dioxide Information Analysis Center, Oak Ridge National Laboratory, Oak Ridge, TN.

Keever, C. 1950. Causes of succession on old fields of the piedmont, North Carolina. *Ecol. Monogr.* 20:231–250.

Keever, C. 1953. Present composition of some stands of the former oak, chestnut forest in the southern Blue Ridge Mountains. *Ecology* 34:44–54.

Kellert, S. R., and E. O. Wilson (eds.). 1993. *The Biophilia Hypothesis.* Island Press, Washington, D.C. 484 pp.

Kellman, M, K. Miyanishi, and P. Hiebert. 1985. Nutrient retention by savanna ecosystems. II. Retention after fire. *J. Ecol.* 73:953–962.

Kellman, M, K. Miyanishi, and P. Hiebert. 1987. Nutrient sequestering by the understory strata of natural *Pinus caribaea* stands subject to prescription burning. *For. Ecol. Manage.* 21:57-73.

Kellogg, L. F. 1939. Site index curves for plantation black walnut in the Central States region. Central States For. Exp. Sta., Res. Note 35. 3 pp.

Kelsall, J. P., E. S. Telfer, and T. D. Wright. 1977. The effects of fire on the ecology of the Boreal Forest, with particular reference to the Canadian north: a review and selected bibliography. Can. Wildlife Serv., Ottawa. Occ. Paper 32. 58 pp.

Kemperman, J. A., and B. V. Barnes. 1976. Clone size in American aspens. *Can J. Bot.* 54:2603–2607.

Kenoyer, L. A. 1933. Forest distribution in southwestern Michigan as interpreted from the original land survey (1826–32). *Papers, Mich. Acad. Sci., Arts, Letters* 19:107–111.

Kerfoot, O. 1968. Mist precipitation on vegetation. *For. Abstracts.* 29(1):8–20.

Keyes, M. R., and C. C. Grier. 1981. Above-and below-ground net production in 40–year-old Douglas-fir stands on low and high productivity sites. *Can. J. For. Res.* 11:599–605.

Khoshoo, T. N. 1959. Polyploidy in gymnosperms. *Evolution* 13:24–39.

Kienitz, H. 1879a. *Vergleichende keimversuche mit Waldbaum-Samen aus klimatisch verschiedenen Orten Mitteleuropas.* Bot. Unters. Herausgegeben von N. J. C. Müller 2.

Kienitz, H. 1879b. *Ueber Formen and Abarten heimischer Waldbäume.* Forstl. Zeitschr. 1.

Kilburn, P. D. 1960. Effects of logging and fire on xerophytic forests in northern Michigan. *Bull. Torrey Bot. Club* 6:402–405.

Kilgore, B. M. 1972. Fire's role in a sequoia forest. *Naturalist* 23:26–37.

Kilgore, B. M. 1973. The ecological role of fire in Sierran conifer forests: its application to national park management. *Quat. Res.* 3:496–513.

Kilgore, B. M. 1975. Restoring fire to national park wilderness. *Am. Forests* 81:16–19.

Kilgore, B. M. 1976a. Fire management in the national parks: an overview. *In Proc. Annual Tall Timbers Fire Ecology Conference and Fire and Land Management Symposium,* 14:45–57. Tall Timbers Res. Sta., Tallahassee, FL.

Kilgore, B. M. 1976b. America's renewable resource potential—1975: the turning point. *Proc. Soc. Amer. For.* 1975:178–188.

Kilgore, B. M. 1976c. From fire control to fire management: an ecological basis for policies. *In Trans. 41st North American Wildlife and Natural Resources Conference.* Wildlife Management Institute, Washington, D.C.

Kilgore, B. M., and R. W. Sando. 1975. Crown-fire potential in a sequoia forest after prescribed burning. *For. Sci.* 21:83–87.

Killingbeck, K. T. 1996. Nutrients in senesced leaves: keys to the search for potential resorption and resorption proficiency. *Ecology* 77:1716–1727

Kirchner, O. von., E. Loew, and C. Schröter. 1927. *Lebensgeschichte der Blütenpflanzen Mitteleuropas.* Vol. 2. Part 1. No. 31/32. Eugen Ulmer, Stuttgart.

Klappa, C. F. 1980. Rhizoliths in terrestrial carbonates: classification. recognition, genesis and significance. *Sedimentology* 27:613–629.

Kleinschmit, J. 1978. Sitka spruce in Germany. *In Proc. IUFRO Joint Meeting of Working Parties, Vol. 2.* British Columbia Ministry of Forests. Vancouver, B.C., Canada.

Kleinschmit, J., and J. Ch. Bastien. 1992. IUFRO's role in Douglas-fir (*Pseudotsuga menziesii* (Mirb.) Franco) tree improvement. *Silvae Genetica* 41:161–173.

Klijn, F., and H. A. Udo de Haes. 1994. A hierarchical approach to ecosystems and its implications for ecological land classification. *Landscape Ecology* 9:89–104.

Klinge, H., and W. A. Rodrigues. 1968a. Litter production in an area of Amazonian terra firme forest. Part I. Litter-fall, organic carbon and total nitrogen contents of litter. *Amazoniana* 1:287–302.

Klinge, H., and W. A. Rodrigues. 1968b. Litter production in an area of Amazonian terra firme forest. Part II. Mineral nutrient content of the litter. *Amazoniana* 1:303–310.

Klinge, H., W. A. Rodrigues, E. Brunig, and E. J. Fittkau. 1975. Biomass and structure in a central Amazon rain forest. *In* F. B. Golley and E. Medina (eds.), *Tropical Ecological Systems: Trends in Terrestrial and Aquatic Research.* Springer-Verlag, New York.

Klinka, K., V. J. Krajina, A. Ceska, and A. M Scagel. 1989. *Indicator Plants of Coastal British Columbia.* Univ. British Columbia Press, Vancouver. 288 pp.

Knight, D. H. 1994. *Mountains and Plains, The Ecology of Wyoming Landscapes.* Yale Univ. Press, New Haven, CT. 338 pp.

Knight, D. H., and O. L. Loucks. 1969. A quantitative analysis of Wisconsin forest vegetation on the basis of plant function and gross morphology. *Ecology* 50:219–234.

Knight, H. 1966. Loss of nitrogen from the forest floor by burning. *For. Chron.* 42:149–152.

Knoepp, J. D., D. P. Turner, and D. T. Tingey. 1993. Effects of ammonium and nitrate on nutrient uptake and activity of nitrogen assimilation enzymes in western hemlock. *For. Ecol. Manage.* 59:179–191.

Knowles, P., and M. C. Grant. 1985. Genetic variation of lodgepole pine over time and microgeographical space. *Can. J. Bot.* 63:722–727.

Knowles, R. 1981. Denitrification. *In* E. A. Paul and J. N. Ladd (eds.), *Soil Biochemistry Vol. 5.* Dekker, New York.

Kohm, K. A., and J. F. Franklin (eds). 1997. *Creating a Forestry for the 21st Century.* Island Press, Washington, D.C. 475 pp.

Kohm, K. A., and J. F. Franklin. 1997. Introduction. *In* K. A. Kohm and J. F. Franklin (eds.), *Creating a Forestry for the 21st Century*, Island Press, Washington, D.C.

Kolb, T. E., and D. A. J. Teulon. 1991. Relationship between sugar maple budburst phenology and pear thrips damage. *Can. J. For Res.* 21:1043–1048.

Kolb, T. E., K. C. Steiner, L. H. McCormick, and T. W. Bowersox. 1990. Growth response of northern red-oak and yellow-poplar seedlings to light, soil moisture and nutrients in relation to ecological strategy. *For. Ecol. Manage.* 38:65–78.

Komarek, E. V. 1962. The use of fire: and historical background. *In Proc. Annual Tall Timbers Fire Ecology Conf.* 1:7–10. Tall Timbers Res. Sta., Tallahassee, FL.

Komarek, E. V. 1971. Principles of fires ecology and fire management in relation to the Alaskan environment. *In* C. W. Slaughter, R. A. Barney, and G. M. Hansen (eds.), *Symp. Proc. Fire in the Northern Environment.* USDA For. Serv. Pacific Northwest For. and Range Exp. Sta., Portland, OR.

Komarek, E. V. 1972. Lightning and fire ecology in Africa. *In Proc. Annual Tall Timbers Fire Ecology Conf.* 11:473–511. Tall Timbers Res. Sta., Tallahassee, FL..

Komarek, E. V. 1973. Ancient fires. *In Proc. Annual Tall Timbers Fire Ecology Conf.* 12:219–240. Tall Timbers Res. Sta., Tallahassee, FL.

Koonce, A. L., and A. González-Cabán. 1990. Social and ecological aspects of fire in Central America. *In* J. G. Goldammer (ed.), *Fire in the Tropical Biota.* (Ecological Studies 84). Springer-Verlag, New York.

Köppen, W. 1931. *Grundriss der Klimakunde.* Walter de Gruyter, Berlin. 388 pp.

Kormanik, P. P., and C. L. Brown. 1969. Origin and development of epicormic branches in sweetgum. USDA For. Serv. Res. Paper SE-54. Southeast For. Exp. Sta., Asheville, NC. 17 pp.

Kormondy, E. J. 1969. *Concepts of Ecology.* Prentice-Hall. Upper Saddle River, NJ. 209 pp.

Korstian, C. F., and T. S. Coile. 1938. Plant competition in forest stands. *Duke Univ. School For. Bull.* 3. 125 pp.

Korstian, C. F., and P. W. Stickel. 1927. The natural replacement of blight-killed chestnut in the hardwood forests of the Northeast. *J. Agr. Res.* 34:631–648.

Koski, V. 1970. A study of pollen dispersal as a mechanism of gene flow in conifers. Comm. *Inst. For. Fenn.* 70:1–78.

Koski, V., and J. R. Tallqvist. 1978. Results of long-time measurements of the quantity of flowering and seed crop of forest trees. *Folia Forestalia* No. 364:1–60

Köstler, J. N., E. Brückner, and H. Bibelriether. 1968. *Untersuchungen zur Morphologie der Waldbäume in Mitteleuropa.* Verlag Paul Parey, Munich. 284 pp.

Kozlowski, T. T. 1949. Light and water in relation to growth and competition of Piedmont forest tree species. *Ecol. Monogr.* 19:207–231.

Kozlowski, T. T. 1971a. *Growth and Development of Trees, I. Seed Germination, Ontogeny, and Shoot Growth.* Academic Press, New York. 443 pp.

Kozlowski, T. T. 1971b. *Growth and Development of Trees, II. Cambial Growth, Root Growth, and Reproductive Growth.* Academic Press, New York. 520 pp.

Kozlowski, T. T., and C. E. Ahlgren (eds.). 1974. *Fire and Ecosystems.* Academic Press, New York. 542 pp.

Kozlowski, T. T., and T. Keller. 1966. Food relations of woody plants. *Bot. Rev.* 32:293–382.

Kozlowski, T. T., P. J. Kramer, and S. G. Pallardy. 1991. *The Physiological Ecology of Woody Plants.* Academic Press, New York. 657 pp.

Kozlowski, T. T., and W. H. Scholtes. 1948. Growth of roots and root hairs of pine and hardwood seedlings in the Piedmont. *J. For.* 46:750–754.

Kozlowski, T. T., and S. G. Pallardy. 1997a. *Physiology of Woody Plants.* 2nd ed. Academic Press, San Diego, CA. 411 pp.

Kozlowski, T. T., and S. G. Pallardy. 1997b. *Growth Control in Woody Plants.* Academic Press, San Diego, CA. 641 pp.

Kramer, P. J. 1943. Amount and duration of growth of various species of tree seedlings. *Plant Physiol.* 18:239–251.

Kramer, P. J. 1957. Some effects of various combinations of day and night temperatures and photoperiod on the growth of loblolly pine seedlings. *For. Sci.* 3:45–55.

Kramer, P. J., and J. S. Boyer. 1995. *Water Relations of Plants and Soils.* Academic Press, New York. 495.

Kramer, P. J., and W. S. Clark. 1947. A comparison of photosynthesis in individual pine needles and entire seedlings at various light intensities. *Plant Physiol.* 22:51–57.

Kramer, P. J., and J. P. Decker. 1944. Relation between light intensity and rate of photosynthesis of loblolly pine and certain hardwoods. *Plant Physiol.* 19:350–358.

Kramer, P. J., W. S. Riley, and T. T. Bannister. 1952. Gas exchange of cypress (*Taxodium distichum*) knees. *Ecology* 33:117–121.

Krammes, J. S. 1965. Seasonal debris movement from steep mountainside slopes in southern California. *USDA Misc. Publ.* 970:85–88.

Krauss, G. A. 1936. Aufgaben der Standortskunde. *Jahresber. deutschen Forstvereins Berlin* pp. 319–329.

Krebs, C. J. 1989. *Ecological Methodology.* Harper and Row, New York. 654 pp.

Krefting, L. W., J. H. Stoeckeler, B. J. Bradle, and W. D. Fitzwater. 1962. Porcupine-timber relationships in the Lake States. *J. For.* 60:325–330.

Kriebel. H. B. 1957. Patterns of genetic variation in sugar maple. *Ohio Agr. Exp. Sta. Res. Bull.* 791. 56 pp.

Kroehler, C. J., and A. E. Linkins. 1988. The root surface phosphatases of *Eriophorum vaginatum*: effects of temperature, pH, substrate concentration, and inorganic phosphorus. *Plant and Soil* 105:3–10.

Küchler, A. W. 1964. The potential natural vegetation of the conterminous United States. *Am. Geogr. Soc. Spec. Publ.* No. 36. 154 pp.

Küchler, A. W. 1967. *Vegetation Mapping.* Ronald Press, New York. 472 pp.

Küppers, M. 1994. Canopy gaps: competitive light interception and economic space filling—a matter of whole-plant allocation. *In* M. M. Caldwell and R. W. Pearcy (eds.), *Exploitation of Environmental Heterogeneity by Plants.* Academic Press, New York.

Kurz, H. and D. Demaree. 1934. Cypress buttresses and knees in relation to water and air. *Ecology* 15:36–41.

Kwesiga, F. R., and J. Grace. 1986. The role of the red/far-red ratio in the response of tropical tree seedlings to shade. *Ann. Bot.* 57:283–290.

Ladefoged, K. 1952. The periodicity of wood formation. *Copenh. Biol. Skr.* 7(3):1–98.

Lahti, T. 1995. Understorey vegetation as an indicator of forest site potential in southern Finland. Finnish Soc. For. Sci., Finnish For. Res. Inst. 68 pp.

Lambert, J. M., and M. B. Dale. 1964. The use of statistics in phytosociology. *Adv. Ecol. Res.* 2:55–99.

Landhäusser, S. M., and R. W. Wein. 1993. Postfire vegetation recovery and tree establishment at the Arctic treeline: climate-change—vegetation-response hypotheses. *J. Ecology* 81:665–672.

Langford, A. N., and M. F. Buell. 1969. Integration, identity and stability in the plant association. *Adv. Ecol. Res.* 6:83–135.

Langlet, O. 1959. A cline or not a cline—a question of Scots pine. *Silvae Genetica* 8:13–22.

Langlet, O. 1971. Two hundred years' genecology. *Taxon* 20:653–721.

Langston, N. 1995. *Forest Dreams, Forest Nightmares, The Paradox of Old Growth in the Inland West.* Univ. Washington Press, Seattle. 368 pp.

Lanner, R. M. 1966. Needed: a new approach to the study of pollen dispersion. *Silvae Genetica* 15:50–52.

Lanner, R. M. 1980. Avian seed dispersal as a factor in the ecology and evolution of limber and whitebark pines. *In* B. P. Dancik and K. O. Higginbotham (eds.), *Proc. Sixth North American For. Biol. Workshop.* Univ. Alberta, Edmonton, Alberta, Canada.

Lanner, R. M. 1981. *The Piñon Pine: A Natural and Cultural History.* Univ. Nevada Press, Reno. 208 pp.

Lanner, R. M. 1982. Adaptations of whitebark pine for seed dispersal by Clark's Nutcracker. *Can. J. For. Res.* 12:391–402.

Lanner, R. M. 1990. Biology, taxonomy, evolution, and geography of stone pines of the world. *In* W. Schmidt and K. McDonald (comps.), *Proc. Symp. Whitebark Pine Ecosystems: Ecology and Management of a High-Mountain Resource,* USDA For. Serv. Gen. Tech. Report INT-270, Intermountain Exp. Sta., Odgen, UT.

Lapin, M. 1990. *The Landscape Ecosystem Groups of the University of Michigan Biological Station: Classification, Mapping, and Analysis of Ecological Diversity.* Master's Thesis, University of Michigan, School of Natural Resources. 140 pp.

Lapin, M., and B. V. Barnes. 1995. Using the landscape ecosystem approach to assess species and ecosystem diversity. *Conserv. Biol.* 9:1148–1158.

Larcher, W. 1969. The effect of environmental and physiological variables on the carbon dioxide gas exchange of trees. *Photosynthetica* 3:167–198.

Larcher, W. 1980. *Physiological Plant Ecology,* 2nd ed. Springer-Verlag, New York. 303 pp.

Larcher, W. 1995. *Physiological Plant Ecology,* 3rd ed. Springer-Verlag, New York. 506 pp.

Larcher, W., and H. Bauer. 1981. Ecological significance of resistance to low temperature. *In* O. L. Lange, P. S. Nobel, C. B. Osmond, and H. Ziegler (eds.), *Physiological Plant Ecology I, Responses to the Physical Environment,* New Series Vol. 12A. Springer-Verlag, New York.

Larsen, J. A. 1925. Natural reproduction after forest fires in northern Idaho. *J. Agr. Res.* 31:1177–1197.

Larsen, J. A. 1980. *The Boreal Ecosystem.* Academic Press, New York. 500 pp.

Larson, M. M. 1967. Effect of temperature on initial development of ponderosa pine seedlings from three sources. *For. Sci.* 13:286–294.

Larson, M. M. 1970. Root regeneration and early growth of red oak seedlings: influence of soil temperature. *For. Sci.* 16:442–446.

Larson, P. R. 1963a. The indirect effect of drought on tracheid diameter in red pine. *For. Sci.* 9:52–62.

Larson, P. R. 1963b. Stem form development of forest trees. *For. Sci. Monogr.* 5. 42 pp.

Larson, P. R. 1964. Some indirect effects of environment on wood formation. *In* M. H. Zimmermann (ed.), *The Formation of Wood in Forest Trees.* Academic Press, New York.

Larsson, S. 1989. Stressful times for the plant stress—insect performance hypothesis. *Oikos* 56:277–283.

Lassoie, J. P. 1982. Physiological processes in Douglas-fir. *In* R. L. Edmonds (ed.), *Analysis of Coniferous Forest Ecosystems in the Western United States.* US/IBP Synthesis Series. Dowden. Hutchinson and Ross, Stroudsburg. PA.

Latham, R. E., and R. E. Ricklefs. 1993a. Global patterns of tree species richness in moist forests: energy-diversity theory does not account for variation in species richness. *Oikos* 67:325–333.

Latham, R. E., and R. E. Ricklefs. 1993b. Continental comparisons of temperate-zone tree species diversity. *In* R. Ricklefs and D. Schluter (eds.), *Species Diversity in Ecological Communities: Historical and Geographical Perspectives.* Univ. Chicago Press, Chicago.

Lavender, D. P., and W. S. Overton. 1972. Thermoperiods and soil temperatures as they affect growth and dormancy of Douglas-fir seedlings of different geographic origin. Oregon State Univ., For. Res. Lab. Res. Paper 13. 26 pp.

Lavoie, N., L-P. Venzina, and H. Margolis. 1992. Absorption and assimilation of nitrate and ammonium ions by jack pine seedlings. *Tree Physiology* 11:171–183.

Lawrence, W. H., and J. H. Rediske. 1962. Fate of sown Douglas-fir seed. *For. Sci.* 8:210–218.

Lawton, J. H., and C. G. Jones. 1995. Linking species and ecosystems: organisms as ecosystem engineers. *In* C. G. Jones and J. H. Lawton (eds.), *Linking Species and Ecosystems.* Chapman and Hall, New York.

Layser, E. F. 1974. Vegetative classification: its application to forestry in the northern Rocky Mountains. *J. For.* 73:354–357.

Leak, W. B. 1975. Age distribution in virgin red spruce and northern hardwoods. *Ecology* 56:1451–1454.

Leak, W. B. 1978. Relationship of species and site index to habitat in the White Mountains of New Hampshire. USDA For. Serv. Res. Paper NE-397. Northeastern For. Exp. Sta., Broomall, PA. 9 pp.

Leak, W. B. 1982. Habitat mapping and interpretation in New England. USDA For. Serv. Res. Paper NE-496. Northeastern For. Exp. Sta. Upper Darby, PA. 10 pp.

Leaphart, C. D., and A. R. Stage. 1971. Climate: a factor in the origin of the pole blight disease of *Pinus monticola* Dougl. *Ecology 52:229–239.*

Ledig, F. T. 1992. Human impacts on genetic diversity in forest ecosystems. *Oikos* 63:87–108.

Ledig, F. T., and J. H. Fryer. 1972. A pocket of variability in *Pinus rigida. Evolution* 26:259–266.

Ledig, F. T., and T. O. Perry. 1969. Net assimilation rate and growth in loblolly pine seedlings. *For. Sci.* 15:431–438.

Lee, D. W. 1983. Unusual strategies of light absorption in rain-forest herbs. *In* T. J. Givnish (ed.), *On the Economy of Plant Form and Function.* Cambridge Univ. Press, Cambridge.

Lee, R., and A. Baumgartner. 1966. The topography and insolation climate of a mountainous forest area. *For. Sci.* 12: 258–267.

Leibundgut, J., and H. Heller. 1960. Photoperiodische Reaktion, Lichtbedarf und Austreiben von Jungpflanzen der Tanne (*Abies alba* Miller). *Beiheft Zeitschr. schweiz. Forstv.* 30:185–198.

Lemieux, G. J. 1965. Soil-vegetation relationships in the northern hardwoods of Quebec. *In* Chester T. Youngberg (ed.), *Forest–Soil Relationships in North America.* Oregon State Univ. Press, Corvallis.

Lennartz, M. R. 1988. The red-cocaded woodpecker: old-growth species in a second-growth landscape. *Natural Areas Journal* 8:160–165

Leopold, A. 1933. *Game Management.* Scribners, New York. 481 pp.

Leopold, A. 1949. *A Sand County Almanac.* Reprinted 1966. Oxford Univ. Press, New York. 189 pp.

Leopold, L. B., M. G. Wolman, and J. P. Miller. 1964. *Fluvial Processes in Geomorphology.* W. H. Freeman, San Francisco. 522 pp.

Lester, D. T. 1967. Variation in cone production of red pine in relation to weather. *Can. J. Bot.* 45:1683–1691.

Levin, D. A. 1976. The chemical defenses of plants to pathogens and herbivores. *Ann. Rev. Ecol. Syst.* 7:121–159.

Levine, J. S. (ed.). 1991a. *Global Biomass Burning, Atmospheric, Climatic, and Biospheric Implications.* MIT Press, Cambridge, MA. 569 pp.

Levine, J. S. 1991b. Introduction: global biomass burning: atmospheric, climatic, and biospheric implications. *In* J. S. Levine (ed.), *Global Biomass Burning, Atmospheric, Climatic, and Biospheric Implications.* MIT Press, Cambridge, MA.

Levins, R. 1969. Dormancy as an adaptive strategy. *Symp. Soc. Exp. Biol.* 23:1–10.

Levitt, J. 1980a. *Responses of Plants to Environmental Stresses. Vol. 1. Chilling, Freezing, and High Temperature Stress.* Academic Press, New York. 497 pp.

Levitt, J. 1980b. *Responses of Plants to Environmental Stresses. Vol. 2. Water, Radiation, Salt, and Other Stresses.* Academic Press, New York. 607 pp.

Lewis, W. H. 1980. Polyploidy in angiosperms: dicotyledons. *In* W. H. Lewis (ed.), *Polyploidy.* Plenum Press, New York.

Li, Hui-Lin. 1953. Present distribution and habitats of the conifers and taxads. *Evolution* 7:245–261.

Lieberman, M., D. Lieberman, and R. Peralta. 1989. Forests are not just Swiss cheese: canopy stereogeometry of non-gaps in tropical forests. *Ecology* 70:550–552.

Liebhold, A. M., W. L. MacDonald, D. Bergdahl, and V. C. Mastro. 1995. Invasion by exotic forest pests: a threat to forest ecosystems. *For. Sci. Monogr.* 30. 49 pp.

Lieth, H. 1973. Primary production in terrestrial ecosystems. *Human Ecology* 1:303–332.

Lieth, H., and M. J. A. Werger (eds.) 1989. *Tropical Rain Forest Ecosystems.* Elsevier, New York. 713 pp.

Ligon, J. D. 1978. Reproductive interdependence of Pinon jays and pinon pines. *Ecol. Monogr.* 48:111–126.

Ligon, J. D., and P. B. Stacey. 1995. Land use, lag times and the detection of demographic change: the case of the acorn woodpecker. *Conserv. Biol.* 10:840–846.

Likens, G. E., F. H. Bormann, R. S. Pierce, J. S. Eaton,1 and N. M. Johnson. 1977. *Biogeochemistry of a Forested Ecosystem.* Springer-Verlag, New York. 146 pp.

Likens, G. E., F. H. Bormann, R. S. Pierce, and W. A. Reiners. 1978. Recovery of a deforested ecosystem. *Science* 199:492–496.

Lilley, J., and C. Webb. 1990. *Climate Warning? Exploring the Answers.* ECA90–ST/1. Environment Council of Alberta, Edmonton, Alberta, Canada. 73 pp.

Lincoln, R. J., G. A. Boxshall, and P. F. Clark. 1982. *A Dictionary of Ecology, Evolution and Systematics.* Cambridge Univ. Press, Cambridge. 298 pp.

Lindberg, S. E., and C. T. Garten. 1988. Sources of sulphur in forest canopy throughfall. *Nature* 336:148–151.

Lindeman, R. L. 1942. The trophic-dynamic aspect of ecology. *Ecology* 23:399–418.

Linder, S. 1971. Photosynthetic action spectra of Scots pine needles of different ages from seedlings grown under different nursery conditions. *Physiol. Plant.* 25:58–63.

Linder, S., and B. Axelsson. 1982. Changes in carbon uptake and allocation patterns as a result of irrigation and fertilization in a young *Pinus sylvestris* stand. *In* R. H. Waring (ed.), *Carbon Uptake and Allocation: Key to Management of Subalpine Forest Ecosystems.* IUFRO Workshop, Forest Resources Laboratory, Oregon State Univ., Corvallis.

Linder, S., J. McDonald, and T. Lohammar. 1981. Effects of nitrogen status and irradiance during cultivation on photosynthesis and respiration of birch seedlings. Energy Forest Project Tech. Rep. 12. Swedish Univ. Agricultural Science, Uppsala, Sweden. 19 pp.

Lindsey, A. A., and L. K. Escobar. 1976. *Eastern Deciduous Forest, II. Beech-Maple Region.* U.S. Dep. Interior, Natl. Park Serv., Nat. Hist. Theme Stud. No. 3. NPS Publ. No. 148. 238 pp.

Lindsey, A. A., and J. A. Newman. 1956. Use of official weather data in spring time temperature analysis of an Indiana phenological record. *Ecology* 37: 812–823.

Lindsey, A. A., R. O. Petty, D. K. Sterling, and W. Van Asdall. 1961. Vegetation and environment along the Wabash and Tippecanoe Rivers. *Ecol. Monogr.* 31:105–156.

Lindsey, A. A., and J. O. Sawyer, Jr. 1970. Vegetation-climate relationships in the eastern United States. *Proc. Indiana Acad. Sci.* 80:210–214.

Linhart, Y. B. 1989. Interactions between genetic and ecological patchiness in forest trees and their dependent species. *In* J. H. Bock and Y. B. Linhart (eds.), *Evolutionary Ecology of Plants.* Westview Press, Boulder, CO.

Linhart, Y. B., J. B. Mitton, K. B. Sturgeon, and M. L. Davis. 1981. Genetic variation in space and time in a population of ponderosa pine. *Heredity* 48:407–426.

Linsser, C. 1867. Die periodischen Erscheinungen des Pflanzenlebens in ihrem Verhältniss zu den Wärmeerscheinungen. *Mém. L'Akad. Imp. d. Sci. de St.-Pétersbourg VII,* Ser. XI, No. 7:1–44.

List, R. J. 1958. *Smithsonian Meteorological Tables.* Smithsonian Inst. Pub. 4014 (rev. ed. 6). 527 pp. (2nd reprinting, 1963).

Little, E. L., Jr. 1979. *Checklist of United States Trees.* USDA For. Serv. Handbook No. 541, Washington D.C. 374 pp.

Little, E. L., Jr., A. Brinkman, and A. L. McComb. 1957. Two natural Iowa hybrid poplars. *For. Sci.* 3:253–262.

Little, S. 1974. Effects of fire on temperate forests: northeastern United States. *In* T. T. Kozlowski and C. E. Ahlgren (eds.), *Fire and Ecosystems.* Academic Press, New York.

Little, S., and E. B. Moore. 1949. The ecological role of prescribed burns in the pine-oak forests of southern New Jersey. *Ecology* 30:223–233.

Lloyd, W. J., and P. E. Lemmon. 1970. Rectifying azimuth (of aspect) in studies of soil-site index relationships *In* C. T. Youngberg and C. B. Davey (eds.), *Tree Growth and Forest Soils.* Oregon State Univ. Press, Corvallis.

Loach. K. 1967. Shade tolerance on tree seedling. I. Leaf photosynthesis and respiration in plants raised under artificial shade. *New Phytol.* 66:607–621.

Loach, K. 1970. Shade tolerance in tree seedlings. II. Growth analysis of plants raised under artificial shade. *New Phytol.* 69:273–292.

Loehle, C. 1986. Phototropism of whole trees: effects of habitat and growth form. *Am. Midl. Nat.* 116:190–196.

Loehle, C., and G. Wein. 1994. Landscape habitat diversity: a multiscale information theory approach. *Ecol. Modelling* 73:311–329.

Logan, K. T. 1965. Growth of tree seedlings as affected by light intensity. I. White birch, yellow birch, sugar maple, and silver maple. Dept. For. Canada, Publ. 1121. 16 pp.

Logan, K. T. 1966a. Growth of tree seedlings as affected by light intensity. II. Red pine, white pine, jack pine and eastern larch. Dept. For. Canada, Publ. 1160. 19 pp.

Logan, K. T. 1966b. Growth of tree seedlings as affected by light intensity. III. Basswood and white elm. Dept. For. Canada, Publ. 1176. 15 pp.

Logan, K. T. 1970. Adaptations of the photosynthetic apparatus of sun- and shade-grown yellow birch (*Betula alleghaniensis* Britt.). *Can. J. Bot.* 48:1681–1688.

Logan, K. T., and G. Krotkov. 1969. Adaptations of the photosynthetic mechanism of sugar maple (*Acer saccharum*) seedlings grown in various light intensities. *Physiol. Plant.* 22:104–116.

Long, J. N. 1980. Productivity of western coniferous forests. *In* R. L. Edmonds (ed.), *Analysis of Coniferous Forest Ecosystems in the Western United States.* US/IBP Synthesis Series, Dowden, Hutchinson and Ross, Stroudsburg, PA.

Lonsdale, W. M. 1990. The self-thinning rule: dead or alive? *Ecology* 71:1373–1388.

Loope, L., M. Duever, A. Herndon, J. Snyder, and D. Jansen. 1994. Hurricane impact on uplands and freshwater swamp forest. *BioScience* 44:238–246.

Lorenz, R. W. 1939. High temperature tolerance of forest trees. *Univ. Minnesota Agr. Exp. Sta. Tech. Bull.* 141. 25 pp.

Lorimer, C. G. 1977. The presettlement forest and natural disturbance cycle of northeastern Maine. *Ecology* 58:139–148.

Lorimer, C. G. 1980. Age structure and disturbance history of a southern Appalachian virgin forest. *Ecology* 61:1169–1184.

Lorimer, C. G. 1984. Development of the red maple understory in northeastern oak forests. *For. Sci.* 30:3–22.

Lorimer, C. G. 1993. Causes of the oak regeneration problem. *In* D. Loftis, D. and C. E. McGee (eds.), *Symp. Proc., Oak Regeneration: Serious Problems, Practical Recommendations.* USDA For. Serv. Gen. Tech. Report SE-84. Southeastern For. Exp. Sta., Asheville, NC.

Lorimer, C. G., L. E. Frelich, and E. V. Nordheim. 1988. Estimating gap origin probabilities for canopy trees. *Ecology* 69:778–785.

Loucks, O. L. 1962. *A Forest Classification for the Maritime Provinces.* For. Res. Br., Can. Dept. For. 167 pp.

Loucks, O. L. 1970. Evolution of diversity, efficiency and community stability. *Amer. Zoologist* 10:17–25.

Lovett, G. M., 1994. Atmospheric deposition of nutrients and pollutants in North America: an ecological perspective. *Ecol. Applications* 4:629–650.

Lovett, G. M., and D. Kinsmann. 1990. Atmospheric pollutant deposition to high-elevation ecosystems. *Atmospheric Env.* 24A:2767–2786.

Lovett, J. C., and S. K. Wasser (eds.). 1993. *Biogeography and Ecology of the Rain Forests of Eastern Africa.* Cambridge Univ. Press, Cambridge. 341 pp.

Lubchenco, J., A. M. Olson. L. B. Brubaker, S. R. Carpenter, M. M. Holland, S. P. Hubbell, S. A. Levin, J. A. MacMahon, P. A. Matson, J. R. Melillo, H. A. Mooney, C. H. Peterson, H. R. Pulliam, L. A. Real, P. J. Regal, and P. G. Risser. 1991. The sustainable biosphere initiative: an ecological research agenda. *Ecology* 72:371–412.

Luckman, B. H. 1986. Reconstruction of Little Ice Age events in the Canadian Rocky Mountains. *Géographie Physique et Quaternaire* 40:17–28.

Lüdi, W. 1945. Besiedlung und Vegetationsentwicklung auf den jungen Seitenmoränen des Grossen Aletschgletschers mit einem Vergleich der Besiedlung im Vorfeld des Rhonegletschers und des Oberen Grindelwaldgletschers. *Bericht über das Geobotanische Forschungsinstitut Rübel.* Zürich, 1944:35–112.

Lugo, A. E., A. Brinson, and S. Brown (eds.). 1990. *Forested Wetlands*. Elsevier, New York. 527 pp.

Lulla, K., and P. Mausel. 1983. Ecological applications of remotely sensed multispectral data. *In* B. F. Richason (ed.), *Introduction to Remote Sensing of the Environment*. Kendall/Hunt, Dubuque, IA.

Lutz, H. J. 1928. Trends and silvicultural significance of upland forest successions in southern New England. *Yale Univ. School For. Bull.* 22. 68 pp.

Lutz, H. J. 1930. The vegetation of Heart's Content, a virgin forest in northwestern Pennsylvania. *Ecology* 11:1–29.

Lutz, H. J. 1934. Ecological relations in the pitch pine plains of southern New Jersey. *Yale School of For. Bull.* 38. 80 pp.

Lutz, H. J. 1940. Disturbance of forest soil resulting from the uprooting of trees. *Yale Univ. School For. Bull.* 45. 37 pp.

Lutz, H. J. 1956. Ecological effects of forest fires in the interior of Alaska. *U.S. Dept. Agr. Tech. Bull.* 1133. 121 pp.

Luvall, J. C., and H. R. Holbo. 1990. Thermal remote sensing methods in landscape ecology. *In* Turner, M. G., and R. H. Gardner (eds.), *Quantitative Methods in Landscape Ecology*. Springer-Verlag. New York.

Lyford, W. H. 1980. Development of the root system of northern red oak (*Quercus rubra* L.). *Harvard Forest Paper* No. 21. 30 pp.

Lyford, W. H., and D. W. MacLean. 1966. Mound and pit microrelief in relation to soil disturbance and tree distribution in New Brunswick, Canada. *Harvard Forest Paper* No. 15. 18 pp.

Lyford, W. R., and B. F. Wilson. 1966. Controlled growth of forest tree roots: technique and application. *Harvard Forest Paper* No. 16. Harvard Univ. Petersham, Mass. 12 pp.

Lyr, H. and G. Hoffmann. 1967. Growth rates and growth periodicity of tree roots. *Int. Rev. For. Res.* 2:181–236.

MacArthur, R. H. 1972. *Geographical Ecology: Patterns in the Distribution of Species*. Harper and Row, New York. 269 pp.

MacArthur, R. H., and J. MacArthur. 1961. On bird species diversity. *Ecology* 42:594–598.

MacDonald, G. M., and L. C. Cwynar. 1985. A fossil pollen based reconstruction of the late Quaternary history of lodgepole pine (*Pinus contorta* ssp. *latifolia*) in the western interior of Canada. *Can. J. For. Res.* 15:1039–1044.

MacDougal, D. T. 1900. Influence of inversions of temperature, ascending and descending currents of air, upon distribution. Biological Lectures Delivered at the Marine Biological Laboratory, Woods Hole 1899. 1900:37–47.

Mackey, H. E., Jr., and N. Sivec. 1973. The present composition of a former oak-chestnut forest in the Allegheny Mountains of western Pennsylvania. *Ecology* 54:915–919.

MacKinnon, A., D. Meidinger, and K. Klinka. 1992. Use of biogeoclimatic ecosystem classification system in British Columbia. *For. Chron.* 68:100–120.

Madgwick, H. A. I., and J. D. Ovington. 1959. The chemical composition of precipitation in adjacent forest and open plots. *Forestry* 32:14–22.

Maguire, W. P. 1955. Radiation, surface temperatures, and seedling survival. *For. Sci.* 1:277–285.

Magurran, A. E. 1988. *Ecological Diversity and Its Measurement*. Princeton Univ. Press, Princeton, NJ. 179 pp.

Maienschein, J. 1994. Pattern and process in early studies of Arizona's San Francisco peaks. *BioScience* 44:479–485.

Major, J. 1951. A functional, factorial approach to plant ecology. *Ecology* 32:392–412.

Major, J. 1969. Historical development of the ecosystem concept. *In* G. M. Van Dyne (ed.), *The Ecosystem Concept in Natural Resource Management*. Academic Press, New York.

Malanson, G. P. 1993. *Riparian Landscapes*. Cambridge Univ. Press, Cambridge. 296 pp.

Marchand, D. E. 1971. Rates and modes of denudation, White Mountains, eastern California. *Amer. J. Sci.* 270:109–135.

Margalef, R. 1972. Homage to Evelyn Hutchinson, or why is there an upper limit to diversity. *Trans. Connect. Acad. Arts Sci.* 44:211–235.

Marion, G. M., and C. H. Black. 1988. Potentially available nitrogen and phosphorus along a chaparral fire cycle chronosequence. *Soil Sci. Soc. Amer. J.* 52:1155–1162.

Marks, P. L. 1974. The role of pin cherry (*Prunus pensylvanica* L.) in the maintenance of stability in northern hardwood ecosystems. *Ecol. Monogr.* 44:73–88.

Marks, P. L. 1975. On the relation between extension growth and successional status of deciduous trees of the northeastern United States. *Bull. Torrey Bot. Club* 102:172–177.

Marks, P. L., and P. A. Harcombe. 1981a. Community diversity of Coastal Plain forests in southern East Texas. *Ecology* 56:1004–1008.

Marks, P. L., and P. A. Harcombe. 1981b. Forest vegetation of the Big Thicket, southeast Texas. *Ecol. Monogr.* 51:287–305.

Marquis, D. A. 1974. The impact of deer browsing on Allegheny hardwood regeneration. USDA For. Serv. Res. Paper NE-308. Northeastern For. Exp. Sta., Upper Darby, PA. 8 pp.

Marquis, D. A. 1975. The Allegheny hardwood forests of Pennsylvania. USDA For. Serv. Gen. Tech. Report NE-15. Northeastern For. Exp. Sta., Upper Darby, PA. 32 pp.

Marquis, D. A. 1981. Effect of deer browsing on timber in Allegheny hardwood forests of northwestern Pennsylvania. USDA For. Serv. Res. Paper NE-475. Northeastern For. Exp. Sta., Upper Darby, PA. 10 pp.

Marr, 1977. The development and movement of tree islands near the upper limit of tree growth in the southern Rocky Mountains. *Ecology* 58:1159–1164.

Marschner, H., M. Häussling, and E. George. 1991. Ammonium and nitrate uptake rates and rhizosphere pH in non-mycorrhizal roots of Norway spruce [*Picea abies* (L.) Karst.]. *Trees* 5:14–21.

Marsh, G. P. 1864. *Man and Nature; or Physical Geography as Modified by Human Action.* Reprinted 1965, Belknap Press of Harvard Univ. Press. Cambridge MA. 472 pp.

Marshall, P. E., and T. T. Kozlowski. 1976. Importance of photosynthetic cotyledons for early growth of woody angiosperms. *Physiol. Plant.* 37:336–340.

Marshall, P. E., and T. T. Kozlowski. 1977. Changes in structure and function of epigeous cotyledons of woody angiosperms during early seedling growth. *Can. J. Bot.* 55:208–215.

Martin, P. S., and B. E. Harrell. 1957. The Pleistocene history of temperate biotas in Mexico and eastern United States. *Ecology* 38:468–480.

Martin, R. E., R. W. Cooper, A. B. Crow, J. A. Cuming, and C. B. Phillips. 1977. Report of task force on prescribed burning. *J. For.* 75:297–301.

Martin, R. E., and D. B. Sapsis. 1992. Fires as agents of biodiversity: pyrodiversity promotes biodiversity. *In* R. R. Harris, D. C. Erman, and H. M. Kerner (eds.), *Symp. Proc., Biodiversity of Northwestern California.* Univ. Calif. Dept. For. and Res. Manage. Berkeley.

Marx, D. H. 1969a. The influence of ectotrophic mycorrhizal fungi on the resistance of pine roots to pathogenic infections. I. Antagonism of mycorrhizal fungi to root pathogenic fungi and soil bacteria. *Phytopathology* 59:153–163.

Marx, D. H. 1969b. The influence of ectotrophic mycorrhizal fungi on the resistance of pine roots to pathogenic infections. II. Production, identification, and biological activity of antibiotics produced by *Leucopaxillus cerealis* var. *piceina. Phytopathology* 59:411–417.

Marx, D. H. 1973. Mycorrhizae and feeder root diseases. *In* G. C. Marks and T. T. Kozlowski (eds.), *Ectomycorrhizae—Their Ecology and Physiology.* Academic Press, New York.

Marx, D. H. 1975. Mycorrhizae and establishment of trees on strip-mined land. *Ohio J. Sci.* 75:288–297.

Marx, D. H., and D. J. Beattie. 1977. Mycorrhizae—promising aid to timber growers. *For. Farmer* 36:6–9.

Maser, C., and Z. Maser. 1988. Interactions among squirrels, mycorrhizal fungi, and coniferous forests in Oregon. *Great Basin Nat.* 48:358–369.

Maser, C., and J. R. Sedell. 1994. *From the Forest to the Sea.* St. Lucie Press, Delray Beach, FL. 200 pp.

Maser, C. R., F. Tarrant, J. M. Trappe, and J. F. Franklin (eds.). 1988. From the forest to the sea: A story of fallen trees. USDA For. Serv. Gen. Tech. Report PNW-229. Washington, D.C. 153 pp.

Maser, C., J. M. Trappe, and R. A. Nussbaum. 1978. Fungal—small mammal interrelationships with emphasis on Oregon conifer forests. *Ecology* 59:799–809.

Mason, D. T. 1915. The lodgepole pine zone in the Rocky Mountains. *USDA Bull.* 154. Washington, D.C. 35 pp.

Mather, K. 1943. Polygenic inheritance and natural selection. *Biol. Rev.* 18:32–64.

Matson, P., P. M. Vitousek, J. T. Ewel, M. J. Mazzarino, and G. P. Robertson. 1987. Nitrogen transformations following tropical forest felling and burning on a volcanic soil. *Ecology* 68:491–502.

Matthews, J. D. 1955. The influence of weather on the frequency of beech mast years in England. *Forestry* 28:107–115.

Matthews, J. D. 1963. Factors affecting the production of seed by forest trees. *For. Abstracts* 24(1):i-xiii.

Matthews, J. A. 1992. *The Ecology of Recently Deglaciated Terrain.* Cambridge Univ. Press. Cambridge. 386 pp.

Mattoon, W. R. 1915. The southern cypress. *USDA Bull.* 272. Washington, D.C. 74 pp.

Mattson, W. J, R. K. Lawrence, R. A. Haack, D. A. Herms, and P-J Charles. 1988. Defensive strategies of woody plants against different insect-feeding guilds in relation to plant ecological strategies and intimacy of association with insects. *In* W. J. Mattson, J. Levieux, and C. Bernard-Dagan (eds.), *Mechanism of Woody Plant Defenses Against Insects.* Springer-Verlag, New York.

Mattson, W. J., and J. M. Scriber. 1987. Nutritional ecology of insect folivores of woody plants: nitrogen, water, fiber, and mineral considerations. *In* F. Slansky Jr. and J. G. Rodriguez (eds.), *Nutritional Ecology of Insects, Mites, and Spiders.* John Wiley, New York.

Mattson, W. J., and N. D. Addy. 1975. Phytophagous insects as regulators of forest primary production. *Science* 190:515–522.

Mattson, W. J., and R. A. Haack. 1987. The role of drought in outbreaks of plant-eating insects. *BioScience* 37:110–118.

Mattson, W. J., D. A. Herms, J. A. Witter, and D. C. Allen. 1991. Woody plant grazing systems: North American outbreak folivores and their host plants. *In* Y. N. Baranchikov, W. J. Mattson, F. P. Hain, and T. L. Payne (eds.), *Forest Insect Guilds: Patterns of Interaction with Host Trees.* USDA For. Serv. Gen. Tech. Report NE-153. Northeastern For. Exp. Sta., Radnor, PA.

Mattson, W. J., P. Niemelä, I. Millers, and Y. Inguanzo. 1994. Immigrant phytophagous insects on woody plants in the United States and Canada: an annotated list. USDA For. Serv. Gen Tech. Report NC-169. North Central For. Exp. Sta., St. Paul, MN. 27 pp.

Mátyás, C., and C. W. Yeatman. 1992. Effect of geographical transfer on growth and survival of jack pine (*Pinus banksiana* Lamb.) populations. *Silvae Genetica* 41:370–376.

Mayfield, H. M. 1962. 1961 decennial census of Kirtland's warbler. *Auk* 79:173–182.

Mayfield, H. M. 1972. Third decennial census of Kirtland's warbler. *Auk* 89:262–268.

McAndrews, J. H. 1966. Postglacial history of prairie, savanna and forest in northwestern Minnesota. *Torreya Bot. Club Mem.* 22 (2). 72 pp.

McBride, J. 1973. Natural replacement of disease-killed elms. *Am. Midl. Nat.* 90:300–306.

McClanahan, T. R. 1986. The effect of a seed source on primary succession in a forest ecosystem. *Vegetatio* 65:175–178.

McCune, B., and T. F. H. Allen. 1985. Forest dynamics in the Bitterroot Canyons, Montana. *Can. J. Bot.* 63:377–383.

McCune, B., and G. Cottam. 1985. The successional status of a southern Wisconsin oak woods. *Ecology* 66:1270–1278.

McDermott, R. E. 1954. Seedling tolerance as a factor in bottomland timber succession. *Mo. Agr. Exp. Sta. Res. Bull.* 557. 11 pp.

McDonnell, M. J., and S. T. A. Pickett. 1993. *Humans as Components of Ecosystems.* Springer-Verlag, New York. 364 pp.

McFarlane, R. W. 1992. *A Stillness in the Pines: The Ecology of the Red-Cocaded Woodpecker.* W. W. Norton, New York. 270 pp.

McIntosh, R. P. 1967. The continuum concept of vegetation. *Bot. Rev.* 33:130–187.

McIntosh, R. P. 1968. Reply. *In* P. Dansereau (ed.), The continuum concept of vegetation: responses. *Bot. Rev.* 34:253–332.

McIntosh, R. P. 1981. Succession and ecological theory. *In* D. C. West, H. H. Shugart and D. B. Botkin (eds.), *Forest Succession: Concepts and Application* . Springer-Verlag, New York.

McIntosh, R. P. 1985. *The Background of Ecology: Concept and Theory*. Cambridge Univ. Press, Cambridge. 383 pp.

McIntosh, R. P. 1993. The continuum continued: John T. Curtis' influence on ecology. *In* J. S. Fralish, R. P. McIntosh, and O. L. Loucks (eds.), *Fifty Years of Wisconsin Plant Ecology*. Univ. Wisconsin. Press, Madison.

McIntyre, L. 1985. Humboldt's way. *National Geographic* 168(3):318–351.

McKevlin, M. R., D. D. Hook, and W. H. McKee, Jr. 1995. Growth and nutrient use efficiency of water tupelo seedlings in flooded and well-drained soil. *Tree Physiol.* 15:753–758.

McKnight, J. S., D. D. Hook, O. G. Langdon, and R. L. Johnson. 1981. Flood tolerance and related characteristics of trees of the bottomland forests of the southern United States. *In* J. R. Clark and J. Benforado (eds.), *Wetlands of Bottomland Hardwood Forests*. Elsevier, Amsterdam.

McMinn, J. W., and R. A. Hardt. 1996. Accumulations of coarse woody debris in southern forests. *In* J. W. McMinn and D. A. Crossley, Jr. (eds.), *Proc. Workshop, Biodiversity and Coarse Woody Debris in Southern Forests*. USDA For. Serv. Gen. Tech. Report SE-94. Southern Res. Sta., Asheville, NC.

McNab, W. H. 1987. Yellow-poplar site quality related to slope type in mountainous terrain. *Northern J. Applied For.* 4:189–192.

McNab, W. H. 1989. Terrain shape index: quantifying effect of minor landforms on tree height. *For Sci.* 35:91–104.

McNab, W. H. 1991. Land classification in the Blue Ridge Province: state-of-the-science report. *In* D. L. Mengel and D. T. Tew (eds.), *Symp. Proc. Ecological Land Classification: Applications to Identify the Productive Potential of Southern Forests*. USDA For. Serv. Gen. Tech. Report SE-68. Southeastern For. Exp. Sta., Asheville, NC.

McNab, W. H. 1993. A topographic index to quantify the effect of mesoscale landform on site productivity. *Can. J. For. Res.* 23:1100–1107.

McNeely, J. A. 1989. *Economics and Biological Diversity: Developing and Using Economic Incentives to Conserve Biological Resources*. Columbia Univ. Press. New York. 232 pp.

Meades, W. J., and B. A. Roberts. 1992. A review of forest site classification activities in Newfoundland and Labrador. *For. Chron.* 68:25–33.

Meeker, D. O., Jr., and D. L. Merkel. 1984. Climax theories and a recommendation for vegetation classification—a viewpoint. *J. Range Manage.* 37:427–430.

Meentemeyer, V. 1978a. Climatic regulation of decomposition rates of organic matter in terrestrial ecosystems. *In* D. C. Adriano and I. L. Brisbin, Jr. (eds.), *Environmental Chemistry and Cycling Processes*. CONF 760429, National Technical Information Service, Springfield, VA.

Meetemeyer, V. 1978b. Macroclimate and lignin control of litter decomposition rates. *Ecology* 59:465–472.

Meeuwig, R. O. 1971. Infiltration and water repellency in granitic soils. USDA For. Serv. Res. Paper INT-111. Intermountain For. and Rge. Exp. Sta., Ogden, UT. 20 pp.

Meffe G. K., and C. R. Carroll. 1997. *Principles of Conservation Biology*, 2nd ed. Sinauer Associates, Sunderland, MA. 673 pp.

Meinecke, E. P. 1928. A camp ground policy. Report, California Dept. Nat. Res., Parks Div., Sacramento, Calif. 16 pp.

Meiners, T. M., D. Wm. Smith, T. L. Sharik, and D. E. Beck. 1984. Soil and plant water stress in an Appalachian oak forest in relation to topography and stand age. *Plant and Soil* 80:171–179.

Melillo, J. M. 1981. Nitrogen cycling in deciduous forests. *In* F. E. Clark and T. Rosswell (eds.), *Terrestrial Nitrogen Cycles*. Ecol. Bull. 33. Swedish Natural Sci. Res. Council, Stockholm.

Melillo, J. M., J. D. Aber, and J. F. Muratore. 1982. Nitrogen and lignin control of hardwood leaf litter decomposition dynamics. *Ecology* 63:621–626.

Melillo, J. M., J. D. Aber, P. A. Steudler, and J. P. Schimel. 1983. Denitrification potentials in a successional sequence of northern hardwood forest stands. *In* R. Hallberg (ed.), *Environmental Biogeochemistry*. Ecol. Bull. 35. Swedish Natural Sci. Res. Council, Stockholm.

Melillo, J. M., A. D. McGuire, D. W. Kicklighter, B. Moore III, C. J. Vorosmarty, and A. L. Schloss. 1993. Climate change and terrestrial net primary productivity. *Nature* 363:234–363.

Menaut, J. C., M. Lepage, and L. Abbadie. 1995. Savannas, woodlands, and dry forests in Africa. *In* Bullock, S. H., Mooney, H. A., and E. Medina (eds.), *Seasonally Dry Tropical Forests*. Cambridge Univ. Press, Cambridge.

Merriam, C. H. 1890. Results of a biological survey of the San Francisco Mountain region and the desert of the Little Colorado, Arizona. *U.S. Dept. Agr. North Amer. Fauna* 3:1–136.

Messier C., and P. Bellefleur. 1988. Light quantity and quality on the forest floor of pioneer and climax stages in a birch-beech-sugar maple stand. *Can. J. For. Res.* 18:615–622.

Metz, L. J., T. Lotti, and R. A. Klawitter. 1961. Some effects of prescribed burning on coastal plain forest soil. USDA For. Serv. Sta. Paper 133. Southeastern For. Exp. Sta., Asheville, NC. 10 pp.

Meyers, R. K., R. Zahner, and S. M. Jones. 1986. Forest habitat regions of South Carolina. Clemson Univ., Dept. Forestry Res. Series No. 42. 31 pp. + map.

Miles, J. 1979. *Vegetation Dynamics*. Chapman and Hall, London. 80 pp.

Miles, J. 1985. The pedogenic effects of different species and vegetation types and the implications of succession. *J. Soil Science* 36:571–584.

Miles, J. 1987. Vegetation succession: past and present perceptions, *In* A. J. Gray, M. J. Crawley, and P. J. Edwards (eds.), *Colonization, Succession and Stability*. Blackwell, Oxford.

Miller, G. T., Jr. 1992. *Living in the Environment: An Introduction to Environmental Science*, 7th ed. Wadsworth, Belmont, CA. 705 pp.

Miller, H. A., and S. H. Lamb. 1985. *Oaks of North America*. Naturegraph Publ., Happy Camp, CA. 327 pp.

Mills, L. S., M. E. Soulé, and D. F. Doak. 1993. The keystone-species concept in ecology and conservation. *BioScience* 43:219–224.

Mirov, N. T. 1967. *The Genus "Pinus."* Ronald Press, New York. 602 pp.

Mitchell, H. L., and R. F. Chandler, Jr. 1939. The nitrogen nutrition and growth of certain deciduous trees of northeastern United Sates. *Black Rock For. Bull.* 11. 94 pp.

Mitton, J. B., and M. C. Grant. 1980. Observations on the ecology and evolution of quaking aspen, *Populus tremuloides*, in the Colorado Front Range. *Amer. J. Bot.* 67:202–209.

Mladenoff, D. J., and F. Stearns. 1993. Eastern hemlock regeneration and deer browsing in the northern Great Lakes region: a re-examination and model simulation. *Conserv. Biology* 7:889–900.

Moehring, D. H., and I. W. Rawls. 1970. Detrimental effects of wet weather logging. *J. For.* 68:166–167.

Moehring, D. H., C. X. Grano, and J. R. Bassett. 1966. Properties of forested loess soils after repeated prescribed burns. USDA For. Serv. Res. Note SO-40. Southern For. Exp. Sta., New Orleans, LA. 4 pp.

Mohn, C. A., and S. Pauley. 1969. Early performance of cottonwood seed sources in Minnesota. *Minn. For. Notes*, 207. 4 pp.

Molina, R., N. Vance, J. F. Weigand, D. Pilz, and M. P. Amaranthus. 1997. Special forest products: integrating social, economic, and biological considerations into ecosystem management. *In* K. A. Kohm and J. F. Franklin (eds.), *Creating a Forestry for the 21st Century*. Island Press, Washington, D.C.

Monk, C. D. 1966. An ecological study of hardwood swamps in north-central Florida. *Ecology* 47:649–654.

Monk, C. D. 1967. Tree species diversity in the eastern deciduous forest with particular reference to north central Florida. *Am. Nat.* 101:173–187.

Monserud, R. A. 1984. Problems with site index: an opinioned review. *In* J. G. Bockheim (ed.), *Symp. Proc. Forest Land Classification: Experience, Problems, Perspectives,* NCR-102 North Central For. Soil Com., Soc. Am. For., USDA For. Serv., USDA Cons. Serv., Madison, WI.

Monserud, R. A., and G. E. Rehfeldt. 1990. Genetic and environmental components of variation of site index in inland Douglas-fir. *For Sci.* 36:1–9.

Mooney, H. A. 1972. The carbon balance of plants. *Ann. Rev. Ecol. Syst.* 3:315–346.

Moosmayer, H. -U. 1957. Zur ertragskundlichen Auswertung der Standortsgliederung im Osteil der Schwäbischen Alb. *Mitt. Vereins forstl. Standortsk. Forstpflz.* 7:3–41.

Mopper, S., and T. G. Whitham. 1992. The plant stress paradox: effects on pinyon sawfly sex ratios and fecundity. *Ecology* 73:515–525.

Mopper, S., J. Maschinski, N. Cobb., and T. G. Whitham. 1991a. A new look at habitat structure: consequences of herbivore-modified plant architecture. *In* S. Bell, E. McCoy, and H. Mushinsky (eds.), *Habitat Complexity: The Physical Arrangement of Objects in Space*. Chapman and Hall, New York.

Mopper, S., J. B. Mitton, T. G. Whitham, N. S. Cobb, and K. M Christensen. 1991b. Genetic differentiation and heterozygosity in pinyon pine associated with resistance to herbivory and environmental stress. *Evolution* 45:989–999.

Mopper, S., T. G. Whitham, and P. W. Price. 1990. Plant phenotype and interspecific competition between insects determine sawfly performance and density. *Ecology* 71:2135–2144.

Morgan, P., and S. C. Bunting. 1990. Fire effects in whitebark pine forests. *In* W. C. Schmidt and K. J. McDonald (comps.), *Symp. Proc. Whitebark Pine Ecosystems: Ecology and Management of a High-Mountain Resource*. USDA For. Serv. Gen Tech. Rep. INT-270, Intermountain Res. Sta., Odgen, UT.

Morgenstsern, E. K. 1996. *Geographic Variation in Forest Trees*. Univ. British Columbia Press, Vancouver, B.C. 209 pp.

Morris, R. F. 1951. The effects of flowering on the foliage production and growth of balsam fir. *For. Chron.* 27:40–57.

Morrison, P. H., and F. J. Swanson. 1990. Fire history and pattern in a Cascade Range landscape. USDA For. Serv. Gen. Tech. Report PNW-GTR-254, Pacific Northwest Res. Sta., Corvallis, OR. 77 pp.

Mosquin, T. 1966. Reproductive specialization as a factor in the evolution of the Canadian flora. *In* R. L. Taylor and R. A. Ludwig (eds.), *The Evolution of Canada's Flora*. Univ. Toronto Press, Toronto.

Mosquin, T., P. G. Whiting, and D. E. McAllister. 1994. *Canada's Biodiversity: the Variety of Life, Its Status, Economic Benefits, Conservation Costs and Unmet Needs*. Canadian Centre for Biodiversity, Canada Museum of Nature, Ottawa. 293 pp.

Moss, C. E. 1913. *Vegetation of the Peak District*. Cambridge Univ. Press, Cambridge. 235 pp.

Moss, E. H. 1932. The vegetation of Alberta IV. The polar association and related vegetation of central Alberta. *J. Ecol.* 20:380–415.

Mount, A. B. 1964. The interdependence of the eucalyptus and forest fires in southern Australia. *Aust. Forestry* 28:166–172.

Mueggler, W. F. 1985. Vegetation associations. *In* N. V. DeByle and R. P. Winokur (eds.), *Aspen: Ecology and Management in the Western United States*. USDA For. Serv. Gen. Tech. Report RM-119. Rocky Mountain For. and Rge. Exp. Sta., Ft. Collins, CO.

Mueggler, W. R., and R. B. Campbell. 1986. Aspen community types of Utah. USDA For. Serv. Res. Paper INT-362. Intermountain Res. Sta., Ogden, UT. 69 pp.

Mueller-Dombois, D. 1987. Natural dieback in forests. *BioScience* 37:575–583.

Mueller-Dombois, D. 1991. The mosaic theory and the spatial dynamics of natural dieback and regeneration in Pacific forests. *In* H. Remmert (ed.), *The Mosaic-Cycle Concept of Ecosystems*. (Ecological Studies Vol. 85). Springer-Verlag, Berlin.

Mueller-Dombois, D., and H. Ellenberg. 1974. *Aims and Methods of Vegetation Ecology*. John Wiley, New York. 547 pp.

Mühlhäusser, G., and S. Müller. 1995. Das südwestdeutsche Verfahren der Forstlichen Standortskartierung. *In* Arbeitsgruppe Biozönosenkunde. 1995. *Ansätze für eine Regionale Biotop und Biozönosenkunde von Baden-Württemberg*. Mitt. Forst. Versuchs u. Forschungsanst. Baden-Württemberg. Freiburg, Germany. 166 pp.

Mühlhäusser, G., W. Hubner, and G. Sturmmer. 1983. Die Forstliche Standortskarte 1:10,000 nach dem Baden-Württembergischen Verfahren. *Mitt. Verein forstl. Standortsk. u. Forstpflz.* 30:3–13.

Muir, J. 1894. *The Mountains of California*. Century, New York. 318 pp.

Muir, P. S., and J. E. Lotan. 1985. Disturbance history and serotiny in *Pinus contorta* in western Montana. *Ecology* 66:1658–1668.

Musselman, R. C., D. T. Lester, and M. S. Adams. 1975. Localized ecotypes of *Thuja occidentalis* L. in Wisconsin. *Ecology* 56:647–655.

Mutch, R. W. 1970. Wildland fires and ecosystems—a hypothesis. *Ecology* 51:1046–1051.

Mutch, R. W. 1976. Fire management and land use planning today: tradition and change in the Forest Service. *Western Wildlands* 3:13–19.

Myers, R. L. 1990. Scrub and high pine. *In* R. L. Myers and J. J. Ewel (eds.), *Ecosystems of Florida*. Univ. Central Florida Press, Orlando.

Myers, R. L., and J. J. Ewel (eds.). 1990. *Ecosystems of Florida*. Univ. Central Florida Press, Orlando. 765 pp.

Nadelhoffer K. J., J. D. Aber, and J. M. Melillo. 1985. Fine roots, net primary production, and soil nitrogen availability: a new hypothesis. *Ecology* 66:1377–1390.

Naess, A. 1986. Intrinsic value: will the defenders of nature please rise? *In* M. E. Soulé (ed.), *Conservation Biology: The Science of Scarcity and Diversity*. Sinauer Associates, Sunderland, MA.

Naveh, Z. 1974. Effects of fire in the Mediterranean region. *In* T. T. Kozlowski and C. E. Ahlgren (eds.), *Fire and Ecosystems*. Academic Press, New York.

Neilson, R. P., and L. H. Wullstein. 1980. Catkin freezing and acorn production in gambel oak in Utah, 1978. *Amer. J. Bot.* 67: 426–428.

Neufeldt, V., and D. B. Guralnik (eds.). 1988. *Webster's New World Dictionary*, Third College Edition. Simon & Schuster, New York. 1574 pp.

Nichols, J. O. 1968. Oak mortality in Pennsylvania, a ten year study. *J. For.* 66: 681–694.

Nienstaedt, H. 1974. Genetic variation in some phenological characteristics of forest trees. *In* H. Lieth (ed.), *Phenology and Seasonality Modeling*. Springer-Verlag, New York.

Nikles, D. G. 1970. Breeding for growth and yield. *Unasylva* 24(2–3):9–22.

Nilsen, E. T. 1985. Seasonal and diurnal leaf movements of *Rhododendron maximum* L. in contrasting irradiance environments. *Oecologia* 65:296–302.

Nilsen, E. T. 1987. Influence of water relations and temperature on leaf movements of Rhododendron species. *Plant Physiology* 83:607–612.

Nilsen, E. T. 1990. Why do *Rhododendron* leaves curl? *Arnoldia* 50:30–35.

Nilsen, E. T. 1992. Thermonastic leaf movements: a synthesis of research with Rhododendron. *Bot. J. Linnean Society* 110:205–233.

Nilsson, T. 1983. *The Pleistocene, Geology and Life in the Quaternary Ice Age*. D. Reidel, Dordrecht. 651 pp.

Nobel, D. I., and R. R. Alexander. 1977. Environmental factors affecting natural regeneration of Engelmann spruce in the central Rocky Mountains. *For. Sci.* 23:420–429.

Noble, I. R., and R. O. Slatyer. 1977. Post-fire succession of plants in Mediterranean ecosystems. *In* H. A. Mooney and C. E. Conrad (eds.), Symp. Proc. *Environmental Consequences of Fire and Fire Management in Mediterranean Ecosystems*. USDA For. Serv. Gen. Tech. Report WO-3. Washington, D.C.

Noble, I. R., and R. O. Slatyer. 1980. The use of vital attributes to predict successional changes in plant communities subject to recurrent disturbances. *Vegetatio* 43:5–21.

Norse, E. A. 1990. *Ancient Forests of the Pacific Northwest*. Island Press. Washington, D.C. 327 pp.

Norton, B. G. 1982. Environmental ethics and nonhuman rights. *Environmental Ethics* 4:17–36.

Norton, D. A., and D. Kelly. 1988. Mast seeding over 33 years by *Dacrydium cupressinum* Lamb. (rimu) (Podocarpaceae) in New Zealand: the importance of economies of scale. *Functional Ecol.* 2:399–408.

Noss, R. F., E. T. LaRoe III, and J. M. Scott. 1995. Endangered ecosystems of the United States: a preliminary assessment of loss and degradation. USDI, Nat. Biol. Serv., Biol. Report 28. Washington D.C. 58 pp.

Nowacki, G. J., and M. D. Abrams. 1991. Community and edaphic analysis of mixed oak forests in the ridge and valley province of central Pennsylvania. *In* L. H. McCormick and K. W. Gottschalk (eds.), *Proc. Eighth Central Hardwood Conference*. USDA, For. Serv. Gen. Tech. Report NE-148. Northeastern For. Exp. Sta., Broomall, PA.

Nowacki, G. J., M. D. Abrams, and C. G. Lorimer. 1990. Composition, structure, and historical development of northern red oak stands along an edaphic gradient in north-central Wisconsin. *For. Sci.* 36:276–292.

Noy-Meir, I., and E. van der Maarel. 1987. Relations between community theory and community analysis in vegetation science: some historical perspectives. *Vegetatio* 69:5–15.

Nye, P. H. 1966. The effect of the nutrient intensity and buffering power of a soil, and the absorbing power, size and root hairs of a root, on nutrient absorption by diffusion. *Plant and Soil* 25:81–105.

Nye, P. H. 1977. The rate-limiting step in plant nutrient absorption from soil. *Soil Science* 123:292–297.

Nye, P. H., and P. B. Tinker. 1977. *Solute Movement in the Soil–Root System.* Blackwell, Oxford. 342 pp.

Primack, R. B. 1993. *Essentials of Conservation Biology.* Sinauer Associates, Sunderland, MA. 564 pp.

O'Dea, M. E., J. C. Zasada, and J. C. Tappeiner III. 1995. Vine maple clone growth and reproduction in managed and unmanaged coastal Oregon Douglas-fir forests. *Ecol. Applications* 5:63–73.

O'Neill, R. V., and D. L. DeAngelis. 1981. Comparative productivity and biomass relations of forest ecosystems. *In* D. E. Reichle (ed.), *Dynamic Properties of Forest Ecosystems,* Int. Biol. Programme 23. Cambridge Univ. Press, Cambridge.

Oberdorfer, E. 1990. *Pflanzensoziologische Exkursionsflora.* Verlag Eugen Ulmer, Stuttgart. 1050 pp.

Odum, E. P. 1971. *Fundamentals of Ecology,* 3rd ed. W. B. Saunders, Philadelphia. 574 pp.

Oechel, W. C., and W. T. Lawrence. 1985. Taiga. *In* B. F. Chabot and H. A. Mooney (eds.), *Physiological Ecology of North American Plant Communities.* Chapman and Hall, New York.

Oldeman, R. A. A. 1990. *Forests: Elements of Silvology.* Springer-Verlag, New York. 624 pp.

Oldemeyer, J. L., A. W. Franzmann, A. L. Brundage, P. D. Arenson, and A. Flynn. 1977. Browse quality and the Kenai moose population. *J. Wildlife Manage.*41:533–542.

Oliver, C. D. 1981. Forest development in North America following major disturbances. *For. Ecol. Manage.* 3:153–168.

Oliver, C. D., and B. C. Larson. 1990. *Forest Stand Dynamics.* McGraw-Hill, New York. 467 pp.

Ollinger, S. V., J. D. Aber, G. M. Lovett, S. E. Millham, R. G. Lathrop, and J. M. Ellis. 1993. A spatial model of atmospheric deposition for the northeastern U.S. *Ecol. Applications* 3:459–472.

Olson, J. S. 1958. Rates of succession and soil changes on southern Lake Michigan sand dunes. *Bot. Gaz.* 119:125–170.

Olson, J. S. 1963. Energy storage and the balance of producers and decomposers in ecological systems. *Ecology* 44:322–331.

Olson, S. R., W. A. Reiners, C. S. Cronan, and G. E. Lang. 1981. The chemistry and flux of throughfall and stemflow in subalpine balsam fir forests. *Holarctic Ecology* 4:291–300.

Omernik, J. M. 1987. Ecoregions of the conterminous United States. *Ann. Assoc. Amer. Geog.* 77:118–125.

Oosting, H. J. 1942. An ecological analysis of the plant communities of Piedmont, North Carolina. *Am. Midl. Nat.* 28:1–126.

Oosting, H. J. 1956. *The Study of Plant Communities,* 2nd ed. Freeman, San Francisco CA. 440 pp.

Osawa, A., and R. B. Allen. 1993. Allometric theory explains self-thinning relationships of mountain beech and red pine. *Ecology* 74:1020–1032.

Osterkamp, W. R., and E. R. Hupp. 1984. Geomorphic and vegetative characteristics along three northern Virginia streams. *Bull. Geol. Soc. Am.* 95:1093–1101.

Ovington, J. D. (ed.). 1983. *Ecosystems of the World, Temperate Broad-Leaved Evergreen Forests, Vol. 10.* Elsevier, New York. 242 pp.

Owens, J. N. 1991. Flowering and seed set. *In* A. S. Raghavendra (ed.), *Physiology of Trees.* John Wiley, New York.

Owens, J. N., M. Molder, and H. Langer. 1977. Bud development in *Picea glauca.* I. Annual growth cycle of vegetative buds and shoot elongation as they relate to date and temperature sums. *Can. J. Bot.* 55:2728–2745.

Paine, R. T. 1969. A note on trophic complexity and community stability. *Am. Nat.* 103:91–93.

Palik, B. J., and K. S. Pregitzer. 1994. White pine seed-tree legacies in an aspen landscape: influences on post-disturbance white pine population structure. *For. Ecol. and Manage.* 67:191–201.

Pallardy, S. G., T. A. Nigh, and H. E. Garrett. 1988. Changes in forest composition in central Missouri: 1968–1982. *Am. Midl. Nat.* 120:380–389.

Pallman, H., and E. Frei. 1943. Beitrag zur Kenntnis der Lokalklimat einiger kennzeichnender Waldgesellschaften des schweizerischen Nationalparkes. *Ergeb. wissens. Untersuch. schweiz. Nationalparkes* 1:437–464.

Park, Y. S., and D. P. Fowler. 1988. Geographic variation of black spruce tested in the maritimes. *Can. J. For. Res.* 18:106–114.

Parker, A. J., and K. C. Parker. 1994. Structural variability of mature lodgepole pine stands on gently sloping terrain in Taylor Park Basin, Colorado. *Can. J. For. Res.* 24: 2020–2029.

Parker, J. 1963. Cold resistance in woody plants. *Bot. Rev.* 29:123–201.

Parks, C. R., N. G. Miller, J. F. Wendel, and K. M. McDougal. 1983. Genetic divergence within the genus *Liriodendron* (Magnoliaceae). *Ann. Missouri Bot. Gard.* 70:658–666.

Pastor, J., J. D. Aber, C.A. McClaugherty, and J. M. Melillo. 1984. Aboveground production and N and P cycling along a nitrogen mineralization gradient on Blackhawk Island, Wisconsin. *Ecology* 65:256–268.

Pastor, J., R. J. Naiman, B. Dewey, and P. McInnes. 1988. Moose, microbes, and the boreal forest. *BioScience* 38:770–777.

Pastor, J., and W. M. Post. 1986. Influence of climate, soil moisture, and succession on forest carbon and nitrogen cycles. *Biogeochemistry* 2:3–27.

Patric, J. H., and P. E. Black. 1968. Potential evapotranspiration and climate in Alaska by Thornthwaite's classification. USDA For. Serv. Res. Paper PNW-71. Pacific Northwest For. and Rge. Exp. Sta., Portland, OR. 28 pp.

Patric, J. H., and J. D. Helvey. 1986. Some effects of grazing on soil and water in the eastern forest. USDA For. Serv. Gen. Tech. Report NE-115, Northeastern For. Exp. Sta., Broomall, PA. 24 pp.

Patterson, J. C. 1976. Soil compaction and its effects upon urban vegetation. *In* F. S. Santamour, Jr., H. D. Gerhold, and S. Little (eds.), *Better Trees for Metropolitan Landscapes: Symposium Proceedings.* USDA For. Serv. Gen. Tech. Report NE-22. Northeastern For. Exp. Sta., Upper Darby, PA.

Patterson, W. A, III, and A. E. Backman. 1988. Fire and disease history of forests. *In* B. Huntley and T. Webb, III (eds.), *Vegetation History.* Kluwer, Dordrecht.

Paul, E. A., and F. E. Clark. 1996. *Soil Microbiology and Biochemistry,* 2nd ed. Academic Press, New York. 340 pp.

Pauley, S. S. 1958. Photoperiodism in relation to tree improvement. *In* K. V. Thimann (ed.), *The Physiology of Forest Trees.* Ronald Press, New York.

Pauley, S. S., and T. O. Perry. 1954. Ecotypic variation of the photoperiodic response in *Populus. J. Arnold Arbor.* 35:167–188.

Pavlik, B. M., P. C. Muick, S. G. Johnson, and M. Popper. 1991. *Oaks of California.* Cachuna Press, Los Olivos, CA. 184 pp.

Pearce, F. 1993. Pinatubo points to vulnerable climate. *New Scientist* 138(1878):7.

Pearce, F. 1996. Lure of the rings. *New Scientist.* 152(2060):38–42.

Pearce, F. 1997. Lightning sparks pollution rethink. *New Scientist.* 153(2066):15.

Pearsall, D. R. 1995. *Landscape Ecosystems of the University of Michigan Biological Station: Ecosystem Diversity and Ground-Cover Diversity.* Ph.D. Thesis. University of Michigan, Ann Arbor. 396 pp.

Pearson, G. A. 1914. A meteorological study of parks and timbered areas in the western yellow-pine forests of Arizona and New Mexico. *Monthly Weather Rev.* 41:1615–1629.

Pearson, G. A. 1936. Some observations on the reaction of pine seedlings to shade. *Ecology* 17:270–276.

Pearson, G. A. 1940. Shade effects in ponderosa pine. *J. For.* 38:778–780.

Peet, R. K. 1974. The measurement of species diversity. *Ann. Rev. Ecol. Syst.* 5:285–307.

Peet, R. K. 1978. Forest vegetation of the Colorado Front Range: patterns of species diversity. *Vegetatio* 37:65–78.

Peet, R. K. 1981. Forest vegetation of the Colorado Front Range: composition and dynamics. *Vegetatio* 45:3–75.

Peet, R. K. 1988. Forests of the Rocky Mountains. *In* M. G. Barbour and W. D. Billings (eds.), *North American Terrestrial Vegetation.* Cambridge Univ. Press, Cambridge.

Peet, R. K. 1992. Community structure and ecosystem function. *In* Glenn-Lewin, D. C., R. K. Peet, and T. T. Veblen (eds.), *Plant Succession: Theory and Prediction.* Chapman and Hall, London.

Peet, R. K., and N. L. Christensen. 1980a. Succession: a population process. *Vegetatio* 43:131–140.

Peet, R. K., and N. L. Christensen. 1980b. Hardwood forest vegetation of the North Carolina piedmont. *Veröff. Geobot. Inst. Eddg. Tech. Hochsch.*, Stift. Rübel, Zürich 69:14–39.

Peet, R. K., and N. L. Christensen. 1987. Competition and tree death. *BioScience* 37: 586–595.

Penfound, W. T. 1952. Southern swamps and marshes. *Bot. Rev.* 18:413–446.

Penning de Vries, F. W. T. 1975. The cost of maintenance processes in plant cells. *Ann. Bot.* 39:77–92.

Perlin, J. 1989. *A Forest Journey, The Role of Wood in the Development of Civilization.* Harvard Univ. Press, Cambridge, MA. 445 pp.

Perry, J. P., Jr. 1991. *The Pines of Mexico and Central America.* Timber Press, Portland, OR. 231 pp.

Perry, P. O., H. E. Sellers, and C. O. Blanchard. 1969. Estimation of photosynthetically active radiation under a forest canopy with chlorophyll extracts and from basal area measurements. *Ecology* 50:39–44.

Perry, T. O. 1964. Soil compaction and loblolly pine growth. USDA For. Serv. *Tree Planters Notes* 69:9.

Perry, T. O. 1971. Dormancy of trees in winter. *Science* 171:29–36.

Perry, T. O., and C. W. Wang. 1960. Genetic variation in the winter chilling requirement for date of dormancy break for *Acer rubrum. Ecology* 41:790–794.

Peterson, C. J., and S. T. A. Pickett. 1995. Forest reorganization: a case study in an old-growth forest catastrophic blowdown. *Ecology* 76:763–774.

Peterson, C. J., W. P. Carson, B. C. McCarthy, and S. T. A. Pickett. 1990. Microsite variation and soil dynamics within newly created treefall pits and mounds. *Oikos* 58:39–46.

Petterssen, S. 1969. *Introduction to Meteorology,* 3rd ed. McGraw-Hill, New York. 333 pp.

Petty, R. O., and M. T. Jackson. 1966. Plant communities. *In* A. A. Lindsey (ed.), *Natural Features of Indiana.* Ind. Acad. Sci., Indianapolis.

Pfister, R. D., and S. F. Arno. 1980. Classifying forest habitats based on potential climax vegetation. *For. Sci.* 26:52–70.

Pfister, R., B. L. Kovalchik, S. F. Arno, and R. C. Presby. 1977. Forest habitat types of Montana. USDA For. Serv. Gen. Tech. Report INT-34. Intermountain For. and Rge. Exp. Sta., Ogden, UT. 174 pp.

Pharis, R. P., and C. G. Kuo. 1977. Physiology of gibberellins in conifers. *Can. J. For. Res.* 7:299–325.

Phillips, J. 1931. The biotic community. *J. Ecol.* 19:1–24.

Phillips, J. 1934–35. Succession, development, the climax, and the complex organism; an analysis of concepts. *J. Ecol.* 22:554–571; 23:210–246.

Phillips, J. 1974. Effects of fire in forest and savanna ecosystems of sub-Saharan Africa. *In.* T. T. Kozlowski and C. E. Ahlgren (eds.), *Fire and Ecosystems.* Academic Press, New York.

Phillips, W. S. 1963. Depth of roots in soil. *Ecology* 44:424.

Phipps, R. L. 1961. Analysis of five years dendrometer data obtained within three deciduous forest communities of Neotoma. Ohio Agr. Exp. Sta. Res. Circ. 105. 34 pp.

Pickett, S. T. A. 1988. Space-for-time substitution as an alternative to long-term studies. *In* G. E. Likens (ed.), *Long-term Studies in Ecology: Approaches and Alternatives.* Springer-Verlag. New York.

Pickett, S. T. A., S. L. Collins, and J. J. Armesto. 1987. Models, mechanisms and pathways of succession. *Bot. Rev.* 53:335–371.

Pickett, S. T. A., and P. S. White (eds.). 1985. *The Ecology of Natural Disturbance and Patch Dynamics.* Academic Press, New York. 472 pp.

Pickup, G. 1990. Remote sensing of landscape processes. *In* R. J. Hobbs and H. A. Mooney (eds.), *Remote Sensing of Biosphere Functioning.* Springer-Verlag, New York.

Pielou, E. C. 1975 . *Ecological Diversity.* John Wiley, Toronto. 363 pp.

Pielou, E. C. 1979. *Biogeography*. John Wiley, New York. 351 pp.

Pielou, E. C. 1991. *After the Ice Age*. Univ. Chicago Press, Chicago. 366 pp.

Pigott, C. D., and J. P. Huntley. 1981. Factors controlling the distribution of *Tilia cordata* at the northern limits of its geographical range. III. Nature and causes of seed sterility. *New Phytol.* 87:817–839.

Pike, L. H., W. C. Denison, D. M. Tracy, M. A. Sherwood, and F. M. Rhoades. 1975. Floristic survey of epiphytic lichens and bryophytes growing on old-growth conifers in western Oregon. *Bryol.* 78:389–402.

Pimm, S. L., G. E. Davis, L. Loope, C. T. Roman, T. J. Smith III, and J. T. Tilmant. 1994. Hurricane Andrew. *BioScience* 44:224–229

Pisek, A., W. Larcher, W. Moser, and I. Pack. 1969. Kardinale Temperaturbereiche der Photosynthese und Grenztemperaturen des Lebens der Blätter verschiedener Spermatophyten. III. Temperaturabhängigkeit und optimaler Temperaturbereich der Netto-Photosynthese. *Flora* (Abt. B) 158:608–630.

Platt, W. J. 1975. The colonization and formation of equilibrium plant species associations on badger disturbances in a tall-grass prairie. *Ecol. Monogr.* 45:285–305.

Polunin, N. 1960. *Introduction to Plant Geography*. McGraw-Hill, New York. 640 pp.

Poole, A. L. 1987. *Southern Beeches*. New Zealand Dept. Sci. and Industrial Res. Information Series No. 162. Sci. Information. Publ. Centre, Wellington, New Zealand. 148 pp.

Poore, M. E. D. 1955. The use of phytosociological methods in ecological investigations. *J. Ecol.* 43:226–269.

Poore, M. E. D. 1968. Studies in Malaysian rain forest. I. The forest of Triassic sediments in Jengka Forest Reserve. *J. Ecol.* 56:143–196.

Poulson, T. L., and W. J. Platt. 1996. Replacement patterns of beech and sugar maple in Warren Woods, Michigan. *Ecology* 77:1234–1253.

Poulson, T. L., and W. B. White. 1969. The cave environment. *Science* 165: 971–981.

Powers, R. F. 1980. Mineralizable soil nitrogen as an index of nitrogen availability to forest trees. *Soil Sci. Soc. Amer. J.* 44:1314–1320.

Power, M E., D. Tilman, J. A. Estes, B. A. Menge, W. J. Bond, L. S. Mills, G. Daily, J. C. Castilla, J. Lubchenco, and R. T. Paine. 1996. Challenges in the quest for keystones. *BioScience* 46:609–620.

Pownall, T. A. 1949. *Topographical Description of the Dominions of the United States of America*. Univ. Pittsburgh Press, Pittsburgh. 235 pp. + maps.

Pregitzer, K. S., B. V. Barnes, and G. E. Lemme. 1983. Relationship of topography to soils and vegetation in an upper Michigan ecosystem. *Soil Sci. Soc. Am. J.* 47:117–123.

Pregitzer, K. S., and B. V. Barnes. 1984. Classification and comparison of the upland hardwood and conifer ecosystems of the Cyrus H. McCormick Experimental Forest. Upper Peninsula, Michigan. *Can. J. For Res.* 14:362–375.

Pregitzer, K. S., and A. L. Friend. 1996. The structure and function of *Populus* root systems. *In* R. F. Stettler, H. D. Bradshaw, Jr., P. E. Heilman, and T. M. Hinckley (eds.), *Biology of Populus and Its Implications for Management and Conservation*. Nat. Res. Council Canada Press, Ottawa.

Pregitzer, K. S., R. L. Hendrick, and R. Fogel. 1992. The demography of fine roots in responses to patches of water and nitrogen. *New Phytol.* 125:575–580.

Pregitzer, K. S., D. R. Zak, P. S. Curtis, M. E. Kubiske, J. A. Teeri, and C. S. Vogel. 1995. Atmospheric CO_2, soil nitrogen, and turnover of fine roots. *New Phytol.* 129:579–585.

Prentice, C., W. Cramer, S. P. Harrison, R. Leemans, R. A. Monserud, and A. M. Solomon. 1992. A global biome model based on plant physiology and dominance, soil properties and climate. *J. Biogeography* 19:117–134.

Priestly, C. H. B., and R. J. Taylor. 1972. On the assessment of surface heat flux and evaporation using large-scale parameters. *Monthly Weather Review* 100:81–92.

Pritchett, W. L., and W. H. Smith. 1970. Fertilizing slash pine on sandy soils of the lower coastal plain. *In* C. T. Youngberg and C. B. Davey (eds.), *Tree Growth and Forest Soils*. Oregon State Univ. Press, Corvallis.

Probst, J. R., and T. R. Crow. 1991. Integrating biological diversity and resource management. *J. For.* 89:12–17.

Probst, J. R., and J. Weinrich. 1993. Relating Kirtland's warbler population to changing landscape composition and structure. *Landscape Ecology* 8:257–271.

Putz, F. E. 1983. Treefall pits and mounds, buried seeds, and the importance of soil disturbance to pioneer trees on Barro Colorado Island, Panama. *Ecology* 64:1069–1074.

Putz, F. E. and R. R. Sharitz. 1991. Hurricane damage to old-growth forest in Congaree Swamp National Monument, South Carolina, U.S.A. *Can. J. For. Res.* 21: 1765–1770.

Pyne, S. J. 1982. *Fire in America*. Princeton Univ. Press, Princeton, NJ. 654 pp.

Pyne, S. J. 1991. *Burning Bush: A Fire History of Australia*. Henry Holt, New York. 520 pp.

Quamme, H. A. 1985. Avoidance of freezing injury in woody plants by deep supercooling. *Acta Hortic.* 168:11–30.

Quattrochi, D. A., and R. E. Pelletier. 1990. Remote sensing for analysis of landscapes: an introduction. *In* M. G. Turner and R. H. Gardner (eds.), *Quantitative Methods in Landscape Ecology*. Springer-Verlag, New York.

Quick, B. E. 1923. A comparative study of the distribution of the climax association in southern Michigan. *Papers Mich. Acad. Sci. Arts Letters* 3:211–244.

Quine, C., M. Coutts, B. Gardiner, and G. Pyatt. 1995. Forests and wind: management to minimize damage. *For. Comm. Bull.* 114. HMSO, London. 27 pp.

Raghavendra, A. S. (ed.). 1991. *Physiology of Trees*. John Wiley, New York. 509 pp.

Raich, J. W., and K. J. Nadelhoffer. 1989. Belowground carbon allocation in forest ecosystems: global trends. *Ecology* 70:1346–1354.

Ralph, C. J. 1985. Habitat association patterns of forest and steppe birds of northern Patagonia, Argentina. *The Condor* 87:471–483.

Ralston, C. W. 1964. Evaluation of forest site productivity. *Int. Rev. For. Res.* 1:171–201.

Raup, D. M. 1986. Biological extinction in Earth history. *Science* 231:1528–1533.

Read, R. A. 1952. Tree species occurrences as influenced by geology and soil on an Ozark north slope. *Ecology* 33:239–246.

Read, R. A. 1980. Genetic variation in seedling progeny of ponderosa pine provenances. *For Sci. Monogr.* 23. 59 pp.

Regal, P. J. 1977. Ecology and evolution of flowering plant dominance. *Science* 196:622–629.

Regan, T. 1981. The nature and possibility of an environmental ethic. *Environmental Ethics* 3:19–34.

Rehfeldt, G. E. 1986a. Adaptive variation in *Pinus ponderosa* from Intermountain Regions, I. Snake and Salmon River Basins. *For. Sci.* 32:79–92.

Rehfeldt, G. E. 1986b. Adaptive variation in *Pinus ponderosa* from intermountain regions. II. Middle Columbia River system. USDA For. Serv. Res. Paper INT-373. Intermountain For. Res., Sta., Odgen, UT. 9 pp.

Rehfeldt, G. E. 1988. Ecological genetics of *Pinus contorta* from the Rocky Mountains (USA): a synthesis. *Silvae Genetica* 37:131–135.

Rehfeldt, G. E. 1989. Ecological adaptation in Douglas-fir (*Pseudotsuga menziesii* var. *glauca*): a synthesis. *For. Ecol. Manage.* 28:203–215.

Rehfeldt, G. E. 1990. Genetic differentiation among populations of *Pinus ponderosa* from the upper Colorado River basin. *Bot. Gaz.* 151:125–137.

Rehfeldt, G. E., A. R. Stage, and R. T. Bingham. 1971. Strobili development in western white pine: periodicity, prediction, and association with weather. *For. Sci.* 17:454–461.

Reich, P. B., C. Uhl, M. B. Walters, and D. S. Ellsworth. 1991. Leaf life-span as a determinant of leaf structure and function among 23 species in Amazonian forest communities. *Oecologia* 86:16–24.

Reich, P. B., M. B. Walters, and D. S. Ellsworth. 1992. Leaf life-span in relation to leaf, plant and stand characteristics among diverse ecosystems. *Ecol. Monogr.* 62:365–392.

Reichard, S. H., and C. W. Hamilton. 1997. Predicting invasions of woody plants introduced into North America. *Conserv. Biol.* 11:193–203.

Reichle, D. E. (ed). 1981. *Dynamic Properties of Forest Ecosystems*. International Biosphere Programme 23, Cambridge Univ. Press, Cambridge. 683 pp.

Reifsnyder, W. 1967. Forest meteorology: the forest energy balance. *Int. Rev. For. Res.* 2:127–179.

Reinartz, J. A., and J. W. Popp. 1987. Structure of clones of northern prickly ash (*Xanthoxylum americanum*). *Amer. J. Bot.* 74:415–428.

Reineke, L. H. 1933. Perfecting a stand-density index for even-aged forests. *J. Agr. Res.* 46:627–638.

Reiners, W. A., and G. E. Lang. 1979. Vegetational patterns and processes in the balsam fir zone, White Mountains, New Hampshire. *Ecology* 60:403–417.

Remington, C. L. 1968. Suture-zones of hybrid interaction between recently joined biotas. *Evolutionary Biology* 2:321–428.

Renkin, R., D. Despain, and D. Clark. 1994. Aspen seedlings following the 1988 Yellowstone fires. *In* E. G. Despain (ed.), *Conf. Proc., Plants and Their Environments.* Tech. Report NPS/NRYELL/NRTR-93/XX. USDI, Nat. Park Service.

Rennie, P. J. 1962. Methods of assessing forest site capacity. *Trans. 7th Inter. Soc. Soil Sci., Comm. IV and V*, pp. 3–18.

Rhoades, D. F. 1976. The anti-herbivore defenses of Larrea. *In* T. J. Mabry, J. H. Hunziker, and D. R. DiFeo (eds.), *The Biology and Chemistry of the Greosote Bush, A Desert Shrub.* Dowden, Hutchinson, and Ross, Stroudsburg, PA.

Richards, N. A., and E. L. Stone. 1964. The application of soil survey to planting site selection: an example from the Allegheny uplands of New York. *J. For.* 62:475–480.

Richards, P. W. 1952. *The Tropical Rain Forest: An Ecological Study.* Cambridge Univ. Press, Cambridge. 450 pp.

Richards, P. W. 1973. The tropical rain forest. *Sci. American* 229:58–67.

Richards, P. W. 1996. *The Tropical Rain Forest: An Ecological Study,* 2nd ed. Cambridge Univ. Press, Cambridge. 575 pp.

Richardson, S. D. 1956. Studies of root growth of *Acer saccharinum* L. IV: The effect of differential shoot and root temperature on root growth. *Proc. K. Ned. Akad. Wet.* 59:428–438.

Ricklefs, R., and D. Schluter. 1993. *Species Diversity in Ecological Communities: Historical and Geographical Perspectives.* Univ. Chicago Press, Chicago. 414 pp.

Riddoch, I., T. Lehto, and J. Grace. 1991. Photosynthesis of tropical tree seedlings in relation to light and nutrient supply. *New Phytol.* 119:137–147.

Ridley, H. N. 1930. *The Dispersal of Plants Throughout the World.* L. Reeve & Co., Ashford, Kent. 744 pp.

Rignot, E. S., J. Way, C. Williams, and L. Viereck. 1994. Radar estimates of aboveground biomass in boreal forests of interior Alaska. *IEEE Trans. Geosci. Remote Sensing* 32:1117–1124.

Robbins, C. S., D. K. Dawson, and B. A. Dowell. 1989. Habitat area requirements of breeding forest birds in the Middle Atlantic states. *Wildl. Monogr.* 103:1–34.

Robertson, G. P. 1982. Factors regulating nitrification in primary and secondary succession. *Ecology* 63:1561–1573.

Robertson, G. P., and J. M. Tiedje. 1984. Denitrification and nitrous oxide production in successional and old-growth Michigan forests. *Soil Sci. Soc. Amer. J.* 48:383–389.

Robertson, G. P., and J. M. Tiedje. 1988. Denitrification in a humid tropical rainforest. *Nature* 336:756–759.

Robertson, G. P., and P. M. Vitousek. 1981. Nitrification potentials in primary and secondary succession. *Ecology* 62:376–386.

Robinson, S. K., R. R. Thompson III, T. M. Donovan, D. R. Whitehead, and J. Faaborg. 1995. Regional forest fragmentation and the nesting success of migratory birds. *Science* 267:1987–1990.

Roe, A. L. 1967. Seed dispersal in a bumper spruce seed year. USDA For. Serv. Res. Paper INT-39. Intermountain For. and Rge. Exp. Sta., Ogden, UT. 10 pp.

Röhrig, E. 1966. Die Wurzelentwicklung der Waldbäume in Abhängigkeit von den ökologischen Verhältnissen. *Forstarchiv* 37:217–229; 237–249.

Röhrig, E. 1991. Temperate deciduous forests in Mexico and Central America. *In* E. Röhrig and B. Ulrich (eds.), *Ecosystems of the World, Temperate Deciduous Forests. Vol. 7.* Elsevier, New York.

Röhrig, E., and B. Ulrich. 1991. *Ecosystems of the World, Temperate Deciduous Forests. Vol. 7.* Elsevier, New York. 635 pp.

Rollet, B. 1990. Leaf morphology. *In* B. Rollet, Ch. Högermann, and I. Roth (eds.), *Stratification of Tropical Forests as Seen in Leaf Structure*, Part 2. Kluwer, Dordrecht.

Romberger, J. A. 1963. Meristems, growth and development in woody plants. *USDA Tech. Bull.* No. 1293. Washington, D.C. 214 pp.

Romme, W. H. 1982. Fire and landscape diversity in subalpine forests of Yellowstone National Park. *Ecol. Monogr.* 52:199–221.

Romme, W. H., and D. G. Despain. 1989. Historical perspective on the Yellowstone fires of 1988. *BioScience* 39:695–699.

Romme, W. H., and D. H. Knight. 1982. Landscape diversity: the concept applied to Yellowstone Park. *BioScience* 32:664–670.

Ronov, A. B., and A. A. Yaroshevsky. 1972. Earth's crust geochemistry *In* R. W. Fairbridge (ed.), *The Encyclopedia of Geochemistry and Environmental Sciences*. Van Nostrand Reinhold. New York.

Rosenzwieg, M. 1968. Net primary productivity of terrestrial communities: prediction from climatological data. *Am. Nat.* 102:67–74.

Roskoski, J. P. 1980. Nitrogen fixation hardwood forests of the northeastern United States. *Plant and Soil* 54:33–44.

Ross, M. S., J. J. O'Brien, and L. J. Flynn. 1992. Ecological site classification of Florida Keys terrestrial habitats. *Biotropica* 24:486–502.

Ross, M. S., T. L. Sharik, and D. Wm. Smith. 1982. Age structural relationships of tree populations in an Appalachian oak forest. *Bull. Torrey Bot. Club* 109:287–298.

Ross, S. D., R. P. Pharis, and W. D. Binder. 1983. Growth regulators and conifers: their physiology and potential uses in forestry. *In* L. G. Nickell (ed.), *Plant Growth Regulating Chemicals, Vol. 2.* CRC Press, Boca Raton, FL.

Roth, I. 1984. *Stratification of Tropical Forests as Seen in Leaf Structure.* W. Junk, Boston. 522 pp.

Roth, I. 1990. Leaf structure of a Venezuelan cloud forest in relation to microclimate. *Encyclopedia of Plant Anatomy,* Vol. 14, Part 1. Gebrüder Borntraeger, Berlin. 244 pp.

Rothstein, D. E., D. R. Zak, and K. S. Pregitzer. 1996. Nitrate deposition in northern hardwood forests and the nitrogen metabolism of *Acer saccharum* marsh. *Oecologia* 108:338–344.

Rousi, M. 1990. Breeding forest trees for resistance to mammalian herbivores—a study based on European white birch. *Acta For. Fenn.* 210:1–20.

Rousi, M., W. J. Mattson, J. Tahvanainen, T. Koike, and I. Uotila. 1996. Growth and hare resistance of birches: testing defense theories. *Oikos* 77:20–30.

Rousi, M., J. Tahvanainen, and I. Uotila. 1989. Inter- and intraspecific variation in the resistance of winter-dormant birch (*Betula* spp.) against browsing by the mountain hare. *Holarctic Ecology* 12:187–192.

Rousi, M., J. Tahvanainen, and I. Uotila. 1991. A mechanism of resistance to hare browsing in winter-dormant European white birch (*Betula pendula*). *Am. Nat.* 137:64–82.

Rowe, J. S. 1956. Uses of undergrowth species in forestry. *Ecology* 37:461–473.

Rowe, J. S. 1960. Can we find a common platform for the different schools of forest type classification? *Silva Fennica* 105:82–88.

Rowe, J. S. 1961a. The level-of-integration concept and ecology. *Ecology* 42:420–427.

Rowe, J. S. 1961b. Critique of some vegetational concepts as applied to forests of northwestern Alberta. *Can. J. Bot.* 39:1007–1017.

Rowe, J. S. 1962. Soil, site and land classification. *For. Chron.* 38:420–432.

Rowe, J. S. 1964. Environmental preconditioning, with special reference to forestry. *Ecology* 45:399–403.

Rowe, J. S. 1966. Phytogeographic zonation: an ecological appreciation. *In* R. L. Taylor and R. A. Ludwig (eds.), *The Evolution of Canada's Flora.* Univ. Toronto Press.

Rowe, J. S. 1969. Plant community as a landscape feature. *In* K. N. H. Greenidge (ed.), *Symp. Proc. Terrestrial Plant Ecology*, Nova Scotia Museum, Halifax.

Rowe, J. S. 1970. Spruce and fire in Northwest Canada and Alaska. *In Proc. Annual Tall Timbers Fire Ecology Conf.* 10:245–254. Tall Timbers Res. Sta., Tallahassee, FL.

Rowe, J. S. 1972. *Forest Regions of Canada.* Can. For. Serv., Dept. Env. Publ. No. 1300, Ottawa. 172 pp + map.

Rowe, J. S. 1983. Concepts of fire effects on plant individuals and species. *In* R. W. Wein and D. A. MacLean (eds.), *The Role of Fire in Northern Circumpolar Ecosystems.* John Wiley, New York.

Rowe, J. S. 1984a. Understanding forest landscapes: what you conceive is what you get. The Leslie L. Schaffer Lectureship in Forest Science. Vancouver, B. C., Canada.

Rowe, J. S. 1984b. Forestland classification: limitations of the use of vegetation. *In* J. G. Bockheim (ed.), *Symp. Proc. Forest Land Classification: Experience, Problems, Perspectives.* NCR-102 North Central For. Soils Comm., Soc. Am. For., USDA For. Serv., and USDA Soil Cons. Serv., Madison, WI.

Rowe, J. S. 1988. Landscape ecology: the ecology of terrain ecosystems. *In* M. R. Moss (ed.), *Landscape Ecology and Management, Proc. First Symp. Can. Soc. Landscape Ecology and Management.* Univ. Guelph., Polyscience Publ., Montreal.

Rowe, J. S. 1989. The importance of conserving systems. *In* M. Hummel (ed.), *Endangered Spaces: The Future for Canada's Wilderness.* Key Porter Books, Toronto.

Rowe, J. S. 1990. *Home Place.* NeWest Publishers, Edmonton, Alberta, Canada. Canadian Parks and Wilderness Society, Henderson Book Series No. 12. 253 pp.

Rowe, J. S. 1992a. Prologue. *For. Chron.* 68:22–24.

Rowe, J. S. 1992b. The ecosystem approach to forestland management. *For. Chron.* 68:222–224.

Rowe, J. S. 1992c. The integration of ecological studies. *Functional Ecology* 6:115–119.

Rowe, J. S. 1994. A new paradigm for forestry. *For. Chronicle* 70:565–568.

Rowe, J. S. 1997. The necessity of protecting ecoscapes. *Global Biodiversity* 7:9–12.

Rowe, J. S., and B. V. Barnes. 1994. Geo-ecosystems and bio-ecosystems. *Ecol. Soc. Amer. Bull.* 75:40–41.

Rowe, J. S., and G. W. Scotter. 1973. Fire in the boreal forest. *Quat. Res.* 3:444–464.

Rowe, J. S., and J. W. Sheard. 1981. Ecological land classification: a survey approach. *Env. Manage.* 5:451–464.

Royall, P. D., P. A. Delcourt, and H. R. Delcourt. 1991. Late Quaternary paleoecology and paleoenvironments of the central Mississippi alluvial valley. *Bull. Geol. Soc. Amer.* 103:157–170.

Rubec. C. D. A. 1992. Thirty years of ecological land surveys in Canada, from 1960 to 1990. *In* G. B. Ingram and M. R. Moss (eds.), *Symp. Proc. Landscape Approaches to Wildlife and Ecosystem Management.* Polysci. Publ. Inc., Morin Heights, Canada.

Ruddiman, W. F., and H. E. Wright, Jr. (eds.). 1987. *North America and Adjacent Oceans During the Last Deglaciation.* Geological Society of America, Boulder, Co. 501 pp.

Rudis, V. A. 1995. Regional forest fragmentation effects on bottomland hardwood community types and resource values. *Landscape Ecology* 10:291–307.

Rudolph, T. D. 1964. Lammas growth and prolepsis in jack pine in the Lake States. *For. Sci. Monogr.* 6. 70 pp.

Rundel, P. W. 1972. Habitat restriction in giant sequoia: the environmental control of grove boundaries. *Am. Midl. Nat.* 87:81–99.

Rundel, P. W. 1981. Fire as an ecological factor. *In* O. L. Lange, P. S. Nobel, C. B. Osmond, and H. Ziegler (eds.), *Physiological Plant Ecology I. Responses to the Physical Environment,* New Series Vol. 12A. Springer-Verlag, New York.

Runge, M., and M. W. Rode. 1991. Effects of soil acidity on plant associations. *In* B. Ulrich and M. E. Sumner (eds.), *Soil Acidity.* Springer-Verlag, Berlin.

Runkle, J. R. 1981. Gap regeneration in some old-growth forests of the eastern United States. *Ecology* 62:1041–1051.

Runkle, J. R. 1982. Patterns of disturbance in some old-growth mesic forests of eastern North America. *Ecology* 63:1533–1546.

Runkle, J. R. 1985. Disturbance regimes in temperate forests. *In* S. T. A. Pickett and P. S. White (eds.), *The Ecology of Natural Disturbance and Patch Dynamics.* Academic Press, New York.

Runkle, J. R. 1990. Gap dynamics in an Ohio *Acer-Fagus* forest and speculations on the geography of disturbance. *Can. J. For. Res.* 20:632–641.

Rusch, D. H., and L. B. Keith. 1971. Ruffed grouse—vegetation relationships in central Alberta. *J. Wildl. Manage.* 35:417–429.

Rushmore, F. M. 1969. Sapsucker damage varies with tree species and seasons. USDA For. Serv. Res. Paper NE-136. Northeastern For. Exp. Sta., Upper Darby, PA. 19 pp.

Ryan, M. G. 1988. *The importance of maintenance respiration by the living cells in sapwood of sub-alpine conifers.* Ph.D. Dissertation. Oregon State Univ., Corvallis. 104 pp.

Ryan, M. G. 1991. Effects of climate change on plant respiration. *Ecol. Applications* 1:157–167.

Ryan, M. G., S. Linder, J. M. Vose, and R. M. Hubbard. 1994. Dark respiration in pines. *Ecol. Bull.* 43:50–63.

Rychert, R., J. Skujins, D. Sorensen, and D. Prochella. 1978. Nitrogen fixation by lichens and free-living microorganisms in deserts. *In* N. E. West and J. Skujins (eds.), *Nitrogen in Desert Ecosystems.* Dowden, Hutchinson, and Ross, Stroudsburg, PA.

Sanchez, P. A. 1976. *Properties and Management of Tropical Soils.* John Wiley, New York. 619 pp.

Sakai, A., and W. Larcher. 1987. *Frost Survival of Plants.* Springer-Verlag, New York. 321 pp.

Sakai, A., and C. J. Weiser. 1973. Freezing resistance of trees in North America with reference to tree regions. *Ecology* 54:118–126.

Salisbury, F. B., and C. W. Ross. 1992. *Plant Physiology,* 4th ed. Wadsworth, Belmont, CA. 682 pp.

Sample, V. A. (ed.). 1994. *Remote Sensing and GIS in Ecosystem Management.* Island Press, Washington, D.C. 400 pp.

Sampson, H. C. 1930. Succession in the swamp forest formation in northern Ohio. *Ohio J. Sci.* 30:340–356.

Sanford, R. L., Jr. 1987. Apogeotropic roots in an Amazon rain forest. *Science* 235:1062–1064.

Sarvas, R. 1962. Investigations on the flowering and seed crop of *Pinus silvestris. Comm. Inst. For. Fenn.* 53:1–198.

Sarvas, R. 1969. Genetical adaptation of forest trees to the heat factor of the climate. Second world consultation on forest tree breeding, Washington, DC. FO-FTB-69–2/15, 11 pp.

Sato, K., and Y. Iwasa. 1993. Modelling of wave regeneration in subalpine *Abies* forests: population dynamics with spatial structure. *Ecology* 74:1538–1550.

Saunders, D. A., R. H. Hobbs, and C. R. Margules. 1991. Biological consequences of ecosystem fragmentation: a review. *Conserv. Biol.* 5:18–32.

Savage, M., and T. W. Swetnam. 1990. Early 19th-century fire decline following sheep pasturing in a Navajo ponderosa pine forest. *Ecology* 71:2374–2378.

Sawyer, J. O., Jr., and A. A. Lindsey. 1964. The Holderidge bioclimatic formations of the eastern and central United States. *Proc. Indiana Acad. Sci.* 72:105–112.

Schaetzl, R. J., D. L. Johnson, S. F. Burns, and T. W. Small. 1989a. Tree uprooting: review of terminology, process, and environmental implications. *Can. J. For. Res.* 19:1–11.

Schaetzl, R. J., S. F. Burns, D. L. Johnson, and T. W. Small. 1989b. Tree uprooting: review of impact on forest ecology. *Vegetatio* 79:165–176.

Schaetzl, R. J., S. F. Burns, T. W. Small, and D. L. Johnson. 1990. Tree uprooting: review of types and patterns of soil disturbance. *Physical Geog.* 11:277–291.

Schaetzl, R. J., and L. R. Follmer. 1990. Longevity of treethrow microtopography: implications for mass wasting. *Geomorphology* 3:113–123.

Schaffalitzky De Muckadell, M. 1959. Investigations on aging of apical meristems in woody plants and its importance in silviculture. *Det Forstl. Forggsv.* i *Danmark* 25:309–455.

Schaffalitzky De Muckadell, M. 1962. Environmental factors in development stages of trees. *In* T. T. Kozlowski (ed.), *Tree Growth.* Ronald Press, New York.

Schier, G. A. 1975. Deterioration of aspen clones in the Middle Rocky Mountains. USDA For. Serv. Res. Paper INT-170. Intermountain For. and Rge. Exp. Sta., Ogden, Utah. 14 pp.

Schimel, D. S., D. C. Coleman, and K. A. Horton. 1985. Organic matter dynamics in paired range-land and cropland toposequences in North Dakota. *Geoderma* 36:201–214.

Schimel, J. P., M. K. Firestone, and K. S. Killham. 1984. Identification of heterotrophic nitrification in a Sierran forest soil. *Applied and Environmental Microbiology* 48:802–806.

Schimper, A. F. W. 1898. *Pflanzen-Geographie auf Physiologischer Grundlage.* Gustav Fischer, Jena. 876 pp.

Schlenker, G. 1960. Zum Problem der Einordnung klimatischer Unterschiede in das System der Waldstandorte Baden-Württembergs. *Mitt. Vereins forstl. Standortsk. Forstpflz.* 9:3–15.

Schlenker, G. 1964. Entwicklung des in Südwestdeutschland angewandten Verfahrens der forstlichen Standortskunde. *In Standort, Wald und Waldwirtschaft in Oberschwaben, "Oberschwäbische Fichtenreviere."* Stuttgart.

Schlesinger, W. H. 1978. On the relative dominance of shrubs in Okefenokee Swamp. *Am. Nat.* 112:949–954.

Schlesinger, W. H., and P. L. Marks. 1977. Mineral cycling and the niche of Spanish moss, *Tillandsia usneoides* L. *Amer. J. Bot.* 64:1254–1262.

Schlesinger, W. H., and W. A. Reiners. 1974. Deposition of water and cations on artificial foliar collectors in fir krummholz of New England Mountains. *Ecology* 55:378–386.

Schlichting, C. D. 1986. The evolution of phenotypic plasticity in plants. *Ann. Rev. Ecol. Syst.* 17:667–693.

Schmidt, H. 1970. *Versuche über die Pollenverteilung in einem Kiefernbestand.* Diss. Forstl. Fak. Univ. Göttingen, Germany.

Schmidt, W. C., and R. C. Shearer. 1971. Ponderosa pine seed—for animals or trees? USDA For. Serv. Res. Paper INT-112. Intermountain For. and Rge. Exp. Sta., Ogden, Utah. 14 pp.

Schmidt, W. C., and W. P. Dufour. 1975. Building a natural area system for Montana. *Western Wildlands* 2:20–29.

Schmidt-Vogt, H. 1977. *Die Fichte.* Verlag Paul Parey, Hamburg. 647 pp.

Schneider, S. H. 1989. The changing climate. *Scientific American* 261:70–79.

Schoenicher, W. 1933. *Deutsche Waldbäume und Waldtypen.* Gustav Fischer Verlag, Jena. 208 pp.

Schoenike, R. R. 1976. Geographical variations in jack pine (*Pinus banksiana*). *Univ. Minnesota. Agr. Exp. Sta. Tech. Bull.* 304. 47 pp.

Schowalter, T. D. 1985. Adaptations of insects to disturbance. *In* S. T. A. Pickett and P. S. White (eds.), *The Ecology of Natural Disturbance and Patch Dynamics.* Academic Press, New York.

Schowalter, T. D. 1989. Canopy arthropod community structure and herbivory in old-growth and regenerating forests in western Oregon. *Can. J. For. Res.* 19:318–322.

Schreiber, K.-F. 1990. The history of landscape ecology in Europe. *In* I. S. Zonneveld and R. T. T. Forman (eds.), *Changing Landscapes: An Ecological Perspective.* Springer-Verlag, New York.

Schüle, W. 1990. Landscape and climate in prehistory: interactions of wildlife, man, and fire. *In* J. G. Goldammer (ed.), *Fire in the Tropical Biota* (Ecological Studies 84). Springer-Verlag, New York.

Schullery, P. 1989. The fires and fire policy. *BioScience* 39:686–694.

Schulze, E. D., M. I. Fuchs, and M. Fuchs. 1977. Spacial distribution of photosynthetic capacity and performance in a montane spruce forest of northern Germany. I. Biomass distribution and daily CO_2 uptake in different crown layers. *Oecologia* 29:43–61.

Schultz, J. C., P. J. Nothnagle, and I. T. Baldwin. 1982. Seasonal and individual variation in leaf quality of two northern hardwoods tree species. *Amer. J. Bot.* 69:753–759.

Schuster, R. L., R. L. Logan, and P. T. Pringle. 1992. Prehistoric rock avalanches in the Olympic Mountains, Washington. *Science* 258:1620–1621.

Schwertmann, U., and R. M. Taylor. 1977. Iron oxides, *In* J. B. Dixon and S. B. Weed (eds.), *Minerals in Soil Environments*, Soil Sci. Soc. Amer., Madison, WI.

Scott, D. A., J. Proctor, and J. Thompson. 1992. Ecological studies on a lowland evergreen rain forest on Maraca Island, Brazil. II. Litter and nutrient cycling. *J. Ecology* 80:705–717.

Scott, R. A., E. S. Barghoorn, and E. B. Leopold. 1960. How old are the angiosperms? *Am. J. Science* 258-A:284–299.

Sebald, O. 1964. Ökologische Artengruppen für den Wuchsbezirk "Oberer Neckar." *Mitt. Vereins forstl. Standortsk. Forstpflz.* 14:60–63.

Secrest, H. C., H. J. MacAloney, and R. C. Lorenz. 1941. Causes of the decadence of hemlock at the Menominee Indian Reservation, Wisconsin. *J. For.* 39:3–12.

Sellers, W. D. 1965. *Physical Climatology.* Univ. Chicago Press, Chicago, 272 pp.

Senn, G. 1923. Über die Ursachen der Brettwurzebildung bei der Pyramiden-Pappel. *Verh. Naturf. Ges. Basel*, 35:405–435.

Shanks, R. E. 1956. Altitudinal and microclimatic relationships of soil temperature under natural vegetation. *Ecology* 37:1–7.

Sharik, T. L., and B. V. Barnes. 1976. Phenology of shoot growth among diverse populations of yellow birch (*Betula alleghaniensis*) and sweet birch (*B. lenta*). *Can. J. Bot.* 54:2122–2129.

Sharik, T. L., P. P. Feret, and R. W. Dyer. 1990. Recovery of the endangered Virginia round-leaf birch (*Betula uber*): a decade of effort. *In* R. S. Mitchell, C. J. Sheviak, and D. J. Leopold (eds.), *Ecosystem Management: Rare Species and Significant Habitats.* Proc. 15th Ann. Natural Areas Conference. New York State Museum Bull. 471.

Shea, K. L. 1990. Genetic variation between and within populations of Engelmann spruce and subalpine fir. *Genome* 33:1–8.

Sheffield, R. M., and M. T. Thompson. 1992. Hurricane Hugo effects on South Carolina's forest resource. USDA For. Serv. Res. Paper SE-284. Southeastern For. Exp. Sta., Asheville, NC. 51 pp.

Shimwell, D. W. 1971. *The Description and Classification of Vegetation.* Univ. Washington Press, Seattle. 322 pp.

Shirley, H. L. 1929. The influence of light intensity and light quality upon the growth of plants. *Amer. J. Bot.* 16:354–390.

Shirley, H. L. 1945a. Light as an ecological factor and its measurement. *Bot. Rev.* 11:497–532.

Shirley, H. L. 1945b. Reproduction of upland conifers in the Lake States as affected by root competition and light. *Am. Midl. Nat.* 33:537–612.

Shiva, V. 1990. Biodiversity, biotechnology, and profit: the need for a peoples' plan to protect biological diversity. *The Ecologist* 20:44–47.

Short, H. L. 1976. Composition and squirrel use of acorns of black and white oak groups. *J. Wildl. Manage.* 40:479–483.

Show, S. B., and E. I. Kotok. 1929. Cover type and fire control in the national forests of northern California. USDA, Dept. Bull. 1495. Washington, D.C. 35 pp.

Shrader-Frechette, K. S., and E. D. McCoy. 1992. Statistics, costs and rationality in ecological inference. *Trends in Evol. and Ecol.* 7:107–145.

Shrader-Frechette, K. S., and E. D. McCoy. 1993. *Method in Ecology;Strategies for Conservation.* Cambridge Univ. Press, Cambridge. 328 pp.

Shumway, J., and W. A. Atkinson. 1978. Predicting nitrogen fertilizer response in unthinned stands of Douglas-fir. *Comm. Soil Sci. Plant Analysis* 9:529–539.

Silen, R. R. 1978. Genetics of Douglas-fir. USDA For. Serv. Res. Paper WO–35, Washington D.C. 34 pp.

Silkworth, D. R., and D. F. Grigal. 1982. Determining and evaluating nutrient losses from whole-tree harvesting of aspen. *Soil Sci. Soc. Amer. J.* 46:626–631.

Silvertown, J. W. 1980. The evolutionary ecology of mast seeding in trees. *Biol. J. Linnean Soc.* 14:235–250.

Silvertown, J. W., and J. L. Doust. 1993. *Introduction to Plant Population Biology.* Blackwell, Oxford. 210 pp.

Silvester, W. B., P. Sollins, T. Verhoeven, and S. P. Cline. 1982. Nitrogen fixation and acetylene reduction in decaying conifer boles: Effects of incubation time, aeration, and moisture content. *Can. J. For. Res.* 12:646–652.

Simard, A. J., D. A. Haines, R. W. Blank, and J. S. Frost. 1983. The Mack Lake fire. USDA For. Serv. Gen. Tech. Report NC-83. North Central For. Exp. Sta. St. Paul, MN. 36 pp.

Simpson, E. H. 1949. Measurement of diversity. *Nature* 163:688.

Simpson, G. G. 1952. How many species? *Evolution* 6:342.

Simpson, T. A. 1990. *Landscape Ecosystems and Cover Types of the Reserve Area and Adjoining Land of the Huron Mountain Club, Marquette Co., Michigan.* Ph.D. Thesis, University of Michigan, Ann Arbor. 384 pp.

Simpson, T. A., P. E. Stuart, and B. V. Barnes. 1990. Landscape ecosystems and cover types of the Reserve Area and adjoining lands of the Huron Mountain Club, Marquette Co., MI. *Huron Mountain Wildlife Foundation Occasional Paper* No. 4. 128 pp.

Sims, R. A., and P. Uhlig. 1992. The current status of forest site classification in Ontario. *For. Chron.* 68:64–77.

Sims, R. A., W. D. Towill, K. A. Baldwin, and G. M. Wickware. 1989. *Field Guide to the Forest Ecosystem Classification for Northwestern Ontario.* Ont. Min. Nat. Res. Toronto, ON. 191 pp.

Singh, T., and J. M. Powell. 1986. Climatic variation and trends in the boreal forest region of western Canada. *Climatic Change* 8:267–278.

Sirén, G. 1955. The development of spruce forest on raw humus sites in northern Finland and its ecology. *Acta For. Fenn.* 62(4):1–408.

Sjörs, H. 1955. Remarks on ecosystems. *Svensk Botanisk Tidskrift* 49:155–169.

Skole, D. L., B. Moore III, and W. H. Chomentowski. Global geographic information systems and databases for vegetation changes studies. *In* A. M. Solomon and H. H. Shugart (eds.), *Vegetation Dynamics & Global Change*. Chapman and Hall, New York.

Slaughter, C. W., R. J. Barnes, and G. M. Hansen (eds.). 1971, *Fire in the Northern Environment—A Symposium*. USDA For. Serv. Pacific Northwest For. and Rge. Exp. Sta., Portland, OR. 275 pp.

Smalley, G. W. 1984. Classification and evaluation of forest sites in the Cumberland Mountains. USDA For. Serv. Gen. Tech. Report SO-50. Southern For. Exp. Sta., New Orleans, LA. 84 pp.

Smalley, G. W. 1991. No more plots; go with what you know: developing a forest land classification system for the Interior Uplands. *In* D. L. Mengel and D. T. Tew (eds.), *Symp. Proc. Ecological Land Classification: Applications to Identify and Productive Potential of Southern Forests*. USDA For. Serv. Gen. Tech. Report SE-68. Southeastern For. Exp. Sta., Asheville, NC.

Smith, C. C. 1970. The coevolution of pine squirrels (*Tamiasciurus*) and conifers. *Ecol. Monogr.* 40:349–371.

Smith, D. D., and D. Follmer. 1972. Food preferences of squirrels. *Ecology* 53:82–91.

Smith, D. M. 1946. Storm damage in New England forests. M.S. thesis, Yale University.

Smith, D. M. 1951. The influence of seedbed conditions on the regeneration of eastern white pine. *Conn. Agr. Exp. Sta. Bull.* 545. 61 pp.

Smith, D. M., B. C. Larson, M. J. Kelty, and P. M. S. Ashton. 1997. *The Practice of Silviculture*. 9th ed. John Wiley, New York. 537 pp.

Smith, H. 1981. Light quality as an ecological factor. *In* J. Grace, E. D. Ford, and P. G. Jarvis (eds.), *Plants and Their Atmospheric Environment*. Blackwell Scientific, Oxford, England.

Smith, R. H. 1966. Resin quality as a factor in the resistance of pines to bark beetles. *In* H. D. Gerhold, R. E. McDermott, E. J. Schreiner, and J. A. Winieski (eds.), *Breeding Pest-Resistant Trees*. Pergamon Press, New York.

Smith, R. H. 1977. Monoterpenes of ponderosa pine xylem resin in western United States. USDA For. Serv., Tech Bull. 1532. 48 pp.

Smith, T. J., III, M. B. Robblee, H. R. Wanless, and T. W. Doyle. 1994. Mangroves, hurricanes, and lightning strikes. *BioScience* 44:256–262.

Smith, W. B., and G. J. Brand. 1983. Allometric biomass equations for 98 species of herbs, shrubs and small trees. USDA For. Serv. Note NC-299. North Central For. Exp. Sta., St. Paul, MN. 8 pp.

Snaydon, R. W., and M. S. Davies. 1972. Rapid population differentiation in a mosaic environment. II. Morphological variation in *Anthoxanthum odoratum*. *Evolution* 26:390–405.

Snaydon, R. W., and M. S. Davies. 1976. Rapid population differentiation in a mosaic environment. IV. Populations of *Anthoxanthum odoratum* at sharp boundaries. *Heredity* 37:9–25.

Snyder, J. D., and R. A. Janke. 1976. Impact of moose browsing on boreal-type forests of Isle Royale National Park. *Am. Midl. Nat.* 95:79–92.

Soares, R. V. 1990. Fire in some tropical and subtropical South American vegetation types: an overview. *In* J. G. Goldammer (ed.), *Fire in the Tropical Biota*. (Ecological Studies 84). Springer-Verlag, New York.

Soil Survey Staff. 1975. *Soil Taxonomy*. USDA Agr. Handbook 436. Washington, D.C. 754 pp.

Solbrig, O. T. 1970. *Principles and Methods of Plant Biosystematics*. Collier-Macmillan, Toronto. 226 pp.

Solbrig, O. T. 1991. The origin and function of biodiversity. *Environment* 33:16–20, 34–38.

Sollins, P., G. P. Robertson, and G. Uehara. 1988. Nutrient mobility in variable-and permanent-charge soils. *Biogeochemistry* 6:181–199.

Sork, V. L., J. Bramble, and S. Owen. 1993. Ecology of mast-fruiting in three species of North American deciduous oaks. *Ecology* 74:528–541.

Soulé, M. E. 1986. *Conservation Biology: The Science of Scarcity and Diversity*. Sinauer Associates, Sunderland, MA. 584 pp.

Sousa, W. P. 1984. The role of disturbance in natural communities. *Ann. Rev. Ecol. Syst.* 15:353–391.

Sowell, J. B. 1985. A predictive model relating North American plant formations and climate. *Vegetatio* 60: 103–111.

Spies, T. A. 1983. *Classification and Analysis of Forest Ecosystems of the Sylvania Recreation Area, Upper Michigan.* Ph.D. Thesis, University of Michigan, Ann Arbor. 321 pp.

Spies, T. A. 1997. Forest stand structure, composition, and function. *In* K. A. Kohm and J. F. Franklin (eds.), *Creating a Forestry for the 21st Century.* Island Press, Washington, D.C.

Spies, T. A., and B. V. Barnes. 1981. A morphological analysis of *Populus alba, P. grandidentata* and their natural hybrids in southeastern Michigan. *Silvae Genetica.* 30:102–106.

Spies, T. A., and B. V. Barnes. 1982. Natural hybridization between *Populus alba* L. and the native aspens in southeastern Michigan. *Can. J. For. Res.* 12:653–660.

Spies, T. A., and B. V. Barnes. 1985a. A multi-factor ecological classification of the northern hardwood and conifer ecosystems of the Sylvania Recreation Area, Upper Peninsula, Michigan. *Can. J. For. Res.* 15:949–960.

Spies, T. A., and B. V. Barnes. 1985b. Ecological species groups of upland northern hardwood-hemlock forest ecosystems of the Sylvania Recreation Area, Upper Peninsula, Michigan. *Can. J. For. Res.* 15:961–972.

Spies, T. A., and J. F. Franklin. 1989. Gap characteristics and vegetation response in tall coniferous forests. *Ecology* 70:543–545.

Spies, T. A., and J. F. Franklin. 1991. The structure of natural young, mature, and old-growth forests in Washington and Oregon. *In* L. F. Ruggiero, K. B. Aubry, A. B. Carey, and M. H. Huff (tech. coords.), *Wildlife and Vegetation of Unmanaged Douglas-fir forests.* USDA For. Serv. PNW-GTR-385. Pacific Northwest Exp. Sta., Portland, OR.

Spies, T. A., J. F. Franklin, and M. Klopsch. 1990. Canopy gaps in Douglas-fir forests of the Cascade Mountains. *Can J. For. Res.* 20:649–658.

Sposito, G. 1989. *The Chemistry of Soils.* Oxford Univ. Press, New York. 277 pp.

Sprugel, D. G. 1976. Dynamic structure of wave-regenerated *Abies balsamea* forests in the northeastern United States. *J. Ecol.* 64:889–911

Sprugel, D. G., and F. H. Bormann. 1981. Natural disturbance and the steady state in high-altitude balsam fir forests. *Science* 211:390–393.

Spurr, S. H. 1945. A new definition of silviculture. *J. For.* 43:44.

Spurr, S. H. 1952a. Origin of the concept of forest succession. *Ecology* 33:426–427.

Spurr, S. H. 1952b. *Forest Inventory.* Ronald Press, New York. 476 pp.

Spurr, S. H. 1954. The forests of Itasca in the nineteenth century as related to fire. *Ecology* 35:21–25.

Spurr, S. H. 1956a. Natural restocking of forests following the 1938 hurricane in central New England. *Ecology* 37:433–451.

Spurr, S. H. 1956b. Forest associations in the Harvard Forest. *Ecol. Monogr.* 26:245–262.

Spurr, S. H. 1957. Local climate in the Harvard Forest. *Ecology* 38:37–46.

Spurr, S. H. 1963. Growth of Douglas-fir in New Zealand. New Zealand For. Serv. For. Res. Inst. Tech. Paper 43. 54 pp.

Spurr, S. H. 1964. *Forest Ecology.* Ronald Press, New York. 352 pp.

Spurr, S. H., and B. V. Barnes. 1980. *Forest Ecology,* 3rd ed. John Wiley, New York. 687 pp.

Squillace, A. E. 1966. Geographic variation in slash pine. *For. Sci. Monogr.* 10. 56 pp.

Staaf, H., and B. Berg. 1982. Accumulation and release of plant nutrients in decomposing Scots pine needle litter. Long-term decomposition in a Scots pine forest II. *Can. J. Bot.* 60:1561–1568.

Stacey, P. B., and W. D. Koenig. 1984. Cooperative breeding in the Acorn Woodpecker. *Sci. American* 252:114–121.

Stage, A. R. 1989. Utility of vegetation-based land classes for predicting forest regeneration and growth. *In* D. E. Ferguson, P. Morgan, F. D. Johnson (comps.), Proceedings—*Land Classifications Based on Vegetation: Applications for Resource Management.* USDA For. Serv. Gen. Tech. Report INT-257. Intermountain Res. Sta., Ogden, UT.

Stahelin, R. 1943. Factors influencing natural restocking of high altitude burns by coniferous trees in the central Rocky Mountains. *Ecology* 24:19–30.

Stark, N. 1968. Seed ecology of *Sequoiadendron giganteum. Madroño* 19:267– 277.

Stark, N. M. 1977. Fire and nutrient cycling in a Douglas-fir/larch forest. *Ecology* 58:16–30

Stebbins, G. L. 1950. *Variation and Evolution in Plants.* Columbia Univ. Press, New York. 643 pp.

Stebbins, G. L. 1966. *Processes of Organic Evolution.* Prentice-Hall, Englewood Cliffs, NJ. 191 pp.

Stebbins, G. L. 1969. The significance of hybridization for plant taxonomy and evolution. *Taxon* 18:26–35.

Stebbins, G. L. 1970. Variation and evolution in plants: progress during the past twenty years. *In* Max K. Hecht and William C. Steere (ed.), *Essays in Evolution and Genetics.* Appleton-Century-Crofts, New York.

Stebbins, G. L. 1976. Seeds, seedlings, and the origin of angiosperms, *In* C. B. Beck (ed.), *Origin and Early Evolution of Angiosperms.* Columbia Univ. Press, New York.

Stebbins, G. L. 1985. Polyploidy, hybridization, and the invasion of new habitats. *Ann. Missouri Bot. Gard.* 72: 824–832.

Steele, B. M., and S. V. Cooper. 1986. Predicting site index and height for selected tree species of northern Idaho. USDA For. Serv. Res. Paper INT-126. Int. For. and Rge. Exp. Sta., Ogden, UT. 16 p.

Steele, M. A., T. Knowles, K. Bridle, and E. L. Simms. 1993. Tannins and partial consumption of acorns: implications for dispersal of oaks by seed predators. *Am. Midl. Nat.* 130:229–238.

Steele, R., and K. Geier-Hayes. 1995. Major Douglas-fir habitat types of central Idaho: a summary of succession and management. USDA For. Serv. Gen. Tech. Rep. INT-GTR-331. Intermountain Res. Sta., Odgen, UT. 23 pp.

Steele, R., R. D. Pfister, R. A. Ryker, and J. A. Kittams. 1981. Forest habitat types of central Idaho. USDA For. Serv. Gen. Tech. Report INT-114. Int. For. Exp. Sta., Ogden, UT. 138 pp.

Steinbrenner, E. C. 1951. Effect of grazing on floristic composition and soil properties of farm woodlands in southern Wisconsin. *J. For.* 49:906–910.

Stephenson, N. L. 1990. Climate control of vegetation distribution: the role of the water balance. *Am. Nat.* 135:649–670.

Stephenson, S. L. 1974. Ecological composition of some former oak-chestnut communities in western Virginia. *Castanea* 39:278–286.

Sterba, H., and R. A. Monserud. 1993. The maximum density concept applied to uneven-aged mixed-species stands. *For. Sci.* 39:432–452.

Stevens, C. L. 1931. Root growth of white pine. *Yale Univ. School For. Bull.* 32. 62 pp.

Stewart, G. H., A. B. Rose, and T. T. Veblen. 1991. Forest development in canopy gaps in old-growth beech (*Nothofagus*) forests, *New Zealand J. Veg. Sci.* 2:679–690.

Stewart, W. N., and G. W. Rothwell. 1993. *Paleobotany and the Evolution of Plants.* Cambridge Univ. Press, Cambridge. 521 pp.

Stoddard, H. L. 1931. *The Bobwhite Quail; Its Habits, Preservation and Increase.* Scribner's, New York. 500 pp.

Stoeckeler, J. H. 1948. The growth of quaking aspen as affected by soil properties and fire. *J. For.* 46:727–737.

Stoeckeler, J. H. 1960. Soil factors affecting the growth of quaking aspen forests in the Lake States. *Univ. Minnesota Agr. Exp. Sta. Tech. Bull.* 233. 48 pp.

Stone, D. M. 1977. Leaf dispersal in a pole-sized maple stand. *Can. J. For. Res.* 7:189–192.

Stone, E. L. 1974. The communal root system of red pine: growth of girdled trees. *For. Sci.* 20:294–305.

Stone, E. L. 1975. Windthrow influences on spatial heterogeneity in a forest soil. *Mitt. Eid. Anst. forstl. Versuchsw.* 51:77–87.

Stone, E. L., and M. H. Stone. 1954. Root collar sprouts in pine. *J. For.* 52: 487–491.

Stott, P. A., J. G. Goldammer, and W. L. Werner. 1990. The role of fire in the tropical lowland deciduous forests of Asia. *In* J. G. Goldammer (ed.), *Fire in the Tropical Biota* (Ecological Studies 84). Springer-Verlag, New York.

Strahler, A. N., and A. H. Strahler. 1989. *Elements of Physical Geography,* 4th ed. John Wiley, New York. 562 pp.

Strong, W. L., E. T. Oswald, and D. J. Downing (eds.). 1990. The Canadian vegetation classification system. First Approx. Ecological Land Classif. Series, No. 25, Env. Canada, Ottawa. 22 pp.

Stuckey, R. L. 1981. Origin and development of the concept of the Prairie Peninsula. *In* R. L. Stuckey and K. J. Reese (eds.), *The Prairie Peninsula—In the "Shadow" of Transeau.* Ohio Biol. Surv. Biol. Notes 15.

Sukachev. V. N., and N. V. Dylis. 1964. *Fundamentals of Forest Biogeocoenology.* Translated by J. M. MacLennan. Oliver and Boyd, Ltd., Edinburgh. 672 pp.

Sutton, R. F. 1969. Form and development of conifer root systems. *Commonw. For. Bur., Tech. Comm.* No. 7. Oxford. 131 pp.

Sutton, R. F., and E. L. Stone Jr. 1974. White grubs: a description for foresters, and an evaluation of their silvicultural significance. Canadian For. Serv. Info. Report 0–X-212. Great Lakes For. Res. Centre. Sault Ste. Marie, Ontario. 21 pp.

Swain, A. M. 1973. A history of fire and vegetation in northeast Minnesota as recorded in lake sediments. *Quat. Res.* 3:383–396.

Swain, A. M. 1978. Environmental changes during the past 2000 years in north-central Wisconsin: analysis of pollen, charcoal, and seeds from varved lake sediments. *Quat. Res.* 10:55–68.

Swank, W. T., and D. A. Crossley (eds.). 1988. *Forest Hydrology and Ecology at Coweeta.* Springer-Verlag, New York. 469 pp.

Swanson, F. J. 1981. Fire and geomorphic processes. *In Proc. Fire Regimes and Ecosystems.* USDA For Serv. Gen. Tech;. Report WO-26. Washington, D.C.

Swanson, F. J., F. J. Franklin, and J. R. Sedell. 1990. Landscape patterns, disturbance, and management in the Pacific Northwest, USA. *In* I. S. Zonneveld and R. T. T. Forman (eds.), *Changing Landscapes: An Ecological Perspective.* Springer-Verlag, New York.

Swanson, F. J., and J. F. Franklin. 1992. New forestry principles from ecosystem analysis of Pacific Northwest forests. *Ecol. Applications* 2:267–274.

Swanson, F. J., R. L. Fredriksen, and F. M. McCorison. 1982a. Material transfer in a western Oregon forested watershed. *In* R. L. Edmonds (ed.), *Analysis of Coniferous Forest Ecosystems in the Western United States.* US/IBP Synthesis Series 14. Hutchinson Ross. Stroudsburg, PA.

Swanson, F. J., S. V. Gregory, J. R. Sedell, and A. G. Campbell. 1982b. Land-water interactions: the riparian zone. *In* R. L. Edmonds (ed.), *Analysis of Coniferous Forest Ecosystems in the Western United States.* US/IBP Synthesis Series 14. Hutchinson Ross. Stroudsburg, PA.

Swanson, F. J., T. K. Kratz, N. Caine, and R. G. Woodmansee. 1988. Landform effects on ecological processes and features. *BioScience* 38:92–98.

Swanson, F. J., and G. W. Lienkaemper. 1978. Physical consequences of large organic debris in Pacific Northwest streams. USDA For. Serv. Gen. Tech. Report PNW-69. Pacific Northwest For. and Rge. Exp. Sat., Portland, OR. 12 pp.

Swanson, F. J., G. W. Lienkaemper, and J. R. Sedell. 1976. History, physical effects, and management implications of large organic debris in western Oregon streams. USDA For. Serv. Gen. Tech. Report PNW-56. Pacific Northwest For. and Rge. Exp. Sta., Portland, OR. 15 pp.

Swanson, F. J., and D. N. Swanston. 1977. Complex mass-movement terrains in the western Cascade Range, Oregon. *Rev. Eng. Geol.* 3:113–124.

Swetnam, T. W. 1993. Fire history and climate change in giant sequoia groves. *Science* 262:885–889.

Taiz, L., and E. Zeiger. 1991. *Plant Physiology.* Benjamin/Cummings, Redwood City, CA. 559 pp.

Tallis, J. H. 1991. *Plant Community History.* Chapman and Hall, New York. 398 pp.

Tansley, A. G. 1920. The classification of vegetation and the concept of development. *J. Ecology* 8:118–144.

Tansley, A. G. 1929. Succession: the concept and its values. *In* B. M. Duggar (ed.), *Proc. International Congress of Plant Sciences*, Ithaca, NY, 1926. George Banta Publ. Co., Menasha, WS.

Tansley, A. G. 1935. The use and abuse of vegetational concepts and terms. *Ecology* 16:284–307.

Tansley, A. G. 1939. *The British Islands and Their Vegetation.* Cambridge Univ. Press, Cambridge. 930 pp.

Tappeiner, J. C., II. 1969. Effect of cone production on branch, needle, and xylem ring growth of sierra Nevada Douglas-fir. *For. Sci.* 15:171–174.

Tappeiner, J. C., D. Lavender, J. Walstad, R. O. Curtis, and D. S. DeBell. 1997 Silvicultural systems and regeneration methods. *In* K. A. Kohm and J. F. Franklin (eds.), *Creating a Forestry for the 21st Century.* Island Press, Washington, DC.

Tappeiner, J., J. Zasada, P. Ryan, and M. Newton. 1991. Salmonberry clonal and populations structure: the basis for a persistent cover. *Ecology* 72:609–618.

Tarrant, R. F. 1956a. Changes in some physical soil properties after a prescribed burn in young ponderosa pine. *J. For.* 54:439–441.

Tarrant, R. F. 1956b. Effect of slash burning on some physical soil properties. *For. Sci.* 2:18–22.

Taub, D. R., and D. Goldberg. 1996. Root system topology of pants from habitats of differing soil resource availability. *Functional Ecology* 10:258–264.

Taylor, A. R. 1974. Forest fire. *In* D. N. Lapedes (ed.), *McGraw-Hill Yearbook of Science and Technology*. McGraw-Hill, London.

Taylor, T. M. C. 1959. The taxonomic relationship between *Picea glauca* (Moench) Voss and *P. engelmannii* Parry. *Madroño* 15:111–115.

Taylor, T. N., and E. L. Taylor. 1993. *The Biology and Evolution of Fossil Plants*. Prentice-Hall, NJ. 982 pp.

Tchebakove, N. M., R. A. Monserud, R. Leemans, and S. Golovanov. 1993. A global vegetation model based on the climatological approach of Budyko. *J. Biogeography* 20:129–144.

Teipner, C. L., E. O. Garton, and L. Nelson, Jr. 1983. Pocket gophers in forest ecosystems. USDA For. Serv. Gen. Tech. Report INT-154. Intermountain For. and Rge. Exp. Sta., Ogden, UT. 53 pp.

Terborgh. J. 1989. *Where Have All the Birds Gone?* Princeton Univ. Press, Princeton, NJ. 207 pp.

Terborgh, J. 1992. *Diversity and the Tropical Rain Forest*. Sci. Amer. Library, HPHLP, New York. 242 pp.

Thilenius, J. F., and D. R. Smith. 1985. Vegetation and soils of an alpine range in the Absaroka Mountains, Wyoming. USDA For. Serv. Gen Tech. Report RM-121. Rocky Mountain For. and Rge. Exp. Sta., Fort Collins, CO. 18 pp.

Thomas, D. A., and H. N. Barber. 1974. Studies on leaf characteristics of a cline of *Eucalyptus urnigera* from Mount Wellington, Tasmania. I. Water repellency and the freezing of leaves. *Australian J. Bot.* 22:501–512.

Thomas, W. L., Jr. (ed). 1956. *Man's Role in Changing the Face of the Earth*. Univ. Chicago Press, Chicago. 1193 pp.

Thompson, D. Q., and R. H. Smith. 1970. The forest primeval in the Northeast—a great myth? *In Proc. Annual Tall Timbers Fire Ecology Conf.* 10:255–265. Tall Timbers Res. Sta., Tallahassee, FL.

Thompson, I. D., and A. U. Mallik. 1989. Moose browsing and allelopathic effects of *Kalmia angustifolia* on balsam fir regeneration in central Newfoundland. *Can. J. For. Res.* 19:524–526.

Thompson, S. C. G., and M. A. Barton. 1994. Ecocentric and anthropocentric attitudes toward the environment. *J. Env. Psychology* 14:149–158.

Thoreau, H. D. 1993. *Faith in a Seed*. Island Press, Washington D.C. 283 pp.

Thornthwaite, C. W. 1948. An approach toward a rational classification of climate. *Geog. Rev.* 38:55–94.

Thornthwaite, C. W., and J. R. Mather. 1955. The water balance. Drexel Inst. Tech., Lab. Climatol., Pub. Climatol., Vol. 8(1). 86 pp.

Tiedemann, A. R., W. P. Clary, and R. J. Barbour. 1987. Underground systems of Gambel oak (*Quercus gambelii*) in central Utah. *Amer. J. Bot.* 74:1065–1071.

Tiedje, J. M. 1988. Ecology of denitrification and dissimilatory nitrate reduction to ammonium. *In* A. J. B. Zehnder (ed.), *Biology of Anaerobic Microorganisms*. John Wiley, New York

Tiedje, J. M., S. Simkins, and P. M. Groffman. 1989. Perspectives on measurement of denitrification in field including recommended protocols for acetylene based methods. *Plant and Soil* 115:261–284.

Tilghman, N. G. 1989. Impacts of white-tailed deer on forest regeneration in northwestern Pennsylvania. *J. Wildl. Manage.* 53:524–532.

Tilman, D. 1985. The resource-ratio hypothesis of plant succession. *Am. Nat.* 125:827–852.

Tilman, D. 1988. *Plant Strategies and the Dynamics and Structure of Plant Communities*. Princeton Univ. Press, Princeton, NJ. 360 pp.

Tinker, D. B., W. H. Romme, W. W. Hargrove, R. H. Gardner, and M. G. Turner. 1994. Landscape-scale heterogeneity in lodgepole pine serotiny. *Can. J. For. Res.* 24:897–903.

Titus, J. H. 1990. Microtopography and woody plant regeneration in a hardwood floodplain swamp in Florida. *Bull. Tor. Bot. Club* 117:429–427.

Tomlinson, P. B. 1983. Tree architecture. *Am. Scientist* 71:141–149.

Tomlinson, P. B. 1986. *The Botany of Mangroves.* Cambridge Univ. Press. Cambridge. 419 pp.

Toumey, J. W. 1947. *Foundations of Silviculture upon an Ecological Basis*, 2nd ed. (rev. by C. F. Korstian). John Wiley, New York. 468 pp.

Toumey, J. W., and R. Kienholz. 1931. Trenched plots under forest canopies. *Yale Univ. School For. Bull.* 30. 31 pp.

Townsend, A. M., S. E. Bentz, and G. R. Johnson. 1995. Variation in response of selected American elm clones to *Ophlostoma ulmi. J. Environ. Hort.* 13:126–128.

Tracey, J. G., 1987. *The vegetation of the Humid Tropical Region of North Queensland.* CSIRO, Melbourne. 124 pp.

Transeau, E. N. 1935. The prairie peninsula. *Ecology* 16:423–437.

Trappe, J. M., and R. D. Fogel. 1977. Ecosystematic function of mycorrhizae. *In* J. K. Marshall (ed.), *The Belowground Ecosystem: A Synthesis of Plant-Associated Processes.* Range Sci. Dep. Sci. Ser. No. 26. Colorado State Univ., Ft. Collins.

Treshow, M. 1970. *Environment and Plant Response.* McGraw-Hill, New York. 422 pp.

Trewartha, G. T. 1968. *An Introduction to Climate*, 4th ed. McGraw-Hill, New York. 408 pp.

Trimble, G. R., Jr., and S. Weitzman. 1956. Site index studies of upland oaks in the northern Appalachians. *For. Sci.* 2:162–173.

Triska, F. J., J. R. Sedell, and S. V. Gregory. 1982. Coniferous forest streams. *In* R. L. Edmonds (ed.), *Analysis of Coniferous Forest Ecosystems in the Western United States.* US/IBP Synthesis Series 14. Hutchinson Ross. Stroudsburg, PA.

Tritton, L. M., and J. W. Hornbeck. 1982. Biomass equations for major tree species of the northeast. USDA For. Serv. Gen. Tech. Report NE-69. Northeastern For. Exp. Sta., Upper Darby, PA. 52 pp.

Troeger, R. 1960. Kiefernprovenienzversuche. I. Teil. Der grosse Kiefernprovenienzversuch im südwürttembergischen Forstbezirk Schussenried. *AFJZ* 131(3–4):49–59.

Troll, C. 1939. Luftbildplan and ökologische Bodenforschung. *Zeitschrift der Gesellschaft für Erdkunde, Berlin.* pp. 241–298.

Troll, C. 1950. Die geographische Landschaft und ihre Erforschung. *Studium Generale* 3:163–181.

Troll, C. 1963a. Landscape ecology and land development with special reference to the tropics. *J. Tropical Geogr.* 17:1–11.

Troll, C. 1963b. Über Landschafts-Sukession. *Arbeiten zur Rheinschen Landeskunde Bonn.* 19:5–12.

Troll, C. 1968. Landschaftsökologie. *In* R. Tüxen (ed.), *Pflanzensoziologie und Landschaftsökologie. Berichte das Internalen Symposiums der Internationalen Vereinigung für Vegetationskunde, Stolzenau/Weser 1963.* W. Junk, The Hague.

Troll, C. 1971. Landscape ecology (geo-ecology) and bio-ceonology—a terminology study. *Geoforum* 8:43–46.

Tubbs, C. H. 1965. Influence of temperature and early spring conditions on sugar maple and yellow birch germination in upper Michigan. USDA For. Serv. Res. Note LS-72. North Central For. Exp. Sta., St. Paul, MN. 2 pp.

Tubbs, G. H. 1969. The influence of light, moisture, and seedbed on yellow birch regeneration. USDA For. Serv. Res. Paper NC-27. North Central For. Exp. Sta., St. Paul, MN. 12 pp.

Tukey, H. B. 1970. The leaching of substances from plants. *Ann. Rev. Plant Physiology* 21:305–324.

Turner, B. L., II, W. C. Clark, R. W. Kates, J. F. Richards, J. T. Mathews, and W. B. Meyer (eds.). 1990. *The Earth as Transformed by Human Action.* Cambridge Univ. Press, New York. 729 pp.

Turner, M G. 1989. Landscape ecology: the effect of pattern on process. *Ann Rev. Ecol. Syst.* 20:171–197.

Turner, M. G., and R. H. Gardner (eds.), 1991. *Quantitative Methods in Landscape Ecology: The Analysis and Interpretation of Landscape Heterogeneity.* Springer-Verlag, New York. 536 pp.

Turreson, G. 1922a. The species and the variety as ecological units. *Hereditas* 3:100–113.

Turreson, G. 1922b. The genotypical response of the plant species to the habitat. *Hereditas* 3:211–350.

Turreson, G. 1923. The scope and import of genecology. *Hereditas* 4:171–176.

U. S. Department of Agriculture. 1974. *Seeds of Woody Plants in the United States*. USDA For. Serv. Agr. Handbook No. 450. Washington, D.C. 883 pp.

U. S. Department of Agriculture. 1990. *Silvics of North America*. USDA For. Serv. Agr. Handbook 654. Washington, D.C. Vol. 1, Conifers, 675 pp; Vol. 2, Hardwoods, 877 pp.

U.S. Corps of Engineers. 1956. *Snow Hydrology. Summary Report of the Snow Investigations*. North Pac. Div., U.S.C.E., Portland, OR. 437 pp.

Udvardy, M. D. F. 1969. Birds of the coniferous forest. *In* R. D. Taber (ed.), *Coniferous Forests of the Northern Rocky Mountains*. Center For. Nat. Res., Missoula, Mont.

Uehara, G., and G. Gillman. 1981. *The Mineralogy, Chemistry and Physics of Tropical Soils with Variable Charge Clays*. Westview Press, Boulder, CO. 170 pp.

Ustin, S. L., M. O. Smith, and J. B. Adams. 1993. Remote sensing of ecological processes: a strategy for developing and testing ecological models using spectral mixture analysis. *In* J. R. Ehleringer and C. B. Field (eds.), *Scaling Physiological Processes: Leaf to Globe*. Academic Press, New York.

Ustin, S. L., R. A. Woodward, and M. G. Barbour. 1984. Relationships between sunfleck dynamics and red fir seedling distribution. *Ecology* 65:1420–1428.

Vaartaja, O. 1954. Temperature and evaporation at and near ground level on certain forest sites. *Can. J. Bot.* 32:760–783.

Van Cleve, K., F. S. Chapin, III, C. T. Dyrness, and L. A. Viereck. 1991. Elemental cycling in taiga forests: state-factor control. *BioScience* 41:78–88.

Van Cleve, K., F. S. Chapin, III, P. W. Flanagan, L. A. Viereck, and C. T. Dyrness (eds.). 1986. *Forest Ecosystems in the Alaskan Taiga*. Springer-Verlag, New York. 230 pp.

Van Cleve, K., L. Oliver, R. Schlentner, L. A. Vierick, and C. T. Dyrness. 1983. Productivity and nutrient cycling in taiga forests. *Can. J. For. Res.* 13: 747–766.

Van der Hammen, T., T. A. Wijmstra, and W. H. Zagwijn. 1971. The floral record of the late Cenozoic of Europe. *In* K. K. Turekian (ed.), *The Late Cenozoic Glacial Ages*. Yale Univ. Press, New Haven, CT.

Van der Maarel, E. 1975. The Braun-Blanquet approach in perspective. *Vegetatio* 30:213–219.

Van der Pijl, L. 1957. The dispersal of plants by bats. *Acta bot. neerl.* 6:291– 315.

Van der Pijl, L. 1972. *Principles of Dispersal in Higher Plants*. Springer-Verlag, New York. 162 pp.

Van Veen, J. A., J. N. Ladd, and M. J. Frissel. 1984. Modelling C & N turnover through the microbial biomass in soil. *Plant and Soil* 76:257–274.

Van Wagner, C. E. 1970a. Fire and red pine. *In Proc. Annual Tall Timbers Fire Ecology Conf.* 10:211–219. Tall Timbers Res. Sta., Tallahassee, FL.

Van Wagner, C. E. 1970b. Temperature gradients in duff and soil during prescribed fires. *Bi-Monthly Research Notes* 26(4):42.

Van Wagner, C. E. 1983. Fire behaviour in northern conifer forests and shrublands. *In* R. W. Wein, and D. A. MacLean (eds.), *The Role of Fire in Northern Circumpolar Ecosystems*. John Wiley, New York.

Vander Wall, S. B., and R. P. Balda. 1977. Coadaptations of the Clark's nutcracker and the piñon pine for efficient seed harvest and dispersal. *Ecol. Monogr.* 47:89–111.

Vankat, J. L. 1979. *The Natural Vegetation of North America*. John Wiley, New York. 486 pp.

Vasilevich, V. I. 1968. Commentary. *In* P. Dansereau (ed.), The continuum concept of vegetation: responses. *Bot. Rev.* 34:253–332.

Veatch, J. O. 1953. *Soils and Land of Michigan*. Michigan State College Press, East Lansing. 241 pp.

Veatch, J. O. 1959. Presettlement forest in Michigan. Michigan State Univ., Dept. Res Development. East Lansing. Map.

Veblen, T. T. 1987. Trees of the trembling earth. *Natural History* 96:43–46.

Veblen, T. T. 1992. Regeneration dynamics. *In* D. C. Glenn-Lewin, R. K. Peet, and T. T. Veblen (eds.), *Plant Succession, Theory and Prediction*. Chapman & Hall, London.

Veblen, T. T., K. S. Hadley, and M S. Reid. 1991a. Disturbance and stand development of a Colorado subalpine forest. *J. Biogeography* 18:707–716.

Veblen, T. T., K. S. Hadley, M. S. Reid, and A. J. Rebertus. 1991b. The response of subalpine forests to spruce beetle outbreak in Colorado. *Ecology* 72:213–231.

Veblen, T. T., and D. C. Lorenz. 1986. Anthropogenic disturbance and recovery patterns in montane forests, Colorado Front Range. *Physical Geog.* 7:1–24.

Veblen, T. T., and D. C. Lorenz. 1988. Recent vegetation changes along the forest/steppe ecotone of northern Patagonia. *Annals Assoc. Amer. Geographers* 78:93–111.

Veblen, T. T., and D. C. Lorenz. 1991. *The Colorado Front Range, A Century of Ecological Change.* Univ. Utah Press, Salt Lake City. 210 pp.

Vegis, A. 1964. Dormancy in higher plants. *Ann. Rev. Plant Phys.* 15:185–224.

Verrall, A. F., and T. W. Graham. 1935. The transmission of *Ceratostomella ulmi* through root grafts. *Phytopathology* 25:1039–1040.

Vézina, P. E. 1961. Variations in total solar radiation in three Norway spruce plantations. *For. Sci.* 7:257–264.

Vézina, P. E., and D. W. K. Boulter. 1966. The spectral composition of near ultraviolet and visible radiation beneath forest canopies. *Can. J. Bot.* 44:1267–1284.

Viereck, L. A. 1970. Forest succession and soil development adjacent to the Chena River in interior Alaska. *Arctic and Alpine Res.* 2:1–26.

Viereck, L. A. 1973. Wildfire in the taiga of Alaska. *Quat. Res.* 3:465–495.

Viereck, L. A. 1982. Effects of fire and firelines on active layer thickness and soil temperatures in interior Alaska. *In* H. M. French (ed.), *Proc. Fourth Can. Permafrost Conference*, Calgary. Natl. Res. Council Can. Ottawa.

Viereck, L. A. 1983. The effects of fire in black spruce ecosystems of Alaska and northern Canada. *In* R. W. Wein and D. A. MacLean (eds.), *The Role of Fire in Northern Circumpolar Ecosystems.* John Wiley, New York.

Viereck, L. A., C. T. Dyrness, and K. Van Cleve. 1984. Potential use of the Alaska vegetation system as an indicator of forest site productivity in interior Alaska. *In* M. Murray (ed.), *Proc. Workshop, Forest Classification at High Latitudes as an Aid to Regeneration.* USDA For. Serv. Gen. Tech. Report PNW-177. Pacific Northwest For. and Rge. Exp. Sta., Portland, OR.

Viereck, L. A., C. T. Dyrness, K. Van Cleve, and M. J. Foote. 1983. Vegetation, soils, and forest productivity in selected forest types in interior Alaska. *Can. J. For. Res.* 13:703–720.

Viereck, L. A., and J. M. Foote. 1970. The status of *Populus balsamifera* and *P. trichocarpa* in Alaska. *Canad. Field-Nat.* 84:169–174.

Viers, S. D. 1980. The influence of fire in coast redwood forests. *In* M. A. Stokes and J. H. Dieterich (eds.), *Proc. Fire History Workshop.* USDA For Serv. Gen Tech. Report. RM-81, Rocky Mountain For. and Rge. Exp. Sta., Fort Collins, CO.

Villiers, T. A. 1972. Seed dormancy. *In* T. T. Kozlowski (ed.), *Seed Biology.* Academic Press, New York.

Vince-Prue, D. 1975. *Photoperiodism in Plants.* McGraw-Hill, New York. 444 pp.

Viosca, P. Jr. 1931. Spontaneous combustion in the marshes of southern Louisiana. *Ecology* 12:439–442.

Viro, P. J. 1974. Effects of forest fire on soil. *In.* T. T. Kozlowski and C. E. Ahlgren (eds.), *Fire and Ecosystems.* Academic Press, New York.

Vitousek, P. M. 1982. Nutrient cycling and nutrient use efficiency. *Am. Nat.* 119:553–572.

Vitousek, P. M. 1984. Litterfall, nutrient cycling, and nutrient limitation in tropical forests. *Ecology* 65:285–298.

Vitousek, P. M., J. R. Gosz, C. C. Grier, J. M. Melillo, and W. A. Reiners. 1982. A comparative analysis of potential nitrification and nitrate mobility in forest ecosystems. *Ecol. Monogr.* 52:155–177.

Vitousek, P. M., and P. A. Matson. 1984. Mechanisms of nitrogen retention in forest ecosystems: a field experiment. *Science* 225:51–52.

Vitousek, P. M., and J. M. Melillo. 1979. Nitrate loss from disturbed forests: patterns and mechanisms. *For. Sci.* 25:605–619.

Vitousek, P. M., H. A. Mooney, J. Lubchenco, J. M. Melillo. 1997. Human domination of Earth's ecosystems. *Science* 277:494–499.

Vitousek, P. M., and W. A. Reiners. 1975. Ecosystem succession and nutrient retention: a hypothesis. *BioScience* 25:376–381.

Vitousek, P. M., and R. L. Sanford, Jr. 1986. Nutrient cycling in moist tropical forests. *Ann. Rev. Ecol. Syst.* 17:137–167.

Vitousek, P. M., L. R. Walker, L. D. Whiteaker, and D. Mueller-Dombois. 1987. Biological invasion by *Myrica faya* alters ecosystem development in Hawaii. *Science* 238:802–804.

Vogelmann, H. W., T. Siccama, D. Leedy, and D. C. Ovitt. 1968. Precipitation from fog moisture in the Green Mountains of Vermont. *Ecology* 49:1205–1207.

Vogelmann, J. E. 1995. Assessment of forest fragmentation in southern New England using remote sensing and geographic information systems technology. *Conserv. Biol.* 9:439–449.

Vogl, R. J. 1964. Vegetational history of the Crex Meadows, a prairie savanna in northwestern Wisconsin. *Am. Midl. Nat.* 72:157–175.

Vogl, R. J. 1970. Fire and the northern Wisconsin pine barrens. *In Proc. Annual Tall Timbers Fire Ecology Conf.* 6:47–96. Tall Timbers Res. Sta., Tallahassee, FL.

Vogl, R. J. 1973. Ecology of knobcone pine in the Santa Ana Mountains, California. *Ecol. Monogr.* 43:125–143.

Vogl, R. J., W. P. Armstrong, K. L. White, and K. L. Cole. 1988. The closed-cone pines and cypresses. *In* M. G. Barbour and J. Major (eds), *Terrestrial Vegetation of California*. CA. Native Plant Soc., Special Publ. No. 9. Berkeley.

Vogl, R. J., and C. Ryder. 1969. Effects of slash burning on conifer reproduction in Montana's Mission Range. *Northwest Sci.* 43:135–147.

Vogt, K. A., C. C. Grier, C. E. Meier, and M. R. Keyes. 1983. Organic matter and nutrient dynamics in forest floors in young and mature *Abies amabilis* stands in western Washington. *Ecol. Monogr.* 53:139–157.

Vogt, K. A., C. C. Grier, and D. J. Vogt. 1986. Production, turnover, and nutrient dynamics of above- and belowground detritus in world forests. *Adv. Ecol. Res.* 15:303–377.

Voigt, G. K. 1968. Variation in nutrient uptake by trees. *In Forest Fertilization*. Tennessee Valley Authority, Muscle Shoals, Ala.

Vose, J. M., B. D. Clinton, and W. T. Swank. 1994. Fire, drought, and forest management influences on pine/hardwood ecosystems in the Southern Appalachians. *In Proc. 12th Int. Conf. on Fire and Forest Meteorology, Jekyll Island, GA.* Soc. Am. For., Bethesda, MD.

Voss, E. G. 1972. *Michigan Flora. I. Gymnosperms and Monocots*. Cranbrook Inst. Sci. and Univ. Mich. Herbarium. Bloomfield Hills, MI. 488 pp.

Voss, E. G., and G. E. Crow. 1976. Across Michigan by covered wagon: a botanical expedition in 1888. *Mich. Botanist* 15:3–70.

Wagner, F. H. 1969. Ecosystem concepts in fish and game management. *In* G. M. Van Dyne (ed.), *The Ecosystem Concept in Natural Resource Management*. Academic Press, New York.

Wagner, W. H., Jr. 1968. Hybridization, taxonomy and evolution. *In* V. H. Heywood (ed.), *Modern Methods in Plant Taxonomy*, Academic Press, New York.

Wagner, W. H., Jr. 1983. Reticulistics: the recognition of hybrids and their role in cladistics and classification. *In* N. I. Platnick and V. A. Funk (eds.), *Advances in Cladistics, Vol. 2*. Columbia Univ. Press, New York.

Wagner, W. H., Jr., and D. J. Schoen. 1976. Shingle oak (*Quercus imbricaria*) and its hybrids in Michigan. *Mich. Botanist* 15:141–155.

Wagner, W. H., Jr., S. R. Taylor, G. Grieve, R. O. Kapp, and W. K. Stewart. 1988. Simple-leaved ashes (*Fraxinus*: Oleaceae) in Michigan. *Mich. Botanist* 27:119–134.

Wahlenberg, W. G. 1949. Forest succession in the southern Piedmont region. *J. For.* 47:713–715.

Waide, J. B., W. H. Caskey, R. L. Todd and L. R. Boring. 1988. Changes in soil nitrogen pools and transformations following forest clearcutting. *In* W. T. Swank, and D. A. Crossley, Jr. (eds.), *Forest Hydrology and Ecology at Coweeta*. (Ecological Studies 66). Springer-Verlag, New York.

Wakeley, P. C. 1954. Planting the southern pines. *USDA For. Serv. Agr. Monogr.* No. 18. Washington, D. C. 233 pp.

Wakeley, P. D., and J. Marrero. 1958. Five-year intercept as site index in southern pine plantations. *J. For.* 56:332–336.

Walker, J., C. H. Thompson, I. F. Fergus, and B. R. Tunstall. 1981. Plant succession and soil development in coastal sand dunes of subtropicaleastern Australia. *In* D. C. West, H. H. Shugart, and D. B. Botkin (eds.), *Forest Succession, Concepts and Application*. Springer-Verlag, New York.

Walker, L. R. 1989. Soil nitrogen changes during primary succession on a floodplain in Alaska, U.S.A. *Arctic and Alpine Res.* 21:341–349.

Walker, L. R., and F. S. Chapin, III. 1986. Physiological controls over seedling growth in primary succession on an Alaskan floodplain. *Ecology* 67:1508– 1523.

Walker, L. R., and F. S. Chapin, III. 1987. Interactions among processes controlling successional change. *Oikos* 50:131–135.

Walker, L. R., J. C. Zasada, and F. S. Chapin, III. 1986. The role of life history processes in primary succession on an Alaskan floodplain. *Ecology* 67:1243–1253.

Wallace, L. L., and E. L. Dunn. 1980. Comparative photosynthesis of three gap phase successional tree species. *Oecologia* 45:331–340.

Walter, H. 1973. *Vegetation of the Earth: In Relation to Climate and the Eco-Physiological Conditions.* Springer-Verlag, New York. 237 pp.

Walters, M. B., E. L. Kruger, and P. B. Reich. 1993. Growth, biomass distribution and CO_2 exchange of northern hardwood seedlings in high and low light: relationships with successional status and shade tolerance. *Oecologia* 94:7–16.

Walters, M. B., and P. B. Reich. 1997. Growth of *Acer saccharum* seedlings in deeply shaded understories of northern Wisconsin: effects of nitrogen and water availability. *Can. J. For. Res.* 27:237–247.

Wani, M. C., H. L. Taylor, M. E. Wall, P. Coggon, and A. T. McPhail. 1971. Plant antitumor agents. VI. The isolation and structure of taxol, a novel antileukemic and antitumor agent from *Taxus brevifolia. J. Amer. Chem. Soc.* 93:2325–2327.

Wardle, D. A. 1992. A comparative assessment of the factors which influence microbial biomass carbon and nitrogen levels in soil. *Biol. Rev.* 67:321–358.

Wardle, P. 1968. Engelmann spruce (*Picea engelmannii* Engel.) at its upper limits in the Front Range, Colorado. *Ecology* 49:483–495.

Wardle, P. 1985. New Zealand timberlines. 3. A synthesis. *New Zealand J. Bot.* 23:263–271.

Wardle, P. 1991. *Vegetation of New Zealand.* Cambridge Univ. Press, Cambridge. 672 pp.

Wareing, P. F. 1959. Problems of juvenility and flowering in trees. *J. Linn. Soc. London, Bot.* 56:282–289.

Wareing, P. F. 1969. Germination and dormancy. *In* M. B. Wilkins (ed.), *The Physiology of Plant Growth and Development.* McGraw-Hill, New York.

Wareing, P. F. 1987. Phase change and vegetative propagation. *In* A. J. Abbott and R. K. Atkin (eds.), *Improving Vegetatively Propagated Crops.* Academic Press, London.

Wareing, P. F., and L. W. Robinson. 1963. Juvenility problems in woody plants. *Rep. Forest Res.* pp. 125–127.

Waring, R. H. 1983. Estimating forest growth and efficiency in relation to canopy leaf area. *Adv. Ecol. Res.* 13:327–354.

Waring, R. H. 1989. Ecosystems: fluxes of matter and energy. *In* J. M. Cherrett (ed.), *Ecological Concepts.* Blackwell, Oxford.

Waring, R. H. 1991. Responses of evergreen trees to multiple stresses. *In* H. A. Mooney, W. E. Winner, and E. J. Pell (eds.), *Response of Plants to Multiple Stresses.* Academic Press, New York.

Waring, R. H., and J. F. Franklin. 1979. The evergreen coniferous forests of the Pacific Northwest. *Science* 204:1380–1386.

Waring, R. H., A. J. S. McDonald, S. Larsson, T. Ericsson, A. Wiren, E. Arwidsson, A. Ericsson, and T. Lohhammar. 1985. Differences in chemical composition of plants grown at constant relative growth rates with stable mineral nutrition. *Oecologia* 66:157–160.

Waring, R. H., and G. B. Pitman. 1985. Modifying lodgepole pine stands to change susceptibility to mountain pine beetle attack. *Ecology* 66:889–897.

Waring, R. H., and W. H. Schlesinger. 1985. *Forest Ecosystems: Concepts and Management.* Academic Press, New York. 340 pp.

Warming, E. 1909. *Oecology of Plants, An Introduction to the Study of Plant Communities.* Clarendon Press, Oxford. 422 pp.

Watt, A. S. 1925. On the ecology of British beechwoods with special reference to their regeneration. Part II: The development and structure of beech communities on the Sussex Downs. *J. Ecology* 13:27–73.

Watt, A. S. 1947. Pattern and process in the plant community. *J. Ecology* 35:1–22.

Watts, W. A. 1970. The full-glacial vegetation of northwestern Georgia. *Ecology* 51:17–33.

Watts, W. A. 1979. Late quaternary vegetation of central Appalachia and the New Jersey Coastal Plain. *Ecol. Monogr.* 49:427–469.

Watts, W. A., and M. Stuiver. 1980. Late Wisconsin climate of northern Florida and the origin of species-rich deciduous forest. *Science* 210:325–327.

Waughman, G. J., R. J. French, and K. Jones. 1981. Nitrogen fixation in some terrestrial environments. *In* W. J. Broughton (ed.), *Nitrogen Fixation, Vol. 1: Ecology*, Clarendon Press, Oxford.

Weaver, H. 1951. Fire as an ecological factor in Southwestern ponderosa pine forests. *J. For.* 49:93–98.

Weaver, H. 1959. Ecological changes in the ponderosa pine forest of the Warm Springs Indian Reservation in Oregon. *J. For.* 57:15–20.

Weaver, H. 1974. Effects of fire on temperate forest: western United States. *In* T. T. Kozlowski and C. E. Ahlgren (eds.), *Fire and Ecosystems*. Academic Press, New York.

Weaver, J. E., and F. E. Clements. 1929. *Plant Ecology.* McGraw-Hill, New York. 520 pp.

Webb, L. J. 1959. A physiognomic classification of Australian rain forests. *J. Ecol.* 47:551–570.

Webb, L. J. 1968. Environmental relationships of the structural types of Australian rain forest vegetation. *Ecology* 49:296–311.

Webb, L. J., J. G. Tracey, and W. T. Williams. 1972. Regeneration and pattern in the subtropical rain forest. *J. Ecology* 6:675–695.

Webb, L. J., J. G. Tracey, W. T. Williams, and G. N. Lance. 1970. Studies in the numerical analysis of complex rain-forest communities. V. A. comparison of the properties of floristic and physiognomic-structure data. *J. Ecol.* 58:203–232.

Webb, S. L. 1986. Potential role of passenger pigeons and other vertebrates in the rapid Holocene migrations of nut trees. *Quat. Res.* 26:367–375.

Webb, T. III. 1988. Eastern North America. *In* B. Huntley and T. Webb, III (eds.),*Vegetation History.* Kluwer, Dordrecht.

Webb, T. III., P. J. Bartlein, S. P. Harrison, and K. H. Anderson. 1993. Vegetation, lake levels, and climate in eastern North America for the past 18,000 years. *In* H. E. Wright, Jr., J. E. Kutzbach, T. Webb III, W. F. Ruddiman, F. A. Street-Perrott, and P. J. Bartlein (eds.), *Global Climates Since the Last Glacial Maximum.* Univ. Minnesota Press, Minneapolis.

Webb, T. III, E. J. Cushing, and H. E. Wright, Jr. 1983. Holocene changes in the vegetation of the Midwest. *In* H. E. Wright, Jr. (ed.), *Late-Quaternary Environments of the United States, Vol. 2, The Holocene.* Univ. Minnesota Press, Minneapolis.

Webb, W. L., W. K. Laurenroth, S. R. Szarek, and R. S. Kinnerson. 1983. Primary production and abiotic controls in forest, grassland, and desert ecosystems in the United States. *Ecology* 59:1239–1247.

Weidman, R. H. 1939. Evidences of racial influence in a 25–year test of ponderosa pine. *J. Agric. Res.* 59:855–887.

Wein, R. W., and D. A. MacLean. 1983. *The Role of Fire in Northern Circumpolar Ecosystems.* John Wiley, New York. 322 pp.

Weiser, C. J. 1970. Cold resistance and injury in woody plants. *Science* 169:1269–1278.

Welbourn, M. L., E. L. Stone, and J. P. Lassoie. 1981. Distribution of net litter inputs with respect to slope position and wind direction. *For. Sci.* 27:651– 659.

Wellner, C. A. 1948. Light intensity related to stand density in mature stands of the western white pine type. *J. For.* 46:16–19.

Wellner, C. A. 1970. Fire history in the northern Rocky Mountains. *In Symp. Proc. The Role of Fire in the Intermountain West.* Intermountain Fire Research Council, Univ. Montana, School of Forestry, Missoula.

Wellner, C. A. 1989. Classification of habitat types in the western United States. *In* D. E. Ferguson, P. Morgan, and F. D. Johnson (comps.), *Proceedings—Land Classifications Based on Vegetation· Applications for Resource Management.* USDA For. Serv. Gen. Tech. Report INT-257. Intermountain Res. Sta., Ogden, UT.

Wells, O. O., G. L. Switzer, and R. C. Schmidtling. 1991. Geographic variation in Mississippi loblolly pine and sweetgum. *Silvae Genetica* 40:105–119.

Wells, O. O., and P. C. Wakeley. 1966. Geographic variation in survival, growth and fusiform rust infection of planted loblolly pine. *For. Sci. Monogr.* 11. 40 pp.

Wells, P. V. 1965. Scarp woodlands, transported grassland soils, and concept of grassland climate in the great plains region. *Science* 148:246–249.

Wells, P. V. 1970. Postglacial vegetational history of the Great Plains. *Science* 167:1574–1582.

Wendel, G. W. 1987. Abundance and distribution of vegetation under four hardwood stands in north-central West Virginia. USDA For. Serv. Res. Paper NE-607. Northeastern For. Exp. Sta., Broomall, PA. 6 pp.

Went, F. W. 1957. *The Experimental Control of Plant Growth*. Ronald Press, New York. 343 pp.

Werner, H. 1962. Untersuchungen über das Wachstum der Hauptholzarten auf den wichtigsten Standortseinheiten der Mittleren Alb. *Mitt. Vereins forstl. Standortsk. forstpflz.* 12:3–52.

Wertz, W. A., and J. F. Arnold. 1972. Land systems inventory. USDA For. Serv., Intermountain Region, Ogden, UT. 12 pp.

Wertz, W. A., and J. F. Arnold. 1975. Land stratification for land-use planning. *In* B. Bernier and C. H. Winget (eds.), *Forest Soils and Forest Land Management*. Laval Univ. Press, Quebec.

West, D. C., H. H. Shugart, and D. B. Botkin. 1981. Introduction. *In* D. C. West, H. H. Shugart, and D. B. Botkin (eds.), *Forest Succession: Concepts and Application*. Springer-Verlag, New York.

West, N. E. 1968. Rodent-influenced establishment of ponderosa pine and bitterbrush seedlings in central Oregon. *Ecology* 49:1009–1011.

Westhoff, V., and E. van der Maarel. 1973. The Braun-Blanquet approach. *In* R. Tüxen (ed.), *Handbook of Vegetation Science*, Part V, *Ordination and Classification of Communities*. Junk, The Hague.

Wharton, C. H., W. M. Kitchens, E. C. Pendleton, and T. W. Sipe. 1982. The ecology of bottomland hardwood swamps of the Southeast: a community profile. U. S. Fish and Wildlife Service, FWS/OBS-81/37. 133 pp.

Wheeler, N. C., and W. B. Critchfield. 1985. The distribution and botanical characteristics of lodgepole pine: biogeographical and management implications. *In* D. M. Baumgartner, R. G. Krebill, J. T. Arnott, and G. F. Weetman (eds.), *Symp. Proc. Lodgepole Pine, The Species and Its Management*. Washington State Univ., Pullman.

Whelan, R. J. 1995. *The Ecology of Fire*. Cambridge Univ. Press. Cambridge. 346 pp.

Whitcomb, R. F., C. S. Robbins, J. F. Lynch, B. L. Whitcomb, M. K. Klimkiewicz, and D. Bystrak. 1981. Effects of forest fragmentation on avifauna of the eastern deciduous forest. *In* R. L. Burgess and D. M. Sharpe (eds.), *Forest Island Dynamics in Man-Dominated Landscapes*. Springer-Verlag, New York.

White, A. S. 1983. The effects of thirteen years of annual prescribed burning on a *Quercus ellipsoidalis* community in Minnesota. *Ecology* 64:1081–1085.

White, A. S. 1986. Prescribed burning for oak savanna restoration in central Minnesota. USDA For. Serv. Res. Paper NC-266. North Central For. Res. Sta., St. Paul, MN. 12 pp.

White, C. S. 1986. Volatile and water-soluble inhibitors of nitrogen mineralization and nitrification in a ponderosa pine ecosystem. *Biology and Fertility of Soils* 2:97–104.

White, C. S. 1988. Nitrification inhibition by monoterpeniods: theoretical mode of action based on molecular structures. *Ecology* 69:1631–1633.

White, J. 1985. The thinning rule and its application to mixtures of plant populations. *In* J. White (ed.), *Studies on Plant Demography*. Academic Press, New York.

White, P. S. 1979. Pattern, process and natural disturbance in vegetation. *Bot. Rev.* 45:229–299.

White, T. C. R. 1969. An index to measure weather-induced stress of trees associated with outbreaks of psyllids in Australia. *Ecology* 50:905–909.

White, T. C. R. 1974. A hypothesis to explain outbreaks of looper caterpillars, with special reference to populations of *Selidosema suavis* in a plantation of *Pinus radiata* in New Zealand. *Oecologia* 16:279–301.

White, T. C. R. 1978. The importance of a relative shortage of food in animal ecology. *Oecologia* 33:71–86.

Whitehead, D. R. 1981. Late-Pleistocene vegetational changes in northeastern North Carolina. *Ecol. Monogr.* 51:451–471.

Whitham, T. G. 1989. Plant hybrid zones as sinks for pests. *Science* 244:1490–1493.

Whitham, T. G., and S. Mopper. 1985. Chronic herbivory: impacts on architecture and sex expression of pinyon pine. *Science* 228:1089–1091.

Whitham, T. G., P. A. Morrow, and B. M. Potts. 1991. Conservation of hybrid plants. *Science* 254:779–780.

Whitham, T. G., P. A. Morrow, and B. M Potts. 1994. Plant hybrid zones as centers of biodiversity: the herbivore community of two endemic Tasmanian eucalypts. *Oecologia* 97:481–490.

Whitham, T. G., and C. N. Slobodchikoff. 1981. Evolution by individuals, plant-herbivore interactions, and mosaics of genetic variability: the adaptive significance of somatic mutations in plants. *Oecologia* 49:287–292.

Whitmore, T. C. 1984. *Tropical Rain Forests of the Far East*. Clarendon Press, Oxford. 352 pp.

Whitmore, T. C. 1991. *An Introduction to Tropical Rain Forests*. Clarendon Press, Oxford. 226 pp.

Whitney, G. G. 1994. *From Coastal Wilderness to Fruited Plain: A History of Environmental Change in Temperate North America From 1500 to the Present*. Cambridge Univ. Press, Cambridge. 451 pp.

Whitney, H. E., and W. C. Johnson. 1984. Ice storms and forest succession in southwestern Virginia. *Bull. Torrey Botanical Club* 111:429–437.

Whittaker, R. H. 1953. A consideration of climax theory: the climax as a population and pattern. *Ecol. Monogr.* 23:41–78.

Whittaker, R. H. 1956. Vegetation of the Great Smoky Mountains. *Ecol. Monogr.* 26:1–80.

Whittaker, R. H. 1960. Vegetation of the Siskiyou Mountains, Oregon and California. *Ecol. Monogr.* 30:279–338.

Whittaker, R. H. 1962. Classification of natural communities. *Bot. Rev.* 28:1–239.

Whittaker, R. H. 1966. Forest dimensions and production in the Great Smoky Mountains. *Ecology* 47:103–121.

Whittaker, R. H. 1967. Gradient analysis of vegetation. *Biol. Rev.* 42:207–264.

Whittaker, R. H. 1972. Evolution and measurement of species diversity. *Taxon* 21:213–251.

Whittaker, R. H. 1975. *Communities and Ecosystems*, 2nd ed. MacMillan, New York. 385 pp.

Whittaker, R. H. 1977. Evolution of species diversity in land communities. *In* M. K. Hecht, W. C. Steere, and B. Wallace (eds.), *Evolutionary Biology*, Vol. 10. Plenum, New York.

Wicken, E. B. 1986. Terrestrial ecozones of Canada. Environment Canada. Ecological Land Classification Series No. 19. Lands Directorate, Ottawa. 26 pp.

Wickware, G., and C. D. A. Rubec. 1989. Ecoregions of Ontario. Ecological Land Classification Series No. 26, Env. Canada, Ottawa. 37 pp.

Wiersma, J. H. 1962. Enkele quantitative aspecten van het exotenvraagstuk. *Ned. Bosbouw Tijdschrift* 34:175–184.

Wiersma, J. H. 1963. A new method of dealing with results of provenance tests. *Silvae Genetica* 12:200–205.

Wilcove, D. S., C. H. McLellan, and A. P. Dobson. 1986. Habitat fragmentation in the Temperate Zone. *In* M. E. Soulé (ed.), *Conservation Biology, The Science of Scarcity and Diversity*. Sinauer Associates, Sunderland, MA.

Wilde, S. A. 1958. *Forest Soils: Their Properties and Relation to Silviculture*. Ronald Press, New York. 537 pp.

Williams, G. J., III, and C. McMillan. 1971. Phenology of six United States provenances of *Liquidambar stryraciflua* under controlled conditions. *Amer. J. Bot.* 58: 24–31.

Williams, M. 1989. *Americans and Their Forests, A Historical Geography*. Cambridge Univ. Press, Cambridge. 599 pp.

Williamson, G. B., and E. M. Black. 1981. High temperature of forest fires under pines as a selective advantage over oaks. *Nature* 293:643–644.

Wilson, A. D. 1990. The effect of grazing on Australian ecosystems. *Proc. Ecol. Soc. Aust.* 16:235–244.

Wilson, B. F. 1966. Development of the shoot system of *Acer rubrum* L. *Harvard Forest Paper* No. 14. Harvard Univ., Petersham, MA. 21 pp.

Wilson, B. F. 1984. *The Growing Tree*. Univ. Massachusetts Press, Amherst. 138 pp.

Wilson, C. C. 1948. Fog and atmospheric carbon dioxide as related to apparent photosynthetic rate of some broadleaf evergreens. *Ecology* 29:507–508.

Wilson, E. O. 1984. *Biophilia*. Harvard Univ. Press, Cambridge, MA. 157 pp.

Wilson, E. O. 1992. *The Diversity of Life*. Belknap Press, Harvard Univ., Cambridge, MA. 424 pp.

Wilson, M. V., and A. Shmida. 1984. Measuring beta diversity with presence-absence data. *J. Ecology* 72:1055–1064.

Wing, S. L., and H.-D. Sues (rapp.) 1993. Mesozoic and early Cenozoic terrestrial ecosystems. *In* A. K. Behrensmeyer, J. D. Damuth, W. A. DiMichele, R. Potts, H.-D. Sues, and S. L. Wing (eds.), 1993. *Terrestrial Ecosystems Through Time*. Univ. Chicago Press, Chicago.

Wistendahl, W. A. 1958. The flood plain of the Raritan River, New Jersey. *Ecol. Monogr.* 28:129–153.

Witter, J. A., and L. A. Waisanen. 1978. The effect of differential flushing times among trembling aspen clones on tortricid caterpillar populations. *Env. Ent.* 7:139–143.

Wofsy, S. P., M. L. Goulden, J. W. Munger, S. -M. Fan, P. S. Bakwin, B. C. Daube, S. L. Bassow, and F. A. Bazzaz. 1993. Net exchange of CO_2 in a mid-latitude forest. *Science* 260:1314–1317.

Wolda, H. 1981. Similarity indices, sample size and diversity. *Oecologia* 50:296–302.

Wolfe, J. A., and E. B. Leopold. 1967. Neogene and early Quaternary vegetation of northwestern North America and northeastern Asia. *In* D. M. Hopkins (ed.), *The Bering Land Bridge*. Stanford Univ. Press, Stanford, CA.

Wolfe, J. N., R. T. Wareham, and H. T. Scofield. 1949. Microclimates and macroclimates of Neotoma, a small valley in central Ohio, *Ohio Biol. Surv. Bull.* 41. 267 pp.

Wood, D. M., and R. del Moral. 1987. Mechanisms of early primary succession in subalpine habitats on Mount St. Helens. *Ecology* 68:780–790.

Wood, T., F. H. Bormann, and G. T. Voigt. 1984. Phosphorus cycling in a northern hardwood forest: Biological and chemical control. *Science* 223:391–393.

Woods, D. B., and N. C. Turner. 1971. Stomatal response to changing light by four tree species of varying shade tolerance. *New Phytol.* 70:77–84.

Woods, F. W. 1953. Disease as a factor in the evolution of forest composition. *J. For.* 51:871–873.

Woods, F. W. 1957. Factors limiting root penetration in deep sands of the southeastern coastal plain. *Ecology* 38:357–359.

Woods, F. W., and R. E. Shanks. 1959. Natural replacement of chestnut by other species in he Great Smoky Mountains National Park. *Ecology* 40:349–361.

Woods, K. D. 1979. Reciprocal replacement and the maintenance of codominance in a beech–maple forest. *Oikos* 33:31–39.

Woods, K. D. 1984. Patterns of tree replacement: canopy effects on understory pattern in hemlock–northern hardwood forests. *Vegetatio* 56:87–107.

Woodward, F. I. 1987. *Climate and Plant Distribution*. Cambridge Univ. Press, Cambridge. 174 pp.

Woodwell, G. M., and D. B. Botkin. 1970. Metabolism of terrestrial ecosystems by gas exchange techniques: The Brookhaven approach. *In* D. Reichle (ed.), *Analysis of Temperate Forest Ecosystems*. (Ecological Studies, Vol. 1) Springer-Verlag, New York.

Worster, D. 1994. *Nature's Economy,* 2nd ed. Cambridge Univ. Press, Cambridge. 505 pp.

Wright, H. E., Jr. 1971. Late Quaternary vegetational history of North America. *In* K. K. Turekian (ed.), *The Late Cenozoic Glacial Ages*. Yale Univ. Press, New Haven, CT.

Wright, H. E., Jr., and M. L. Heinselman (eds.). 1973. The ecological role of fire in natural conifer forests of western and northern North America. *Quat. Res.* 3:317–513.

Wright, H. E., Jr., J. E. Kutzbach, T. Webb III, W. F. Ruddiman, F. A. Street-Perrott, and P. J. Bartlein (eds.). 1993. *Global Climates Since the Last Glacial Maximum*. Univ. Minnesota Press. Minneapolis. 569 pp.

Wright, J. W. 1955. Species crossability in spruce in relation to distribution and taxonomy. *For. Sci.* 1:319–352.

Wright, J. W. 1970. Genetics of eastern white pine (*Pinus strobus* L.). USDA For. Serv. Res. Paper WO-9. Washington, D.C. 16 pp.

Wright, J. W. 1976. *Introduction to Forest Genetics*. Academic Press, New York. 463 pp.

Wright, J. W., S. S. Pauley, R. B. Polk, J. J. Jokela, and R. A. Read. 1966. Performance of Scotch pine varieties in the North Central Region. *Silvae Genetica* 15:101–110.

Wright, J. W., L. F. Wilson, and W. Randall. 1967. Differences among Scotch pine varieties in susceptibility to European pine sawfly. *For. Sci.* 13:175–181.

Wright, R. A., R. W. Wein, and B. P. Dancik. 1992. Population differentiation in seedling root size between adjacent stands of jack pine. *For. Sci.* 38:777–785.

Wuerthner, G. 1988. *Yellowstone and the Fires of Change.* Haggis House Publ, Salt Lake City, UT.

Yaffee, S. L. 1994. *The Wisdom of the Spotted Owl.* Island Press. Washington, D.C. 430 pp.

Yaffee, S. L., A. F. Phillips, I. C. Frentz, P. W. Hardy, S. M. Maleki, and B. E. Thorpe. 1996. *Ecosystem Management in the United States: An Assessment of Current Experience.* Island Press. Washington, D.C. 352 pp.

Yahner, R. H. 1988. Changes in wildlife communities near edges. *Conserv. Biol.* 2:333–339.

Yahner, R. H. 1995. *Eastern Deciduous Forest, Ecology and Wildlife Conservation.* Univ. Minnesota Press. Minneapolis. 220 pp.

Yocom, H. A. 1968. Shortleaf pine seed dispersal. *J. For.* 66:422.

Youngberg, C. T. 1959. The influence of soil conditions, following tractor logging, on the growth of planted Douglas-fir seedlings. *Proc. Soil Sci. Soc. Amer.* 23:76–78.

Youngberg, C. T., and A. G. Wollum. 1976. Nitrogen accretion in developing *Ceonothus velutinus* stands. *Soil Sci. Soc. Amer. Proc.* 40: 109–112.

Youngblood, A. P., and R. L. Mauk. 1985. Coniferous forest habitat types of central and southern Utah. USDA For. Serv. Gen. Tech. Report INT-187. Intermountain For. Exp. Sta., Ogden, UT. 89 pp.

Zahner, R. 1956. Evaluating summer water deficiencies. USDA For. Serv. Occ. Paper 150. Southern For. Exp. Sta., New Orleans, LA. 18 pp.

Zahner, R. 1958. Site-quality relationships of pine forests in southern Arkansas and northern Louisiana. *For. Sci.* 4:162–176.

Zahner, R. 1968. Water deficits and growth of trees. *In* T. T. Kozlowski (ed.), *Water Deficits and Plant Growth. II.* Academic Press, New York.

Zahner, R., and J. R. Donnelly. 1967. Refining correlations of rainfall and radial growth in young red pine. *Ecology* 48:525–530.

Zahner, R., and A. R. Stage. 1966. A procedure for calculating daily moisture stress and its utility in regressions of tree growth on weather. *Ecology* 47:64–74.

Zahner, R., and F. W. Whitmore. 1960. Early growth of radically thinned loblolly pine. *J. For.* 58:628–634.

Zahner, R., and N. A. Crawford. 1965. The clonal concept in aspen site relations. *In* C. T. Youngberg and C. B. Davey (eds.), *Tree Growth and Forest Soils.* Oregon State Univ. Press, Corvallis.

Zak, B. 1965. Aphids feeding on mycorrhizae Douglas-fir. *For. Sci.* 11:410–411.

Zak, D. R., P. M. Groffman, K. S. Pregitzer, S. Christensen, and J. M. Tiedje. 1990. The vernal dam: plant-microbe competition for nitrogen in northern hardwood forests. *Ecology* 71:651–656.

Zak, D. R., G. E. Host, and K. S. Pregitzer. 1989. Regional variability in nitrogen mineralization, nitrification, and overstory biomass in northern Lower Michigan. *Can. J. For. Res.* 19:1521–1526.

Zak, D. R., K. S. Pregitzer, and G. E. Host. 1986. Landscape variation in nitrogen mineralization and nitrification. *Can. J. For. Res.* 16:1258–1263.

Zak, D. R., and K. S. Pregitzer. 1990. Spatial and temporal variability of nitrogen cycling in northern Lower Michigan. *For. Sci.* 36:367–380.

Zak, D. R., D. Tilman, R. R. Parmenter, C. W. Rice, F. M. Fisher, J. Vose, D. Milchunas, and C. W. Martin. 1994. Plant production and soil microorgansims in late-successional ecosystems: a continental-scale study. *Ecology* 75:2333–2347.

Zasada, J. 1985. Production, dispersal, and germination, and first year seedling survival of white spruce and birch in the Rosie Creek burn. *In* G. P. Juday and C. T. Dyrness (eds.), *Early Results of the Rosie Creek Research Project—1984.* Agr. and For. Exp. Sta., Univ. Alaska, Fairbanks, AK. Misc. Publ. 85-2.

Zasada, J. C., T. L. Sharik, and M. Nygren. 1992. The reproductive process in boreal forest trees. *In* H. H. Shugart, R. Leemans, and G. B. Bonan (eds.), *A Systems Analysis of the Global Boreal Forest.* Cambridge Univ. Press, New York.

Zasada, J. C., K. Van Cleve, R. Werner, J. A., McQueen, and E. Nyland. 1977. Forest biology and management in high latitude North American forest. *In Forest Lands at Latitudes North of 60 Degrees.* School Agr. Land Res. Mange., Univ. Alaska, Fairbanks.

Zimmermann, M. H. 1971. Transport in the xylem. *In* M. H. Zimmermann and C. L. Brown, *Trees: Structure and Function.* Springer-Verlag, New York.

Zimmermann, M. H., and C. L. Brown. 1971. *Trees: Structure and Function.* Springer-Verlag, New York. 336 pp.

Zobel, B., and J. Talbert. 1984. *Applied Forest Tree Improvement.* John Wiley, New York. 505 pp.

Zobel, D. B. 1969. Factors affecting the distribution of *Pinus pungens,* an Appalachian endemic. *Ecol. Monogr.* 39:303–333.

Zogg, G. P., and B. V. Barnes. 1995. Ecological classification and analysis of wetland ecosystems, northern Lower Michigan. *Can. J. For. Res.* 25:1865–1875.

Zoltai, S. C. 1965. Forest sites of site regions 5S and 4S, Northwestern Ontario. Vol. I. Ontario Dept. Lands and For. Res. Rept. 65. 121 pp.

Zonneveld, I. S. 1990. Scope and concepts of landscape ecology as an emerging science. *In* I. S. Zonneveld and R. T. T. Forman (eds.), *Changing Landscapes: An Ecological Perspective.* Springer-Verlag, New York.

Zonneveld, I. S., and R. T. T. Forman (eds.). 1990. *Changing Landscapes: An Ecological Perspective.* Springer-Verlag, New York. 286 pp.

Zou, X., C. Theiss, and B. V. Barnes. 1992. Pattern of Kirtland's warbler occurrence in relation to the landscape structure of its summer habitat in northern Lower Michigan. *Landscape Ecology.* 6:221–231.

SCIENTIFIC NAMES OF TREES AND SHRUBS

Acacia	*Acacia* spp.
Ailanthus; tree-of-heaven	*Ailanthus altissima* (Mill.) Swingle
Alaska-cedar	*Chamaecyparis nootkatensis* (D. Don) Spach
Alder, European or black	*Alnus glutinosa* (L.) Gaertn.
Alder, green or Sitka	*Alnus crispa* (Ait.) Pursh.
Alder, red or Oregon	*Alnus rubra* (Bong.)
Alder, Sitka	*Alnus sinuata* (Reg.) Rydb.
Alder, speckled	*Alnus rugosa* (Du Roi) Spreng.
Apple, common	*Malus pumila* Miller
Apple, wild crab	*Malus coronaria* (L.) Miller
Arborvitae	*Thuja occidentalis* L.
Ash, black	*Fraxinus nigra* Marsh.
Ash, blue	*Fraxinus quadrangulata* Michx.
Ash, European	*Fraxinus excelsior* L.
Ash, mountain (Australia)	*Eucalyptus regnans* F. Muell.
Ash, red or green	*Fraxinus pennsylvanica* Marsh.
Ash, white	*Fraxinus americana* L.
Ash, prickly-	*Zanthoxylum americanum* Mill.
Aspen, bigtooth	*Populus grandidentata* Michx.
Aspen, European	*Populus tremula* L.
Aspen, trembling or quaking	*Populus tremuloides* Michx.
Australian pine	*Casuarina* spp.
Baldcypress	*Taxodium distichum* (L.) Rich.
Balsa	*Ochroma pyramidale* (Cav.) Urban
Basswood	*Tilia americana* L.
Beech, American	*Fagus grandifolia* Ehrh.
Beech, Antarctic	*Nothofagus antarctica* (Forst.) Oerst.
Beech, blue-	*Carpinus caroliniana* Walt.
Beech, European	*Fagus sylvatica* L.
Beech, Oriental	*Fagus orientalis* Lipsky

Bigtree; giant sequoia	*Sequoiadendron giganteum* (Lindl.) Buchholz
Birch, sweet or black	*Betula lenta* L.
Birch, bog, swamp, or low	*Betula pumila* L.
Birch, European or silver	*Betula pendula* Roth.
Birch, gray	*Betula populifolila* Marsh.
Birch, moor or pubescent	*Betula pubescens* Ehrh.
Birch, paper or white	*Betula papyrifera* Marsh.
Birch, river or red	*Betula nigra* L.
Birch, Virginia round-leaf	*Betula uber* (Ashe) Fern.
Birch, yellow	*Betula alleghaniensis* Britton
Blackgum	*Nyssa sylvatica* Marsh.
Boxelder	*Acer negundo* L.
Brazilian pepper tree	*Schinus terebinthifolius* Raddi
Buckeye, Ohio	*Aesculus glabra* Willd.
Buckeye, painted	*Aesculus sylvatica* Bartr.
Buckeye, red	*Aesculus pavia* L.
Buckeye, yellow	*Aesculus octandra* Marsh.
Buckthorn, alder	*Rhamnus alnifolia* L'Héritier
Buckthorn, common	*Rhamnus cathartica* L.
Buckthorn, glossy	*Rhamnus frangula* L.
Butternut	*Juglans cinerea* L.
Catalpa, northern	*Catalpa speciosa* Warder
Catalpa, southern	*Catalpa bignonioides* Walter
Ceanothus, redstem	*Ceanothus sanguineus* Pursh.
Cedar, Alaska-	*Chamaecyparis nootkatensis* (D. Don) Spach
Cedar, Deodar	*Cedrus deodara* (Roxb.) Loud.
Cedar, eastern red-	*Juniperus virginiana* L.
Cedar, incense	*Libocedrus decurrens* Torr.
Cedar, Lebanon	*Cedrus libani* Loud.
Cedar, northern white-	*Thuja occidentalis* L.
Cedar, Port Orford	*Chamaecyparis lawsoniana* (A. Murr.) Parl.
Cedar, Atlantic white-	*Chamaecyparis thyoides* (L.) B. S. P.
Cedar, eastern red-	*Juniperus virginiana* L.
Cedar, western red-	*Thuja plicata* Donn
Cherry, black	*Prunus serotina* Ehrh.
Cherry, choke	*Prunus virginiana* L.
Cherry, pin or fire	*Prunus pensylvanica* L. f.

Cherry, sweet, European bird	*Prunus padus* L.
Chestnut, American	*Castanea dentata* (Marsh.) Borkh.
Chinese fir	*Cunninghamia lanceolata* (Lamb.) Hook.
Chinquapin, Ozark	*Castanea ozarkensis* Ashe
Coconut	*Cocos nucifera* L.
Cottonwood, black	*Populus trichocarpa* Torr. & Gray
Cottonwood, eastern	*Populus deltoides* Bartr.
Cottonwood, European; black poplar	*Populus nigra* L.
Cottonwood, Fremont	*Populus fremontii* Wats.
Cottonwood, narrowleaf	*Populus angustifolia* James
Cottonwood, plains	*Populus deltoides var. occidentalis* Rydb.
Creeping strawberry-bush	*Euonymus obovata* Nutt.
Creosote bush	*Larrea* spp.
Cypress, Arizona	*Cupressus arizonica* Greens
Cypress, Mexican	*Cupressus lusitanica* Miller
Cypress pine	*Callitris* spp.
Dawn redwood	*Metasequoia glyptostroboides* H. H. Hu & Cheng
Dogwood, alternate-leaf	*Cornus alternifolia* L.
Dogwood, flowering	*Cornus florida* L.
Dogwood, gray	*Cornus foemina* Miller
Dogwood, red-osier	*Cornus stolonifera* Michx.
Dogwood, roundleaf	*Cornus rugosa* Lamarck
Dogwood, silky	*Cornus amomum* Miller
Douglas-fir	*Pseudotsuga menziesii* (Mirb.) Franco
Elm, American	*Ulmus americana* L.
Elm, rock	*Ulmus thomasii* Sarg.
Elm, Siberian	*Ulmus pumila* L.
Elm, slippery	*Ulmus rubra* Mühl.
Elm, winged	*Ulmus alata* Michx.
Eucalyptus	*Eucalyptus* spp.
Eucalyptus, alpine ash	*Eucalyptus delegatensis* B. T. Baker
Eucalyptus, blackbutt	*Eucalyptus pilularis* Smith
Eucalyptus, flooded gum	*Eucalyptus grandis* Hill ex Maiden
Eucalyptus, jarrah	*Eucalyptus marginata* Donn ex Smith
Eucalyptus, karri	*Eucalyptus diversicolor* F. Muell.
Eucalyptus, messmate stringybark	*Eucalyptus obliqua* L'Hérit.
Eucalyptus, mountain ash	*Eucalyptus regnans* F. Muell.
Fetterbush	*Leucothoe* spp.
Fir, subalpine or alpine	*Abies lasiocarpa* (Hook.) Nutt.

Fir, balsam	*Abies balsamea* (L.) Mill.
Fir, European silver	*Abies alba* Mill.
Fir, Fraser	*Abies fraseri* (Pursh.) Poir.
Fir, grand or lowland white	*Abies grandis* (Dougl.) Lindl.
Fir, noble	*Abies procera* Rehd.
Fir, Pacific silver	*Abies amabilis* (Dougl.) Forbes
Fir, red	*Abies magnifica* A. Murr.
Fir, west Himalayan	*Abies pindrow* Royle
Fir, white	*Abies concolor* (Gord. & Glend.) Lindl.
Giant sequoia	*Sequoiadendron giganteum* (Lindl.) Buchholz
Ginkgo	*Ginkgo biloba* L.
Gum, southern blue	*Eucalyptus globulus* Labill.
Hackberry	*Celtis occidentalis* L.
Hackberry, dwarf	*Celtis tenuifolia* Nuttall
Haw, black	*Viburnum prunifolium* L.
Haw, dotted	*Crataegus punctata* Jacquin
Hawthorn	*Crataegus* spp.
Hazel, beaked	*Corylus cornuta* Marsh.
Hemlock, eastern	*Tsuga canadensis* (L.) Carr.
Hemlock, mountain	*Tsuga mertensiana* (Bong.) Carr.
Hemlock, western	*Tsuga heterophylla* (Raf.) Sarg.
Hickory, bitternut	*Carya cordiformis* (Wangenh.) K. Koch
Hickory, mockernut	*Carya tomentosa* Nutt.
Hickory, pignut	*Carya glabra* (Mill.) Sweet
Hickory, shagbark	*Carya ovata* (Mill.) K. Koch
Hickory, shellbark	*Carya laciniosa* (Michaux f.) G. Don
Hickory, water	*Carya aquatica* (Michx. f.) Nutt.
Hobble-bush	*Viburnum alnifolium* Marsh.
Holly, American	*Ilex opaca* Ait.
Hop-hornbeam; ironwood	*Ostrya virginiana* (Mill.) K. Koch
Hornbeam	*Carpinus betulus* L.
Horsechestnut	*Aesculus hippcastanum* L.
Horsechestnut, red	*Aesculus* ✕*carnea* Hayne
Ironwood; hop-hornbeam	*Ostrya virginiana* (Mill.) K. Koch
Juniper, alligator	*Juniperus deppeana* Steud.
Juniper, common	*Juniperus communis* L.
Juniper, ground	*Juniperus communis* L. var. depressa Pursh.
Juniper, one-seed	*Juniperus monosperma* (Engelm.) Sarg.

Juniper, Rocky Mountain	*Juniperus scopulorum* Sarg.
Juniper, Utah	*Juniperus osteosperma* (Torr.) Little
Juniper, western	*Juniperus occidentalis* Hook.
Kahikatea	*Podocarpus dacrydioides* A. Rich.
Kauri	*Agathis australis* Hort. ex Lindl.
Kentucky coffeetree	*Gymnocladus dioicus* (L.) K. Koch
Kudzu	*Pueraria lobata* (Willd.) Ohwi
Larch, eastern; tamarack	*Larix laricina* (Du Roi) K. Koch
Larch, European	*Larix decidua* Mill.
Larch, Japanese	*Larix leptolepis* (Sieb. & Zucc.) Gord.
Larch, subalpine	*Larix lyallii* Parl.
Larch, western	*Larix occidentalis* Nutt.
Laurel, mountain	*Kalmia latifolia* L.
Lime–see Linden	
Linden, Large-leaved or Summer	*Tilia platyphyllos* Scop.
Linden, Small-leaved European or Winter	*Tilia cordata* Mill.
Locust, black	*Robinia pseudoacacia* L.
Locust, honey	*Gleditsia triacanthos* L.
Madrone, Pacific	*Arbutus menziesii* Pursh.
Magnolia, Fraser	*Magnolia fraseri* Walt.
Magnolia, southern or evergreen	*Magnolia grandiflora* L.
Magnolia, sweetbay	*Magnolia virginiana* L.
Magnolia, umbrella	*Magnolia tripetala* L.
Mahogany, West Indies	*Swietenia mahagoni* Jacq.
Maidenhair tree	*Ginkgo biloba* L.
Mangrove, black	*Avicennia nitida* Jacq.
Mangrove, red	*Rhizophora mangle* L.
Maple, Amur	*Acer ginnala* Maxim.
Maple, ash-leaf	*Acer negundo* L.
Maple, bigleaf	*Acer macrophyllum* Pursh.
Maple, bigtooth	*Acer grandidentatum* Nutt.
Maple, black	*Acer nigrum* Michaux f.
Maple, field	*Acer campestre* L.
Maple, Japanese	*Acer palmatum* Thunb.
Maple, mountain	*Acer spicatum* Lam.
Maple, Norway	*Acer platanoides* L.
Maple, red	*Acer rubrum* L.
Maple, silver	*Acer saccharinum* L.
Maple, striped	*Acer pensylvanicum* L.
Maple, sugar	*Acer saccharum* Marsh.

Maple, sycamore	*Acer pseudoplatanus* L.
Maple, vine	*Acer circinatum* Pursh.
Melaleuca	*Melaleuca quinquenervia* (Cav.) Blake
Mesquite	*Prosopis juliflora* (Sw.) DC.
Mexican cypress	*Cupressus lusitanica* Miller
Mimosa	*Albizzia julibrissin* Durazz.
Mountain-ash, American	*Sorbus americana* Marsh.
Mountain-ash, showy	*Sorbus decora* (Sarg.) Schneid.
Mountain laurel	*Kalmia latifolia* L.
Mulberry, red	*Morus rubra* L.
Mulberry, white	*Morus alba* L.
Nannyberry	*Viburnum lentago* L.
Oak, bear	*Quercus ilicifolia* Wangenh.
Oak, black	*Quercus velutina* Lam.
Oak, blackjack	*Quercus marilandica* Muenchh.
Oak, bur	*Quercus macrocarpa* Michx.
Oak, California black	*Quercus kelloggii* Newb.
Oak, California live	*Quercus agrifolia* Née
Oak, canyon live	*Quercus chrysolepis* Liebm.
Oak, cherrybark	*Quercus pagoda Raf.* (syn. *Q. falcata* var. *pagodifolia* Ill.)
Oak, chestnut or rock chestnut	*Quercus prinus* L.
Oak, chinquapin or yellow	*Quercus muehlenbergii* Engelm.
Oak, cork	*Quercus suber* L.
Oak, diamond leaf	*Quercus laurifolia* Michx.
Oak, durmast	*Quercus petraea* (Mattuschka) Lieblein.
Oak, dwarf chestnut	*Quercus prinoides* Willd.
Oak, Emory	*Quercus emoryi* Torr.
Oak, English, European, or common	*Quercus robur* L.
Oak, holm	*Quercus ilex* L.
Oak, interior live	*Quercus wislizenii* A. DC.
Oak, live	*Quercus virginiana* Mill.
Oak, mongolian	*Quercus mongolica* Fisch. et Turcz.
Oak, northern pin	*Quercus ellipsoidalis* E. J. Hill
Oak, Oregon	*Quercus garryana* Dougl.
Oak, overcup	*Quercus lyrata* Walt.
Oak, pin	*Quercus palustris* Muenchh.
Oak, post	*Quercus stellata* Wangenh.
Oak, pubescent	*Quercus pubescens* Willd.
Oak, northern red	*Quercus rubra* L.

Oak, sand live	*Quercus geminata* Small. [syn. *Q. virginiana* var. *maritima* (Michx.) Sarg.]
Oak, shingle	*Quercus imbricaria* Michx.
Oak, southern red	*Quercus falcata* Michx.
Oak, swamp chestnut	*Quercus michauxii* Nutt.
Oak, swamp laurel	*Quercus laurifolia* Michx.
Oak, swamp white	*Quercus bicolor* Willd.
Oak, turkey	*Quercus laevis* Walt.
Oak, white	*Quercus alba* L.
Oak, willow	*Quercus phellos* L.
Osage-orange	*Maclura pomifera* (Raf.) C. K. Schneid.
Pacific silver fir	*Abies amabilis* (Dougl.) Forbes
Palm, cabbage	*Sabal palmetto* (Walt.) Lodd.
Palmetto, cabbage	*Sabal palmetto* (Walt.) Lodd.
Palmetto, saw	*Serenos repens* (Bartr.) Small
Paperbark tree	*Melaleuca quinquenervia* (Cav.) Blake
Pawpaw	*Asimina triloba* (L.) Dunal
Pepper tree, Brazilian	*Schinus terebinthifolius* Raddi
Pine, aleppo	*Pinus halepensis* Mill.
Pine, bishop	*Pinus muricata* D. Don
Pine, black or Austrian	*Pinus nigra* Arnold
Pine, bristlecone	*Pinus aristata* Engelm.
Pine, Canary Island	*Pinus canariensis* Smith
Pine, Caribbean	*Pinus caribaea* Morelet
Pine, chir	*Pinus roxburghii* Sarg.
Pine, Calabrian	*Pinus brutia* Ten.
Pine, cypress	*Callitris* spp.
Pine, digger	*Pinus sabiniana* Dougl.
Pine, eastern white	*Pinus strobus* L.
Pine, erectcone or Calabrian	*Pinus brutia* Ten.
Pine, Formosa white	*Pinus formosana* Hayata
Pine, Himalayan white	*Pinus griffithii* McClelland
Pine, Italian stone	*Pinus pinea* L.
Pine, Jack	*Pinus banksiana* Lamb.
Pine, Japanese black	*Pinus thunbergii* Part.
Pine, Japanese red	*Pinus densiflora* Sieb. & Zucc.
Pine, Japanese white	*Pinus parviflora* Sieb. & Zucc.
Pine, Jeffrey	*Pinus jeffreyi* Grev. & Balf.
Pine, knobcone	*Pinus attenuata* Lemmon
Pine, limber	*Pinus flexilis* James

Pine, loblolly	*Pinus taeda* L.
Pine, lodgepole	*Pinus contorta* Dougl.
Coastal lodgepole pine	*Pinus contorta* ssp. *contorta*
Mendocino White Plains lodgepole pine	*Pinus contorta* ssp. *bolanderi*
Rocky Mountain-Intermountain lodgepole pine	*Pinus contorta* ssp. *latifolia*
Sierra–Cascade lodgepole pine	*Pinus contorta* ssp. *murrayana*
Pine, longleaf	*Pinus palustris* Mill.
Pine, Macedonia white	*Pinus peuce* Griseb.
Pine, maritime	*Pinus pinaster* Ait.
Pine, Merkus	*Pinus merkusii* Jungh et De Vriese
Pine, Mexican pinyon (piñon)	*Pinus cembroides* Zucc.
Pine, Mexican white	*Pinus ayacahuite* Ehrenb.
Pine, Monterey or radiata	*Pinus radiata* D. Don
Pine, mugo or mountain	*Pinus mugo* Turra. [*P. montana* Mill.]
Pine, patula	*Pinus patula* Schl. & Cham.
Pine, pinyon (piñon)	*Pinus edulis* Engelm. or *P. monophylla* Torr. & Frem. or *P. cembroides* Zucc. or *P. quadrifolia* Parl. ex Sudw.
Pine, pitch	*Pinus rigida* Mill.
Pine, ponderosa	*Pinus ponderosa* Laws.
Pine, red	*Pinus resinosa* Ait.
Pine, sand	*Pinus clausa* (Chapm.) Vasey
Pine, Scots or Scotch	*Pinus sylvestris* L.
Pine, shortleaf	*Pinus echinata* Mill.
Pine, slash	*Pinus elliottii* Engelm. var. *elliottii*
Pine, sugar	*Pinus lambertiana* Dougl.
Pine, stone or Swiss stone	*Pinus cembra* L.
Pine, table mountain	*Pinus pungens* Lamb.
Pine, Torrey	*Pinus torreyana* Parry ex Carr.
Pine, Virginia	*Pinus virginiana* Mill.
Pine, western white	*Pinus monticola* Dougl.
Pine, whitebark	*Pinus albicaulis* Engelm.
Planetree, oriental	*Platanus orientalis* L.
Plum, Allegheny	*Prunus alleghaniensis* Porter
Pondcypress	*Taxodium distichum* var. *imbricarium* (Nutt.) Croon (syn = *T. ascendens* Brong.)
Poplar, balsam	*Populus balsamifera* L.
Poplar, black hybrid	*Populus* ×*euramericana*

Poplar, Lombardy	*Populus nigra* 'Italica' Muenchh.
Poplar, European white	*Populus alba* L.
Prickley-ash	*Zanthoxylum americanum* Mill.
Torreya, Florida	*Torreya taxifolia* Arn.
Redbud, eastern	*Cercis canadensis* L.
Redwood	*Sequoia sempervirens* (D. Don) Endl.
Redwood, dawn	*Metasequoia glyptostroboides* H. H. Hu & Cheng
Rimu	*Dacrydium cupressinum* Soland. ex Forst.
Rhododendron, rosebay	*Rhododendron maximum* L.
Sagebrush	*Artemisia tridentata* Nutt.
Saguaro; giant cactus	*Cereus giganteus* Engelm.
Sassafras	*Sassafras albidum* (Nutt.) Nees
Sequoia, giant; bigtree	*Sequoiadendron giganteum* (Lindl.) Buchholz
Serviceberry, downy	*Amelanchier arborea* (Michaux f.) Fernald
Serviceberry, smooth	*Amelanchier laevis* Wicg.
Sheepberry	*Viburnum lentago* L.
Silver bell; Carolina silverbell	*Halesia carolina* L.
Sourwood	*Oxydendrum arboreum* (L.) DC.
Spruce, black	*Picea mariana* (Mill.) B. S. P.
Spruce, Colorado blue	*Picea pungens* Engelm.
Spruce, Engelmann	*Picea engelmannii* Parry
Spruce, Norway or European	*Picea abies* (L.) Karst.
Spruce, red	*Picea rubens* Sarg.
Spruce, Siberian	*Picea obovata* Ledeb.
Spruce, Sitka	*Picea sitchensis* (Bong.) Carr.
Spruce, white	*Picea glauca* (Moench) Voss
Sugarberry	*Celtis laevigata* Willd.
Sugi	*Cryptomeria japonica* (Linn. f.) D. Don
Sumac, smooth	*Rhus glabra* L.
Sweetgum	*Liquidambar styraciflua* L.
Sycamore, American	*Platanus occidentalis* L.
Tamarack; eastern larch	*Larix laricina* (Du Roi) K. Koch
Tamarix, five-stamen	*Tamarix pentandra* Pall.
Teak	*Tectona grandis* L.f.
Totara	*Podocarpus totara* G. Benn. ex D. Don
Tree-of-heaven; ailanthus	*Ailanthus altissima* (Mill.) Swingle
Tuliptree; yellow-poplar	*Liriodendron tulipifera* L.

Tupelo, water or swamp	*Nyssa aquatica* L.
Virginia creeper	*Parthenocissus quinquefolia* (L.) Planchon
Walnut, black	*Juglans nigra* L.
Willow, black	*Salix nigra* Marsh.
Willow, crack	*Salix fragilis* L.
Willow, peachleaf	*Salix amygdaloides* Andersson
Willow, sandbar	*Salix interior* Rowlee
Willow, weeping	*Salix babylonica* L.
Witch-hazel	*Hamamelis virginiana* L.
Yellow-poplar; tuliptree	*Liriodendron tulipifera* L.
Yew, Canada	*Taxus canadensis* Marshall
Yew, English	*Taxus baccata* L.
Yew, Florida	*Taxus floridana* Nutt.
Yew, Japanese	*Taxus cuspidata* Sieb. & Zucc.
Yew, Pacific	*Taxus brevifolia* Nutt.

INDEX